Biochemical and Biotechnological Applications of Electrospray Ionization Mass Spectrometry

ACS SYMPOSIUM SERIES **619**

Biochemical and Biotechnological Applications of Electrospray Ionization Mass Spectrometry

A. Peter Snyder, EDITOR
U.S. Army Edgewood Research, Development and Engineering Center

Developed from a symposium sponsored
by the Division of Analytical Chemistry
at the 209th National Meeting
of the American Chemical Society,
Anaheim, California,
April 2–6, 1995

American Chemical Society, Washington, DC 1995

Library of Congress Cataloging-in-Publication Data

Biochemical and biotechnological applications of electrospray ionization mass spectrometry / A. Peter Snyder, editor.

p. cm.—(ACS symposium series, ISSN 0097–6156; 619)

"Developed from a symposium sponsored by the Division of Analytical Chemistry at the 209th National Meeting of the American Chemical Society, Anaheim, California, April 2–6, 1995."

Includes bibliographical references and indexes.

ISBN 0–8412–3378–0

1. Mass spectrometry—Congresses. 2. Ionization—Congresses. 3. Biomolecules—analysis—Congresses.

I. Snyder, A. Peter, 1952– . II. American Chemical Society. Division of Analytical Chemistry. III. American Chemical Society. Meeting (209th: 1995: Anaheim, Calif.) IV. Series.
[DNLM: 1. Spectrum Analysis, Mass. 2. Chromatography, Liquid. QC 454.M3 B614 1995]

QP519.9.M3B46 1995
574.19'285—dc20
DNLM/DLC
for Library of Congress 95–50381
CIP

This book is printed on acid-free, recycled paper.

PRINTED IN THE UNITED STATES OF AMERICA

1995 Advisory Board

ACS Symposium Series

Foreword

THE ACS SYMPOSIUM SERIES was first published in 1974 to provide a mechanism for publishing symposia quickly in book form. The purpose of this series is to publish comprehensive books developed from symposia, which are usually "snapshots in time" of the current research being done on a topic, plus some review material on the topic. For this reason, it is necessary that the papers be published as quickly as possible.

Before a symposium-based book is put under contract, the proposed table of contents is reviewed for appropriateness to the topic and for comprehensiveness of the collection. Some papers are excluded at this point, and others are added to round out the scope of the volume. In addition, a draft of each paper is peer-reviewed prior to final acceptance or rejection. This anonymous review process is supervised by the organizer(s) of the symposium, who become the editor(s) of the book. The authors then revise their papers according to the recommendations of both the reviewers and the editors, prepare camera-ready copy, and submit the final papers to the editors, who check that all necessary revisions have been made.

As a rule, only original research papers and original review papers are included in the volumes. Verbatim reproductions of previously published papers are not accepted.

Contents

Preface xi

1. **Electrospray: A Popular Ionization Technique for Mass Spectrometry** 1
   A. Peter Snyder

2. **Analytical Characteristics of the Electrospray Ionization Process** 21
   Thomas Covey

3. **Electrospray Ion Formation: Desorption Versus Desertion** 60
   J. B. Fenn, J. Rosell, T. Nohmi, S. Shen, and F. J. Banks, Jr.

4. **Liquid Chromatography with Electrospray Ionization Tandem Mass Spectrometry: Profiling Carbohydrates in Whole Bacterial Cell Hydrolysates** 81
   Gavin E. Black and Alvin Fox

5. **Identification of a Cytoplasmic Peptidoglycan Precursor in Antibiotic-Resistant Bacteria** 106
   Mike S. Lee, Kevin J. Volk, Jinping Liu, Michael J. Pucci, and Sandra Handwerger

6. **Structural Characterization of Prokaryotic Glycans and Oligosaccharides** 130
   Vernon N. Reinhold and Bruce B. Reinhold

7. **Structural Characterization of Lipopolysaccharides from *Pseudomonas aeruginosa* Using Capillary Electrophoresis Electrospray Ionization Mass Spectrometry and Tandem Mass Spectrometry** 149
   S. Auriola, P. Thibault, I. Sadovskaya, E. Altman, H. Masoud, and J. C. Richards

8. **Determining Structures and Functions of Surface Glycolipids in Pathogenic *Haemophilus* Bacteria by Electrospray Ionization Mass Spectrometry** 166
Bradford W. Gibson, Nancy J. Phillips, William Melaugh, and Jeffrey J. Engstrom

9. **Electrospray Ionization Mass Spectrometry for Structural Characterization of the Lipid A Component in Bacterial Endotoxins** 185
Richard B. Cole

10. **Microcolumn Liquid Chromatography–Electrospray Ionization Tandem Mass Spectrometry: Analysis of Immunological Samples** 207
Ashley L. McCormack, Jimmy K. Eng, Paul C. DeRoos, Alexander Y. Rudensky, and John R. Yates, III

11. **Identification of Proteins from Two-Dimensional Electrophoresis Gels by Peptide Mass Fingerprinting** 226
David P. Arnott, William J. Henzel, and John T. Stults

12. **Lipid Metabolism of the Mosquito Pathogen *Lagenidium giganteum* and Its Hosts: Glycerophospholipids and Hydroxy Fatty Acids** 244
James L. Kerwin

13. **Characterization of Polyunsaturated Phospholipid Remodeling in Mammalian Cells by High-Performance Liquid Chromatography–Electrospray Ionization Mass Spectrometry** 267
Hee-Yong Kim, Tao-Chin Lin Wang, and Yee-Chung Ma

14. **Application of Electrospray Ionization Mass Spectrometry to the Analysis of Oligodeoxynucleotides** 281
Charles R. Iden, Robert A. Rieger, M. Cecilia Torres, and LeRoy B. Martin

15. **Noncovalent Complexes of Nucleic Acids and Proteins Studied by Electrospray Ionization Mass Spectrometry** 294
Richard D. Smith, Xueheng Cheng, Brenda L. Schwartz, Ruidan Chen, and Steven A. Hofstadler

16. **Analysis of Xanomeline, a Potential Drug for Alzheimer's Disease, by Electrospray Ionization Tandem Mass Spectrometry** 315
Todd A. Gillespie, Thomas J. Lindsay, J. David Cornpropst, Peter L. Bonate, Theresa G. Skaggs, Allyn F. DeLong, and Lisa A. Shipley

17. **High-Performance Liquid Chromatography with Atmospheric Pressure Ionization Tandem Mass Spectrometry as a Tool in Quantitative Bioanalytical Chemistry** 330
John D. Gilbert, Timothy V. Olah, and Debra A. McLoughlin

18. **Analysis of Diarrhetic Shellfish Poisoning Toxins and Metabolites in Plankton and Shellfish by Ion-Spray Liquid Chromatography–Mass Spectrometry** 351
M. A. Quilliam and N. W. Ross

19. **Identification of Active-Site Residues in Glycosidases** 365
Stephen G. Withers

20. **Electrospray Ionization Mass Spectrometric Investigation of Signal Transduction Pathways: Determination of Sites of Inducible Protein Phosphorylation in Activated T Cells** 381
Julian D. Watts, Michael Affolter, Danielle L. Krebs, Ronald L. Wange, Lawrence E. Samelson, and Ruedi Aebersold

21. **Characterization of Recombinant Glycoproteins from Chinese Hamster Ovary Cells** 408
Michael F. Rohde, Viswanatham Katta, Patricia Derby, and Robert S. Rush

22. **Complications in the Determination of Molecular Weights of Proteins and Peptides Using Electrospray Ionization Mass Spectrometry** 424
Catherine Fenselau and Michele Kelly

23. **Use of Hyphenated Liquid-Phase Analyses–Mass Spectrometric Approaches for the Characterization of Glycoproteins Derived from Recombinant DNA** 432
A. Apffel, J. Chakel, S. Udiavar, W. S. Hancock, C. Souders, and E. Pungor, Jr.

24. From Protein Primary Sequence to the Gamut of Covalent Modifications Using Mass Spectrometry **472**
A. L. Burlingame, K. F. Medzihradszky, K. R. Clauser, S. C. Hall, D. A. Maltby, and F. C. Walls

25. Analysis of Biomolecules Using Electrospray Ionization–Ion-Trap Mass Spectrometry and Laser Photodissociation **512**
James L. Stephenson, Jr., Matthew M. Booth, Stephen M. Boué, John R. Eyler, and Richard A. Yost

26. Atmospheric Pressure Ionization Liquid Chromatography–Mass Spectrometry for Environmental Analysis **565**
Robert D. Voyksner

INDEXES

Author Index **585**

Affiliation Index **586**

Subject Index **586**

Preface

ELECTROSPRAY IONIZATION MASS SPECTROMETRY (ESI-MS) has grown as an analytical technique in exponential proportion over the past 5 or 6 years. During this time, a void has developed in the documentation of applications of ESI-MS for the many faceted and significant areas of the bio-related sciences. This book was developed from a series of symposia that addressed applications. It represents a focal point for the first collection of application-driven, problem-solving approaches on a multitude of contemporary analytical, biochemical, biological, biotechnological, environmental, immunological, microbiological, and pharmaceutical issues with the very simple sample introduction and ionization analytical technique of ESI-MS.

ESI-MS is a technique that can provide significant information to society-relevant, biologically related problems in the commercial sector; organism detection and detoxification; immunology; high-profile human diseases from the host, organism, and environmental points of view; and drug therapy fields. In short, the books says "Use me! Look what you get when you combine an amazingly simple and user-friendly sample introduction system with an easy-to-operate yet very sophisticated piece of analysis hardware. (Yes, unbelievably, there are relatively easy-to-operate, state-of-the-art, mass spectrometers in the commercial domain.) See for yourself how seemingly complex biological samples can be handled with ease."

Thus one may ask how can meaningful, even ground-breaking, information be obtained in an area as large as bio-related science with a probe that sounds too good to be true from an operator or technician's point of view. The answer is twofold. This book testifies to the fact that ESI faithfully transfers the sample, including noncovalently bound analytes, to a very powerful, but straightforward in design, MS analyzer system. A wealth of data and information can be obtained, even with as little as picomoles–femtomoles of sample, that quite often cannot be realized by other analytical systems. These two characteristics make for an ideal partnership in the exploration of biologically related phenomena and substances, the latter of which can have from very low to very high molecular weights in the hundreds of thousands to millions of daltons.

In this book, scientists from various avenues of scientific life provide their insights on the interpretation of ESI-MS investigations. Thus, the power of this book is in the planning and methods used to resolve the data from investigations of various biological processes into useful information. This concept, presented by an interdisciplinary group of scientists, should appeal to beginners and seasoned practitioners. Such issues as "What can I expect and what shall I look out for in *my* experiments" are addressed throughout the book. This makes for very useful reading by any account.

Acknowledgments

The seed of this project was planted by Richard R. Smardzewski. I listened to his ideas, and the result is the book you are holding in your hands—an accounting of the power of ESI-MS challenged with one of the most complex entities that Mother Nature has to offer: biological molecules.

I express my sincere appreciation to the scientists who contributed their time in reviewing the manuscripts contained herein: Ruth H. Angeletti, Timothy R. Baker, Gerald W. Becker, Ian A. Blair, James H. Bourell, Brian T. Chait, Stephen H. Chan, Swapan K. Chowdhury, Mark J. Cole, Catherine A. Costello, Robert J. Cotter, Pamela F. Crain, Jack B. Cunniff, Deanne M. Dulik, Gerald C. DiDonato, Gottfried J. Feistner, Douglas A. Gage, Marie E. Grace, Michael L. Gross, Sohrab Habibi-Goudarzi, Steven R. Hagen, Herman van Halbeek, Kathleen A. Harrison, Michael J. Hayes, Jack D. Henion, Ronald A. Hites, Feng-Yin Hsieh, T. William Hutchens, Constance M. John, Lawrence S. Kaminsky, Donald V. Kenny, Jeffrey P. Kiplinger, Steven E. Klohr, Michael D. Knierman, Wilson B. Knight, Roger A. Laine, S. Randolph Long, Joseph A. Loo, James A. McCloskey, Juan Fernandez de la Mora, Melanie M. C. G. Peters, James C. Richards, Kenneth L. Rinehart, Patrick A. Schindler, Jeffrey Shabanowitz, Jhobe Steadman, Justin G. Stroh, Andy J. Tomlinson, Frantisek Turecek, Paul Vouros, Eric Watson, Craig M. Whitehouse, and Ronald N. Zuckermann. A tremendous amount of clerical and administrative assistance and organizational concepts were provided by Linda G. Jarvis, and I thank her for her accomplishments. Also, I recognize Barbara E. Pralle for her organizational planning and support of this volume at the American Chemical Society.

A. PETER SNYDER
U.S. Army Edgewood Research,
Development and Engineering Center
Aberdeen Proving Ground, MD 21010–5423

August 22, 1995

Chapter 1

Electrospray: A Popular Ionization Technique for Mass Spectrometry

A. Peter Snyder

U.S. Army Edgewood Research, Development and Engineering Center, Aberdeen Proving Ground, MD 21010–5423

A comparison of major sample ionization techniques for mass spectrometry (MS) is reviewed in a quantitative and qualitative fashion. The Proceedings of the American Society for Mass Spectrometry (ASMS) for the years 1979-1995 were consulted with respect to ionization methods. Trends were established and comparisons noted for the total number and percentage of papers with selected ionization techniques. Electrospray ionization (ESI) and its derivatives are noted in particular, and the total number as well as percentage of ESI papers presented at the 1994 and 1995 ASMS conferences exceeded that of all ionization techniques including the ubiquitous electron ionization method. Descriptions of the contents of the book and overviews of the scientific fields are presented where analytical, biological, biotechnological, biochemical, environmental, immunological, microbiological and pharmaceutical applications of ESI-tandem mass spectrometry occupy central roles. The chapters are grouped around the following topics: tutorial and mechanisms of ESI, the internal and surface molecules and macromolecules of bacterial cells, non-covalent biomolecule association and interactions, nucleic acids, drugs, drug metabolites, marine toxins, man-made environmental contaminants, enzyme active sites, immunological processes, protein identification with database analysis, recombinant and post-translational protein investigations, and complementary protein structure and function information from ESI and matrix-assisted laser desorption/ionization (MALDI).

If only Malcolm Dole could have known. That is, the great potential of a relatively simple technique known as electrospray ionization (ESI) was just that for more than a decade. But that is what the stuff of science is about. We only need to look in our own backyard for examples. Witness the length of time that it took to advance the

Published 1996 American Chemical Society

technique of gas chromatography (GC), from a very basic column system using, not mega, but literally macro-bore, rigid columns in the early 1950s (1,2), to the ubiquitous capillary variety of the eighties which are standard today. These changes in column usage occurred over a thirty year gestation and maturation period. Even more sobering is the time span for evolution of mass spectrometry (MS). The formative years for MS were 1900-1940 (3) to provide for commercially-available computerized quadrupole and sector mass spectrometers. In the forties and fifties, MS could be noted for its basic characterization concerning different analyzer geometries. In the late fifties, the beginnings of fundamental yet formulation-building applications on chemical compounds were taking effect, and it wasn't until the mid-to-late eighties that MS was taken from the oscilloscope to the personal computer. A few publications and presentations on electrospray ionization (ESI), mainly by Dole, can be found from the late sixties to the early eighties (4-10). This literature found ESI interfaced to ion mobility spectrometry (IMS) - a low resolution, poor man's, but cheap, pseudo-mass spectrometer which operates at atmospheric pressure. Even at his retirement in 1984, Dole employed IMS as a detector for ESI (11, 12). At that time he was attempting to use ESI to deliver large compounds into an IMS analyzer. Thus, a superior sample ionization technique was interfaced to a low resolution device, however, an important advantage here is that both the analyzer and sample introduction device conveniently operate at atmospheric pressure. But where was mass spectrometry with respect to ESI? Around that time, John Fenn must have had a higher form of resolution in mind, because he, along with Yamashita and Whitehouse (13-15), conducted experiments to show the usefulness of Dole's ESI sample transfer and ionization method to a mass spectrometer. The enticing aspect of ESI and MS as partners in science is that no heat, vacuum or sophisticated interfaces are needed. Just some pressure is required. Pressure pushes the liquid-containing analyte past the tip of a metal needle that has applied kilovolt voltages. The high electric field density places many charges on and/or in the many tiny droplets of liquid spraying from the ESI needle tip. The transition from highly charged, tiny droplets to gas phase ions doesn't seem like a big deal, but it is. It is a highly controversial subject. A number of theories have been offered and Chapter 3 is a very timely report on the subject. John Fenn has attempted to place the various theories into a rational perspective.

The philosophy in preparing this book was that electrospray is ready to be adopted by the entire biological community. In practice, this philosophy is envisioned as two intertwined goals in that the contents of this book would appeal to the uninitiated as well as to the seasoned practitioner. A strength of this book is in the many applications of ESI-MS from analytical, biochemical, biological, biotechnological, environmental, immunological, microbiological, and pharmaceutical perspectives as observed from a varied group of people. Biologists, biochemists, microbiologists, chemists and their analytical counterparts are all represented.

A Survey of Electrospray Ionization

In order to appreciate the scientific experience of these participants, an introduction to the practice of electrospray ionization and the major parameters that affect ESI performance appeared to be a useful addition. Tom Covey was one of the pioneers of the electrospray technique when in the laboratory of Jack Henion, an outer sheath of nebulization gas was added coaxially near the tip of the sample introduction capillary, thus, nebulization-assisted electrospray, or ionspray, was born. Tom provides an overview in Chapter 2 of the ESI process, what makes it tick, and the operational and logistic choices an operator has, and together, both Chapters 2 and 3 provide the reader with a concise report on the operational and theoretical aspects of ESI-MS. The vast array of problems that can be addressed by the relatively simple technique of ESI-MS make it a powerful, even revolutionary tool.

A barometer for the previous statement can be argued as that of the Proceedings of the American Society for Mass Spectrometry Conferences on Mass Spectrometry and Allied Topics (ASMS Proceedings) (16). The ASMS Proceedings provides the mass spectrometry practitioner with the latest in research, applications and hardware development. These Proceedings can also provide a competitive database for sample introduction and ionization techniques for mass spectrometry systems. Hence, important clues can be obtained as to the trends of sample introduction/ionization techniques in terms of which have succeeded in a relative sense and for how long. By noting the trends, it may be possible to predict, postulate or at least to observe which relatively new ideas are worth watching and investigating. The ASMS Proceedings were consulted by noting every presentation, and some papers used multiple techniques.

Table 1 describes the various abbreviations and names for each sample introduction/ionization technique found in the pages of the 27th-43rd ASMS Proceedings. Some techniques are known by a variety of names and acronyms. In certain instances, a number of different techniques have been combined into one category. For purposes of clarity, one abbreviation was chosen to represent each category, and that abbreviation is presented first. Provision for an explanation of each technique is outside the scope of this Introduction, however, the reader can refer to the appended references.

Figures 1 and 2 represent trends of the seven most commonly used methods of sample ionization. Figure 1 plots the year vs. the total number of papers in the respective ASMS Proceedings featuring each ionization method, and Figure 2 plots the year vs. the percentage of papers of each of the ionization methods with respect to the total number of ASMS Proceedings papers in the respective year. Table 2 presents a survey of the number of papers containing other types of sample introduction and ionization techniques for MS.

In Figures 1 and 2, ESI, MALDI and LIMS refer to the categories; EI refers to the EI category plus that part of the following categories of papers which use the EI technique: Py methods, SID, GC-EI, PB and SFC. In Figures 1 and 2, CI refers to the CI category plus that part of the following categories which use the CI technique: Py methods, heated nebulizer APCI, DCI, GC-CI, PB and SFC. In Figures 1 and 2, the FAB and SIMS categories have been combined, but are tabulated separately in Table 2. The SFC and Py categories in Table 2 represent the total number of ASMS Proceedings papers for each sample introduction method, regardless of ionization technique.

Over the analyzed time span, it appears that the number of LIMS investigations have approximately doubled (Figure 1), yet Figure 2 shows that LIMS has not grown in overall stature compared to fifteen years ago. Thus, it can be assumed that since 1979, LIMS has retained a relatively constant percentage of investigators as the MS community has grown, while nevertheless, doubling in size. LIMS realizes a relatively

Table 1. Glossary of sample ionization/introduction techniques to mass spectrometry analyzers.

Abbreviation	Nomenclature	Reference
ESI, ESP, ES, ESPI	electrospray ionization	17, 18
ISP	ionspray	17, 18
IE	ion evaporation	17, 18
	ultraspray	19
API	atmospheric pressure ionization	18
APCI	atmospheric pressure chemical ionization	18
	high pressure ionization	18
	radioactive 63-Ni energetic (beta) electron emission source	18, 20
	heated nebulizer APCI	18
APS	atmospheric pressure spray ionization	21
MALDI	matrix-assisted laser desorption/ ionization	22, 23
	aerosol MALDI, laser spray ionization	22
FD	field desorption	24
FI	field ionization	24

Continued on next page

Table 1. *Continued*

Abbreviation	Nomenclature	Reference
EI	electron ionization	23
NRMS	neutralization-reionization MS	25
SIFT	selected ion flow tube	26
FA	flowing afterglow	26
EB flow	electron bombardment flow	27
EI-PD	EI-photodissociation	28
EI-PID	EI-photon-induced dissociation	28
EC	electron capture	
	electron attachment	29, 30
	The following differ in the method of sample introduction to an EI source	
HPLC	High pressure liquid chromatography-moving belt	
DLI	direct liquid introduction	
	Townshend discharge	31
	Knudson cell	32
MIMS	membrane inlet MS	33
CRIMS	chemical reaction interface MS	34
	supersonic molecular beam	35
LD	laser desorption	36
		37
		38
CI	chemical ionization	23
	The following differ in the method of sample introduction to a CI source.	
HPLC	High pressure liquid chromatography-moving belt	31
DLI	direct liquid introduction	32
LD	laser desorption	39
GC-EI-MS		29, 30
GC-CI-MS		29, 30

Continued on next page

Table 1. *Continued*

Abbreviation	Nomenclature	Reference
Py[1]	pyrolysis	40
TD	thermal desorption	41
TGA	thermogravimetric analysis	41
DIP	direct insertion probe, direct probe, solids probe	42
DEP	direct exposure probe	42
DCI[2]	desorption chemical ionization	42
SID[3]	surface-induced dissociation	43
PB[4]	particle beam, Thermabeam	44
MAGIC-LC	Monodisperse Aerosol Generation Interface for Combining LC with MS	44
FAB	fast atom bombardment	23, 45, 46
Cf-FAB	continuous flow FAB	45, 46
SIMS	secondary ion MS	23, 47, 48
LSIMS	liquid SIMS	46, 47
	gaseous SIMS	46, 47
FIB	fast ion bombardment	46, 47
PIB	pulsed ion bombardment	46, 47
LSI	liquid secondary ionization	46, 47
LIMS	liquid ion MS	46, 47
LMI	liquid metal ionization	46, 47
Cs^+	cesium SIMS	46, 47
Ga^+	gallium SIMS	46, 47
	Internal Ion Impact Ionization	46, 47
LIMS	laser ionization MS	38
RIMS	resonance ionization MS	38
REMPI	resonance-enhanced multiphoton ionization	38
MPI/MUPI/MPRI	multiphoton resonance ionization	38
LAMMA	laser microprobe mass analysis	38
LD-LI	laser desorption or laser ablation-ionization	23, 38
PD/PI	photodissociation/photoionization	23, 38
PIPICO	photoion-photoion coincidence spectrometry	49
PEPICO/PIPECO	photoelectron photoion coincidence spectrometry	50, 51
PD	plasma desorption, ^{252}Cf	23, 52

Continued on next page

Table 1. *Continued*

Abbreviation	Nomenclature	Reference
GD	glow discharge	23, 53
HCP	hollow cathode discharge	53
	Penning ionization	53
ASGDI	atmospheric sampling glow discharge ionization	54
MIP	microwave-induced plasma	55
SSMS	spark source MS	56
ICP	inductively coupled plasma	23, 57
TSP	thermospray	58, 59
	plasmaspray	58, 59
LIMS	liquid ionization MS	60
EHD, EHDI	electroheterodynamic ionization	61
TIMS	thermal ionization MS	62
AMS	accelerator mass spectrometry	63
K^+IDS	potassium ionization of desorbed species	64
SFC or SCF[5]	supercritical fluid chromatography	65

[1]These methods are heating or sample introduction methods, not ionization techniques. EI and CI dominate as ionization techniques.

[2]This is the analogue of the EI-DEP method, because the sample is heated on a probe in a torr pressure CI source.

[3]This is a sample introduction method for EI.

[4]These are sample introduction techniques for EI or CI.

[5]Sample introduction method for ionization methods including EI, CI, TSP, APCI, LD and photodissociation.

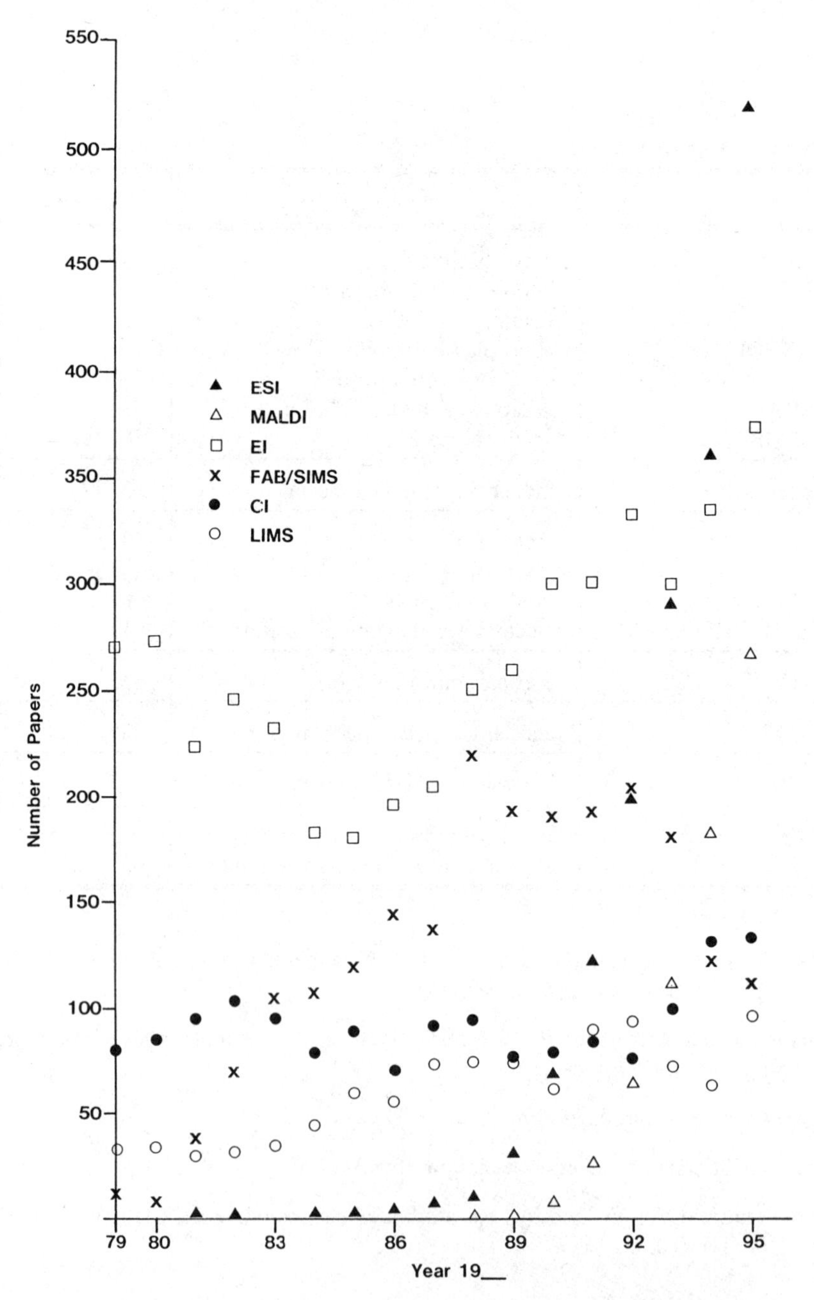

Figure 1. *Calendar year vs. number of papers in the ASMS Proceedings for the most common methods of sample ionization for MS analysis.*

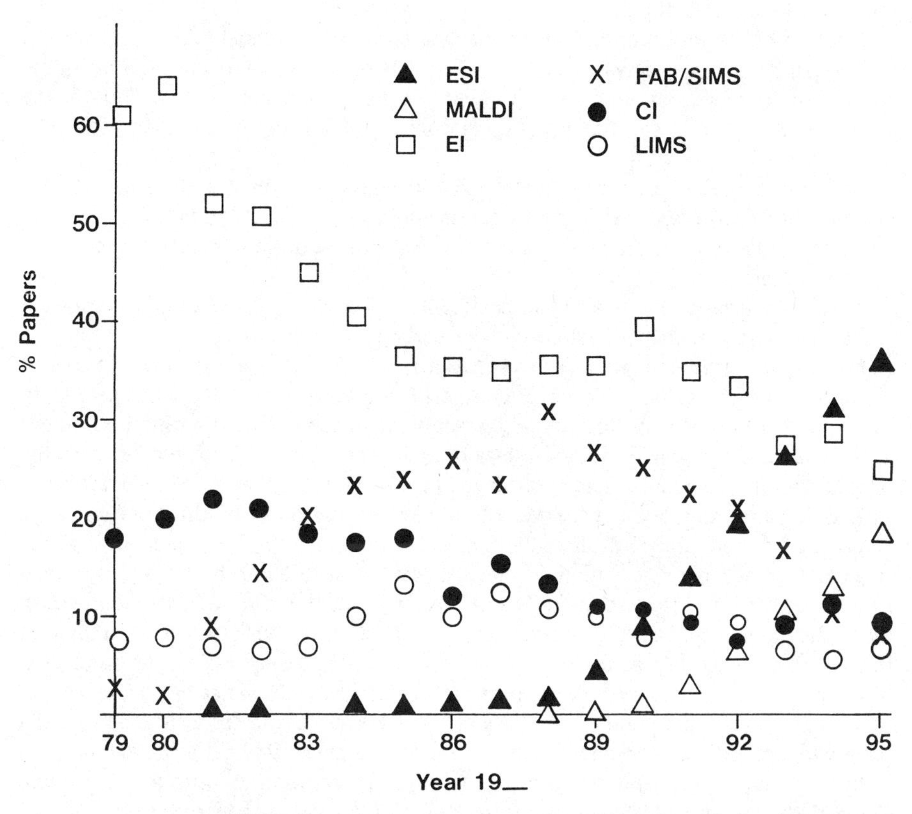

Figure 2. Calendar year vs. the percentage of papers in the ASMS Proceedings for the most common methods of sample ionization for MS analysis.

low percentage of the MS ionization techniques (Figure 2) averaging 6.6% in 1995 and 9% over the past seventeen years. However, that should be considered a healthy percentage of the entire pool of MS ionization techniques.

In the 1979-1988 time period the FAB/SIMS desorption techniques steadily gained in popularity, slowly leveled off in the early nineties, and then took a sudden drop in usage in 1994 (Figure 1). Figure 2 shows essentially that the relative use of FAB/SIMS took a somewhat bell-shaped curve where the techniques peaked in the late 1980s. Nevertheless, in 1995, the desorption techniques showed an 8% incidence of use in all reported MS investigations.

The use of CI ionization techniques has varied in approximately three cycles from 70-100 papers per year between 1979-1993, and 1994 had a significant increase in the number of CI reports (Figure 1). However, Figure 2 shows a steady decline in the overall percentage of presentations, ranging from approximately 20% in the early 1980s to 9% in 1995.

MALDI is a relatively recent phenomenon in mass spectrometry and since 1990 it launched into a steep, geometric increase in the number of reports (Figure 1). Likewise, as Figure 2 attests, its share in the amount of papers rose at an impressive rate to that of approximately 19% in 1995.

Electron ionization has had a commanding lead over the years. EI clearly dominated as the ionization method of choice in the early to mid 1980s (Figure 1) despite a significant drop in the number of investigations. Then EI took a turn for the better in the mid-1980s and sustained a consistent increase up until 1993, all the while leading in the number of reports as the method for sample ionization. However, Figure 2 shows another analysis in that EI posted a steady downward trend. As stated, EI was the front-runner in total number and percentage of investigations in the ASMS Conferences. ESI, the star of this book, unseated EI in 1994 in the areas of this survey. It is surprising to note that ESI had the longest gestation period (acceptance? public relations?) of all major ionization techniques. An ESI-IMS report by Dole (10) as well as a paper on ion evaporation (66) were presented at the 1981 ASMS Conference. Since then, approximately 30 papers were presented up to 1988; this is an eight year time period. In 1983, there were no scientific reports using ESI as an ionization technique. ES (electrospray) did appear in one report by Beavis *et al.*, however, not as an ionization technique (67). ES was used as a sample transfer technique to a flat plate which interfaced to a SIMS source. However, since 1989, ESI was taking hold in the MS community, and it was promoted in a geometric fashion up to the present. Figure 2 shows that ESI has climbed from obscurity to a very impressive 35.8% of total ASMS presentations in 1995. However, even GC, a current workhorse of many a laboratory, followed a similar track record in the literature (68). Its literary gestation spanned the 1940s until 1955 when it started a geometric increase in the rate of publications. EI, in contrast, has steadily decreased in its share of scientific investigations in the ASMS Proceedings from a grand high (and probably never to be reached again by any MS method) of 64% in 1980 (nearly two-thirds of all papers) to 25.8% in 1995. A further analysis (Figure 2) reveals some interesting trends and noteworthy points. 1993 saw ESI and EI sharing essentially the same percentage of investigations at 26.5% and 27.3 %, respectively. In 1994, ESI (30.9%) jumped ahead of EI (28.7%) and increased its share in 1995. These relative and absolute percentage levels of sample ionization methods in the MS user community were matched only once, in 1988,

Table 2. Number of papers of the relatively lesser used ionization and sample introduction techniques in the ASMS Proceedings in the years 1979-1995.

	79	80	81	82	83	84	85	86	87	88	89	90	91	92	93	94	95
FD	31	29	16	17	14	9	7	5	6	6	8	1	4	4	7	2	7
TIMS	9	6	9	18	13	8	10	8	5	8	9	4	3	0	2	3	2
TSP	2	2	4	7	11	21	37	55	47	60	43	55	46	37	33	19	15
AMS	0	0	0	0	0	0	0	0	0	0	0	0	0	0	4	0	1
K+IDS	0	0	0	0	0	0	0	1	2	3	2	1	1	0	1	1	0
GD	10	8	9	7	13	6	6	3	7	9	19	20	15	19	23	20	25
ICP	1	0	2	4	4	2	1	8	7	1	14	5	5	8	11	9	9
SFC	0	0	0	0	0	0	0	1	3	2	2	1	5	0	1	1	5
Py	18	22	18	17	21	10	17	20	19	25	21	26	24	14	24	24	29
FAB	0	0	21	35	71	73	81	109	101	161	140	142	136	123	100	69	69
SIMS	12	8	18	25	34	34	38	35	37	58	53	49	57	81	81	54	47
Total number papers	443	430	433	485	517	452	496	555	590	703	732	760	862	994	1099	1169	1444

between FAB/SIMS (31.3%) and EI (35.7%). The relatively significant drop in percentage of investigations by EI in the 1980s could have been mostly due to the popularity of the FAB/SIMS desorption techniques. EI dropped from 60% to a 35% level while FAB/SIMS rose from a few percent to 25-30%. Thus it appears that FAB/SIMS pioneered the way as new and more versatile methods for sample ionization at the expense of EI. Then, FAB/SIMS decreased in relative popularity, and this appears to coincide with the relative increase in ESI and MALDI techniques. Even though FAB/SIMS usage decreased in the late 1980s and early 1990s, the popularity of the ESI and MALDI techniques appears to have not only made up for the loss of FAB/SIMS investigations, but they also caused further inroads in the overall share of the EI research programs. In this time frame, EI decreased from 35.5% in 1989 to 25.8% in 1995.

Biological Applications of Electrospray Ionization Mass Spectrometry

As the survey indicates, ESI has become one of the leading sample introduction/ionization techniques for MS, and the literature shows that it has made a major impact on the bio-related areas of pure and applied science and technology. Thus, it appeared desirable to gather a series of scientific contributions which presented timely biological applications with ESI-MS-MS.

We start with the bacterial cell, a complex entity unto itself and entirely too formidable to be considered an analyte for an ESI-MS analysis. Thus, monomeric, oligomeric and polymeric components must be considered in an elucidation of structure as it relates to function as well as detection and characterization of the microbial entity. Chapters 4 and 5 present chemical and physical methods in the characterization of the peptidoglycan- associated sugar monomers and glycopeptide oligomers, respectively. Black and Fox show in Chapter 4 that it is possible to devise methods that can become a part of a comprehensive identification scheme for bacteria. This is based on the neutral and amino-sugar profiles of hydrolyzates of bacterial cells, and a large portion of the sugars derive from the the cell wall peptidoglycan and lipopolysaccharide layers. Comparisons are made on time-consuming derivatization methods with GC-MS and analyses consisting of ESI-MS/MS procedures on underivatized, direct hydrolyzates. Chapter 5 also discusses the peptidoglycan layer, however, from a structural perspective. Lee *et al.* analyze the glycopeptide components of peptidoglycan, because that is where antibiotic action resides. By probing antibiotic-resistant bacteria, such as *E. faecalis* and *Leuconostoc*, alterations were discovered in the peptide portion of the peptidoglycan from that of the native molecular architecture. These structural studies led to the development of an assay to determine the binding affinity of antibiotics such as vancomycin with the native and altered glycopeptide components of peptidoglycan.

Chapters 6-10 are concerned with the surface of the organism where there exists complex arrays of biochemical entities. Organisms must communicate and interact in some fashion with other organisms, cells from a host body, and protein antibodies from host cells. Structure-function relationships enter into these complex recognition events by the bacterial cell with host components during infection. Reinhold and Reinhold (Chapter 6) explore the complex bacterial surface motif which includes polysaccharides and lipopolysaccharides. These surface entities contribute to toxic host responses, and as such, biochemical extraction and selected modifications of these macromolecules provide important information on their composition. This chapter also is an excellent primer in the understanding of these two classes of bacterial macromolecules. Thibault *et al.* (Chapter 7) tackle this problem through the determination of the "bare" oligosaccharide nucleus of the lipopolysaccharide (LPS) of *Pseudomonas aeruginosa* with subsequent determination of the various pendant groups and modified carbohydrate residues. An appreciation to detail is presented in the reconstruction of the intact LPS species. Gibson *et al.* (Chapter 8) concentrate on finding the select lipooligosaccharides (LOS), among many surface species, that are relevant to the disease processes in *Haemophilus*. To aid in this determination, various forms of chemically-processed LOS were investigated. Cole (Chapter 9) uses acid and base chemical procedures to determine the decomposition pathways of lipid A in the LPS endotoxin structure. Various chemical reaction parameters are implemented and yield chemical detoxification procedures which can be correlated with degrees of reduced toxicity for the lipid A fragments from *Enterobacter agglomerans* and *Salmonella minnesota* organisms with ESI-MS and PDMS. Chapter 10 discusses a class of surface biomarker that is relatively low in molecular weight and high in the number of different isomers and species. The major histocompatibility complex (MHC) consists of a broad class of glycoprotein surface immune markers which bind an extremely diverse array of peptides, and the latter act as antigenic determinants for cells. Yates *et al.* present studies on mutant cell surface markers in order to identify differences in the peptide arrays from that of the native cell. These differences can provide clues as to the cellular mechanisms and structure-antigenic activity of the MHC species and are important steps toward the goal of developing drugs to combat diseases. Chapter 10 also offers a scheme for the identification of these antigenic small peptides based on a comparison of the raw ESI-product ion mass spectrum, a synthetic mass spectrum, and a search of a database containing known protein sequences. A different, automated method of protein identification is given in Chapter 11. The hallmark of these automated methods is that the investigator is relieved of having to manually interpret a complex peptide fragment mass spectrum. This process can be very time consuming, and expert knowledge is usually required for unraveling series of peptide ions and being alert to the fact that not all potential peptide ion sequences in a protein may be present in a mass spectrum. Stults *et al.* present a scheme for protein identification with ESI-MS/MS and two-dimensional gel electrophoresis (Chapter 11) by digesting a series of proteins with the same enzyme and sorting the mass spectral sequences. As few as three or four peptides from a complex protein digest can be used to differentiate one protein from another, and this technique gives rise to peptide mass fingerprinting. Complex protein extracts from normal and enlarged heart cells, the latter of which leads to congestive heart failure, is illustrative of this identification method.

Lipids represent an important class of compounds in cellular systems. The control of the mosquito insect pest is usually accomplished by chemical pesticides, however, it may be possible to control mosquito presence in certain environmental situations by the introduction of fungal species. Lipids could play an important role, because they are required for successful invasion of the mosquito by the fungal parasite. Kerwin provides an analysis on the *Lagenidium giganteum* water mold that is a parasite on mosquito larva (Chapter 12). Both need phospholipids and eicosanoids, yet both do not synthesize or produce these necessary metabolic compounds. Thus ESI-MS/MS was used to study the metabolic profile of these compounds during water fungal mold infection under controlled diet and growth conditions. Chapter 13 explores the major classes of phospholipid in glia cells and rodent brain membrane tissue. Kim *et al.* conduct investigations with respect to the qualitative and quantitative aspects of the dynamics of phospholipid turnover and maintenance in these cells using ESI-MS/MS.

All cellular functions ultimately focus on DNA, the biological molecule which encodes life as we know it. ESI-MS has created in-roads into the structure and function aspects of nucleic acids by tackling such complex molecules as high molecular weight single and double stranded DNA (69). Both transfer ribonucleic acid (t-RNA) (70) and, incredibly, the 110 megadalton DNA from coliphage T4 (71) have yielded to molecular weight studies. Polymerase chain reaction, or PCR, is a technology which amplifies or increases the amount and concentration of a specific sequence of DNA which can be present in trace levels. Only one DNA strand containing a specific sequence of nucleotides is required. PCR has received recent coverage in the popular and scientific news because its forensic utility has signficant use in a judicial court of law (72, 73). Analysis is usually accomplished by staining the DNA and performing gel electrophoresis. ESI-MS/MS has recently been shown to be able to analyze for PCR gene products (74). In order for the reader to appreciate these concepts, Iden (Chapter 14) presents a primer on the basics of an understanding of the primary sequences of DNA oligomers with ESI-MS/MS. The four primary DNA bases can also be found in modified states, and strategies for their characterization are also presented.

Biochemical enzyme reactions (refer to Chapter 19) as well as electrostatic, non-covalent biomolecule interactions are the lifeblood for cellular function and recognition events. The latter interaction finds immense importance in cellular functions such as immunological antigen-antibody events, the protective histone protein coating of DNA, drug-biomolecule interactions, enzyme processes requiring co-factors, and the MHC recognition events (Chapters 10 and 20). The non-covalent biomolecule association phenomenon is a new field for ESI-MS/MS and is explored by Smith *et al.* (Chapter 15) using DNA. Evidence for three types of non-covalent complex interactions is shown for oligonucleotides with their complementary strands, multimeric proteins, and complexes between an oligonucleotide strand and a small drug molecule.

Drug therapy for combatting human disease and poisoning agents are topics which focus on a wide variety of chemical compounds. Disease and poisoning agents can be very debilitating, and as such, ESI-MS/MS has provided great strides in the understanding of the biochemical origins of these assaults on the quality of human life. Gillespie (Chapter 16) takes us on a tour of a systematic analysis of the muscarinic M_1 drug agonists of Alzheimer's disease. This disease is primarily responsible for memory loss in the aged population, and has received a heightened awareness in the public over the last few years. Biotransformation products from the drug analyte are observed. However, urine is shown to provide effective roadblocks in the separation of drug metabolites due to the overwhelming amount of endogenous interferences, and methods are suggested to alleviate this problem.

Gilbert then provides a treatise (Chapter 17) on the investigations of drug analyses in human subjects from a quantitative perspective by ESI-MS/MS. Advantages/disadvantages of LC-MS and LC-MS/MS are presented with respect to sensitivity and selectivity, and comparisons to the established radioimmunoassay and HPLC-ultraviolet detection methods for quantitative drug determination in mammalian fluids are discussed. Cross validation methods are shown to be useful in correlating drug test results from the different detection methods, and isotope addition to samples highlights the quantitative aspect. Quilliam provides a quantitative and qualitative presentation of the poisoning agents in plankton shellfish (Chapter 18). An interesting observation was that different toxin preparation/extraction procedures resulted in different mass spectral responses. In addition, different culture ages produced different toxin levels which suggested endogenous esterase action over time on the toxins. ISP-MS/MS was used to resolve these phenomena, and this investigation presents numerous, almost seemingly random, problems in bio-analysis scenarios that can be solved in a systematic fashion.

The biochemical dynamics of cells is the next subject for discussion in Chapters 19-22. A deliberate analysis of events in the cell such as enzymatic processes, complex interactions of a killer T cell with host tissue, and post-translational products of regulatory enzymes presents a unique evolving picture in the depth of ESI-MS/MS as applied to the elucidation of biological proceses.

Withers starts our journey by investigating the active sites of *Bacillus subtilis* and *Cellulomonas fimi* glycosidases and human lysosomal glucocerebrosidase in Chapter 19. Covalent labeling with active site inhibitors and affinity labels was done, and subsequent peptic digests allowed ESI-MS/MS to accurately map out which amino acid residues are involved. The relative ease of this procedure has obvious implications for enzymes of unknown structures.

Complex intracellular signal transduction pathways are essential for various biochemical processes. Aebersold shows in Chapter 20 how one can track tyrosine phosphorylation pathways on protein tyrosine kinases and T cell receptors. These series of complex immunochemical reactions are essential for successful T cell recognition of surface peptides complexed to the MHC molecule described by Yates, *et al.* in Chapter 10. Attempts to unravel the phosphorylation pathways are accomplished by chemical, enzymatic and immobilized metal affinity chromatography - HPLC - ESI-MS/MS methods.

Chapter 21 presents the analysis of recombinant proteins with a slightly different objective than just straightforward structure characterization. It is well known that bacterial cell-produced recombinant protein can sometimes be different with respect to the native- produced protein from mammalian cells *in vivo*. Salient features between the mammalian-derived protein may not be expressed in the bacterial-derived recombinant. Some of these features include proper disulfide pairs and post-translational carbohydrate attachment sites and sequences. Mammalian Chinese hamster assay cells produce recombinant proteins that are usually faithful replicates of the host mammalian cell protein, and Rohde *et al.* attempt to confirm this with ESI-MS/MS experiments on glycosylated proteins.

Another topic that has much public awareness is the human immunodeficiency virus (HIV), because it is a disease state that is eventually fatal and no cure exists. Fenselau (Chapter 22) investigates the polypeptides produced from the human and bovine immunodeficiency viruses, HIV and BIV, respectively, with respect to their molecular weight distributions. Cellular processing and post-translational modifications incurred by the polypeptides are investigated with respect to protein multiplicity by mutations during viral replication.

ESI and MALDI team up in the next two chapters and provide the core technologies in a suite of methods so as to provide structural characterization for complex proteins and the heterogeneous glycoprotein biomolecule. Chapter 23 finds Hancock *et al.*

exploiting both mass spectrometry methods, as well as LC and capillary electrophoresis, and they are found to be complementary in the understanding of the structure and integrity of the complex single chain plasminogen activator glycoprotein. Burlingame and associates (Chapter 24) present a classical approach at solving complex biochemical structural problems by using an array of modern analytical and biochemical techniques. Polyacrylamide gel electrophoresis, select enzyme digests, HPLC, ESI-MS/MS, MALDI time-of-flight MS, LSIMS and low resolution quadrupole and high resolution electric/magnetic sector mass spectrometers comprise the tools. These biochemical and analytical methods provide levels of information that, when taken together, provide a relatively complete picture of complex biochemical structures, including the isomer and stereoisomer issues. The signal recognition particle ribonucleoprotein, human melanoma A375 cell proteins, and bovine fetuin glycoproteins are investigated by the techniques. This chapter serves as a reminder that in many cases various scientific avenues are necessary in order to provide a comprehensive, informative analysis, and depending on the degree of complexity of a substance, some scientific systems provide more complete information than others.

Chapter 25 is more fundamental in nature and finds molecules such as proteins/peptides, oligosaccharides and single-stranded RNA as the targets of investigation by ESI-ion trap MS analysis with photodissociation as the collision-induced dissociation (CID) route. Stephenson *et al.* show that this method has advantages over the conventional ion-molecule CID procedure, and photodissociation CID is used to search for salient mass spectral details in the characterization of these small molecules.

Of great importance is the interface between man and the environment. Environmental factors can take the form of the contamination of soil, water and air with a host of harmful or irritant chemicals and by-products through inadvertant or deliberate means. Short as well as long term exposure and the effects of a host of man-made chemicals on humans are a significant part of the Environmental Protection Agency's responsibilities. GC/MS is a premier technology in the field screening and determination of many environmental contaminants. Voyksner (Chapter 26) provides descriptions of analyses concerning environmental contaminants such as dyes, pesticides, herbicides, amines, oxygenated compounds and hydrocarbons, and it appears that LC-ESI-MS/MS is a powerful tool in the discovery and identification of a host of environmental contaminants. Especially for these important classes of compounds, the ESI-MS technique provides information that often cannot directly be obtained from the conventional method of GC/MS.

The Future of Electrospray Ionization Mass Spectrometry

This is a tough question. Despite its success and predictions that could be made (75), one must keep in mind the phenomenon of the FAB/SIMS desorption technologies as quantitatively documented in Figures 1 and 2. One could have easily suggested that 1989 should have been the year for FAB/SIMS to take the lead from EI in the percentage of total ionization techniques.

In 1988, ESI was a mere shadow of its present utility when FAB/SIMS were knocking on the door of EI. Thus, what of the future of ESI? Clues to this question should be found in what current methods are "shadows" or latest technologies for possible future increased exploitation. MALDI already has passed through its latent period and this occurred a few years after FAB/SIMS had their peak (Figure 2). MALDI is a sample desorption/ionization method that is constantly gaining in popularity, virtually at the same rate as ESI (Figure 2). Other than MALDI, the ASMS Proceedings provide no tangible evidence of other techniques on the horizon that could yield success stories like FAB/SIMS, MALDI or ESI.

However, on an optimistic note, there is no reason to suspect that the popularity of ESI-MS will decline in the near future. ESI has firmly established itself as an easy to use, atmospheric pressure sample introduction and ionization system (75). It can handle virtually any analyte in a gentle manner and in almost all cases faithfully transforms a neutral of interest into nothing less than an ion of equal interest. Interpretation of ESI-mass spectra for the most part is straightforward, and even if it is not, the resulting clues and information more than make up for any tedious data analysis. This can be said because information generated by ESI-MS, as well as MALDI, may not be evident or even be impossible to observe with other ionization techniques (76). ESI-MS is exquisitely sensitive as is evident with the recent advent of micro-spray technology (Chapter 2).

A revolutionary method can manifest itself by having an impact on pre-existing methods, and usually it suppresses or supplants these methods. The ESI technique invented by Dole a quarter of a century ago and reapplied by Fenn appears to be revolutionary, at least with respect to its acceptance and broad array of applications in many fields of science.

Thus, the significant advantages that drive the large body of work at the present equally fuel the increasing amount of research that is trying to solve major disadvantages of ESI, such as ion signal suppression by dissolved salts and the unambiguous interpretation of the charge states of CID product ions from a multiply charged precursor ion. The future looks good for electrospray ionization!

Literature Cited

1. James, A.T.; Martin, A.J.P. *Biochemical J.* **1952,** *50*, 679-690.
2. James, A.T.; Martin, A.J.P. *Analyst* **1952,** *77*, 915-932.
3. Brunnee, C. *Intl. J. Mass Spectrom. Ion Physics* **1982,** *45*, 51-86.
4. Dole, M.; Mack, L.L.; Hines, R.L.; Mobley, R.C.; Ferguson, L.D.; Alice, M.B. *J. Chem. Phys.* **1968,** *49*, 2240-2249.
5. Mack, L.L.; Kralik, P.; Rheude, A.; Dole, M. *J. Chem. Phys.* **1970,** *52*, 4977-4986.
6. Clegg, G.A.; Dole, M. *Biopolymers* **1971,** *10*, 821-826.
7. Gienic, J.; Cox, H.L. Jr.; Teer, D.; Dole, M. Proc. 20th Am. Soc. Mass Spectrom. Conference on Mass Spectrometry and Allied Topics, Dallas, TX, **1972,** 276-280.
8. Teer, D.; Dole, M. *J. Polymer Sci.* **1975,** *13*, 985-995.
9. Dole, M.; Gupta, C.V.; Mack, L.L.; Nakamae, K. *Polymer Preprints* **1977,** *18*, 188-193.
10. Nakamae, K.; Kumar, V.; Dole, M. Proc. 29th Am. Soc. Mass Spectrom. Conference on Mass Spectrometry and Allied Topics, Minneapolis, MN, **1981,** 517-518.

11. Gieniec, J.; Mack, L.L.; Nakamae, K.; Gupta, C.; Kumar, V.; Dole, M. *Biomed. Mass Spectrom.* **1984,** *11*, 259-268.
12. Dole, M. Proc. 33rd Am. Soc. Mass Spectrom. Conference on Mass Spectrometry and Allied Topics, San Diego, CA, **1985,** 196-197.
13. Yamashita, M.; Fenn, J.B. *J. Phys. Chem.* **1984,** *88,* 4451-4459.
14. Yamashita, M.; Fenn, J.B. *J. Phys. Chem.* **1984,** *88,* 4671-4675.
15. Whitehouse, C.M.; Dreyer, R.N.; Yamashita, M.; Fenn, J.B. *Anal. Chem.* **1985,** *57,* 675-679.
16. Niessen, W.M.A.; Tinke, A.P. *J. Chromatogr.* **1995,** *703*, 37-57.
17. Iribarne, J.V.; Dziedzic, P.J.; Thomson, B.A. *Intl. J. Mass Spectrom. Ion Physics* **1983,** *50,* 331-347.
18. Huang, E.C.; Wachs, T.; Conboy, J.J.; Henion, J.D. *Anal. Chem.* **1990,** *62,* 713A-725A.
19. Banks, J.F., Jr.; Shen, S.; Whitehouse, C.M.; Fenn, J.B. *Anal. Chem.* **1994,** *66,* 406-414.
20. *Ion Mobility Spectrometry;* Eiceman, G.A.; Karpas, Z., Eds., CRC Press: Boca Raton, FL, 1994.
21. Hirabayashi, A.; Takada, Y.; Kambara, H.; Umemura, Y.; Ohta, H.; Ito, H; Kuchitsu, K. *Intl. J. Mass Spectrom. Ion Processes* **1992**, *120,* 207-216.
22. Murray, K.K.; Lewis, T.M.; Beeson, M.D.; Russell, D.H. *Anal. Chem.* **1994,** *66,* 1601-1609.
23. Busch, K.L. *J. Mass Spectrom.* **1995**, *30*, 233-240.
24. Schulten, H-R.; In *Soft Ionization Biological Mass Spectrometry*; Morris, H.R., Ed.; Heyden: London, UK, 1981; pp 6-38.
25. Goldberg, N.; Schwarz, H. *Acc. Chem. Res.* **1994,** *27,* 347-352.
26. Brickhouse, M.D.; Chyall, L.J.; Sunderlin, L.S.; Squires, R.R. *Rapid Commun. Mass Spectrom.* **1993,** *7,* 386-391.
27. Redman, E.W.; Johri, K.K.; Morton, T.H. *J. Am. Chem. Soc.* **1985,** *107,* 780-784.
28. Dunbar, R.C.; Chen, J.H.; So, H.Y.; Asamoto, B. *J. Chem. Phys.* **1987,** *86,* 2081-2086.
29. *Chromatography Today*; Poole, C.F.; Poole, S.K., Eds.; Elsevier: Amsterdam, Netherlands, 1991.
30. Evershed, R.P. In *Gas Chromatography, A Practical Approach*; Baugh, P.J., Ed.; Oxford University Press: Oxford, England, Chapter 11, 1993.
31. Privett, O.S.; Erdahl, W.L. *Chem. Phys. Lipids.* **1978**, *21,* 361-387.
32. Dedieu, M.; Juin, C.; Arpino, P.J.; Bounine, J.P.; Guiochon, G. *J. Chromatogr.* **1982,** *251,* 203-213.
33. Hunt, D.F.; McEwen, C.N.; Harvey, T.M. *Anal. Chem.* **1975,** *47,* 1730-1734.
34. Chupka, W.A.; Inghram, M.G. *J. Phys. Chem.* **1955,** *59,* 100-104.
35. Bauer, S.; Solyom, D. *Anal. Chem.* **1994,** *66,* 4422-4431.
36. Chace, D.H.; Abramson, F.P. *Anal. Chem.* **1989,** *61,* 2724-2730.
37. Dagan, S.; Amirav, A. *Intl. J. Mass Spectrom. Ion Processes* **1994,** *133,* 187-210.
38. Cotter, R.J. *Anal. Chem.* **1984,** *56,* 485A-504A.
39. Amster, I.J.; Land, D.P.; Hemminger, J.C.; McIver, R.T., Jr.; *Anal. Chem.* **1989,** *61,* 184-186.

40. *Analytical Pyrolysis: Techniques and Applications*; Voorhees, K.J., Ed.; Butterworths: London, UK, 1984.
41. Pavlath, A.E.; Gregorski, K.S. *J. Anal. Appl. Pyrolysis* **1985,** *8,* 41-48.
42. Cotter, R.J. *Anal. Chem.* **1980,** *52,* 1589A-1606A.
43. Cooks, R.G.; Ast, T.; Pradeep, T. *Accounts Chem. Res.* **1994,** *27,* 316-323.
44. Willoughby, R.C.; Browner, R.F. *Anal. Chem.* **1984,** *56,* 2626-2631.
45. Caprioli, R.M. *Biochemistry* **1988,** *27,* 513-521.
46. Hafok-Peters, Ch.; Maurer-Fogy, I.; Schmid, E.R. *Biomed. Environ. Mass Spectrom.* **1990,** *19,* 159-163.
47. Aberth, W.H.; Burlingame, A.L. *Anal. Chem.* **1988,** *60,* 1426-1428.
48. Monegier, B.; Clerc, F.F.; Van Dorsselaer, A.; Vuilhorgne, M.: Green, B.: Cartwright, T. *BioPharm* **1990,** *3*, 26-35.
49. Price, S.D.; Eland, J.H.D.; Fournier, P.G.; Fournier, J.; Millie, P. *J. Chem. Phys.* **1988,** *88,* 1511-1515.
50. Eland, J.H.D. *Intl. J. Mass Spectrom. Ion Physics* **1972,** *8,* 143-151.
51. Danby, C.J.; Eland, J.H.D. *Intl. J. Mass Spectrom. Ion Physics* **1972,** *8,* 153-161.
52. Macfarlane, R.D. *Anal. Chem.* **1983,** *55,* 1247A-1264A.
53. Coburn, J.W.; Harrison, W.W. *Appl. Spectroscopy Rev.* **1981**, 17, 95-164.
54. McLuckey, S.A.; Glish, G.L.; Asano, K.G.; Grant, B.C. *Anal. Chem.* **1988,** *60,* 2220-2227.
55. Fujii, T. *Chem. Phys. Lett.* **1992**, *191,* 162-168.
56. Bacon, J.R.; Ure, A.M. *Analyst* **1984,** *109,* 1229-1254.
57. Houk, R.S. *Accounts Chem. Res.* **1994,** *27,* 333-339.
58. Tomer, K.B.; Parker, C.E. *J. Chromatogr.* **1989,** *492,* 189-221.
59. Bowers, L.D. *Clin. Chem.* **1989,** *35,* 1282-1287.
60. Tsuchiya, M.; Kuwabara, H. *Anal. Chem.* **1984,** *56,* 14-19.
61. Evans, C.A., Jr.; Hendricks, C.D. *Rev. Sci. Instrum.* **1972,** *43*, 1527-1530.
62. Aggarwal, S.K.; Saxena, M.K.; Shah, P.M.; Kumar, S.; Jairaman, U.; Jain, H.C. *Intl. J. Mass Spectrom. Ion Processes* **1994,** *139,* 111-126.
63. Liu, Y.; Guo, Z.; Liu, X.; Qu, T.; Xie, *J. Pure Appl. Chem.* **1994,** *66,* 305-334.
64. Bombick, D.; Pinkston, J.D.; Allison, J. *Anal. Chem.* **1984,** *56,* 396-402.
65. Tyrefores, L.N.; Moulder, R.X.; Markides, K.E. *Anal. Chem.* **1993,** *65,* 2835-2840.
66. Iribarne, J.V.; Dziedzic, P.J.; Thomson, B.A. Proc. 29th Am. Soc. Mass Spectrom. Conference on Mass Spectrometry and Allied Topics, Minneapolis, MN, **1981,** 519-520.
67. Beavis, R.; Ens, W.; Standing, K.G.; Westmore, J.B.; Schiller, P.W. Proc. 31st Am. Soc. Mass Spectrom. Conference on Mass Spectrometry and Allied Topics, Boston, MA, **1983,** 679-680.
68. *Gas-Liquid Chromatography: Theory and Practice*; Nogare, S.D.; Juvet, R.S., Jr., Eds., John Wiley and Sons, New York, NY, 1962.
69. Little, D.P.; Thannhauser, T.W.; McLafferty, F.W. *Proc. Natl. Acad. Sci.* **1995,** *92*, 2318-2322.
70. Limbach, P.A.; Crain, P.F.; McCloskey, J.A. *J. Am. Soc. Mass Spectrom.* **1995,** *6*, 27-39.

71. Chen, R.; Cheng, X.; Mitchell, D.W.; Hofstadler, S.A.; Wu, Q.; Rockwood, A.L.; Sherman, M.G.; Smith, R.D. *Anal. Chem.* **1995**, *67*, 1159-1163.
72. *DNA in the Courtroom, A Trial Watcher's Guide*; Coleman, H.; Sewnson, E., Genelex Corp., Seattle, WA 1994.
73. Cohen, J. *Science*, **1995**, *268*, 22-23.
74. Doktycz, M.J.; Hurst, G.B.; Habibi-Goudarzi, S.; McLuckey, S.A.; Tang, K.; Chen, C.H.; Uziel, M.; Jacobson, K.B.; Woychik, R.P.; Buchanan, M.V. *Anal. Biochem.* **1995**, in press.
75. Borman, S. *Chem. Eng. News*, **1995**, *73*, 23-32.
76. Siuzdak, G. *Proc. Natl. Acad. Sci,* **1994,** *91,* 11290-11297.

RECEIVED November 9, 1995

Chapter 2

Analytical Characteristics of the Electrospray Ionization Process

Thomas Covey

Perkin Elmer-Sciex, 71 Four Valley Drive, Concord, Ontario L4K 4V8, Canada

Electrospray mass spectrometry has gained prominence as the most versatile detector for high performance liquid chromatography and capillary electrophoresis. Key issues of primary concern when considering this technique for any particular application are sensitivity, dynamic range, mobile phase flow rates and composition, appropriate HPLC columns and pumps, stream splitting, concentration sensitivity, and types of molecules amenable to ionization. Particular attention is given to electrospray ionizer designs operational from the nanoliter/minute to mL/min flow range. Applications demonstrating the general applicability of the technique span from highly acidic and basic biopolymers to non-polar neutral species and inorganic ions.

Outline

A. Introduction

B. Flow Rates
1. Low Flows-nanoliters/min
2. Medium Flows-microliters/min
3. High Flows-milliliters/min
4. Post Column Stream Splitting & Concentration vs Mass Sensitivity

C. Sensitivity, Linearity & Dynamic Range

D. Molecules Amenable to Ionization
1. Species Charged in Solution
2. Polar Neutral Species
3. Non-Polar Neutral Species
4. Inorganic anions and cations

E. Conclusions

0097–6156/95/0619–0021$16.75/0

A. Introduction

Electrospray Ionization (ESI) is the most versatile ionization technique in existence today. The only absolute prerequisite for ionization is that the analytes of interest be soluble in some solvent. Since the commercial inception of ESI in 1989 as a device for coupling liquid sample introduction techniques to mass spectrometry (MS), its use has quickly become widespread and is now considered the most generally useful ionization technique in mass spectrometry. The utility of ESI-MS as a sensitive, specific, and versatile detector for liquid sample introduction techniques has launched it into prominence as one of the most important detectors for high pressure liquid chromatography (HPLC) and capillary zone electrophoresis (CZE).

A very large amount of work has gone into understanding the fundamentals of the ESI ionization process, optimization of analytical methods utilizing ESI, and the design and development of instrumentation. It is the purpose of this chapter to focus on the second category, the analytical characteristics of ESI which are important to know when considering solutions to any particular problem. The versatility of the technique presents one with a vast array of analytical scenarios each of which has its own particular requirements, however, there are some basic principles which are of common concern to nearly all analytical situations utilizing ESI-MS. These are sensitivity, dynamic range, mobile phase flow rates and composition, and types of molecules amenable to ESI ionization. To a large degree these variables are a function of the sample introduction and chromatographic tools employed. They also are affected by elements of the ESP ionizer and ion source design which will be discussed. This chapter is not meant to be a literature review, however many of the cited articles can be considered landmark references.

Discussion of the fundamentals of the ionization process is left to one of the original pioneers of ESI-MS, John Fenn (chapter 3). A tremendous amount of work has occurred in the area of ion optics and vacuum system design over the past few years (1-3). For the purpose of this overview suffice it to say that electrospray instrumentation is currently entering into what may be referred to as second generation systems. Mass spectrometers designed from the ground up, with the expressed purpose of optimizing performance with ESI ionization, are supplanting instruments that were modifications of previous systems, whose original intent was to provide optimal performance for other forms of ionization. The vast majority of the systems in place today are quadrupole based mass analyzers which have proven to be robust and productive. This is partly because of the natural compatibility the quadrupole system has with ESI. Vacuum requirements are relatively lax and accelerating voltages are low making the coupling of the atmospheric ion source to the quadrupole mass analyzer a simpler task. However, the coupling of virtually every other mass analyzer has been shown by independent researchers and commercial instrumentation companies. Every type of analyzer brings with it distinct advantages and disadvantages which must be carefully considered in light of the type of analytical problems to be solved.

B. Flow Rates

The effects of liquid flow, and thus HPLC column diameter on the performance of ESI-MS, is the single most important consideration when approaching solutions to analytical problems with ESI. ESI is commonly viewed as a liquid phase ionization technique that requires very low liquid flow rates and thus very small diameter HPLC columns. Embedded in this view is an element of truth, but mostly misperception. It is true that the highest absolute sensitivity (greatest signal for a fixed sample amount injected) is obtained with the lowest possible flows and column diameters. This is a consequence of both the increasing concentration of eluting peaks from smaller column diameters and an increased efficiency of the ionization process at the lower liquid flows.

However, with the proper technique and hardware any column diameter can be coupled to ESI. The decreasing ionization efficiency at the higher flows can be countered to some degree by measures that enhance the spray process and increase the rate of solvent evaporation from the highly charged droplets. This includes the use of pneumatic nebulizers and thermal input as discussed below. The decreasing concentration of eluting peaks in larger columns is an unalterable reality which will decrease sensitivity. In general it can be said that to maximize the overall practicality of an analytical method based on ESI-MS detection, the largest possible column diameter is chosen that provides sufficient sensitivity to solve the problem at hand. The practicality of the technique is primarily related to the difficulty of the chromatographic separation step and has far less to do with complications in the operation of the ESI ion source at different flow rates. The proper choice of column diameter, mobile phase flow, and thus HPLC pump hardware depends on the characteristics of the samples and analytical problem to be solved combined with a knowledge of the relationship between ESI sensitivity and these chromatographic considerations. It should be noted that flow injection experiments are a poor way to obtain an estimate of the relative sensitivities of different columns at various flows because they totally discount the effects of sample dilution (or concentration) as it passes through columns of different diameters.

As a general starting point Figure 1 shows the differences in concentration of a fixed amount of analyte eluting from different column diameters operating at their respective optimum flow rates. Superimposed is the relative column capacities of the different columns. The predicted sensitivity difference between the 4.6mm and .18mm column, considering only the difference in sample concentration as it elutes from the column, is approximately a factor of 600 (falls with the square of the column diameter). In this regard, electrospray behaves in a fashion similar to what one would observe with a UV detector having a zero dead volume flow cell and optimum light path length (practically not achievable). When the amount of sample is limited, and sensitivity is of primary concern, it is logical to use as small a column diameter as practical. Sequencing proteins isolated from 2-D gels is a good example of this where the use of 1 mm and packed capillary columns operating in the sub-μL/min flow range is the norm. However, column loading capacity increases with increasing column diameter in a fashion similar to the sensitivity increases observed with decreasing column diameter (Figure 1). Compensation for the signal loss incurred with larger column diameters is naturally achieved by simply injecting more sample, when sample is not limited. For applications such as quality control and fermentation monitoring, the logical choice is standard 4.6 mm columns operating at ≈ 1 mL/min flow rates. Contaminants at concentration levels of .01% of the major component are readily observed by simply injecting more sample. In the field of qualitative drug metabolite identification, high through-put quantitation of plasma drug levels for pharmacokinetic purposes, and environmental analyses both sensitivity and column loading are important. A compromise is often struck in these areas with the use of intermediate size 2.1 mm columns operating between 100 - 300 μl/min liquid flows.

In order to maximize the ionization efficiency for the full range of mobile phase flows, an ESI ion source needs to accommodate a range of 10^6, from nanoliter to mL/min. To do this and maintain optimum performance across the entire range, different approaches have been used for the sprayer design. Flow rates can be roughly divided into three ranges each of which has its own special considerations and will be discussed separately. These ranges are: a) nanoliter to 1 μL/min, b) 1 μL/min to ≈200 μL/min, and c) 200 μL to 1 mL/min. The issue of post-column flow splitting will be

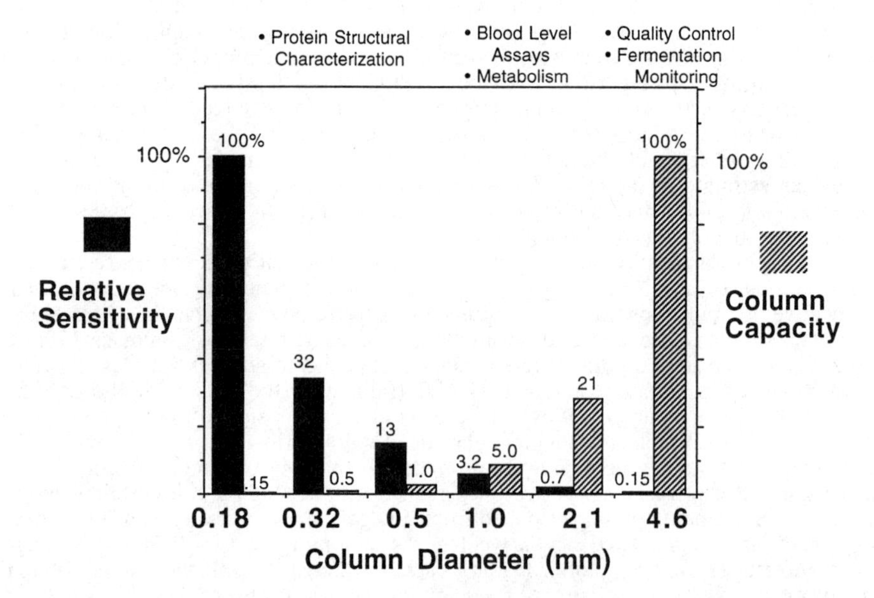

Figure 1. Plot of the relative sensitivity vs column internal diameter (left axis) and relative column capacity vs. column internal diameter (right axis). Relative sensitivities are calculated based on the difference in concentration of an eluting peak from each column. Relative column capacities are calculated based on the differences in cross sectional area of the columns.

elaborated on separately as an alternative means of accommodating high mobile phase flow rates and as a means for utilizing alternative detectors in parallel with the mass spectrometer. Flow splitting highlights the issue of concentration versus mass sensitivity which is a central issue regarding the observed behavior of the electrospray response.

During the formation of ions by the ion evaporation mechanism there are four distinct steps that occur in succession: 1. liquid charging, 2. droplet formation, 3. droplet evaporation and 4. ion emission from the liquid. Of these four steps the most important ones to consider when optimizing an electrospray device for a particular flow rate range are the droplet formation step and the droplet evaporation step. Charging of the liquid occurs quite naturally by simply applying a high voltage to the liquid through an appropriate electrical contact such as a metalized tube. Provided the charge polarity and solution chemistry is correct for a given analyte ion, emission of the ions from the droplets is merely a consequence of doing the first three steps properly. To obtain optimum sensitivity across the nanoliter to milliliter range different techniques are used to optimize the droplet formation and evaporation processes which will be discussed below. Figure 2a-c shows three variations of the electrospray nebulizer modified to maximize performance in three flow ranges spanning the nanoliter to milliliter/min range.

1. Low Flow Range-1 nL to < 1µL/min. The most recent advances in ESP ionization technology have occurred in this area. Production of stable signals at these flows is of interest for purposes of coupling to micro-chromatographic and electrophoretic techniques such as packed capillary HPLC and CZE, thereby obtaining sensitivity gains from the very high concentration eluting peaks from these systems (4-7). The interest in low flow rate is also being driven by the ability to achieve long acquisition times on small amounts of samples without the use of chromatography (8). Substantial enhancements in sensitivity are observed by taking advantage of signal averaging techniques over long periods of time. In the field of peptide chemistry femtomole to attomole sensitivities have been reported (4,8-10).

To achieve stable ESI flows in the nanoliter flow range small sprayer apertures are required. The sprayer diameter establishes the optimum flow rate range, lower flow rates requiring smaller diameters. When the linear velocity of liquid through the sprayer is too high large coarse droplets will emerge. Taken to the extreme, a constant stream of liquid will emerge with no droplet formation. When the linear velocity of liquid through the sprayer is too low for a particular diameter aperture a stable Taylor cone required to produce a constant and uniform stream of charged droplets cannot be maintained and unstable signals result.

One embodiment of this approach is to use a "pure" ESI device (no pneumatic nebulization) with no liquid being actively delivered by a separate pumping mechanism (Figure 3) (8). Finely drawn borosilicate glass pipettes with tip apertures on the order of 1-3 microns are used with gold vapor plated on the surface to obtain an electrical contact. Samples of ≈ 1µL volume are deposited into these nanosprayers and the tips are inserted into the electrospray source. When the voltage is applied the liquid is drawn from the tip through a Taylor cone due to the electrostatic forces from the applied voltage. Through this mechanism the liquid flows unassisted by pumps at a rate regulated by the tip diameter, solvent viscosity, surface tension, and electrolyte concentration, and falls generally around 25 nL/min. The name NanoES has been recently coined to describe these devices (9,10). Solvents containing 100% H_2O have not been problematic with these sprayers unlike spraying with unassisted electrospray at the µL/min flows using larger diameter tips. Extremely efficient sample utilization can be achieved with these nanosprayers. In a typical analytical scenario one microliter of sample can be loaded into such a capillary tip and sprayed for

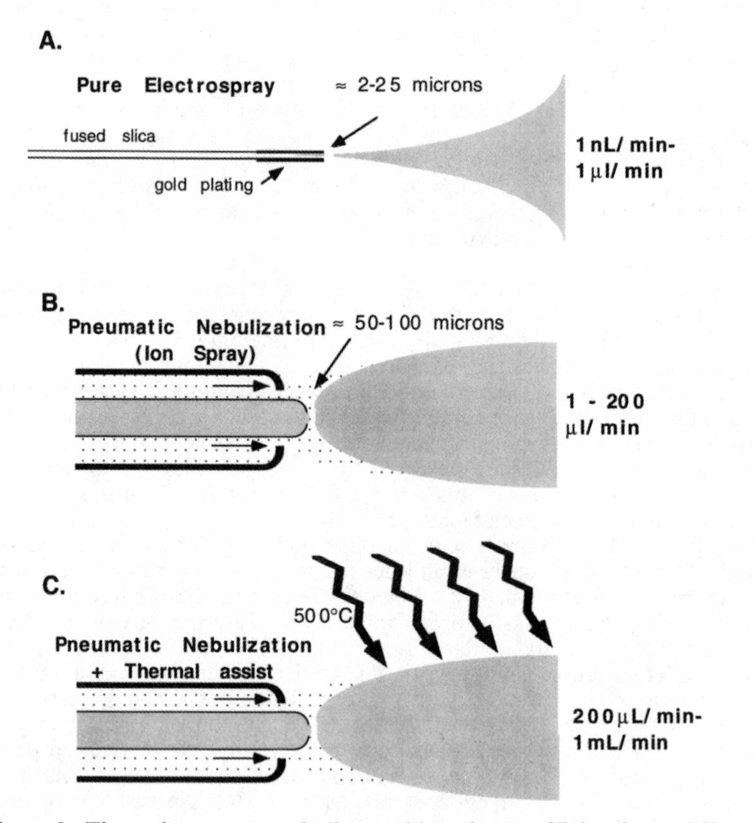

Figure 2. Three electrospray nebulizers with optimum efficiencies at different flow rates.

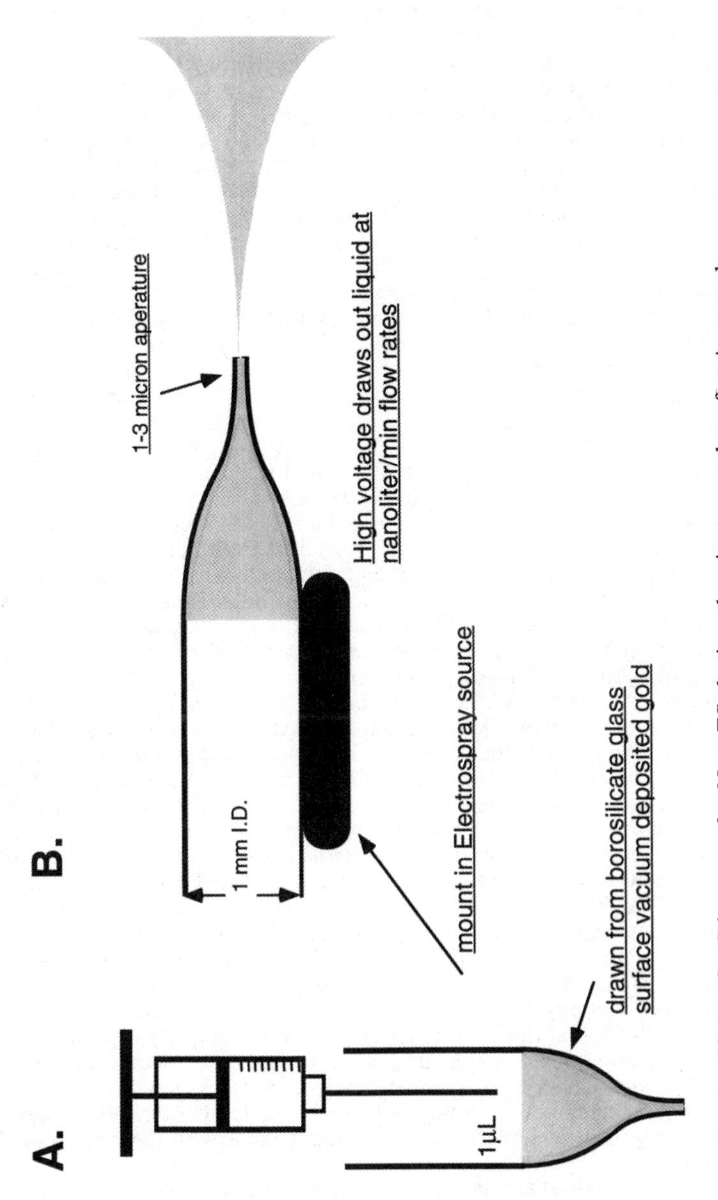

Figure 3. Diagram of a NanoES device showing procedure for A. sample loading and B. spraying.

approximately 26 minutes (Figure 4a). The technique is of particular advantage where scan averaging can be employed over long periods of time to obtain significant signal to noise enhancement. When background chemical noise is substantially reduced by using MS/MS techniques, long acquisition times can be advantageous in maximizing the signal to noise of fragment ion spectra.

With such small apertures (1-3 μ) one can expect clogging to be a problem. However the nanosprayers described above are surprisingly robust and plug relatively infrequently. This can be attributed to the fact that liquid is not being forced through the tip by an external pump. The slow drawing action of the electrospray process will allow particles to lodge loosely in the taper of the tip but seldom do they jam tightly enough to stop the flow. Of course samples must be reasonably clean but, as an example, procedures have been developed for digesting and extracting proteins from 2-D polyacrylamide gels and analyzing the extracts without plugging problems (9,10). Filling these tips with chromatographic packing material, in an attempt to accomplish "in situ" sample clean-up, can lead to plugging problems.

Figure 5a-c is an example of the advantages such long acquisition times can afford. A full scan of a dilute peptide solution shows fairly low signal to noise above the chemical background (Figure 5a). Because of the chemical background slow scanning or signal averaging will not improve the signal to noise. Signal averaging was used to improve the signal to noise of a precursor (parent) ion scan for the m/z 175 fragment (Figure 5b). Tryptic peptides all have C-terminal residues consisting of lysine or arginine which give characteristic immonium fragment ions at 145 and 175, respectively. Once the tryptic molecular ion is confirmed with reasonable confidence from the precursor ion scan an additional 75 nanoliters of the sample was used to sum several full scan product ion spectra (Figure 5c). The entire three step experiment requires ≈ 3.7 fmoles, or 158 nanoliters in approximately 5 minutes, and sufficient sample remains to acquire successive product ion scans on several other components. In this case the sample was a single component but, with samples sufficiently de-buffered, the above approach could be used to obtain product ion spectra on many of the components of entire tryptic digests of proteins.

For purposes of packed capillary HPLC and CZE it is usually beneficial to use reduced diameter sprayers whose flow rate optimum matches the flow being delivered from the column. Using fused silica tubing with inside diameters between 2-100 μ, a transfer line from the column to the sprayer can be chosen whose stable ESP flow characteristics are correct for the mobile phase flow. The end of the transfer line (or CZE column) becomes the sprayer tip by depositing a stabilized gold coating for electrical contact (5) as shown in Figure 2a. Alternatively the voltage can be applied upstream of the sprayer tip at a metal junction between the fused silica transfer line and the fused silica used for the sprayer aperture (4). A 100μ ID sprayer tube gives optimum stability and sensitivity around 4 μL/min, 20μID around 200nL, 10μ around 50nL/min, and 5μ around 10nl/min. Depending on the mobile phase conditions the use of a nebulization gas may be beneficial. Several groups have used this tip metalization approach for direct coupling of CZE to ESP (5, 7, 11,12). It also serves well as a discrete sample introduction device without on-line chromatographic separation in a fashion similar to that outlined with the μSprayer. In this case however the sample is pumped to the sprayer with a syringe pump that will deliver these low flow rates. With forced flow, plugging and back-pressure can be problematic. For this reason the 20 μ ID sprayers have gained acceptance as the smallest size with the most practical implementation.

It is important to understand the nature of the sensitivity enhancements observed when using these reduced aperture sprayers. The absolute signal that is obtained from any particular concentration of analyte is roughly equivalent in the various flow regimes. For instance, when one is operating with a few micron diameter aperture at

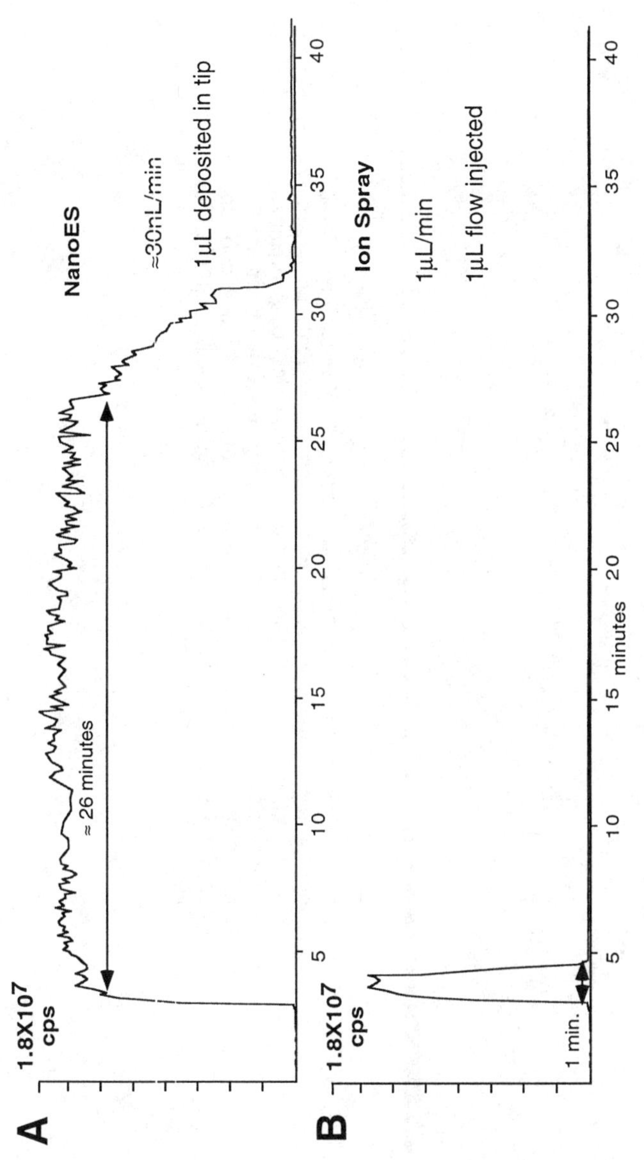

Figure 4. Comparison of signal from 1 μL of a 1 pmole/μl sample of glufibrinopeptide B with two different electrospray introduction techniques. A. One μL deposited into a NanoES (Figure 2) and allowed to electrospray at its own natural flow rate. B. One μl flow injected into a pumped stream at 1μL/min to a Ion Spray nebulizer (Figure 2b)

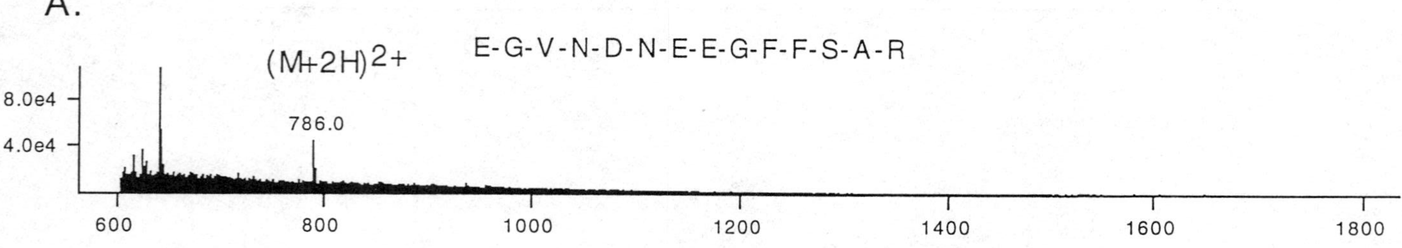

Figure 5. A) A single 15 second mass spectrum of a 25fmole/µl solution of glufibrinopeptide B from the NanoES devise. Flow rate ≈ 30nL/min. Of the one microliter loaded 7nL was consumed or 175 attomoles. B) A precursor ion scan for m/z 175 on the same sample scan averaging for 2.5 minutes (10 scans) and consuming 75 nL or 1.8 fmole. C) Full product ion scan on the same sample with averaging for 2.5 minutes (10 scans) consuming 75 nL or 1.8 fmole.

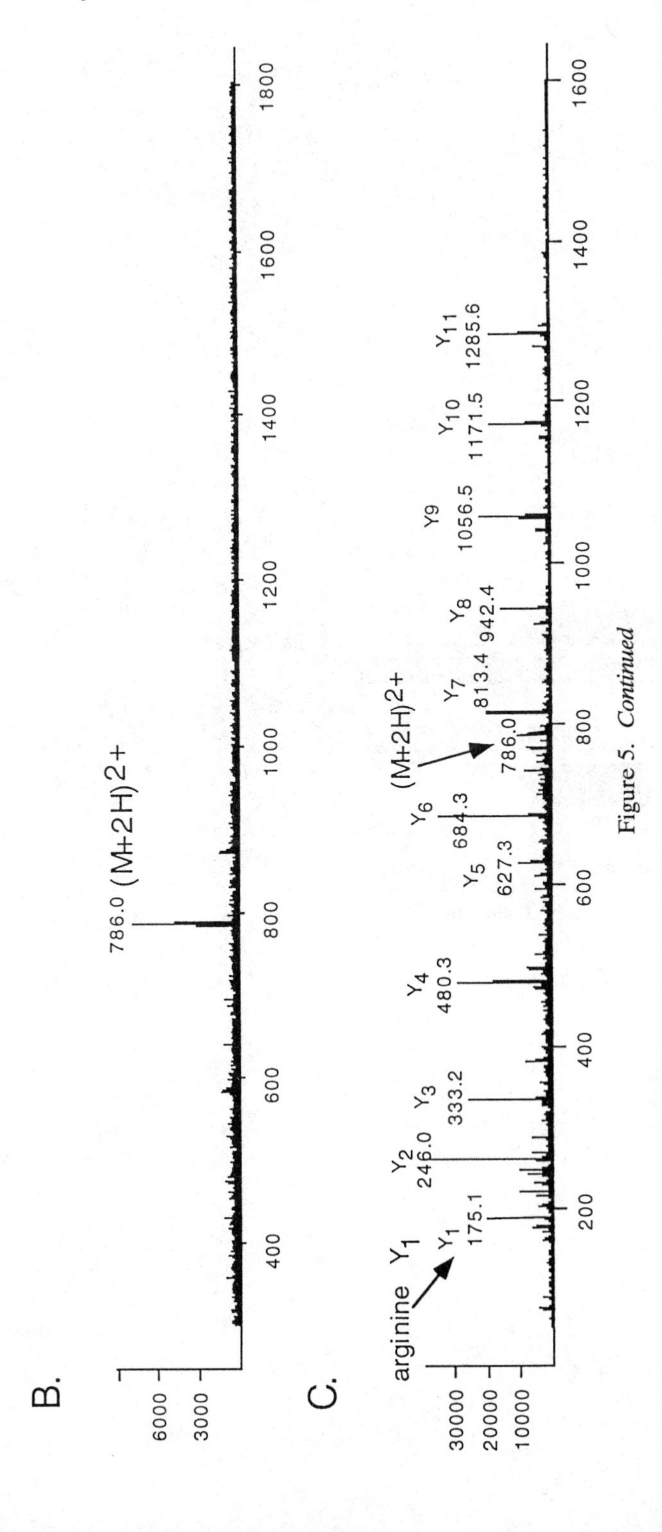

Figure 5. *Continued*

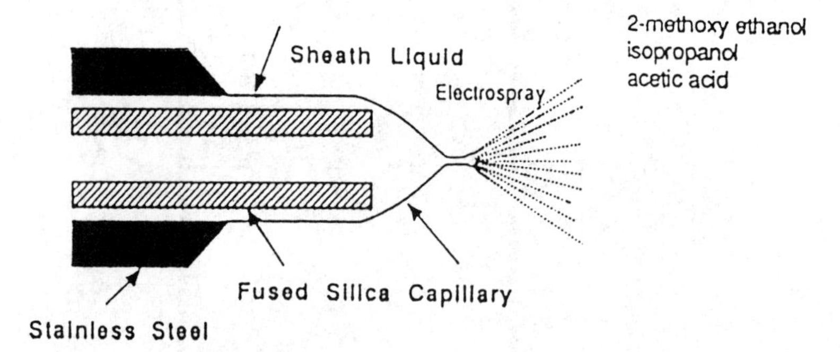

Figure 6. Diagram of the co-axial flow sprayer configuration.

nanoliter flows the signal obtained in each spectrum is similar to that when operating at a few μL/min flows with more conventional sprayers of ≈ 100 micron diameter. Figure 4 illustrates this point. The ion current trace from a 1 μL sample in a NanoES tip is shown to last ≈ 26 minutes in Fig 4a. A sample injected into a conventional pneumatically assisted ESI source (Ion Spray) at a flow of 1μL/min lasts for ≈ one minute, but the same signal is obtained in any individual spectrum (Figure 4b). Clearly the efficiency of utilization of the analyte molecules is ≈ 20X greater with the nanoES tip. With the same number of molecules dissolved in the 1 μL analyzed, a factor of 26 more of them were detected utilizing the nanosprayer versus the Ion Spray. Several explanations for this phenomena have been put forth and are presented in the section on mass versus concentration sensitivity.

An alternate method of coupling chromatographic separations requiring sub μL/min flows to ESI has been to add a co-axial make-up flow of solvent to raise the flow into a regime where more standard pneumatically assisted ESI works well, i.e. in the low μL/min range. Figure 6 is a diagram of the co-axial approach. An abundance of applications using this approach for coupling CZE and packed capillary LC requiring sub-μL/min flows have been published (13,14). One advantage of this approach is that non-ideal mobile phases containing non-volatile buffers and high aqueous content can be diluted with more favorable spray solvents. The sensitivity is somewhat compromised by the dilution of the sample with the make-up flow and for this reason the majority of the sub-μL flow chromatographic applications are being supplanted by the reduced spray aperture approach discussed above.

A separate but important issue involves the delivery of sub-μL/min flows to capillary HPLC columns with reproducible gradients. Current HPLC pump technology has not been designed with these flow rates in mind. The most popular approach is to form the gradients at a standard flow rate and split off the desired flow before the injector using a balanced backpressure (15,16).

2. Intermediate Flow Range 1μL-200μL. It was recognized early on that, for efficient coupling to liquid chromatographic techniques requiring in excess of 1μL/min, great benefit could be derived by separating the droplet charging process from the droplet spraying process (19). Utilizing the electric field alone to both disperse the liquid into a spray and charge the droplets did not prove sufficiently reliable for routine operation at these higher flows, especially if solvents with high aqueous concentrations were involved. Another form of energy input was required to stabilize the spraying process. The pneumatically assisted ESI evolved from a combination of earlier pure electrospray devices (17) and ion evaporation devices (18) and thus the term "Ion Spray" was coined (19). By mechanically shearing the liquid at the sprayer tip, pneumatic nebulizers have proven to effectively generate small droplets from high flows and high aqueous solutions without adjustment to account for changing mobile phase condition. Over the years pneumatic nebulization has proven to be the most simple and practical approach to solving this problem.

The vast majority of the applications involving electrospray have been done using pneumatically assisted devices. Columns from 0.5 to 2 mm diameter offer high analyte concentrations without excessively sacrificing the ruggedness and reliability of the chromatographic system. Where both high sensitivity and good column loading are required, such as in qualitative and quantitative drug metabolism, 1.0 and especially 2.1 mm columns are most frequently used. Assays quantitating plasma levels of pharmaceutical compounds to the low pg/mL levels are being done by many groups and the low ng/mL range is relatively routine. Tandem mass spectrometry (MS/MS) techniques employing multiple reaction monitoring (MRM) have gained widespread acceptance as the method of choice for high throughput quantitative analyses because of the exceptional selectivity. In an assay developed for FK506, the ammoniated

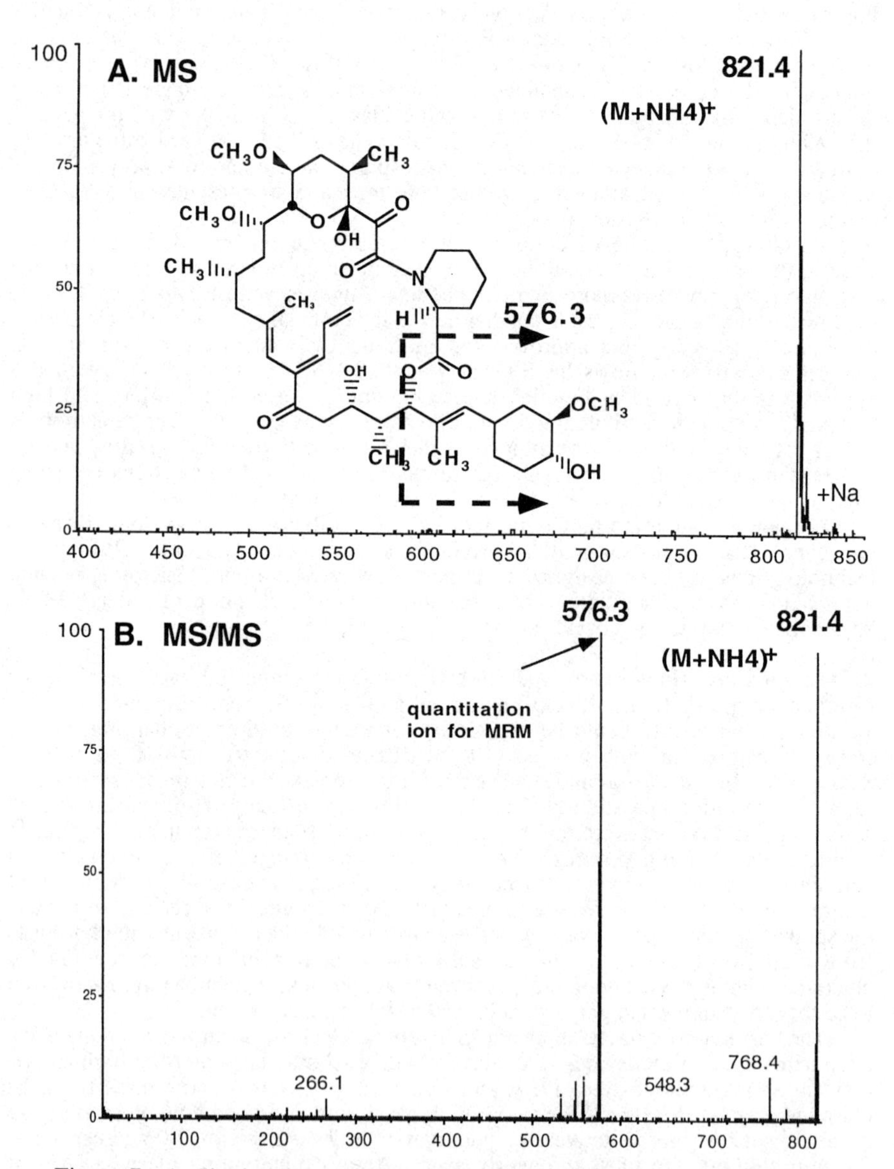

Figure 7. A. Single MS spectrum of FK 506 dissolved in 20mM NH_4OAc. B. Product ion spectrum of ammoniated molecular ion with 20 eV collision energy (argon).

molecular ion (Figure 7a) is used as the parent with a fragment at m/z 576 (Figure 7b) serving as the quantitation product ion. Virtually no background signal is observed at the 50 pg/mL level using MRM and fast 2.1 mm chromatography with retention times ≈ 2 minutes per sample (Figure 8). Chemical analogs are frequently used as internal standards primarily because they are more conveniently procured than stable isotopes and provide acceptable results even when they are chromatographically separated by several minutes. Because of the reliability of intermediate bore columns, sample throughputs greater than 200/day are not uncommonly encountered in the quantitative analysis field. Some laboratories have reported in excess of 600 samples per day (20). In the environmental field the same considerations of striking a balance between column loading, sensitivity, and through-put exist which suggests that this intermediate flow chromatography regime will continue to gain importance as more environmental problems are solved with this technique.

Because of the exquisite selectivity of MRM techniques there is the temptation to reduce the chromatographic run times to a point where almost no chromatographic retention is realized. Limits to the speed are imposed by the phenomena of ionization suppression. Analytes co-eluting with several orders of magnitude difference in concentration will exhibit severe reduction in response. This is presumably due to a competition for desorption sites on the surface of the droplet. Some very useful chromatographic techniques can be employed to minimize this problem. Among them is the use of ballistic gradients which also allow for injection volumes much greater than one would normally expect for small diameter columns. Figure 9 is an example of the implementation of such a gradient with 1 mm columns to produce a 10 pg/mL assay with retention times < 1 minute and a time between injections of 2 minutes. Clearly these types of applications challenge current HPLC pump and autosampler technology and hopefully will provide impetus for development in this area.

Most applications for peptide mapping do not require high throughput or high column loading but high sensitivity is always a concern. Use of 1 mm columns provides sensitivities in the 1-100 pmole range which is sufficient for many applications. Of particular practical value is the ability to post-column split the effluent for the purpose of purification for further analysis. A typical splitting set-up is shown in Figure 10 with the splitter constructed from a standard 1/16" "tee" fitting. It is a relatively simple task to adjust the split ratio using different lengths of 50-100 μ ID fused silica for the split line to the fraction collector and to the sprayer. A flow of 40 μL/min (typical for 1 mm columns) can be easily diverted such that >90% of the sample is conserved (21,22). Because ESI ionization exhibits the peculiar but advantageous phenomenon of concentration sensitivity, no signal is lost in stream splitting situations where the majority of the sample is shunted away from the mass spectrometer. This issue is discussed further under the topic of stream splitting and concentration sensitivity.

Figure 11b is an example of a HPLC/MS total ion current trace (TIC) of a peptide map with 1 mm chromatography of a tryptic digest of a protein isolated from 2-D gel electrophoresis. The UV trace (Figure 11a) was monitored simultaneously with the mass spectrometer on the split line and used as a manual indicator for fraction collection of the individual peaks. Approximately 15 pmoles were injected with ≈ 13.5 pmoles (in ≈36μL) recovered for further analysis by Edman sequencing or MS/MS analysis. A mass spectrum of peak 21 (Figure 12) indicates a mixture of at least 5 co-eluting components which poses a serious problem for Edman sequencing. At this point a single microliter from the 36 μL recovered would be sufficient to obtain MS/MS spectra if it were placed into the nanoSprayer system discussed above.

3. High Flow Range-200μL-2mL/min. In the high flow rate range droplet evaporation rates can seriously limit ionization efficiency and thus sensitivity. The

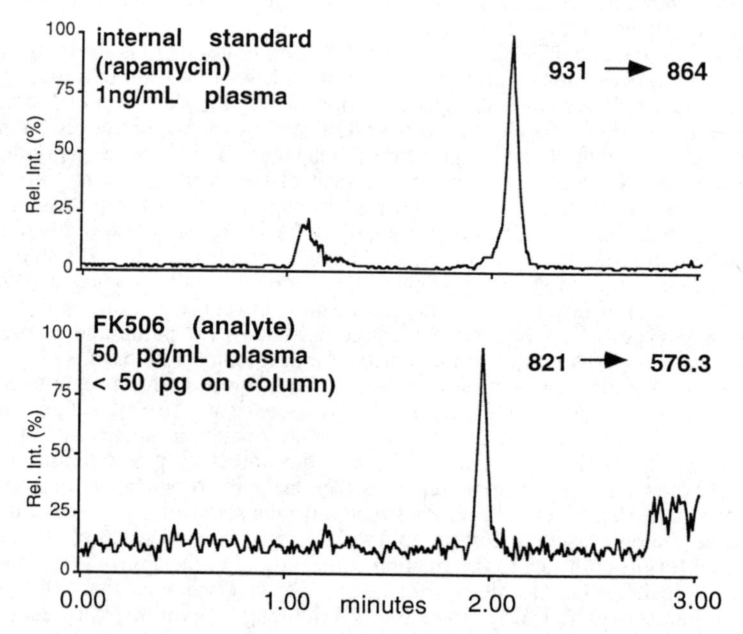

Figure 8. Lower limit of quantitation for FK506 in plasma. Chromatography was done with a 2.1 X 100 mm BDS Hypersil C18 @ 40°C and a flow of 100 μL/min of 80/20 CH_3CN/H_2O 20mM NH_4OAc with no split and a 50 μL injection volume. The assay CV was < 12% from 50pg to 10ng/mL plasma and calibration was by simple linear regression weighted 1/x; intercept =0.0060, slope =0.3731, correlation coefficient. =0.9935.

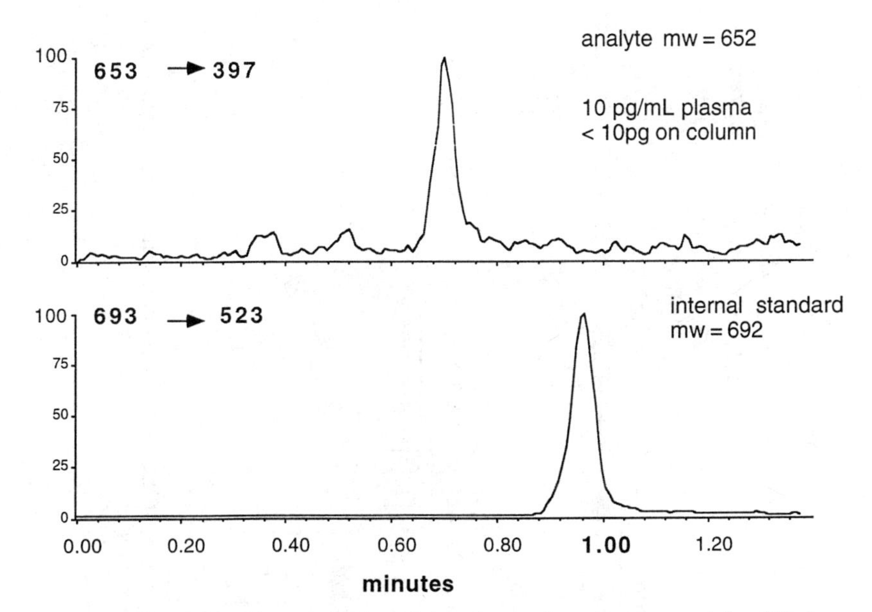

Figure 9. Limit of quantitation for a drug of molecular weight 652 by HPLC/MS/MS MRM using ballistic gradients on short 1 mm columns. Column: 1 X 50 mm Hypersil C18 @ 50μL/min. Injection volume 20 μL. Gradient from 60%-90% CH_3CN in 90 seconds with a re-equilibration time of 30 seconds. Method validated from 10pg/mL to 20 ng/mL plasma, CV = 9.9%.

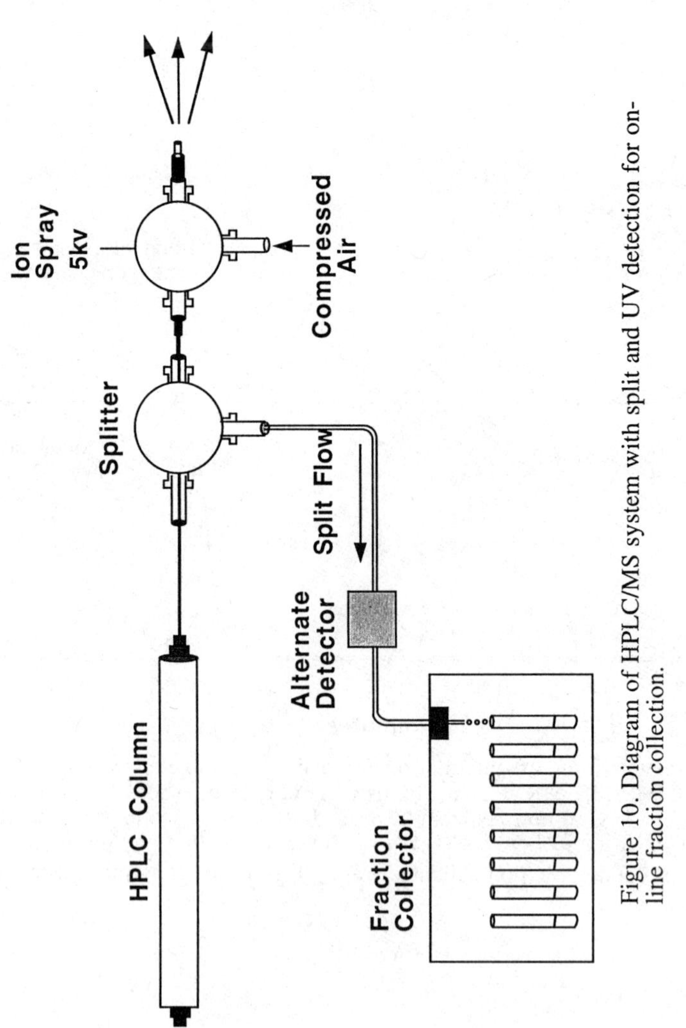

Figure 10. Diagram of HPLC/MS system with split and UV detection for on-line fraction collection.

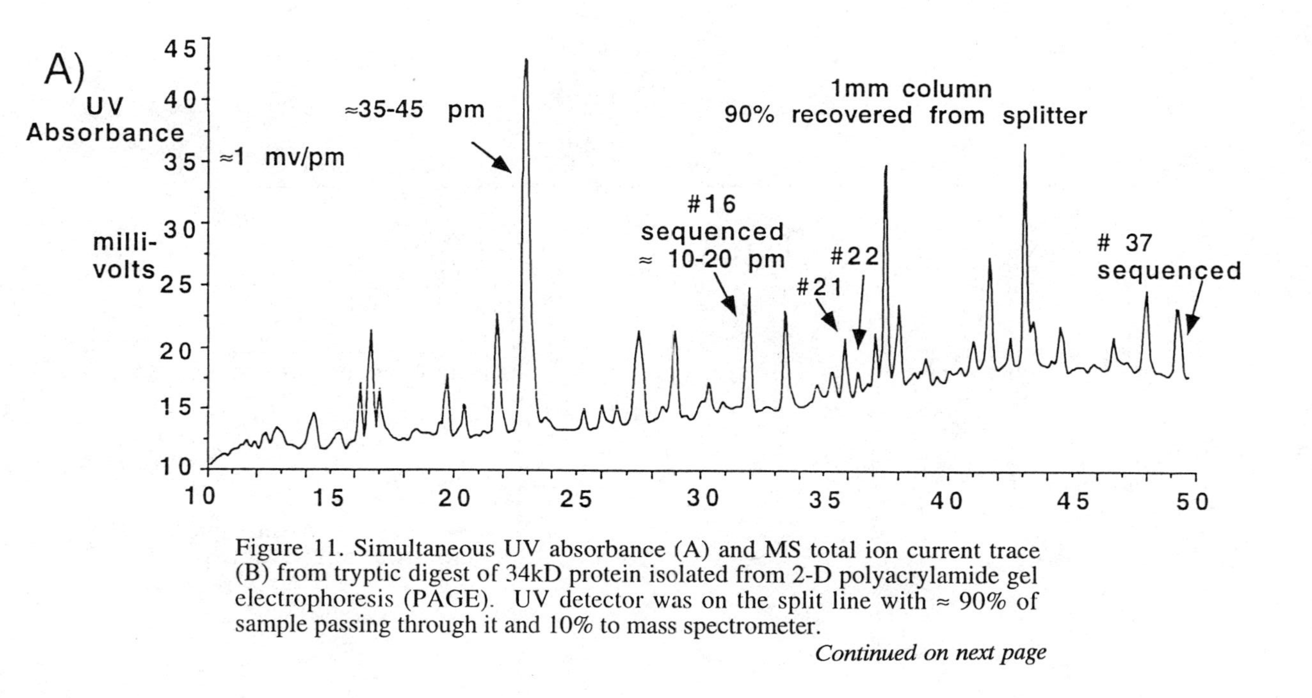

Figure 11. Simultaneous UV absorbance (A) and MS total ion current trace (B) from tryptic digest of 34kD protein isolated from 2-D polyacrylamide gel electrophoresis (PAGE). UV detector was on the split line with ≈ 90% of sample passing through it and 10% to mass spectrometer.

Continued on next page

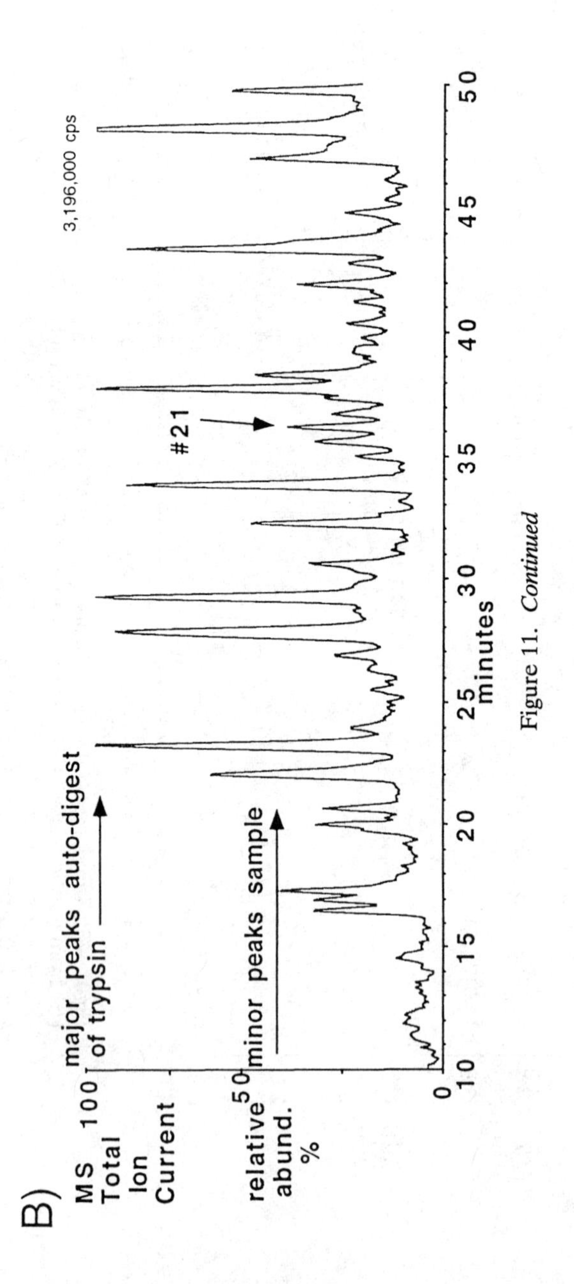

Figure 11. *Continued*

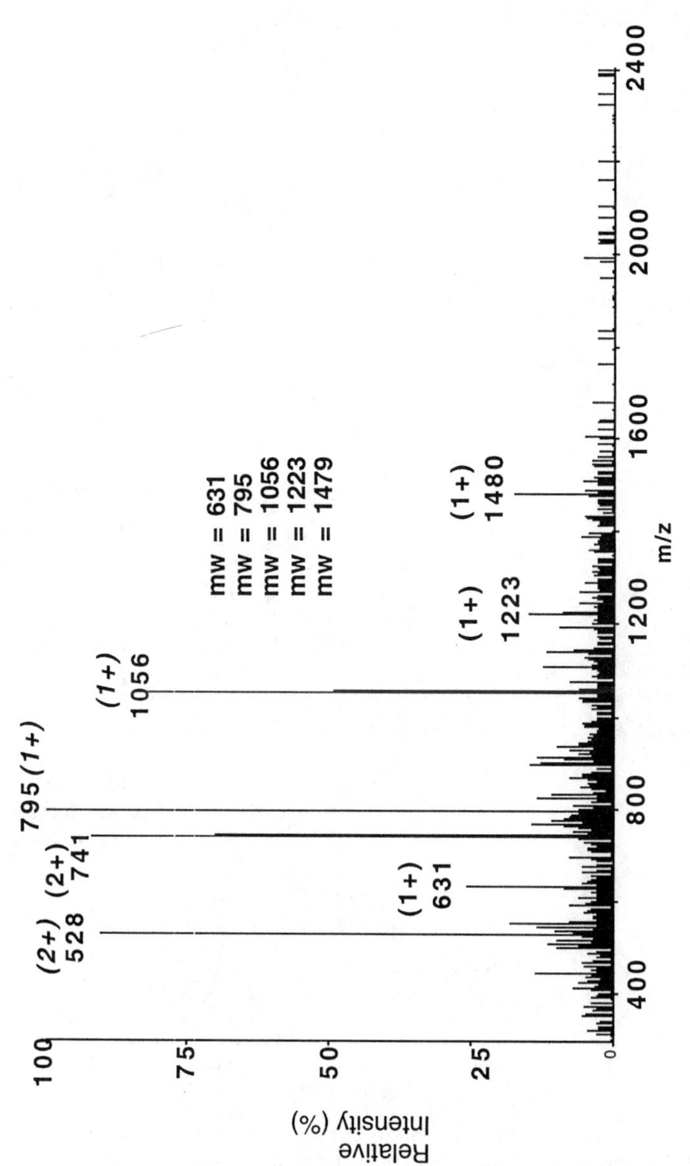

Figure 12. Spectrum from peak 21 in Figure 11 showing the doubly and singly charged ions of 5 components.

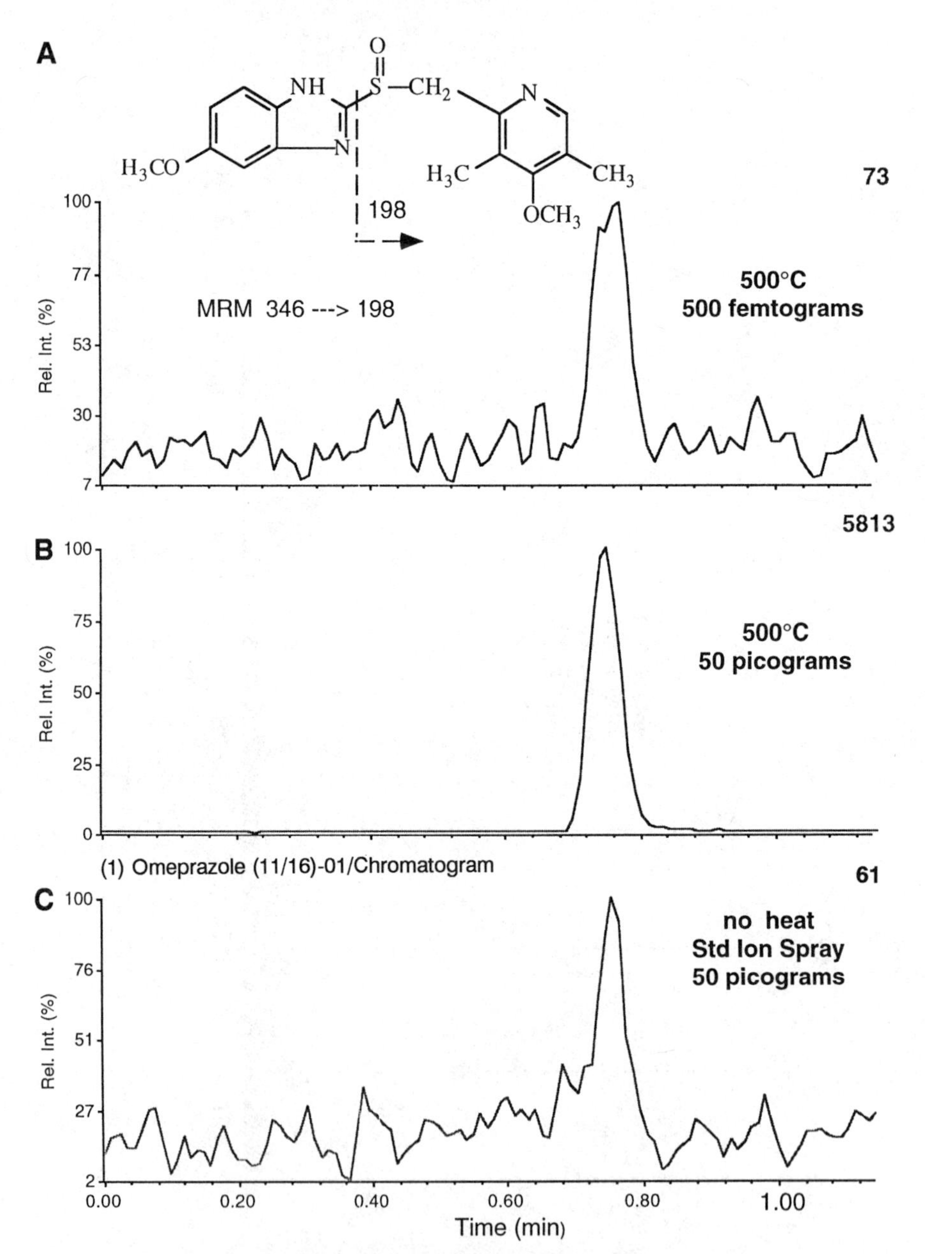

Figure 13. Comparison of omeprazole MRM signal from a 4.6 mm column at 1 mL/min (no split) under different ion source temperatures. A) detection limit with 500°C gas stream injected into the pneumatically nebulized electrospray, B) same as A) with 50 pg injected, C) no heating of the pneumatically nebulized electrospray. Chromatographic conditions: 1/1 CH_3CN/H_2O, 2 mM NH_4OAc, 4.6 X 100 mm C18 column. Peak heights in counts/minute in upper left of chromatograms.

difficulty arises when trying to produce the sub-micron sized droplets required for ion emission from large liquid volumes during the transit time from spray tip to vacuum entrance orifice (less than 1 millisecond). Additional thermal energy input into the sprayed droplets has proven effective for enhancing the evaporation rates (23). Evaporation rates at the lower flows are not limiting so almost no effects are observed at <10 µL/min, and they do not begin to become very significant until the high flow rate range is reached. An example is shown in Figures 13 and 14 with the anti-ulcer drug omeprazole. Using a heated gas stream of 500°C intersecting the spray from a standard pneumatically assisted electrospray device, a signal enhancement of 95X is observed from the compound eluting from a 4.6 mm column at 1 mL/min flow (Figure 13). The identical experiment with a 1 mm column at 40 µL/min shows an enhancement of only 8X (Figure 14). Note that even though the gain is less the sensitivity is still nearly 20X greater due to the higher analyte concentration eluting from the 1 mm column. In addition to flow, solvent composition and analyte polarity can affect the extent of the gain are observed. For instance, high aqueous solvent compositions benefit to a larger degree from the injection of heat than do high organic containing mobile phases. An assay for the anti-tissue rejection drug cyclosporine (Figure 15) with 2.1 mm columns showed a sensitivity increase of ≈ 4X at elevated temperatures when eluting in a mobile phase containing ≈ 80% acetonitrile (Figure 16 A&B).

Different commercial approaches have been employed for heating the sprayed droplets including; 1. injecting a heated gas stream into the spray, 2. heating the droplets as they are inhaled into the vacuum through a long hot steel capillary, and 3. by spraying the droplets through heated channels in the atmospheric region of the ion source. The same principles apply for all of them except that temperature parameters are adjusted in different fashions. In order to rapidly desolvate, significant amounts of heat are required where local temperatures around the droplets need to exceed several hundred degrees centigrade. In such an environment thermal degradation of organic molecules would seem likely. However because of the intense cooling of the liquid that occurs during the solvent evaporation process, droplet temperatures do not rise above ≈ 50°C and thermal degradation of even the most labile compounds has not been observed.

In application areas such as quality control where sample is not limited, larger bore columns are almost exclusively used with ESP. In these cases identifying contaminants of a sample at a concentration of .01% of the major component can be done with standard 4.6 mm conventional chromatography using the approaches outlined above or by simply utilizing post column stream splitting to maintain as high an ionization efficiency as possible.

4. Stream Splitting & Mass vs. Concentration Sensitivity. Any discussion of ESP and the relationship between sensitivity and flow rates must begin with an understanding of the concepts of mass and concentration sensitive detectors. A concentration sensitive detector is one where detector response is a function of the concentration of analyte in the mobile phase. The UV detector is a classical case following the well known Beer Lambert relationship between concentration and absorbance. The smaller the column diameter the higher the sensitivity because the analyte elutes in a more concentrated plug for the total sample amount injected. If the chromatographic peak widths are approximately the same width in time with the different column diameters (theoretically they should be) then the elution volumes will be exponentially smaller for the smaller diameter columns. The concentration increases as an inverse function of the square of the column diameter, thus, a 1 mm column provides peaks that have 16 times greater concentration than a 4 mm column provided that the peak symmetries and widths are identical (25).

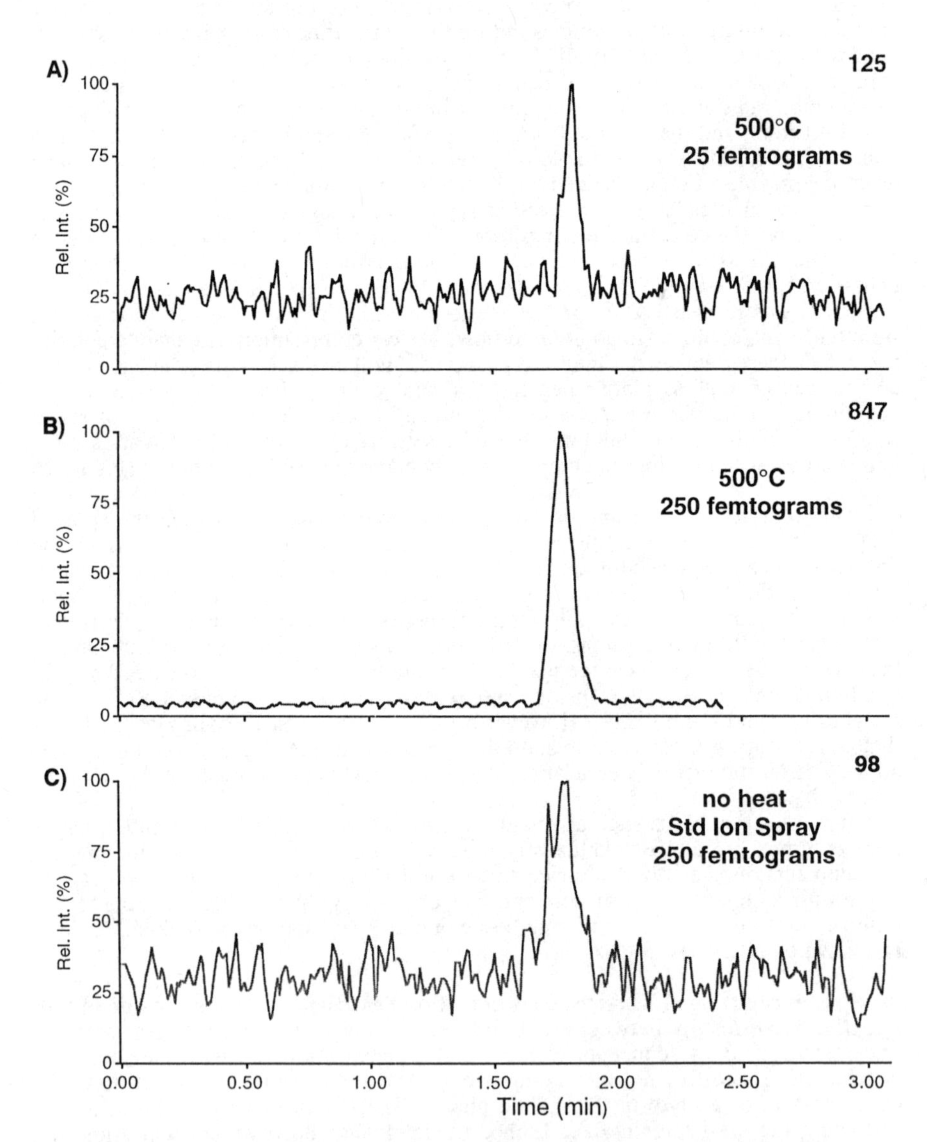

Figure 14. Comparison of omeprazole MRM signal from a 1.0 mm column at 50 μL/min (no split) under different ion source temperatures. A) detection limit with 500°C gas stream injected into the pneumatically nebulized electrospray, B) same as A) with 250 fg injected C) no heating of the pneumatically nebulized electrospray. Chromatographic conditions: 1/1 CH_3CN/H_2O, 2 mM NH_4OAc, 1.0 X 100 mm C18 column. Peak heights in counts/minute in upper left of chromatograms.

Mass sensitive detectors are those whose response is a function of the absolute mass of analyte passing through it irregardless of its concentration in a solvent or gas. Classical mass sensitive detectors are radioisotope detectors and mass spectrometers operating with electron impact or chemical ionization sources. The atmospheric pressure chemical ionization inlet for HPLC coupling to mass spectrometers is another example of a purely mass sensitive detector. With mass sensitive detectors no sensitivity gains are achieved with smaller columns or flows, therefore, they are seldom used with anything but 4.6 mm columns.

Unlike other mass spectrometer ion sources, ESI gives the appearance of behaving like a concentration sensitive device. A dramatic example of this is observed when using a post-column split and diverting a percentage of the column effluent away from the ion source. Figure 17 shows the results of a splitting experiment with a drug of molecular weight 581 (t_r=1.2 min) and an internal standard (t_r=1.7 min) chromatographed through a 2.1 mm column and subjected to various split ratios. Virtually no difference in response is observed even though the amount of analyte entering the source varies by > 25X between the maximum split and a no split condition. This result illustrates the concentration sensitive-like response since the response is not a function of the absolute mass of material entering the source, only the concentration. In this case the organic content of the mobile phase was high assuring good nebulization efficiency across the range of flows introduced into the source and no heat was required to compensate.

A second line of evidence comes from comparing the response obtained from the injection of known amounts of analyte on columns of different diameters run at their respective flows and with ESP ionizers that have been optimized for maximum performance at each respective flow. The sensitivity differences follow what one would expect from a concentration sensitive detector. That is they vary proportional to the square of the column diameter. An example of this is the omeprazole data shown in Figures 13 and 14.

Experiments with the reduced aperture low flow rate systems described above also support the concept of ESI as an ionization process that gives the appearance of concentration-sensitive detection. When a sample is continuously introduced into the mass spectrometer at flows that can differ by > 100X, the signal level remains constant while the length of the available acquisition time increases with decreasing flow.

Two explanations for this apparent concentration sensitivity phenomena exist. They are grounded in the concept of the ionization process as being fundamentally mass sensitive where the apparent concentration-sensitive response is the result of two competing processes which balance each other. No obvious means for distinguishing which explanation is correct has been proposed.

The first envisions a balance between ionization efficiency and mass flux into the electrospray needle. As the flow rate is decreased the efficiency of the ion desorption process increases in direct proportion to the decreased mass flux into the sprayer. In this scenario the rate of ionization increases as the flow is lowered but an increase in signal is not observed because of the proportional drop in mass flux at the lower flow rate. Thus the signal remains constant. The alternative explanation using a mass sensitive model views the ionization efficiency as remaining constant at different flows, but the fraction of the available ions passing through the orifice at the atmospheric to vacuum interface is increasing as the flow is lowered. This physical model involves the repulsion of droplets of like charge. At higher flows the larger number of charged droplets results in a scattering of these ion emission particles into regions far enough away from the atmosphere to vacuum orifice that they don't pass into the mass spectrometer. Again a balance is struck, this time between the efficiency of ion transfer into the MS and mass flux to the sprayer such that little change in signal occurs.

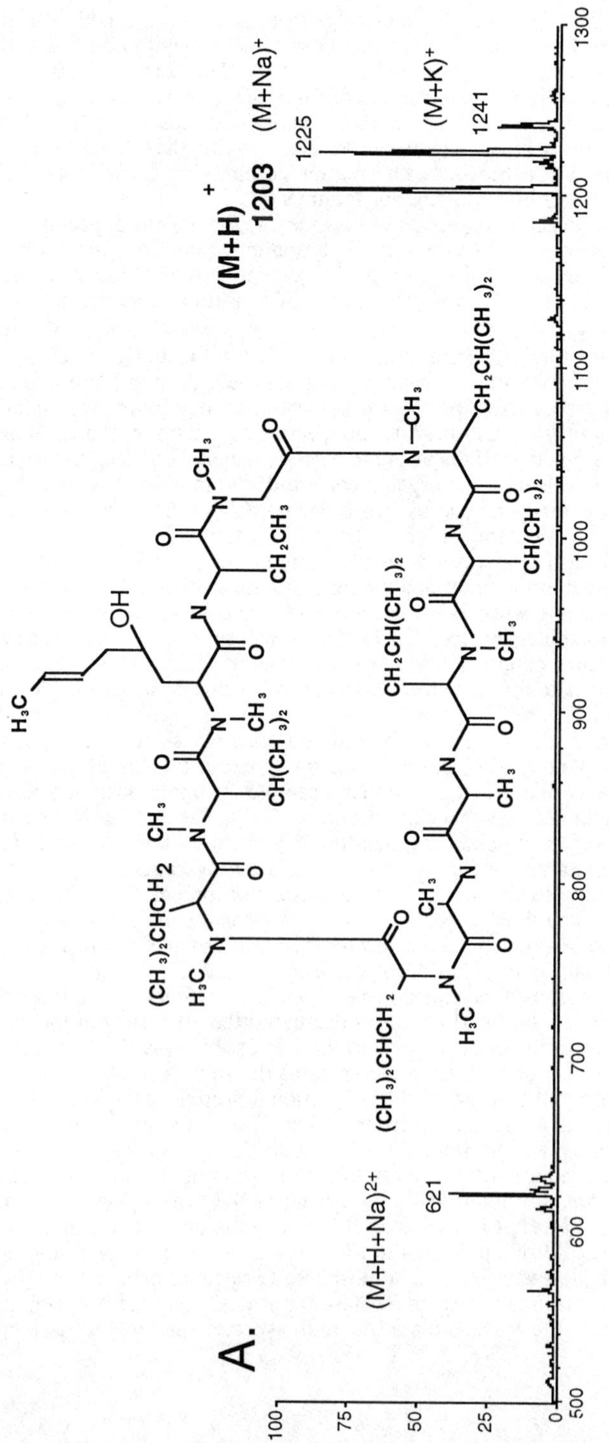

Figure 15. A) Ion Evaporation spectrum of cyclosporine. B) Product ion spectrum of cyclosporine at 40eV collision energy.

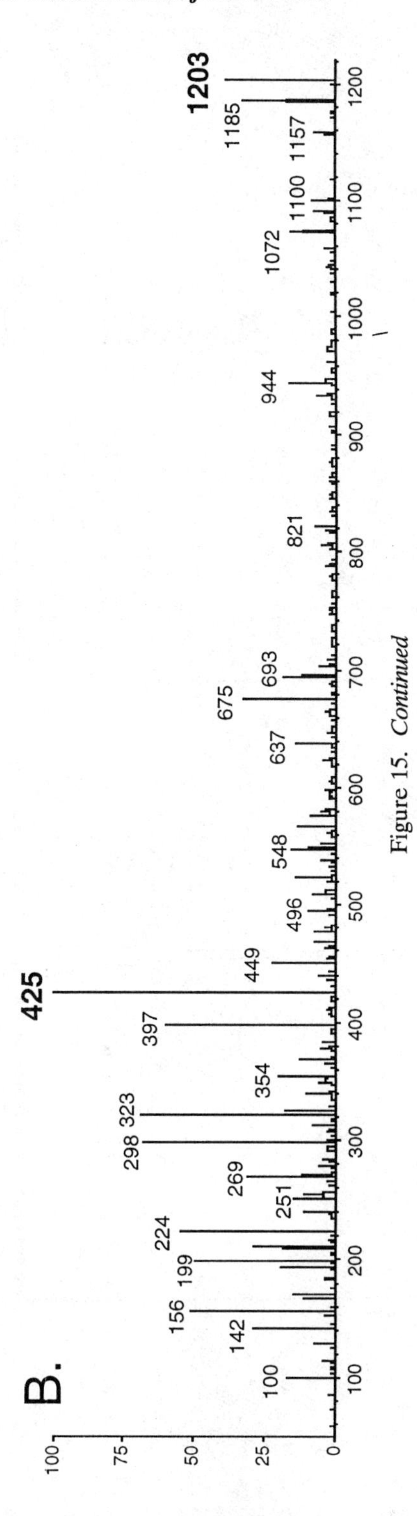

Figure 15. *Continued*

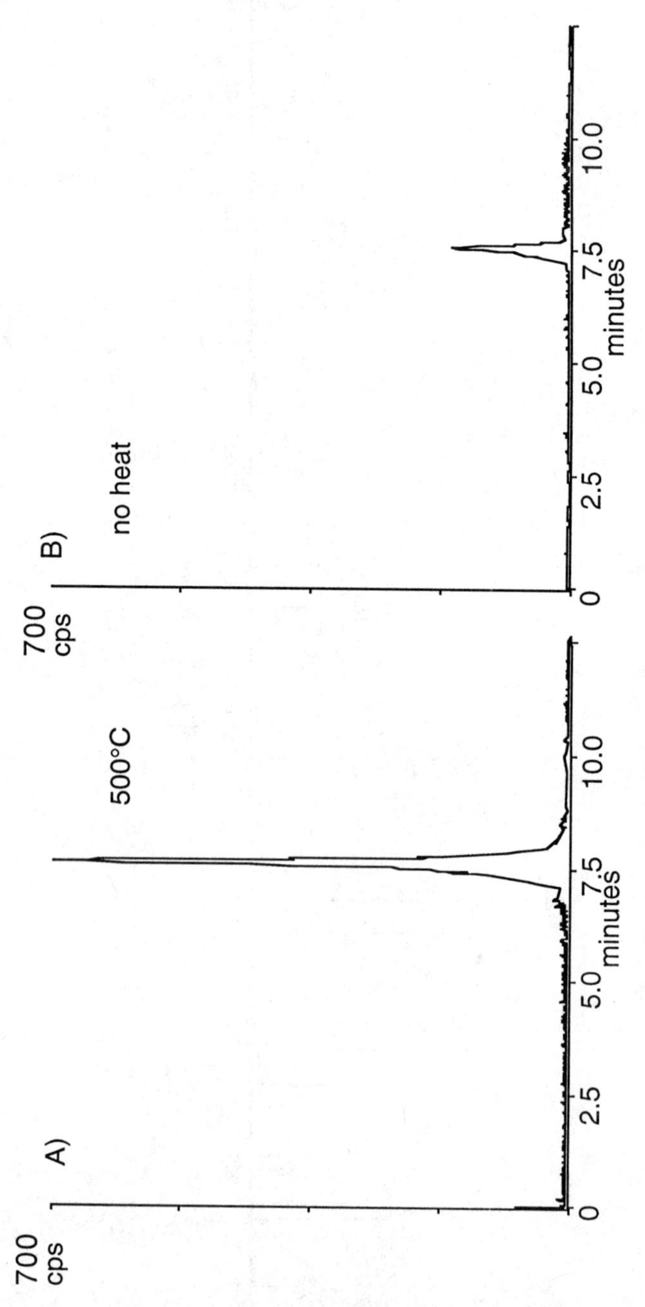

Figure 16. Comparison of MRM signal from 10 pg cyclosporine (A) with and (B) a 500°C orthogonal heat source. Column: 2.1X100mm C18 BDS Hypersil @ 60°C. Gradient: from 50-100% CH_3CN in 10 min. (0.1% formic acid). Ions monitored: 1203--->425.

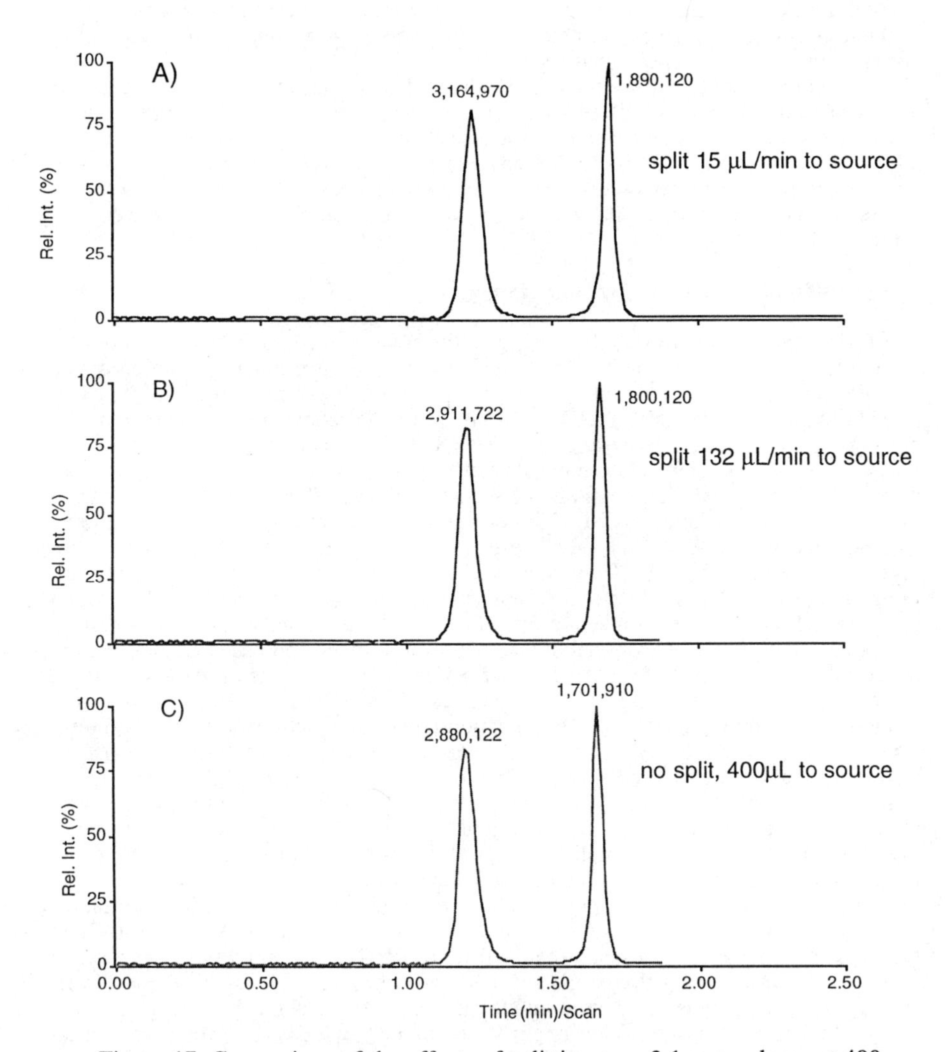

Figure 17. Comparison of the effects of splitting on a 2.1 mm column at 400 μL/min mobile phase, 15/85 H_2O/CH_3CN, 5mM NH_4OAc + 0.2% formic acid. A) 15 μL/min split to ion source, B) 132 μL/min split to ion source, C) 400 μL/min split to ion source. The integrated peak areas are displayed above each peak. The analytes (581 and 662 molecular weight drugs) were monitored by selected ion monitoring.

From an analytical point of view it is essentially irrelevant which model one adapts to explain the observed behavior of electrospray. The important point is that it gives the appearance of a concentration sensitive detector. The implications are 1) smaller columns mean higher absolute sensitivity, 2) lower flows mean longer acquisition times on a given amount of sample, and 3) post column stream splitting does not result in significant signal loss.

Post-column splitters have proven to be reliable, easy to set-up, and do not add to a system's complexity. Good reasons for using splitters are that they offer unique advantages for simultaneous sample purification, allow the simultaneous use of alternative detectors in parallel to the mass spectrometer, and provide lower flows to the sprayer for high ionization efficiency. An additional benefit is that their use keeps the ion source cleaner since only a small fraction of the sample actually enters the ion source.

C. Sensitivity, Linearity & Dynamic Range

ESI is generally considered an exquisitely sensitive technique but, as should be clear from the discussion on flow rate, the sensitivity is very much a function of the column diameter, flow rate, and solvent composition. Examples from the fields of peptide chemistry, and drug metabolism are presented above in the context of the chromatographic system most appropriate to that area. The chemical nature of the analyte may also strongly determine its lower limit of detection. The single most important property of a compound is the surface activity. This logically follows from the proposed mechanisms of ESI which envision it as a surface ionization technique. Compounds that have surface active properties (detergents for instance) give the highest response with ESI. They can also be the most effective at suppressing the ionization of other less surface active molecules, presumably by shielding the droplet exterior. The ability of a molecule to ionize in solution is also important. Acidic compounds are primarily observed in the negative ion mode and bases are observed in the positive mode. Aside from the familiar acid/base mechanisms of ionization, there are a variety of ways a molecule can pick up charge such as adduction to a salt (ammonium, sodium, acetate) or via oxidation/reduction reactions at the metal/liquid interface in the sprayer. Although different chemical structures will give rise to different response factors, compounds in a mixture that are reasonably similar will ionize to a degree that will reflect their relative concentrations. An example of this would be a peptide and its phosphorylated version. Only in cases of extremely divergent chemical properties will the spectrum of a mixture provide a distorted view of the relative concentrations of the individual components.

For the purposes of developing quantitative target analysis with ESI the issue of linearity and dynamic range is of key importance. In the pharmaceutical industry ESI (along side atmospheric pressure chemical ionization) is the quantitative method of choice for characterizing drugs and metabolites in biological samples at low levels (low pg/mL plasma) with very high sample throughput (several hundred/day).

One of the earliest reported applications of ESP with HPLC/MS involved an assessment of the dynamic range for purposes of quantitative determination of sulfate conjugates of drug metabolites in biological samples (26). A linear dynamic range of 2.5 orders of magnitude was reported in those early experiments on sulfate conjugates of anabolic steroids showing a significant leveling of the curve at high concentrations. Since the earliest quantitative work cited, reports of 10^3 to 10^4 dynamic ranges have been regularly reported. The expansion of the range, since the earlier reports, has been achieved in two ways. Fitting the calibration curves to weighted or quadratic models significantly the improves dynamic range of the assay by extending the high concentration end of the curve into non-linear areas. The statistical fit which best mimics the ESP process is the weighted quadratic (1/X). The other way to increase the dynamic range is to decrease the lower limit of quantitation. Assay sensitivities have increased significantly over the past few years due to improvements to the ion source and analyzers of triple quadrupole systems. Although the upper end of the dynamic range appears to be fixed by the fundamentals of the ionization process, lower concentration detection limits have the effect of extending the range.

A detailed examination of the high concentration role-off phenomena was conducted which characterized the upper limit of the sample ion signal observed at approximately 10^{-5} M concentration, regardless of the analyte (27,28). This is roughly equivalent to between tens to hundreds of nanograms injected on-column of a < 1000 amu compound, depending of course on molecular weight, column diameters and flow rates. This phenomena is characteristic of the ionization process itself and has been consistently observed on all types of ESI systems. It appears that this plateauing effect is due to a competition between analyte molecules for sites on the surface of the ion emitting droplet. Thus future extension of the dynamic range of ESI will probably only occur by lowering the detection limit with improvements in instrumentation sensitivity.

D. Molecules Amenable to Ionization

1. Species Charged in Solution. Although there is some controversy over the precise mechanism of ion formation, one thing is clear. Molecules that are inherently charged in solution by virtue of their chemical structure, by weak associations with other charged species, or by chemical reactions occurring in the solution eventually leave the liquid phase and become gas phase ions in the atmospheric region of the ion source. As a rule-of-thumb, acidic compounds are observed best in the negative ion mode under neutral to slightly basic solution conditions. Basic compounds are best observed in the positive ions mode under slightly acidic conditions. Examples abound for proteins, nucleotides, and a variety of acidic and basic small molecular weight compounds. Three additional examples are reported below which illustrate the versatility of the technique and suggest future application areas that undoubtedly will be implemented on a much broader basis than they are today.

2. Polar Neutral Species. As a rule of thumb, neutral molecules that retain any propensity for hydrogen bonding will form ions by adduction to ammonium or alkali metal ions. Carbohydrates are a good example. This is one reason why ammonium acetate is so popular as a mobile phase additive, because it facilitates the ionization of polar neutral species in solution. A good all purpose "cocktail" is 1-10 mM NH_4OAc plus 0.1% acetic or formic acid in aqueous containing mobile phases. Ammonium hydroxide can be substituted for the acids to enhance the formation of negative ions. Another reason this volatile buffer is used is to substitute for inorganic buffers which tend to suppress ionization and form extensive background cluster ions.

Normal phase chromatographic systems can be very useful for obtaining selective separations on many types of neutral species. If doped with small amounts of ammonium ions, adduction will occur. Figure 18 is the TIC from a normal phase chromatogram of Triton X-100 oligomers; a separation very difficult to achieve with reverse phase chromatography. Figure 19 shows the spectrum of one of the ammonium adducted oligomers with a small amount of synthetic by-product. The presence of a small amount of a chlorinated solvent is an important safety consideration to suppress ignition of the spray.

3. Non-Polar Neutral Species. Non-polar substances have been considered to be relatively unionizable by electrospray because they do not normally exist in solution in

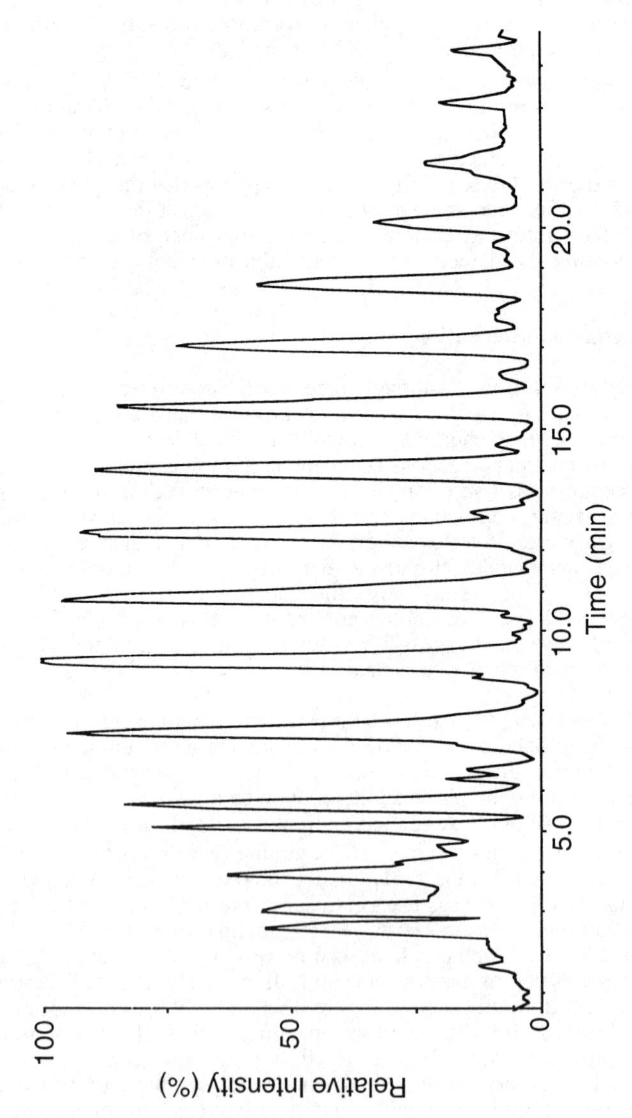

Figure 18. TIC of a normal phase separation of Triton X - 100 oligomers on a 4.6 X 100 mm cyano column; Mobile phase: solvent A 95/5 hexane/dichloromethane; solvent B 5/4/1 hexane/ dichloromethane/methanol containing 1mM NH_4OAc. Gradient: 100%A to 100%B in 30 minutes.

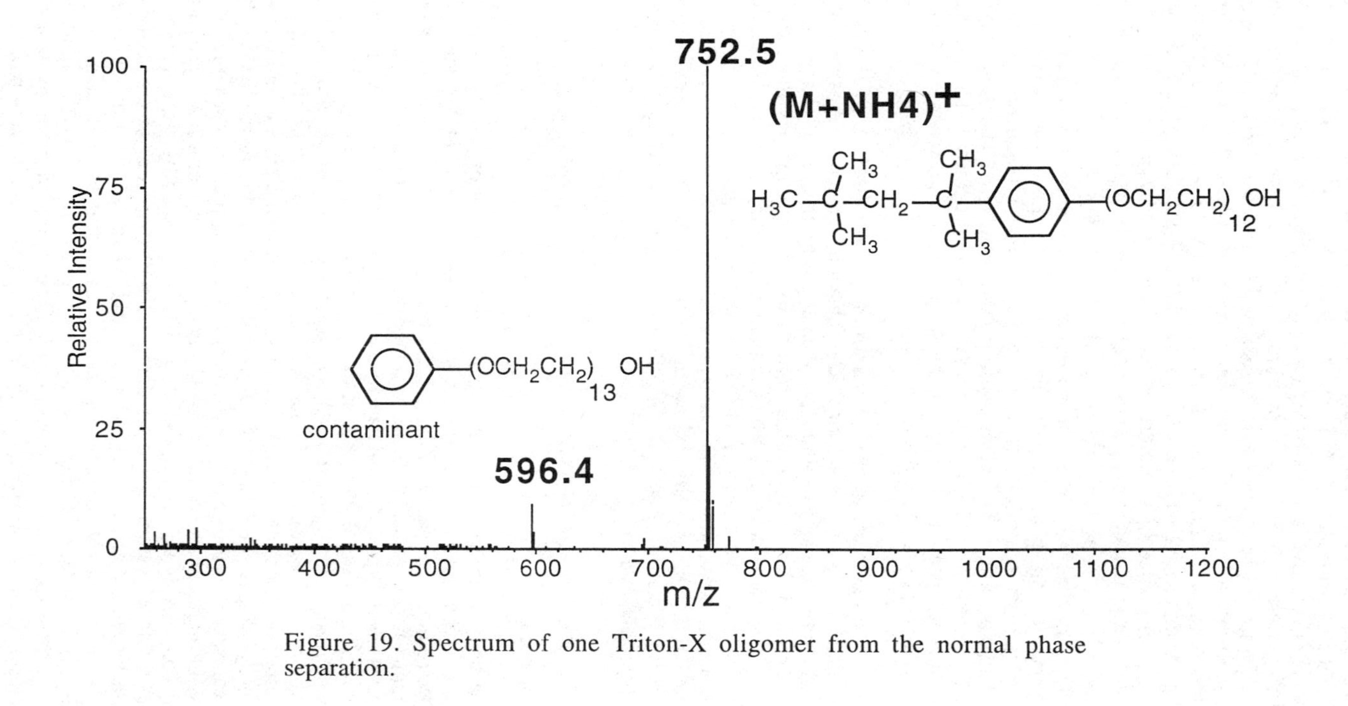

Figure 19. Spectrum of one Triton-X oligomer from the normal phase separation.

an ionized form and have little tendency to form adducts to other ions by hydrogen bonding. It has been subsequently shown that electrochemical oxidation/reduction reactions will occur in the electrospray tip to generate solution phase ions from highly non-polar species, such as polycyclic aromatic hydrocarbons (PAHs') and fullerenes, indicating future environmental applications of the technique (29,30). Charge transfer reagents are typically used to drive the reaction in non-polar solvents toward the production of radical cation molecular ions. Figure 20a shows the electron impact (70eV) spectrum of the PAH benzo(a)pyrene. When benzo(a)pyrene is introduced into an electrospray source in a mixture of toluene and methylene chloride (in this case there is no charge transfer reagent), a mixed ionization process occurs (Figure 20b). Oxidation in the electrospray tip forms the $M^{+\bullet}$. The origin of the $(M+H)^+$ is apparently from chemical ionization caused by some corona discharge on the sprayer tip. The product ion spectrum spectrum from a triple quadrupole mass spectrometer (Figure 20c) is structurally rich and suggests some advantages of collisionally activated dissociation over electron impact ionization. The tendency of PAH compounds to ionize by chemical ionization has been demonstrated earlier and it is suggested that atmospheric pressure chemical ionization is the method of choice for the analysis of complex PAH samples (31).

4. Inorganic anions and cations. Some of the earliest discussions of applications of ESI showed data on elemental species (32). This area has only been sparsely exploited to date but interest is rapidly climbing, especially for purposes of elemental speciation (33-35). The total ion current trace from an ion chromatography/mass spectrometry detection of a series of anionic species is shown in Figure 21. Of particular interest is the ability of molecular anions of various species to fragment in the free jet expansion of the electrospray ion source. For instance, the molecular anion of perchlorate can be fragmented to the chlorine atoms showing successive losses of molecular oxygen with a high degree of control (Figure 22). Collision energy is imparted with the ion source declustering potentials.

E. Conclusions

Electrospray has become an indispensable tool in the organic mass spectrometry laboratory. Indications are that it will also become important in the field of inorganic mass spectrometry. Because of the general utility of the technique, it has rapidly left the domain of the mass spectrometry laboratory and infiltrated chromatography laboratories of all types. Significant future improvements to the technique will, to a large degree, be a result of the input of the chromatography field. There is good reason now to pay serious attention to the development of rugged microflow pumps, autosamplers, and columns without losing track of the continued utility of the conventional higher flow systems. High speed autosamplers and integrated column switching systems for on-line purification will be required to match the very high daily sample capacity the electrospray mass spectrometer has proven to accommodate. Very high resolution integrated orthogonal chromatographic systems may play an important role in deconvoluting the complex mixtures being generated by combinatorial chemical methods. These systems must be designed with the goal of mass spectrometry detection where electrospray is the only light on the horizon at the current time. Beyond its use as a chromatographic detector and structural elucidation technique, there are glimpses that the electrospray mass spectrometer may serve as a bioassay tool. This idea has arisen as a result of the observations of non-covalent complexes between enzymes and receptors and their associated substrates and ligands with

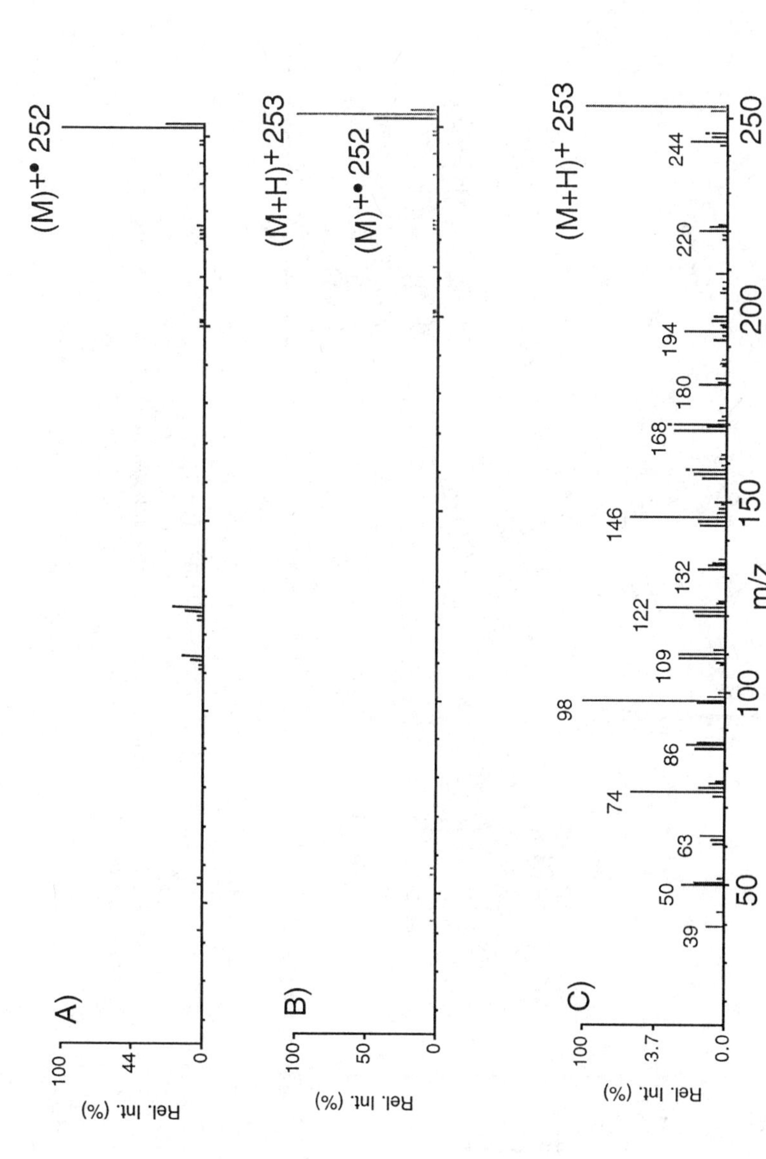

Figure 20. A) 70eV electron impact spectrum of benzo(a)pyrene, B) ESI spectrum, and C) product ion spectrum of 253 $(M+H)^+$ from electrospray ionization.

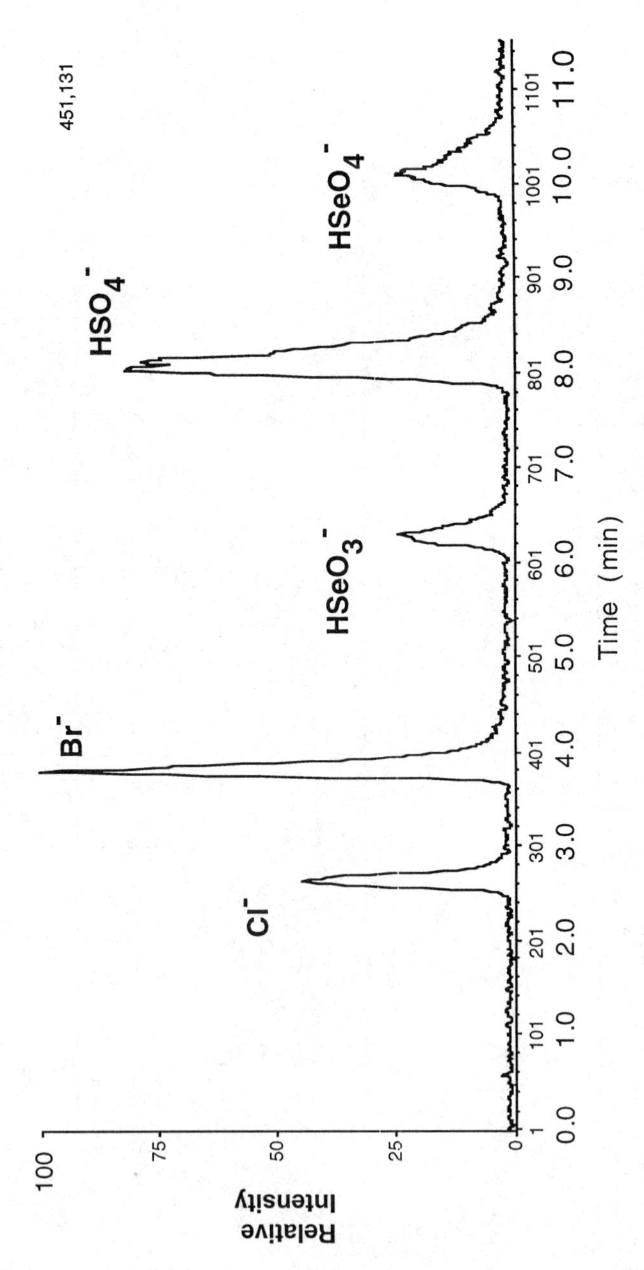

Figure 21. TIC trace from an ion chromatography/MS analysis of a mixture of inorganic anions.

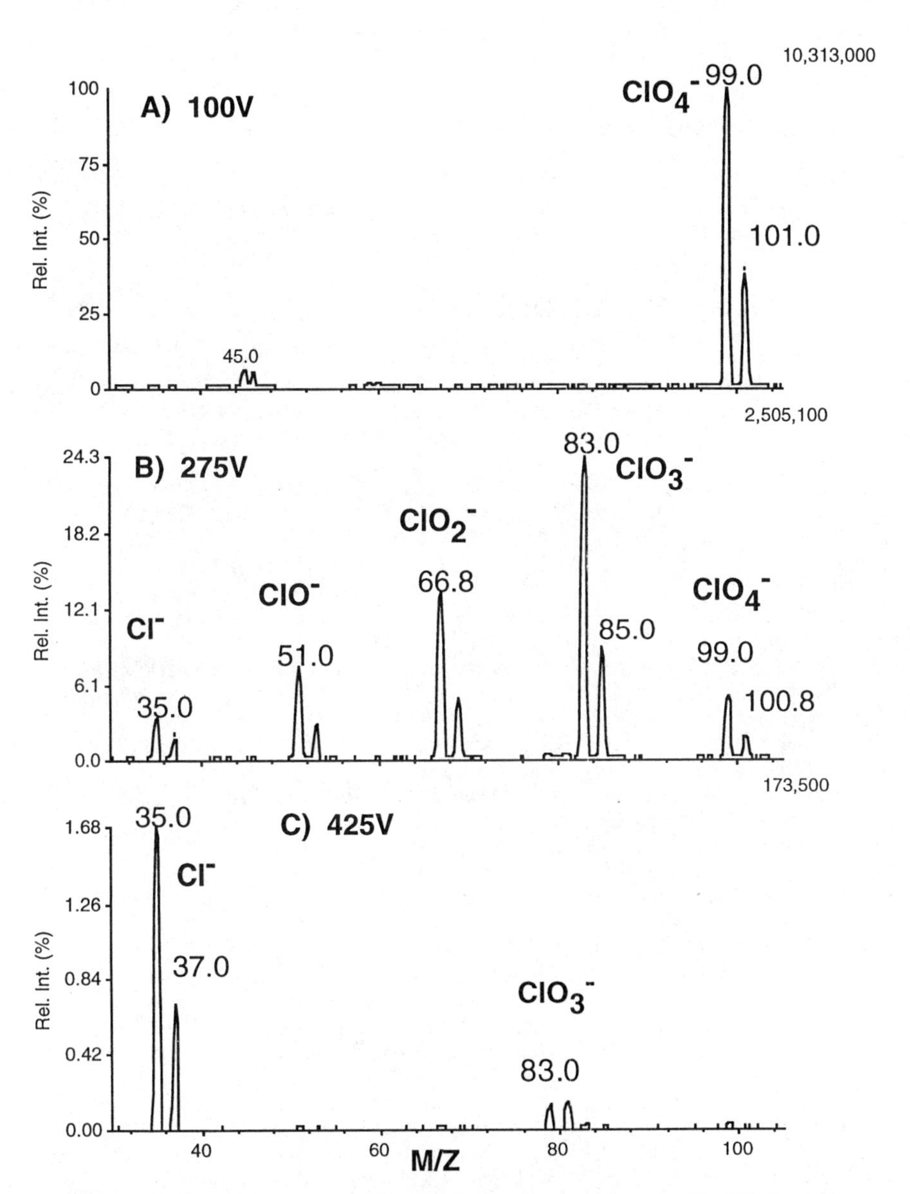

Figure 22. Ion evaporation spectra of perchlorate taken at A) 100V source declustering potential, B) 275V source declustering potential, and C) 425V source declustering potential.

electrospray mass spectrometry (36-38). It is not inconceivable that the future may bring electrospray systems of many flavors, each designed and dedicated for each of the many application areas in which it has proven such great utility.

References

1. Bruins, A.P. *Mass Spectrom. Reviews* **1991,** *10* , 53.
2. Douglas, D. *J. Am. Soc. Mass Spectrom* **1992**, *3,* 398.
3. Bruins, A.P. *Trends in Anal.Chem.* **1994**, *13* , 37.
4. Emmett, M.R.; Caproli, R.M. *J. Am. Soc. Mass Spectrom.* **1994**, *5*, 605.
5. Kriger, M.S.; Cook, K.D.; Ramsey, R.S. *Anal. Chem.* **1995**, *67*, 385.
6. Smith, R.D.; Wahl, J.H.; Goodlett, D.R.; Hofstadler, S.A. *Anal. Chem.* **1993**, *65*, 574A.
7. Hofstadler, S.A.; Wahl, J.H.; Bruce, J.E.; Smith, R.D. *J. Ame.Chem. Soc.* **1993**, *115*, 6983.
8. Wilm, M.; Mann, M. *Intl. J. Mass Spectrom. Ion Physics* **1994**, *136,* 167.
9. Wilm, M.; Shevchenko,A.; Houthaeve,T.; Mann,M. in *Abstracts of the 43rd ASMS Conference on Mass Spectrom. and Allied Topics,* Atlanta **1995** p. 93.
10. Wilm, M.; Shevchenko, A.; Houthaeve,T.; Mann, M. *Protein Science* 1995, submitted.
11. Kelly, J.F.; Locke,S.J.; Ramaley,L.; Thibeault,P. J. in *Abstracts of the 43rd ASMS Conference on Mass Spectrom. and Allied Topics,* Atlanta **1995,** p. 12.
12. Wahl, J. H.; Smith, R. D. *J. Chromatogr.* **1994**, *659*, 217.
13. Smith, R. D.; Udseth, H. R.; Barinaga, C. J.; Edmonds, C. *J. Chromatogr.* **1991**, *559*, 197.
14. Pleasance, S.; Thibault, P.; *In Capillary Electrophoresis: Theory, and Practice*; P. Camillieri; CRC Press, Boca Raton, Fl. 1993, p. 311.
15. Chervet, J.P.; Meijvogel, C. J.; Ursem, M.; Salzmann, J.P. *LC/GC* **1992**, *10*, 140.
16. Balogh, M.P.; Stacey, C.C. in *Abstracts of the 38th ASMS Conference on Mass Spectrom. and Allied Topics,* Tuscon **1990**, p. 1709.
17. Whitehouse, C.M.; Dreyer, R.N.; Yamashita, M.; Fenn, J.B. *Anal.Chem.* **1985**, *57*, 675.
18. Thomson, B.A.; Dziedzic, P.J.; Iribarne, J.V. *Anal. Chem.* **1982**, *54*, 2219.
19. Bruins, A.P.; Covey, T.R.; Henion, J.D. *Anal. Chem.* **1987**, *59,* 2642.
20. Personel communication Phoenix International
21. Covey, T.R.; Huang, E.C.; Henion, J.D., *Anal. Chem*.. **1991**, *63*, 1193.
22. Hess, D.; Covey, T.R.; Winz, R.; Brownsey, R.W.; Abersold, R. *Protein Science* **1993,** 2, 1342-1351.
23. Ikonomou, M.G.; Kebarle, P. *J. Am. Soc. Mass Spectrom.* **1994**, 5, 791.
24. Ling,V.; Guzzetta,A.W.; Canova-Davis,E.; Stults, J.T.;Hancock,W.S.; Covey,T.R.; Shushan, B.I. *Anal. Chem*.. **1991,** *63*, 2909.
25. Hopfgartner,G.; Bean, K.; Henion, J.; Henry, R. *J. Chromatogr.* **1993**, *647*, 51.
26. Weidolf, L.O.G.; Lee, E.D.; Henion, J.D.*Biomed. Environ. Mass Spectrom.* **1988** *15*, 283.
27. Kostianinen, R.; Bruins, A.P. *Rapid Commun. Mass Spectrom.* **1994,** *2*, 549.
28. Raffaelli, A.; Bruins, A.P. *Rapid Commun. Mass Spectrom.* **1991**, 5, 269.
29. VanBirkel, G.J.; McLuckey, S.A.; Glish, G.L., *Anal. Chem*.. **1992,** *64*, 1586.
30. Anacleto, J.F.; Quilliam, M.A.; Boyd, R.K.; Howard, J.B.; Lafleur, A.L.; Yadav, T. *Rapid Commun. Mass Spectrom.*, **1993**, *7*, 229.
31. Anacleto, J.F.; Perrault, H.; Boyd, B.; Pleasance, S.; Quilliam, M.; Sim, P. G.; Howard, J. B.; Makarovsky, J.; LaFleur, A. L., *Rapid Commun. Mass Spectrom.* **1996**, *6*, 214.

32. Iribarne, J.V.; Dziedzic, P.J.;Thomson, B.C. *Intl. Mass Spectrom. Ion Physics* **1983**, *50* , 331.
33. Agnes, G.D.; Horlick, G., *Applied Spectroscopy,* **1992,** 46, 401.
34. Cheng, Z.L.; Siu, K.W.M.; Guevremont, R.; Berman, S.S., *Org. Mass Spectrom.* **1992**, *27*, 1370.
35. Corr, J.J.; Douglas, D.J. in *Abstracts of the 41st ASMS Conference on Mass Spectrom. and Allied Topics,* San Francisco **1993** p. 202.
36. Ganem, B.; Li,Y.; Henion,J.D. *J. Am. Chem Soc.* **1991**, *113*, 6294.
37. Ganguly, A.K.; Praminik, B.N.; Tsarbopoulos, A.; Covey, T.R.; Huang, E.; Fuhrman, S.A. *J. Am. Chem .Soc..* **1992**, *114,* 6559.
38. Smith, R.D.; Light-Wahl, K.J. *Biol. Mass Spectrom.* **1993**, *22*, 493.

RECEIVED September 13, 1995

Chapter 3

Electrospray Ion Formation: Desorption Versus Desertion

J. B. Fenn[1], J. Rosell[1], T. Nohmi[2], S. Shen[3], and F. J. Banks, Jr.[3]

[1]Department of Chemistry, Virginia Commonwealth University, Richmond, VA 23284–2006
[2]Nohmi Bosai Ltd., Tokyo 102, Japan
[3]Analytica of Branford, Inc., 29 Business Park Drive, Branford, CT 06405

There are two widely debated mechanisms for the formation of gas phase ions from solute species in charged droplets. One is embodied in the Charged Residue Model (CRM) of Malcolm Dole. The other forms the basis of the Ion Desorption Model (IDM) of Iribarne and Thomson. Careful consideration of a variety of selected experimental results leads to a conclusion that the IDM provides a more satisfactory explanation of the ionization process for all species except perhaps very large molecules with molecular weights of a million or more.

Malcolm Dole - A Daring Doer

Electrically charged droplets were first proposed as a source of ions for mass spectrometry (MS) in 1968 by Malcolm Dole and his colleagues (*1*). Interested in trying to determine molecular weight distributions of synthetic polymers they assembled an apparatus designed to stage the following scenario: A dilute solution of the polymer analyte in a volatile solvent is introduced through a small tube into an electrospray (ES) source chamber through which dry bath gas (nitrogen) flows at atmospheric pressure. A potential difference of a few kilovolts between the tube and the chamber walls produces an intense electrostatic field at the tube exit and disperses the emerging solution into a fine spray of charged droplets. As the droplets lose solvent by evaporation their charge density increases until the Rayleigh limit is reached at which Coulomb forces overcome surface tension and the droplet breaks up into small droplets which repeat that sequence. Dole argued that if initial concentrations of solute polymer are low enough, a succession of such "Coulomb explosions" would ultimately lead to droplets so small that each would contain only one molecule of analyte. As the last of the solvent evaporates that residual molecule would retain some of the droplet charge, thereby becoming a free macroion. A portion of the resulting dispersion of macroions in bath gas could then be introduced into a vacuum system by way of a free jet expansion from an orifice in the source chamber. Once in the vacuum system they could be "weighed" by an appropriate mass

0097–6156/95/0619–0060$12.25/0

analyzer. Dole clearly recognized and stressed the vital role of the bath gas in providing the enthalpy necessary to vaporize solvent from the droplets. Neglect of that requirement was the fatal flaw in so-called "Electrohydrodynamic Ionization" (EHDI) which electrosprays analyte solution directly into vacuum. After much frustration in trying to develop EHDI into a useful source of ions for mass spectrometry, most investigators have abandoned the technique. Its status was reviewed by Cook in 1986 (*2*).

That first paper of Dole et al also presented some experimental results purporting to show the validity of the scenario they had envisioned. The apparatus did seem to produce free ions from polystyrene oligomers with molecular weights in the range from 50,000 to 500,000. Several subsequent papers reported similar results with other large molecular species (*3-5*). In those first experiments the method of determining the mass of the ions consisted in measuring the retarding potential required to prevent ions from reaching a Faraday cup and contributing to the read-out current after they had been accelerated during free jet expansion into vacuum from the ES chamber at atmospheric pressure. In later experiments ion mobility was used to characterize the ions. Dole's interpretation of his observations now seems open to question, largely because of shortcomings in his methods of interrogating the ions. Moreover, he did not recognize the probability of, or take any steps to avoid, resolvation of those ions during the free jet expansion into vacuum. Nor did he realize that the velocity of the macroions was substantially less than the velocity of the carrier gas because of slip effects during that expansion. Thus, from the vantage point of hindsight it seems clear that his experiments did not provide convincing evidence for the scenario that spawned them. Even so, it cannot be denied that Dole took the first steps on the path to ES ionization as it is widely practiced today. That technique produces ions of large molecules by experimental procedures that are nearly the same as those he described. Moreover, his model for the mechanism of ion formation is still preferred by many investigators.

The Truth According to Iribarne and Thomson

In spite of Dole's provocative experiments the use of charged droplets to produce ions was pretty much ignored by other investigators until 1979 when Iribarne and Thomson found evidence from measurements of mobility that small ions were produced during evaporation of charged water droplets (*6*). The motivation for their studies was far removed from a concern for the needs of mass spectrometry. Instead it stemmed from their interest in charged droplets as a possible source of ions in the atmosphere. They proposed a model for such ion formation based on the idea that on charged droplets that were small enough, evaporation could make the surface field sufficiently intense to lift solute ions from the droplet into the ambient gas before the Rayleigh limit was reached. In a 1979 paper they reported mass spectrometric identification of the ions whose presence they had earlier inferred from the mobility measurements (*7*). They also provided a more complete analysis of their model of the ion formation process and christened it "Atmospheric Pressure Ion Evaporation" (APIE). That term literally covers what is now a widely accepted mechanism of ion formation from charged droplets no matter how those droplets are produced. It will be referred to hereinafter as the Ion Desorption Model or IDM. (We choose to refer to ion departure from the droplet as "desorption" rather than "evaporation" in order to avoid confusion due to frequent use of the latter term to refer to the departure of solvent molecules from the droplet.) The operational technique of the Iribarne-Thomson experiments will be referred to as Aerospray Ionization (ASI) in recognition of its use of aerodynamic forces to nebulize the sample liquid. A third paper from the Iribarne group in 1983 presented mass spectrometric evidence that a wide variety of solutes could be evaporated (desorbed) as ions from charged droplets (*8*).

The Last Becomes First - Marvin Vestal's Thermospray

One of history's many ironies is that in spite of the noteworthy findings of the Dole and Iribarne groups, the everyday use of charged droplets as ion sources was actually born of very different parentage. That birth came in 1978, all of ten years after Dole's first paper, when Marvin Vestal and his colleagues introduced Thermospray Ionization (TSI) (*9-12*). Moreover, TSI had no apparent roots in the work of Dole and Iribarne. Indeed, it seems to have been discovered and reduced to practice before the role of charged droplets in its ion production was realized. Whatever its ancestry, for most of the next decade TSI was the method of choice for interfacing Liquid Chromatography with Mass Spectrometry and must be recognized as the first widely-practiced technique to use charged droplets as a source of analyte ions for mass analysis.

To be remembered is that in Vestal's TSI as well as in the first experiments of Iribarne and Thomson with ASI, charging of the droplets is due to statistical fluctuations in the distribution of cations and anions among those droplets when an ion-bearing liquid is nebulized. Conservation of charge requires that the total amount of charge carried on positively charged droplets must equal that carried by negatively charged droplets. Therefore, the numbers of negatively and positively charged droplets are approximately the same. The difference between ASI and TSI is in how the nebulization is carried out. ASI uses pneumatic atomization with a separate gas. TSI passes the sample solution through a heated tube wherein rapid vaporization of most of the solvent and expansion of the resulting vapor provides the gas dynamic forces needed for atomization of the remaining liquid. Interestingly enough, Iribarne and Thomson introduced an induction or polarizing electrode at high voltage near the nebulizing region and found that the resulting droplets all had the same polarity as well as much higher charge/mass ratios (*7*). This use of an electric field to polarize the emerging liquid seems reminiscent of what goes on in ESI. The difference is that in spite of the addition of a polarizing electrode the nebulization work in ASI is still performed by gas dynamic forces. In ESI most or all of that work is done by electrostatic forces. As a result, ESI droplets always have higher charge/mass ratios and provide higher analytical sensitivity than those produced by ASI or TSI, with or without the assistance of a polarizing field.

Ions Without Droplet Evaporation?

The usual assumption in ASI, TSI and ESI is that the evaporation of solvent from charged droplets plays the key role in producing the free ions. It is noteworthy in passing, therefore, that Guevremont et al found that the distribution of charge states for ES ions from protein solutions was strikingly similar to the most probable distribution of charge states for protein ions within that solution as calculated from pKa values for the basic groups of those proteins (*13*). They concluded that the ES ions could be identified with the solution ions, not only in the case of proteins but for all other solutes as well. This simplistic idea seems to have found wide acceptance. It is now taken as a given in many papers and has led to some startling conclusions. For example, the Siu group has suggested that ES ions are not formed by the evaporation of charged droplets but are emitted directly from the tip of the Taylor cone at the exit of the electrospray needle (*14,15*). This possibility seems somewhat remote to anyone who has deliberately or inadvertently interfered with the flow of counter-current drying gas in an Analytica source, or lowered the temperature of the heated tube in a Chait-type source. Without these sources of enthalpy for droplet evaporation the MS signal vanishes.

In our view the evidence in support of the idea that ES ions are simply the solute ions in neutral bulk sample solution is at best tenuous while the evidence against it is formidable. In the first place one cannot explain the emission of a free ion from a neutral droplet without either violating the principle of charge conservation or ignoring the Coulomb force that would perforce drive such an ion from an initially neutral droplet back to the now-charged droplet from which it came. The only charges available to produce free gas-phase ions from a droplet of solution, are the excess charges the droplet carries when it is formed from the jet of liquid emerging from the Taylor cone at the exit of the spray "needle." Those excess charges are all at the droplet surface and have no accompanying counterion, as do all the solute ions in the sub-surface bulk solution which is neutral overall. Those omnipresent counter ions are frequently ignored in attempts to account for the marked differences often found between ion distributions in ESMS spectra and those "known" to obtain in the droplet liquid.

Some particularly revealing examples of such differences were reported by Catherine Fenselau and her colleagues (*16*). They found, for example, that the distribution of charge states in a positive ion ESMS spectrum for a methanol-water solution of equine myoglobin at a pH of 3.5 ranged from 8 to 21. A negative ion spectrum <u>for the same solution</u> showed a remarkably similar distribution of charge states ranging from 9 to 17. The authors reckoned that only one of every 3100 molecules in that solution had as much as one negative charge. (Even that negative charge also had a cation nearby!) The number of ions in solution with more than one negative charge was indeed negligible, and yet the spectrum showed ions with up to 17 negative charges! Clearly, those charges must have come from the anionic charges that were in excess on the droplet surface, not with those in the neutral bulk liquid below the droplet surface. Each of those sub-surface anions always had a cation nearby. The excess anions on the surface resulted from the loss of their cation partners which were driven to the spray needle (cathode) by the high field at the needle tip. In our view this result from the Fenselau group effectively demolishes any basis for the still-too-frequent assumption that ES ions can be identified with the ions in bulk solution that overall is neutral.

There are many other examples of discrepancies between the charge distributions of particular solute species when they are ions in the droplet solution and when they have left the droplet to become free ions in the gas phase. Space limitations preclude a more complete review of these examples and a more extensive critique of the notion that there is a one-to-one correspondence between the free ions from charged droplets and the solute ions in neutral bulk solutions. To our knowledge none of the papers that assume this correspondence, or attempt to support it, have ever presented a mechanistic

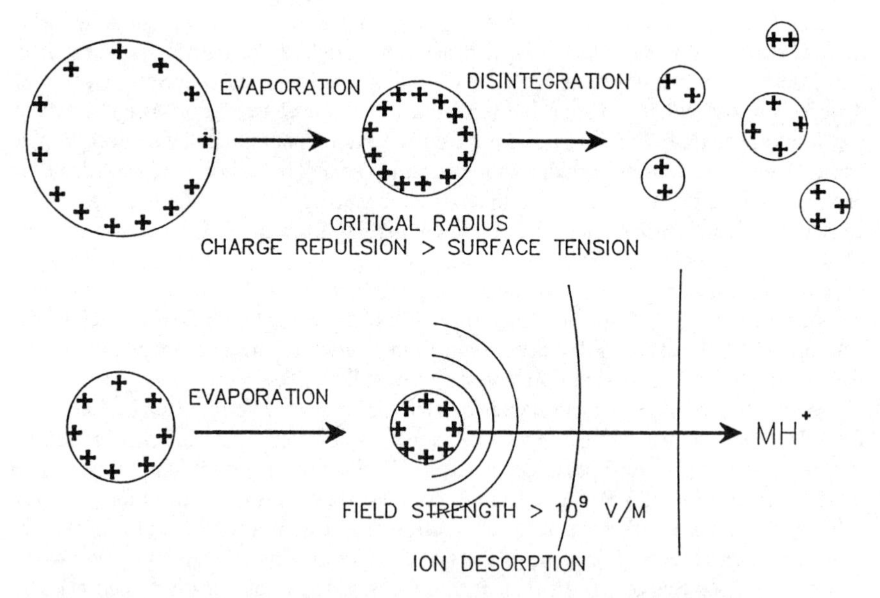

Figure 1. Schematic representation of the Charged Residue Model (CRM) and the Ion Desorption Model (IDM) for ion formation. The upper series depicts a sequence of evaporation fission steps that by the CRM leads to ultimate droplets containing only one solute molecule. The lower series shows an intermediate stage at which a droplet's surface charge density is below the Rayleigh stability limit but high enough to provide a surface field sufficiently intense to desorb a solute molecule as an ion in accord with the IDM.

explanation of how the solute ions in bulk liquid escape from the droplet (and their counterions) while maintaining the equilibrium charge distribution they supposedly have while in neutral solution. Thus, in the remainder of this report we will not labor the issue further but for convenience will simply assume that ions from charged droplets are formed either by the CRM or the IDM. Our more immediate objective then becomes to determine which of these two mechanisms can better explain the experimental results to be considered.

Distinguishing Between the CRM and the IDM

Figure 1 is an attempt at a schematic portrayal of the two models. At first glance the distinction between their ionization mechanisms seems clear and obvious. In Dole's CRM, represented in the upper sequence of sketches, evaporation of solvent molecules from a charged droplet steadily decreases its size, thereby increasing its surface charge density. The droplet continues to shrink until it reaches the Rayleigh limit at which the Coulomb repulsion due to the charge overcomes the surface tension. The resulting instability breaks up the parent droplet into a hatch of offspring droplets, each of which continues to evaporate until it too reaches the Rayleigh limit. This sequence continues until the offspring droplets ultimately become so small that they contain only one analyte molecule. That molecule retains some of the droplet charge to become an ion as the last solvent molecules evaporate. The key feature of this scenario is that the solvent molecules are separated from the lone solute molecule in the ultimate droplet by evaporating individually, leaving behind a residue comprising the single analyte molecule together with all other non-volatile material, if any, in that ultimate droplet. Implicit in this perspective is either (a) the absence of any other solute molecules in the ultimate droplet or (b) an instability in the charged residue such that at least one of its component non-volatile solute molecules (the analyte) escapes from the others, carrying with it at least some of the droplet charge.

In the IDM, as envisaged by Iribarne and Thomson, the charged droplet commences the same sequence of evaporation and Coulomb fission steps as in the Dole CRM scenario. However, the IDM holds that at some intermediate stage, before offspring droplets are so small that they contain only one analyte molecule, they are small enough so that at a surface charge density below the Rayleigh limit the surface field is sufficiently intense to overcome solvation forces and to lift a charged analyte molecule (ion) from the droplet surface into the ambient bath gas. This departure from the CRM scenario is shown schematically in the lower sequence of sketches in Figure 1. Generally presumed, but not always explicitly stated, is that the droplet is much larger than the departing ion and contains many other solute molecules. Thus, analyte ions become separate entities in the gas phase by desorbing individually, one at a time, from the aggregation of solvent and other solute molecules constituting the droplet. In this perspective the fundamental difference between the CRM and the IDM is in how an analyte molecule with a charge becomes separated from all its companions in the droplet. In the CRM the separation comes about by repetitive subdivision of the initial droplet by Coulomb forces until the resulting particles of liquid are so small that they contain only one solute molecule. From that point on the solvent molecules evaporate individually from that ultimate droplet one at a time, until, in the words of the children's game song "the cheeze stands alone" as a solute ion with some of the droplet's charge. In the IDM the isolation comes about by desorption of a single analyte ion from a droplet that contains many other solute and solvent molecules. In other words does an ion escape from the solvent molecules or do the solvent molecules escape from the ion?

The Horns of the Dilemma

Unfortunately, as is too often the case, the situation is not that simple. There is yet no convincing experimental evidence on the extent of solvation of free solute ions in their nascent state as they leave the droplet. They encounter a wide variety of conditions which can change their solvation between the time they are formed and the time they are weighed in a mass analyzer. It seems highly likely that unless their parent molecules are inherently and strongly solvophobic, these ions would desorb in aggregation with some number of solvent molecules as they leave the droplet. (Clearly, species that are too highly solvophobic could not be solutes in the first place!) Thus there arises a semantic dilemma. How many solvent molecules can be attached to a departing ion before the resulting aggregate can be considered as the product droplet of a Rayleigh instability in the perspective of the CRM? In other words can desorption of a single solvated ion from a parent droplet be accurately characterized as a Coulomb explosion that produces a single tiny droplet from that parent droplet? Is "evaporative" loss of solvent molecules from a solvated ion entirely equivalent to the evaporation of solvent from an ultimate droplet of the CRM perspective? Conversely, how many solvent molecules can a desorbing ion take in tow before it should be regarded as a droplet resulting from Coulomb fission?

Consideration of questions like these has led some investigators (*17,18*) to suggest that there may be no real difference between the CRM and the IDM. They argue that the desorption of a solvated ion from a droplet is essentially the same phenomenon as a Coulomb explosion of that droplet that leads to one or more ultimate product "droplets" that are neither more nor less than highly solvated ions. Unfortunately, it is not yet possible to probe directly the plausibility of this hypothesis because the present state of experimental art does not allow an investigator to measure the numbers of charges and solvent molecules in the nascent state of what might be either the ultimate droplet of the CRM or the nascent desorbed ion of the IDM.

Resolving the Dilemma by Results

There are two characteristics of ES ions, at the time of their mass analysis, that might provide insight on whether their formation was by way of the IDM or the CRM, namely their relative abundance and the number of their charges. It will emerge in what follows that these properties, both of which are readily observable, do indeed shed light on the issue of IDM vs. CRM. For example, a key feature of the IDM is the strong dependence of ion desorption rate, and therefore the measured abundance (peak height in the mass spectrum), on the work required to remove an ion from a charged droplet, i.e. the difference in free energy between an ion on the droplet solution and that same ion when it is in the gas phase. Clearly, this free energy depends upon the properties of the solvent, of the analyte molecule, and of the ion that it forms. On the other hand, the abundance of ions formed by a CRM process must be determined essentially by the number and rate of formation of ultimate droplets from parent droplets by Coulomb fission, i.e. on the amount of work required to fragment a parent droplet into ultimate droplets. That work requirement will be determined largely by the properties of the solvent, primarily its surface tension and polarizability. It follows that for a particular solvent, if relatively small differences in structure or composition of analyte molecules of comparable size make substantial differences in measured abundance of the corresponding ions, one could reasonably conclude that the IDM governs the ion formation. Unfortunately, the influence of solvent properties alone cannot be so readily elucidated. Solvents with different surface tensions, for example, are also likely to show different solvation energies for a particular analyte molecule or ion. Thus, changing the solvent might make a large difference in charge state and/or abundance of the ions of a particular analyte species. However, in the absence of other information, one could not know whether to attribute such changes to a difference in the number of droplets formed, which would mean the CRM was at work, or to changes in the solvation energy of the analyte ion, which would mean the IDM was in control.

The other characteristic of an ion that can be revealing is the number of its adduct charges as indicated by the mass analysis. For example, evidence from a number of experiments with droplets having diameters above the diffraction limit so that they could be interrogated optically, i.e. larger than about half a micron, shows that the small offspring droplets formed by a Coulomb explosion have a substantially higher charge/mass ratio than the parent droplet (*19,20*). If one makes the reasonable assumption that similar behaviour is exhibited in the submicron size range, then it follows that the ultimate droplets of the CRM are likely to have a plurality of charges. Therefore, if analyte species capable of retaining two charges, for example, show up in the mass analyzer with only one charge, one could reasonably conclude that those singly charged ions were produced by the IDM scenario. The only way the CRM can produce singly charged ions from species capable of retaining two or more charges is by way of singly charged droplets. Moreover, because the CRM assumes that one ultimate droplet produces only one ion, the measured relative abundance of analyte ions is also a measure of the number of ultimate droplets that are formed. By the same token, if the CRM is to account for observed abundances of ES ions, the liquid dispersion and droplet fission processes must produce at least as many droplets as ions.

It can be argued, as noted earlier, that the desorption of a singly charged ion with solvent ligands by the IDM is exactly equivalent to a Coulomb fission that produces an offspring droplet so small that it contains only one analyte molecule and a single charge. It is certainly true that Coulomb repulsion is the driving force in either case. In our view, however the production of a singly charged ion by such a process is much more in the spirit of the IDM than of the CRM. Moreover, as will be discussed later, the apparent similarity between the two perspectives becomes tenuous when one considers the large molecules that show a broad range of charge state distributions.

On the basis of these considerations, and their logical extensions, we suggest that if mass analysis of a population of ES ions shows any of the following features, the formation of the ions in question cannot be accounted for by the CRM. Consequently, it must attributed to the IDM.

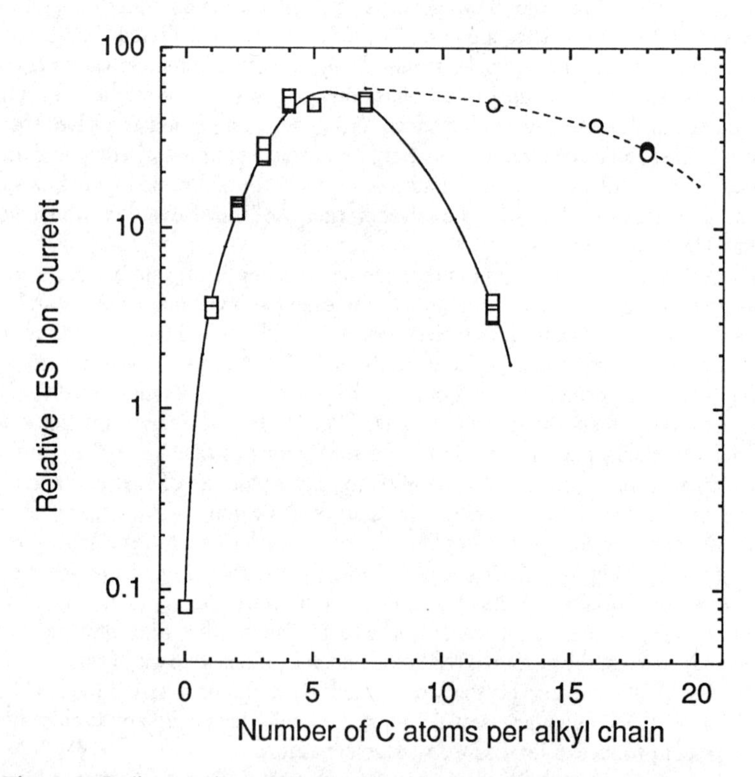

Figure 2. Ordinate values show the relative magnitudes of mass selected ion currents for analytes comprising alkyl ammonium bromides. Abscissae values indicate the number of carbon atoms in each of the four side chains for each analyte. The solid curve is for a solvent of 50:50 methanol-water. The dashed curve is for 50:25:25 methanol-water-n-butanol.

[A] Ions for which the number of charges is less than the number of charges their parent molecule can retain, and less than the number of charges likely to have been on the ultimate droplets from which those ions could have come. In other words, an ultimate droplet with n or more charges containing a molecule that can hold n or more charges cannot by the CRM give rise to an ion with fewer than n charges. (We assume here that charges are not removed from an ion after it leaves the droplet, e.g. by charge exchange with gas phase molecules or by charge stripping in energetic collisions).

[B] Ions for which, other things being equal, a small change in structure or composition of their parent molecule makes a big difference in their abundance.

[C] The absence of ions for which parent molecules are known to be present in the sample solution in detectable abundance under the conditions of the experiment.

The Experimental Evidence

In the perspective of the preceding discussion we will now examine the results of a variety of experiments with particular attention to the criteria A, B and C by which we believe it is possible to determine whether the CRM or the IDM provides a better accounting for the observations.

Tetra Alkyl Ammonium Ions. One of the first and most strikingly unequivocal demonstrations that ES could produce intact ions from non-volatile solute species was an experiment with a solution of seven quaternary ammonium and phosphonium halides in 50-50 methanol-water at concentrations of from 2 to 10 ppm. The mass spectrum showed a clean peak for each species, a very high signal/noise ratio, and no evidence of fragmentation (*21*). Since then these tetra alkyl compounds have been widely used as mass markers for calibration of analyzer scales in the m/z range below about 500. They desolvate easily, do not readily form clusters, are always singly charged and show a linear dependence of free-ion abundance on initial solution concentration over a range of 4 or more decades. We will now examine what they can tell us about CRM vs. IEM.

The ordinate values along the solid curve in Figure 2 show the relative abundances of ES ions obtained with 10 different tetra alkyl ammonium bromides, each at a concentrations of 10 micromoles/liter in 50-50 methanol-water infused at a flow rate of 3 uL/min. The abscissa values are the numbers of carbon atoms in each of the four identical n-alkyl chains on each molecule. As the number C of carbon atoms increases, the ion abundance also goes up, at first very rapidly and then more slowly. Thus, from $C = 0$ (ammonium ions) to $C = 1$, the increase is by a factor of almost 40. That factor goes down to 5 between $C = 1$ and $C = 2$ and reaches zero somewhere between $C = 4$ and $C = 7$. Above $C = 7$ the factor is less than unity so that the abundance at $C = 12$ is about the same it was for $C = 1$. Above $C = 12$ the abundance nosedives.

To explain this fairly complex behaviour in terms of the CRM poses a formidable task. In the first place that model would require a forty-fold increase in the number of ultimate droplets to account for the forty-fold increase in ion abundance as each H in ammonia becomes a methyl group. Another five-fold increase in number of droplets would be required to explain the additional increase in ion abundance when those methyl groups (C = 1) become ethyl groups (C = 2) To account for the overall differences in relative abundances the CRM would require the formation of 1000 times as many ultimate droplets when the analyte species comprised tetrabutyl ammonium ions as when they comprised plain ammonium ions! At the outset it is difficult to understand how these relatively small changes in the size and nature of analyte molecules at micromolar concentrations could make so much difference in the number of ultimate droplets that are produced. The task becomes even more forbidding in the face of the results obtained by our former colleagues, Prof. A. Gomez and his student K. Tang at Yale who examined some of these same sprays by Phase-Doppler Anemometry. No significant differences in the size and number density of the droplets could be "seen" in the sprays as the alkyl side chains went from methyl to amyl, i.e. C = 1 to C = 5. They could not "see" the droplets when the diameters went below about one micron. Thus, the increase in numbers of ultimate droplets required to explain the ion abundances would have had to occur late in the sequence of Coulomb fissions. How that could happen is not at all clear.

Then there is the added difficulty of explaining the marked decrease in ion populations when the number of carbon atoms in the alkyl groups rises above six or seven. Why should the ultimate droplets become much less numerous, and therefore bigger, with increasing chain length of the solute alkyl groups, especially when the surface activity of these groups would be expected to lower surface tension? Moreover, as the points along the dashed curve in Figure 2 show, there was a very much smaller decrease in ion abundance when the solvent was changed from 50-50 methanol-water to 50-25-25 n-propanol-methanol-water. In terms of the CRM this smaller decrease implies that in propanol-methanol-water an increase in the alkyl chain length from C = 7 to C = 12 or more has little or no effect on the number of ultimate droplets formed. In methanol-water the same increase in chain length decreases the abundance of ultimate droplets by a factor of 15 or more. Such behaviour is indeed a "puzzlement."

It is fair to ask whether the IDM can provide a more satisfactory explanation of these observations. We think the answer is yes. That model predicts an exponential dependence of ion abundance on the work required to remove an ion from the droplet solution. As the number of carbon atoms in the side chains increases the quaternary ammonium cations become increasingly solvophobic so their escaping tendency increases and the work required to remove them from the droplet decreases. Indeed, when the number of side-chain carbon atoms gets above seven the solvophobicity becomes so great that solubility of the analyte nosedives. Thus, the concentration of analyte cations at the droplet surface can never get above the saturation value which is pretty low for chains with 12 or more carbon atoms. Consequently, the ion abundance, which depends directly upon the concentration of analyte at the droplet surface, plummets. Adding n-propanol to the solvent greatly increases the solubility of these tetra-alkyl ammonium ions with many carbons in the alkyl groups so the ion abundance climbs back almost to the maximum level shown in methanol-water when C = 4 in this homologous series. One of the reviewers of this paper pointed out that the marked decrease in ion abundance when C is greater than seven may be associated with micelle formation by the solute. That possibility is certainly something to be considered. However, the key point is that the decrease in signal is due to the decrease in concentration of monomeric solute molecules at the surface. From the standpoint of ion formation it would not seem to matter whether that decrease was due to micelle formation or to more conventional precipitation. In sum, the IDM seems able to explain Figure 2 with much less strain on one's credulity than the CRM calls for.

A Glimpse of Metal Cation Behaviour. A number of investigators have found that ES ionization can produce free cations of various metals but the yields have always been much lower than those encountered for many of the organic analyte species that have been studied. Having come to realize the importance of solvophobicity in determining ion abundance we reasoned that an appropriate choice of solvent might increase the yield of these very hydrophilic species. Accordingly, a few years ago we decided to carry out some preliminary experiments with a non-aqueous solvent. We were delighted to find much higher abundances of several metal cations than we had ever obtained with methanol-water, our workhorse solvent. Figure 3 shows a result that is particularly germane to the present discussion. It was obtained by S. Fuerstenau in our laboratory with a solution comprising acetonitrile in which a few small crystals of copper sulfate had been rinsed for a few seconds (*22*). Water was added in varying amounts to this solution so that the abscissa value for each point represents the percentage of water in the solution over and above any small but unknown amount that might have been in the acetonitrile to begin with, or was associated with the small amount of copper sulfate that dissolved. The observed ions showed no solvation because a potential difference of 200 V was applied to the free jet between the source orifice and the first skimmer. The resulting field accelerated the ions relative to the neutral bath gas molecules, bringing about collisions with sufficient energy to strip off any solvent ligands.

The noteworthy feature of Figure 3 is the remarkable decrease in ion abundance as the water content of the solution rises. Adding 0.1 per cent of water to the solution caused a signal loss of about 15 per cent. At a water content of 10 per cent the signal is only a quarter of its original value. According to the CRM these decreases can only be explained by equivalent decreases in the number of ultimate droplets. It is hard to understand how such small amounts of water in the acetonitrile solvent could cause such a big decrease in the number of ultimate droplets. In our view these results are more readily understood in terms of the affinity of copper ions for water molecules. It is to be expected that as the number of hydration ligands on a copper ion increases, the work required to remove the resulting cluster from a droplet would increase, given the high solubility of water in acetonitrile.

The IDM calls for an exponential dependence of the rate of ion desorption on the removal work. Thus, it would seem to offer a more reasonable accounting for these results. Another noteworthy feature of this experiment is that the ES ions of copper have one charge rather than two, even though they were divalent in the source material, cupric sulfate. Blades et al have explained this valence reduction by the fact that the second ionization potential of copper is higher than the first ionization potential for most solvents and ligands (*23*).

A Tale of Two Peptides. Gramicidin S (GMCD) and Cyclosporin A (CYSP) are cyclic peptides of nearly the same size, having molecular weights respectively of 1141 and 1202. Many ES mass spectra of these species have been made in our laboratory and others under a wide variety of conditions. For example, a solution of GMCD at a concentration of 0.01 gm/L in 50:50 methanol-water shows a dominant peak due to doubly charged parent molecules which has 15 times the amplitude of the peak due to singly charged molecules. In a mass spectrum of CYSP under the same conditions the peak for singly charged ions is dominant, having twice the height of the peak for doubly charged ions. These results clearly show that both species can retain two charges. The CRM would require a singly charged ultimate droplet to produce a singly charged ion under these circumstances. Thus the immediate question, which was asked earlier, is: "How can ultimate droplets be produced with a single charge when all the experimental evidence indicates that the clutch of droplets produced by a Coulomb explosion generally contains about fifteen per cent of the parent droplet charge but only two or three per cent or less of its mass. The answer is: only if the amount of solvent in an ultimate "droplet" is very small, in fact so small that the droplet is essentially a solvated ion that has desorbed from a much larger droplet in accordance with the IDM. Another obvious question is then: "How can some of the ions be doubly charged while others are singly charged?" The CRM can explain this result only by assuming that in the case of GMCD most of the ultimate droplets have two or more charges and only a few are singly charged. In the case of CYSP, on the other hand, the singly charged droplets must outnumber those with two or more charges by a substantial margin. It is very difficult to understand how a six per cent difference in the molecular weight of two cyclic peptides in the same solvent and under the same conditions could cause such a large shift in the ratio of singly charged droplets to multiply charged droplets.

The IDM offers an explanation that seems more acceptable than the CRM can provide. That explanation is straightforward but a bit intricate and has been set forth previously (24). In brief, GMCD ions have more charges than CYSP ions because GMCD has more affinity for the solvent than does CYSP, as is shown, for example, by its much shorter retention times in reverse-phase liquid chromatography. Therefore, more lift is required to remove a GMCD ion from the droplet surface. That increased lift becomes available only late in the droplet evaporation process when the surface field increases as the surface charges get closer together. Indeed they get close enough for a GMCD molecule to span the distance between two of them and thus become doubly charged. The CYSP molecules, being much more solvophobic, can escape the droplet with much less lift. Consequently, most of them leave earlier during the droplet evaporation process while the charges are so far apart that a CYSP molecule cannot reach two of them at the same time.

This picture is confirmed by what happens when the concentration of CYSP is greatly decreased. As required by the IDM and confirmed by experiment, the rate of analyte ion desorption is directly proportional to the surface concentration of the parent analyte molecules. Decreasing that concentration thus decreases the rate of ion desorption and therefore, the rate at which the droplet's charge is depleted. For most situations the total charge on the droplet when ion desorption begins is essentially independent of the analyte concentration. It follows that decreasing the initial concentration of analyte must result in an increase of charge density on the droplet, and thus a decrease in the distance between charges, at all stages of solvent evaporation from the droplet. Consequently, the distance between charges will become as small as the distance between charge sites on the analyte molecule at ever earlier stages of droplet evaporation as the initial analyte concentration decreases. It follows that decreasing the analyte concentration must increase the fraction of the droplet's evaporation history during which doubly charges ions can desorb. Therefore, the abundance ratio of doubly charged ions to singly charged ions must increase with decreasing analyte concentration in the sample solution. By the same kind of arguments one concludes that increasing the concentration of GMCD in the sample solution might increase the ratio of singly charged ions to those with two charges. That conclusion is indeed confirmed by experiment. In sum, the ESMS behaviour of these two cyclic peptides is much more consonant with the IDM than with the CRM.

The Predicament Posed by Proteins. Our comparison of the CRM with the IDM has thus far been limited to relatively simple ions having only one or two charges. Even so it has emerged that to explain the experimental observations one must identify and describe a fairly intricate sequence of events. At the risk of becoming entangled in even greater complexities we will now try to elucidate what goes on during the formation of ions with the extensive charge and peak multiplicity that are the "trademark" of ES ions made from large biomolecules like proteins. Figure 4 shows a spectrum for cytochrome C that displays these features. That particular spectrum is of some historic interest because it was one of the first 8 spectra that were ever obtained by ESMS for proteins, an octet that ignited what has sometimes been called the "Electrospray Revolution." The characteristic feature of these spectra is a sequence of several peaks such that the ions of each peak differ only by one adduct charge from the ions of immediately adjacent peaks. It is this coherence that makes possible the accurate determination of molecular weight values for these large molecules with relatively modest mass analyzers.

An important point made earlier in the discussion is that the number of charges on an ion formed from an ultimate droplet of the CRM must be equal to the number it is able to retain or the number on the droplet it occupies, whichever is smaller. The spectrum of Figure 4 clearly shows that cytochrome C molecules can retain any number of charges from 12 to 22. Consequently, to account for the spectrum of Figure 4 by the CRM one would have to presume the production of a population of ultimate droplets in which there are some droplets with each of all possible numbers of charges from 12 to 22. Moreover, the number distribution of those droplets in that range of charge states would have to parallel exactly the distribution of peak heights shown in Figure 4 for the various charge states of the analyte ions. Other analyte species would require other such distributions. For example, ES ions of insulin can have 4, 5, or 6 charges, whereas those of alcohol dehydrogenase can have any number from 32 to 46. How could the initial droplets produced in electrospray dispersion possibly "anticipate" what charge state distributions of droplets they will have to provide for the analyte ions that they contain? Even if they somehow gain that information a priori, how can the subsequent sequence of evaporation and fission steps be programmed or controlled so as to produce the required distributions? That anticipation and control would have to be very sophisticated indeed to account for the further complication implicit in the discovery by Chowdhury et al (*25*) that ES ions of cytochrome C can have at least three different charge state distributions depending on the pH of the initial solution. They attributed these differences to the conformation of the molecules. Similar effects have been found with other species.

The IDM can much more easily explain the great variety of charge distributions found in proteins and other large molecules than can the CRM. The underlying rationale is the same as that used above to explain the behaviour of cyclosporin and gramicidin S (*24*).

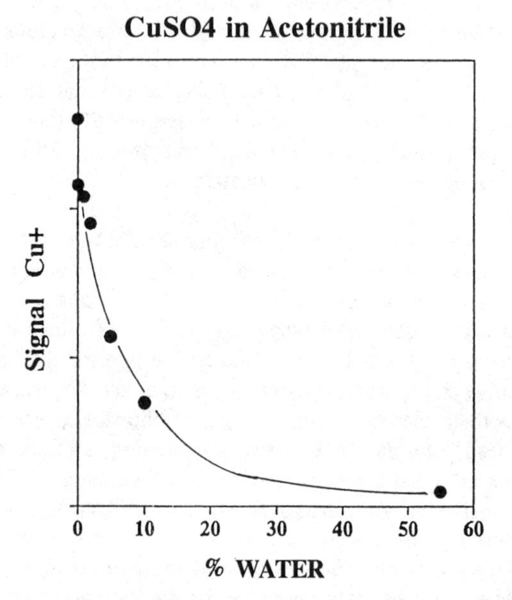

Figure 3. Copper ion intensity vs. water content of an electrosprayed acetonitrile solution. Addition of even a small quantity of water to a solution of $CuSO_4$ in acetonitrile leads to a marked decrease in signal.

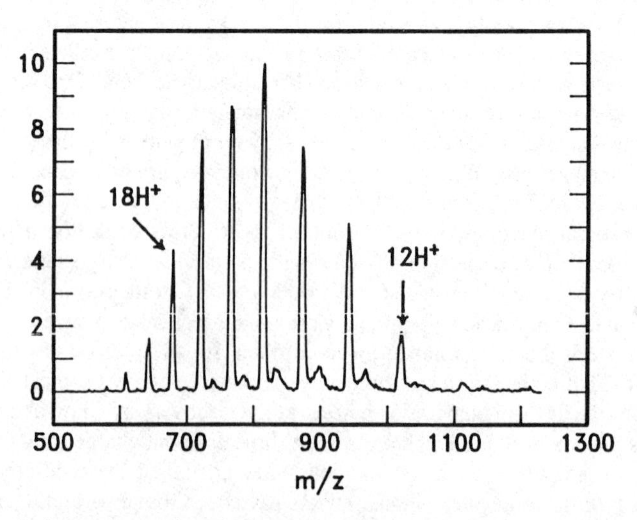

Figure 4. A representative electrospray mass spectrum for cytochrome C in 50:50 methanol–water. The labels on two of the peaks indicate the number of adduct protons on the ions of those peaks. Reproduced with permission from Reference 29. Copyright 1989 American Association of the Advancement of Science.

It assumes that a droplet's charges, protons in the case of the acid solutions that work best with most proteins, are distributed over the droplet surface in an array having as nearly as possible the equidistant spacing that corresponds to the minimum in the electrostatic repulsive energy for the system of charges on the droplet surface. As the droplet shrinks by evaporation the distance between the protons decreases. When a protein molecule approaches the surface in such an attitude that one of its basic groups comes near a surface proton, that group attaches to that proton and becomes pinned to its location. Brownian motion causes the rest of the molecule to thrash around till another basic group attaches to a nearby proton that is within reach. In due course some of the other basic groups on the molecule also attach themselves to surface protons. The number of protons to which the molecule as a whole can attach is clearly determined by the extent of the match between the spatial distribution of the charges and the number and spatial distribution of the attachment sites on the molecule. To be remembered is that the droplet's *excess* charges are locked into an equidistant array on its surface by Coulomb forces. Moreover, the distance between them decreases as evaporation shrinks the droplet. Concomitantly, the "lifting field" at the droplet surface steadily increases as the charges get closer together. The total lifting force felt by the droplet is the product of the total number of attached charges by the strength of the field. At the stage of droplet evaporation when that force becomes strong enough to overcome the solvation forces that anchor the ion to the droplet, the ion desorbs into the ambient gas.

It will be convenient to summarize briefly some salient features of the scenario just outlined along with the consequences that relate to the charge states found on desorbed ions:

[A] At any stage of droplet evaporation the factors determining the rate of analyte ion departure include the concentration of analyte molecules on the surface, the field strength, the total number of charges to which a molecule is attached and the work required to remove the ion from the droplet surface.

[B] The number of charges to which an analyte molecule can be attached is determined by (a) the number of its prospective charge sites along with their spacing, as determined by its size and shape (conformation) and (b) the charge spacing on the droplet surface. As droplet evaporation progresses the distance between charges decreases so the number within attachment-reach of a prospective site increases, as does the surface field. Thus, ions leaving a droplet at an early stage of droplet evaporation depart at a slower rate and with fewer charges than those leaving at a later stage.

[C] Because each droplet contains a limited number of both charges and analyte molecules, it follows that the overall distribution of ions among the various charge states as shown by a mass spectrum will depend on the charge and analyte content of the droplet when ion desorption starts as well as on the rate of solvent evaporation. If that rate is slow, the proportion of low charge states in the analyte mass spectrum will be higher than if the solvent evaporation rate is high.

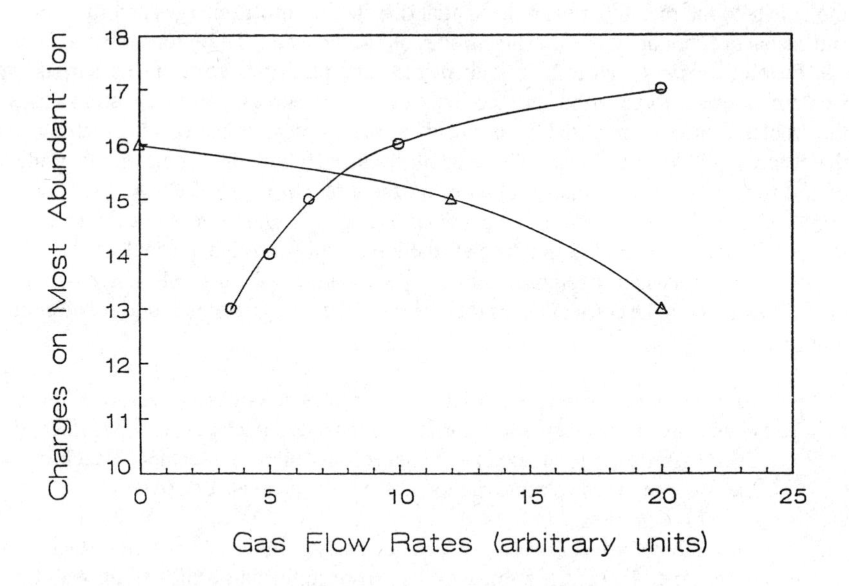

Figure 5. The ordinate values show the number of charges on the most abundant ions in an ES spectrum of cytochrome C in 50:50 methanol–water. The extent of evaporation when ions are sampled increases from left to right for the circle points and decreases for the triangles. Reproduced with permission from Reference 25. Copyright 1993 Elsevier (New York).

The Importance of Droplet Evaporation Rate. An experiment previously described (*25*) confirms the above conclusions on how charge state relates to when an ion desorbs. In that experiment the ES source used ultrasonic vibration of the sprayer to obtain satisfactory nebulization of sample liquid at high flow rates. The piezo-electric vibrator was bathed in a flow of cooling gas that issued from the sprayer in an annular flow around the injection needle, concurrent with the charged droplets drifting "down" the electric field toward the entrance of the sampling tube that passed some of the resulting mixture of bath gas and ES ions into the vacuum system for mass analysis. As in all Analytica sources, dry heated bath gas was introduced annularly around that sampling tube countercurrent to the trajectories of the ions and charged droplets as well as to the flow of cooling gas from the sprayer. These opposing flows of cooling gas and drying gas met in a "stagnation plane" between the sprayer and the sampling tube at an axial position determined by the relative magnitudes of the two flows. On the upstream (sprayer) side of that stagnation plane the droplets and ions were immersed in cooling gas. On the downstream (sampling tube) side they were in drying gas. The electric field between the sprayer and the sampling tube was always strong enough to drive the droplets and ions toward the sampling tube even when they were immersed in bath gas that was trying to drag them in the opposite direction. Two kinds of experiments were carried out:

(i) For a fixed flow rate of cooling gas the volume flow rate of countercurrent drying gas was increased. The result was a decrease in the axial velocity of the charged species (droplets and/or ions) toward the orifice leading into the vacuum system, thus allowing more time for desolvation. In addition the fraction of travel time spent in warm dry bath gas and, therefore, the rate of evaporation, was increased. Both of these effects meant that the sample of ions entering the vacuum system for mass analysis was biased toward those ions leaving the droplets at later stages of the droplet evaporation. Therefore, increasing the drying gas flow should have increased the average number of charges per ion and thus a shift of peaks toward lower m/z values. The curve through the circle points shows an increase in the number of charges on the most abundant ion with increasing flow of drying gas, confirming our speculation.

(ii) The flow rate of cooling gas was increased while the flow rate of drying gas remained constant. The result was to decrease the time available for droplet evaporation before the ions entered the vacuum system. Moreover, during that shorter time the droplet's evaporation rate was also decreased because it spent more of that available time in cooling gas which had a lower temperature and a higher content of solvent vapor than did the drying gas. Consequently, as the flow rate of cooling gas increased, the ions reaching the analyzer should have been increasingly representative of those that desorbed at earlier stages of the droplet evaporation and thus have lower m/z values on average. The curve through the triangle points in Figure 5 shows a decrease in the number of charges on the most abundant ion, thus again confirming a prediction of the IDM model.

In our view these results constitute persuasive evidence that in populations of multiply charged ions produced from charged droplets the charge state distributions depend strongly upon the time dependence of evaporation rate during the droplet lifetime. There seems to be no way that the CRM can account for these variations in charge state distributions. The IEM both predicts and explains them.

Summary and Concluding Remarks

Two principal models have been proposed to account for the formation of free ions in the gas phase from solute species in charged droplets of solution: (a) the Charged Residue Model (CRM) of Malcolm Dole and his colleagues and (b) the Ion Desorption Model (IDM) of Iribarne and Thomson. In this report we have first undertaken to provide a consistent, clear and unequivocal distinction between these two models. Then, in the perspective of these models, we have examined several sets of experimental results, obtained with different kinds of species under varying experimental conditions. In every case the IDM seemed better able than the CRM to provide a credible explanation of the observations.

One final piece of evidence should be mentioned. In a paper just accepted for publication in *J. Chem. Phys.* Loscertales and Fernandez de la Mora present values of droplet surface field during ion desoption that they have deduced from mobility measurement of the droplet residues (*26*). Their results include inferences on the dependence of surface field on the nature of solute and solvent as well as on rate constants for the ion desorption process. The story they tell is overall so consistently consonant with the IDM, and so incomprehensible in terms of the CRM, that it comprises compelling evidence for the former's validity in the cases they studied.

The question remains as to whether there are any situations in which the CRM might apply. The only one we have encountered emerged in some experiments in our laboratory on poly (ethylene glycols) (PEGs) having molecular weights up to five million (*27*). There it was found that in order to obtain a stable spray with the largest oligomers, the sample solution had to be so dilute that there was only one solute molecule in most of the initial droplets. Those droplets were 2.8 um in diameter and the molecules comprised chains 40 or more microns in length. Clearly, the molecule in each droplet had to be coiled up so that the droplet resembled a piece of composite material like fiberglass. Consequently, the droplet could not very well undergo a Coulomb explosion nor could the molecule very well leave the droplet as an ion by the same kind of desorption mechanism that would apply to the much smaller solute molecules we have been considering. We believe instead that the solvent molecules evaporated from the droplet individually, one at a time, all the way to "dryness." As the surface charge density approached the Rayleigh limit, the effective surface tension became too small to maintain the droplet shape at the minimum surface/volume ratio associated with a sphere. Consequently, the Coulomb repulsion stretched that droplet into a long thin column comprising a long linear molecule sheathed in a layer of solvent molecules that continued their evaporation to leave a linear multiply-charged ion of PEG with an m/z value around 1200.

The number of those charges found on the largest oligomers provided strong evidence that the ion formation occurred as we have described it. An earlier paper on ESMS of PEGs presented an electrostatic repulsion model by which one could compute the maximum number of charges that a PEG ion of any size could hold (*28*). In all experiments with smaller oligomers the observed ions had at most only about two thirds of the maximum number charges, according to that model. At the time we guessed this deficit might be due to the possibility that the ions desorbed before they had attached as many charges as they could hold. That guess now seems to have been a good one in view of our present understanding, set forth above, on how droplet evaporation leads to desorption in accordance with the IDM. The noteworthy observation for the very large PEG oligomers was that they retained the full number of charges allowed by the electrostatic repulsion model. We guess again that the reason for this charge saturation is that those very large oligomers could not leave the droplet by desorption. The initial droplet had about ten times as many charges as the molecule could hold so there was always plenty of charge around as the solvent molecules evaporated. Therefore, the molecule retained a charge on as many sites as the electrostatic repulsion model allowed.

In sum, the available evidence indicates that the Ion Desorption Model of Iribarne and Thomson applies to the formation of ions from charged droplets in almost all experiments. The Charged Residue model of Malcolm Dole seems applicable only when the solute ions have linear dimensions substantially larger than those of the droplet that contains them.

Acknowledgments

This work was supported in part by NIH Grant R01 GM31660-09 and in part by NSF grant MCB-9118224. We thank Professor Alessandro Gomez and Keqi Tang at Yale University for their measurements of droplet size distributions in electrosprays of some of our sample solutions. We are grateful to Professor Juan F. de la Mora for many provocative and helpful discussions.

Literature Cited

1. Dole, M.; Mach, L. L.; Hines, R. L.; Mobley, R. C.; Ferguson, L. P.; Alice, M. B. *J. Chem. Phys.* **1968**, *49*, 2240.
2. Cook, K. D. *Mass Spectrom. Rev.* **1986**, *5*, 467.
3. Mach, L. L.; Kralik, P.; Rheude, A.; Dole, M. *J. Chem. Phys.* **1970**, *52*, 4977.
4. Clegg, G. A.; Dole, M. *Biopolymers* **1971**, *10*, 821.
5. Teer, D.; Dole, M. *J. Poly. Sci.* **1975**, *13*, 985.
6. Iribarne, J. V.; Thomson, B. A. *J. Chem. Phys.* **1976**, *64*, 2287.
7. Thomson, B. A.; Iribarne, J. V. *J. Chem. Phys.* **1979**, *71*, 4451.
8. Iribarne, J. V.; Dziedzic, P. *J. Int. J. Mass Spectrom. Ion Phys.* **1983**, 50, 331.
9. Blakley, C. R.; McAdams, M. J.; Vestal, M. L. *J. Chromat.* **1977**, *158*, 264.
10. Blakley, C. R.; Carmody, J. J.; Vestal, M. L. *Anal. Chem.* **1980**, *52*, 1636.
11. Blakley, C. R.; Carmody, J. J.; Vestal, M. L. *Clin. Chem.* **1980**, *26*, 1467.
12. Blakley, C. R.; Carmody, J. J.; Vestal, M. L. *J. Am. Chem. Soc.* **1980**, 102, 5931.
13. Guevremont, R; Siu, K. W. M.; LeBlanc, J. C. Y.; Berman, W. W. *J. Am. Soc. Mass Spectrom.* **1992**, *3*, 216.
14. Siu, K. W. M.; Guevremont, R.; LeBlanc, J. C. Y.; O'Brien, R. T.; Berman. S. S. *Org. Mass Spectrom.* **1993**, *28*, 579.
15. Guevremont, R.; LeBlanc, J. C. Y. *Org. Mass Spectrom.* (in press).
16. Kelly, M. A; Vestling, M. M; Fenselau, C; Smith, P. B. *Org. Mass Spectrom.* **1992**, *27*, 1143.
17. Schmelzeisen-Redeker, G.; Buttering, L.; Roellgen, F. W. *Int. J. Mass Spectrom. Ion Processes* **1989**, *90*, 139.

18. Kebarle, P.; Tang, L. *Anal. Chem.* **1993**, *65*, 973A.
19. Taflin, D. C.; Ward, T. L.; Davis, E. J. *Langmuir* **1989**, *5*, 376.
20. Davis, E. J., Bridges, M. A. *J. Aerosol Sci.* **1994**, *25*, 1179.
21. Yamashita, M.; Fenn, J. B. *J. Phys. Chem.* **1984**, 4451.
22. Fuerstenau, S., *Ph.D. Thesis*, **1994**, Yale University.
23. Blades, A. T.; Jayweera, P.; Ikonomou, M. G.; Kebarle, P. *J. Chem. Phys.* **1990** *92*, 5900.
24. Chowdhury, S. K.; Katta, V.; Chait, B. *J. Am. Chem. Soc.* **1990,** *12*, 9012.
25. Fenn, J. B. *J. Am. Soc. Mass Spectrom.* **1993,** *4*, 524.
26. Loscertales, I. G.; Fernandez de la Mora, J. "Experiments on the Kinetics of Field-Evaporation of Small Ions from Droplets" *J. Chem. Phys.* **1995**, in press.
27. Nohmi, T.; Fenn, J. B. *J. Am. Chem. Soc.* **1992,** *114*, 3245.
28. Wong, S. F.; Meng, C. K.; Fenn, J. B. *J. Phys. Chem.* **1988**, *92*, 546.
29. Fenn, J. B.; Mann, M.; Meng, C. K.; Wong, S. F.; Whitehouse, C. M. *Science* **1989,** *246,* 64.

RECEIVED November 13, 1995

Chapter 4

Liquid Chromatography with Electrospray Ionization Tandem Mass Spectrometry

Profiling Carbohydrates in Whole Bacterial Cell Hydrolysates

Gavin E. Black and Alvin Fox[1]

Department of Microbiology and Immunology, School of Medicine, University of South Carolina, Columbia, SC 29208

Carbohydrates can serve as chemical markers that allow for the taxonomic differentiation of bacteria. Gas chromatography-mass spectrometry (GC-MS) is a proven technique for profiling neutral and amino sugars (as alditol acetates) in bacterial cell hydrolysates. The chromatograms display low background and mass spectra are readily interpretable. Unfortunately, the required derivatization is time consuming and not applicable to all sugars of interest. This review concerns the profiling of underivatized sugars using electrospray ionization tandem mass spectrometry with either direct injection (MS-MS) or on-line liquid chromatography (LC-MS-MS). Both MS-MS and LC-MS-MS can facilitate the rapid identification of carbohydrates over conventional GC-MS. However, unlike MS-MS, LC-MS-MS can readily discriminate sugar isomers. Sugar standards can be chromatographically resolved and thus be analyzed by the rather non-specific pulsed amperometric detector. However, chromatograms obtained for bacterial whole cell hydrolysates are complicated and require the specificity of MS for analysis. For MS analysis, following high pH anion exchange chromatography, a cation suppressor is used to remove sodium hydroxide from the eluent. Amino sugars are removed in the suppressor. Thus analyses are focussed on profiling of acidic and neutral sugars. The LC mobile phase generates considerable background ions on MS analysis, thus MS-MS is vital to obtain product ion spectra for sugar identification. LC-MS-MS instrumentation is more expensive than GC-MS and profiles different groups of sugars. However, sample preparation for LC-MS-MS and MS-MS is simpler in comparison to GC-MS.

Electrospray (ES) is a well established technique allowing mass spectrometry (MS) analysis of polar compounds from aqueous solution without derivatization. For this reason, ES is exquisitely suited for on-line analysis in conjunction with liquid chromatography (LC). The potential of LC-MS for analysis of sugar monomers has been

[1]Corresponding author

0097–6156/95/0619–0081$13.25/0

demonstrated (1-3), however their analysis in more complex matrices requires further investigation. High pH anion exchange LC (HPAEC) with pulsed amperometric detection (PAD) is suitable for profiling of sugar monomers (4). Due to the simplicity and sensitivity of LC-PAD it has rapidly become one of the most common techniques for chromatographic analysis of carbohydrates. However, PAD is non-selective and is unsuitable for complex biological matrices (5, 6). For complex samples, the specificity offered by MS is essential.

Microbial identification and taxonomic differentiation are traditionally accomplished by culture-based methods to determine physiological characteristics. Alternatively, identification can be achieved by determining the presence of unique sugars (chemical markers) as alditol acetate derivatives utilizing GC-MS analysis (7-9). The alditol acetate procedure allows derivatization of neutral and amino sugars but not acidic sugars. This limitation arises from the fact that hydroxyl and amino groups are amenable to acylation but carboxylic acid functions are not.

There is considerable diversity of carbohydrates among bacterial species, many of which are readily identifiable by GC-MS. Sugars exist in a multitude of stereoisomers. These stereoisomers cannot be readily distinguished on grounds of their molecular mass but often have different chromatographic retention times (10). Thus chromatographic characteristics are extremely important in sugar identification. An excellent review detailing all known monosaccharide components of bacterial polysaccharides (85 in total) reported in the literature before 1989 has been published. As noted by Lindberg "only a limited number of all the bacterial families and tribes have been investigated for their cell wall or extracellular polysaccharides or both" (11).

A brief overview of some aspects of sugar chemistry provides a perspective on variability in sugar structure. Further information on sugar structure is provided elsewhere (12). Sugars are polyhydroxyl compounds and their backbones commonly contain from 3-7 carbons. Each internal carbon acts as an optical center. For example, for a simple hexose (e.g. glucose) there are 4 optical centers or 4^2 isomers. Monosaccharides commonly exist in both aldose and ketose forms (carbonyl function in either position 1 or 2 respectively). Replacement of hydroxyl and/or aldehyde functions with amino or carboxylic acid functions leads to aminosugars, alduronic acids (COOH in position 6), aldonic acids (COOH in position 1) and aldaric acids (COOH in position 1 and 6). The primary and secondary hydroxyl groups may be reduced to give methyl or deoxysugars respectively and dehydration of acidic or neutral sugars leads to lactone and anhydro sugars respectively. Derivatization of the functional groups (acylation, methylation) is also commonly observed.

As an example of taxonomic discrimination provided by carbohydrate markers, muramic acid (MA, (3-O-lactyl glucosamine) is a sugar common to almost all eubacteria (as a component of cell wall peptidoglycan) while not present in non-bacterial matter, including fungi (13,14). In a wider sense, neutral and aminosugar profiles allow discrimination among Gram negative bacteria and Gram positive bacteria. For example, aminodideoxyhexoses are useful for distinguishing among the *Legionellaceae*. Fucosamine is characteristic of the genus *Legionella* whilst quinovosamine is found in the genus *Tatlockia* (15,16). O-methylated sugars are useful in discriminating various *Bacillus* species. Thus 2-O-methyl and 3-O-methyl-rhamnose are found in the spores of *B. cereus* but only the 3-O-methyl isomer in B. *anthracis* (17) .

Profiling of carbohydrates by GC-MS of the corresponding alditol acetates has become routine in our laboratory. It is noteworthy that chemotaxonomic speciation provided by carbohydrate profiles agrees closely with molecular differentiation techniques such as ribosomal RNA sequencing (18) and PCR amplification of ribosomal RNA spacer regions (19). Unfortunately, GC-MS analysis requires derivatization which is both labor intensive and time consuming.

Simple and rapid analysis is achievable by ES-MS-MS without prior chromatography. However, suppression from sample matrix components can adversely affect detection limits (20). Furthermore, much chemotaxonomic information resides in the differentiation of sugar isomers (15-19). Such isomers are unlikely to be readily discriminated by MS-MS analysis. On the other hand, sugar isomers can be distinguished on the grounds of different chromatographic retention times (16,17). For example, *B. anthracis* and *B. cereus* can often be distinguished by the presence of galactose although both species commonly contain glucose (17). Due to interferences, unambiguous identification of sugars in complex matrices requires high detection specificity. The latter may be provided by LC-MS, or better yet, LC-MS-MS. Because LC-MS-MS does not require derivatization it offers an appealing alternative or complementary technique to GC-MS profiling

Excellent chromatography of native sugars can be obtained using HPAEC with pulsed amperometric detection (PAD) (21). However, PAD is not a selective detector, (5, 6) and for detection of sugars in complex matrices (e.g. bacterial whole cell hydrolysates) MS is required. Unfortunately, the high ionic strength solutions necessary for this type of LC separation are not well suited to MS (2,3). This limitation may be overcome by using on-line suppressors (replacing NaOH with H_2O) but the price paid for it is that positively charged chemotaxonomic markers, such as aminosugars, are lost in the process. The purpose of this review is to describe the current status of carbohydrate profiling by LC-ES-MS, ES-MS-MS and LC-ES-MS-MS. Reviews on the current status of GC-MS can be found elsewhere (7-9).

Direct injection mass spectrometry and tandem mass spectrometry

Electrospray ionization followed by MS-MS analysis has been used for the analysis of underivatized carbohydrates from whole cell hydrolysates (20). MA was first released by hydrolysis in 2N sulfuric acid. Acid was then removed by extraction with *N,N* dioctylmethylamine in chloroform followed by hydrophobic clean-up using C-18 columns. No derivatization was performed and sample processing time was reduced to ~4 hr compared to a 50 hr preparation time for GC-MS. Additionally, the elimination of chromatography reduced instrumental analysis time from 60 min to 2 min per sample. For compounds present at low levels, however, signal suppression by components of the sample matrix adversely affected sensitivity, often making product ion spectra unobservable.

For aminosugar analysis, positive ion ES produced abundant molecular ions $(M+H)^+$ with little fragmentation. Acidic sugar analysis, using negative ion ES, produced primarily deprotonated molecular ions $(M-H)^-$. Figure 1 shows ES MS spectra of

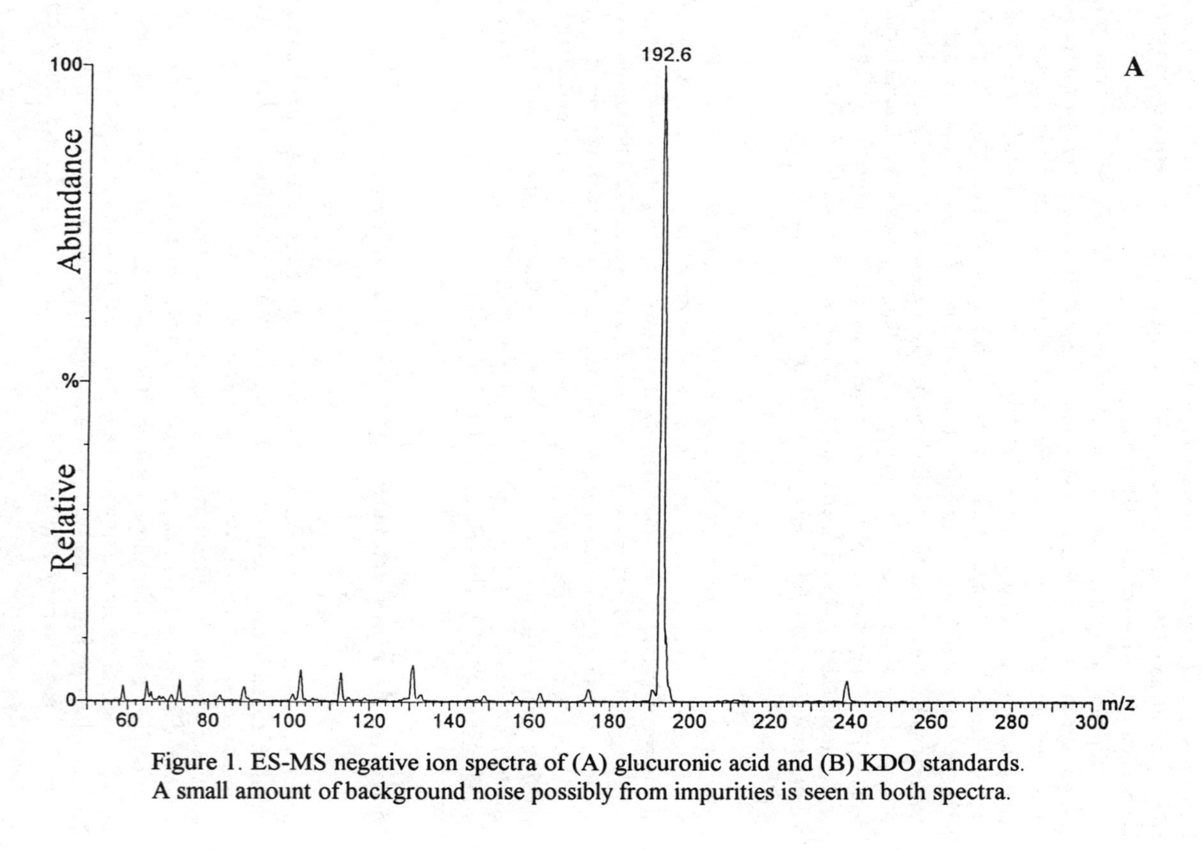

Figure 1. ES-MS negative ion spectra of (A) glucuronic acid and (B) KDO standards. A small amount of background noise possibly from impurities is seen in both spectra.

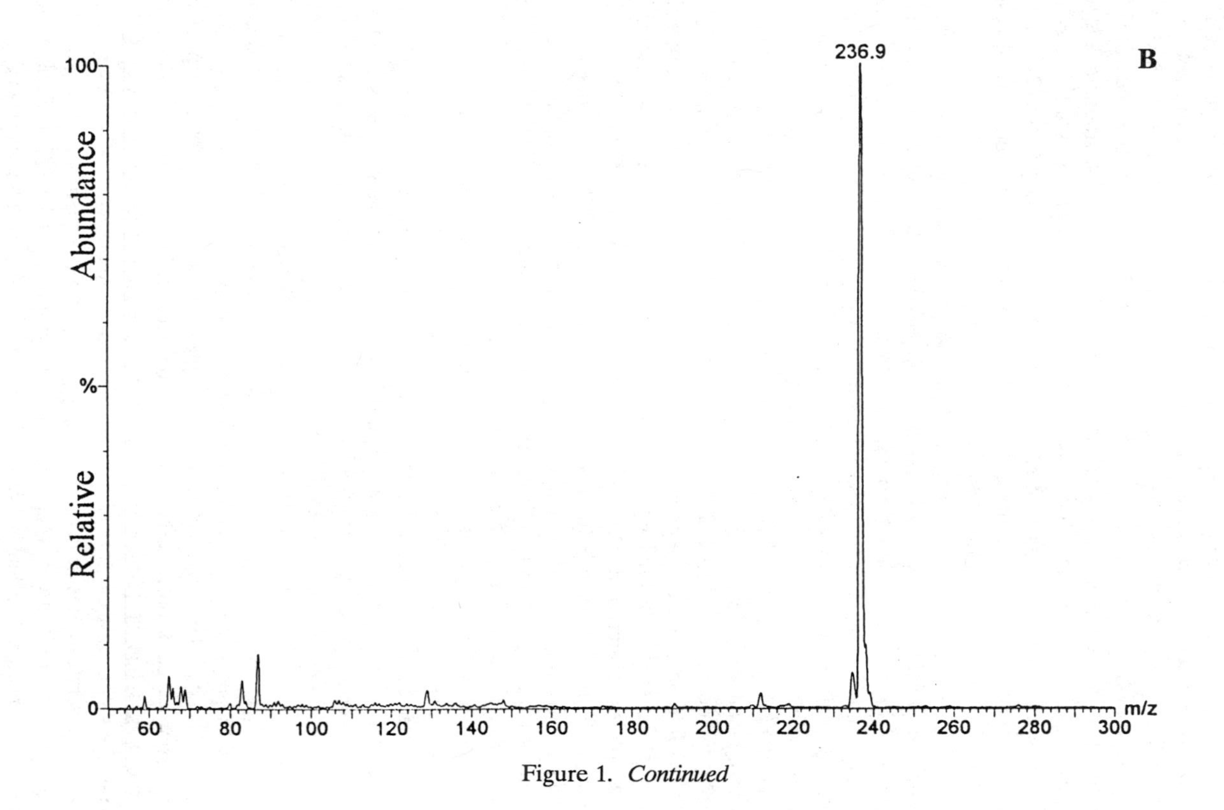

Figure 1. *Continued*

glucuronic acid (GlcUA, mol.wt. 194) and 3-deoxy-D-manno-octulosonic acid (KDO, molecular wight 238) as examples of acidic sugars. These sugars both form predominantly deprotonated molecular ions (m/z 193 for GlcUA, m/z 237 for KDO).

Neutral sugars ionize weakly under both positive ion and negative ion ES conditions. However, neutral sugars can form adducts under both positive and negative ion conditions. Neutral sugars, including glucose and fucose, have been detected using positive ion ES as their Na^+, NH_4^+ and Li^+ adducts (2,3). It has been observed for plasmaspray (2) and thermospray (3) analysis, that formation of ammonium adducts were necessary for sugar ionization. The negative ion mode has been used for detection of low levels of anhydroglucitol in serum in form of chlorine adduct ions (22). In the presence of acetate, neutral sugars generate $[M+AcO]^-$ and $[M-H]^-$ under negative ion ES-LC-MS conditions (unpublished data).

For compounds containing both carboxyl and amino groups, such as phospholipids (23) mycotoxins (24) and MA positive and negative ionization can be used to increase selectivity. Additional specificity is provided by collision-induced dissociation (CID). For example, in the positive ion mode, MA produced a prominent protonated molecular ion at m/z 252 $[M+H]^+$ (20) whereas in the negative ion mode the deprotonated ion 250 $[M-H]^-$ was observed (9). The precursor ion m/z 250 generated an ion at m/z 89, presumably from the loss of a lactate anion on CID (9). MA is known to lose lactic acid under alkaline conditions and this fact has become the basis for a spectrometric assay for MA (25,26).

Multiple cellular components in whole cell hydrolysates produce a complex mixture of signals. Thus full scan or selected ion monitoring does not allow definitive detection of chemical markers. When chemical markers are present at high concentrations, it is possible to obtain product ion spectra. However, the increased sensitivity of multiple reaction monitoring (MRM) is often essential. In this instance, a characteristic ion (selected in the first mass analyzer) produces specific fragments after CID, which are monitored in the second mass analyzer. MS-MS produces lower ion intensities than MS analysis, due to poor transmission between the two mass analyzers, but improved signal to noise ratio leads to lower detection limits.

As an example of MS-MS analysis of whole cell hydrolysates, monitoring for ions characteristic of MA did not allow discrimination of bacteria and fungi when ES-MS was used. However, monitoring for product ions of MA allowed ready differentiation of bacteria and fungi (9,20). The acidic sugar, KDO, is present in the lipopolysaccharides (LPS) of most, but not all, Gram negative bacteria. MS-MS analysis of authentic KDO produced an identical fragment spectrum to KDO from a hydrolysate of the Gram negative bacterium *Legionella pneumophila* (see Figure 2).

When complex samples containing multiple compounds enter the ES source, ion suppression is a commonly observed phenomenon with some compounds ionizing better than others. Thus, despite the sensitivity of ES, the limit of detection may be greatly diminished as compared to pure compounds. For example, GlcUA is present in the hyaluronic acid capsules of certain group A and group C streptococci (27), we were unable to detect GlcUA in streptococcal hydrolysates using ES-MS-MS with direct injection. It required on-line LC-MS to detect this sugar (see following text).

Liquid chromatography-mass spectrometry

Many LC systems for sugar analysis have been described but few provide both high resolution separations and column durability (28, 29). One column that displays these characteristics is an HPAEC column (PA-1) developed by Dionex. This column has been primarily used in conjunction with a PAD detector (4,21,28). As noted above, it may be of advantage to combine HPAEC with ES-MS or ES-MS-MS. However this combination is not straightforward (2,3). To ionize carbohydrates for HPAEC, high concentrations of NaOH (18-100 mM) are necessary. For acidic sugars (including KDO and GlcUA), a gradient of Na^+AcO^- was needed for displacement from the anion exchange resin (28).

A cation suppressor exchanges the sodium ions in solution with hydrogen ions, thus replacing sodium hydroxide with water. A 2 mm anion self-regenerating suppressor from Dionex (Sunnyvale, CA) was used in our studies. This anode-based electrolysis system produces H^+ and O_2 from H_2O. The generated H^+ ions replace Na^+ (present in the mobile phase in the form of Na^+AcO^-). Removal of Na^+ thus generates acetic acid, which is tolerated by ES ionization. Using a 2 mm column, the flow rate can be reduced to 100 μl/min. Whilst not entirely optimal for chromatographic resolution this flow rate of concentrated NaOH is readily handled by the ion suppression system.

The potential of HPAEC LC-MS, was demonstrated previously using thermospray (2) and ES (3) for the analysis of sugar standards. Amino-containing compounds (eg. aminosugars) do not pass through the suppressor. In complex samples which contain a large amount of protein, high concentrations of amino acids are generated. The presence of the on-line suppressor effectively removes these amino acids, thus acting as an on-line clean-up procedure. Previous LC-MS studies have not addressed the analysis of acidic sugars from bacterial cell hydrolysates. As mentioned above, acidic sugars are particularly difficult to analyze by GC-MS due to their carboxyl groups which require additional derivatization. Certain alternative columns can be used at neutral pH (eliminating the need for an ion suppressor) which makes them more compatible with MS detection, but chromatographic resolution can be compromised. Previous LC-MS studies have relied upon post-column addition of reagents that promote adduct ion formation (3,22). Instead the desired adduct ions may also be formed without the need of post-column reagents by selecting an appropriate LC buffer. For example, we have used Na^+AcO^- gradients to form the molecular acetate adduct ions (unpublished). Deoxyhexoses (dHex), including rhamnose and fucose, have a molecular weight of 164 and under negative ion ES these sugars formed both acetate adduct ions ($[dHex+AcO]^-$ = m/z 223) and deprotonated molecular ions ($[dHex-H]^-$ = m/z 163). Similarly glucose (mol.wt. 180) and ribose (mol.wt. 150), both commonly found in bacteria, produced acetate adduct ions $[M+AcO]^-$; m/z 239 and m/z 209 respectively and deprotonated molecular ions, $[M-H]^-$; at m/z 179 and m/z 149 respectively. Sometimes dimers are observed e.g. $[Hex-Hex]^-$ at m/z 359. Acidic compounds formed primarily deprotonated molecular ions; gluconic acid (GlcOA, used as an internal standard) which has a molecular weight of 196 formed an ion of m/z 195, N-acetyl neuraminic acid (mol. wt. 309) observed at m/z 308, KDO (mol. wt. 238) at m/z 237 and glucuronic acid (GlcUA, mol. wt. 194) at m/z 193.

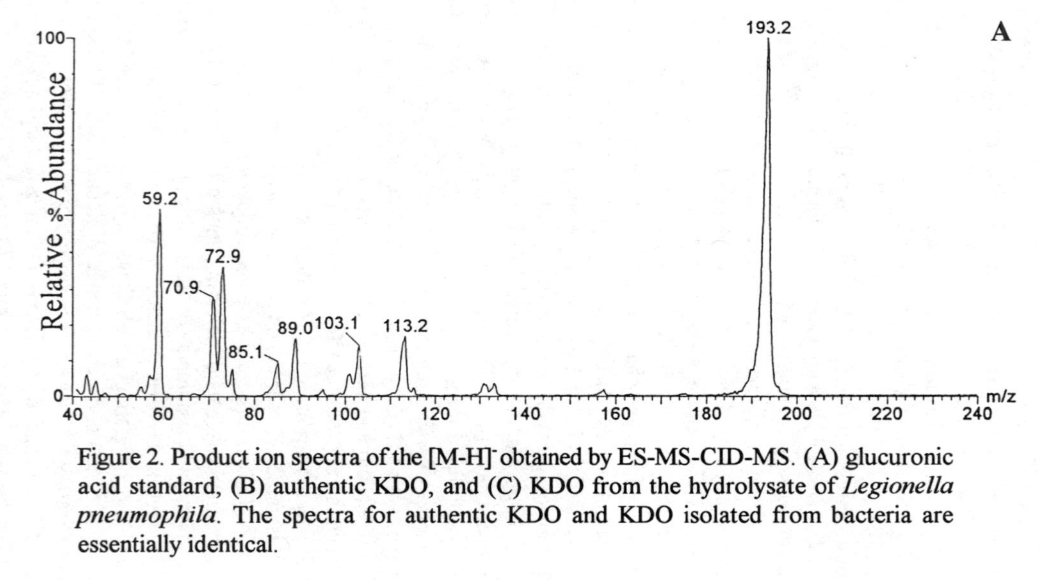

Figure 2. Product ion spectra of the $[M-H]^-$ obtained by ES-MS-CID-MS. (A) glucuronic acid standard, (B) authentic KDO, and (C) KDO from the hydrolysate of *Legionella pneumophila*. The spectra for authentic KDO and KDO isolated from bacteria are essentially identical.

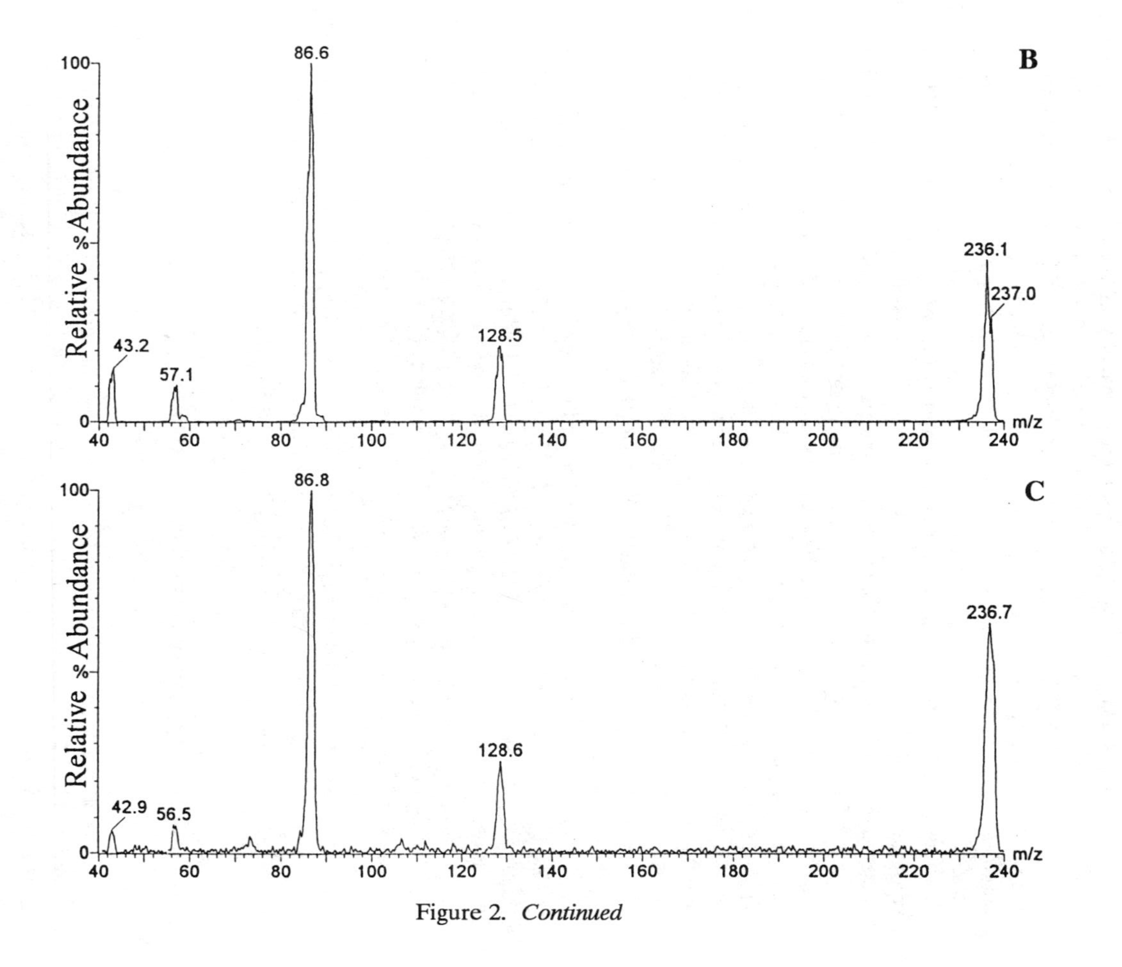

Figure 2. *Continued*

In total ion chromatograms, peaks characteristic of sugars were observed on a high baseline or background signal generated by components of the mobile phase (e.g. AcO^-, m/z 59). In the SIR mode data acquisition was focussed only on the deprotonated molecular ion (for neutral and acidic sugars) or acetate adduct ions (neutral sugars) rather than full scans. Chromatograms showed signals for all the compounds in the standard mixture (compare Figures 3 and 4) except for the amino sugars. For example, as seen in the LC-MS SIR chromatogram monitoring for m/z 250 $[M-H^-]$ showed the absence of MA although a peak for this compound is readily seen in the LC-PAD chromatogram. As noted above, amino-containing compounds are unable to pass through the cation permeable suppressor.

Bacterial hydrolysates presented considerably more complicated LC-PAD chromatograms (Figure 5) than standards (Figure 3). Even when set to optimize sugar detection, the PAD is not a specific detector. There are many compounds present in high concentrations in the bacterial cell which may contribute significant background response. However, LC-MS SIR chromatograms allowed ready identification of neutral and acidic sugars in whole cell hydrolysates. For example, as shown in Figure 6, rhamnose, glucose, ribose and GlcUA acid were readily detected in *S. zooepidemicus*. GC-MS analysis of alditol acetates of this organism confirmed the presence of these neutral sugars. Additionally GC-MS analysis showed the presence of aminosugars including MA, glucosamine, and galactosamine (data not shown). Thus LC-MS and GC-MS together produce complementary sugar profiles allowing identification of amino, neutral, and acidic sugars.

Carbohydrates were also generated by hydrolysis of whole cells of the Gram negative organism, *Legionella pneumophila.* LC-MS demonstrated the presence of rhamnose, hexoses, ribose, and KDO (see Figure 7). Under these chromatographic conditions mannose and glucose were not resolved. Previous GC-MS studies showed the presence of the neutral sugars rhamnose, mannose, glucose, and ribose, in addition to the aminosugars MA, glucosamine and quinovosamine (16). Similarly, GC-MS has previously demonstrated the presence of KDO in *L. pneumophila* (32). To determine the sensitivity of this technique, spiking experiments were performed. GlcUA acid (not present in *L. pneumophila*) corresponding to 0.01 through 10.0% of the bacterial dry weight in each sample was added to cell hydrolysates. The lowest level of detection, using SIR, corresponded to 0.05% (data not shown).

Liquid chromatography-tandem mass spectrometry

The common ionization techniques for GC-MS are electron ionization (EI) and chemical ionization (CI). The former is used for structural identification and produces fragment mass spectra, whereas CI produces intact molecular ions and is helpful in establishing molecular weight. As noted above, ES produces primarily intact, molecular ions, although fragmentation can be accomplished by varying ionization conditions. In the positive ion mode aminosugars gain a proton, whilst in negative ion mode acidic sugars lose a proton. However, due to the high abundance of low mass ions from the LC eluent, in-source production of fragment ions from the small molecules can be difficult

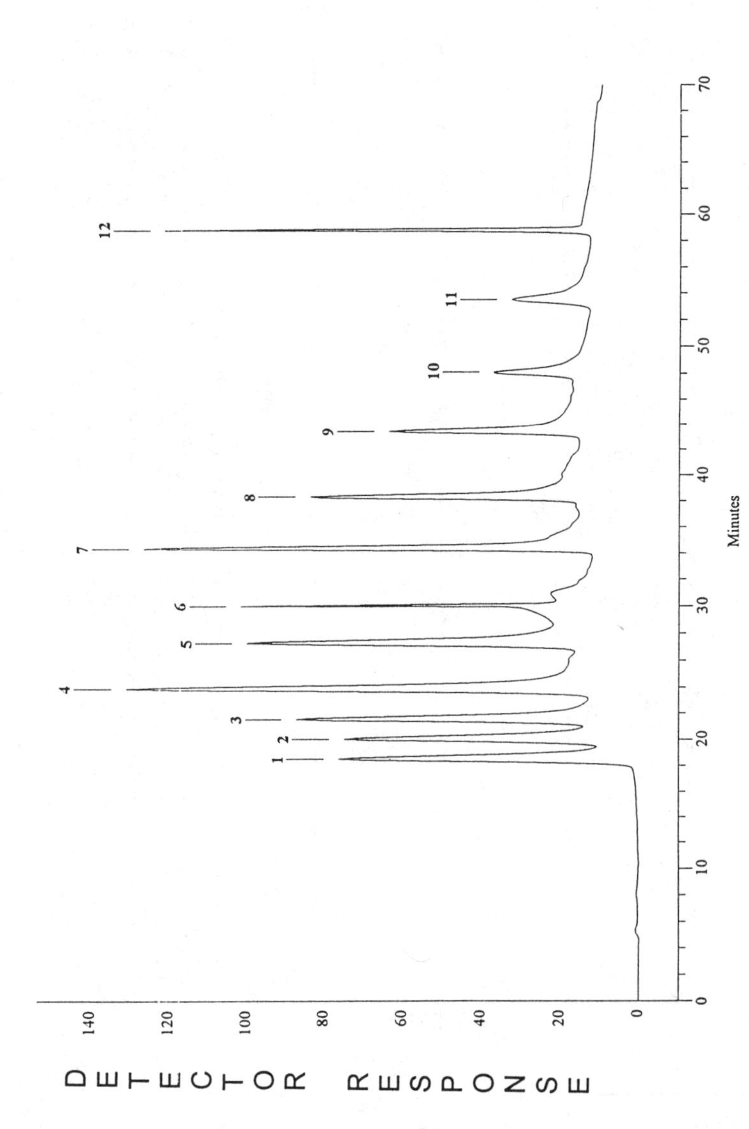

Figure 3. PAD chromatogram of a standard mixture of common bacterial sugars: 1, rhamnose; 2, galactosamine; 3, glucosamine; 4, hexose (mannose or glucose); 5, ribose; 6, acetate signal; 7, gluconic acid (internal standard); 8, N-acetyl neuraminic acid; 9, ketodeoxyoctonic acid; 10, muramic acid; 11, galacturonic acid; 12, glucuronic acid.

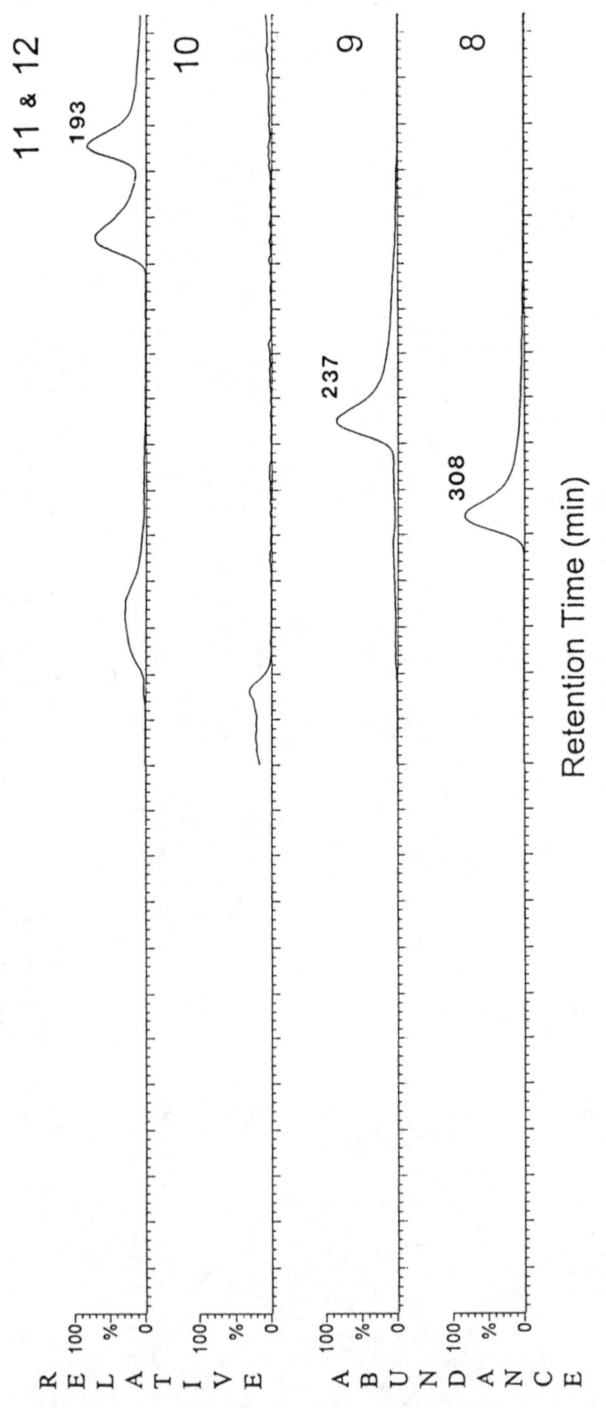

Figure 4. SIR chromatograms of the same mixture as in Fig.3. The numbers on the right hand side of each chromatogram are the same compounds listed in Figure 3. Note compound 10 (muramic acid, an aminosugar) which elutes after KDO (compound 9) is not detected. The numbers above each peak refer to the m/z ratios monitored at the respective retention time.

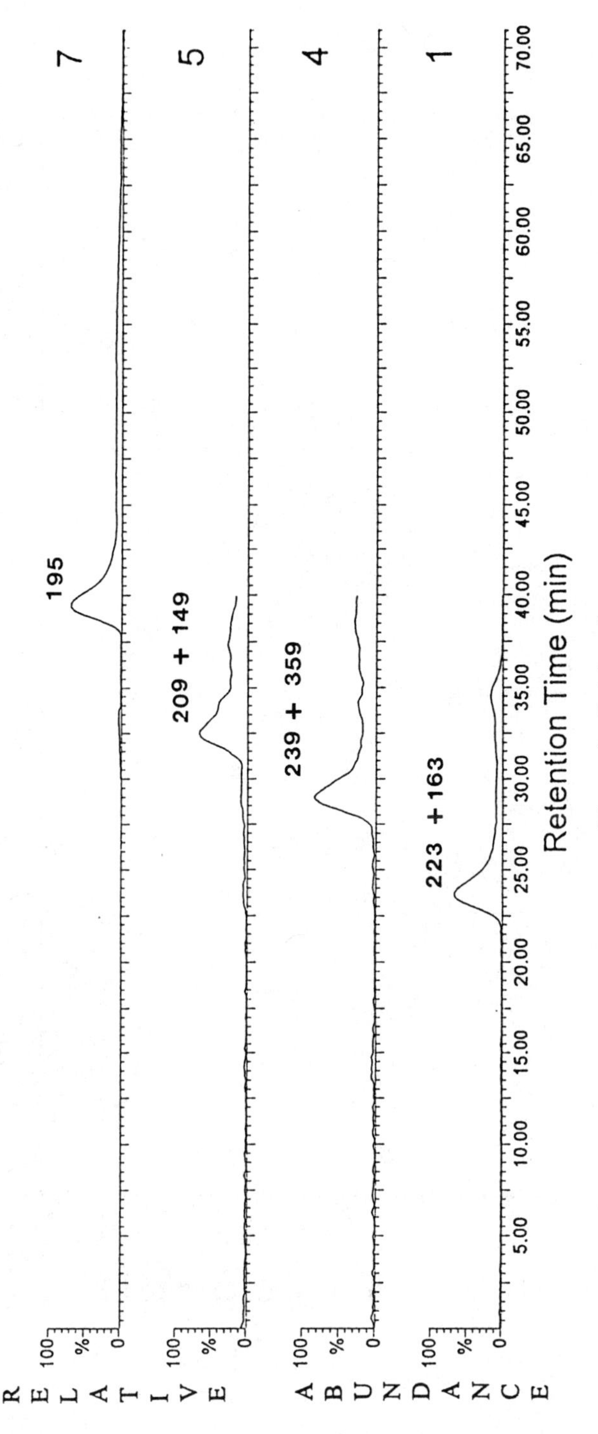

Figure 4. *Continued*

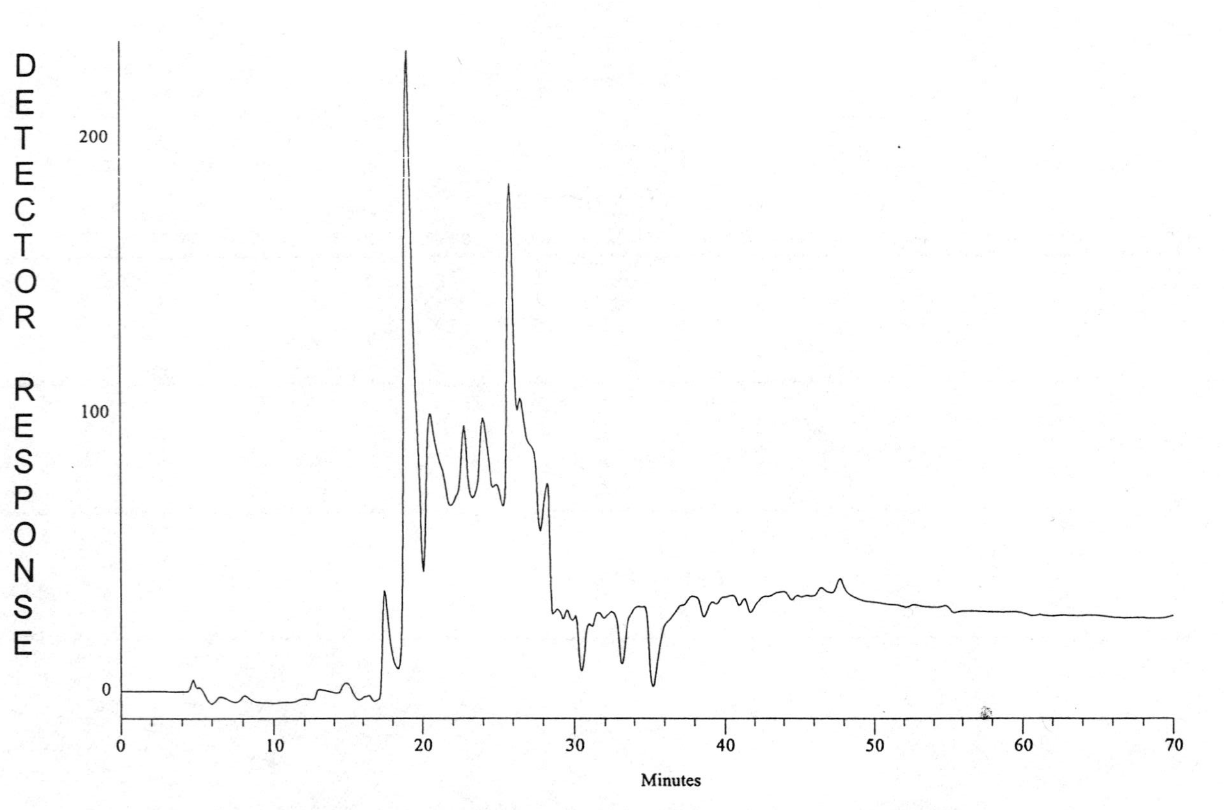

Figure 5. PAD chromatogram of *Streptococcus zooepidemicus* hydrolysate.

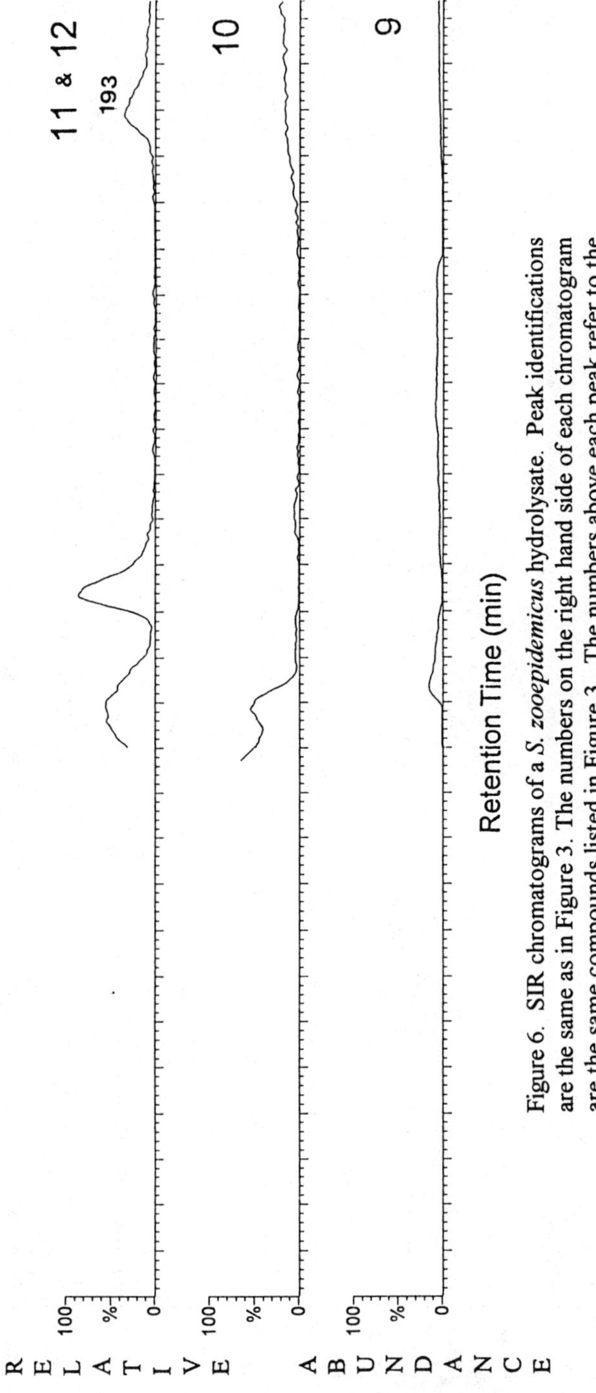

Figure 6. SIR chromatograms of a *S. zooepidemicus* hydrolysate. Peak identifications are the same as in Figure 3. The numbers on the right hand side of each chromatogram are the same compounds listed in Figure 3. The numbers above each peak refer to the m/z ratios monitored at the respective retention time.

Continued on next page

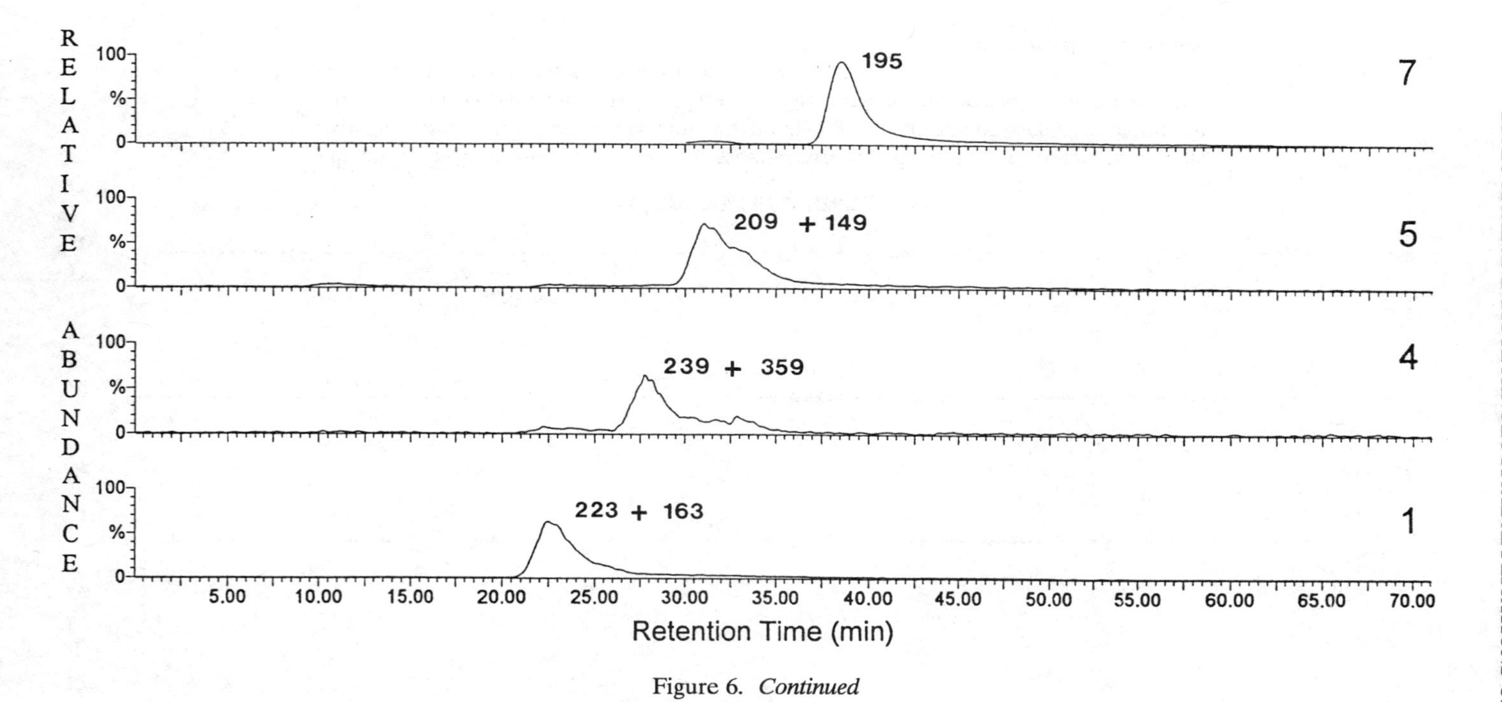

Figure 6. *Continued*

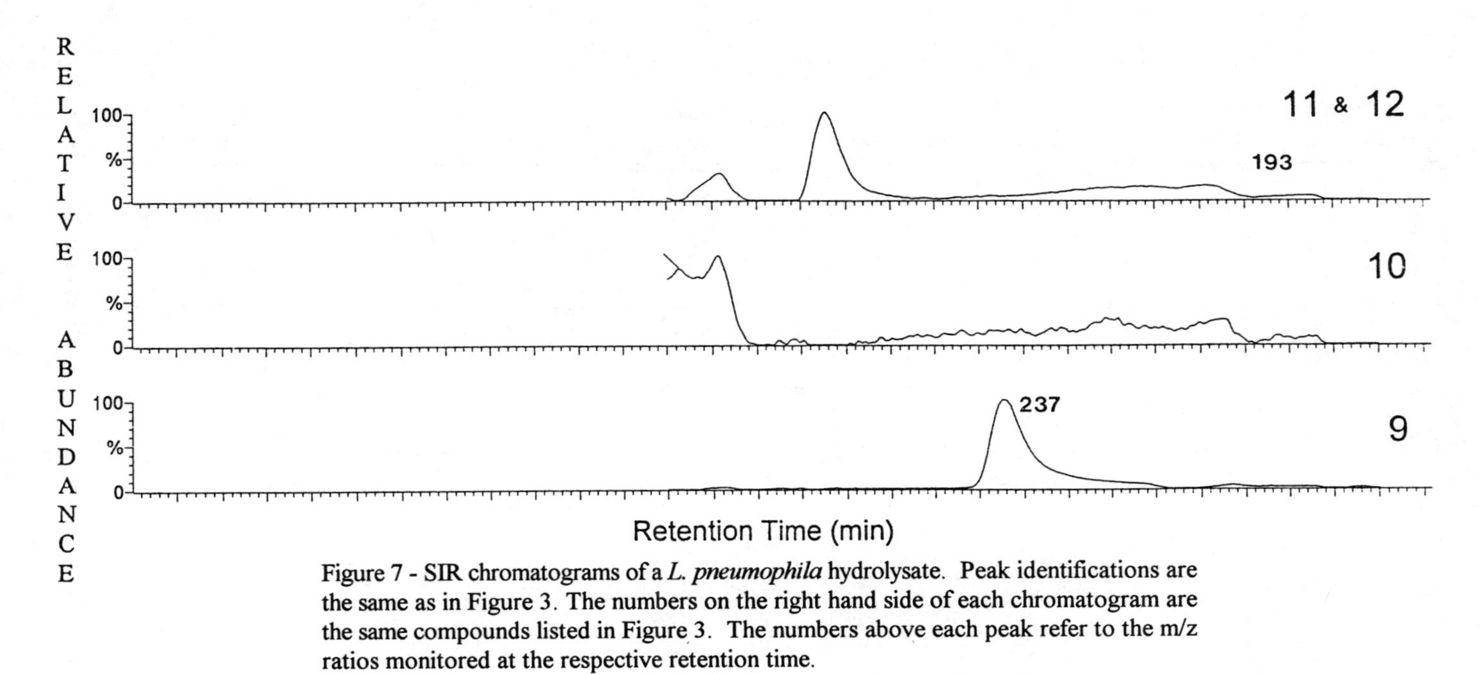

Figure 7 - SIR chromatograms of a *L. pneumophila* hydrolysate. Peak identifications are the same as in Figure 3. The numbers on the right hand side of each chromatogram are the same compounds listed in Figure 3. The numbers above each peak refer to the m/z ratios monitored at the respective retention time.

Continued on next page

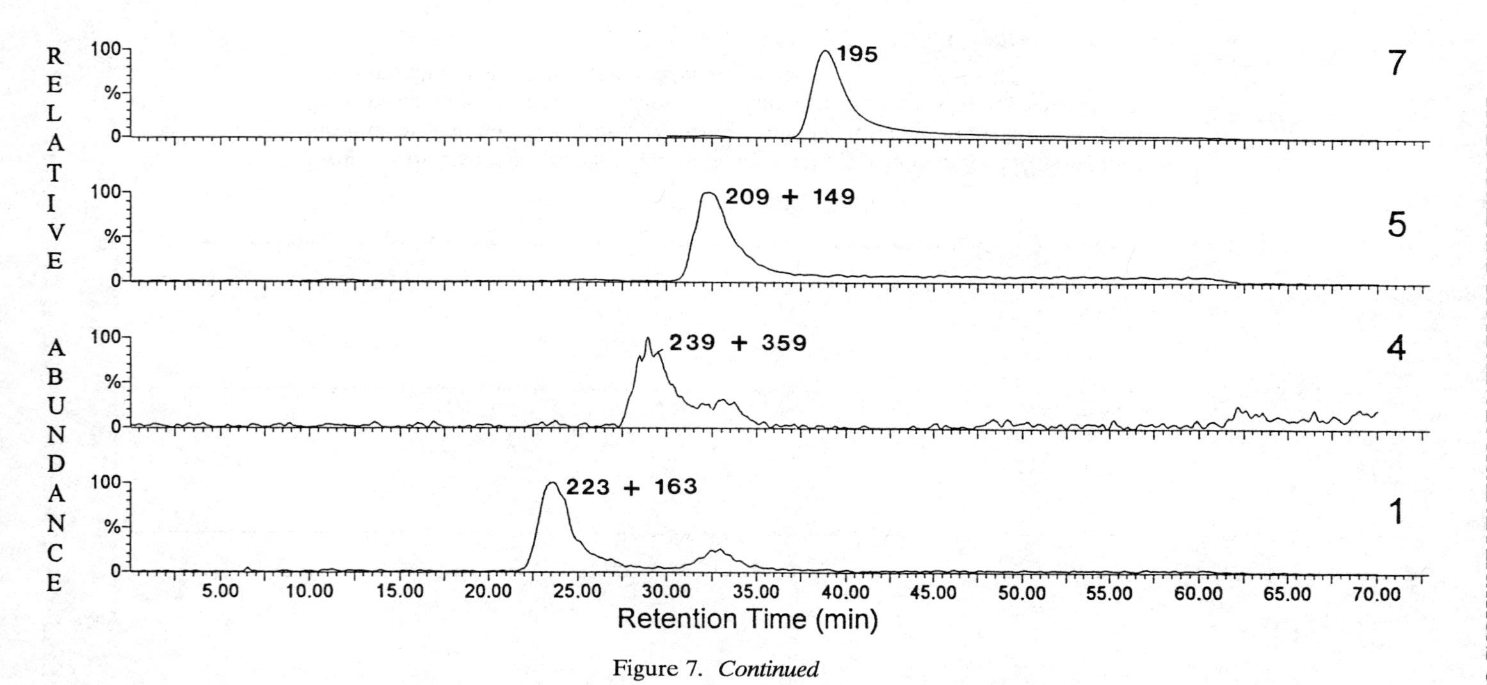

Figure 7. *Continued*

to distinguish from background. Instead, MS-MS allowed selective transmission of the highest abundance molecular (or adduct) ion followed by generation of product ion spectra which were substantially free from background noise. Detection of the molecular ion in the first mass analyzer in conjunction with product ions in the second analyzer is analogous to performing separate CI and EI analyses for GC-MS.

The neutral sugar ribose is present at high levels in bacterial cells, principally as a component in nucleic acids. Figure 8 compares product ion spectra obtained by LC-MS-MS of authentic ribose and ribose present in a hydrolysate of *L. pneumophila*. Both product ion spectra of m/z 209 $[M-AcO]^-$ contained m/z 149, m/z 89, m/z 71, and m/z 59 (see Figure 8). Acidic compounds did not form adducts but formed exclusively their deprotonated molecular ions. Product ion spectra of KDO in a hydrolysate *of L. pneumophila* were also identical to fragments produced by a KDO standard (Figure 9).

Current status of LC-MS-MS, MS-MS and GC-MS for profiling sugars in bacterial hydrolysates

Sample preparation for LC-MS-MS or MS-MS analysis of carbohydrates in whole cell hydrolysates requires no derivatization. The entire procedure consists of three steps (acid hydrolysis, removal of the acid, and solid phase extraction with C-18 columns) and takes a few hours (20). In contrast, sample preparation for GC-MS analysis of bacterial carbohydrates is a multi-step procedure which currently requires three days (7-9).

Sugar analysis using HPAEC-LC is performed using concentrated sodium hydroxide in the mobile phase. Standard mass spectrometers are not designed to handle high concentrations of NaOH present in the LC eluent. Thus a cation permeable ion suppressor is placed after the LC column and prior to the MS. The ion suppressor exchanges hydrogen for sodium ions; thus replacing sodium hydroxide with water. Other cationic species (including basic and amphoteric compounds) are removed by the suppressor and this serves as an on-line clean-up step (e.g. in removing amino acids and peptides) (2,3).

In GC-MS analysis, total ion chromatograms of neutral and aminosugars (as alditol acetate derivatives) from bacterial cell hydrolysates display low background in the total ion chromatogram. Full spectra also display low background and are readily interpreted. In LC-MS analysis, there is considerable generation of signal from the components of the mobile phase. By selection of a parent ion in the first mass analyzer, product ion mass spectra free of contribution from the mobile phase are observed. However, LC-MS-MS (triple quadropole) instrumentation is considerably more expensive than GC-MS instruments. Commercial benchtop GC-MS instruments have been available since the 1980's whereas user-friendly LC-MS and LC-MS-MS instruments have only become available in the 1990's.

Further development of LC-MS-MS instruments for routine use (including simplification of software and further integration of the control of LC, MS-MS and autosamplers) is vital. The recent availability of modestly priced ion trap instruments that can be operated in the LC-MS-MS mode may contribute to rapid changes in both

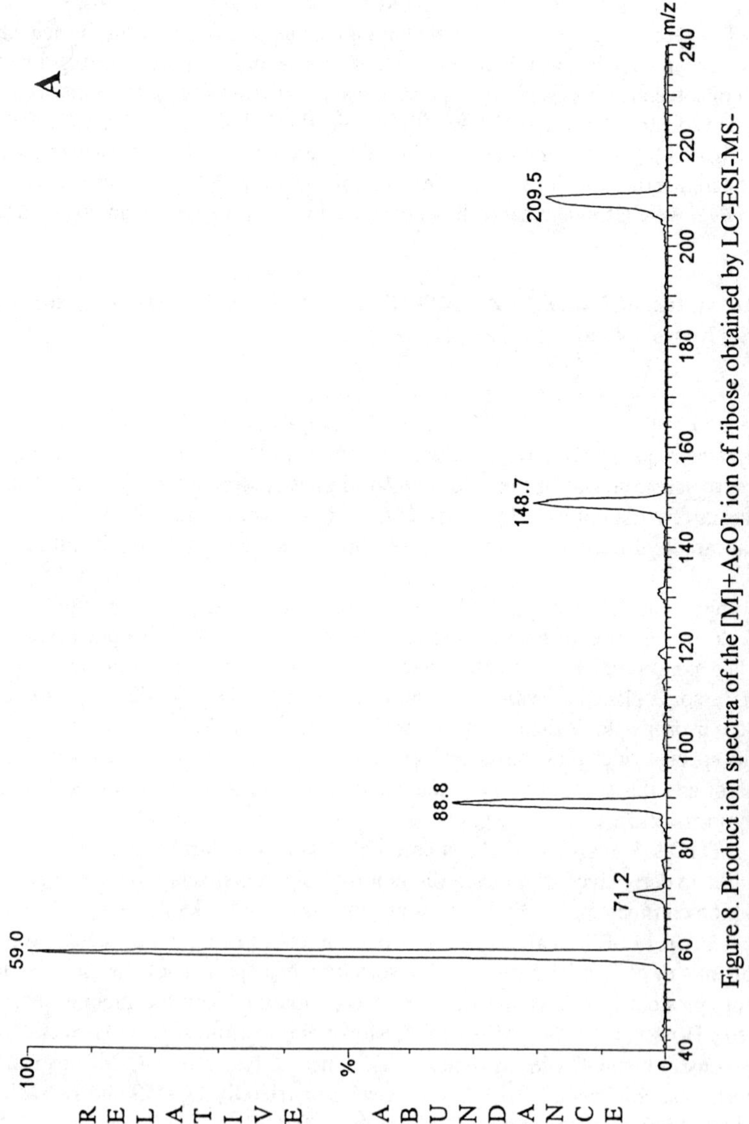

Figure 8. Product ion spectra of the $[M]+AcO]^-$ ion of ribose obtained by LC-ESI-MS-CID-MS (A) authentic ribose and (B) hydrolysate of *L. pneumophila.*

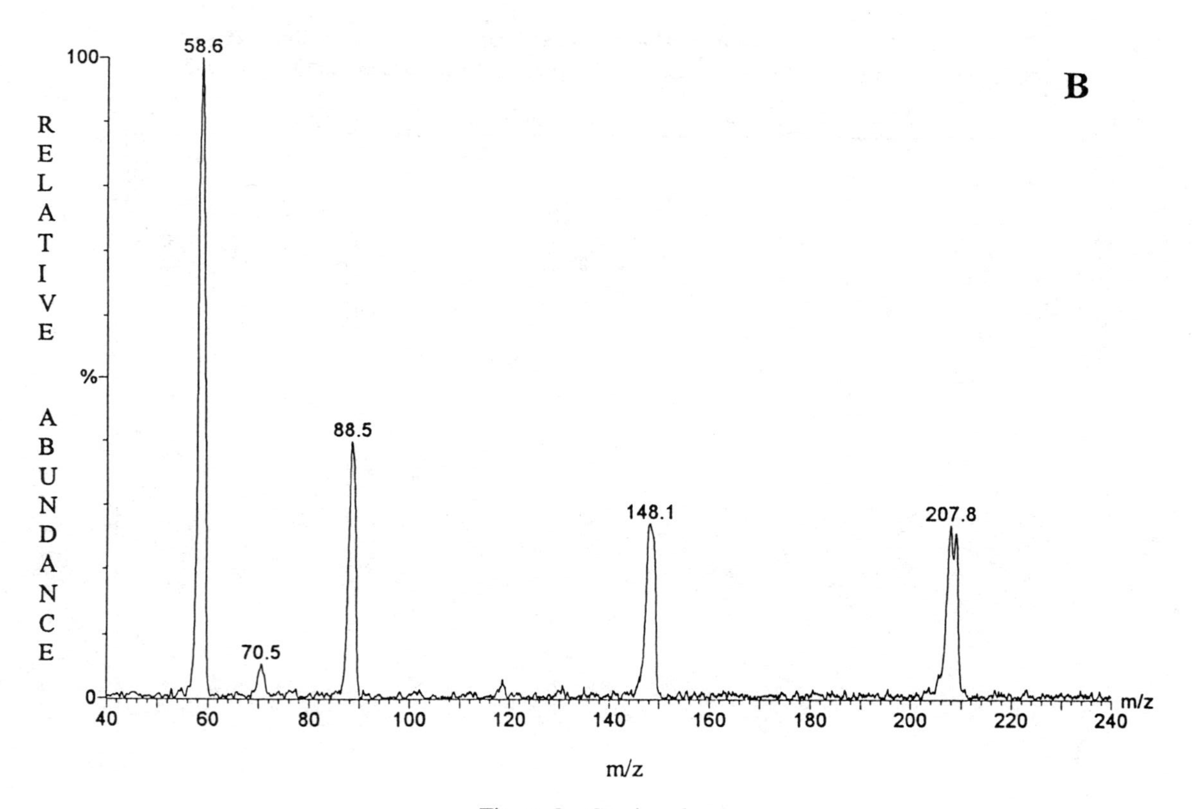

Figure 8. *Continued*

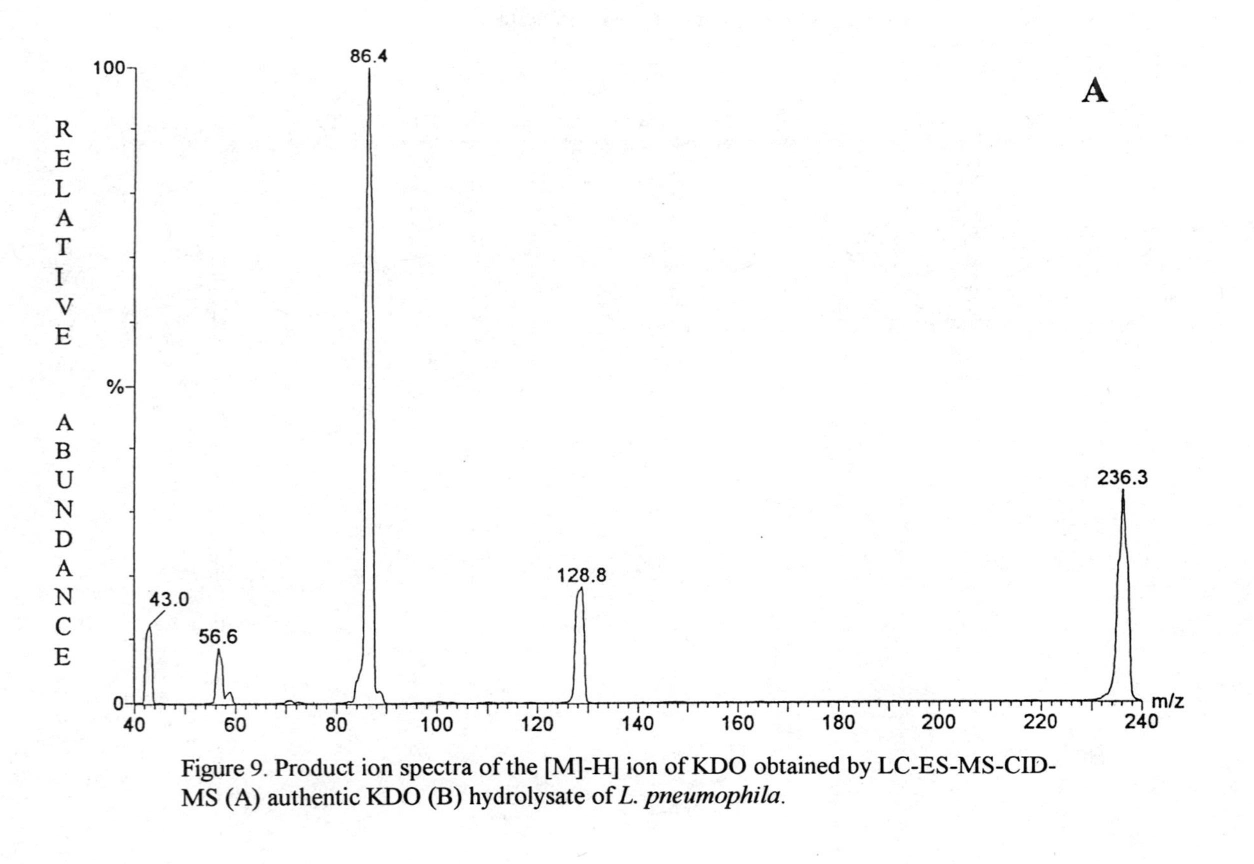

Figure 9. Product ion spectra of the [M]-H] ion of KDO obtained by LC-ES-MS-CID-MS (A) authentic KDO (B) hydrolysate of *L. pneumophila*.

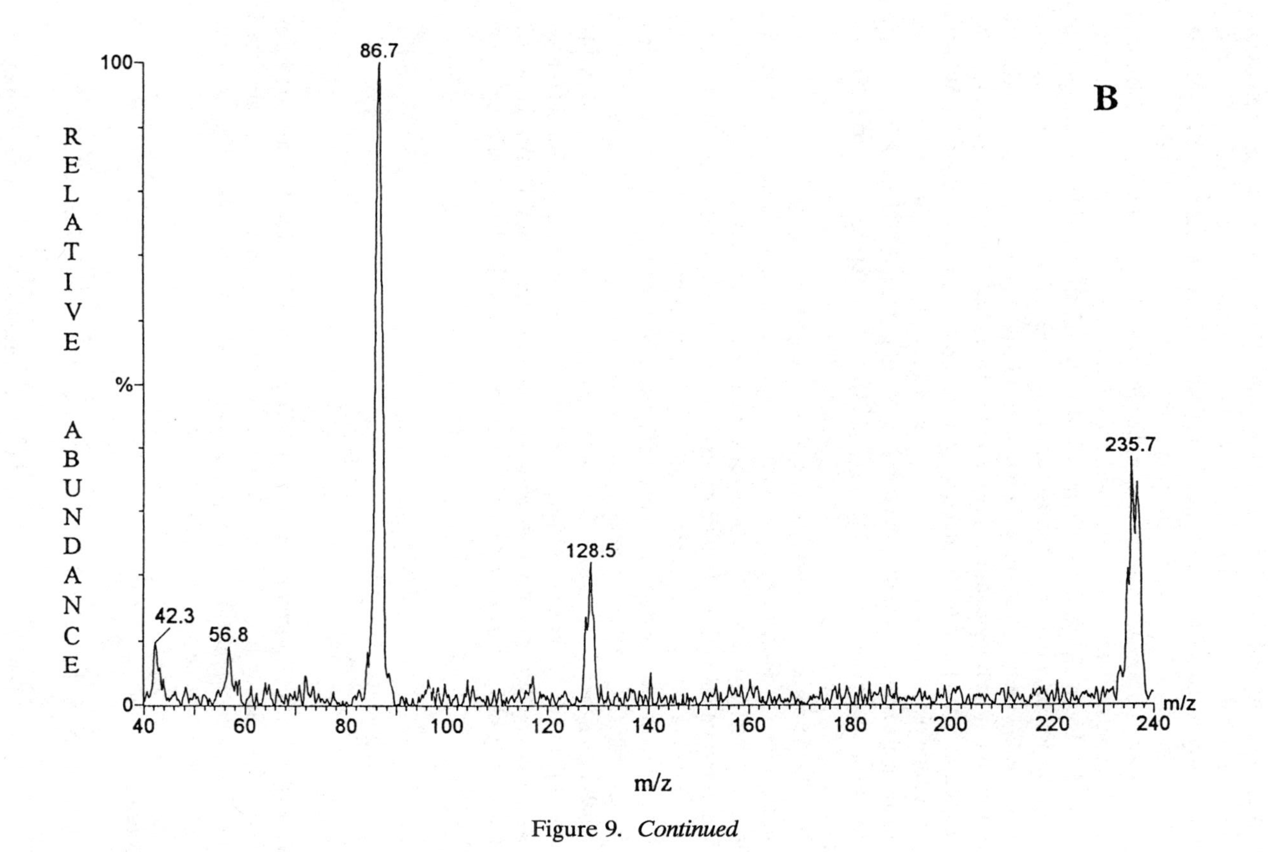

Figure 9. *Continued*

ease of use and cost of analysis. Analysis using the current configuration of LC-MS-MS instruments still involves some compromise in terms of optimal chromatography, mass spectrometry and the ion suppression system. For example, in the current work optimal flow rates for chromatography occurs at around 250 μl/ minute. However, the ion suppressor performs best around 100 μl/ minute. Operating the LC-MS-MS at flow rates of 100 μl/ min is not optimal for chromatographic resolution.

As noted above, sample preparation for ES-MS-MS is identical to LC-MS-MS and also avoids derivatization. A typical MS-MS analysis takes one or two minutes versus 30-50 min for an LC-MS or GC-MS run. These considerations would be extremely important if carbohydrate profiling were adapted for routine bacterial identification. Many sugars found in microorganisms only provide chemotaxonomic information if chromatographically-resolved from their respective isomers. The presence of galactosamine differentiates *B. anthracis* from *B. cereus*. Without chromatographic resolution it would be extremely difficult to discriminate galactosamine from the mannosamine present in both species (17,19). Regarding the analysis of sugar isomers, GC-MS and LC-MS-MS are superior to direct injection MS-MS.

In the current ES investigations, the concentration of sugars was quite high, around 0.05-1.0 % dry weight of the sample. In addition, sample size was not limited, allowing analysis of 5-10 mg of sample. However in other GC-MS studies, trace detection of sugars in complex matrices has displayed considerably greater sensitivity (33). MA has been detected in serum from animals injected with cell wall monomer (muramyl dipeptide) at concentrations of 100 ng/ml (1 part in 10 million) using GC-MS (34). As a component of naturally present bacteria, MA was detected at levels of 4 ng/mg (1 part in 250,000) in house dust using GC-MS and GC-MS-MS. Thus, current GC-MS and GC-MS-MS sensitivity for detection of sugars in complex matrices is superior by several orders of magnitude to LC-MS-MS (14). However, the sensitivity of LC-MS-MS for the task of bacterial identification appears more than adequate.

Acknowledgments

We wish to thank Christopher Maple, Jim Thayer, and Nebojsa Avdalovic (Dionex Corporation, Sunnyvale, CA) for advice in determining an LC configuration and chromatographic conditions appropriate for LC-MS-MS analysis. This included selection of a suitable on-line ion suppressor.

Literature Cited

1. Esteban, N.V.; Liberato, D.J.; Sidbury, J.B.; Yergey, A.L. *Anal. Chem.* **1987**, *59*, 1674.
2. Simpson, R.C.; Fenselau, C.C.; Hardy, M.R.; Townsend, R.R.; Lee, Y.C.; Cotter, R.J. *Anal. Chem.* **1990**, *62*, 248.
3. Conboy, J.J.; Henion, J. *Biol. Mass Spectrom.* **1992**, *21*, 397.

4. Rocklin, R.D.; Pohl, C.A. *J. Liq. Chrom.* **1983**, *6*, 1577.
5. Anumula, K.R. *Anal. Biochem.* **1994**, *220*, 275.
6. Peelen, G.O.H.; de Jong, J.G.N.; Wevers, R.A. *Anal. Biochem.* **1991**, *198*, 334.
7. Fox, A.; Morgan, S.L.; Gilbart, J. In *Analysis of Carbohydrates by GLC and MS;* Biermann, C.J.; McGinnis, G.D., Eds.; CRC Press: Boca Raton, FL, **1989**, 87.
8. Fox, A.; Black, G.E. In *Mass Spectrometry for the Characterization of Microorganisms*; Fenselau, C., Ed.; American Chemical Society: Washington, DC, **1994**, 107.
9. Black, G.E.; Fox, A. *J. Chromatogr.*, in press
10. Higgins M.; Bly S.; Morgan S.L.; Fox A. Anal. Chem. **1994** *66*, 2656
11. Lindberg, B. Adv. Carbohyd. Chem. Biochem. **1990** *48*, 279.
12. Fox A.; Gilbart J.; Morgan S.L. Analytical Microbiology Methods: chromatography and mass spectrometry. Fox A.; Larsson L.; Morgan S.L.; Odham G., Eds. Plenum, NY, NY.**1990.** 71.
13. Fox, A.; Rogers, J.C.; Gilbart, J.; Morgan, S.; Davis, C.H.; Knight, S.; Wyrick, P.B. *Infect. Immun.* **1990**, *58*, 835.
14. Fox, A.; Wright, L.; Fox, K. *J. Microbiol. Meth.*, **1995**, *22*, 11.
15. Walla, M.D.; Lau, P.Y.; Morgan, S.L.; Fox, A.; Brown, A. *J. Chromatogr.* **1984**, *288*, 399.
16. Fox, A.; Rogers, J.C.; Fox, K.F.; Schnitzer, G.; Morgan, S.L.; Brown, A.; Aono, R. *J. Clin. Microbiol.* **1990**, *28*, 546.
17. Fox, A.; Black, G.E.; Fox, K.; Rostovtseva, S. *J. Clin. Microbiol.* **1993**, *31*, 887.
18. Fox, K.F.; Brown, A.; Fox, A.; Schnitzer, G. *System. Appl. Microbiol.* **1991**, *14*, 52.
19. Wunschel, D.; Fox, K.F.; Black, G.E.; Fox, A. *System. Appl. Microbiol.*, **1995**, *17*, 625
20. Black, G.E.; Fox, A.; Fox, K.; Snyder, A.P.; Smith, P.B.W. *Anal. Chem.* **1994**, *66*, 4171.
21. Lee, Y.C. *Anal. Biochem.* **1990**, *189*, 151.
22. Niwa, T.; Dewald, L.; Sone, J.; Miyazaki, T.; Kajita, M. Clin. Chem. **1994**, *40*, 260.
23. Smith, P.B.W.; Snyder, A.P.; Harden, C.S. *Anal. Chem.*, **1995**, 67, 1824-1830.
24. Caldas, E.D.; Jones, A.D.; Winter, C.K.; Ward, B.; Gilchrist, D.G. *Anal. Chem.* **1995**, *67*, 196.
25. Hadžija, O. *Anal. Biochem.* **1974**, *60*, 512.
26. Tipper, D.J. *Biochem.* **1968**, *7*, 1441.
27. Fillit, H.M.; McCarty, M.; Blake, M. *J. Exp. Med.* **1986**, *164*, 762.
28. Clarke, A.J.; Sarabia, V.; Keenleyside, W.; MacLachlan, P.R.; Whitfield, C. *Anal. Biochem.* **1991**, *199*, 68.
29. Hicks K. B. Adv. Carb. Chem. Biochem. **1988**, *46*, 17.
30. Herbreteau B.; Lafosse M.; Morin-Allory L.; Dreux M. Chromatographia **1992** 33, 325.
31. Elmroth, I.; Larsson, L.; Westerdahl, G.; Odham, G. *J. Chromatogr.* **1992**, *598*, 43.
32. Sonesson, A.; Jantzen, E.; Bryn, K.; Larsson, L.; Eng., J. *Arch. Microbiol.* **1989**, *153*, 72.
33. Fox A.; Schwab J.H.; Cochran T. Infect. Immun. **1980,** *29*, 526.
34. Fox A.; Fox K. Infect. Immun. **1991,** *59*, 1202.

RECEIVED August 25, 1995

Chapter 5

Identification of a Cytoplasmic Peptidoglycan Precursor in Antibiotic-Resistant Bacteria

Mike S. Lee[1], Kevin J. Volk[2], Jinping Liu[2], Michael J. Pucci[3], and Sandra Handwerger[4]

[1]Analytical Research and Development, Pharmaceutical Research Institute, Bristol-Myers Squibb, Princeton, NJ 08543–4000
[2]Analytical Research and Development and [3]Anti-Infective Microbiology, Pharmaceutical Research Institute, Bristol-Myers Squibb, Wallingford, CT 06492
[4]Laboratory of Microbiology, Rockefeller University, New York, NY 10021

Research in the area of bacterial strains resistant to antibiotic therapies has gained much attention since the recent emergence of vancomycin resistant bacteria. Vancomycin is a glycopeptide antibiotic which functions by binding directly to the D-Ala-D-Ala terminus of peptidoglycan precursors thereby inhibiting cross-linking by the transpeptidase enzyme. We have developed an analytical approach which utilizes sensitive and selective electrospray LC/MS techniques to determine the molecular weight of cytoplasmic precursors present in samples harvested from vancomycin resistant cells. Substructure analysis strategies utilizing LC/MS/MS protocols provide on-line structure identification by comparison of the mass spectrometric characteristics of unknown precursors with the substructural "template" of the standard pentapeptide precursor. Differences between resulting product ion spectra are indicative of altered and/or modified substructures. Using a novel application of affinity capillary electrophoresis (ACE) techniques, we have also obtained binding information for these precursors. Rapid screening of a variety of peptidoglycan molecules has been performed and resulting binding constants were obtained from Scatchard plots. The affinity of peptidoglycan precursors to vancomycin can be related to structure and provide a basis for proposing possible mechanisms of action. The results of these findings are significant with respect to revealing the mechanism(s) of bacteria resistance as structural evidence was obtained which confirmed the presence of a peptidoglycan precursor terminating in lactate rather than alanine. This structural information serves as a valuable platform for understanding mechanisms of resistance, and hopefully, the design of novel and unique therapies for intervention.

0097–6156/95/0619–0106$13.00/0

Vancomycin is used extensively throughout the world for treatment of infections due to methicillin-resistant *Staphylococcus aureus* (MRSA) and Gram-positive organisms in patients allergic to ß-lactam antibiotics (1). Treatments utilizing vancomycin have been quite attractive and successful. This is due, in part, to its selectivity for the D-Ala-D-Ala terminus of peptidoglycan precursors thought to be ubiquitous in many bacterial cell walls. However, these treatments often represent our last line of defense to fight infectious disease.

Research involving the glycopeptide family of antibiotics has gained much attention since the emergence of vancomycin resistant bacteria (2-4). Reports of transmissible high-level resistance to vancomycin in *Enterococcus faecalis* have been associated with the production of a 38 kDa membrane protein VanA (5-8). This protein is responsible for the synthesis of a number of mixed dipeptides, D-Ala-X, which could be incorporated into a peptidoglycan precursor resulting in structural alteration and possible reduction of its affinity to vancomycin (9). This situation serves to highlight the serious and potentially disastrous implications of a microbe's seemingly endless capability for adaptation and subsequent development of resistance.

Vancomycins, like other bacterial agents such as penicillins and cephalosporins, target the bacterial cell wall and enzymes called transpeptidases which are involved with cell wall synthesis. The peptidoglycan molecule is the basic building block in the bacterial cell wall and consists of both carbohydrate and peptide substructures shown in Figure 1. The carbohydrate portion of the molecule is made up of the two sugar units N-acetyl-glucosamine and N-acetyl-muramic acid while the peptide consists of five amino acids L-Ala-D-Glu-(L-Lys or m-Dap)-D-Ala-D-Ala where m-Dap is meso-diaminopimelic acid. A requisite for cell wall synthesis involves the cross-linking of peptidyl substructures on adjacent glycan strands by transpeptidation between an amine group on one strand and the penultimate D-alanine of a D-Ala-D-Ala terminus on an adjacent strand (Figure 2). The ß-lactam antibiotics (penicillins and cephalosporins) work by binding to a transpeptidase or an ensemble of transpeptidases which catalyze this cross-linking process (10). Vancomycin is a glycopeptide antibiotic and inhibits peptidoglycan synthesis by binding directly to the D-Ala-D-Ala terminus thereby inhibiting cross-linking by the transpeptidase (11-12).

The structure of vancomycin features a modified heptapeptide with cross-linked tyrosine residues substituted with a sugar and aminosugar (Figure 3). The core structure containing the seven amino acids is biologically active while the sugar substituents do not affect antibiotic activity *in vitro*. It is the heptapeptide core which binds to stem pentapeptides terminating in D-Ala-D-Ala within the bacterial cell wall layer resulting in inhibition of peptidoglycan synthesis (13-17). Inhibition of this critical process during cell wall synthesis results in the accumulation of lipid intermediates and the peptidoglycan precursor uridine diphosphate-N-acetyl-muramyl-pentapeptide (UDP-N-acetyl-muramyl-pentapeptide) in the cytoplasm.

Recent studies performed in our laboratory focus on the application of mass spectrometry (MS) based techniques for the structure profile analysis of vancomycin resistant bacteria. New advances in coupling liquid chromatography with electrospray/mass spectrometry have resulted in new areas of research and biological applications (18-19). Our analytical approach utilizes sensitive and selective micro-liquid chromatography/mass spectrometry (micro-LC/MS) profiling techniques to determine the molecular weight (MW) of cytoplasmic

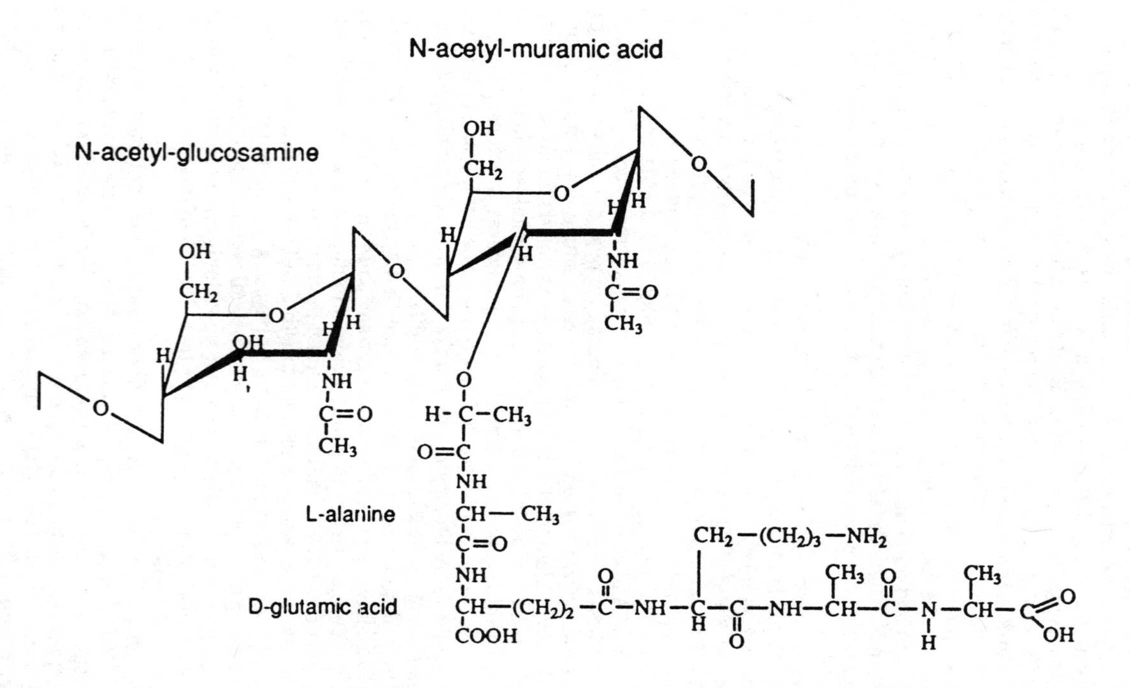

Figure 1. A peptidoglycan molecule consisting of carbohydrate and peptide substructures.

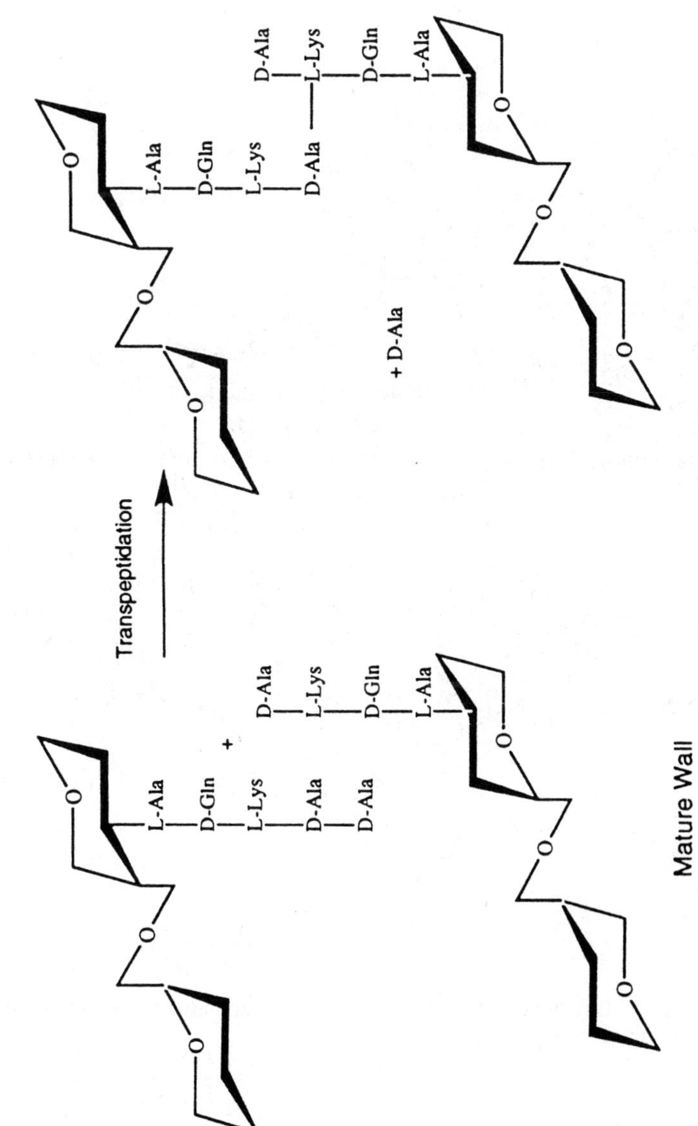

Figure 2. Representation of the cross-linking reaction during bacterial cell wall synthesis.

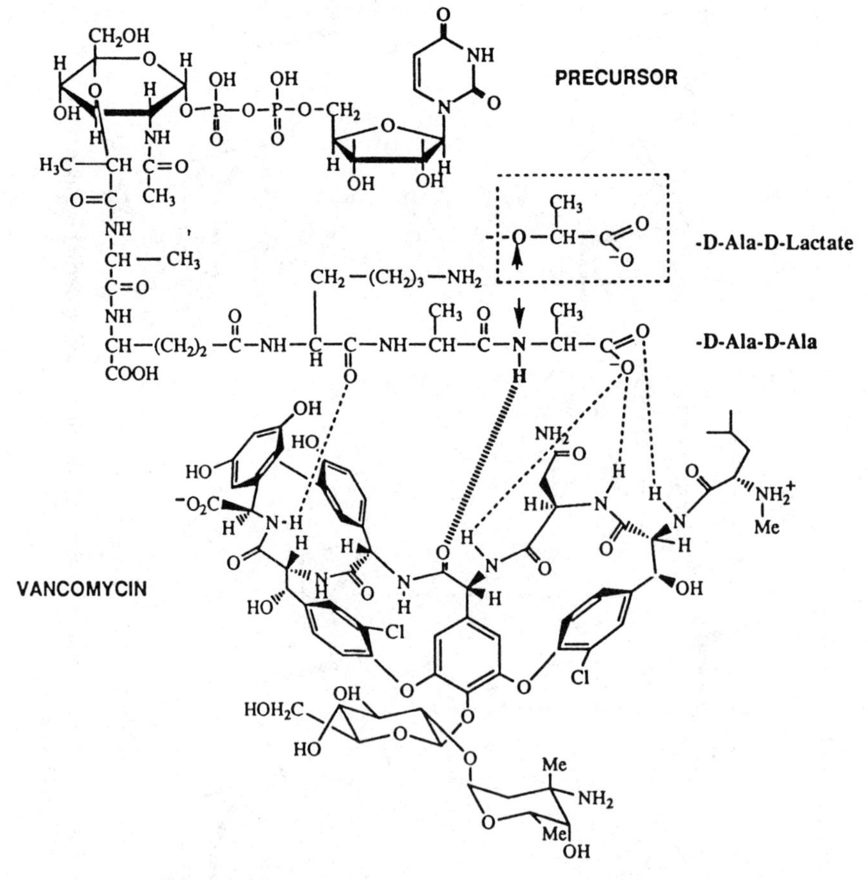

Figure 3. Illustration of the binding between vancomycin and a normal peptidoglycan precursor.

peptidoglycan precursors present in samples harvested from vancomycin resistant cells. Substructure analysis strategies utilizing tandem mass spectrometry (MS/MS) protocols provide on-line structure identification by comparison of the mass spectrometric characteristics of unknown precursors with the substructural "template" of the standard pentapeptide precursor present in non-resistant bacteria. This approach is based on the premise that the precursors of interest would be expected to retain much of the substructural characteristics of the standard peptidoglycan precursor and would therefore be expected to undergo similar MS/MS fragmentations. Identical MS/MS product ions and neutral losses are correlated and provide direct evidence for common substructures. Differences between resulting MS/MS product ion spectra are indicative of altered and/or modified substructures. This MS/MS substructure analysis strategy has been delineated in previous studies involving the rapid identification of drug metabolites (20) Recently, integrated LC/MS and LC/MS/MS profiling approaches have been utilized in our laboratories with studies involving the structure profile analysis of natural products (21-23) contained in complex matrices. Here we describe details of our recent findings (24,25) which reveal the presence of a peptidoglycan precursor terminating in D-Ala-lactate rather than D-Ala-D-Ala.

We also feature our subsequent work dealing with the novel application of affinity capillary electrophoresis (ACE) techniques for the rapid generation of binding or molecular recognition profiles (26). Rapid screening of a variety of peptidoglycan molecules are performed and binding constants are obtained from Scatchard plots (29). The affinity of peptidoglycan precursors to vancomycin is related to structure and provides a basis for proposing possible mechanisms of action.

The analytical methodology described in this chapter highlights the powerful capabilities of nebulizer assisted electrospray ionization (ESI) techniques, particularly for the sensitive analysis of challenging and difficult to analyze samples. When incorporated within a micro-LC/MS environment, this arrangement affords excellent opportunities for application with sample limited conditions such as the structure profile analysis of peptidoglycan precursors. Sensitivity and speed of analysis are other factors worthwhile noting since peptidoglycan precursors are transient intermediates present in very low amounts. The results obtained from these micro-LC/MS structure profiles are fed into subsequent ACE binding assays.

The purpose of such an integrated effort is to provide a rapid assessment of structure and binding. This combined information can serve as a valuable platform for revealing, predicting or postulating mechanisms of resistance, and hopefully, the design of novel and unique therapies for intervention.

Experimental

Cytoplasmic pools of UDP-linked peptidoglycan precursors were extracted as previously described (24,25). Briefly, *Enterococcus faecalis* 221 was obtained by introduction of the glycopeptide resistance plasmid phKK100 (27) into the susceptible *Enterococcus faecalis* JH2-2. All bacterial cultures including *Leuconostoc mesenteroides* VR1 and *Lactobacillus casei* ATCC 7469 were grown to midlogarithmic phase and bacitracin was added to a final concentration of 100 µg/mL to accumulate precursors. After 1 h, cultures were chilled rapidly, and cells were harvested by centrifugation and extracted with cold trichloroacetic acid (final concentration 5%) for 30 min. The supernatant fluid was separated by gel

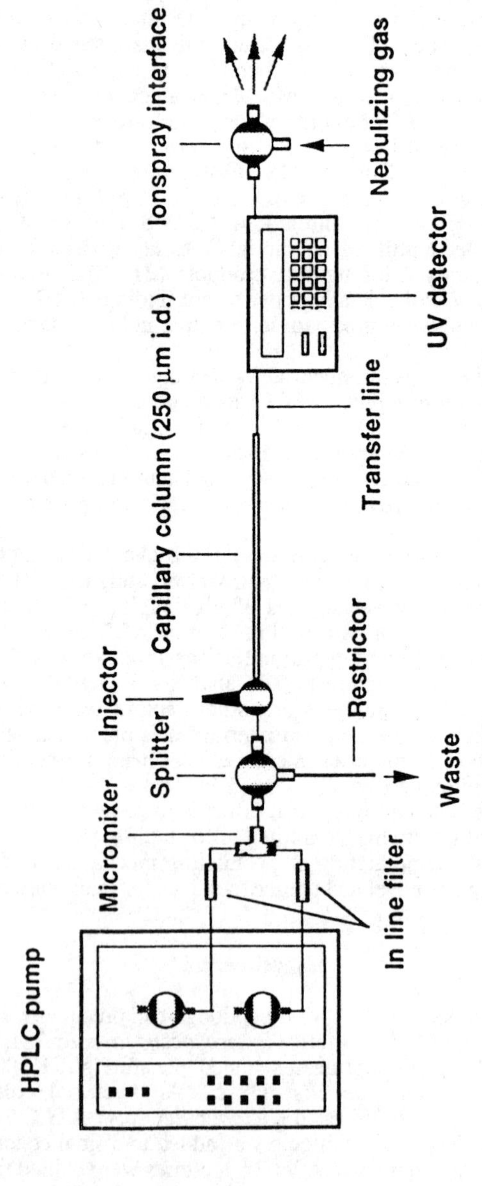

Figure 4. Schematic diagram of the microcolumn-LC//MS system.

filtration (Sephadex G-25) with water elution. Hexosamine-containing fractions were identified by the assay of Ghuysen et al. (28), and they were pooled and lyophilized.

Precursor extracts were chromatographically separated on-line with a Beckman System Gold high-performance liquid chromatography (Fullerton, CA, U.S.A.) and a Sciex API triple quadrupole mass spectrometer (Sciex, Thornhill, Ontario, Canada) equipped with a nebulizer assisted electrospray LC/MS interface (Figure 4). The conventional HPLC system was modified for performing micro-LC at low flow rates. Solvent gradients were directly delivered into a micromixer obtained from Upchurch Scientific, Inc. (Oak Harbor, WA, U.S.A.) at flow rates of 0.2-0.4 mL/min. A precolumn splitting device was used to obtain appropriate output flow rates (approximately 3 µL/min) for packed capillary columns. The split ratio was easily regulated by adjusting the length of restriction line where a fused silica capillary with 50 µm i.d. was used. The capillary columns (36 cm x 250 µm i.d., 375 µm o.d.) were packed in-house with C_{18}, 5-µm particle of 300 Å pore size, from Vydac (Hesperia, CA, U.S.A.) using an ISCO mLC-500 pump (Lincoln, Nebraska, U.S.A). The columns were directly connected into a Valco microinjector with 100 nL and 500 nL internal loops. The transfer line from the column outlet consisted of fused silica capillary with 50 µm i.d. and 190 µm o.d. on-line connected to a UV detector and mass spectrometer. An ABI Model 785A UV detector (Applied Biosystem Inc., Foster City, CA, U.S.A.) equipped with Z-shape capillary flow cell obtained from LCpacking (San Francisco, CA, U.S.A.) was used in this study. The mobile phase used in this system consisted of 50 mM ammonium formate at pH=6.75 (solvent A) and 30% solvent A/70% acetonitrile (solvent B). Solvent gradient (0% - 20% solvent B over 20 minutes) was performed for all separations.

The low eluent flow from the micro-LC system was coupled directly to the mass spectrometer equipped with an articulated nebulizer assisted electrospray interface with a voltage of +5300V. The mass spectrometer was scanned from m/z 400-1500 with a step size of 0.4 u and an acquisition time of 2.6s per scan. MS/MS experiments were performed on the doubly charged MH_2^{2+} ions of the precursors with a collision energy of 45 eV and an argon collision gas thickness of 400×10^{12} atoms/cm^2.

Experiments involving ACE were performed with a Beckman P/ACE Model 2100 CE system (Fullerton, CA, U.S.A.). An untreated fused silica capillary (Polymicro Technology, Phoenix, AZ, U.S.A.), 57.3 cm X 50 µm i.d., was used within a capillary chamber maintained at 25°C. The resulting components were monitored using the UV detector which was set at 254 nm.

All binding assays were performed using an open-tubular ACE system described previously (26). The running buffer consisted of 0.2 M glycine and 0.03 M Tris at pH=8.30. A 25 µM vancomycin stock solution was used as ligand substrate added into the running buffer. A neutral marker, mesityl oxide (0.2 mg/mL), was utilized as a reference for the measurement of relative migration times reported (29). A 10 µL aliquot of each peptidoglycan precursor was mixed with 10 µL of mesityl oxide and used directly for injection without further sample purification. Pressure injection was utilized with a duration time of 20 seconds, which represents about 20 nL of sample volume injected. A series of running buffers containing appropriate concentrations of vancomycin ligand was used to measure migration shifting with precursors. The relationship between ligand concentration and migration time can be illustrated using Scatchard analysis affording the determination of binding constants or dissociation constants.

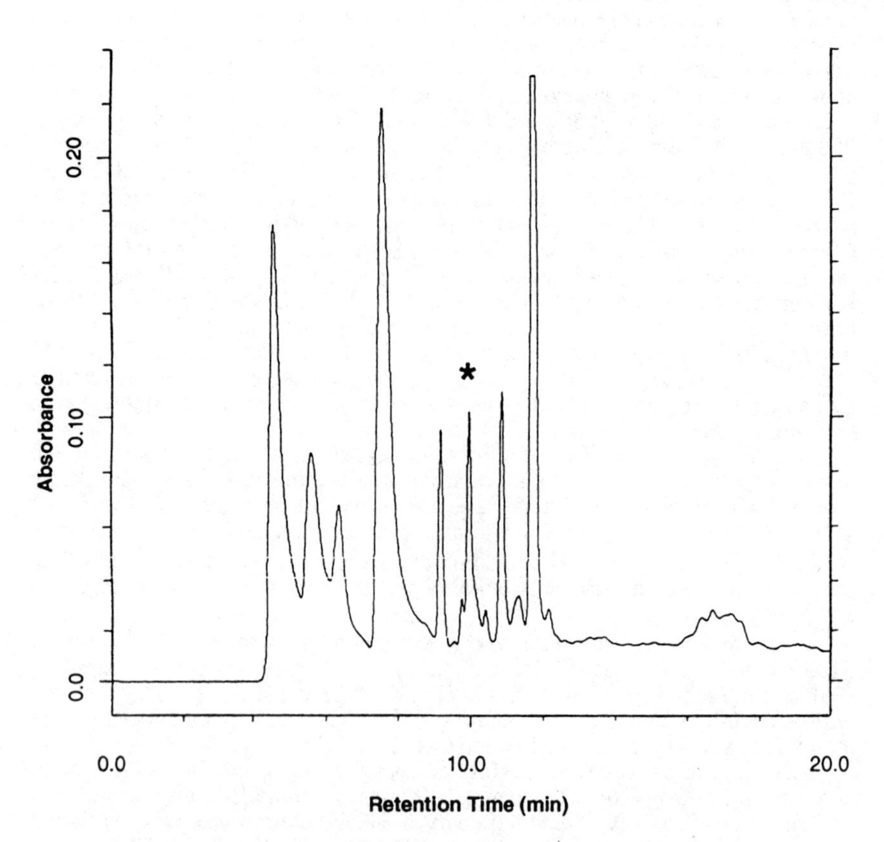

Figure 5. Microcolumn-LC/UV chromatogram of the precursor extract obtained from *Enterococcus faecalis*. The asterisk denotes the peak corresponding to the modified peptidoglycan precursor.

Structure Analysis

Mixtures of peptidoglycan precursors extracted from vancomycin resistant bacteria were characterized using micro-LC/MS profiling techniques. As discussed previously, vancomycin acts by binding to the D-Ala-D-Ala terminus of the stem pentapeptides present in bacterial peptidoglycan. Although it was considered highly unlikely that vancomycin resistance would develop since the D-Ala-D-Ala terminus is ubiquitous among bacterial species, vancomycin resistant *Enterococcus faecalis* strains have been identified. In addition, other species such as *Leuconostoc mesenteroides* and *Lactobacillus casei* are known to be intrinsically resistant to high levels of glycopeptide antibiotics such as vancomycin.

Based on the development of resistant strains, it was postulated that vancomycin resistant bacteria could produce a stem pentapeptide terminating in something other than D-Alanine. The initial goal of these studies was to develop analytical methodologies to profile and characterize precursor extracts from various types of bacteria. To determine whether the mechanism of acquired resistance in specific strains of *Enterococcus faecalis*, and intrinsically resistant *Leuconostoc mesenteroides* and *Lactobacillus casei* was similar, cytoplasmic precursors from resistant and non-resistant strains of *Enterococcus faecalis* were obtained and profiled on-line using micro-LC/MS and compared to those of *Leuconostoc mesenteroides* and *Lactobacillus casei.* A tripeptide precursor, UDP-N-acetyl-muramyl-L-Ala-D-Glu-L-Lys (MW 1107), isolated by inhibition of murein synthesis in the presence of D-cycloserine and the normal UDP-linked pentapeptide precursor, UDP-N-acetyl-muramyl-L-Ala-D-Glu-L-Lys-D-Ala-D-Ala (MW 1149), isolated from *Staphylococcus aureus*, were used as reference materials and substructural "templates".

The micro-LC separation of precursor extracts from *Enterococcus faecalis, Leuconostoc mesenteroides,* and *Lactobacillus casei* each demonstrated a peak which eluted later than the normal pentapeptide precursor from *Staphylococcus aureus*. Figure 5 illustrates the micro-LC/UV chromatogram of the precursor extract from *Enterococcus faecalis* and indicates the modified precursor is a component in the mixture.

Full scan mass spectra containing both singly charged and doubly charged ions were generated from the on-line analysis of the standard and precursors. The standard pentapeptide precursor isolated from *Staphylococcus aureus* has a molecular weight of 1149 Da. The corresponding full scan mass spectrum of the standard pentapeptide precursor is shown in Figure 6 and illustrates the ions corresponding to the protonated (MH^+) and doubly charged (MH_2^{2+}) molecular ions accompanied by a fragment ion at m/z 746 corresponding to a neutral loss of the uridine diphosphate (UDP) substructure (404 Da) from the MH^+ species. This full scan fragmentation information is specific and can be used diagnostically to identify peptidoglycan precursors in complex mixtures and indicate modifications of the N-acetyl-muramyl-peptide substructure. The molecular weight analysis performed on extracts from *Enterococcus faecalis* and *Lactobacillus casei* indicated a modified precursor containing the UDP substructure with a MW of 1150 Da. Micro-LC/MS analysis of the extract obtained from *Leuconostoc mesenteroides* indicated a precursor with a MW of 1221 Da.

Substructural "templates" were generated for the tripeptide and normal pentapeptide precursors using micro-LC/MS/MS techniques. The association of specific MS/MS product ions with unique substructures or fragment ions of known precursors provides a basis for the interpretation of the MS/MS substructural data

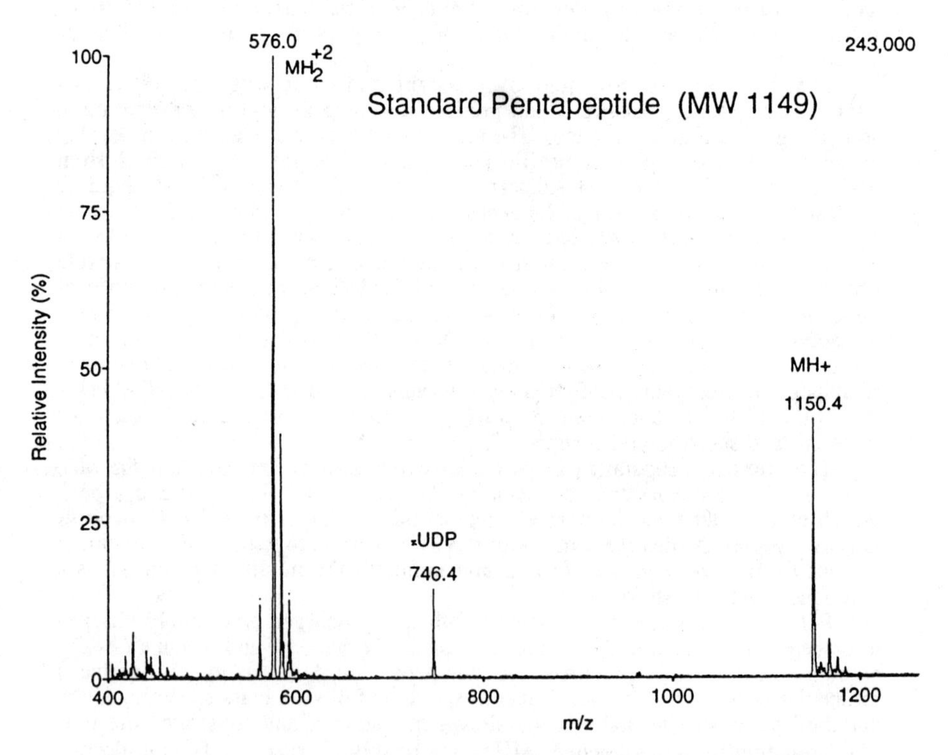

Figure 6. A typical full scan mass spectrum of the standard pentapeptide precursor isolated from *Staphylococcus aureus*. This spectrum was obtained from a standard averaged over 8 seconds.

for the modified precursors. The MS/MS product ion spectrum corresponding to the MH_2^{2+} species of the standard pentapeptide precursor is shown in Figure 7. Neutral loss of the UDP substructure is a facile fragmentation route resulting in the singly and doubly charged product ions at m/z 746 and m/z 374, respectively. Cleavage of the muramyl-peptide ether linkage also yields a product ion which provides MW information for the peptide portion of the peptidoglycan precursor. Several important fragmentations occur in the peptidic backbone which helps to delineate the portion of the precursor which is modified. One common pathway involves cleavage of the Glu-Lys amide bond to yield a Y series product ion at m/z 289 (Lys-Ala-Ala substructure) and cleavage of the Ala-Glu amide bond to yield a Y series product ion at m/z 418 (Glu-Lys-Ala-Ala substructure). A second important fragmentation pathway involves the neutral loss of the C-terminal Ala-Ala substructure from the product ion at m/z 418 to produce a product ion at m/z 258. The neutral loss of the C-terminal and penultimate amino acids, such as Ala-Ala in the standard pentapeptide precursor, is a common fragmentation theme for all of the peptidoglycan precursors studied.

Previous studies in other laboratories have indicated key proteins (VanH and VanA) involved in peptidoglycan synthesis which are modified in resistant bacteria and appear to have altered substrate specificity (9). Hydroxy acids such as D-2-hydroxybutyrate were proposed as potential substrates in these studies. In studies performed in our laboratories, we have utilized LC/MS/MS methodologies to provide detailed information about modified precursors which indicate hydroxybutyrate is not incorporated into the modified peptidoglycan precursors we have characterized. The MW (1150 Da) and product ion spectra of the precursors from *Enterococcus faecalis* and *Leuconostoc mesenteroides* were identical. The MS/MS product ion spectra of the doubly charged MH_2^{2+} species of the precursors from *Leuconostoc mesenteroides*, *Lactobacillus casei*, and *Enterococcus faecalis* were consistent with substructures of modified pentapeptide and modified tetrapeptide precursors. Comparison of the MS/MS product ion spectra of the precursors from *Enterococcus faecalis*, *Lactobacillus casei* (Figure 8) and *Leuconostoc mesenteroides* (Figure 9) to the fragmentation "template" of the normal pentapeptide obtained from *Staphylococcus aureus* (Figure 7) indicated an alteration of an amino acid near the C-terminus for each precursor. As discussed previously, a key fragmentation route common to each precursor involved cleavage of the Glu-Lys amide bond. Y-series fragment ions were observed consistent with Lys-Ala-Ala (m/z 289), Lys-Ala-Lactate (m/z 290), and Lys-(Ala)-Ala-Lactate (m/z 361) substructures for the standard pentapeptide precursor (*Staphylococcus aureus*), modified tetrapeptide precursor (*Lactobacillus casei* and *Enterococcus faecalis*), and modified pentapeptide precursor (*Leuconostoc mesenteroides*), respectively. Cytoplasmic hexapeptide precursors in bacterial species with interpeptide bridges are present in some *Leuconostoc mesenteroides* and *Lactobacillus casei* species. In these organisms, the first amino acid of the interpeptide bridge is added to the e-amino group of the L-Lys at the level of the UDP-linked cytoplasmic precursor rather than at the lipid intermediate level (30). Consecutive MS/MS fragmentation resulting in the neutral loss of the Ala-Lactate substructure from the Y-series product ions (Glu-Lys-Ala-Lactate and Glu-Lys-(Ala)-Ala-Lactate) provide additional complementary evidence for the proposed structural modification.

Although not fully utilized in these studies, MS/MS screening techniques such as constant neutral loss or precursor scans may be employed on a wider scale to evaluate other species of bacteria suspected of producing modified precursors. For

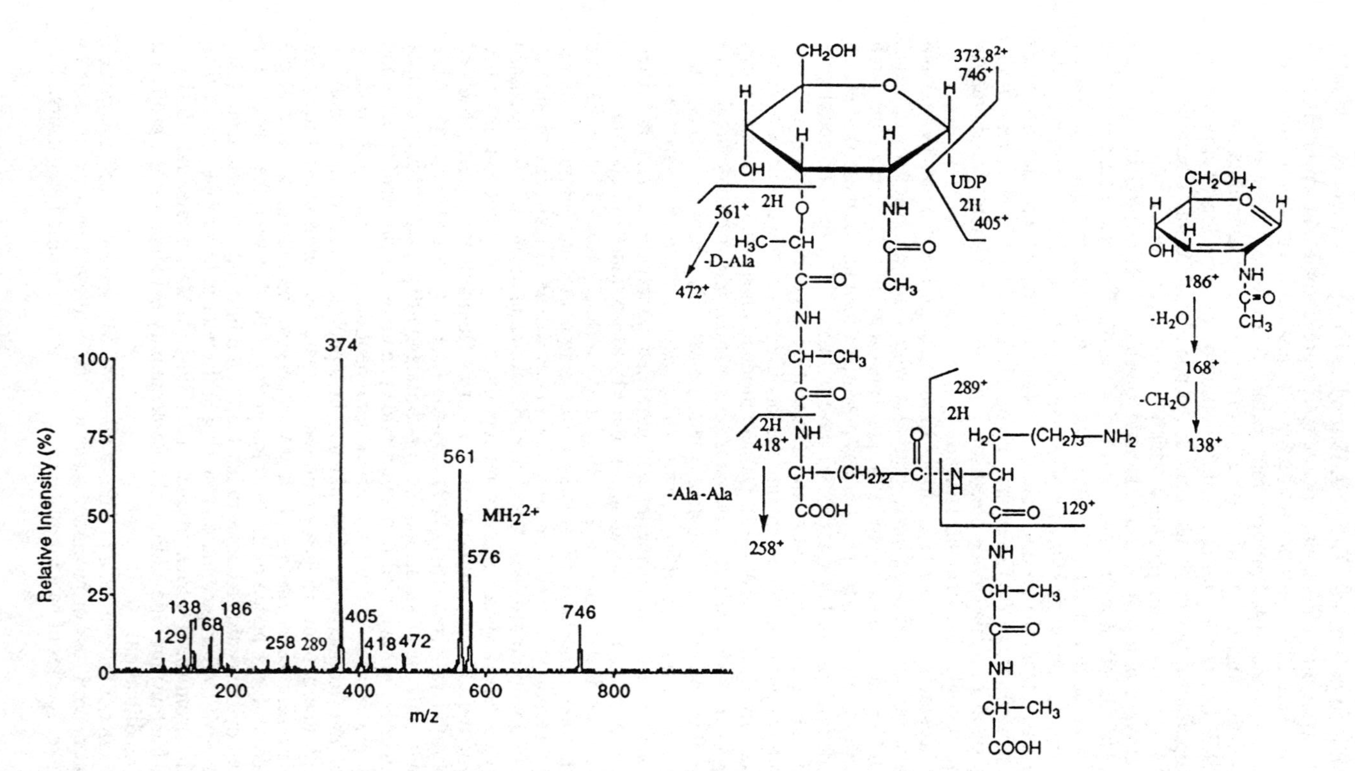

Figure 7. The MS/MS product ion spectrum corresponding to the MH_2^{2+} species of the standard pentapeptide precursor isolated from *Staphylococcus aureus*.

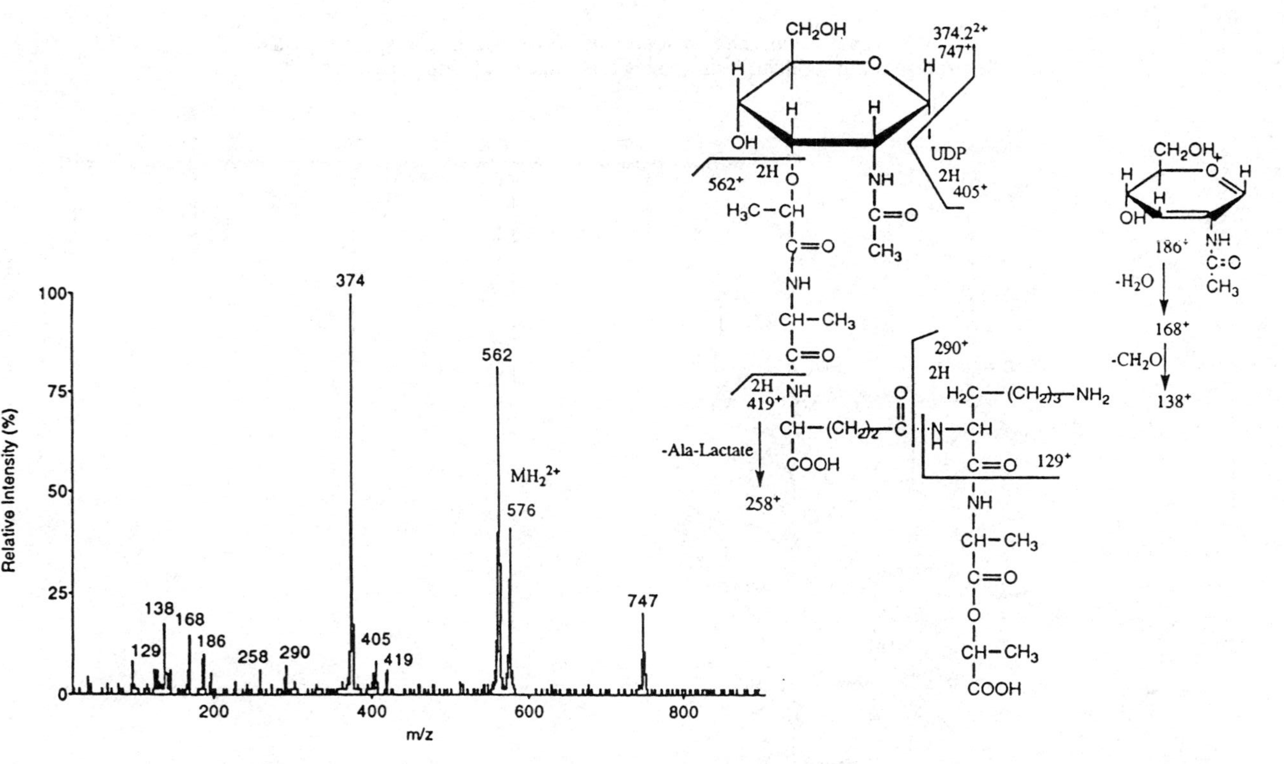

Figure 8. The MS/MS product ion spectrum corresponding to the MH_2^{2+} species of the precursors from *Lactobacillus casei*.

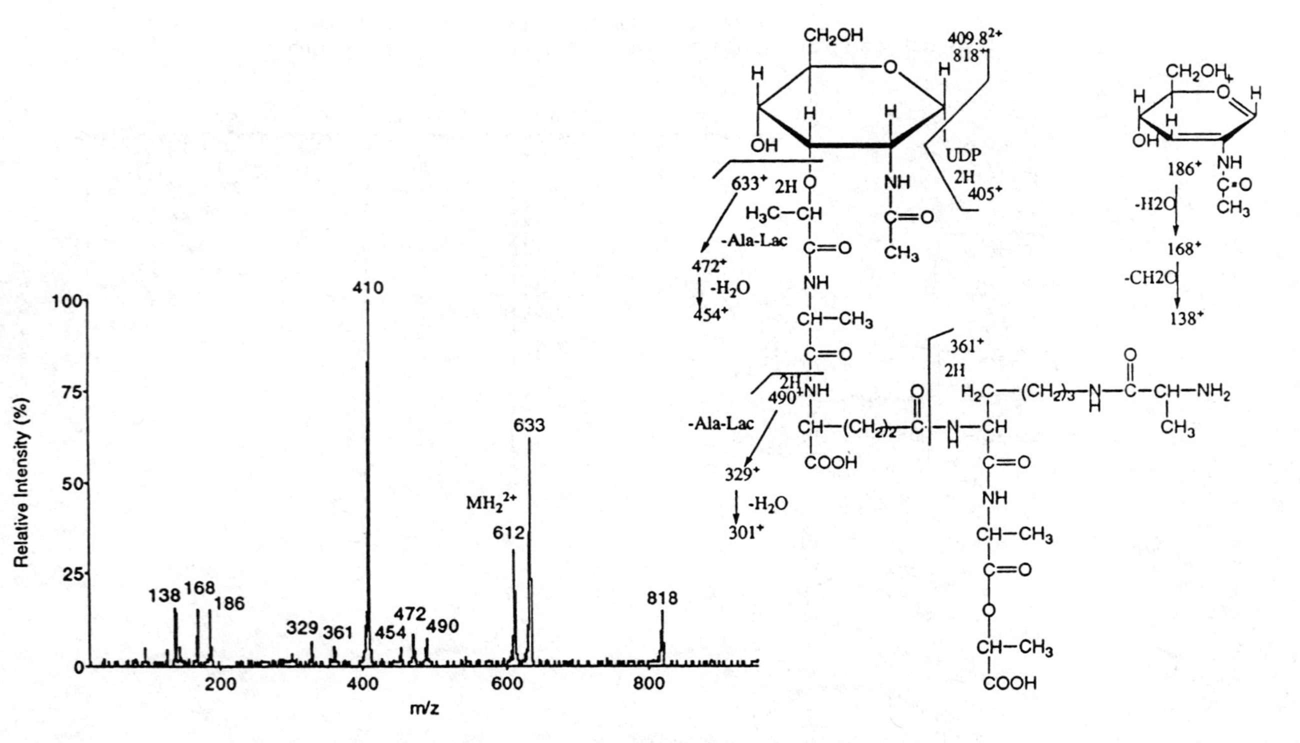

Figure 9. The MS/MS product ion spectrum corresponding to the MH_2^{2+} species of the precursors from *Leuconostoc mesenteroides*.

example, a constant neutral loss scan corresponding to the UDP substructure may be used to screen complex samples for the presence of peptidoglycan precursors.

Binding Studies

The identification of a peptidoglycan precursor terminating in Ala-Lactate resulted in efforts directed toward establishing a general binding assay to further confirm these findings. A novel ACE approach was developed and incorporated into our investigations to provide a rapid assessment of the binding affinity of vancomycin to these precursors and provide additional evidence for the identification of altered peptidoglycan precursors.

The normal pentapeptide precursor isolated from *Staphylococcus aureus* (UDP-N-acetyl-muramyl-L-Ala-D-Glu-L-Lys-D-Ala-D-Ala) was used as a model system in this portion of our studies. The proposed structure of this precursor together with its binding to vancomycin is displayed in Figure 3. Based on the proposed hydrogen bond interaction of vancomycin and the terminus of the precursor, a modification of the amide linkage to an ester linkage as indicated by the arrow in Figure 3 could disrupt the binding process (9). Affinity interaction between the precursor and vancomycin is mediated by a series of five hydrogen bonds between the dipeptide backbone and the glycopeptide. It is known that the amide linkage and the free terminal carboxyl group are essential for vancomycin binding (13-17). Modification or replacement of these functional groups (by other structural moieties) may significantly reduce their interaction with vancomycin (30). The binding behavior of this normal precursor with vancomycin serves as a binding profile "template" for subsequent studies involving other precursors.

Vancomycin was selected as the substrate ligand (added directly into the operating buffer) while peptidoglycan precursors were used as the probe for monitoring migration shifting. A commonly used biological buffer consisting of 0.2 M glycine/0.03 M Tris (pH=8.30) was selected. Two typical electropherograms obtained under conditions with and without vancomycin in the buffer is shown in Figure 10. The neutral marker, mesityl oxide, has the same migration time (2.9 min) under both conditions. Upon the addition of 25 μM vancomycin to the buffer, the precursor peak shifts towards the reference peak corresponding to a shift in migration time from 4.3 to 3.9 min. The complexation between the precursor and vancomycin results in a reduction of net charge and the complex is thus "dragged" towards the cathode (detection window) somewhat faster than the unbound component due to the electroosmotic flow generated under present experimental conditions. The observed change in mobility of the complex indicates that the normal pentapeptide precursor (UDP-N-acetyl-muramyl-L-Ala-D-Glu-L-Lys-D-Ala-D-Ala) from *Staphylococcus aureus*, is bound to vancomycin. As expected, the formed complex has a different mobility from the original free precursor.

A single component appears in both the micro-LC chromatogram and the CE electropherogram for the normal peptidoglycan precursor from *Staphylococcus aureus*. The major components present in the extracts correspond to precursors, and similarities in structure are inferred from migration times. It is important to note that the precursors studied here are prepared from cytoplasmic extracts and pre-purified by gel filtration prior to other analyses.

A series of buffers containing various concentrations of vancomycin ranging from 0 to 200 μM were prepared for binding constant determination. The corresponding series of ACE binding profiles as a function of ligand concentration are shown in Figure 11. An increase of vancomycin concentration results in gradual migration shifting of the precursor towards the reference peak. Binding

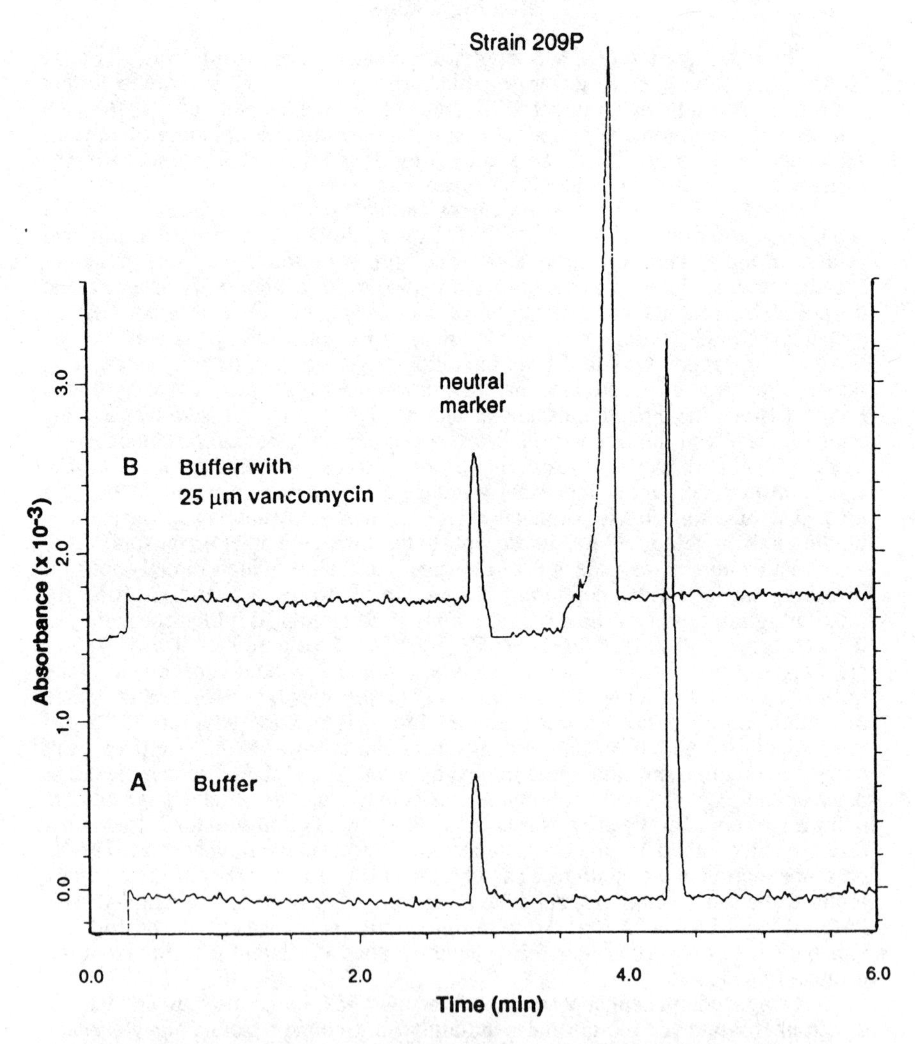

Figure 10. Evaluation of vancomycin binding to pentapeptide precursors from *Staphylococcus aureus* by affinity capillary electrophoresis. 1 - neutral marker; 2 - UDP-N-acetyl-muramyl-L-ala-D-glu-L-lys-D-ala-D-ala. Buffer in upper trace (A): 0.2 M glycine/0.03 M Tris (pH=8.30) and additional 25 μM vancomycin, and in lower trace (B): 0.2 M glycine/0.03 M Tris (pH=8.30) without vancomycin. Sample injection: 20 s; Operating voltage: 25 kV (7 μA).

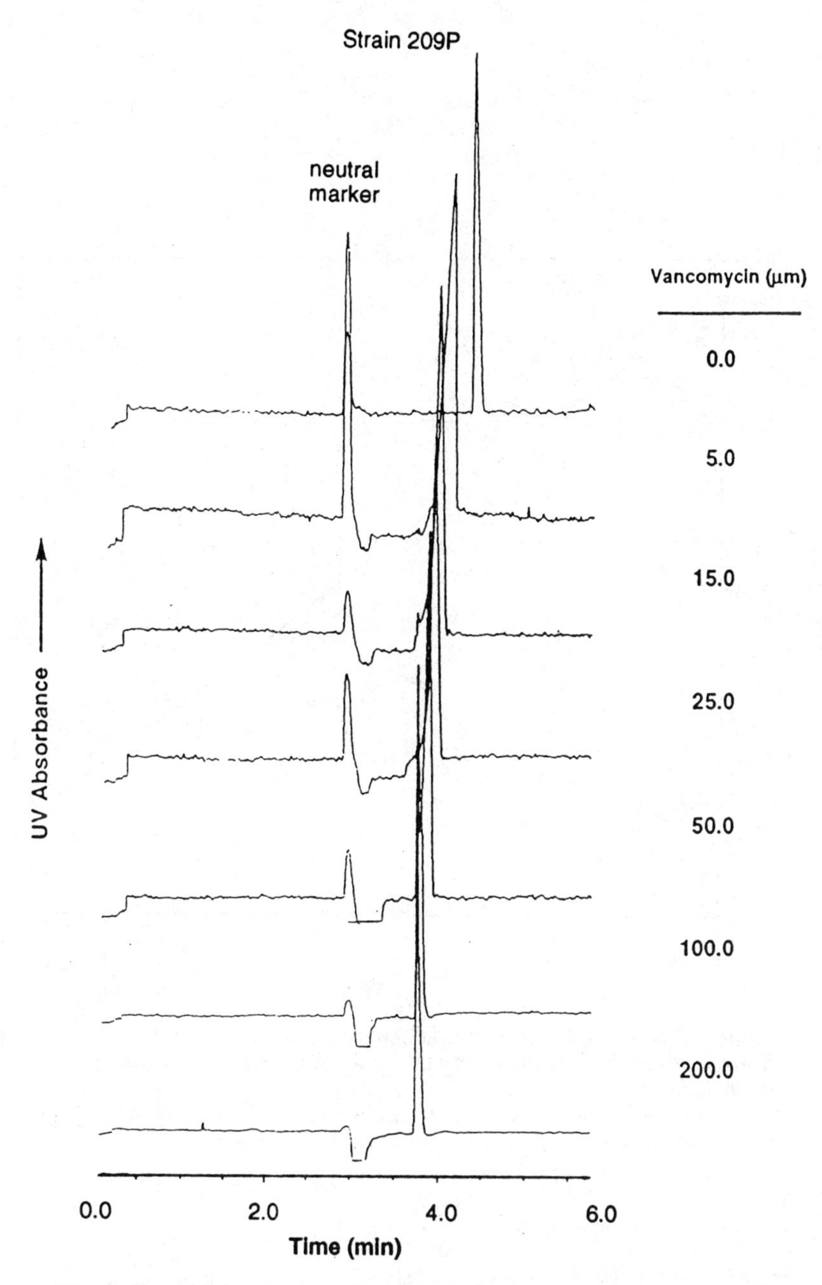

Figure 11. Affinity capillary electrophoresis of pentapeptide precursors from *Staphylococcus aureus* using 0.2 M glycine/0.03 M Tris (pH=8.30) containing vancomycin with concentrations ranging from 0 to 200 μM. Instrumental conditions are the same as in Figure 10.

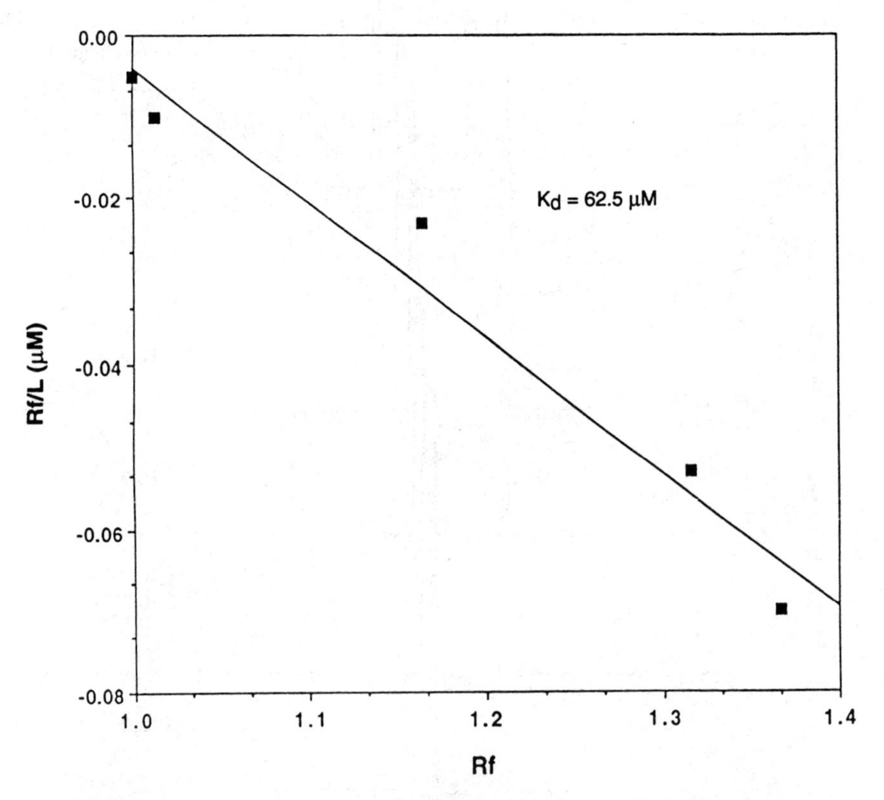

Figure 12. Scatchard plot from data obtained in Figure 11 for the determination of the binding constant for the peptidoglycan precursor from *Staphylococcus aureus* .
$R_f = dDt/ dDt_S$, dDt_S is the relative migration shift of receptor at saturating concentration of ligand, [L] is the concentration of ligand.

saturation was observed at a concentration of about 200 μM vancomycin. Peak broadening was observed at some intermediate concentrations. This condition is generally caused by the retardation of migrating molecules due to their frequent interactions with the ligand.

A receptor-ligand binding interaction in biological studies is generally a thermodynamic equilibrium process. Relaxation times between free receptor and its saturated complex with the ligand appear to take longer in the range of intermediate substrate concentrations, reflecting the slow process of equilibration between species with different migration times, which results in slightly broader peaks. The precursor peak becomes sharper at the saturating concentrations of the ligand. Similar results were also reported by simulation studies of this phenomenon (29). A possible explanation for decreased peak height of the neutral marker may be due to the evaporation of mesityl oxide which is miscible with common organic solvent and evaporates slowly. It should be emphasized that this change has no effect on the measurement of the relative migration time as long as the migration time of the neutral marker remains unchanged and measurable. Based on some assumptions dealing with interaction, equilibrium and surface absorption, equations relating the migration time to binding constant were proposed (29,32). The general equation for Scatchard analysis in ACE studies was derived as $Rf/[L] = K_b - K_b Rf$, where $Rf = dDt/dDt_S$, dDt_S is the relative migration shift of receptor at saturating concentration of ligand, [L] is the concentration of ligand, and K_b is the binding constant. A Scatchard plot derived from these experiments (Figure 12) displays linearity with a correlation coefficient of 0.97. The binding constant, K_b, measured is $1.6 \times 10^5\ M^{-1}$ and the dissociation constant, K_d was found to be 6.25 μM, which compares well with those obtained from other assays for structurally similar compounds (9,11).

Binding profiles were also obtained for precursors isolated from *Leuconostoc mesenteroides* and *Lactobacillus casei* using the method established for vancomycin-peptidoglycan precursor binding discussed above. Figure 13 displays two electropherograms obtained from *Leuconostoc mesenteroides* under the same conditions as previously described. The cell extract was directly injected without further purification. No apparent migration shifting occurs under these conditions which indicates the bioaffinity of this precursor to vancomycin has been reduced to a great extent, suggesting a structural modification at the terminal amino acid. These findings appear to be consistent with our earlier micro-LC/MS/MS studies which revealed a modified terminus corresponding to Ala-Lactate. A similar lack of migration shift was also observed for the precursor from *Lactobacillus casei*, as shown in Figure 14, supporting an altered C-terminus structure.

The amide hydrogen of the C-terminal D-Ala-D-Ala has been reported to be a crucial binding site involved in the formation of hydrogen bonds, which binds specifically to the carbonyl oxygen of the (*p*-hydroxyphenyl) glycine residue located at the "core" center of vancomycin (17). The modified structure resulted from the incorporation of a depsipeptide terminating in D-lactate which no longer possesses this amide linkage, and therefore, loses its capacity to hydrogen-bond and its affinity to vancomycin. This substitution of the amide NH for oxygen has been reported to result in at least 1000-fold lower binding (9,11), which is not detectable by this assay.

Conclusions and Future Prospects

Our structural studies dealing with abnormal precursors provide useful information aimed at developing a better understanding of molecular recognition and potential mechanisms of resistance. These findings along with previously

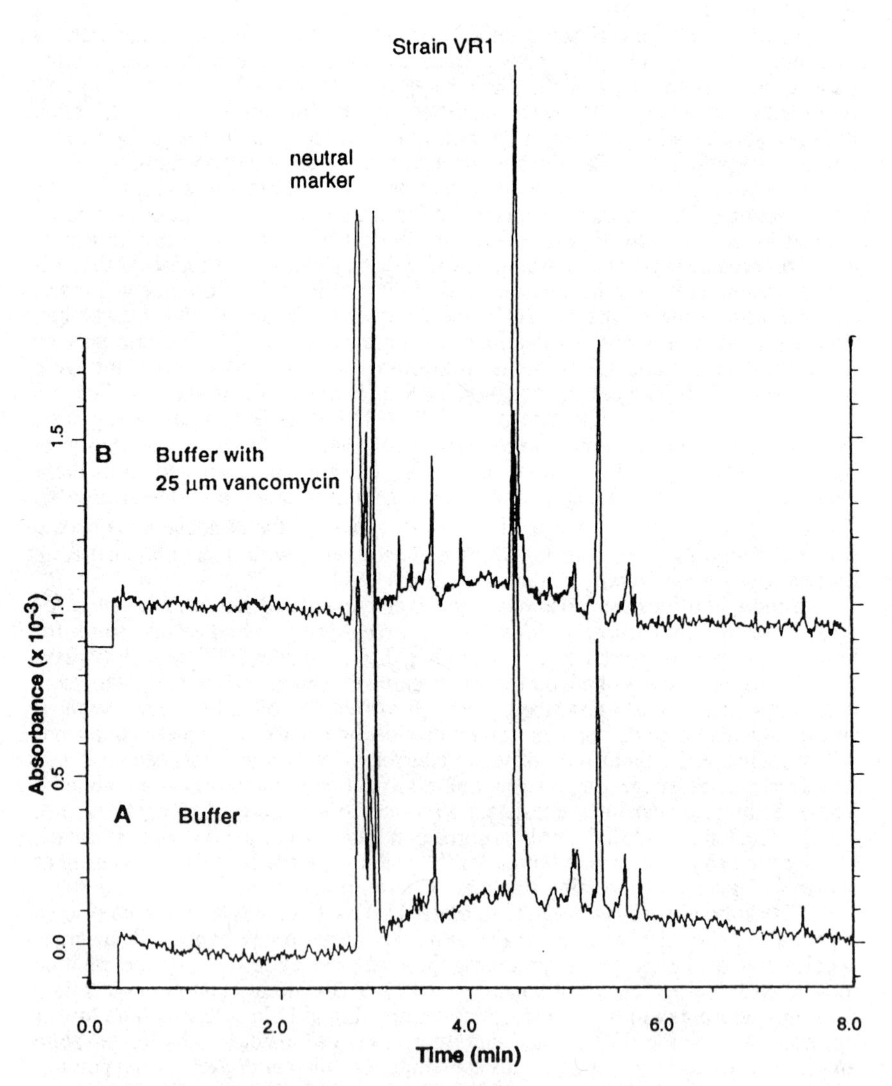

Figure 13. Evaluation of vancomycin binding to pentapeptide precursors from *Leuconostoc mesenteroides* by affinity capillary electrophoresis. 1 - neutral marker; 2 - UDP-N-acetyl-muramyl-L-ala-D-glu-L-lys-(L-ala)-D-ala-D-lactate. Instrumental conditions are the same as in Figure 10.

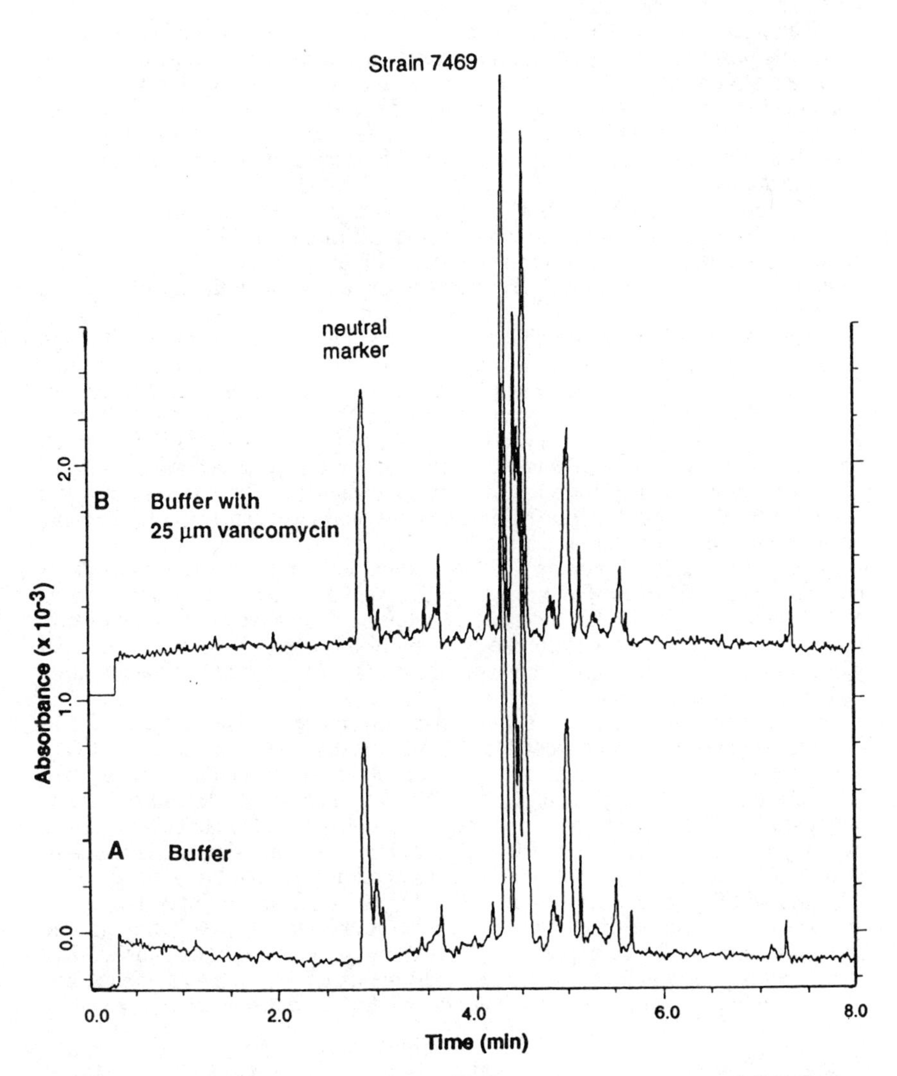

Figure 14. Evaluation of vancomycin binding to pentapeptide precursors from *Lactobacillus casei* by affinity capillary electrophoresis. 1 - neutral marker; 2 - UDP-N-acetyl-muramyl-L-ala-D-glu-L-lys-D-ala-D-lactate. Instrumental conditions are the same as in Figure 10.

described biological data (9,24-26,33) demonstrate that the mechanism of glycopeptide resistance found in *Enterococcus faecalis*, *Leuconostoc mesenteroides*, and *Lactobacillus casei* involves utilization of a peptidoglycan precursor terminating in lactate rather than alanine.

The micro-LC and CE based techniques described in this chapter represent extremely powerful approaches for the separation, analysis and characterization of complex biological mixtures. The high separation efficiency of micro-LC in combination with MS/MS provides a rapid and systematic assessment of MW and structure profile. Subsequent ACE assays generate binding profiles for a variety of precursors. Certainly, the strategies described here are suitable for a wide variety of applications.

Micro-LC/MS methods offer distinct advantages over conventional LC/MS systems. First, reduced column diameters result in enhanced mass sensitivity. Also, high linear mobile phase velocities can be generated facilitating fast separations. These analytical features are quite useful with sample-limited applications.

The novel ACE approach affords many useful advantages dealing with high separation efficiency, high sensitivity and an extremely rapid analysis time. Similar to other CE based techniques, ACE also offers advantages of reproducibility and ease of automation. Perhaps the most attractive feature of ACE is its unique ability to profile receptor-ligand interactions with only small amounts of non-radiolabeled sample. High protein purity or accurate values of concentration are not requisites for ACE binding analyses. These factors are quite attractive for the profile analysis of simultaneous binding events (receptor-ligands) occurring in the same solution.

The advantages of modern LC/MS techniques for the profile analysis of complex mixtures are evident in this work as well as other fundamental studies and applications. To continue and further delineate these advantages would likely be redundant and unnecessary for the purposes of this book. However, we should pause and recognize the significance of the current quality of LC/MS technology and instrumentation available today.

Continued exploitation of the analytical advantages offered by the LC/MS interface (nebulizer assisted electrospray) will result in numerous and significant advances across scientific disciplines. As in the past, we should follow these advances closely while "stopping" along the way to participate, share and learn. We realize that application of this valuable technology has led our investigations of today to places, destinations and collaborations that were simply visions 5-10 years ago. The extension of analytical methodologies involving the LC/MS interface will likely continue beyond micro-LC or CE. The anticipation of newer and improved technologies will no doubt culminate in novel and unique applications. The future appears bright, signaling a new beginning for LC/MS based techniques which appear ready to assume expanding roles in science and play an integral part in providing a faster rate of discovery than ever experienced.

References

1. Wilhelm, M.P. *Mayo Clin. Proc.* **1991**, 66, 165-1170.
2. Courvalin, P. *Antimicrob. Agents Chemother.* **1990**, 34, 2291-2296.
3. Wright, G.D.; Walsh, C.T. *Acc. Chem. Res.* **1992**, 25, 468-473.

4. Fan, C.; Moews, P.C.; Walsh, C.T.; Knox, J.R. *Science* **1994**, 266, 439-443.
5. Johnson, A.P.; Uttley, A.H.C.; Woodford, N.; George, R.C. *Clin. Microbiol. Rev.* **1990**, 3, 280-291.
6. Leclercq. R.; Derlot, E.; Duval, J.; Courvalin, P. *N. Engl. J. Med.* **1988**, 319, 157-161.
7. Nicas, T.I.; Wu, C.Y.E.; Hobbs, J.N.; Preston, D.A.; Allen, N.E. *Antimicrob. Agents Chemother.* **1989**, 33, 1121-1124.
8. Shlaes, D.M.; Bouvet A.; Devin, C.; Shlaes, J.H. *Antimicrob. Agents Chemother.* **1989**, 33, 198-203.
9. Bugg, T.D.H.; Wright, G.D.; Dutka-Malen, S.; Arthur, M.; Courvalin, P.; Walsh, C.T. *Biochemistry* **1991**, 30, 10408-10415.
10. Waxman, D.J. *Annu. Rev. Biochem.* **1983**, 52, 825-869.
11. Nieto, M.; Perkins, H.R. *Biochem. J.* **1971**, 123, 789-803.
12. Sheldrick, G.M.; Jones, P.G.; Kennard, O.; Williams, H.; Smith, G.A. *Nature* **1978**, 271, 223-225.
13. Reynolds, P.E. *Biochem. Biophys. Acta* **1961**, 52, 403-405.
14. Perkins, H.R. *Biochem. J.* **1969**, 111, 195-205.
15. Williams, D.H.; Butcher, D.W. *J. Am. Chem. Soc.* **1981**, 103, 5697-5700.
16. Williams, D.H. *Acc. Chem. Res.* **1984**, 17, 364-369.
17. Williamson, M. P.; Williams, D. H.; Hammond, S. J. *Tetrahedron* **1984**, 40, 569-577.
18. Caprioli, R.M.; Emmett, M.R.; *J. Am. Soc Mass Spec.*, **1994**, 5, 605-613.
19. Covey, T.R. ; Huang, E.C.; Pramanik, B.N.; Tsarbopoulos, A.; Reichert, P.; Ganguly, A.K.; Trotta, P.P.; Nagabhushan, T.L. *J. Am. Soc Mass Spec.*, **1993**, 4, 624-630.
20. Lee, M.S.; Yost, R.A. *Biomed. Mass Spectrom.* **1988**, 15, 193-204.
21. Lee, M.S.; Hook, D.J.; Kerns, E.H.; Volk, K.J.; Rosenberg, I.E. *Biol. Mass Spectrom.* **1993**, 22, 84-88.
22. Kerns, E.H.; Volk, K.J.; Hill, S.E.; Lee, M.S. *J. Nat. Prod.*, **1994**, 57, 1391-1403.
23. Kerns, E.H.; Volk, K.J.; Hill, S.E.; Lee, M.S. *J. Nat. Prod.*, **1994**, Submitted.
24. Handwerger, S.; Pucci, M.J.; Volk, K.J.; Liu, J.; Lee, M.S. *J. Bacteriol.* **1992**, 174, 5982-5984.
25. Handwerger, S.; Pucci, M.J.; Volk, K.J.; Liu, J.; Lee, M.S. *J. Bacteriol.* **1994**, 176, 260-264.
26. Liu, J.; Volk, K.J.; Lee, M.S.; Pucci, M.J.; Handwerger, S. *Anal. Chem.* **1994**, 66, 2412-2416.
27. Handwerger, S.; Pucci, M.J.; Kolokathis, A. *Antimicrob. Agents Chemother.* **1990**, 34, 358-360.
28. Ghuysen, J.M.; Tipper, D.J.; Strominger, J.L. *Methods Enzymol.* **1966**, 8, 684-699.
29. Avila, L.Z.; Chu, Y.-H.; Blossey, E.C.; Whitesides, G.M. *J. Med. Chem.* **1993**, 36, 126-133.
30. Plapp, R.; Strominger, J.L.; *J. Biol. Chem.* **1970**, 245, 3667-3674.
31. Borna, J.C.J.; Williams, D.H. *Annu. Rev. Microbiol.* **1984**, 38, 339-357.
32. Honda, S.; Taga, A.; Suzuki, K.; Suzuki, S.; Kakehi, D. *J. Chromatogr.* **1992**, 597, 377-382.
33. Orberg, P.K.; Sandine, W.E.; *Appl. Environ. Microbiol.* **1984**, 48, 1129-1133.

RECEIVED July 14, 1995

Chapter 6

Structural Characterization of Prokaryotic Glycans and Oligosaccharides

Vernon N. Reinhold and Bruce B. Reinhold

Mass Spectrometry Resource Center, Boston University School of Medicine, Boston, MA 02118

The diversity of oligosaccharides involved in cellular function (1) poses intriguing questions in glycobiology and a greater understanding of structure is a biological priority. Complexities of structure, however, make a detailed pursuit of these features a difficult problem. Electrospray ionization mass spectrometry has offered an improved instrumental approach to these materials by providing for the first time a quantitative measure of molecular distribution (glycoforms) with an understanding of glycotype at exceptional fidelity. Coupled with collisional activation and tandem analyzers, contemporary instruments can bring solutions to the details of linkage and branching at sensitivities appropriate to the questions posed. To best apply this technology, chemical modification of the native sample is generally beneficial. Alkylation or acylation of the hydroxyl groups dramatically enhances sensitivity while other derivatization strategies provide mass shifts to the molecular ion or its fragments that reflect specific structural features. These procedures, and their attributes, are discussed in the context of understanding structural detail and solving unknown problems.

This chapter describes the structural solution of three prokaryotic carbohydrate components using electrospray ionization and collision induced dissociation (CID). In two of these, as in many oligosaccharide applications, it was necessary to chemically modify the sample prior to its analysis. A crucial demand of the derivatization protocol is that it be practical, in the picomolar range, and possess low background to preserve the high sensitivity of electrospray-mass spectrometry (ES-MS). CID of these samples uncovers linkage or branching information that is not observed by direct (underivatized) mass analysis. Along with greatly increased sensitivity, derivatization also converts very dissimilar molecules into a homogeneous family with similar solubility properties. For example, acidic (Neu5Ac, uronic acid), 2-

0097–6156/95/0619–0130$12.00/0

acetamido hexoses (GlcNAc, GalNAc), and neutral residues (Glc, Gal, Man) are transformed by methylation into a homogeneous family of lipophilic analytes that extract well into organic solvents, and readily ionize by metal cation adduction. This feature minimizes the opportunity for sample disproportionation, a significant problem with ionization techniques that require a matrix. Spectral identification relies on piecing together the fragments consistent with the chemical modifications and established structural motifs. In keeping with the editor's attempt to provide a book for the uninitiated as well as the seasoned practitioner, we have tried to provide a detailed illustration of these strategies by restricting applications to three divergent problems. Although the applications are limited, the advantage is a detailed look at the structural features arising from this approach. Application to other carbohydrate problems is both implicit in what is discussed here and a matter for the references.

Although we are committed practitioners of ES within the domain of carbohydrate structure, we make no claim to a definitive strategy, only that these general procedures have greatly aided our characterization of numerous carbohydrate materials using the smallest amounts of sample. Sensitivity and working with complex mixtures are well-known strengths of mass spectrometry, however, no single methodology exists, nor can be foreseen, that will resolve the manifold structural problems inherent in an oligosaccharide sample. Degradative procedures and determining residue and linkage compositions remain important elements in the solution of an unknown. Features such as the anomeric and monomeric configuration are difficult to obtain by MS and easily resolved by NMR (given an adequate amount of sample) or enzymatic procedures. Thus, a detailed carbohydrate characterization still demands multiple techniques providing orthogonal information.

Profile Analysis By ES-MS. Mass spectrometry of carbohydrates has been largely based on ballistic ionization strategies, e.g., fast atom bombardment (FAB), or secondary ion (SI), which exhibit poor ionization efficiencies and allow considerable ion source fragmentation. For the purposes of carbohydrate analysis, ES is the most effective and generally applicable technique currently available for transforming these molecules into gas-phase ions (2). For carbohydrate samples, ES-MS has provided a direct measure of molecular distribution with an absence of degradation. Since most glycosylated molecules appear with variable degrees of polymerization, this measure has profound consequence in sample characterization. Glycan heterogeneity reflects the temporal expression of cellular activity or alterations in the environment (3). In cell adhesion, membrane surface glycoconjugates serve as platform molecules for fucose attachment to develop selectin ligands. The degree and location of these modifications on the oligosaccharide's peripheral antennae determine its functional activity, and an accurate intact molecular representation is crucial. Isolation strategies or ionization techniques that misrepresent this information through desialylation, variations in surface activity, or mass discrimination during ionization, greatly compromises structure-function understanding.

High molecular weight polysaccharides, glycoconjugates, glycoproteins and glycopeptides are difficult to analyze by electrospray directly. There are probably many reasons for this difficulty. Protein glycoconjugates might appear in high charge states with an accumulation of glycan heterogeneity and adducts that challenge the resolution limits of many mass spectrometers. With these samples, the ES-MS spectrum is a large unresolved "hump". High performance liquid chromatography (HPLC) with MS interfacing is often effective in extending the molecular weight limits probably by removing many extraneous adducts. For polysaccharides the large solvation energy of an extended and highly hydrophilic molecule may limit the desorption kinetics. A glycoprotein's glycan distribution may be assessed, however, with smaller glycopeptides or following deglycosylation and derivatization (e.g., methylation) of the released glycan. For polysaccharides, enzymatic or chemical degradation can be utilized to obtain these smaller sub-fractions, analogous to overlapping strategies for peptide sequencing. Derivatization provides an approximate 100-fold improvement in sensitivity when infused with metal ion salts in methanol/water solutions. Using this strategy, samples greater than 10 kDa have been profiled for molecular weight distributions. Unfortunately, detailed sequence information by CID of these multiply-charged ions are not observed. Smaller glycans (≈5 kDa) do yield abundant product ions (CID-MS/MS) with two levels of structural information. Single and multiple glycosidic bond rupture provide sequence and branching detail, and two-bond cross-ring cleavage yields linkage information (4).

In summary, derivatization brings together several important features for effective ES-MS analysis. Two of the most important are the increases observed in sensitivity and the exposure of linkage information. Derivatization also offers easy sample clean-up by organic extraction; a single adduct type in the ES-MS, e.g. sodium which is appropriate for quantitative profiling; and, the procedures are easily adaptable for introducing and preparing deuterated analogs for structural confirmation. Described below are specific applications using these general strategies.

Collision Induced Dissociation at Low Energy. CID is normally a two step process, where a fraction of the ion's kinetic energy is transferred into internal rotational and vibrational energy followed by an ergodic evolution on the potential surface and finally, unimolecular dissociation. Low energy (eV) CID with a triple quadrupole involves multiple collisions and potentially large, highly structured ions. The kinetics of energy redistribution amongst the various internal degrees of freedom may be limited by weak couplings (phase space bottlenecks) so that parts of the molecule are effectively at higher temperatures than other parts, i.e., on the dissociation time scale the phase space evolution is not ergodic. As an example, we have observed extensive fragmentation of the carbohydrate with some glycopeptides and glycosyl-phosphatidylinositol (GPI) anchors together with minimal fragmentation of the peptide's amide bonds (5). For other glycoconjugates such clear discrepancies are not observed. The failure, in the first case, of the peptide portion of the molecule to serve as an energy sink for the carbohydrate may reflect the fact that in some of these structures the carbohydrate is exposed with a large collision cross-section and absorbs most of the excitation energy. The energy is then only slowly transported through an isolated bond to the peptide portion of the molecule. As noted below, lipid A, consisting of several carboxyl esters, exhibits a markedly nonergodic distribution of CID products. The rate and localization of energy deposition and the extent of redistribution before dissociation are complex but central mechanistic issues in CID of high mass ions. The kinetics of ion heating compared with the dissociation kinetics tend to restrict dissociation to the lowest energy pathways in the triple quadrupole. Although the heating rate is slow (compared with keV collisions), the residence time in the collision cell is comparatively long and ions with many degrees of freedom will eventually acquire enough energy to dissociate. Thus, molecular ions over a large mass range will fragment with good yields but are restricted to the lowest energy pathways, which are generally the glycosidic cleavages for carbohydrates. This provides abundant product ion spectra to characterize carbohydrate samples. Thus, CID fragments can define three important components of structural information, *(i)* sequence and *(ii)* branching detail from glycosidic cleavage; and, *(iii)* linkage position from cross-ring cleavages. Collisional activation of glycosidic bonds is most effective in defining sequence/branching information, or, more precisely, the topology between monomeric units. Linkage information seems to require higher energy fragmentation pathways (6) and is often restricted by the very labile glycosidic cleavages on the reducing side of Neu5Ac and HexNAc residues. When these residues are not present ring-opening cleavages are easily observed.

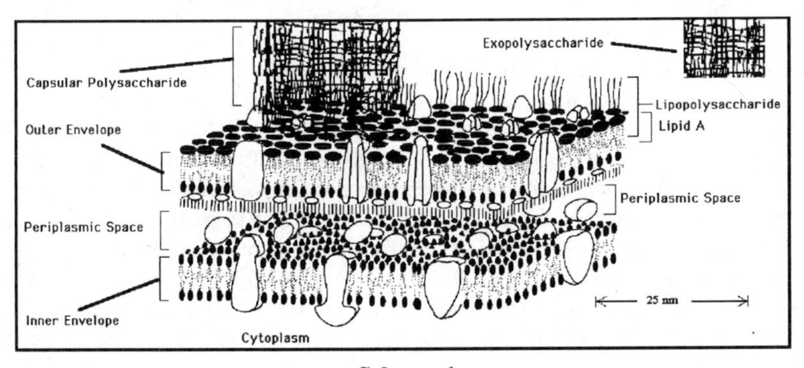

Scheme 1

Prokaryotic Cell Surface Components. In the characterization of glycans, oligosaccharides, and other glycoconjugates, the prokaryotic cell wall provides many challenges. A schematic representation of the Gram-negative bacterial cell wall is shown in Scheme 1. The cell-surface exopolysaccharides (EPS) are highly hydrated molecules that form the outermost barrier between the cell and its surroundings. These polymers protect the cell against phagocytosis, phage attachment, and adverse environmental conditions. They are also involved in binding, detoxification of metal ions, and appear to function as adhesive agents for the attachment to inert or biological surfaces. The association between microorganisms and the tissues of both animals and plants has been a major research topic for many years and continues to grows in importance for the control of health in animals, and understanding fundamental processes in plants.

Examples of this can be found in bacterial EPSs and plant interactions. The genus Rhizobia has been intensively studied in order to understand the symbiotic processes that lead to nitrogen fixation. In a deleterious interaction, endotoxins that cause Gram negative sepsis originate from lipopolysaccharides (LPS) that consist of a highly variable polysaccharide linked to an acylated glucosamine disaccharide. The latter disaccharide serves to anchor the extracellular polysaccharide and this specific moiety is considered to be the etiological agent of sepsis (Scheme 8). Peptidoglycans separate the inner and outer membranes and play very important roles in maintaining bacterial cell integrity. Cyclic (1,2)-β-glucans are found in the periplasm, the space between inner and outer membrane, and are present in high amounts when the osmolarity of the medium is low (7, 8). The subtleties of carbohydrate involvement are diverse and understanding function is a matter of structural detail.

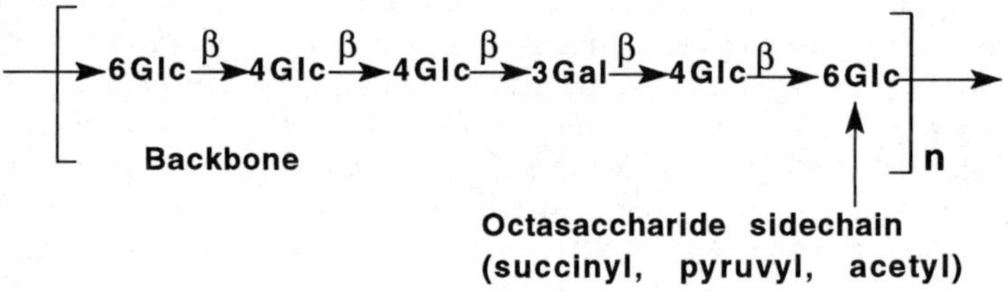

$$\left[\rightarrow 6Glc \xrightarrow{\beta} 4Glc \xrightarrow{\beta} 4Glc \xrightarrow{\beta} 3Gal \xrightarrow{\beta} 4Glc \xrightarrow{\beta} 6Glc \right]_n \rightarrow$$

Backbone

Octasaccharide sidechain (succinyl, pyruvyl, acetyl)

Scheme 2

Oligosaccharides. Root hair infection of leguminous plants by nitrogen fixing bacteria starts with recognition and cell surface attachment and is followed by a series of complex processes involving the formation of an infection thread, nodule formation, invasion of the nodule by the bacterium and finally, nitrogen fixation. In the interplay of chemical signals between guest and host species it has been shown that glycoconjugates play both a generic and specific role (9-11). EPS from *R. meliloti* (called succinoglycan) induce the formation of nodules, but mutants to this structure are unable to invade. In a similar way, mutant variations in the EPS structure also block this plant/host relationship. The features of EPS that induce nodule formation and the subsequent step of invasion has motivated interest in its detailed understanding.

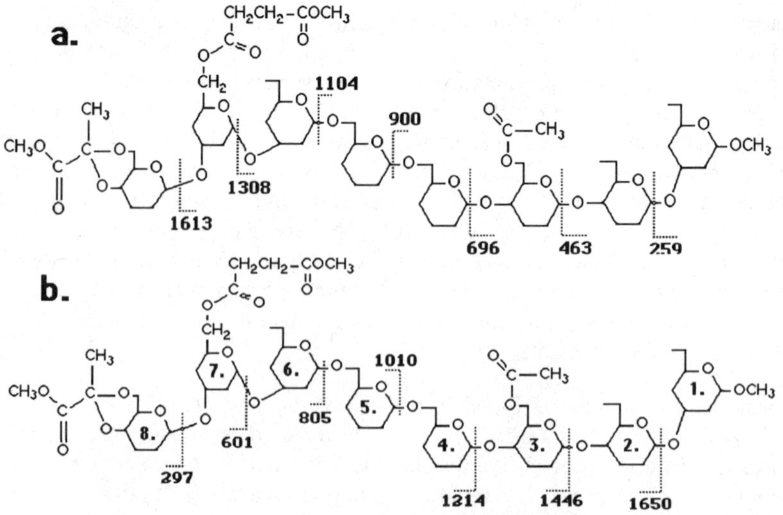

Scheme 3

Early studies had reported the succinoglycan to contain glucose, galactose, pyruvic acid, and O-acetyl groups in the proportion 7:1:1:1 (12). The inter-residue linkages were examined by methylation analysis and O-acetyl groups were reported at the 6 carbon of 3 and 4-linked glucose residues (12). At this time the succinate modification was not recognized and the assignment of the O-acetyl group to a 3-linked glucose was in error. Evidence for a repeating unit (Scheme 2) of EPS was reported and the linkage sequence in the main chain was determined by a modified Smith degradation (13). Most of these structural details for *R. meliloti* strain Rm 1021 were subsequently confirmed as glycosyl sequencing methodology developed (14). Polymer degradation with β-D-glycanases and methylation analysis demonstrated that the exopolysaccharides from *R. meliloti, Alcaligenes faecalis* var. myxogenes, *Agrobacterium radiobacter, Agrobacterium rhizogenes* and *Agrobacterium umefaciens* were identical except for the acetyl or succinyl substituents on a side chain (15).

The side chain of succinoglycan in *R. meliloti* and *Agrobacterium sp.* was found to consist of a octasaccharide containing a pyruvate, an acetate and a succinate group (15) but the exact locations of these substituents on the side-chain were not identified (Scheme 2). The subunit was prepared by digestion with a succinoglycan depolymerase from *Cytophaga arvensicola* (16) and the products purified by gel filtration chromatography on Bio-Gel P4. Permethylation under neutral conditions (to preserve any acyl modifications) followed by ES-MS identified a major ion at *m/z* 955. This corresponds to a molecular weight of 1864 Da adducted with two sodium ions (955 = (1864+23+23)/2) and can be equated to eight methylated hexose residues with a pyruvyl, acetyl, and a succinyl group. CID of the doubly charged ion gave the spectrum in Figure 1 and a sequence for the octamer structure as shown in (Scheme 3) (17).

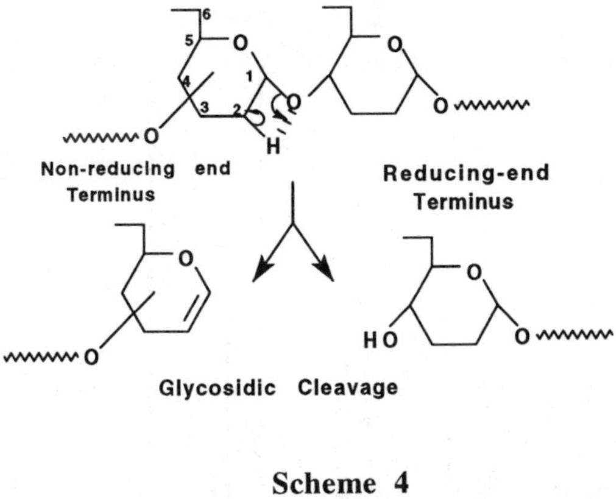

Non-reducing end Terminus

Reducing-end Terminus

Glycosidic Cleavage

Scheme 4

This spectrum provided features common to methylated oligosaccharides with a combination of glycosidic and cross-ring cleavages for an understanding of sequence and linkage detail. Most fragments were detected as single sodium ion adducts with only a few retaining both sodium cations. Glycosidic fragments result from cleavage of the C_1-O bond (Scheme 4) with a hydrogen transfer to the "exposed" oxygen. This simple elimination yields three sets of informative sodiated fragments; a set containing the reducing-end with one unblocked hydroxyl, a non-reducing set with unsaturation, and internal fragments (not shown; losses from opposing termini) (Scheme 4). Substituent groups on specific residues (succinyl, pyruvyl, acyl) were

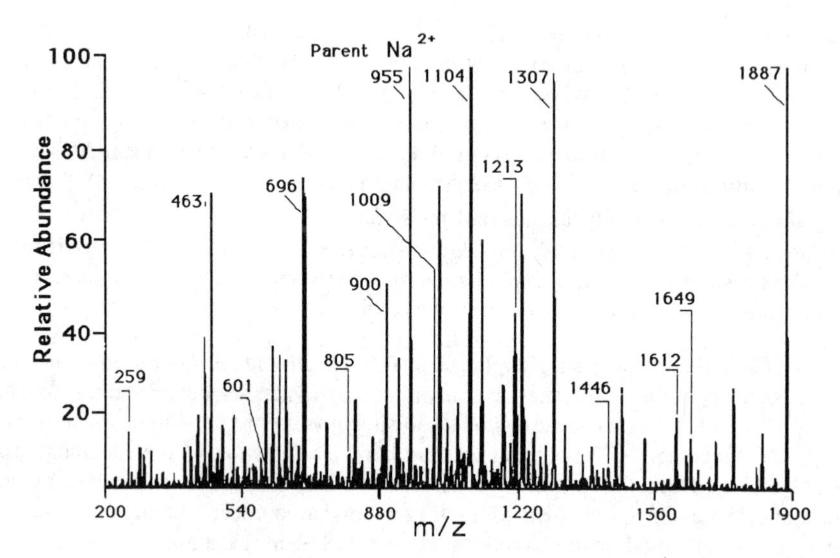

Figure 1. ES-CID-MS/MS spectrum of the neutral methylated *R. meliloti* side chain octasaccharide. Doubly charged precursor ion *m/z* 955 selected for collision. Glycosidic cleavage ions labeled.

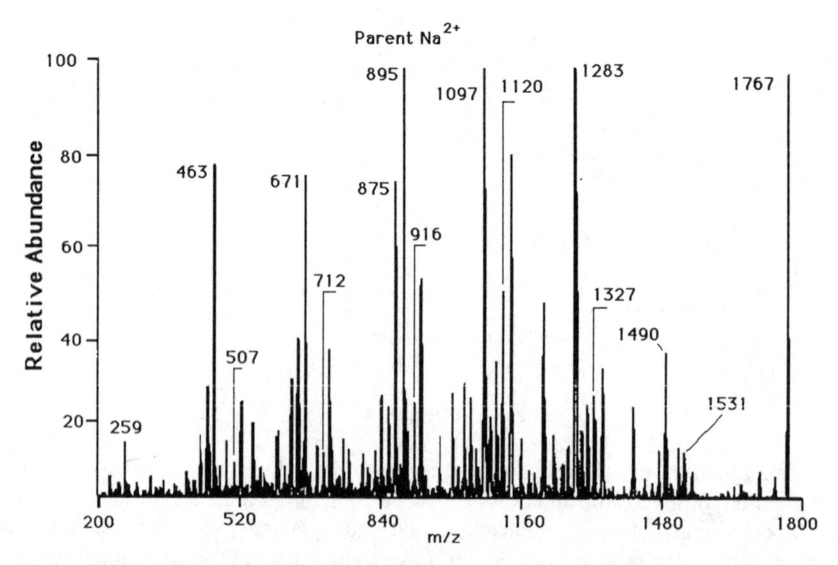

Figure 2. ES-CID-MS/MS spectrum of the *R. meliloti* side chain octasaccharide following basic CD_3-remethylation. Doubly charged precursor ion *m/z* 895 selected for collision. Remethylation exchanges the succinyl and acetyl esters with perdeuterated methyl ethers.

identified by the mass differences between the glycosidic cleavage ions in both the reducing (Scheme 3a.) and non-reducing fragments (Scheme 3b). Prominent reducing-end fragments, *m/z* 259, 463, 696, 900, 1104, 1308, and 1613, clearly position the acetyl group on the third residue from the reducing terminus, while the succinyl and pyruvyl moieties occur on the penultimate and terminal residues.

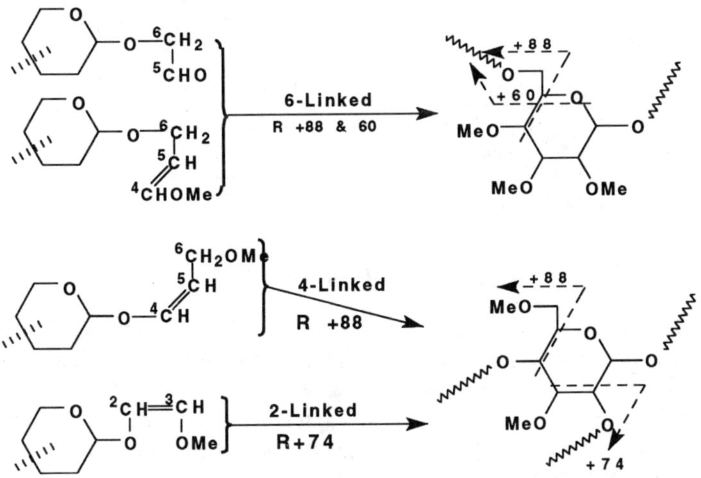

(R = unsaturated monomeric unit, Scheme 4)

Non-Reducing Termini

Scheme 5

Fragments that define linkage do not involve rupture of glycosidic bonds, but are derived from cleavage across the pyranose ring. This rupture results in pieces of the ring attached to non-reducing and reducing fragments and both are observed. Remnants linked to non-reducing terminal and internal fragments differ with linkage position and occur as satellite peaks above their respective glycosidic ions, e.g., 2-O-linkages, +74 Da increment; 4-O-linkages, +88 Da increment; 6-O-linkages, two ions +60 and +88 Da increments, and for 3-O-linkages, no related ions (Scheme 5). All glycosidic and linkage fragments are monosodiated.

For reasons not understood CID fails to identify the linkage to a reducing-end residue, as well as to most moieties that posses acyl substituents. Thus, linkage to the acetylated residues were not observed in the spectrum. The exact position of acyl substituents were assigned by an additional methylation step using perdeuterated methyliodide under basic conditions. This strategy replaces the ester residues with a CD_3-methyl analog while methyl ethers, prepared initially as a consequence of neutral methylation, remain unaltered. The spectrum (Fig. 2) showed cross-ring fragments attached to residues 6, 5, 4, and 3 (Scheme 3b). The absence of any cross-ring fragments attached to residues 8, and 7 strongly suggest each to be 3-linked. The position of O-acetylation on residue 3 was determined from the single +91 u increment, (Scheme 6, Figure 3). In contrast to the complicated procedures that marked the elucidation of earlier succinoglycan structures, recent advances in MS, especially sample ionization and CID, brought extensive structural characterization with orders of magnitude less material.

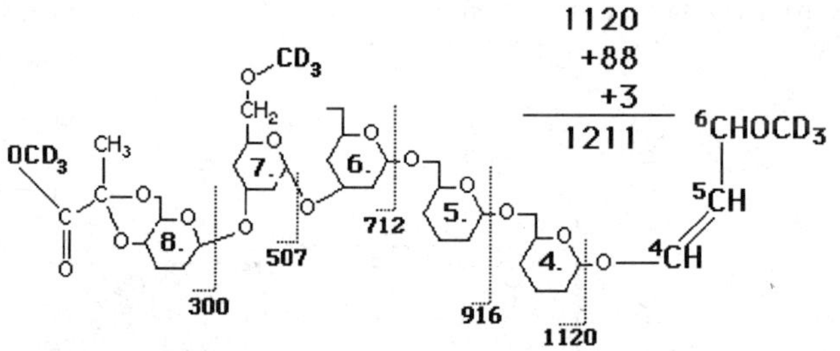

Scheme 6

Protein Glycans. A detailed characterization of protein N and O linked glycans must be approached in the absence of the protein. Certainly some information can be realized by ES-MS of glycopeptides and small glycoproteins, but for the reasons discussed above, ES-MS of deglycosylated and derivatized samples has proven to be more effective. Deglycosylation techniques have vastly improved and releasing enzymes (amidases and endoglycosidases) (18, 19) of excellent specificity and purity are commercially available. It is not unusual to find some glycans that are inaccessible to the enzyme, and often a combination of proteolysis, denaturization or reduction and alkylation can provide quantitative release. Chemical release, using hydrazinolysis or basic reductive elimination for N and O-linked glycans, respectively, are also improving but at present are unsatisfactory for an accurate representation of glycans. This question of structural fidelity should be pursued rigorously with any new methodology, especially for a measure of glycoform distribution where a correct answer is not genomically imprinted. (For an excellent treatise on glycoprotein techniques see (20)).

Protein glycosylation in bacteria, in contrast to eukaryotes, is poorly understood, and our knowledge rests mostly with carbohydrate oligomers from the cell wall and lipopolysaccharides. The first reports of a single N-linked, and multiple O-linked, carbohydrates were in the envelope protein of *Halobacterium salinarium,* (21). More recent studies have reported the presence of N-linked carbohydrates in bacterial cell-wall proteins (22, 23) and O-linked glycans in Thr/Pro-rich regions of cellulase complexes of the cellulolytic bacteria *Clostridium thermocellum* and *Bacteriodes cellulosolvens,* (24).

Recently, a study of a *Flavobacterium* species suggested protein glycosylation different from what had been previously observed in prokaryotes. This Gram-negative bacterium secretes at least eight major proteins including two proteases and four oligosaccharide chain-cleaving enzymes, including one amidase, termed PNGase2 F, and three endoglycosidases designated Endo F1, F2, and F3 (25). Mass spectrometry confirmed that Endo F2 and Endo F3 were post-translationally modified with carbohydrate during secretion by a possible linkage to the hydroxyl group of serine

(18). Further evidence suggested the O-linked carbohydrate on P40, Endo F2, and Endo F3 to be identical (26). Prokaryotic extracellular glycoproteins like Endo F2, Endo F3, and P40 that have O-linked oligosaccharides at specific consensus sites have never been described before.

Carbohydrate composition analysis had shown the presence of glucuronic acid which was further supported by a 42 u increment (3 X 14) to the glycan following methyl esterification (26). Uronyl residues are destroyed under the basic conditions frequently utilized to cause glycan release. To offset this, the carboxyl groups were first esterified and then reduced to the corresponding alcohol. Composition and linkage analyses following these steps identified two new components, 2-deoxy-2-acetamidoglucose and 2-O-methylglucose, indicating three uronyl precursors to be glucuronyl analogs, (glucuronic acid, 2-deoxy-2-acetamidoglucuronic acid and 2-O-methylglucuronic acid). From these results, the combined composition analysis suggested three hexoses, (mannose, glucose, and 2-O-methylglucose), one deoxyhexose, (2-O-methylrhamnose), and the three glucuronyl analogs for a total of seven unique structures. A repeat of the composition analysis following reductive elimination from the peptide detected the unique presence of mannitol and an absence of mannose, providing evidence for O-linkage to the peptide. Linkage analysis, aided by CD_3-methylation, indicated two termini, (2-O-methylmannose, 2-O-methylrhamnose), and a 4-O-linked glucosyl residue. Interestingly, no dibranched structures were detected.

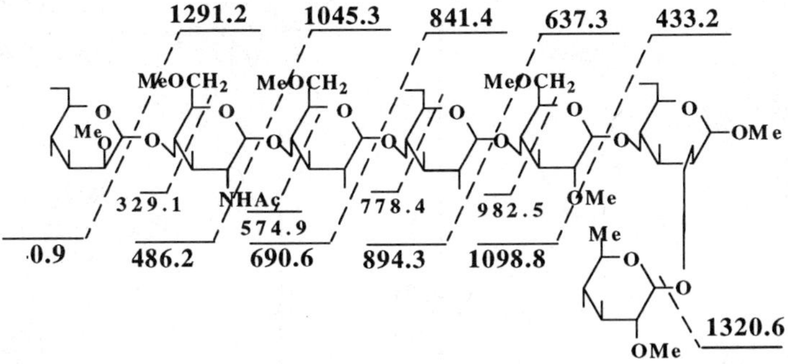

Scheme 7

Reduction of the uronyl esters within the oligomer had three important advantages. It evades the problem of alkali instability, yields a hexose conformer for characterization, and the hexose product was suitable for linkage analysis by CID. The esterified glycopeptide was reduced to the primary alcohol and released by both reductive elimination as the alditol and methylation as the methyl glycoside. Each sample provided a major doubly charged parent ion when analyzed by ES-MS and an absence of base degradation. CID analysis of the analog released by methylation, *m/z* 765.6^{2+}, provided the spectrum in Figure 4.

As expected, a major glycosidic cleavage dominated the spectrum which occurred adjacent to the acetamidohexose residue leaving a non-reducing terminal disaccharide fragment, *m/z* 486.2, and its pentasaccharide counterpart, *m/z* 1045.3.

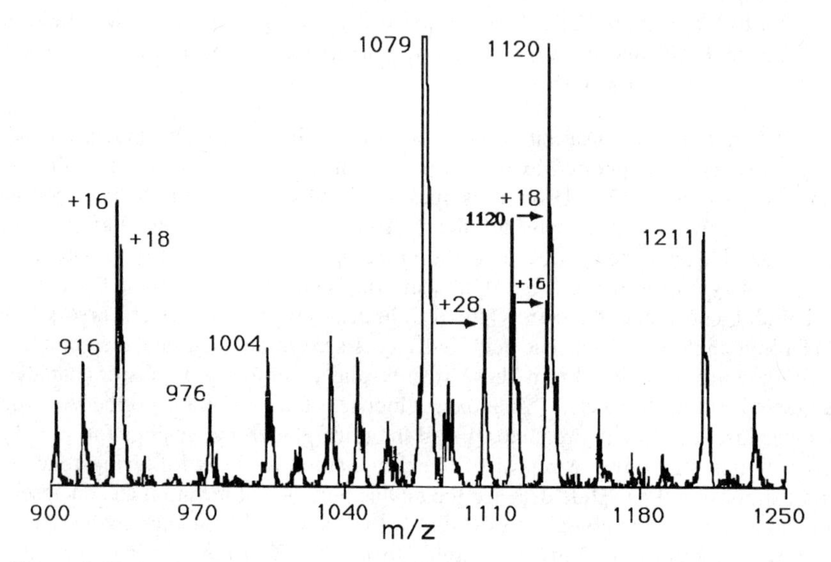

Figure 3. Expanded view of the *m/z* 900-1250 region of Figure 2 illustrating ring-opening fragments used to identify 6-linkage (Scheme 6, *m/z* 1211, R + 88 u).

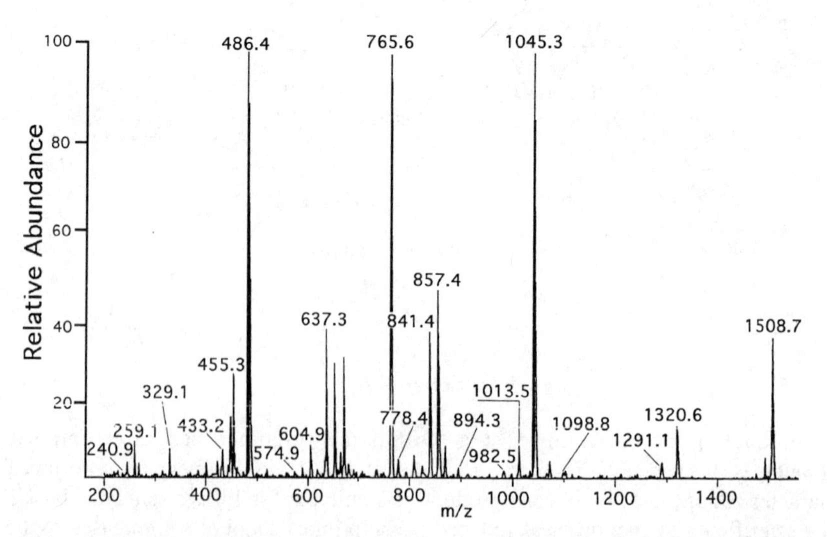

Figure 4. ES-CID-MS/MS spectrum of *F. meningosepticum* P40 glycan prepared with uronyl carboxyl groups reduced. Precursor ion selected *m/z* 765.6.$^{2+}$ Prepared from the glycopeptide by uronyl ester reduction and released by basic methylation.

Smaller fragments defined the sequence from both termini and four cross-ring fragments identify 4-O-linkages for each of four internal residues, *m/z* 329.1, 574.9, 778.4, and 982.5 (Scheme 7). A similar fragment to identify the penultimate GlcA-Man linkage was not expected or detected for the terminally positioned residue.

In contrast to these more typical structures, an O-glycosyl tyrosine linkage has been reported where the linking sugar moiety may be a glucosyl moiety, as in the insect larvae glycogenin (27, 28), and more recently, a galactosyl residue has been discovered in the crystalline surface layer glycoproteins of *Clostridium thermohydrosulfuricus* S102-70 (29). The present report introduces considerable novelty into prokaryotic glycosylation with a heptasaccharide chain linked through an O-mannosyl serine linkage, and each sugar residue is different (30). Although the data in this area is limited, it appears the conserved glycan motifs established in mammalian glycosylation will be less constrained in bacterial species.

As evident throughout this text, electrospray ionization (2), can ionize biopolymers with minimal internal energy and with an absence of fragmentation. Although much of the ES work has been focused on analyzing positive ions adducted with protons or metal ions, molecules containing one or more acidic sites may be susceptible to deprotonation with the formation of negative ions. As an example, Cole and coworkers (31) (see elsewhere in this volume) have demonstrated that the acidity of the phosphate group on lipid A provides improved ionization efficiency in the negative ion mode and we summarize some of our studies in this regard using a preparation from *Salmonella minnesota* Re595 (32).

Lipid A. Endotoxins play important roles in the interaction of Gram-negative bacteria with higher organisms and represent immunoreactive surface antigens. Set free by multiplying or disintegrating bacteria, endotoxins exhibit a broad spectrum of biological activities (pyrogenicity, hypotension, lethal shock), and because of these properties, they are held responsible for certain events which results in the activation of cells leading to endotoxic shock. Moreover, they are immunostimulators causing activation of B lymphocytes, granulocytes and mononuclear cells. These characteristics have stimulated research into their chemical nature and primary structure to understand, at a molecular level, the steps involved in endotoxin action.

Structural studies of endotoxins derived from various bacterial families appear to share a common architecture consisting of a polysaccharide and a covalently bound lipid component, (lipid A), and hence, the name lipopolysaccharides (LPS). The polysaccharide portion consists of two domains, the outer O-polysaccharide chain and an inner core oligosaccharide, which differs considerably in chemical structure (e.g., *S. minnesota*, Scheme 8). The lipid A moiety represents the endotoxic center of LPS and its structure and toxicity varies among different bacterial species. Endeavors to understand the diverse structure-function relationships are difficult but, from an immunological standpoint, profoundly important.

Previous studies of *Salmonella minnesota* Re595 monophosphoryl lipid A (MLA) have described the β(1→6) linked disaccharide, (2-amino-2-deoxy-D-glucopyranose), to consist of free hydroxyl groups at C1, C4 and C6' with a phosphate residue located at the C4' position. The C2, C3, C2', and C3' groups are esterified with β-hydroxymyristic acid which on positions C2, C2', C3' is further esterified with a palmitic, lauric, and myristic acids, respectively, MW = 1955, (33). The spectrum of this MLA preparation is presented in Figure 5. A series of molecular weight-related, deprotonated anions in eight clusters were detected with one or two major components within each group. The major, high mass ion, (*m/z* 1954.6), supports the above heptaacyl structure (Scheme 8), and a careful assessment of adjacent peaks indicates alkane heterogeneity above and below this component, e.g., *m/z* 1982.3 (9%), 1926.6 (13%), and 1898.6 (4%). The percent values are relative to the base ion *m/z* 1954.6.

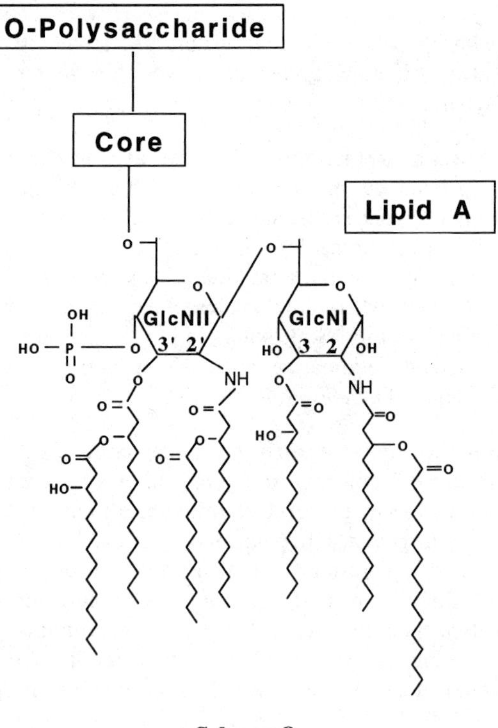

Scheme 8

The next four ion clusters, in decreasing mass, can be related to this heptaacyl MLA by the respective absence of 1, 2, 3, and 4 acyl groups. The remaining three clusters are represented by monosaccharide structures. Thus, all major ions in the spectrum can be related through the presence or absence of acyl residues or esterified amine monosaccharide residues. The second ion cluster, centered around *m/z* 1716.4, illustrates the effects of losing one acyl group from each of the four positions, combined with alkane heterogeneity. The ions *m/z* 1772.3, 1744.2, 1728.0 and 1716.4 account for the respective absence of lauroyl, myristoyl, hydroxymyristoyl, and palmitoyl groups, while the minor alkane heterogeneity can only be observed at *m/z* 1700.6 and 1688.2. The remaining alkane peaks are hidden under other major ions and would require higher resolution for identification. The third cluster, centered around *m/z* 1505.9, becomes more complex with the major ions accounted for by an absence of two acyl groups combined with the usual alkane heterogeneity. Double acyl absence from any combination of the four positions should in principle provide six ions, and all six are detected. These details were observed with an expanded mass range for the combined absence of lauroyl and myristoyl, *m/z* 1562.0; lauroyl and hydroxymyristoyl, *m/z* 1546.2; lauroyl and palmitoyl, *m/z* 1534.1; myristoyl and hydroxymyristoyl, *m/z* 1518.0; myristoyl and palmitoyl, *m/z* 1505.9; and palmitoyl and hydroxymyristoyl, *m/z* 1489.9. The two lowest mass ions, *m/z* 1462.0 and 1477.9, can be accounted for as alkane heterogeneity. Their abundance is considerably diminished from the ratios measured for the heptaacyl MLA suggesting the absent acyl groups, myristoyl and palmitoyl residues, contribute significantly to the heterogeneity. Interestingly, the alkane heterogeneity was absent from the lower mass ions composed only of hydroxymyristoyl groups. This suggests that the alkane heterogeneity is associated only with the three branched esters, lauroyl, myristoyl, and palmitoyl.

The fourth cluster, adjacent to the ion *m/z* 1279.7, shows all the expected components for an absence of three acyl groups: lauroyl, hydroxymyristoyl, and myristoyl, *m/z* 1336.2; lauroyl, palmitoyl, and myristoyl, *m/z* 1323.7; lauroyl, palmitoyl, and hydroxymyristoyl, *m/z* 1307.9; and, myristoyl, hydroxymyristoyl, and palmitoyl, *m/z* 1279.7. This latter ion is isobaric with the double acyl loss of myristylhydroxymyristoyl and palmitoyl. A single 28 Da lower ion (9%) can be observed, *m/z* 1261.5, which represents the heterogeneity contributed by a lauroyl residue. In the next ion cluster, represented by the absence of four acyl residues, all structures appear to collapse into the single ion, *m/z* 1097.5. The three remaining hydroxymyristyl groups could in principle give rise to four structural isomers, each missing a single hydroxymyristoyl group from one of four possible positions. Considering the greater stability of acyl amides there are probably only two structures with hydroxymyristoyl groups at C2, C3, C2' and C2, C2', C3'. Close examination of the ion profile, *m/z* 1097.8, indicates a shoulder which on scale expansion resolved into the ion *m/z* 1102.7. This is consistent with a fully acylated GNII structure, the first of several monosaccharide ions observed at lower mass.

Lipid A Structural Analysis by Collision Induced Dissociation using MS/MS. The highest mass observed in the analysis of the *Salmonella minnesota* Re595 MLA sample was the heptaacyl component, *m/z* 1954.6 (Figure 5). Selection of that ion followed by CID and MS/MS provided a series of ion clusters related to the number of residues eliminated (data not shown). Since the acyl residues are of different mass, each cluster expresses the heterogeneity of remaining groups and their abundance. The first hexaacyl cluster of four ions, *m/z* 1726.6 (68%), *m/z* 1710.3 (16%), *m/z* 1698.3 (12%), and *m/z* 1754.3 (4%), represent the respective loss of myristic, hydroxymyristic, palmitic and lauric acids. Interestingly, the elimination of myristic acid, as measured by product ion intensity after CID, was several times more abundant than palmitic acid. And the lauroyl ester was barely detectable. Since the detailed position of each acyl group has been established in *Salmonella minnesota* Re595, the mass would define the ion structure (Scheme 8). Even though all acyl groups possess chemically similar linkage, their position appears to have a profound influence on their lability. As noted with the heptaacyl analog, myristic acid elimination appears to be highly favored and the lauroyl moiety is largely unaffected. Selection of the ion representing the hexaacyl MLA (*m/z* 1716.1, Figure 5) followed by CID and MS/MS provided a series of ion clusters again related to the number of

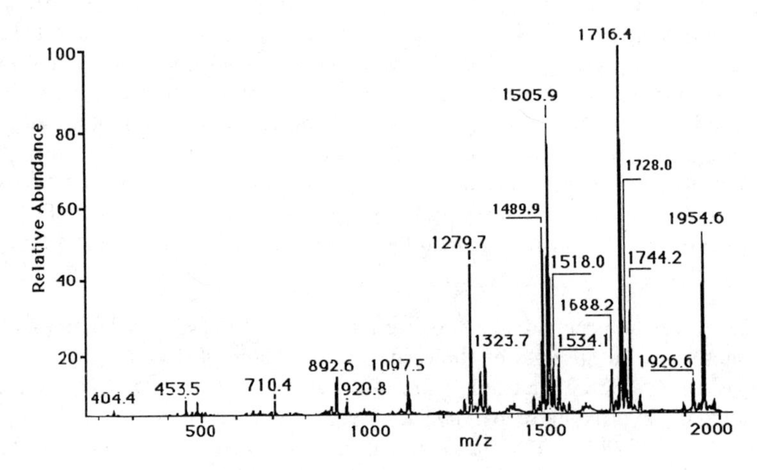

Figure 5. Lipid A; Negative ES-MS profile spectrum derived from MLA homologs of *Salmonella minnesota* Re595

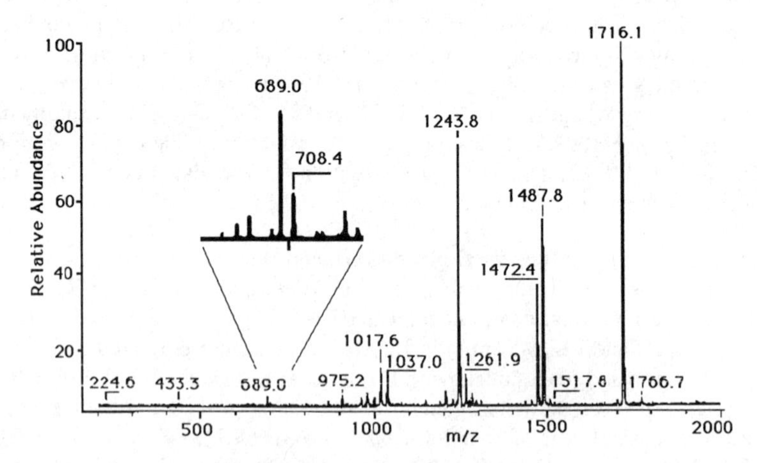

Figure 6. CID spectrum for the $[M-1]^-$ ions associated with the hexaacyl MLA structure (*m/z* 1716) derived from *Salmonella minnesota* Re595 lipopolysaccharide. Insert, glycosidic cleavage fragment accounted for as GNII which underwent either elimination (*m/z* 689.0) or acylium cleavage (*m/z* 708.4) at the lauroyl residue.

residues eliminated (Figure 6). The elimination of either one branched residue or two separate acyl residues can account for the fragments *m/z* 1261.9 and 1243.8, respectively, (Scheme 9). A combined loss of three acyl groups provided two major fragments, *m/z* 1017.6 and *m/z* 1037.0, where the former ion can be derived by a double elimination of one branched (myristylhydroxymyristoyl) and one unbranched (hydroxymyristoyl) residue. The fragment at *m/z* 1037.0 is 18 Da higher in mass and suggests one or both of the groups underwent partial acylium cleavage and not elimination. Two other fragments are of considerable interest which appear to be related to glycosidic cleavage. They were observed in the CID spectra of the hepta-, hexa-, and pentaacyl analogs and may represent the non-reducing terminus of the disaccharide, GNII. The glycosidic oxygen was retained with fragments which may be accounted for as GNII which underwent either elimination (*m/z* 689.0) or acylium cleavage (*m/z* 708.4) at the lauroyl residue (Fig. 6, insert).

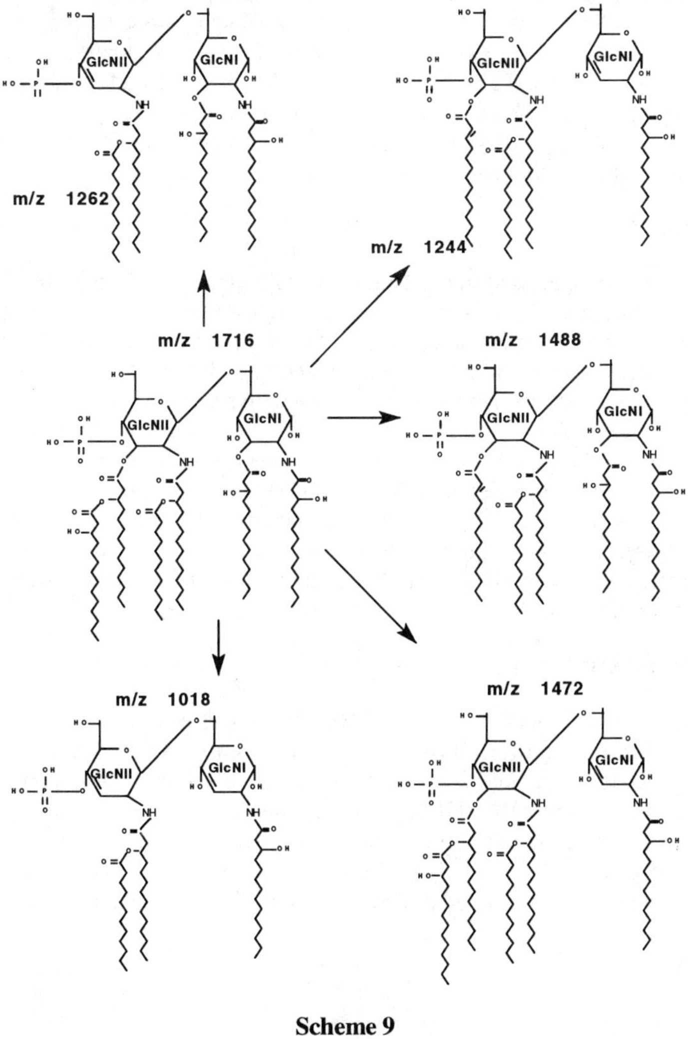

m/z 1262

m/z 1244

m/z 1716

m/z 1488

m/z 1018

m/z 1472

Scheme 9

Conclusions

Instrumental advances are progressing at a rapid pace and nowhere are the effects more apparent than in carbohydrate mass spectrometry. Problems with sample gasification, for decades a major obstacle, have now been overcome providing for the first time molecular ion profiles in the absence of fragmentation and fidelity in glycoform distributions with the potential for quantification in a single analysis. These techniques are being matched with developments in MS analyzers where ion trap technology promises unprecedented gains in sensitivity and resolution.

Utilizing this technology to unravel biopolymer problems is not always straightforward and frequently demands modifications in both instrumental operation and analyte preparation. Such is the case for carbohydrate samples which exhibit improved electrospray properties from methanol solutions when prepared as metal adducted lipophilic analytes. Molecular weight profiling by these techniques provides a measure of composition to assess glycotype (complex, hybrid, high mannose), and their concentration (glycoform distribution). Further detailed studies may proceed on any selected glycan by CID and MS/MS. As we proceed to study glycoconjugates of greater complexity, comprising epitopes of diverse function and vanishing concentration, these techniques may be helpful.

Experimental

The mass spectrometer used in this study was a Finnigan-MAT TSQ-700 (Finnigan-MAT Corp., San Jose, CA) equipped with an electrospray ion source. Samples were dissolved in methanol:water solutions (70:30 v/v) containing 0.5 mM sodium acetate and analyzed by syringe pump flow injected at a rate of 0.75 µl/min directly into the electrospray chamber through a stainless steel hypodermic needle. The voltage difference between the needle tip and the source electrode was -3.5 keV. A unique feature of ES is the generation of multiply charged ions from a single molecular species. ES data was analyzed directly. For CID studies, multiply charged parent ions were selectively transmitted by the first mass analyzer and directed into the collision cell containing argon at roughly 2 mtorr with acceleration voltages of 30 - 40 V, hence kinetic energies of 60 - 80 eV. For negative ion studies oxygen was used as a sheath gas for electron scavenging.

Abbreviations Used

MLA, monophosphoryl lipid A; LPS, lipopolysaccharides; Neu5Ac, N-acetylneuraminic acid; GlcA, glucuronic acid; GlcNAc, 2-acetamido-2-deoxy-glucose; GalNAc, 2-acetamido-2-deoxy-galactose; Glc, glucose; Gal, galactose, Man, mannose; GNI, reducing-end residue of 2-amino-2-deoxy-glucopyranose disaccharide; GNII, non-reducing-end residue of 2-amino-2-deoxy glucopyranose disaccharide; ES-MS, electrospray mass spectrometry; FAB, fast atom bombardment; CID, collision induced dissociation; MS/MS, tandem MS.

Acknowledgements

The authors gratefully acknowledge our collaborators, Drs. Walker and Plummer, for identifying these problems and supplying sample to complete their structure. We also acknowledge support and collaboration from Susan Goelz and Werner Meier of Biogen Corp. This work has been supported in parat by NSF grants (MCB9400633) and NIH grant (R01 GM4578).

References

(1) Varki, A. *Glycobiology* **1993,** *3*, 97-130.
(2) Fenn, J. B.; Mann, M.; Meng, C. K.; Wong, S. F.; Whitehouse, C. M. *Science* **1989,** *246*, 64-71.
(3) Goochee, C. F. *Develop. In Biol.Standard.* **1992,** *76*, 95-104.
(4) Reinhold, V. N.; Reinhold, B. B.; Costello, C. E. *Anal. Chem.* **1995,** *67*, 1775-1784.
(5) Redman, C. A.; Green, B. N.; Thomasoates, J. E.; Reinhold, V. N.; Ferguson, M. A. J. *Glycoconjugate J.* **1994,** *11*, 187-193.
(6) Lemoine, J.; Fournet, B.; Despeyroux, D.; Jennings, K. R.; Rosenberg, R.; de Hoffmann, E. *J. Am. Soc. Mass. Spectrom.* **1993,** *4*, 197-203.
(7) Miller, K. J.; Kennedy, E. P.; Reinhold, V. N. *Science* **1986,** *231*, 48-51.
(8) Miller, K. J.; Reinhold, V. N.; Weissborn, A. C.; Kennedy, E. P. *Biochim. Biophy. Acta* **1987,** *901*, 112-118.
(9) Spaink, H. P. *Antonie van Leeuwenhoek* **1994,** *65*, 81-98.
(10) Spaink, H. P.; Sheeley, D. M.; van, B. A.; Glushka, J.; York, W. S.; Tak, T.; Geiger, O.; Kennedy, E. P.; Reinhold, V. N.; Lugtenberg, B. J. *Nature* **1991,** *354*, 125-130.
(11) Halverson, L. J.; Stacey, G. *Microbiol. Rev.* **1986,** *56*, 321-326.
(12) Bjorndal, H.; Erbing, C.; Lindberg, B.; Fahraeus, G.; Ljunggren, H. *Acta Chem. Scand.* **1971,** *25*, 1281-1286.
(13) Jansson, P. E.; Kenne, L.; Lindberg, B.; Ljunggren, H.; Lonngren, J.; Ruden, U.; Svensson, S. *J. Am. Chem. Soc.* **1977,** *99*, 3812-3815.
(14) Aman, P.; Franzen, L. E.; Darvill, J. E.; Albersheim, P. *Carbohydr. Res.* **1982,** *95*, 263-281.
(15) Hisamatsu, M.; Abe, J.; Amemura, A.; Harada, T. *Agric. Biol. Chem.* **1980,** *44*, 1049-1055.
(16) Amemura, A.; Moorl, K. *Biochim. Biophys. Acta* **1974,** *334*; 398-409.
(17) Reinhold, B. B.; Chan, S.-Y.; Reuber, L.; Walker, G. C.; Reinhold, V. N. *J. Bact.* **1994,** *176*, 1997-2002.
(18) Tarentino, A. L.; Plummer, T. H. In *Guide to Techniques in Glycobiology*; Lennarz, W. J., Hart, G. W., Eds.; Methods In Enzymology; Academic Press Inc: San Diego, CA 92101-4495, 1994; Vol. 230, pp 44-57.
(19) Tarentino, A. L.; Trimble, R. B.; Plummer, T. H. Jr. *Meth. Cell Biol.* **1989,** *32*, 111-139.
(20) *Guide to Techniques in Glycobiology*; Lennarz, W. J.; Hart, G. W., Eds.; Methods In Enzymology; Academic Press Inc: San Diego, CA 92101-4495, 1994; Vol. 230.

(21) Mescher, M. F.; Strominger, J. L. *J. Biol. Chem.* **1976,** *251*, 2005-2014.
(22) Erickson, P. R.; Herzberg, M. C. *J. Biol. Chem.* **1993,** *268*, 23780-23783.
(23) Karcher, U.; Schroder, H.; Haslinger, E.; Allmaier, G.; Schreiner, R.; Wieland, F.; Haselbeck, A.; Konig, H. *J. Biol. Chem.* **1993,** *268*, 26821-26826.
(24) Gerwig, G. J.; Kamerling, J. P.; Vliegenthart, J. F. G.; Morag, E.; Lamed, R.; Bayer, E. A. *J. Biol. Chem.* **1993,** *268*, 26956-26960.
(25) Plummer, T. H. Jr.; Tarentino, A. L. *Glycobiology* **1991,** *1*, 257-263.
(26) Plummer, T. H. J.; Tarentino, T. L.; Hauer, C. R. *J. Biol. Chem.* **1995,** *270*, 13192-13196.
(27) Herscovics, A.; Orlean, P. *FASEB J.* **1993,** *7*, 540-550.
(28) Krusius, T.; Reinhold, V. N.; Margolis, R. K.; Margolis, R. U. *Biochem. J.* **1987,** *245*, 229-234.
(29) Spiro, R. G.; Bhoyroo, V. D. *J. Biol. Chem.* **1980,** *255*, 5347-5354.
(30) Reinhold, B. B.; Hauer, C. R.; Plummer, T. H.; Reinhold, V. N. *J. Biol. Chem.* **1995,** *270*, 13197-13203.
(31) Harrata, A. K.; Domelsmith, L. N.; Cole, R. B. *J. Biol. Chem.* **1993,** *22*, 59-67.
(32) Chan, S.; Reinhold, V. N. *Anal. Biochem.* **1994,** *219*, 63-73.
(33) Johnson, R. S.; Her, G. R.; Grabarek, J.; Hawiger, J.; Reinhold, V. N. *J. Biol. Chem.* **1990,** *265*, 8108-81016.

RECEIVED October 11, 1995

Chapter 7

Structural Characterization of Lipopolysaccharides from *Pseudomonas aeruginosa* Using Capillary Electrophoresis Electrospray Ionization Mass Spectrometry and Tandem Mass Spectrometry

S. Auriola[1,3], P. Thibault[1,4], I. Sadovskaya[2], E. Altman[2], H. Masoud[2], and J. C. Richards[2]

[1]Institute for Marine Biosciences, National Research Council, 1411 Oxford Street, Halifax, Nova Scotia B3H 3Z1, Canada
[2]Institute for Biological Sciences, National Research Council, 100 Sussex Drive, Ottawa, Ontario K1A 0R6, Canada

The analysis of underivatized lipopolysaccharides (LPS) arising from mild hydrazinolysis and acid hydrolysis of endotoxins from *Pseudomonas aeruginosa* serotype O6 was achieved using on-line capillary electrophoresis-electrospray mass spectrometry (CE-ESMS). This technique also provided unparalleled resolution of the different core oligosaccharides obtained from the complete N,O-deacylation and dephosphorylation of the native endotoxins. Electrophoretic conditions, enabling the separation of anionic and cationic analytes, were developed to determine possible sites of heterogeneity on either the core or the O-chain structures. Structural characterization of underivatized LPS and glycans identified in the ion electropherograms was achieved using tandem mass spectrometry under low collision energy conditions.

The gram-negative bacterium *Pseudomonas aeruginosa* is an opportunistic pathogen among debilitated burn-wounded, immunocompromised individuals, and patients suffering from cystic fibrosis and cancer (*1-3*). As in most pathogenic gram-negative bacteria, the major virulence factors of *P. aeruginosa* are associated with lipopolysaccharides (LPS) or endotoxins, which are major components of the outer membrane of the bacterial cell walls (*4,5*). The structure of LPS found in *P. aeruginosa* is characterized by a common molecular architecture similar to that of *Enterobacteriaceae* (*6,7*), and consists of three distinct regions: the O-specific chain, the core, and the lipid A. The first two regions comprise the hydrophilic polysaccharide chain bonded to the lipid A, a glucosamine disaccharide containing both N and O-linked fatty acids. In some bacteria, the O-specific chain is either absent or truncated as a result of genetic mutation or as a given characteristic of the bacterial strains. These modified O-specific chains are known as rough-type LPS. The term lipooligosaccharides (LOS) has also been to refer to those LPS which lack the repeating O-antigen, but contain instead a more variable and branched core oligosaccharide.

[3]Current address: Faculty of Pharmacy, University of Kuopio, Finland
[4]Corresponding author

0097–6156/95/0619–0149$12.00/0

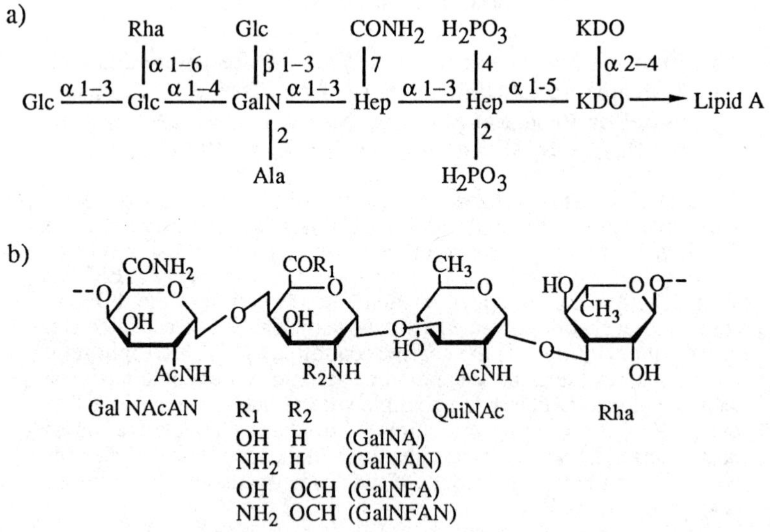

Figure 1: Structures of core oligosaccharide (a) and O-chain (b) from *P. aeruginosa* serotype O6, adapted from references (*18*) and (*7*), respectively. The lipid A structure is presented in Figure 5.

The differences among the specific compositions and structures of the O-chain polysaccharides form the basis of differentiation amongst the 20 serologically distinguishable strains of *P. aeruginosa* (*8,9*). Among these strains, the serotype O6 is the most prevalently encountered in isolates from clinical sources (*8*). The presence and the length of the O-chain influences various cell surface processes, including antibiotic susceptibility (*10,11*), bacteriophage recognition (*12,13*), and virulence and sensitivity to bactericidal action of normal human serum (*4,14*). In a large number of strains of *P. aeruginosa*, only partial or tentative structures of the core regions are known (*15-17*), and detailed structural analysis encompassing LPS heterogeneity is still fragmentary. Characterization of O-deacylated lipooligosaccharides (LOS) from serotype O6 core-deficient mutant strains R5 and A28 from *P. aeruginosa* was achieved recently using ^{31}P, ^{1}H and ^{13}C nuclear magnetic resonance (NMR) spectroscopy (*18,19*). Based on these investigations, a structural model for the complete core oligosaccharide from *P. aeruginosa* is presented in Figure 1 (See Table I for identifier). For wild-type strains of *P. aeruginosa* the core region is extended by further additions of outer O-chains. In all strains belonging to serotype O6, the O-chain (Figure 1) is comprised of complex tetrasaccharide repeating units composed of L-rhamnose, N-acetyl-D-quinovosamine (QuiNAc), N-formyl-D-galactosaminuronic acid (GalNFA) or its corresponding amide form (GalNFAN), and N-acetyl-D-galactosaminuronamide (GalNAcAN)(*7*).

Mass spectrometry has also played a pivotal role in the structural characterization of LPS from different pathogenic microorganisms. For example, liquid secondary ion mass spectrometry (LSIMS) facilitated the characterization of LOS from *Neisseria gonorrheae* and *Haemophilus spp.* (*20*), and fast atom bombardment mass spectrometry (FABMS) was successfully applied to the analysis of lipid A from *P. aeruginosa* (*21*) . More recently, electrospray mass spectrometry (ESMS) has been used to elucidate the structures of O-deacylated LOS from *Haemophilus*, *Neisseria*, and *Salmonella* strains (22), and for the analyses of lipid A from *Salmonella minnesota* Re595, *Shigella flexneri* (*23*), and *Enterobacter agglomerans* (*24*).

To date most mass spectrometric investigations have been performed directly on samples purified by either gel or ion exchange chromatography. However, the lack of on-line separation techniques often precludes the determination of sample heterogeneity associated with substituents attached to the oligosaccharide backbone (i.e. fatty acids, phosphates, acetyl groups, substituted carbohydrate chains). To this end, significant efforts have been devoted by our group to develop electrophoretic separation techniques compatible with electrospray mass spectrometry. This report presents the application of capillary electrophoresis-electrospray mass spectrometry (CE-ESMS) to the analysis of LPS and glycans arising from acid hydrolysis, mild hydrazinolysis, and complete deacylation and dephosphorylation of intact LPS from wild strains of *P. aeruginosa* serotype O6. Structural characterization of the different components identified in CE-ESMS separations was achieved using tandem mass spectrometry (MS-MS) analyses of selected precursor ions. The potential of the combined approach is evaluated for the characterization of O-deacylated, acid hydrolysed LPS and for lipid A samples derived from the native LPS.

Methods

Bacterial growth and isolation of LPS. *P. aeruginosa* serotype O6 wild-type (ATCC 33354), was obtained from the American Type Culture Collection (Rockville, MD, USA), and was grown on tryptic soy or agar plates (Difco Laboratories, Detroit, MI). The LPS fraction was isolated by 95% hot phenol-water extraction and purified by ultracentrifugation (105,000 x g, 4 h, 4°C) as described previously(*25, 26*).

Preparation and purification of O-deacylated LPS. LPS were O-deacylated using anhydrous hydrazine according to a procedure described previously (*18*). Briefly, 15 mg of LPS was incubated with 1 mL of anhydrous hydrazine (Sigma Chemicals, St-Louis, MO., USA) for 30 min. at 37°C. The samples were cooled to 0°C, and 1 mL of chilled acetone was added dropwise to destroy excess hydrazine and to precipitate the O-deacylated LPS. The O-deacylated LPS were then isolated by low-speed centrifugation (5000 rpm, 10 min) and purified by gel filtration using a Sephadex G-50 column (90 x 1 cm I.D.). Elution of the final product was achieved using 20 mM pyridinium acetate (pH 5.6) at a flow rate of 0.4 mL/min with refractive index detection. Low molecular weight O-deacylated LPS fractions were collected. After lyophilization the O-deacylated LPS were converted to the sodium salt using Rexyn 101 (H+ form) ion-exchange resin followed by neutralization with dilute sodium hydroxide. The purified material was obtained following lyophilization.

Preparation and purification of N,O-deacylated and dephosphorylated LPS. Complete deacylation and dephosphorylation were achieved according to procedures described previously (*26*). For small-scale preparations, approximately 5 mg of O-deacylated LPS was incubated in 100µL of 48% hydrofluoric acid at 0°C for 48 hours. The dephosphorylated LPS was then reduced by sodium borohydride (Aldrich Chemicals, Milwaukee, WI, USA) at 30°C for 3 hours. Complete deacylation was achieved by treating the previously reduced LPS in 200 µL of anhydrous hydrazine for 7 days at 85°C. After removal of hydrazine in vacuo over sulfuric acid, the core oligosaccharides were separated on a polyethylenimine cellulose anion exchange resin (medium mesh, 1.06 mequiv/g, Sigma) and eluted with water (20 mL). The backbone oligosaccharide was further purified on a Bio-Gel P-2 column (140 cm x 2.6 cm I.D.) with 50 mM pyridinium acetate (pH 4.5).

Preparation of lipid A and backbone oligosaccharide. Purified LPS (10 mg) was heated in 1 mL of 1 % aqueous acetic acid for 1 h at 100°C, the solution was cooled (4°C), and the precipitated lipid A was removed by centrifugation (500 rpm, 30 min), resuspended in water and lyophilized. The supernatant solution was lyophilized, and the oligosaccharide was purified on a Sephadex G-50 column as described above.

CE-ESMS analysis. A Crystal model 310 CE instrument (ATI Unicam, Boston, MA, USA) was coupled to an API/III+ mass spectrometer (Perkin Elmer/SCIEX, Concord, ON, Canada) via a coaxial interface configuration (*27*). The sheath solution was delivered at 7 µL/min to the back tee of the CE-ESMS interface and was comprised of 0.2 % formic acid (v:v) in 25 % aqueous methanol in positive ion mode, and of 10 mM ammonium formate pH 8.5 in 50 % aqueous methanol for negative ion detection. Fused silica columns of 1.0 m x 50 µm I.D. were obtained from Polymicro Technologies (Tucson, AZ, USA). Anionic separations were typically achieved at 30 kV, using 50 mM ammonium formate, pH 8.5. Cationic separations were obtained by polarity reversal on Polybrene coated capillaries (*26*) using a 1M formic acid buffer and an effective voltage of 20 kV across the capillary. Mass spectral acquisition was performed using dwell times of 3.5 msec per 1 Da in the full mass scan mode. A Macintosh Quadra 950 computer was used for instrument control, data acquisition and data processing.

Tandem mass spectrometry. All MS-MS analyses were conducted using the API/III+ triple quadrupole mass spectrometer described above. Flow injection (2 µL injection) was typically used to introduce the sample to the mass spectrometer, except for on-line CE-MS-MS experiments. LPS and glycans were dissolved in deionized water with either 1% acetic acid (v:v) for positive ion analysis or 1mM ammonium hydroxide for negative ion mode, whereas lipid A samples were solubilized in chloroform: methanol (2:1). Collisional activation of selected precursor ions in the rf-only quadrupole collision cell was achieved using argon at a collision thickness of $3.5 x 10^{15}$ atoms·cm^{-2}. Collision energies of typically 50eV were used for both negative and positive ion MS-MS experiments . Tandem mass spectra were acquired using a dwell time of 1 msec per step of 0.1 Da. The resolution of the third quadrupole was set to unit mass (25% valley definition). Combined CE-MS-MS analyses were performed in multiple reaction monitoring (MRM) acquisition mode by selecting the desired precursor ions using the first quadrupole while stepping the third quadrupole to transmit specific fragment ions. A total of up to 11 pairs of precursor/fragment ions were recorded simultaneously using a dwell time of 50 msec per channel monitored.

Results

General strategy for the structural analysis of underivatized LPS and oligosaccharides. Mild hydrazinolysis of the intact LPS is commonly used to obtain water-soluble O-deacylated LPS amenable to negative ion ESMS analysis. Alternatively, the relatively labile nature of the glycosidic bond between the 3-deoxy-D-manno-2-octulosonic acid (KDO) and the glucosamine of the lipid A can be exploited to produce a hydrophilic oligosaccharide and an insoluble lipid A using mild acid hydrolysis. These relatively mild reactions lead to LPS or glycans that usually comprise both O-chain and core oligosaccharide. These procedures also yield hydrophilic products, often containing ionic functionalities such as phosphate groups and/or acidic KDO residues imparting a net negative charge to the molecule, thus favoring their analyses by combined CE-ESMS. However, structural characterization of these underivatized compounds using tandem mass spectrometry is relatively tedious due to the overall complexity of the sample and relatively small numbers of fragment ions obtained in the low energy collision regime. A more successful approach to the elucidation of the core structure is to fully deacylate and dephosphorylate the lipopolysaccharides to produce oligosaccharides which can then be analyzed using both positive and negative ion ESMS. The heterogeneity in the carbohydrate moiety associated with either the native LPS or from the chemical treatment yield a distribution of molecules differing by the addition or subtraction of saccharide residues. The difference in molecular weights of the LPS-derived molecules thus provide characteristic masses which can be related to the known carbohydrate residues (Table I). The molecular weight of the oligosaccharide may be calculated by summing the masses of the different residues (Table I) and terminal group (typically H, OH). The characterization of products arising from each of the different preparation steps will be described separately in the following sections.

Characterization of core oligosaccharides. Fractions arising from the complete deacylation and dephosphorylation were subjected to CE-ESMS analysis using cationic separation on a Polybrene coated column. A narrow mass range (m/z 600-1300) was scanned rapidly to provide adequate temporal and mass spectral resolution of the different oligosaccharides. The total ion electropherogram is shown in Figure 2a, along with the two dimensional depiction of ion intensity as a function of both m/z and time for the same analysis (Figure 2b). The contour profile shown in Figure 2b provides an overall view of the separation and illustrates best the complexity of this fraction. This representation clearly shows a series of at least five diagonal lines between 16 and 20 min, corresponding to doubly- and triply-protonated oligosaccharide molecules. The series labelled 4 consists of truncated core glycans, the highest member of which is observed at m/z 1017 (molecular weight of 2032 Da). This glycan, labeled oligosaccharide A, corresponds to the fully deacylated and

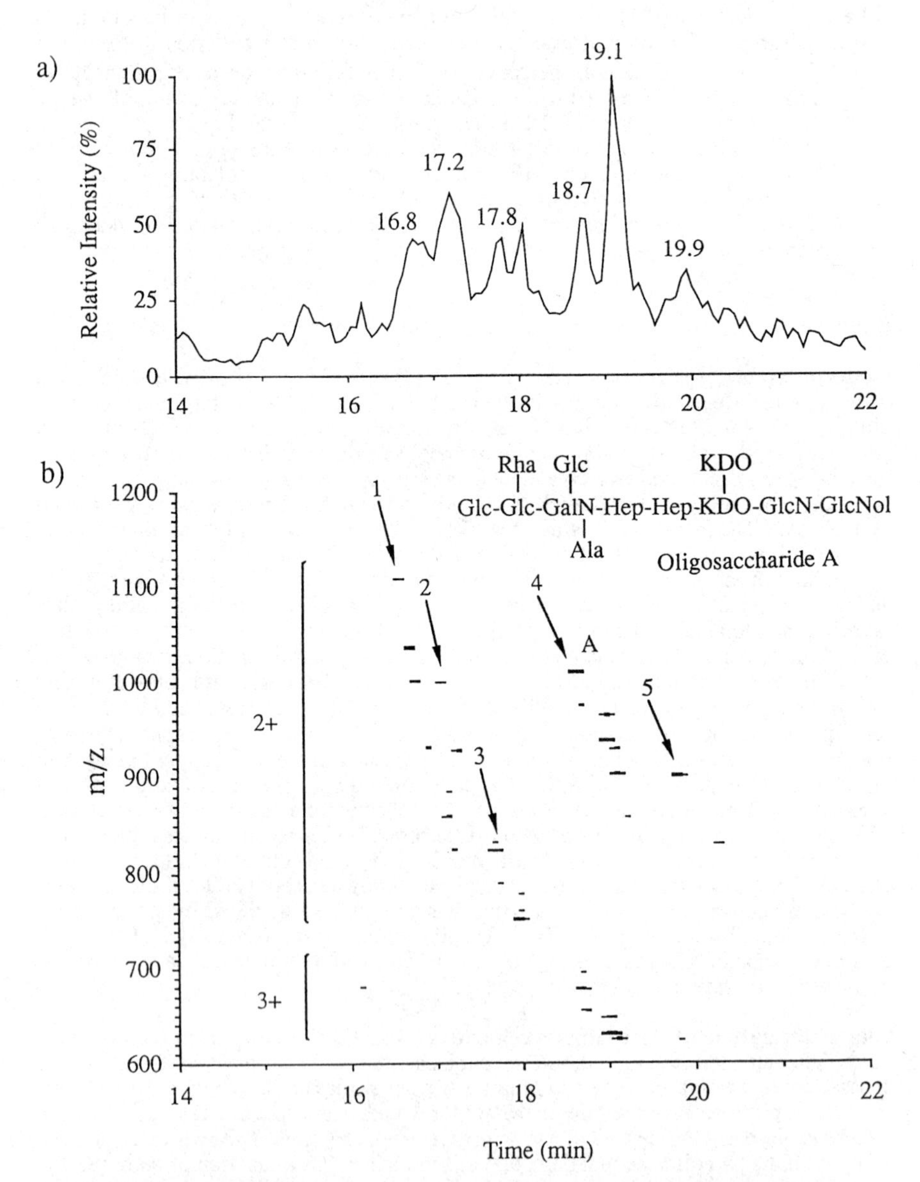

Figure 2: CE-ESMS analysis (positive ion) of oligosaccharides arising from complete N,O deacylation, and dephosphorylation of LPS from *P. aeruginosa*. a) Total ion electropherogram for m/z 600-1300. b) Contour profile of m/z *vs*. time showing characteristic heterogeneity patterns as indicated by the parallel diagonal lines. CE conditions: Polybrene coated fused silica capillary (100 cm x 50 µm I.D.), electrolyte: 1.0 M formic acid, -20 kV, 20 ng of the oligosaccharide mixture injected.

dephosphorylated core oligosaccharide having a glucosaminitol (GlcNol) at the reducing end but lacking the carbamoyl substituent on the Hep residue. In each series, the diagonal line reflects the concurrent changes of both electrophoretic mobilities and molecular weights of truncated core oligosaccharides lacking alanine (Ala), rhamnose (Rha), and/or glucose (Glc) residues. For example, series 1 is characterized by doubly-protonated ions at m/z 982 (18.8 min), 944 (19.1 min), and 908 (19.2 min), and consist of a truncated oligosaccharide A devoid of Ala, Rha, and Ala plus Rha, respectively. Series 1 differs from series 4 by the addition of a 3-hydroxy lauric acid (HL) on oligosaccharide A, presumably on the glucosamine residue (GlcN). Both series 1 and 4 are accompanied by parallel diagonal lines corresponding to oligosaccharides of higher mobilities (series 2 and 5), arising from the elimination of a KDO residue. Finally the removal of a KDO residue from the core, and of a GlcN residue from the lipid A disaccharide, affords oligosaccharides of intermediate mobilities indicated in series 3. The proposed compositions of the most abundant glycans found in this sample are conveniently summarized in Table II.

The structures of components identified in the CE-ESMS analysis were confirmed using direct MS-MS analyses. Selected examples of MS-MS spectra of core oligosaccharides from N,O deacylated and dephosphorylated lipopolysaccharides are shown in Figures 3 a and b for precursor ions at m/z 1017 and 907.5, respectively. For sensitivity reasons, the MS-MS spectra were obtained by flow injection analyses rather than by CE-MS-MS. This latter type of analysis was impractical in view of the small sample loading (ng injection sizes) and the requirement for fast scanning acquisition required to obtain a sufficient number of data points across the CE peaks. Furthermore, the resolution of the first quadrupole was reduced to 2-3 Da at 50% peak width to provide better sensitivity, and to facilitate the identification of multiply-charged fragment ions appearing as isotopic multiplets.

As a result of the experimental conditions selected, the MS-MS spectrum of the ion at m/z 907.5 is actually a composite of the doubly-protonated precursor ions identified at m/z 908 (19.2 min) and m/z 907 (19.9 min) in Figure 2 b. The first component (19.2 min) corresponds to structure A (Figure 2) lacking both an Ala and a Rha residue, while the latter oligosaccharide (19.9 min) is devoid of a KDO residue. The product ion spectra of these $[M+2H]^{2+}$ oligosaccharide ions were mostly dominated by fragment ions corresponding to cleavage of the glycosidic bond with the formation of complementary singly-charged fragment pairs, B and Y ions (*28*). The most prominent series was observed for the dissociation of the KDO-GlcN bond, giving rise to m/z 343 and its complementary B fragment ion (m/z 1689 and 1470 in Figures 3 a and b, respectively). It is noteworthy that the characteristic Y ion at m/z 343 was shifted to m/z 541 for precursor ions of glycans from series 1 and 2 (data not shown), consistent with the addition of an N-linked HL fatty acid. Major fragmentations are identified in Figures 3 a and b.

Interestingly, the glycosidic bond between the two KDO residues is more susceptible to cleavage, and the abundance of the corresponding fragment ion can be used as a diagnostic tool to identify oligosaccharides having KDO side chains. This typical feature is illustrated in Figure 4 for the CE-MS-MS analysis of precursor ions at m/z 907.5 where 10 multiple reaction monitoring (MRM) transitions were acquired simultaneously. The coincidences of given sets of MRM transitions were used to establish connectivities between fragment ions in the MS-MS spectrum of Figure 3b. For example, cleavage of the KDO-GlcN bond, affording a B fragment ion (m/z 1470 in Figure 3b), was observed only for oligosaccharides lacking a side chain KDO residue. In contrast, oligosaccharides having a side chain KDO gave rise to concomitant losses of KDO and of the lipid A disaccharide (B fragment ion at m/z 1252 in Figure 3b). This observation attests to the enhanced stabilization effect of the side chain KDO on the adjacent GlcN residue. Rationalization of the fragmentations observed in the different MS-MS spectra were consistent with the structure of oligosaccharide A proposed previously (*19*)

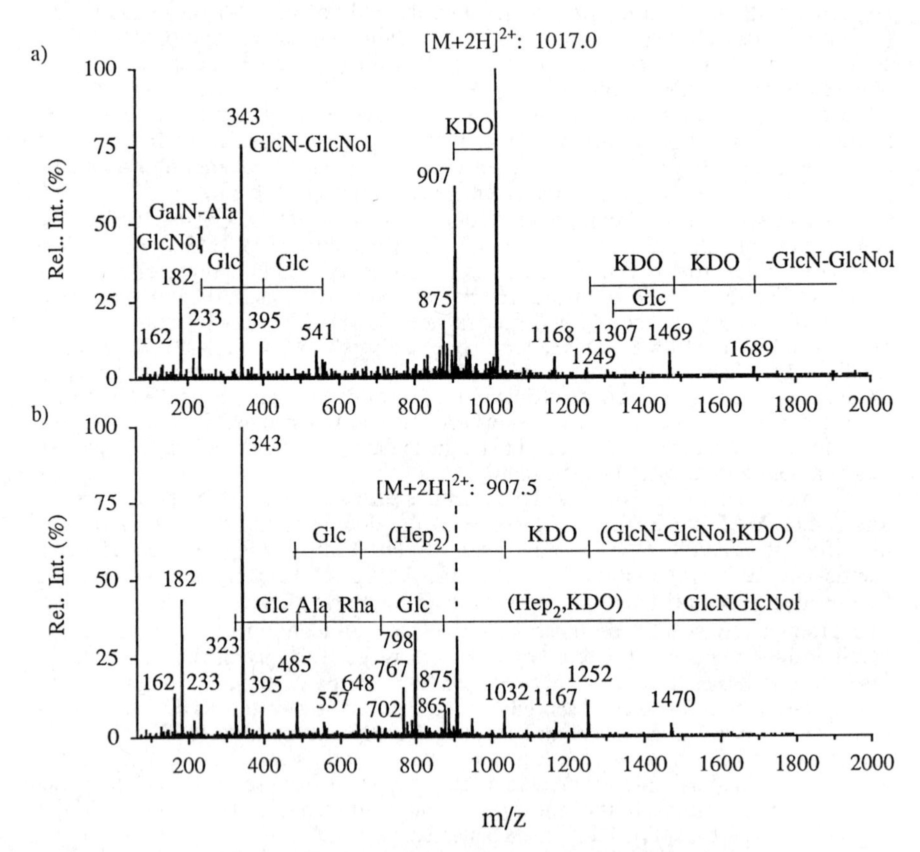

Figure 3: Tandem mass spectra of selected precursor ions of N,O deacylated and dephosphorylated oligosaccharides from *P. aeruginosa*. Product ions for $[M+2H]^{2+}$ precursors at m/z 1017 (a) and 907.5 (b). Conditions: Flow injection of 2 µg of the unseparated oligosaccharide mixture, collision energy 50 eV (laboratory frame of reference), argon target gas, 3.5 x10^{15} atoms·cm^2. Mass assignment is given to the nearest integer.

Table I: Masses of residues found in *P. aeruginosa* serotype O6

Residue	*Mass*	
	Monoisotopic	*Chemical average*[a]
Ala	71.0371	71.0788
Rha	146.0579	146.1430
GlcN, GalN	161.0688	161.1577
Glc	162.0528	162.1424
GlcNol	163.0845	163.1735
GalNAN	174.0640	174.1564
GalNA	175.0481	175.1411
QuiNAc	187.0845	187.1955
Hep	192.0634	192.1687
GalNFAN	202.0590	202.1668
GalNFA	203.0430	203.1516
GalNAcAN	216.0746	216.1936
KDO	220.0583	220.1791

[a] Calculated using H = 1.0079, C = 12.0110, N = 14.0067, O = 15.9994

Table II: Principal components found in sample of N, O-deacylated and dephosphorylated LPS analyzed by CE-ESMS (Figure 2)

Time (min)	*Mr (obs.)*[a]	*Mr (calc.)*	*Assignment*[b]
16.6	2230	2231	A + HL
16.7	2084	2085	A + HL - Rha
16.8	2014	2014	A + HL - (Rha, Ala)
17.1	2010	2011	A + HL - KDO
17.3	1864	1864	A + HL - KDO -Rha
17.7	1650	1651	A - KDO - GlcN
18.0	1504	1505	A - KDO - GlcN- Rha
18.7	2032	2032	A
18.8	1961	1962	A - Ala
19.1	1886	1887	A- Rha
19.2	1812	1813	A - (Rha, Ala)
19.9	1815	1816	A - KDO
20.4	1670	1670	A- KDO - Rha

[a] Mass assignment within ± 1Da.

[b] Oligosaccharide A is composed of: Rha_1, Glc_3, $GalN_1$, $GlcN_1$, GlcNol, Ala, Hep_2, KDO_2. HL denotes hydroxy lauric acid.

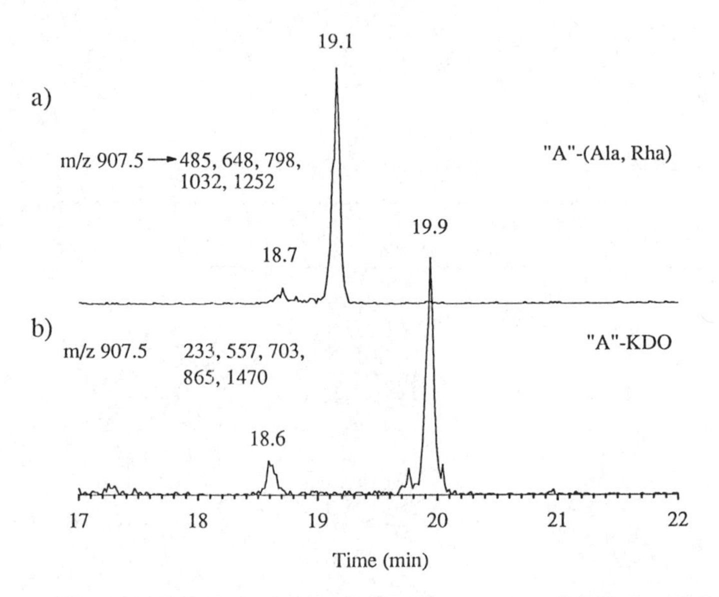

Figure 4: CE-MS-MS analysis of $[M+2H]^{2+}$ precursors at m/z 907.5 using MRM acquisition mode (10 MRM transitions monitored). The sums of the ion currents corresponding to selected transitions are shown in a) and b). Conditions as for Figure 2 except argon introduced in rf-only quadrupole at gas thickness of 3.5 $x10^{15}$ atoms·cm^2, collision energy 50 eV (laboratory frame of reference).

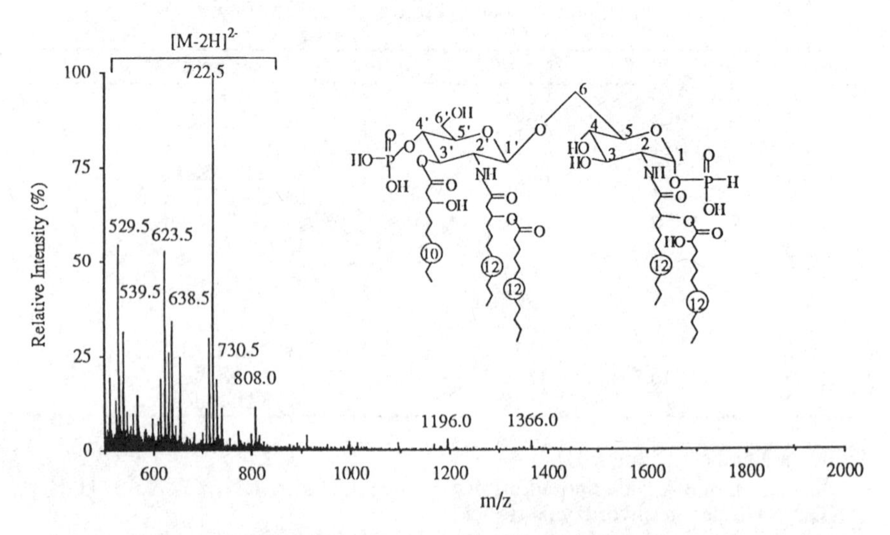

Figure 5: Negative-ion ESMS mass spectrum of lipid A from *P. aeruginosa.* Inset shows the lipid A disaccharide structure with pendant acyl substituents and phosphate groups based on previous investigations (*21*). Spectrum obtained by continuous infusion of a 0.1 mg/mL solution of lipid A fraction dissolved in chloroform: methanol (2:1).

Analysis of the lipid A disaccharide. The negative ion ESMS mass spectrum of lipid A from *P. aeruginosa* serotype O6 (Figure 5) shows predominantly doubly deprotonated molecules $[M\text{-}2H]^{2-}$. The most abundant ion was observed at m/z 722.5 whose product ion spectrum (not shown) was characterized by consecutive losses of the pendant O-linked fatty acids and phosphate groups from the precursor ion. The presence of different acyl substituents was also confirmed by the observation of low mass fragment ions (not shown) such as those at m/z 187, 199, and 215 for 3-hydroxy capric (HC), lauric (L), and hydroxy lauric (HL) acids, respectively. Fragment ions observed in the product ion spectrum of the ion at m/z 722.5 (not shown) were consistent with a pentaacyl lipid A (inset Figure 5) containing two phosphomonoester groups, two N-linked HL acyl moieties, and three O-linked fatty acids (HC, L, and HL), as observed previously for the lipid A of *P. aeruginosa* serotype O5 (*21*). The mass spectrum of the lipid A sample also shows other ions of lower intensities, such as those at m/z 529.5, 623.5, and 638.5 (Figure 5). These ions correspond to truncated lipid A molecules which have lost different acyl chains (HC, HL or both) during the mild acid hydrolysis reaction (*23,24*). Interestingly, the lipid A from *P. aeruginosa* appears to show very little heterogeneity in the substitution of phosphate groups (pyrophosphate or phosphoethanolamine), in contrast to lipid A isolated from *Moraxella catarrhalis* (*29*).

Heterogeneity of core and O-chain of *P. aeruginosa* serotype O6. At pH above 4, the O-deacylated LPS have an overall net negative charge and can be separated under anionic conditions using uncoated capillaries. Optimized separation conditions providing shorter analysis times, and favoring the formation of anionic species with minimal mass spectral interferences, were obtained using a 50 mM ammonium formate buffer adjusted to pH 8.5. The CE-ESMS analysis of O-deacylated LPS performed under full mass scan acquisition is presented in Figure 6. The mass spectra extracted for each peak were characterized by abundant $[M\text{-}3H]^{-3}$ and $[M\text{-}4H]^{-4}$ ions, from which individual molecular weight profiles could be obtained (shown as insets in Figure 6).

For each ionic species identified in the CE-ESMS analysis, tentative composition assignments could be made based on known structural features (*7, 18,19*) and on additional information obtained from MS-MS experiments (*vide infra*). The isotope-averaged molecular weights, migration times, and proposed assignments are conveniently summarized in Table III. It is interesting to note that the migration order is consistent with predicted changes in electrophoretic mobilities based on substitution of ionic functionalities and variation of molecular weight. The electrophoretic mobility is expected to be more significantly affected by the presence of highly ionizable functionalities (phosphate and carboxylate groups). The lighter O-deacylated LPS members devoid of an O-chain, which have the highest mobilities, are observed between 18.0-18.3 min in Figure 6. The heterogeneity of the O-chain arises mostly from substitution on the galactosamine uronic acid residue giving rise to either an acid (GalNA) or an amide (GalNAN) on the O-chain, whereas residues Rha, QuiNAc and GalNAcAN remain unchanged. Each of these two sets of peaks is accompanied by a series of closely related LPS spaced by 28 Da, corresponding to their N-formyl analogs (7). Additional heterogeneity is also observed in the numbers of phosphate groups appended to the core structure. Such substitution is thought to arise through the formation of pyrophosphate groups, possibly at the Hep residue, since the incremental number of phosphate groups does not yield a significant increase in mobility.

Hydrophilic oligosaccharides obtained following mild acid hydrolysis were also subjected to CE-ESMS to confirm previous structural assignments. Figure 7 shows the total ion electropherogram (m/z 700-1400) for this analysis together with the reconstructed molecular weight profiles calculated from the $[M\text{-}2H]^{2-}$ and $[M\text{-}3H]^{3-}$

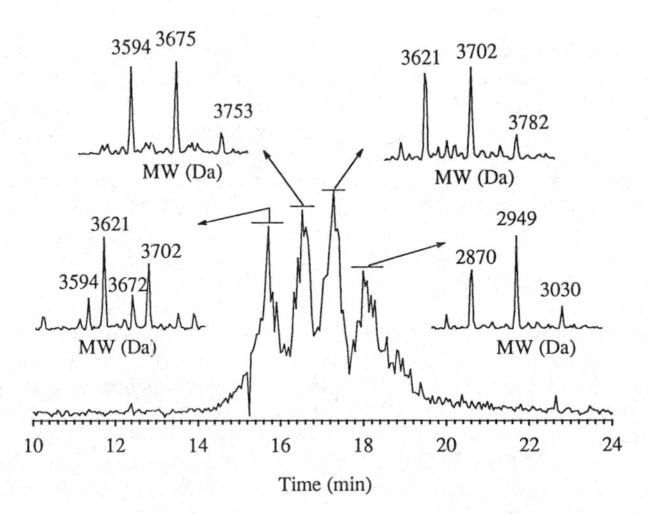

Figure 6: Total ion electropherogram obtained for the CE-ESMS analysis (m/z 700-1400) of anionic O-deacylated LPS from *P. aeruginosa*. Reconstructed molecular weight profiles, calculated from the multiply charged ions observed for each CE peak are shown as insets. Conditions: Uncoated fused silica capillary (100 cm x 50 μm I.D.), electrolyte: 50 mM ammonium formate pH 8.5, 20 ng of O-deacylated LPS injected.

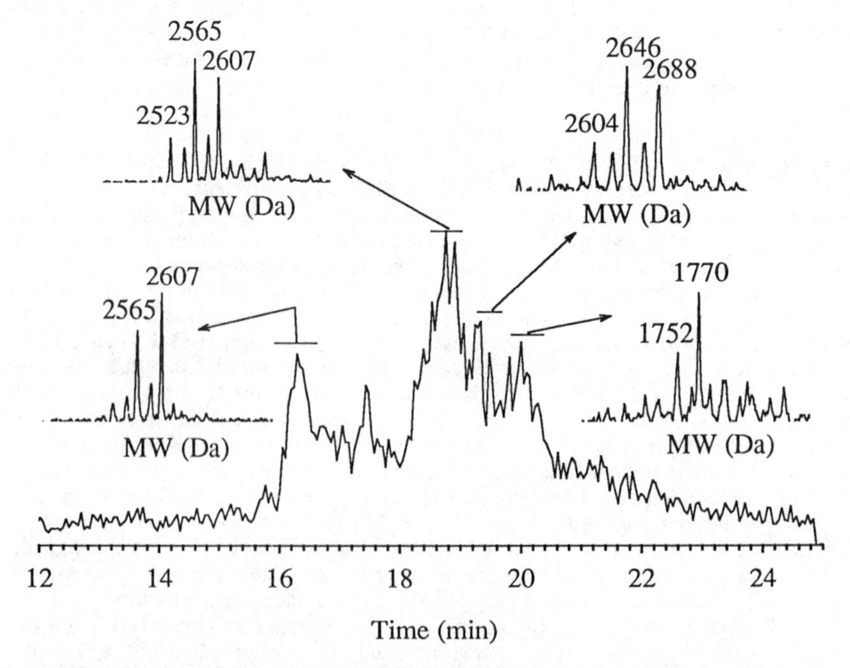

Figure 7: Total ion electropherogram obtained for the CE-ESMS analysis (m/z 700-1400) of anionic oligosaccharides arising from the mild acid hydrolysis of LPS from *P. aeruginosa*. Reconstructed molecular weight profiles calculated from the multiply charged ions observed for each CE peak, are shown as insets. Conditions as for Figure 6.

Table III: Assignment of O-deacylated LPS analyzed by CE-ESMS (Figure 6)

Time (min)	Mr (obs.)[b]	Mr (calc.)	Proposed composition[a] HPO_3	O-chain modification GalNFA	GalNFAN	GalNA	GalNAN
15.3	3594	3593	1	-	-	-	1
15.5	3675	3673	2	-	-	-	1
15.6	3753	3753	3	-	-	-	1
15.7	3621	3621	1	-	1	-	-
15.8	3702	3701	2	-	1	-	-
15.9	3783	3781	3	-	1	-	-
16.5	3594	3594	1	-	-	1	-
16.7	3675	3674	2	-	-	1	-
16.8	3753	3754	3	-	-	1	-
17.2	3621	3622	1	1	-	-	-
17.4	3702	3702	2	1	-	-	-
17.5	3783	3782	3	1	-	-	-
18.0	2868	2869[c]	1	-	-	-	-
18.2	2949	2949[c]	2	-	-	-	-
18.3	3030	3029[c]	3	-	-	-	-

[a] Appended to an invariant structure composed of: Rha, QuiNAc, GalNAcAN, Glc_3, $GalN_1$, Rha_1, Ala, Hep_2, $CONH_2$, $(H_2PO_3)_4$, KDO_2 $GlcN_2$, (hydroxy lauric acid)$_2$

[b] Mass assignment within ± 1Da.

[c] Appended to a core glycolipid containing Glc_3, $GalN_1$, Rha_1, Ala, Hep_2, $CONH_2$, $(H_2PO_3)_4$, KDO_2 $GlcN_2$, (hydroxy lauric acid)$_2$

ions observed for each peak. In contrast to the previous analysis (Figure 6), oligosaccharides obtained from the acetic acid digestion appear to exhibit a more heterogenous distribution of substituents (acetyl, formyl or phosphate groups) attached to the core or O-chain structure. Mass spectra extracted for each peak were characterized by oligosaccharides of similar electrophoretic mobilities but differing by 28 and 42 Da, consistent with the addition or removal of formyl and acetyl groups, respectively. All oligosaccharides formed during the acetic acid hydrolysis also lacked the side chain KDO, and had at most 3 phosphate groups bonded on the core structure. Glycans devoid of O-chain (Mr: 1752, 1770) were observed at approximately 20 min in Figure 7. Interestingly, two groups of peaks at 16.3 and 18.7 min (Figure 7) have the same molecular weights, a situation also encountered for O-deacylated LPS (Figure 6 and Table III).

Further characterization of these closely related compounds was achieved using tandem mass spectrometry. The MS-MS spectra of m/z 1281.5 (Mr: 2565) and 1303 (Mr: 2607) are presented in Figures 8 a and 8 b, respectively. Fragmentations observed in both spectra are relatively sparse and are dominated by small fragment ions formed by cleavage of glycosidic bonds near acidic sites such as KDO and phosphorylated Hep residues. An abundant fragment ion at m/z 299 suggests the presence of a phosphorylated KDO residue. Similar fragment ions have been observed previously for LPS derived oligosaccharides from *Haemophilus* species (*20*). However, recent detailed chemical and NMR structural studies provide no evidence for the presence of phosphorylated KDO residues in the inner core region of *P. aeruginosa* LPS (*18,19*). The occurence of the fragment ion at m/z 299 in the MS-MS spectrum shown in Figure 8 may be explained by either a side reaction product of the mild acid hydrolysis used in the preparation of the oligosaccharide or by a rearrangement reaction involving a transfer of a phosphate group to the KDO residue during the collisional activation. The most revealing feature concerning these closely related glycans was the

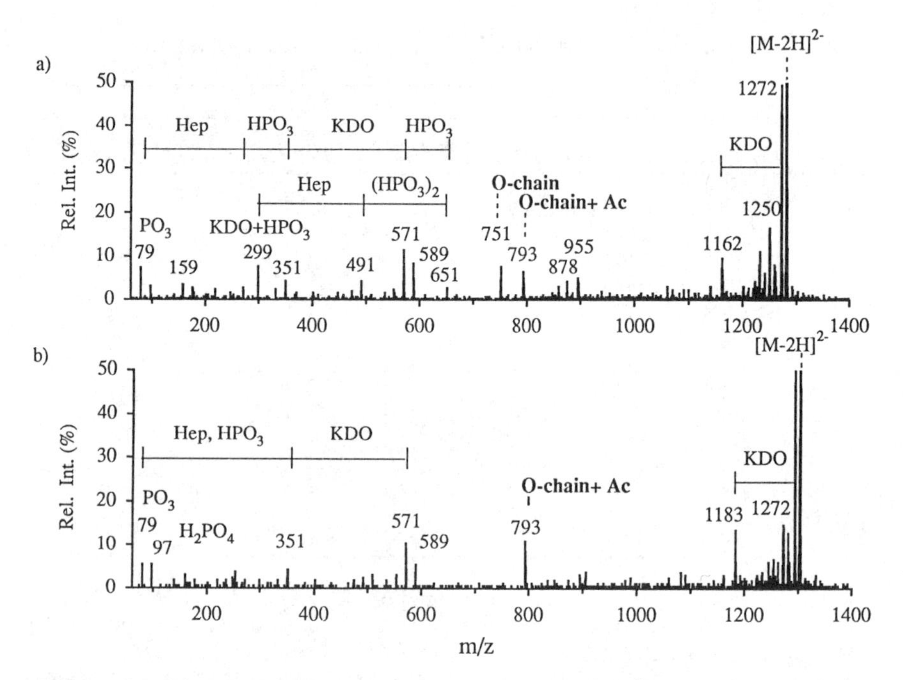

Figure 8: Tandem mass spectra of selected $[M-2H]^{2-}$ precursor ions of oligosaccharides arising from the mild acid hydrolysis of LPS from *P. aeruginosa*. Product ions for m/z 1281.5 (a) and 1303.0 (b). Conditions: Flow injection of 2 µg of the unseparated oligosaccharide mixture, collision energy 50 eV (laboratory frame of reference), argon target gas, 3.5 $\times 10^{15}$ atoms·cm^2.

observation of two fragment ions at m/z 751 and 793 corresponding to the tetrasaccharide O-chain (GalNAcAN-GalNFA-QuiNAc-Rha) and its acetylated analog. The observation of both fragments m/z 751 and 793, for the precursor ion at m/z 1281, could be explained by heterogeneity of O-acetylation in the O-chain. Acetylation of hydroxyl groups from the O-chain GalNAcAN and GalNFA residues has been described previously for different strains of *P. aeruginosa* (7) and could possibly explain the present observation.

In a separate series of experiments, a set of 11 fragment ions identified in the product ion spectrum of m/z 1281.5 (Figure 8 a) were further analyzed by combined CE-MS-MS. The corresponding electropherograms are presented in Figure 9 a and 9 b. Oligosaccharides observed at 16.5 and 19.4 min in Figure 9 have similar core structures with a triphosphorylated Hep residue. However, fragment ions at m/z 751 and 793 were observed only for the oligosaccharide at 19.4 min, suggesting that the earlier component has an amide GalNFAN rather than a GalNFA residue. It is important to note that such fragment ions arise only for O-chain glycans having acidic residues (GalNFA, GalNA), as the carboxylic acid functionality offers a localization site for the negative charge giving rise to fewer product ions compared to the amide. This is further supported by the fact that these two closely compounds have different electrophoretic mobilities, with the more acidic oligosaccharide migrating last. Although a molecular weight difference of 1 Da is expected between the amide and acid analogs, the sensitivity under full mass scan acquisition conditions prevented sufficiently precise mass measurements.

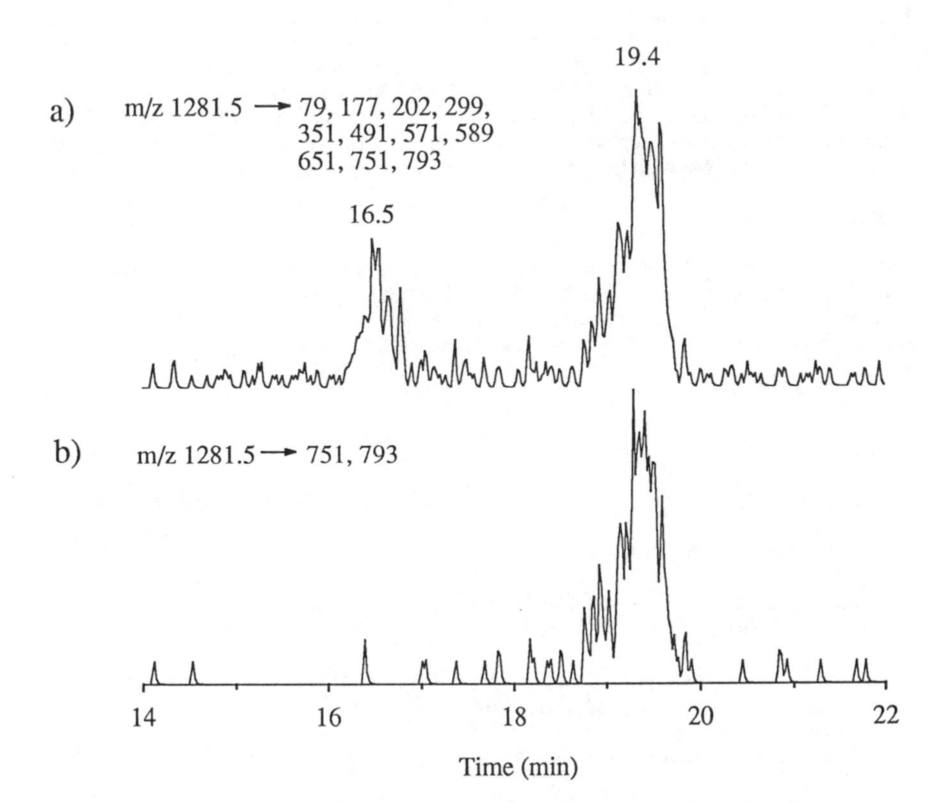

Figure 9: CE-MS-MS analysis of $[M\text{-}2H]^{2-}$ precursor ion at m/z 1281.5 using MRM acquisition mode. a) Total ion electropherogram for 11 reaction channels. The sum of the ion currents corresponding to transitions giving rise to fragment ions m/z 751 and 793 is shown in b). Conditions as for Figure 6 except argon introduced in rf-only quadrupole at gas thickness of 3.5×10^{15} atoms·cm^2, collision energy 50 eV (laboratory frame of reference).

Summary

Combined CE-ESMS analyses of oligosaccharides and glycolipids, arising from mild treatment of intact LPS, provides an efficient analytical tool for monitoring the heterogeneity of substituents appended to the O-chain (acetyl and formyl groups) or to the core structure (phosphate and pyrophosphate groups). Characterization of the core structure of *P. aeruginosa* serotype O6 was facilitated by tandem mass spectrometric experiments on fully deacylated and dephosphorylated oligosaccharides, affording fragment ions corresponding to sequential cleavages of glycosidic bonds. More importantly, the formation of characteristic O-chain fragment ions was used in CE-MS-MS experiments to differentiate between closely related glycans differing only by the incorporation of an amide (GalNFAN and GalNAN) or an acidic residue (GalNFAN and GalNAN) in the O-chain structure. Such an approach could also be used to distinguish different serotypes from similar pathogenic bacteria, thus providing a valuable correlation between the LPS structure and the immunochemical response.

Acknowledgments

The technical assistance of Doug Griffith in the preparation of the bacterial cultures used this study.is greatfully acknowledged. Financial support was provided by the Canadian Bacterial Disease Network of Centers of Excellence. S.A. acknowledges support from the Academy of Finland, and the Finnish Cultural Foundation.

NRCC#: 38096

Literature Cited

1. Bodey, G.P.; Bolivar, R.; Fainstein, V.; Jadaja, L. *Rev. Infect. Dis.* **1983**, *5*, 279-313.
2. Botzenhart, K.; Wolz, C.; Döring, G.*Antibiot. Chemother.* (Basel) **1991**, *44*, 8-12.
3. Pennington, J.E. *J. Infect. Dis.* **1974**, *130*, S159-S162.
4. Crys, S.J.,Jr.; Pitt, T.L.; Fürer, E.; Germanier, R. *Infect. Immun.* **1984,** *44*, 508-513.
5. Kimaru, A.; Hansen, E. *Infect. Immun.* **1986,** *51*, 69-79.
6. Wilkinson, S.G. *Rev. Infect. Dis.* **1983**, *5*, S941-S949.
7. Knirel, Y.A. *Crit. Rev. Microbiol.* **1990**, *17*, 273-304.
8. Liu, P.V.; Matsumoto, H.; Kusama, H.; Bergan, T., *Int. J. Syst. Bacteriol.* **1983**, *33*, 256-264.
9. Liu, P.V.; Wang, S. *J. Clin. Microbiol.* **1990**, *28*, 922-925.
10. Angus, B.L.; Fyfe, J.M.; Hancock, R.E.W. *J. Gen. Microbiol.* **1986**, *133*, 2905-2914.
11. Godfrey, A.J.; Shahrabadi, M.S.; Bryan, L.E. *Antimicrob. Agents Chemother.* **1986**, *30*, 802-805.
12. Bergan, T.; Midtvedt, T., *Acta Pathol. Microbiol. Scand. Sect. B*, **1975**, *83*, 1-9
13. Kuzio, J.; Kropinski, A.M. *J. Bacteriol.* **1983**, *155*, 203-213.
14. Engels, W.; Endert, J.; Kamps M.A.F., van Boven, C.P.A., *Infect. Immun.* **1985**, *49*, 182-189.
15. Drewry, D.T.; Symes, K.C.; Gray, G.W.; Wilkinson, S.G. *Biochem. J.*, **1975**, *149*, 93-106.
16. Kropinski, A.M.; Chan, L.C.; Milazzo, F.H. *Can. J. Microbiol.* **1979**, *25*, 390-398.

17. Rowe, S.N.; Meadow, P.M. *Eur. J. Biochem.* **1983**, *132*, 326-337.
18. Masoud, H.; Altman, E.; Richards, J.C.; Lam, J.S. *Biochemistry* **1994**, *33*, 10568-10578.
19. Masoud, H.; Sadovskaya, I.; de Kievit, T.; Altman, E.; Richards, J.C.; Lam, J.S., *J. Bacteriol.*, **1995**, in press.
20. Gibson, B.W.; Phillips, N.J.; John, C.M.; Melaugh, W. in *Mass Spectrometry for the Characterization of Microorganisms*; Fenseleau, C., Ed.; ACS symposium series: Washington, DC, 1994, Vol. 541; pp 185-202.
21. Karunaratne, D.N.; Richards, J.C.; Hancock, R.E.W. *Arch. Biochem. Biophys* **1992**, *299*, 368-376.
22. Gibson, B.W.; Melaugh, W.; Phillips, N.J.; Apicella, M.A.; Campagnari, A.A.; McLeod Griffis, J. *J. Bacteriol.* **1993**, *175*, 2702-2712.
23. Chan, S.; Reinhold, V.N. *Anal. Biochem.* **1994**, *218*, 63-73.
24. Harrata, A.K.; Domelsmith, L.N.; Cole, R.B., *Biol. Mass Spectrom.* **1993**, *22*, 59-67.
25. Westphal, O.; Jann, K. *Methods Carbohydr. Chem.* **1965**, *5*, 83-91.
26. Masoud, H.; Perry, M.B.; Brisson, J.R.; Uhrin, D.; Richards, J.C. *Can. J. Chem.* **1994**,*72*, 1466-1477
27. Kelly, J.F.; Locke, S.J.; Ramaley, L.; Thibault, P., *J. Chromatogr.* **1995**, in press.
28. Domon, B.; Costello, C.E. *Glycoconjugate J.* **1988**, *5*, 397-409.
29. Masoud, H.; Perry, M.B.; Richards, J.C. *Eur. J. Biochem.* **1994**, *220*, 209-216.

RECEIVED October 30, 1995

Chapter 8

Determining Structures and Functions of Surface Glycolipids in Pathogenic *Haemophilus* Bacteria by Electrospray Ionization Mass Spectrometry

Bradford W. Gibson, Nancy J. Phillips, William Melaugh, and Jeffrey J. Engstrom

Department of Pharmaceutical Chemistry, School of Pharmacy, University of California, San Francisco, CA 94143–0446

An abundance of immunochemical and biological data implicates bacterial glycolipids, or lipooligosaccharides (LOS), in the diseases caused by pathogenic strains of *Haemophilus* and *Neisseria* species. These LOS consist of a lipid A attached to a variable and sometimes highly branched oligosaccharide region. There are various reasons why electrospray ionization mass spectrometry (ESI-MS) is an ideal tool for investigating the structure/function of bacterial LOS. First, LOS are biosynthesized as heterogeneous populations that cannot be readily separated, and mass spectrometry is well-suited for mixture analysis. For example, ESI-MS and ESI-tandem mass spectrometry have been used to determine the molecular weights and partial structures of LOS containing as many as twenty distinct components isolated from a single bacterial strain. Second, in the determination of the structure/function relationship of biological molecules, one would like to focus on LOS most relevant to the actual disease states. To conduct structural characterization studies under biologically relevant conditions requires working with very small amounts of sample, consistent with the sensitivity of mass spectrometry. In this chapter, we will present examples which demonstrate the application of ESI-MS in characterizing LOS from pathogenic strains of *Haemophilus*.

Pathogenic bacteria from *Neisseria* and *Haemophilus* are responsible for a large number of human diseases, including meningitis, gonorrhea, pneumonia, otitis media and chancroid. Over the last ten years, an abundance of immunochemical, chemical and biological data has been reported that has begun to show a clear link between the outer-membrane glycolipids of these bacteria and specific pathogenic processes contributing to these human diseases. These glycolipids, or lipooligosaccharides (LOS), consist of a largely conserved lipid A region linked to a highly variable, branched and complex set of oligosaccharide moieties (see Figure 1). The lipid A is made up of a hexaacyl-substituted glucosamine disaccharide containing two phosphates that is attached to the oligosaccharide region through an acidic sugar, 2-keto-3-deoxy-mannooctulosonic acid (KDO). The LOS from *Haemophilus* and *Neisseria* differ from most enteric bacterial surface glycolipids, or lipopolysaccharides (LPS), by their much smaller size ($M_r \approx 3500\text{-}6000$), variable composition, antigenic diversity, high cytotoxicity, and lack of repeating terminal structures.

0097–6156/95/0619–0166$12.00/0

The lack of repeating terminal saccharides in LOS is the most dramatic difference between LOS and LPS, although the remaining oligosaccharide regions of LOS represent an even more important structural difference when compared to the analogous core oligosaccharides of LPS. The LPS structures of the enteric bacteria such as *Salmonella typhimurium* and *Escherichia coli* (*1,2*), for example, contain relatively conserved core oligosaccharide regions that are extended by the addition of an O-antigen, a polysaccharide made up of repeating oligosaccharide units, neither of which bear much resemblance to mammalian sugar sequences. In contrast, LOS structures from *H. influenzae*, *H. ducreyi* and *N. gonorrhoeae* have been shown to contain sugar sequences that resemble those present in human glycosphingolipids, such as lactose, N-acetyllactosamine, lacto-*N*-neotetraose, N-acetyl sialyllactosamine (NeuAcα2→3Galβ1→4GlcNAc) and the P^k antigen, Galα1→4Galβ1→4Glc (for review, see (*3*)). In *H. ducreyi,* LOS itself has been shown to be highly cytotoxic in animal models, forming necrotic skin lesions that resemble the chancroid ulcers found in the human disease (*4*). Moreover, in a process termed 'phase variation' (*5,6*), LOS populations can change in structures and/or in relative proportions, making these non-enteric bacterial pathogens highly adaptive to the changing and hostile environment typical of that encountered in the human host.

One of the first goals towards elucidating the roles that LOS play in host-pathogen interactions is to determine the precise LOS species expressed by these bacteria, preferably under the conditions of the disease process itself. Elucidating LOS structures expressed *in vivo* will require major advances in both the detection and sample preparation protocols over what is currently available. While there are a number of techniques used by chemists to characterize the structures of LOS and LPS, virtually all of these methods suffer serious shortcomings in regards to determining the precise heterogeneity of LOS glycoforms. Most of these techniques, for example, require chemical treatment of LOS, such as mild acid hydrolysis to generate a water soluble oligosaccharide fraction and a chloroform/methanol soluble lipid A fraction. Indeed, the vast majority of chemical and structural studies of LOS and LPS has been on these oligosaccharide and lipid A fractions. Although mild acid hydrolysis of LOS and LPS is designed specifically to cleave at the labile glycosidic bond between the core KDO and lipid A moiety (see Figure 1), this procedure can also remove other acid labile moieties such as sialic acid or result in the β-elimination of specific phosphate groups on KDO (*7*). Furthermore, heterogeneity can be artificially generated in the lipid A portion as well, giving rise to monophosphoryl lipid A species and/or lipid A species lacking one or more *O*-acyl substituted fatty acids moieties. Even treatment with aqueous HF under conditions designed to selectively remove phosphate groups can produce artifactual LOS species by cleaving glycosidic bonds. Given this situation, we have sought to develop mass spectrometric-based procedures that will provide an accurate as possible assessment of LOS heterogeneity without introducing artifacts or losing important structural information due to chemical degradation of specific functional groups.

While we have used several mass spectrometric techniques to analyze LOS and their oligosaccharide and lipid A moieties, there are a number of reasons why electrospray ionization mass spectrometry (ESI-MS) is well-suited to investigating the structure/function of bacterial LOS. First, ESI-MS is an excellent technique for complex mixture analysis under conditions where one sees a somewhat restricted number of charge states per molecular species. Second, given the presence of several phosphate groups on the LOS, the fixed negative charges ensure excellent ionization in the negative ion mode and within a mass range where they can be readily observed as their major doubly or triply charged species. Third, the low background and large dynamic range afforded by ESI-MS allow for relatively minor species to be detected even in the presence of LOS species of considerably greater abundance. Fourth, in the determination of the structure/function relationship of biological molecules, one would like to focus on LOS species most relevant to the actual disease states. ESI-MS

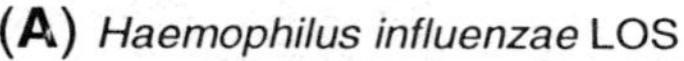

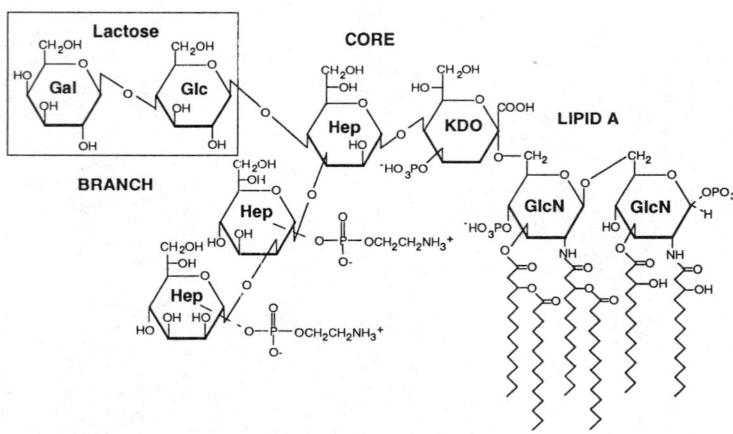

Figure 1. LOS structures found in (A) *Haemophilus influenzae* strain 2019 (*7*) and (B) *Haemophilus ducreyi* strain 35000 (*19,20*). Note the terminal sugars in these LOS constitute lactose or N-acetyl sialyllactosamine, structures common to many human glycoconjugates. The position of the phosphoethanolamine (PEA) group is not known but has been presumed to be linked to one of the core heptoses. However, more recent unpublished data suggests an additional PEA group is lost during mild acid treatment and may be linked to the phosphate of KDO. Abbreviations for sugars used in the figure and throughout the text are as follows: Gal is galactose, Glc is glucose; GlcN or GlcNAc is N-acetyl-glucosamine; Heptose is L-*glycero*-D-*manno* heptose, KDO is D-*manno*-2-keto-3-deoxy octulosonic acid; NeuAc is N-acetylneuraminic acid (or silaic acid); KDO(P) is 4-phosphoKDO. Hep* refers to an unusual heptose, D-*glycero*-D-*manno* heptose.

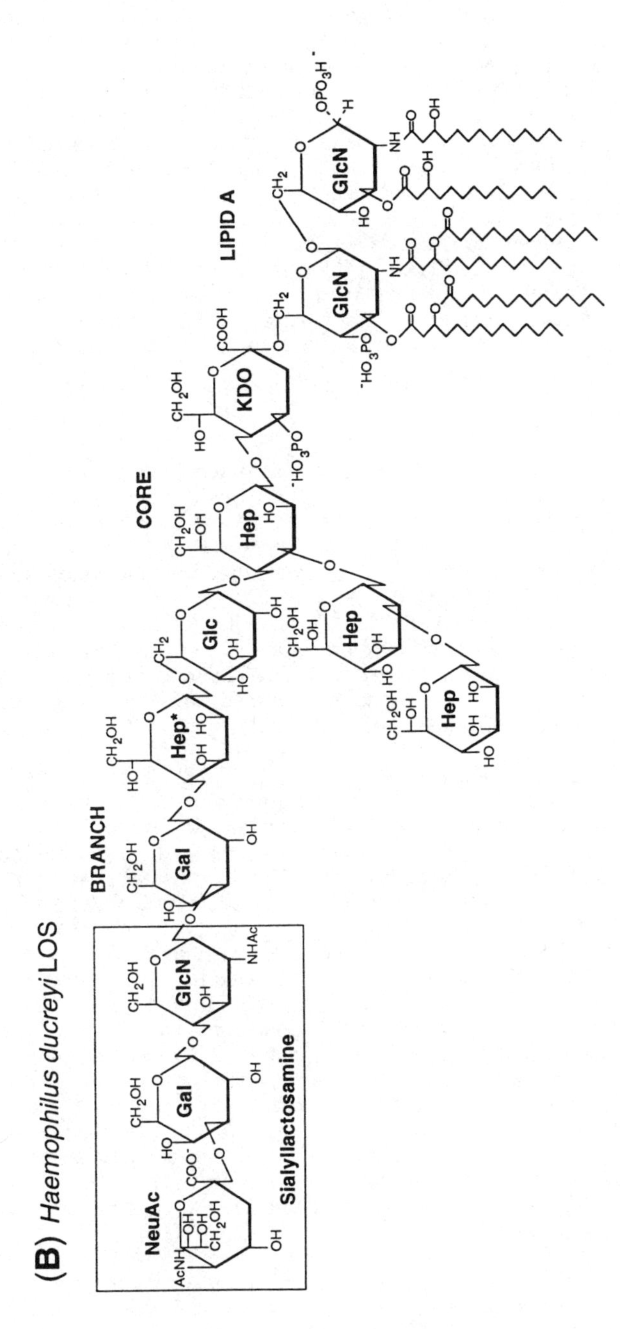

Figure 1. *Continued*

appears to have the necessary high degree of sensitivity to conduct structural characterization studies under biologically relevant conditions that may require working with very small amounts of sample, perhaps as little as a few femtomoles. Recent advances in ESI-MS sample introduction techniques such as the nanoliter spray methods (*8*) (see Chapter 2 of this volume) look particularly encouraging in this regard.

In this chapter, we will discuss some of our most recent efforts to analyze LOS by ESI-MS, either as fully intact species or as their *O*-deacylated derivatives. For a discussion on the ESI approaches for the analysis of lipid A, one should see Chapter 9 of this volume and two recent papers (*9,10*). Our discussion, however, will emphasize methods for determining the structures of the variable oligosaccharide portions from pathogenic *Haemophilus* LOS and will be primarily limited to molecular mass determinations of LOS in complex mixtures (*11*). A few examples, however, will be presented that address strategies for limited structure determination of LOS by using ESI-tandem mass spectrometry or treatment with glycosidases in combination with ESI-MS.

Methods.

Isolation and Purification of LOS and *O*-deacylated LOS. LOS from *H. ducreyi* strains 35000 and 188-2 were isolated using a modified phenol/water extraction procedure of Westphal and Jahn (*12*). The LOS from *H. influenzae* strain A2 was prepared by the procedure of Darveau and Hancock (*13*). For mass spectrometric analysis, small amounts of LOS were *O*-deacylated according to a modified procedure of Helander *et al.* (*14*). Briefly, 0.5-1 mg of LOS was incubated with 200 μL of anhydrous hydrazine for 20 min at 37°C. The samples were then cooled to -20°C and chilled acetone was added dropwise to precipitate the *O*-deacylated LOS which was then centrifuged at 12,000 x *g* for 20 min. The supernatant was removed and the pellet washed again with cold acetone and centrifuged. Precipitated *O*-deacylated LOS was then resuspended in 500 μL of water and lyophilized.

Electrospray Ionization Mass Spectrometry (ESI-MS). For negative-ion ESI-MS analysis, either a VG-Fisons Platform quadrupole or a VG-Fisons Bio-Q triple quadrupole mass spectrometer with an electrospray ion source was used to mass analyze unmodified LOS and *O*-deacylated LOS. For the high resolution analysis of *O*-deacylated LOS from *H. influenzae* strain A2, a VG-Fisons AutoSpec magnetic sector instrument operating at a resolving power of 1000-3000 Δm/m was used with an electrospray ion source.

For these ESI-MS experiments, the *O*-deacylated LOS samples were first dissolved in water to make a 1 μg/μL solution. One μL of this solution was mixed with 4 μL of the running solvent and the 5 μL was injected via a Rheodyne injector into a constant stream of H_2O/CH_3CN (3:1) containing 1% acetic acid running at 10 μL/min. For the analysis of unmodified *H. ducreyi* 35000 LOS, a similar procedure was followed except that the samples were first dissolved in H_2O/CH_3CN/triethylamine (1/1/1, v/v) to make a 1 μg/μL solution, and the running solvent was H_2O/CH_3CN (1/1, v/v) containing 1% acetic acid. It was important to prepare fresh solutions of LOS prior to the run, since degradation occurred if the sample was allowed to remain in solution for prolonged periods of time. Mass calibration was carried out with an external horse heart myoglobin reference using the supplied VG-Fisons software.

Electrospray tandem mass spectrometric studies of *O*-deacylated LOS were performed on a VG-Fisons BioQ triple quadrupole mass spectrometer. Conditions were similar to those mentioned above for ESI-MS of *O*-deacylated LOS. For the tandem experiments argon was used as the collision gas. The argon gas pressure in the collision quadrupole chamber was 10^{-2} torr and the collision energy was 17V.

Results and Discussions.

Over the last five years we have determined the structures of LOS from several pathogenic species of *Neisseria* and *Haemophilus* using liquid secondary ion mass spectrometry (LSIMS), tandem mass spectrometry, methylation analysis and NMR techniques (*7,15-20*). More recently, we have developed ESI-MS techniques as an integral part of our LOS structural studies (*11*). Other groups have reported structures and/or mass spectrometric methods for the analysis of the largely invariant lipid A regions from these same organisms (*14,21,22*). From these data, we have constructed a structural model for LOS that begins to explain why these organisms are effective at surviving in the body and colonizing specific host tissues (see Figure 1) (*17*). For example, these bacteria mimic host carbohydrate structures in their LOS, and may provide the bacteria a means either to attach to and/or invade host tissues, or to evade the immune system. Recent reports have shown that sialylation of lactosamine-containing LOS from *Neisseria* confers serum resistance and inhibits opsonization and killing by human neutrophils (*23-26*).

As mentioned earlier, the analysis of LOS by ESI-MS techniques has generally required the conversion of LOS to their *O*-deacylated forms by hydrazine treatment (see Figure 2). This simple one-step procedure removes four of the six fatty acyl moieties on the lipid A, producing a water soluble LOS derivative that now contains only two N-linked fatty acids (β-hydroxy myristic acid) and three free phosphates (two on lipid A and one on KDO for *Haemophilus* species). Although further treatment with aqueous HF can produce LOS derivatives that are now amenable to MS techniques that prefer less anionic forms for optimum ionization efficiency such as LSIMS or matrix-assisted laser desorption ionization (MALDI) mass spectrometry, HF treatment produces some degree of glycosidic bond cleavage which can yield truncated LOS species that can confound the determination of the true LOS-glycoform population. Therefore, our results and discussion will be limited to the analysis of *O*-deacylated and intact LOS species by ESI-MS. A summary of chromatographic and mass spectrometric strategies used and/or are under investigation for analyzing intact LOS, *O*-deacylated LOS, and *O*-deacylated and dephosphorylated LOS, is summarized in Figure 2.

Determination of sialic acid in the LOS of *H. ducreyi* strain 35000. One of the first ESI-MS spectra taken of *O*-deacylated LOS was from *H. ducreyi* strain 35000 (*11*). Before this sample had been analyzed, we had previously determined the major oligosaccharide structure present in this LOS mixture after mild acid treatment using LSIMS and high-energy CID experiments (*19*), and later NMR (*20*):

Galβ1→4GlcNAcβ1→3Galβ1→4Hepα1→6Glcβ1→4Hepα1→*anhydro*KDO
3
↑
Hepα1→2Hepα1

However, we were also aware that SDS-PAGE analysis of the total LOS pool had shown the presence of at least 3-4 additional glycoforms with both higher and lower molecular weights (*27*). The negative-ion ESI-MS spectrum of this *O*-deacylated LOS mixture shown in Figure 3, therefore, provided the first accurate molecular weight data that allowed us to properly assess the chemical nature of these less abundant glycoforms (*11,20*). In fact, it was relatively straightforward to make preliminary composition assignments of these additional LOS species based on simple extension or truncation of the oligosaccharide region of the major LOS-**B** form shown above. For example, a lower mass species (LOS-**A**) was present corresponding to the loss of a single hexose (Hex) residue, and three of the four higher mass species could be

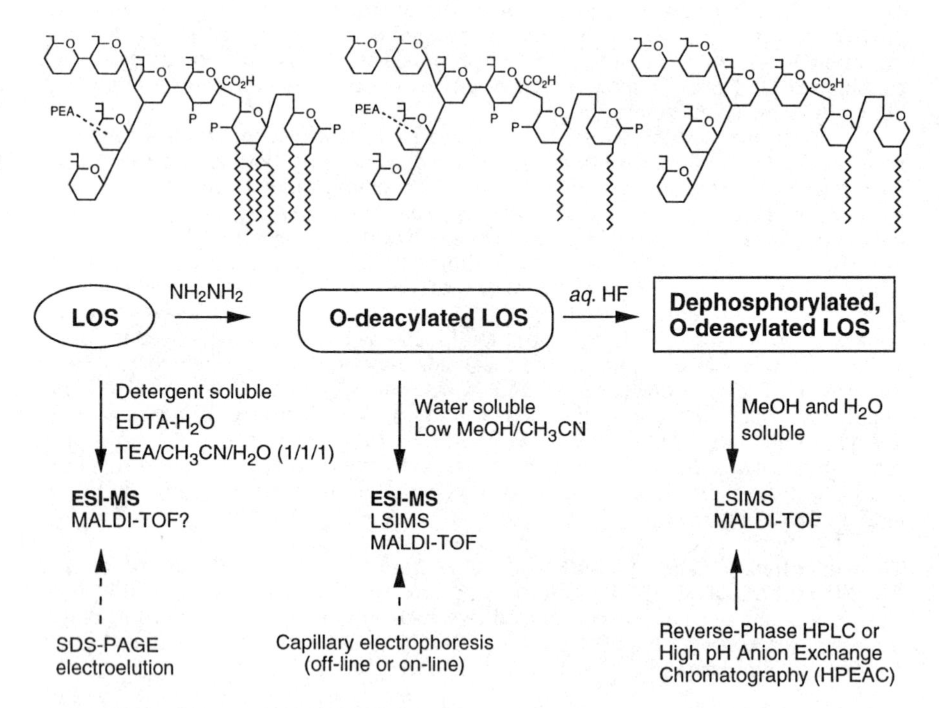

Figure 2. Analytical scheme for the preparation of LOS derivatives. Dashed arrows identify chromatographic interfaces to ESI that are currently under development in our group for LOS analysis. SDS-PAGE refers to sodium dodecylsulfate polyacrylamide gel electrophoresis.

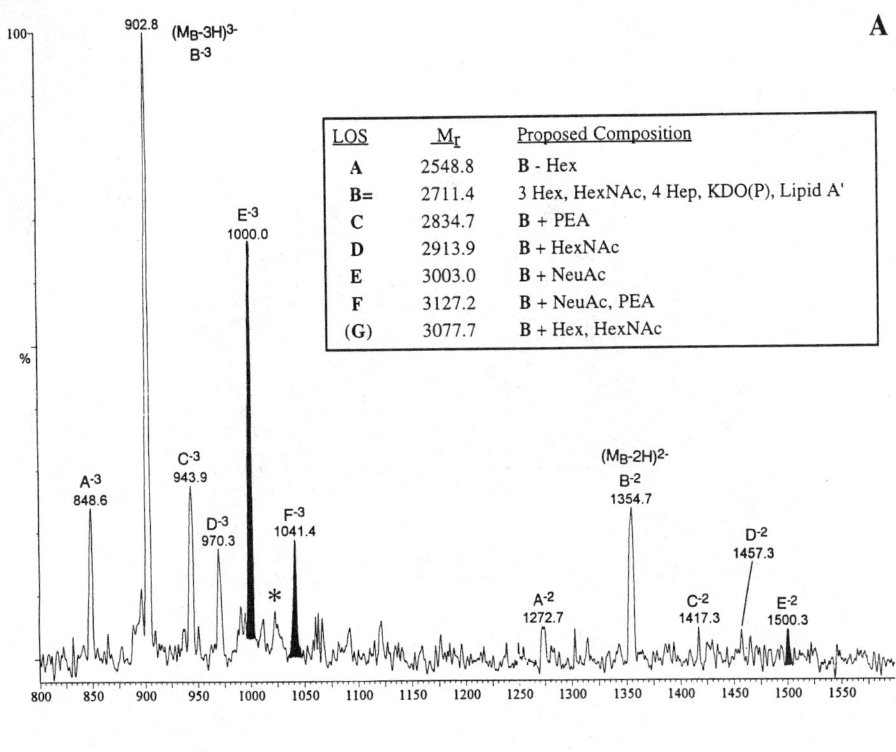

LOS	M_r	Proposed Composition
A	2548.8	B - Hex
B=	2711.4	3 Hex, HexNAc, 4 Hep, KDO(P), Lipid A'
C	2834.7	B + PEA
D	2913.9	B + HexNAc
E	3003.0	B + NeuAc
F	3127.2	B + NeuAc, PEA
(G)	3077.7	B + Hex, HexNAc

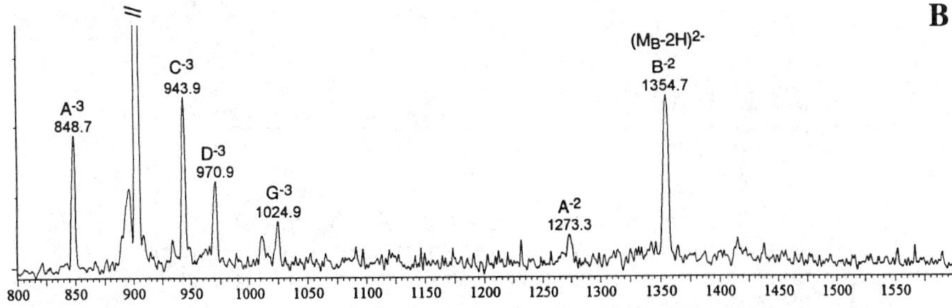

Figure 3. Negative-ion ESI-MS spectra before (A) and after (B) treatment with neuraminidase to remove sialic acid (NeuAc) moieties from specific *O*-deacylated LOS glycoforms isolated from *H. ducreyi* strain 35000. Adapted from ref. (*20*).

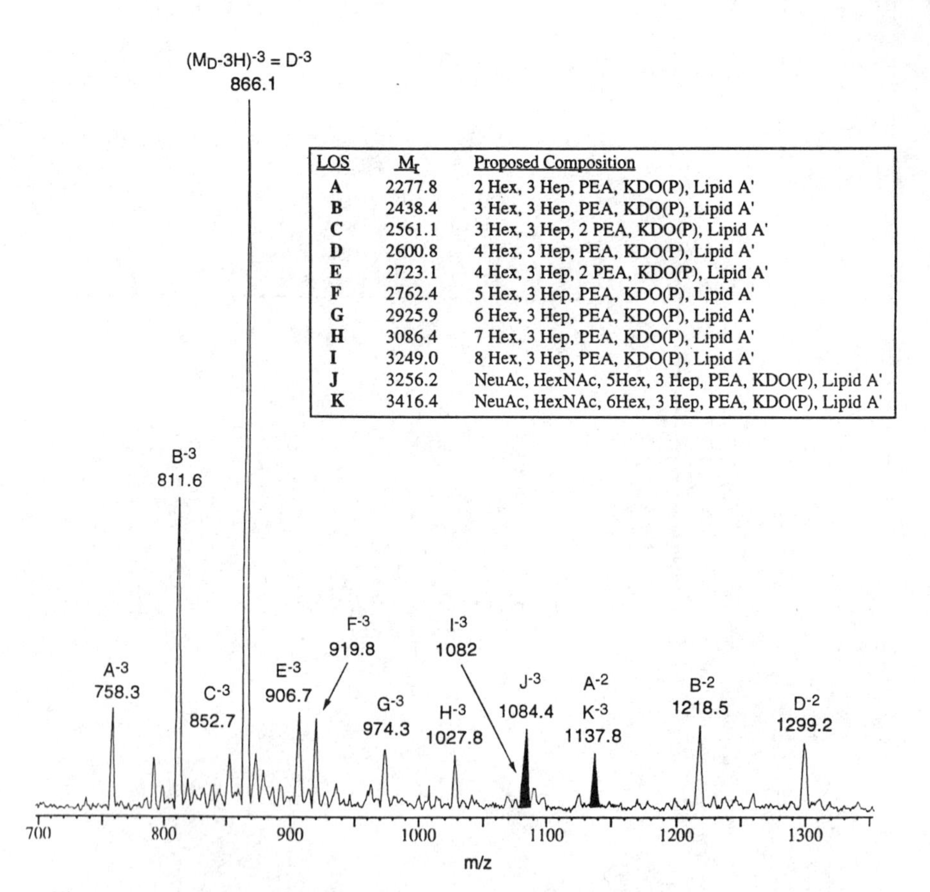

LOS	M_r	Proposed Composition
A	2277.8	2 Hex, 3 Hep, PEA, KDO(P), Lipid A'
B	2438.4	3 Hex, 3 Hep, PEA, KDO(P), Lipid A'
C	2561.1	3 Hex, 3 Hep, 2 PEA, KDO(P), Lipid A'
D	2600.8	4 Hex, 3 Hep, PEA, KDO(P), Lipid A'
E	2723.1	4 Hex, 3 Hep, 2 PEA, KDO(P), Lipid A'
F	2762.4	5 Hex, 3 Hep, PEA, KDO(P), Lipid A'
G	2925.9	6 Hex, 3 Hep, PEA, KDO(P), Lipid A'
H	3086.4	7 Hex, 3 Hep, PEA, KDO(P), Lipid A'
I	3249.0	8 Hex, 3 Hep, PEA, KDO(P), Lipid A'
J	3256.2	NeuAc, HexNAc, 5Hex, 3 Hep, PEA, KDO(P), Lipid A'
K	3416.4	NeuAc, HexNAc, 6Hex, 3 Hep, PEA, KDO(P), Lipid A'

Figure 4. Negative ion ESI-MS analysis of LOS from *H. influenzae* strain A2 from a VG-Fisons BioQ quadrupole mass spectrometer at a resolution of ≈500 (M/ΔM). Note the lack of separation of the components LOS-**I** and LOS-**J** at ≈1082 and 1084.4, respectively (adapted after ref. (*18*)).

assigned as extensions of the major LOS-**B** form by the addition of PEA (LOS-**C**), N-acetylhexosamine (HexNAc) (LOS-**D**), or Hex+HexNAc (LOS-**F**). However, the second most abundant species (LOS-**E**) had an unexpected mass 291 Da larger, which was recognized as the mass of sialic acid (N-acetylneuraminic acid, or NeuAc). Even though this species constituted a major percentage of the LOS species based on relative ion abundances, it went undetected in the previous oligosaccharide experiments presumably due to degradation during the mild acid treatment step. To confirm this tentative assignment, we simply treated this LOS mixture with neuraminidase to enzymatically remove the terminal α-linked sialic acid and re-ran the sample by ESI-MS. As shown in Figure 3b, the peaks assigned as LOS-**E** and LOS-**F** at *m/z* 1000 and 1041.4, respectively, which had been tentatively identified as sialic acid-containing LOS components disappeared, thus supporting the presence of a terminal sialic acid in these two LOS species.

Analysis of LOS glycoforms from *H. influenzae* strain A2. Based on our initial results with the *H. ducreyi* 35000 LOS, a second study was carried out to evaluate a much more complex mixture of LOS species from a type b strain of *H. influenzae*. As with the previous example, the LOS from *H. influenzae* A2 had been shown to be highly heterogeneous by SDS-PAGE analysis (≈8 or more bands) (*28*), and contained perhaps twice as many distinct LOS species as seen in the *H. ducreyi* strain 35000. The negative ion ESI-MS spectrum was consistent with this initial assessment by SDS-PAGE, and ten distinct molecular species could be identified either as their doubly and/or triply charged ions, $(M\text{-}2H)^{2-}$ and $(M\text{-}3H)^{3-}$ (see Figure 4). Based in part on LSIMS analysis of individual oligosaccharide masses determined from the hydrolyzed sample (*18*), compositional assignments were made based on the observed molecular weights, and a general structural formula could be constructed as follows, where the number of Hex moieties (x+y) on the two branches varied from 2-8:

$$
\begin{array}{l}
(\text{Hex})_x\rightarrow\text{Hep}\rightarrow\text{KDO(P)}\rightarrow O\text{-deacyl-Lipid A}\\
\qquad\qquad\;\uparrow\\
(\text{Hex})_y\rightarrow\text{Hep--PEA}\\
\qquad\qquad\;\uparrow\\
\qquad\qquad\text{Hep}
\end{array}
$$

In addition, a second set of LOS species were conjectured to exist based on the observed masses. These LOS species start out by assembling five hexose residues (x+y = 5), followed by the addition of HexNAc and sialic acid, presumably as a sialylated N-acetyl lactosamine structure. Although these tentative assignments could be made relatively easily, uncertainties and/or discrepancies in the observed masses compared to the expected mass values of these LOS were as large as ±1-2 Da, making these interpretations somewhat in doubt. Indeed, had it not been for the oligosaccharide data obtained previously from LSIMS analysis, it would have been impossible to make some of these interpretations. For example, for the less abundant LOS species, the precise charge state(s) could not be determined directly from the data due to the lack of a confirming doubly charged ion. This problem was particularly noted in the *m/z* 1100-1200 region where the triply charged peaks from the higher mass LOS (LOS-**I**, -**J**, and -**K**) begin to overlap with the doubly charged peaks from the lower mass components (LOS-**A** and -**B**), which was especially problematic for the $(M\text{-}3H)^{3-}$ ion for LOS-**K** and the $(M\text{-}2H)^{2-}$ ion for LOS-**A**. Furthermore, there was one ion series as noted above that seemed to indicate the presence of a divergent biosynthetic pathway leading to the synthesis of two N-acetyl sialyllactosamine terminating structures, LOS-**J** and -**K**. But in these cases, there was no evidence for the unsialylated N-acetyl lactosamine LOS species. This suggests that unlike that observed in *H. ducreyi* LOS, once N-acetyl lactosamine was formed as the presumed sialic acid acceptor, it was completely modified to the sialylated species,

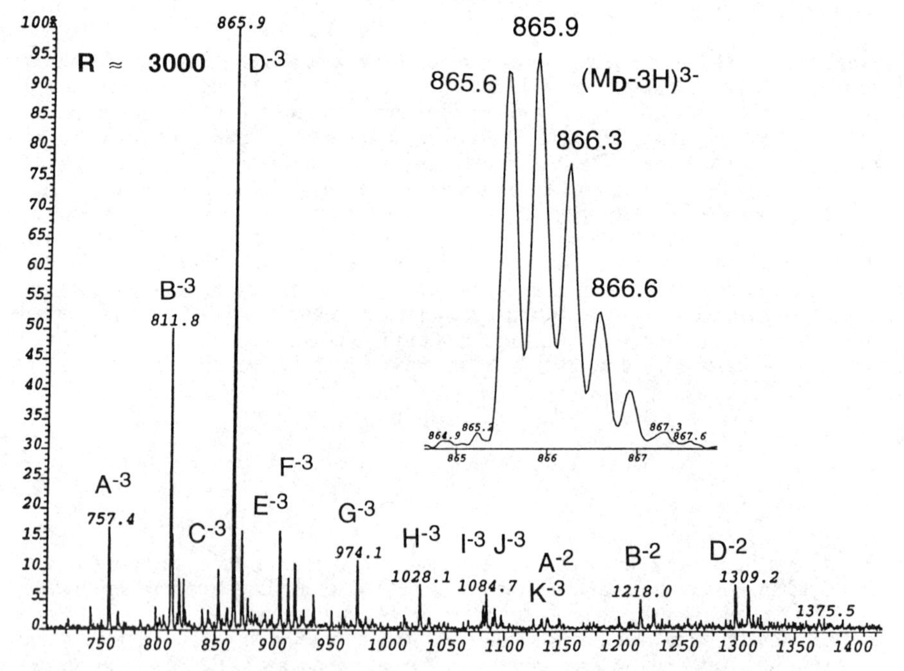

Figure 5. Negative-ion ESI-MS analysis of *H. influenzae* A2 LOS from a magnetic sector VG/Fisons Autospec mass spectrometer at a resolution of 1000 and 3000 M/ΔM (inset). Note the separation of the individual isotopes in LOS-**C** at the higher mass resolving power.

NeuAc→Galβ1→4GlcNAc. Clearly, limitations of this relatively low resolving power quadrupole analyzer (R = ≈500 M/ΔM) were compromising our abilities to make accurate molecular weight and composition assignments.

In an attempt to get around the limitations of mass accuracy and resolution, we investigated the utility of performing these experiments on a sector instrument with higher resolving power. And as we will see, improvements obtained on the AutoSpec magnetic sector instrument were impressive and offered several advantages (see Figure 5). First, the increased mass resolving power of the ESI-magnetic sector instrument provided much better separation of closely related LOS components. For example, the ion at *m/z* 866 can now be clearly assigned as a triply charged ion (see Figure 5, inset). In addition, the peaks at *m/z* 1082.2 and 1084.7 which were not fully separated in the previous quadrupole ESI-MS experiment (Figure 4) (*18*) are fully resolved at both the 1000 and 3000 M/ΔM resolution settings. Moreover, the higher resolving power reveals the isotope pattern in these two molecular ions and allows one to assign charge states in a completely unambiguous manner. This is especially important for LOS analyses since we generally see only one (and sometimes two) charge state per LOS component under negative-ion ESI conditions (primarily z = -3, corresponding to the number of phosphates) (*11*). The lack of a confirming ion series, therefore, can make it difficult to precisely determine the molecular weights as well as make accurate component identifications based on observations of single peaks in a complex spectrum. Therefore, better resolution MS data from the VG-Fisons AutoSpec instrument provided a higher degree of mass precision (and in this case, mass accuracy) in the assignments of the LOS components. For example, LOS-**I** and -**J** can be completely resolved and their masses determined to within ±0.1-0.2 Da of their ^{12}C monoisotopic M_r (see Figure 6, Table I).

Table I. Masses of *Haemophilus influenzae* A2 O-deacylated LOS at ≈3000 M/ΔM Resolution Under Negative-ion ESI-MS Conditions

Observed ^{12}C mass	*Calculated ^{12}C mass*	*ΔM*	*Proposed LOS Compositions*
A= 2275.3	2275.78	0.4	2Hex,3Hep,PEA,KDO(P),LA[a]
B= 2437.5	2437.84	0.3	3Hex,3Hep,PEA,KDO(P),LA
C= 2559.8	2599.89	0.1	3Hex,3Hep,2 PEA,KDO,LA
D= 2599.8	2599.89	0.1	4Hex,3Hep,PEA,KDO(P),LA
E= 2723.0	2722.90	0.1	4Hex,3Hep,2 PEA,KDO(P),LA
F= 2762.0	2761.94	<0.1	5Hex,3Hep,PEA,KDO(P),LA
G= 2924.3	2924.00	0.3	6Hex,3Hep,PEA,KDO(P),LA
H= 3086.2	3086.05	0.15	7Hex,3Hep,PEA,KDO(P),LA
I= 3248.3	3248.10	0.2	8Hex,3Hep,PEA,KDO(P),LA
J= 3256.2	3256.12	<0.1	NeuAc,HexNAc,5Hex,3Hep,PEA,KDO(P),LA
K= 3416.4	3416.19	0.2	NeuAc,HexNAc,6Hex,3Hep,PEA,KDO(P),LA

[a]LA stands for *O*-deacylated diphosphoryl lipid A.

Mass accuracy is an important factor as we often attempt to assign saccharide compositions directly from mass spectrometric data. In most cases, we employ a computer algorithm that searches possible monosaccharide compositions and variations in the number of phosphate moieties present against the experimental (observed) molecular weights (see Table II). An overall mass accuracy of ±0.1-0.2 Da (or 0.03-0.06 *m/z* units) means that we can now drastically narrow the number of

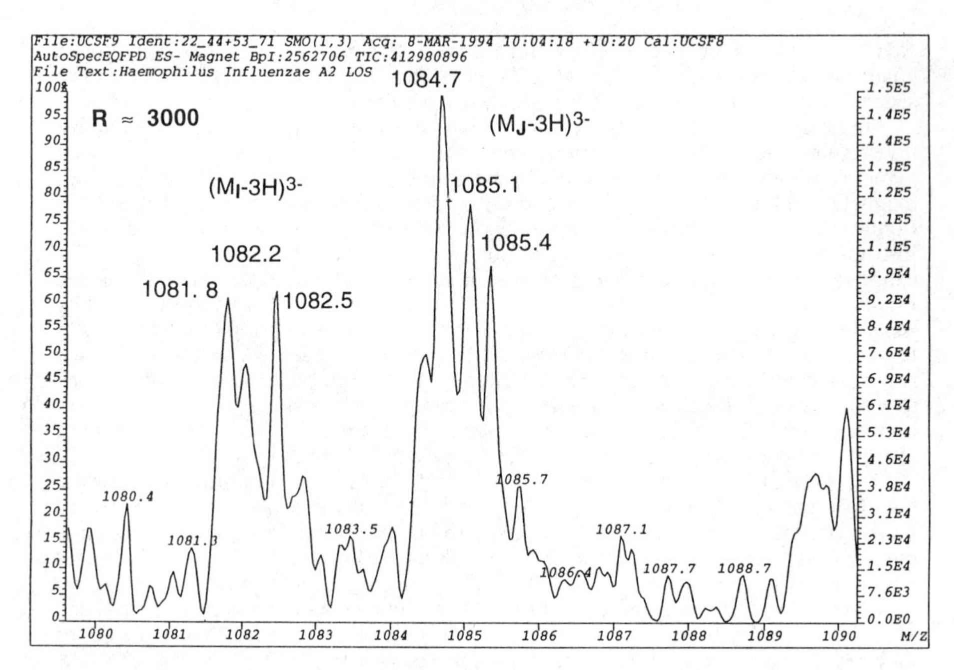

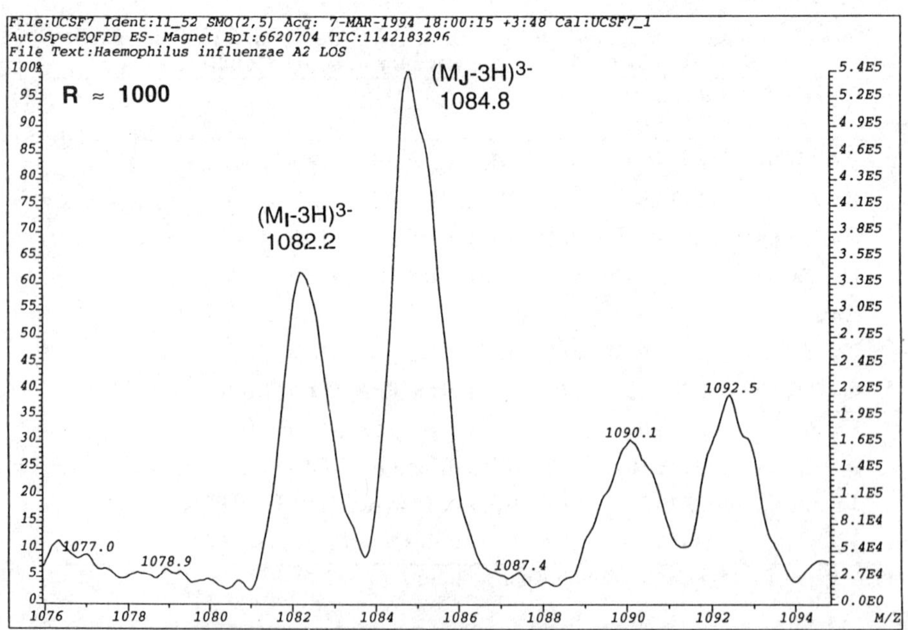

Figure 6. Expansion of the *m/z* 1080-1090 region shown in Figure 5 of *O*-deacylated LOS from *H. influenzae* strain A2 taken by a VG AutoSpec mass spectrometer at a resolution of ≈3000 and ≈1000 M/ΔM. The higher resolving power of 3000 allows the charge states to be determined directly, although both settings resolve the overlapping LOS-**I** and -**J** components.

possible LOS compositions compared to mass data obtained from instruments operating at lower resolving power (see Table II). For example, two common constituents of LOS, HexNAc (203.079) and phosphate-PEA (202.975), both have the same nominal mass of 203, but differ by ≈ 0.1 Da. Therefore, by obtaining ESI mass spectrometric data from LOS mixtures at these higher resolving powers, e.g., between 5,000-10,000 M/ΔM, we can expect to improve our mass accuracy to better than ±0.01 *m/z* , allowing us to better define generic saccharide compositions from the molecular weight data alone.

Table II. Gretta Carbos[a] Composition Search Output for $M_{ave.}$ = 3257.9 with ±1.0 and ±0.1 Da Mass Accuracy

Looking for Mass = 3257.87 with Tolerance +/- 1.0 Da:

No.	Mass	NeuAc	Hexose	HexNAc	Heptose	PEA	P	KDO(P)	O-DPLA
1	3257.48	1	2	1	3	3	3	1	1
2	3257.65	1	2	2	3	2	2	1	1
3	3257.82	1	2	3	3	1	1	1	1
4	3257.98	1	2	4	3	0	0	1	1
5	3257.71	1	5	0	3	2	1	1	1
6	**3257.87**	**1**	**5**	**1**	**3**	**1**	**0**	**1**	**1**

Looking for Mass = 3257.87 with Tolerance +/- 0.1 Da:

No.	Mass	NeuAc	Hexose	HexNAc	Heptose	PEA	P	KDO(P)	O-DPLA
1	3257.82	1	2	3	3	1	1	1	1
2	**3257.87**	**1**	**5**	**1**	**3**	**1**	**0**	**1**	**1**

[a]Program developed by W. Hines, UCSF. Abbreviations are as follows: NeuAc, N-acetylneuraminic acid (sialic acid); Hexose, Galactose and/or Glucose; Heptose, L-glycero-D-manno heptose; PEA, phosphoethanolamine; P, phosphate, KDO(P), 4-phospho-3-deoxy-2-keto octulosonic; O-DPLA, *O*-deacylated diphosphoryl lipid A.

Collision induced dissociation (CID) of *O*-deacylated LOS. To investigate what types of structural information can be obtained by tandem electrospray, we carried out a CID experiment of the major doubly charged *O*-deacylated LOS peak from a variant strain 188-2 of *H. ducreyi* (*29*). The base peak at *m/z* 821.4 for the triply charged molecular ion $(M\text{-}3H)^{3-}$ was selected for low energy collision induced dissociation (CID), and has been previously determined to consist of a diphosphoryl *O*-deacylated lipid A moiety attached to a typical Hep_3-phosphorylated KDO core with a trisaccharide branch:

Hex→Hep→Hex→Hep,PEA→KDO(P)-Lipid A
↑
Hep→Hep

The ESI-MS/MS spectrum for this major LOS-glycoform is shown in Figure 7. Besides the triply charged parent ion at *m/z* 821.4 there is also a peak at *m/z* 756.8 which corresponds to a triply charged ion formed by the loss of a terminal heptose (-192 Da/3) from the parent ion. The peak at *m/z* 951.9 represents the singly charged deprotonated molecular ion $(M\text{-}H)^-$ for lipid A and its doubly charged ion is evident at *m/z* 474.3. The peak at *m/z* 1416.4 is the singly charged Y-type ion representing the loss of lipid A, H_3PO_4 and H_2O from the molecular ion. The peak at *m/z* 1372.6 represents decarboxylation of the ion at *m/z* 1416.4.

The preliminary data from the ESI-MS/MS experiment indicate that this method gives little structural information on the oligosaccharide region and cannot be

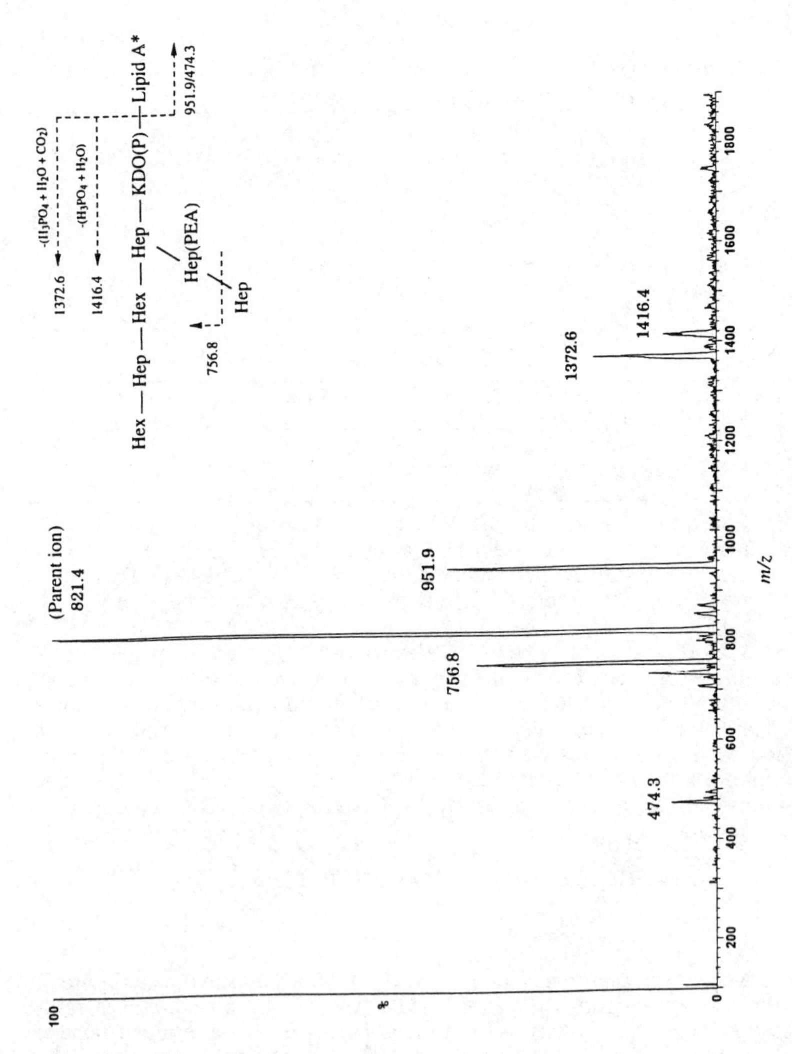

Figure 7. Negative-ion ESI-MS/MS of *m/z* 821.4 from the *O*-deacylated LOS mixture obtained from *H. ducreyi* strain 188-2 taken on a VG/Fisons BioQ triple quadrupole mass spectrometer. Lipid A* refers to *O*-deacylated diphosphoryl lipid A.

considered a suitable alternative to LSIMS high energy CID. However, the presence of the singly charged ion $(M-H)^-$ at *m/z* 951.9 for the *O*-deacylated lipid A suggests that this could be used as a fast method for obtaining the molecular weight of lipid A without LOS hydrolysis. This could save valuable time when screening different strains in order to determine if the LOS warrant further structural analysis by pointing to differences in the lipid A and/or oligosaccharide portion of the *O*-deacylated LOS molecules.

Intact LOS analysis by ESI-MS. Methods for analysis of intact unmodified LOS have proved difficult to develop due to the insolubility of LOS in water or suitable organic solvents without the aid of detergents. However, LOS can be solubilized in triethylamine, and if triethylamine is combined with a mixture of water and acetonitrile, a solution of LOS can be prepared that is amenable to ESI-MS analysis (see Figure 2). For example, the unmodified LOS from *H. ducreyi* strain 35000 was prepared in this manner just prior to ESI-MS analysis (see Figure 8).

This spectrum is considerably more complex than its *O*-deacylated counterpart (Figure 3) and not as readily interpretable. Nonetheless, one observes two series of ions in the spectrum that differ by 226 Da, the mass of a hydroxymyristoyl group. The higher charged series shows a triply charged molecular ion $(M-3H)^{3-}$ at *m/z* 1234.7, its doubly charged counterpart at *m/z* 1852.6, as well as several extended analogs in each charge state. This triply charged molecular ion corresponds to an intact LOS species with a molecular mass of 3706.5 Da which is 123 Da (the mass of a PEA moiety) more than expected for the mass of the major LOS. The ions at *m/z* 1275.6, 1332.0 and 1372.7 represent triply charged LOS species which contain an additional PEA, sialic acid, and PEA plus sialic acid, respectively, relative to the LOS species with M_r 3706.5. Similarly, peaks at *m/z* 1914.2, 1998.3 and 2059.7 denote the same LOS species as their doubly charged ions. The lower mass series represents an LOS species containing one less hydroxymyristoyl moiety (-226 Da) on the lipid A moiety and has a triply charged molecular ion at *m/z* 1158.4 and a corresponding doubly charged ion at *m/z* 1739.3. The ions at *m/z* 1200.1 and 1800.6 signify the presence of additional LOS components containing an extra PEA as the triply and doubly charged ions, respectively. The main problem with this latter interpretation is that there was no additional evidence for an LOS species containing two PEA groups in the ESI-MS spectrum of *O*-deacylated LOS from *H. ducreyi* strain 35000 (Figure 3, Table inset). Furthermore, it appears that significant degradation has occurred in the *O*-linked fatty acid of the lipid A region, producing unwanted artifacts such as the LOS species missing a hydroxymyristoyl group that make a proper determination of the biologically significant LOS-glycoforms difficult. Clearly, the ESI-MS/MS data shown here are preliminary and one can hope that more efficient ways of solubilizing LOS can be found that may improve the utility of ESI-MS for intact LOS analysis.

Summary.

Electrospray ionization mass spectrometry has been successfully used to evaluate the molecular heterogeneity of bacterial lipooligosaccharides. The primary goal of these present studies was to determine the molecular masses of these glycolipids and, to a lesser extent, to allow some sequence or structural assignments to be made. The conversion of LOS to water soluble *O*-deacylated forms is relatively straightforward and has the advantage of not creating artifactual species through unwanted chemical degradation. In addition, *O*-deacylation appears to convert the lipid A to a N-linked diacyl form that is invariant, therefore allowing us to concentrate on the heterogeneity present in the oligosaccharide region of LOS. It also appears that higher resolution data obtained by ESI on the magnetic sector instrument is important in that it allows better separation of LOS mixtures that have closely related masses, better overall mass accuracies and unambiguous determination of the charge states. The issue of charge

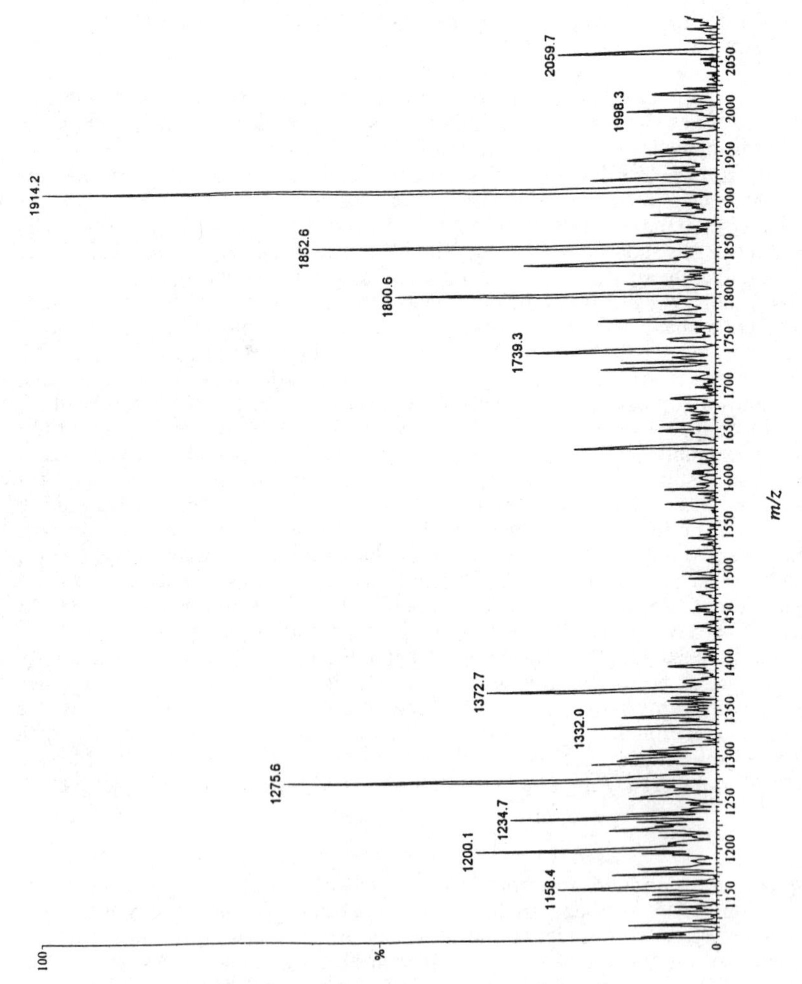

Figure 8. Negative-ion ESI-MS spectrum of intact LOS from *H. ducreyi* 35000 taken on a VG/Fisons quadrupole mass spectrometer.

state determination is especially important because *O*-deacylated LOS often give only one charge state based primarily on the number of phosphates.

It is also worth commenting on the relatively large dynamic range for LOS mixture analysis, since it is one of the major advantages of ESI-MS. It is crucial to be able to identify even lower abundance LOS components than are observed in the ESI-MS spectra of the LOS from *H. influenzae* A2 or *H. ducreyi* 35000. For example, we have previously shown that LOS structures present in wild-type strains of *N. gonorrhoeae* contain a biologically important epitope (the P^k antigen, Galα1→4Galβ1→4Glc) between 0.1-0.2% of the total LOS glycoforms when expressed under *in vitro* conditions (*16*). However, under *in vivo* conditions of urinary tract and genital infection, the LOS glycoform(s) containing this P^k-epitope may be selectively expressed at higher levels and critical for establishing infection in these tissues. Therefore, we clearly need a very large dynamic range as well as high detection efficiencies to identify the full range of LOS glycoforms expressed under both *in vitro* and *in vivo* conditions so we can better correlate specific LOS structures to specific biological and pathological functions.

As we investigated the analysis of intact LOS by ESI-MS, these data proved to be somewhat less useful. Some additional heterogeneity was observed in the ESI-MS spectra of the intact LOS as compared to the *O*-deacylated spectra due primarily to the different acylation states of the lipid A. This is likely to be the result of chemical degradation, as prolonged exposure of LOS to the solubilizing triethylamine-containing solution increased the abundances of lower mass species. Nonetheless, the data obtained from the intact LOS did provide useful structural information and confirmed the existence of the hexacyl lipid A species as well as the other glycoforms previously observed in the *O*-deacylated LOS spectra.

Lastly, obtaining additional structural (i.e., sequence and branching) information on these LOS species is highly desirable, but has generally been obtained from experiments carried out on the separately isolated oligosaccharides, primarily by high-energy CID techniques (*7,17-19,29*). The low energy CID data obtained by ESI-MS/MS analysis in this study were somewhat disappointing, as most fragment ions appeared to result from charge-localized cleavages near or at the lipid A moiety. As previously shown by our group (*7,15,17-19*), high energy CID of oligosaccharides yields very extensive fragmentation, a good deal of which arises from charge remote fragmentation, allowing for a complete sequence determination to be made. We hope in the near future to carry out medium-to-high energy CID experiments on these *O*-deacylated LOS using a magnetic sector or a time-of-flight mass analyzer that would have the required ionization and detection efficiencies suitable for LOS compounds between 2000-5000 Da in mass. In an alternative approach, glycosidase treatment of LOS can provide very useful terminal sequence information, such as for the identification of LOS species containing sialic acid. Clearly other glycosidic enzymes can be used, and it may also be possible to employ immobilized lectins or monoclonal antibodies in methods based on selective binding for identifying LOS with other terminal epitopes or structures, such as terminal N-acetylglucosamine, galactose or lactose.

Acknowledgments.

This work was supported by grants from the National Institutes of Health (AI24016 and AI31254; BWG) and the University of California University AIDS Research Program (BWG). We also acknowledge support for the Mass Spectrometry Facility through grants from the National Center for Research Resources (RR 01614; A.L. Burlingame, Director and BWG, Deputy Director) and NIEHS (ES04705; ALB).

References.

1. Wilkinson, S. G. In *Surface Carbohydrates of the Prokaryotic Cell*; I. Sunderland, Ed.; Academic Press: New York, 1977; pp 97-175.
2. Freudenberg, M. A.; Galanos, C. *Int. Rev. Immunol.* **1990**, *6*, 207-221.
3. Mandrell, R. E.; Apicella, M. A. *Immunobiology* **1993**, *187*, 387-402.
4. Campagnari, A. A.; Wild, L. M.; Griffiths, G. E.; Karalus, R. J.; Wirth, M. A.; Spinola, S. M. *Infect Immun* **1991**, *59*, 2601-8.
5. Maskell, D. J.; Szabo, M.; Butler, P. D.; Williams, A. E.; Moxon, E. R. *Res. Microbiol.* **1991**, *142*, 719-724.
6. Weiser, J. N.; Love, J.; Moxon, E. R. *Cell* **1989**, *59*, 657-665.
7. Phillips, N. J.; Apicella, M. A.; Griffiss, J. M.; Gibson, B. W. *Biochemistry* **1992**, *31*, 4515-4526.
8. Andren, P. E.; Emmett, M. R.; Caprioli, R. M. *J. Amer. Soc. Mass Spectrom.* **1995**, *5*, 867-869.
9. Harrata, A. K.; Domelsmith, L. D.; Cole, R. B. *Biol. Mass Spectrom.* **1993**, *22*, 59-67.
10. Chan, S.; Reinhold, V. N. *Anal. Chem.* **1994**, *218*, 63-73.
11. Gibson, B. W.; W., M.; Phillips, N. J.; Apicella, M. A.; Campagnari, A. A.; Griffiss, J. M. *J. Bacteriol.* **1993**, *175*, 2702-2712.
12. Westphal, O.; Jahn, K. In *Methods in Carbohydrate Chemistry*; R. L. Whistler, Ed.; Academic Press Inc.: New York, 1965; Vol. 5; pp 83.
13. Darveau, R. P.; Hancock, R. E. W. *J. Bacteriol.* **1983**, *155*, 831-838.
14. Helander, I. M.; Lindner, B.; Brade, H.; Altmann, K.; Lindberg, A. A.; Rietschel, E. T.; Zähringer, U. *Eur. J. Biochem.* **1988**, *177*, 483-492.
15. Gibson, B. W.; Webb, J. W.; Yamasaki, R.; Fisher, S. J.; Burlingame, A. L.; Mandrell, R. E.; Schneider, H.; Griffiss, J. M. *Proc. Natl. Acad. Sci. USA* **1989**, *86*, 17-21.
16. John, C. M.; Griffiss, J. M.; Apicella, M. A.; Mandrell, R. E.; Gibson, B. W. *J. Biol. Chem.* **1991**, *266*, 19303-19311.
17. Phillips, N. J.; John, C. M.; Reinders, L. G.; Gibson, B. W.; Apicella, M. A.; Griffiss, J. M. *Biomed. Environ. Mass Spectrom.* **1990**, *19*, 731-745.
18. Phillips, N. J.; Apicella, M. A.; Griffiss, J. M.; Gibson, B. W. *Biochemistry* **1993**, *32*, 2003-2012.
19. Melaugh, W.; Phillips, N. J.; Campagnari, A. A.; Karalus, R.; Gibson, B. W. *J. Biol. Chem.* **1992**, *267*, 13434-13439.
20. Melaugh, W.; Phillips, N. J.; Campagnari, A. A.; Tullius, M. V.; Gibson, B. W. *Biochemistry* **1994**, *33*, 13070-13078.
21. Takayama, K.; Qureshi, N.; Hyver, K.; Honovich, J.; Cotter, R. J.; Mascagni, P.; Schneider, H. *J. Biol. Chem.* **1986**, *261*, 10624-10631.
22. Cotter, R. J.; Honovich, J.; Qureshi, N.; Takayama, K. *Biomed. Environ. Mass Spectrom.* **1987**, *14*, 591-598.
23. McNeil, M.; Darvill, A. G.; Aman, P.; Franzen, P.; Albersheim, P. *Meth. Enz.* **1982**, *83*, 3-45.
24. Rest, R. F.; Frangipane, J. V. *Infect. Immun.* **1992**, *60*, 989-997.
25. Estabrook, M. M.; Christopher, N. C.; Griffiss, J. M.; Baker, C. J.; Mandrell, R. E. *J. Infect. Dis.* **1992**, *166*, 1079-1088.
26. Kim, J.; Zhou, D.; Mandrell, R. E.; Griffiss, J. M. *Infect. Immun.* **1992**, *60*, 4439-4442
27. Campagnari, A. A.; Spinola, S. M.; Lesse, A. J.; Abu Kwaik, Y.; Mandrell, R. E.; Apicella, M. A. *Microbial Pathogen.* **1990**, *8*, 353-362.
28. McLaughin, R.; Spinola, S. M.; Apicella, M. A. *J. Bacteriol.* **1992**, *174*, 6455-6459.
29. Campagnari, A.; Karalus, R.; Apicella, M. A.; Melaugh, W.; Lesse, A. J.; Gibson, B. W. *Infect. Immun.* **1994**, *62*, 2379-2386.

RECEIVED September 13, 1995

Chapter 9

Electrospray Ionization Mass Spectrometry for Structural Characterization of the Lipid A Component in Bacterial Endotoxins

Richard B. Cole

Department of Chemistry, University of New Orleans–Lakefront, New Orleans, LA 70148

Electrospray ionization mass spectrometry (ESMS) represents a recently developed technology for the direct analysis of glycolipids at high sensitivity. This report investigates phosphoglycolipids known as 'lipid A', originating from lipopolysaccharides of the gram-negative bacteria *Enterobacter agglomerans* (EA) and *Salmonella minnesota* (SM). Lipid A is the lipophilic component of the outermost bacterial cell wall. ESMS was found to be more than one order of magnitude more sensitive for lipid A analysis in the positive ion mode, and more than two orders of magnitude more sensitive in the negative ion mode as compared to plasma desorption mass spectrometry (PDMS). Due to the extremely "soft" nature of the ionization process, negative ion ESMS readily produces a profile of the heterogeneity of a crude lipid A preparation by displaying singly charged deprotonated molecules of the various monophosphoryl lipid A forms, and singly or doubly deprotonated molecules corresponding to diphosphoryl forms. Lipid A samples from chemically dephosphorylated lipopolysaccharides produced "distal adduct ions" in the positive ion mode which appear in the form of sodium cations attached to the non-reducing monoglucosamine. These ions are analogs of oxonium ions observed in PDMS. Although fragmentation of ions generated via ESMS is typically quite low, to augment obtained structural information, formed ions can be subjected to collision-induced dissociations (CID), either "pre-analyzer" (in-source) on a single analyzer instrument or in the collision cell of a tandem mass spectrometer. Among the CID processes which can occur are dehydrations, acylium cleavages, carboxylic acid eliminations, and glycosidic cleavages. ESMS has been used to characterize the structures of lipid A species which have undergone hydrolysis either by acid or base treatment. Both acid and base treatments are known to reduce the 'endotoxin level' of bacterial species as measured by the *Limulus* amebocyte lysate (LAL) assay. Base treatment is effective at eliminating pulmonary toxicity.

Bacterial endotoxins are lipopolysaccharides (LPS) which can play numerous roles to influence biological activity in humans as well as other organisms. Induced

0097–6156/95/0619–0185$12.50/0

{O-Specific Chain} - {Outer Core} - {Inner Core - $(Kdo)_n$} - {Lipid A}

Core Region

Figure 1. Schematic representation of the component parts of enterobacterial lipopolysaccharides.

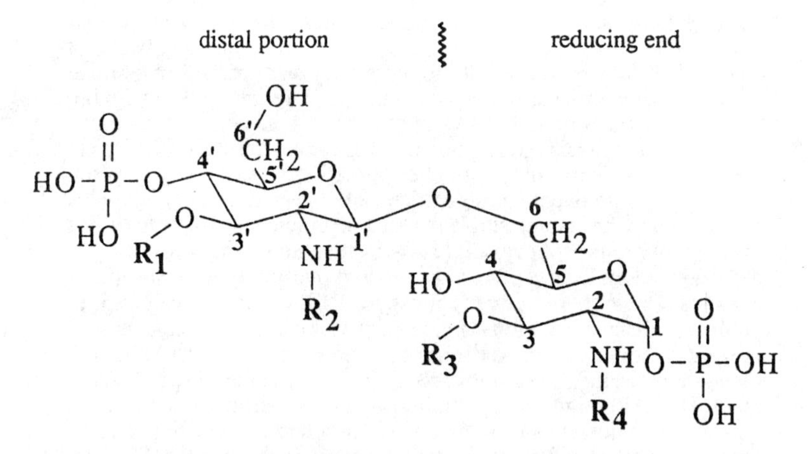

Figure 2. General structure, including stereochemistry, of the common type of diphosphoryl, diglucosamine backbone of lipid A shared by *Salmonella minnesota* and *Escherichia coli.*

effects can be quite harmful, ranging from promotion of fever, diarrhea, and shock, to lethal toxicity (1,2). Ironically, the same endotoxins can reportedly enhance the body's resistance to infections and cancer (2). The bacterial cell wall can be coarsely divided into three primary regions (3) as shown in Figure 1. The O-specific chain is composed of polysaccharides (generally glycosyl residues) linked in a manner which tends to be highly diverse when comparing different bacterial species. The O-specific chain is attached to the core region which can be subdivided into "outer" and "inner" cores. These cores consist of polysaccharides which have a rather high degree of structural uniformity when comparing different species within a bacterial genus. The outer surface of the bacterial cell is composed of lipopolysaccharides, with the extreme outer lipophilic layer (referred to as lipid A) linked to the core region by several KDO (2-keto-3-deoxyoctonic acid) moieties. The 6' position of the lipid A portion (Fig. 2) is attached to the KDO moiety. The ability to completely assign the structures of lipid A species is an essential step toward understanding the complex and multi-faceted functions of bacterial endotoxins.

It has been established in various toxicity studies that the toxic activity of bacterial lipopolysaccharides (LPS) originates in the lipid A portion (4-6). Because these endotoxins are common components of organic dusts that cause or contribute to symptoms associated with organic dust diseases, an understanding of the complex relationship between chemical structure and biological function can only arise from a detailed chemical characterization of the lipopolysaccharide. The possibility to completely characterize lipopolysaccharide structures has only rather recently become a reality, due to the advent and refinement of new analytical approaches (7,8). Lipid A is the hydrophobic portion of the endotoxin. With certain known exceptions, lipid A from different sources often share a common basic structure consisting of a mono- or di-phosphorylated 1,6-β-linked D-glucosamine disaccharide with several fatty acyl side chains attached via amide and ester linkages (Fig. 2). The amide linkages are usually found at the 2 and 2' positions, while ester linkages are located at the 3 and 3' positions, with phosphate groups at the 1 and 4' positions of the disaccharide backbone. Additional fatty acyl groups may also be attached via ester linkages at sites α or β to the carbonyl of either the amide- or ester-linked side chains. The structures of several lipid As including *Salmonella typhimurium* G30/21, *Salmonella minnesota* R595, *E. coli* and *Rhodobacter sphaeroides* ATCC17023 (9) have been definitively established using a combination of analytical approaches.

Mass spectrometry (MS) has played a vital role in the characterization of these and other varieties of bacterial lipid A species. The earliest use of mass spectrometry involved gas chromatography-electron ionization MS analysis of fatty acid side chains which had been chemically removed from their lipid A anchor. With the advent of the particle-induced desorption techniques, such as fast atom bombardment (10,11), laser desorption (12-14), and plasma desorption (PD) (15-17), the possibility to desorb whole lipid A species into the gas phase became a reality. In general, these techniques produce singly charged ions representative of intact analyte molecules. Fast atom bombardment (FAB) was the first particle-induced desorption technique to be used for direct analysis of heterogeneous lipid A preparations (10), wherein deprotonated molecules (negative charge localized on the phosphate group) were observed in the negative ion mode. Deprotonated molecules were also generated via laser desorption MS, and later, plasma desorption MS, both used in conjunction with linear time-of-flight mass analyzers. However, in the latter studies involving desorption from solid targets, the presence of phosphate groups, which provided a facile mechanism for creating charged molecules, inhibited efficient desorption from the target. Desorption ability was improved significantly via the chemical removal of the phosphate moiety from the 4' position of the lipid A molecule. In the process, of course, the propensity to form negatively charged ions was virtually eliminated. Dephosphorylated ions had to be detected in the positive ion mode, and they appeared mostly as sodium adducts of the various lipid A species. It seems that in removing the phosphate group, a gain in desorption efficiency came at the expense of some loss in ionization efficiency. Nonetheless, later PDMS work (18) demonstrated that good signal-to-noise ratio mass spectra could be obtained directly from the free acid form of lipid A (monophosphoryl variety from *E. coli*).

More recently, electrospray ionization mass spectrometry (ESMS) has proven its utility for structural studies of lipid A. Electrospray ionization provides efficient production of intact multiply charged and singly charged ions from a wide variety of polar species and has been increasingly used for analyses of biopolymers (19,20). The earliest reports of the use of ESMS for lipid A analysis (21-26) established that the technique is highly suited to lipid A structural studies. The extreme "softness" of the ionization process yields minimal gas-phase fragmentations to allow viewing of the profile of forms of lipid A produced during the hydrolysis steps taken to liberate lipid A from the LPS (24). Structural information can be augmented when "pre-analyzer" collision-induced dissociation (CID) is employed to fragment ES-desorbed species before mass analysis on a conventional mass spectrometer (24). The ability to assign daughter ions to a specific lipid A parent ion is vastly improved via the use of tandem mass spectrometry (27).

We have been working (17,21,22,24,26) on mass spectrometry studies of LPS produced by *Enterobacter agglomerans* ATCC 27996 (abbreviated hereafter as EA), a gram-negative bacterium often found in field cotton. Such endotoxins in cotton dust are known to contribute to respiratory diseases such as byssinosis (28). The Pittsburgh animal test model (29) has established that LPS from *Salmonella minnesota* (30) and *Enterobacter agglomerans* (31) can be deleterious to normal pulmonary functioning. Toxic activity is believed to originate in the lipid A moiety. This paper summarizes the capabilities of ESMS for investigating the structures of lipid A compounds, while also explaining how the technique can aid in the investigation of decomposition pathways intended to destroy the toxic activity of the molecule.

Experimental

Lipopolysaccharides (LPS) from *Enterobacter agglomerans* ATCC 27996 (abbreviated as EA) were used to generate lipid A used in this study. The hot water/phenol method (32) was used to extract LPS from *E. agglomerans* cells. The methodology employed to culture the organism and isolate LPS has been described previously (17). Purification of the LPS was performed using a variation (17) of the alcohol precipitation method of Kato and coworkers (33).

All lipid A samples referred to in the 'PDMS vs. ESMS' section were prepared by mild acid hydrolysis of LPS according to a previously published procedure (34,35). Dephosphorylated lipid A samples were prepared by treating the LPS with 48% hydrofluoric acid at 4°C for 42 hours. Following dialysis, the dephosphorylated LPS was acid hydrolyzed with 0.1 M HCl at 110°C for two hours (36), to yield dephosphorylated lipid A. Monophosphoryl lipid A from *Salmonella minnesota* R595 was purchased from Sigma Chemical Company.

^{252}Cf-plasma desorption mass spectrometry experiments were performed on a Bio-Ion Nordic 20 (Applied Biosystems, Uppsala, Sweden). In plasma desorption experiments, approximately 6 μg of lipid A in 4:1 chloroform:methanol was placed on an aluminized mylar foil. Displayed plasma desorption mass spectra were signal averaged for approximately one hour. In the positive ion mode, an acceleration energy of 18 keV was employed, while 14 keV acceleration was used in the negative ion mode.

Electrospray ionization mass spectrometry experiments were performed on a Vestec model 201 (PerSeptive Biosystems, Vestec Corp., Houston, TX) single quadrupole mass spectrometer capable of analyzing ions up to m/z 2000. For positive ion experiments, dissolved NaCl (saturated in methanol, 2% of final volume) was added to promote sodium adduct formation. The electrospray capillary (needle) was operated between 1.5 and 3.5 kV, while the nozzle and collimator voltages were maintained at 280 V and 10 V, respectively. Negative voltages of approximately the same magnitude were employed for negative ion operation.

PDMS vs. ESMS for lipid A characterization

For direct analyses of lipid A preparations from *E. agglomerans* (abbreviated as EA), plasma desorption (17) and electrospray ionization (24) can both be extremely useful methods. Plasma desorption is an inherently pulsed ionization method which makes it readily compatible with time-of-flight mass analyzers. The electrospray technique, on the other hand, continuously generates ions at atmospheric pressure which makes it more suitable for coupling to scanning mass analyzers such as the quadrupole mass filter. It is possible, however, to convert ions formed via electrospray into a discrete package suitable for time-of-flight analysis (37).

In preparation for both positive ion PDMS and positive ion ESMS, lipid A preparations were taken through a dephosphorylation step involving treatment with 48% hydrofluoric acid at 4 °C for 42 hrs. While it is true that, in our hands, monophosphoryl compounds could be detected by PDMS in the positive ion mode from crude lipid A preparations which had not undergone dephosphorylation in PDMS, the signals were not highly intense. Using chloroform:methanol solvent, comparable crude lipid A samples were difficult to detect via ESMS. However, after removal of the phosphate moiety, intense positive ion signals due to cation (sodium) attachment to neutral lipid A molecules were readily observed by both techniques. Figures 3a and 3b display the positive ion mass spectra of dephosphorylated lipid A produced by *E. agglomerans* obtained using PDMS and ESMS, respectively. Figure 3a presents PDMS data acquired at 18 keV acceleration energy which was signal averaged for approximately one hour. The ES mass spectrum in Figure 3b was acquired over approximately 10 minutes. Prior to the ESMS experiment, a small amount of sodium chloride was added to promote sodium cation attachment to the neutral dephosphorylated lipid A molecules.

In the ES mass spectrum (Figure 3b), all peaks were found to correspond to sodium-containing ions (24). The peaks observed in the plasma desorption mass spectrum (Figure 3a) have also been assigned (17) to sodium-containing ions except for those at m/z 1008 and 1024. These latter peaks correspond to the oxonium ions of the type shown in Figure 4a (m/z 1008). This form of oxonium ion has also been observed from lipid A preparations analyzed by FAB (11) and PDMS (14). These oxonium ions (no sodium attachment) were not observed in ESMS. However, analogous ions in the form of sodium-adducts of hydroxylated varieties of the distal

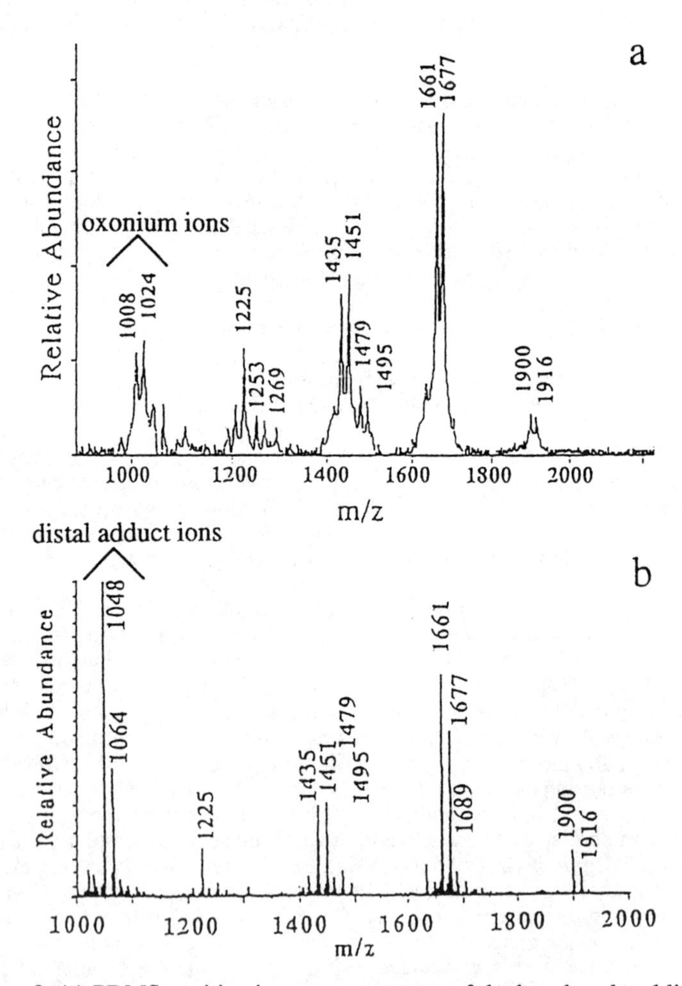

Figure 3. (a) PDMS positive ion mass spectrum of dephosphorylated lipid A from *Enterobacter agglomerans* including oxonium ions at lower mass. (b) ESMS positive ion mass spectrum of dephosphorylated lipid A from *E. agglomerans* including lower mass distal adduct ions.

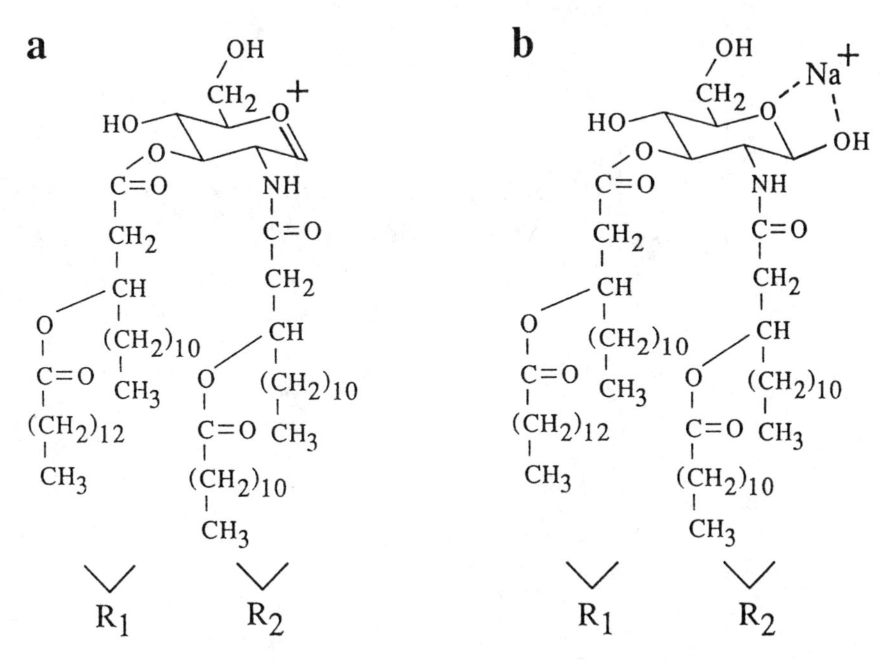

Figure 4. (a) Structure of the m/z 1008 oxonium ion generated via PDMS which appears in Figure 3a. (b) Structure of the m/z 1048 distal adduct ion produced during the ESMS experiment which appears in Figure 3b.

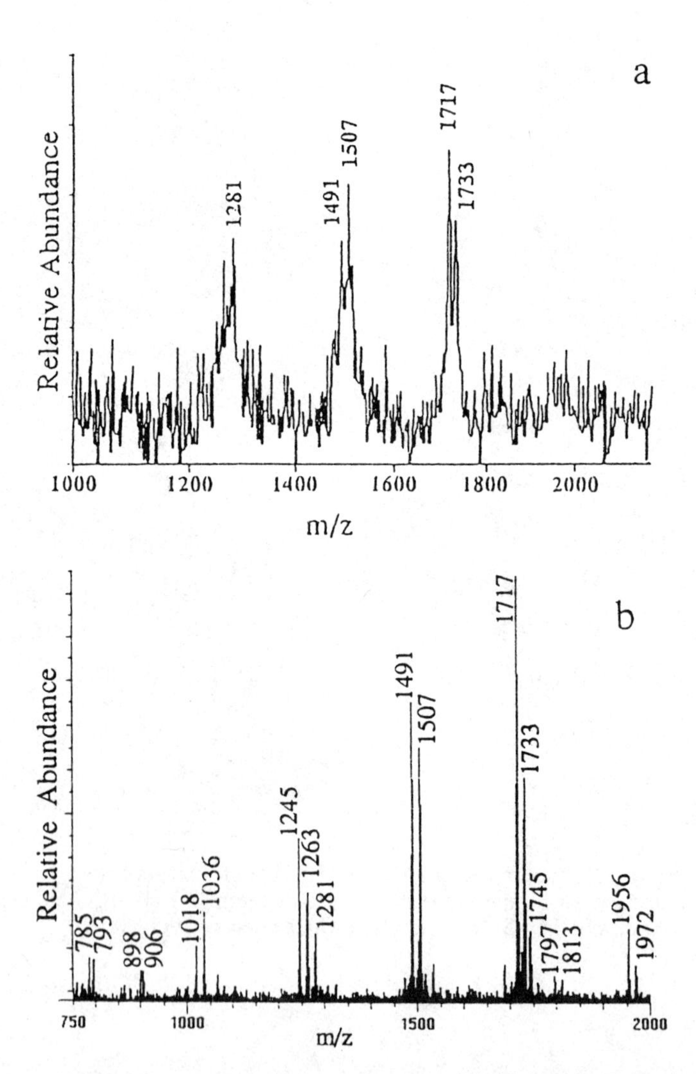

Figure 5. (a) PDMS negative ion mass spectrum of lipid A from *Enterobacter agglomerans*. (b) ESMS negative ion mass spectrum of lipid A from *E. agglomerans*.

lipid A portion (Figure 4b, m/z 1048), referred to here as distal adduct ions (appearing at m/z 1048 and 1064, Fig. 3b) were formed in ESMS (26).

Turning now to the negative ion mode, plasma desorption (-14 kV acceleration voltage) and electrospray mass spectra of monophosphoryl lipid A from *E. agglomerans* are shown in Figures 5a and 5b, respectively. Deprotonation to yield singly charged anions occurred in rather high yield with each technique. Data acquisition for PDMS involved signal averaging for about one hour, whereas the ESMS data were averaged for only 7 minutes. The mass analyzers were calibrated to measure average mass values and mass spectral assignments appear in Table 1. Separate ESMS experiments (24) performed under minimal fragmentation conditions (mass spectrum not shown) revealed that the peaks at m/z 1972, 1956, 1745, 1733, 1717, 1507, 1491, and 1281 correspond to eight different intact lipid A molecules present in the sample mixture prepared from hydrolysis of LPS from *E. agglomerans*.

The mass spectrum shown in Figure 5b was obtained at a skimmer potential which produces a moderate level of fragmentation via pre-analyzer collision-induced dissociations. All ions have been assigned to lipid A species wherein the negative charge is localized on the phosphate moiety, and any gas-phase fragmentations are charge-remote fragmentations. Gas-phase decompositions occurring by pre-analyzer CID take the form of dehydrations (losses of H_2O) (24) possibly involving the phosphate moiety (e.g., cyclization). Decomposition of ester-linked acyl side chains may also be occurring via a hydrogen transfer from the departing long chain acyl group to the ester oxygen, thus forming a hydroxyl group. In this scenario, the departing fatty acyl group would be lost either as a ketene, or as a cyclic or unsaturated ketone (24). The above types of gas-phase decomposition have also been proposed to occur in the collision cell of a triple quadrupole mass spectrometer (27). In the latter study, additional gas-phase decomposition pathways were also reported including losses of acyl side chains in the form of neutral carboxylic acids. This elimination of a long chain carboxylic acid left a site of unsaturation (double bond) either on another fatty acyl group originally bound to the departing chain, or on one of the glucosamine rings. Gas-phase cleavage of the glycosidic oxygen was also reported (27) during CID with proposed formation of a carbonyl group at the 1' position.

All peaks observed in negative ion PDMS (Fig. 5a) were also observed in negative ion ESMS (Fig. 5b). Fewer peaks were observable via PDMS, although PDMS is known to impart enough energy in the ionization process to produce some fragmentation (38). For example, it is likely that the oxonium ions observed in Figure 4a represent fragments formed on the mylar PDMS target. Unique to the ES mass spectra are peaks corresponding to doubly charged diphosphoryl species which appear at m/z 906, 898, 793, and 785. Doubly charged negative ions are formed by deprotonation of both acidic phosphoryl sites (see Fig. 2). The relative abundances of doubly and singly charged ions have been found to vary dramatically with the solvent employed for ESMS (23,25). This effect was attributed to the variable solution dielectric which caused different degrees of association/dissociation of anions and positive counterions. Previous reports (23,25) revealed that solvents with high dielectric constants increased the abundances of doubly charged species relative to singly charged species in ES mass spectra. A combination of results obtained by PDMS (17) and later ESMS (24) led to the conclusion that the structure of hepta-acyl lipid A from *E. agglomerans* is that shown in Figure 6. This lipid A from *E. agglomerans* thus resembles the known structure of lipid A from *S. minnesota* (39,40).

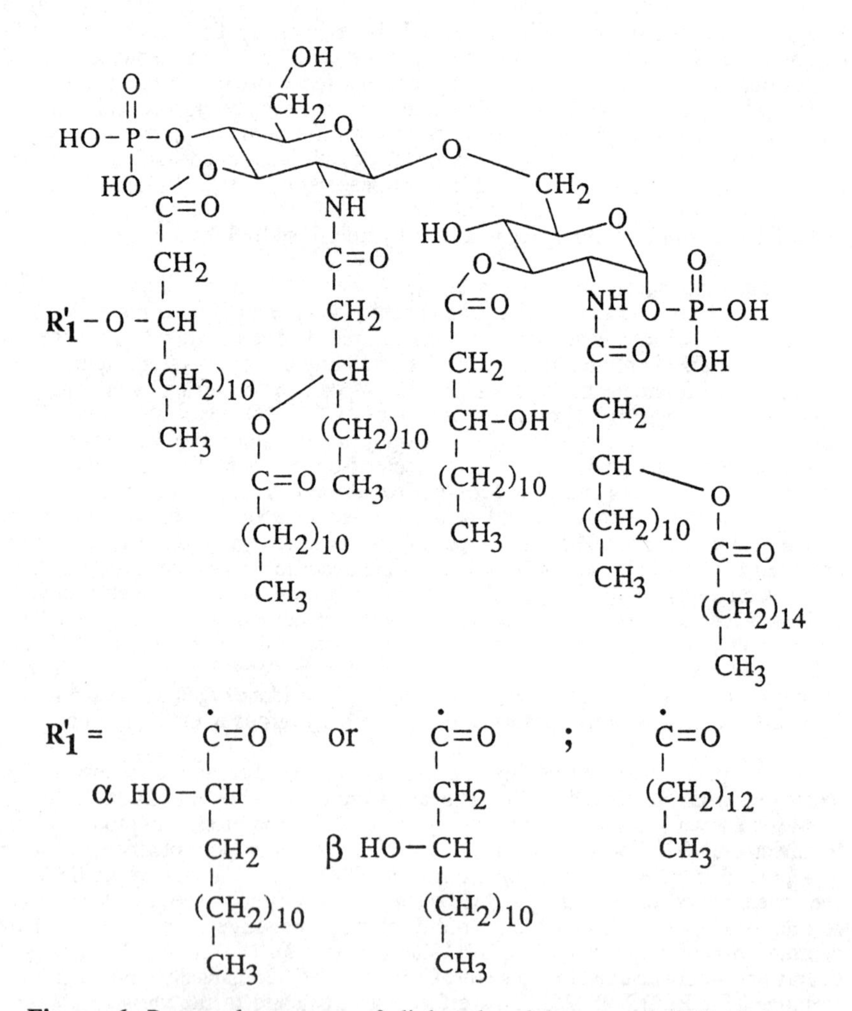

Figure 6. Proposed structure of diphosphoryl hepta-acyl lipid A from *Enterobacter agglomerans*. The R'_1 acyl chain may be present either in the form of a myristoyl (majority, bottom right) or a hydroxymyristoyl (α- or β-hydroxyl, bottom left) group.

Table 1.

Interpretation of negative ion ES mass spectra of lipid A from *E. agglomerans* and *S. minnesota*

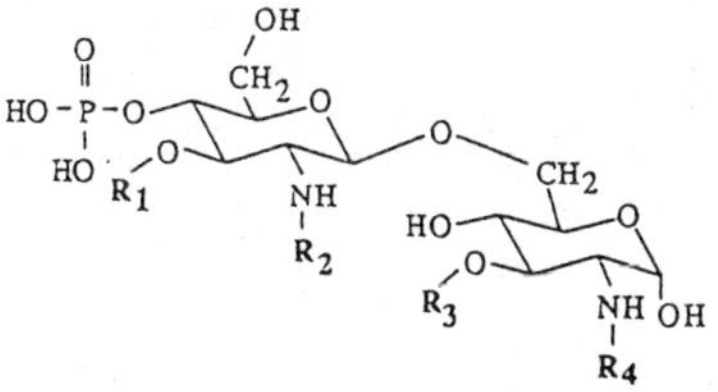

Lipid A backbone with different acyl-linked side chains

m/z	empirical formula	R_1	R_2	R_3	R_4
1972	$[C_{110}H_{206}N_2O_{24}P]^-$	HM-OM·	L-OM·	HM·	P-OM·
1956	$[C_{110}H_{206}N_2O_{23}P]^-$	M-OM·	L-OM·	HM·	P-OM·
1928	$[C_{108}H_{202}N_2O_{23}P]^-$	less C_2H_4 relative to m/z 1956			
1813*	$[C_{94}H_{177}N_2O_{26}P_2]^-$	HM-OM·	L-OM·	HM·	HM·
1797*	$[C_{94}H_{177}N_2O_{25}P_2]^-$	M-OM·	L-OM·	HM·	HM·
1745	$[C_{96}H_{180}N_2O_{22}P]^-$	HM·	L-OM·	HM·	P-OM·
1733	$[C_{94}H_{176}N_2O_{23}P]^-$	HM-OM·	L-OM·	HM·	HM·
1717	$[C_{94}H_{176}N_2O_{22}P]^-$	M-OM·	L-OM·	HM·	HM·
1689	$[C_{92}H_{172}N_2O_{22}P]^-$	less C_2H_4 relative to m/z 1717			
1587*	$[C_{80}H_{151}N_2O_{24}P_2]^-$	HM·	L-OM·	HM·	HM·
1571*	$[C_{80}H_{151}N_2O_{23}P_2]^-$	M-OM·	L-OM·	H·	HM·
1507	$[C_{80}H_{150}N_2O_{21}P]^-$	HM·	L-OM·	HM·	HM·
1491	$[C_{80}H_{150}N_2O_{20}P]^-$	M-OM·	L-OM·	H·	HM·
1361*	$[C_{66}H_{125}N_2O_{22}P_2]^-$	▲HM·	L-OM·	H·	HM·
1281	$[C_{66}H_{124}N_2O_{19}P]^-$	▲HM·	L-OM·	H·	HM·
1263	$[C_{66}H_{122}N_2O_{18}P]^-$	less H_2O relative to m/z 1281			
1245	$[C_{66}H_{120}N_2O_{17}P]^-$	less $2H_2O$ relative to m/z 1281			
1178*	$[C_{54}H_{103}N_2O_{21}P_2]^-$	▲HM·	HM·	H·	HM·
1054	$[C_{52}H_{98}N_2O_{17}P]^-$	H·	L-OM·	H·	HM·
1036	$[C_{52}H_{96}N_2O_{16}P]^-$	less H_2O relative to m/z 1054			
1018	$[C_{52}H_{94}N_2O_{15}P]^-$	less $2H_2O$ relative to m/z 1054			
906*	$[C_{94}H_{177}N_2O_{26}P_2]^{2-}$	HM-OM·	L-OM·	HM·	HM·
898*	$[C_{94}H_{177}N_2O_{25}P_2]^{2-}$	M-OM·	L-OM·	HM·	HM·
793*	$[C_{80}H_{150}N_2O_{24}P_2]^{2-}$	HM·	L-OM·	HM·	HM·
785*	$[C_{80}H_{150}N_2O_{23}P_2]^{2-}$	M-OM·	L-OM·	H·	HM·

Legend

P= palmitoyl ($-C_{16}$); M= myristoyl ($-C_{14}$); HM= hydroxymyristoyl ($-C_{14}OH$); L= lauroyl ($-C_{12}$); OM=oxymyristoyl ($-C_{14}O$); H=hydrogen

* Ions representative of diphosphoryl lipid A; the second phosphoryl group (not shown) is located in the place of the (1-position) axial hydroxyl group on the extreme right of the figure.

▲ HM· could also be located on R_3 instead of R_1

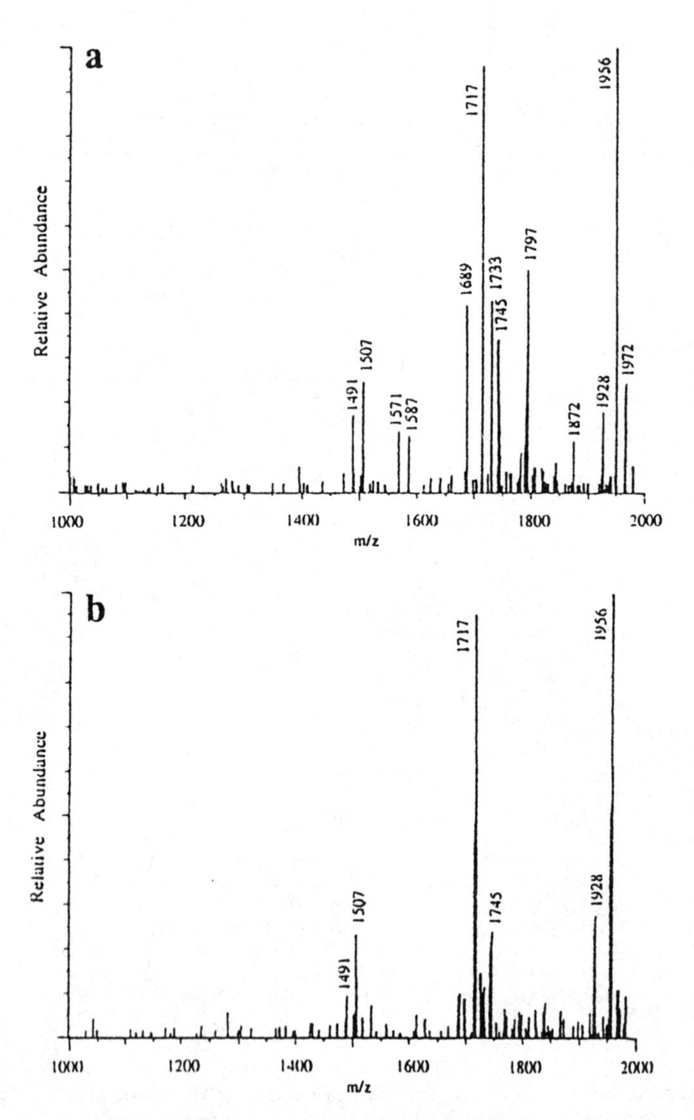

Figure 7. Negative ion ES mass spectra of lipid A obtained under minimal hydrolysis conditions from (a) *Enterobacter agglomerans*; and (b) *Salmonella minnesota*.

A comparison of the analysis times employed, and the signal-to-noise ratios obtained, using electrospray ionization and plasma desorption in both the positive and negative ion modes reveals that sensitivity is better in both cases when ESMS is employed. An improvement in signal-to-noise ratio of 10- to 20-fold was obtained for analysis of dephosphorylated lipid A (positive ion mode). An even larger improvement (100- to 200-fold) was realized for negative ion analyses of monophosphoryl lipid A. It should be emphasized that this comparison was made using typical operating conditions (PDMS and ESMS) which were not rigorously optimized in either case.

Mass resolution for each technique, of course, depends upon the specific mass analyzer employed. In PDMS where a linear time-of-flight mass analyzer was employed, obtained peaks were significantly broader than those obtained via ESMS where a quadrupole mass analyzer was used. For the analysis of lipid A, both analyzers could adequately resolve peaks which were separated by a minimum of 16 m/z units (singly charged ions, excluding peaks attributable to elemental isotopes). The quadrupole analyzer used in this study has an upper mass range limit of m/z 2000 which could preclude the detection of higher mass lipid A species in certain cases. For this study, however, the PD mass spectra show no peaks between m/z 2000 and m/z 7000.

In summarizing the comparison of the utility of PDMS and ESMS for the analysis of lipid A molecules originating from endotoxins, the techniques can be complementary. While ESMS has clear advantages in terms of sensitivity, PDMS can provide structurally-informative fragmentations which may not be observed in ESMS, owing largely to the "soft" nature of the ES technique which inherently produces little fragmentation. However, decompositions can be promoted in ESMS by the use of pre-analyzer collision-induced dissociation. This ability to control the level of decomposition in ESMS has added to the versatility of the technique. Control of fragmentation behavior in PDMS is much less direct. PD time-of-flight instruments often have the advantage of being able to detect singly charged species at much higher m/z values than ES quadrupole instruments.

Acid Hydrolysis of Lipid A

Lipid A is usually prepared via mild acid hydrolysis of LPS to selectively hydrolyze the KDO-glucosamine bond which links the core region to lipid A. To release lipid A from LPS in a gentle manner (i.e., without disrupting other structural features of the molecule), 1% acetic acid (pH=2.8) at 100 °C was chosen as a standard condition for lipid A preparation from *E. agglomerans* (abbreviated as EA) LPS. The same conditions were used to liberate lipid A from *S. minnesota* R595 (abbreviated as SM). As mentioned above and shown in Figure 6, Lipid A (EA) was previously found (17,24) to have a structure similar to that of lipid A (SM). *Salmonella minnesota* hepta-acyl lipid A structures identical to that shown in Figure 6 have been reported (39,40), including the same form of heterogeneity at the R'_1 side chain in one report (39). Lipid As from *E. agglomerans* and *S. minnesota* generated in previous studies (17,24) were obtained from the respective lipopolysaccharides using harsher conditions, (e.g., lower pH, higher temperature, and/or longer treatment times) which provoked decompositions of the lipid A side chains. Figures 7a and 7b show the negative ion ES mass spectra of lipid A from *E. agglomerans* and *S. minnesota*, respectively, obtained under mild conditions (1% acetic acid (pH=2.8) at 100 °C); mass spectral assignments are given in Table 1. These mass spectra contrast significantly with those of comparable lipid As produced under harsher

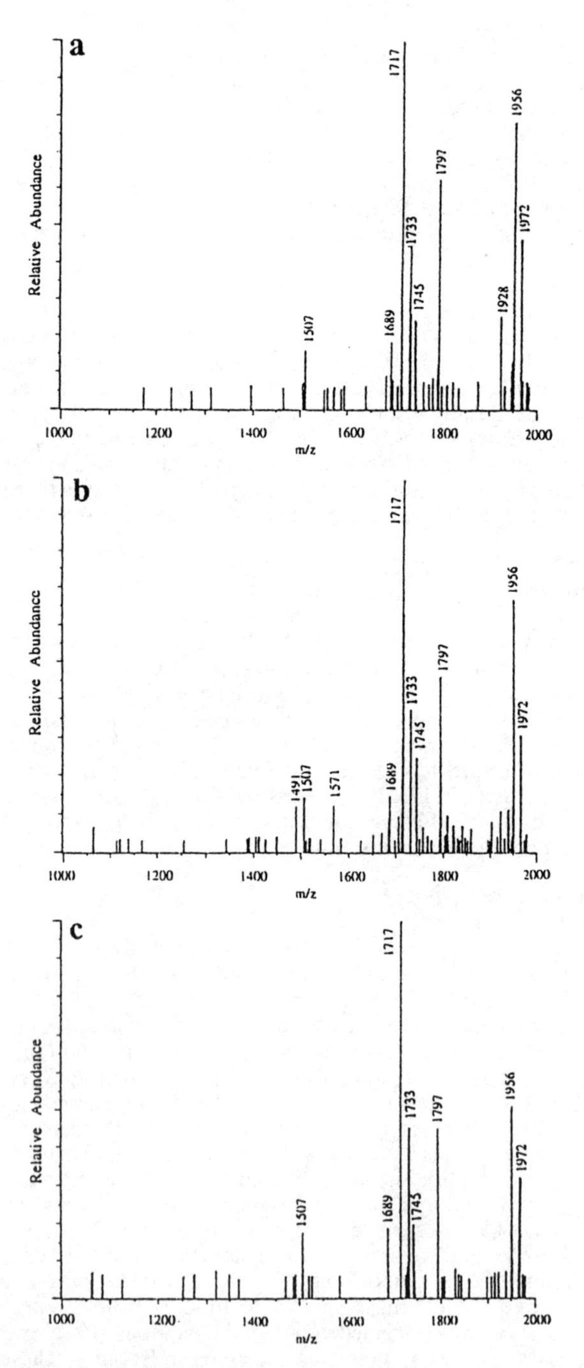

Figure 8. Negative ion ES mass spectra obtained from *Enterobacter agglomerans* lipid A subjected to mild acid treatment (1% acetic acid, 100 °C) for: (a) 30 min; (b) 60 min; and (c) 120 min.

conditions (Fig. 5b) in that a much lower degree of acyl side chain hydrolysis is observed in Figures 7a and 7b.

In comparing the two mass spectra shown in Figs. 7a and 7b, it is evident that the EA lipid A has undergone a higher level of hydrolysis than the SM lipid A. For example, the relative intensities of monophosphoryl mass spectral peaks at m/z 1745, 1507, and 1491 are higher in Fig. 7a than in Fig. 7b. Additional hydrolysis can be attributed to a longer treatment time required to release lipid A (EA) from its LPS (30-35 min) than was needed for lipid A (SM) from its source material (5-8 min). Use of the minimal hydrolysis condition to generate lipid A primarily yields lipid As from *E. agglomerans* and *S. minnesota* in hepta-acyl (m/z 1956) and hexa-acyl (m/z 1717) forms. Raising the acid concentration employed for lipid A preparation produced lipid A in increasingly hydrolyzed forms (causing loss of progressively larger numbers of acyl chains) (41). In addition, prolonging exposure (even under mild conditions: 1% acetic acid at 100 °C) will also progressively diminish the relative intensity of the fully acylated (hepta-acyl) m/z 1956 peak forming, in large part, the hexa-acyl counterpart corresponding to loss of the palmitoyl group at R'_4.

Negative ion electrospray ionization mass spectra corresponding to lipid A (EA) and lipid A (SM) which have undergone exposure to the mild acid conditions for different lengths of time are shown in Figs. 8 and 9, respectively. A pseudo-first order (excess acid) rate constant for the hydrolysis of the R'_4 palmitoyl group was determined to be 3.3×10^{-3} min^{-1} for *E. agglomerans*, and 8.4×10^{-3} min^{-1} for *S. minnesota* at 100 °C (41). The existence of diphosphoryl lipid A components in lipid A (EA) which can undergo competitive hydrolyses may be contributing to the difference in the k' values for lipid A (EA) and lipid A (SM). It is worth noting that the diphosphoryl lipid A peak at m/z 1797 corresponding to the hexa-acyl species which has lost the palmitoyl group at R'_4 (Table 1) is the most prominent diphosphoryl peak present in Figure 8a, b, and c. This serves as evidence to establish that even in the presence of a second phosphoryl group at the 1 position (rather near to the R'_4 palmitoyl group, Figure 2), the palmitoyl group at R'_4 is still the acyl side chain most susceptible to acid hydrolysis. This is significant because the native LPS will contain a phosphoryl group at the 1 position. Thus, the presence of this added phosphoryl group does not alter the fact that, relative to the other acyl chains, the R'_4 palmitoyl chain is the most readily hydrolyzed in the presence of acid.

Base Hydrolysis of Lipid A

Ethanolic base treatment (0.05 N NaOH in 95% EtOH) of *Enterobacter agglomerans* LPS was found by Domelsmith and coworkers (35,42) to reduce pulmonary toxicity more satisfactorily than other solvent systems (e.g. HCl, H_3PO_4, K_2CO_3, DMSO (dimethylsulfoxide), DMF (dimethylformamide), and DEG (diethyleneglycol), also in 95% EtOH). In order to preserve inorganic salts in the cotton fiber, nonaqueous solutions were employed in those studies geared toward deactivating bacterial endotoxins in cotton dust. Such salts may improve the ability to process cotton fibers by helping to dissipate static charges (43,44). Prior to base treatment in our study, lipid A "starting material" was obtained from LPS (EA) via the mild acid hydrolysis conditions described above. Base treatment was initiated by dissolving the lipid A sample in 500 µl of $CHCl_3$, and combining it with 500 µl of 0.05 M NaOH in 95% EtOH. After treatment times of various durations, samples run in parallel were cooled and neutralized by addition of 500 µl of 1% acetic acid. The post-treatment products were extracted and analyzed by ESMS.

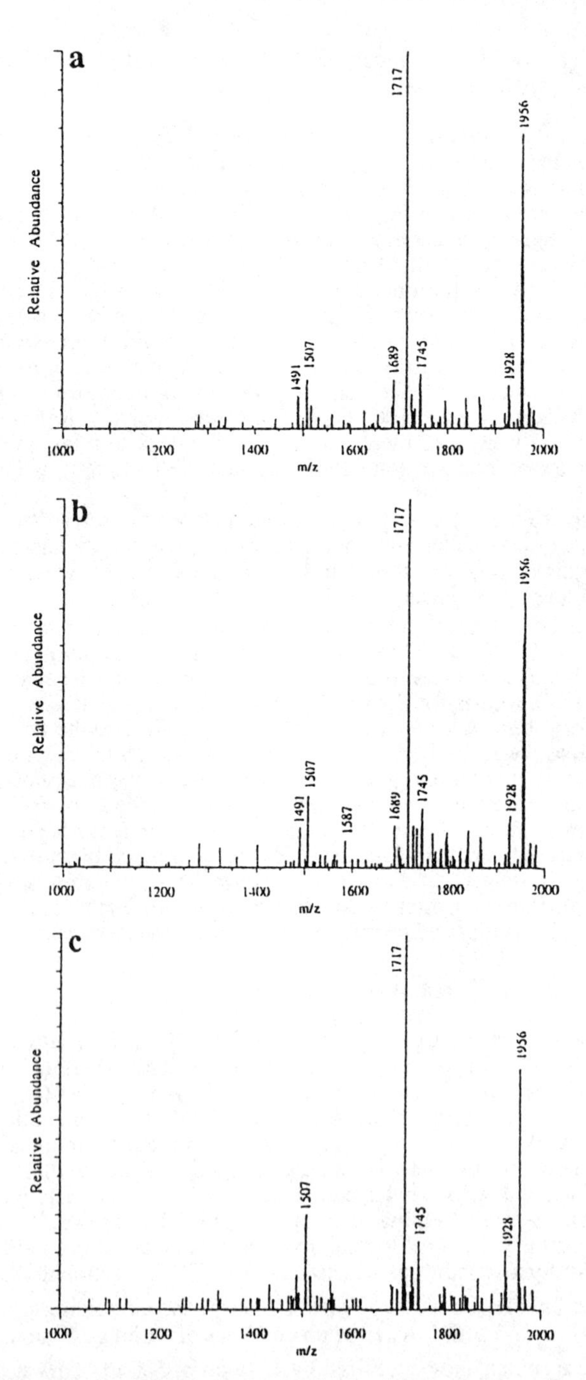

Figure 9. Negative ion ES mass spectra obtained from *Salmonella minnesota* lipid A subjected to mild acid treatment (1% acetic acid, 100 °C) for: (a) 10 min; (b) 20 min; and (c) 30 min.

ES mass spectra of the products of 0, 15, 30, and 60 min base treatment appear in Fig. 10; assignments for observed peaks appear in Table 1. These mass spectra clearly reflect changes attributable to progressive base hydrolysis of lipid A (EA). Base-induced hydrolysis of ester-linked acyl chains from the parent monophosphoryl hepta-acyl or hexa-acyl lipid A (EA) yields ions which appear at m/z 1507, 1491, 1281, and 1054. The lipid A starting material (Fig. 10a) contained some diphosphoryl hexa-acyl lipid A (m/z 1797) which was apparently rather resistant to breakdown (e.g., see Fig. 10b,c), but eventually underwent decomposition after 60 min of base treatment (Fig. 10d). The peaks at m/z 1587, 1571, 1361, and 1178 were formed by base hydrolysis of various ester-linked acyl chains from parent diphosphoryl lipid A (Table 1). Under the employed conditions, base hydrolysis of ester side chains proceeded much faster than acid hydrolysis.

Under more lengthy base treatment conditions (e.g., one hour), ions corresponding to intact lipid A species which had lost only acyl side chains were no longer observed; in fact, no ions above m/z 900 were detected (Fig. 10d). It is possible that the lengthy one hour treatment resulted in a breakage of the disaccharide lipid A backbone wherein the ether linkage between the distal portion and the reducing end was ruptured (41). It is also possible that the phosphate group itself (bearing the charge) had been hydrolyzed from the rest of the molecule. In this case, lipid A decomposition products (now dephosphorylated) would no longer be detectable by negative ion ESMS.

Conclusion

This report has summarized several aspects of the utility and effectiveness of electrospray ionization for the investigation of lipid A species originating from bacterial endotoxins. Under the employed experimental conditions, ESMS has proven to be more than one order of magnitude more sensitive in the positive ion mode (dephosphorylated lipid A analysis) and more than two orders of magnitude more sensitive in the negative ion mode (monophosphoryl lipid A analysis) as compared to PDMS. The formation of distal adduct ions in ESMS, and the formation of oxonium ions in PDMS, provide information which is extremely useful in elucidating the structures of unknown lipid A forms. The extreme "softness" of the ESMS technique is advantageous for direct determinations of the composition of heterogenous mixtures such as crude lipid A provided that signal suppression from competing analytes (45,46) is not too severe. Under appropriate experimental conditions, the contribution of gas-phase fragmentations to ES mass spectra is minimal. Structural information can be augmented by promoting decompositions via the use of collision-induced dissociation. Although not pursued here, the solution introduction feature makes the electrospray technique particularly suited to coupling with solution-based separations techniques such as liquid chromatography or capillary electrophoresis. Due to the higher energy imparted to analyte species in PDMS, without prior chromatographic separation, gas-phase fragmentation products (e.g., acyl side chain losses) may be indistinguishable from identical ions formed via solution-phase hydrolysis.

The ability to observe intact species by ESMS has been exploited to monitor the products of lipid A hydrolysis upon acid and base treatment. Among several possible detoxification approaches, acid and base treatments of LPS are known to be effective at reducing the 'endotoxin level' (43) as measured by the *Limulus* amebocyte lysate (LAL) assay. ES mass spectra reveal that during acid hydrolysis of lipid A, ester-linked side chains were progressively removed. The palmitoyl group at R'_4 was

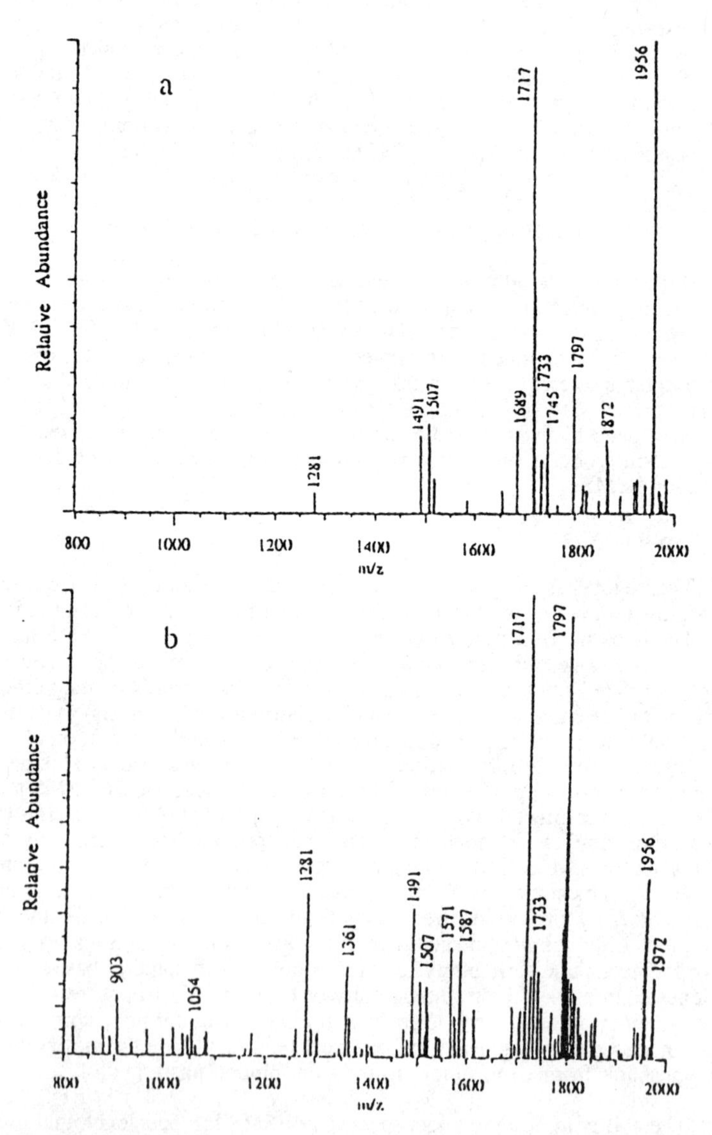

Figure 10. Negative ion ES mass spectra obtained from *Enterobacter agglomerans* lipid A subjected to base treatment (0.05 N NaOH in 95% ethanol, 65 °C) for: (a) 0 min; (b) 15 min; (c) 30 min; and (d) 60 min.

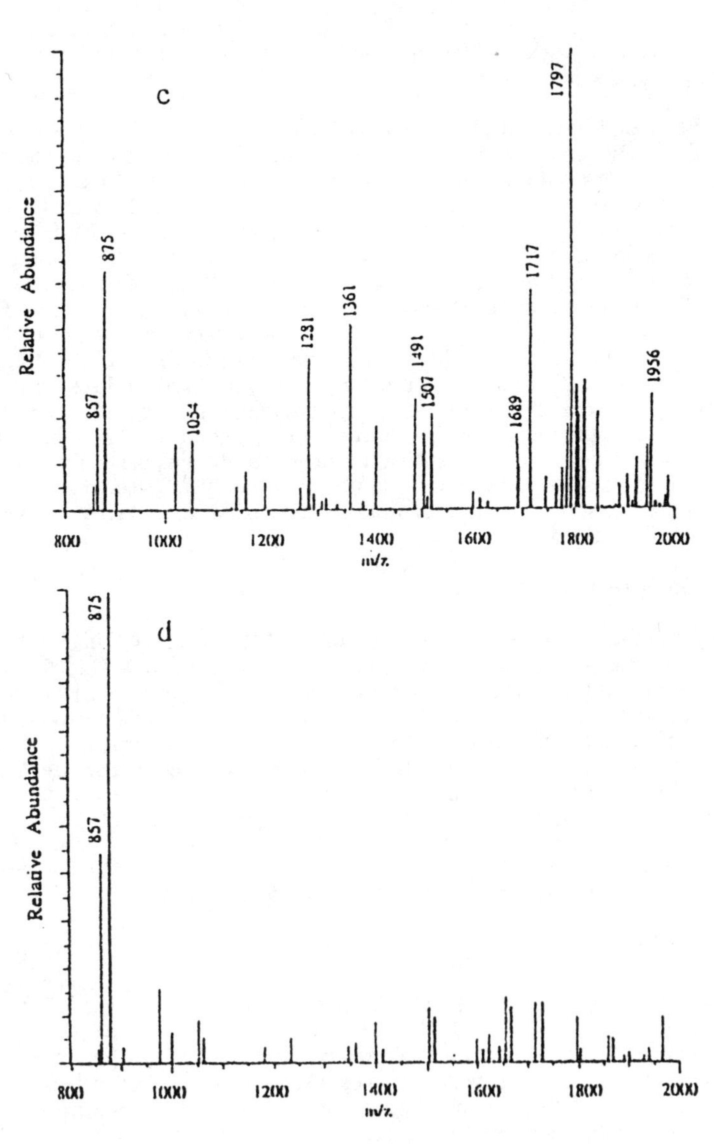

Figure 10. *Continued*

found to be the most readily removed and its hydrolysis in the presence of excess acid conformed to pseudo-first order reaction kinetics. An approximate order of acid hydrolysis of the ester side chains was determined (41) to be: R'_4 (palmitoyl) > R'_1 (myristoyl or hydroxymyristoyl) > R_3 (myristoyl group at position 3) > R_1 (oxymyristoyl group at position 3') > R'_2 (lauroyl). Upon acid treatment, it is likely that the mechanism responsible for diminution of the 'endotoxin level' is the removal of the acyl linked side chains.

The use of base hydrolysis (0.05 N NaOH in 95% ethanol, 65 °C) was even more effective at removing acyl side chains. This treatment was also consistently effective in lowering the endotoxin level, as measured by the LAL assay, which serves as additional evidence that the removal of side chains is the underlying factor responsible for lowering of the endotoxin level. This base treatment of LPS (EA) was also effective in reducing the pulmonary toxicity of cotton dust, as monitored by changes in respiratory frequency and tidal volume in a guinea pig animal model (35,42,47). The mechanism for eliminating the pulmonary toxicity from LPS (EA) via base treatment likely goes beyond hydrolysis of the acyl side chains, possibly involving breakage of the glycosidic linkage of the disaccharide backbone of lipid A to produce a monoglucosamine structure. Monoglucosamines are often considerably less toxic than their parent diglucosamine structures (2). For example, lipid X, which is the phosphorylated form of the reducing end monoglucosamine (48), is known to be nontoxic. Breakage of the glycosidic linkage of lipid A may be responsible for the elimination of pulmonary toxicity in base treated samples, however, the minimum structure of lipid A responsible for toxicity has yet to be unambiguously defined.

Acknowledgments

The author gratefully acknowledges the vital participation of his former group members Mr. Yan Wang in all of the hydrolysis studies, and Dr. A. Kamel Harrata in the early structural characterization work employing ES mass spectrometry. The invaluable assistance of Dr. Linda N. Domelsmith was an essential component to the early structural characterization work. Dr. Anthony J. DeLucca of the Southern Regional Research Center of the U.S.D.A. maintained the *Enterobacter agglomerans* culture and kindly provided the EA LPS used in all experimental work. Ms. Connie M. David supplied valuable assistance in acquiring the displayed PD mass spectra obtained in the laboratory of Professor Roger A. Laine at Louisiana State University. The support of Dr. Mary-Alice Rousselle and Dr. William Franklin of the Southern Regional Research Center of the U.S.D.A. is also greatly appreciated.

Literature Cited

1. Galanos, C.; Roppel, J.; Weckesser, J.; Rietschel, E.T.; Mayer, H., *Infect. Immunity*, **1977**, 16, 407-412.
2. Rietschel, E.T.; Brade, H., *Scientific American*, **1992**, August, pp. 54-61.
3. Luderitz, O.; Freudenberg, M.A.; Galanos, C.; Lehmann, V.; Rietschel, E.T.; Shaw, D.H., **1982**, Microbial Membrane Lipids, Vol. 17 (S. Razin and S. Rottem, Eds.) Academic Press, New York, pp. 79-151.
4. Homma, J.Y.; Tanamoto, K., **1984**, Bacterial Endotoxins: Chemical Biological, and Clinical Aspects (Homma, J.Y.; Kanegasaki, S.; Luderitz, O.; Shiba, T.; Westphal, O., Eds.) Verlag Chemie, Weinheim, pp. 159-172.

5. Shiba, T.; Kusumoto, S., **1984**, Handbook of Endotoxin, Vol. 1: Chemistry of Endotoxin, (Rietschel, E.T., ed.), Elsevier Science Publishers, Amsterdam, pp. 284-307.
6. Yasuda, T.; Kanegasaki, S.; Tsumita, T.; Tadakuma, T.; Homma, J.Y.; Inage, M.; Kusumoto, S.; Shiba, T., *European J. Biochem.*, **1982**, 124, 405-407.
7. Caroff, M.; Deprun, C.; Karibian, D.; Szabo, L., *J. Biol. Chem.*, **1991**, 266, 18543-18549.
8. John, C.M.; McLeod Griffiss, J.; Apicella, M.A.; Mandrell, R.E.; Gibson, B.W., *J. Biol. Chem.*, **1991**, 266, 19303-19311.
9. Rietschel, E.T.; Brade, L.; Lindner, B.; Zahringer, U., **1992**, Bacterial Endotoxic Lipopolysacharides, Vol.1 (Morrison, D.C. and Ryan, J.L., Eds.) CRC Press, pp. 3-43.
10. Qureshi, N.; Takayama, K.; Ribi, E., *J. Biol. Chem.*, **1982**, 257,11808-11815.
11. Qureshi, N.; Takayama, K.; Heller, D.; Fenselau, C., *J. Biol. Chem.*, **1983**, 258, 12947-12951.
12. Seydel, U.; Lindner, B., **1983**, Springer Ser. Chem. Phys., 25, 240-244.
13. Seydel, U.; Lindner, B.; Wollenweber, H.-W.; Rietschel, E.T., *European J. Biochem.*, **1984**, 145, 505-509.
14. Cotter, R.J.; Honovich, J.P.; Qureshi, N.; Takayama, K., *Biomed. Environ. Mass Spectrom.*, **1987**, 14, 591-598.
15. Karibian, D.; Deprun, C.; Szabo, L.; LeBeyec, Y.; Caroff, M., *Int. J. Mass Spectrom. Ion Proc.*, **1991**, 111, 273-286.
16. Cotter, R.J., *Anal. Chem.*, **1988**, 60, 781A-793A.
17. Cole, R.B.; Domelsmith, L.N.; David, C.M.; Laine, R.A.; Delucca, A.J., *Rapid Commun. Mass Spectrom.*, **1992**, 6, 616-622.
18. Wang, R.; Chen, L.; Cotter, R.J.; Qureshi, N.; Takayama, K.J., *Microbiol. Methods*, **1992**, 15, 151-166.
19. Fenn, J.B.; Mann, M.; Meng, C.K.; Wong, S.F.; Whitehouse, C.M. *Mass Spectrom. Revs.*, **1990**, 9, 37-70.
20. Smith, R.D.; Loo, J.A.; Ogorzalek Loo, R.R.; Busman, M., Udseth, H.R. *Mass Spectrom. Revs.* **1991**, 10, 359-451.
21. Harrata, A.K.; Domelsmith, L.N.; and Cole, R.B., **1991**, Book of Abstracts, 12th Int. Mass Spectrom. Conf. (26-30 August 1991, Amsterdam, Holland) p. 287.
22. Cole, R.B.; Harrata, A.K.; Domelsmith, L.N., **1992**, Proceedings of the 40th ASMS Conf. on Mass Spectrom. and Allied Topics, (Washington, D.C. May 31 - June 5, 1992) pp. 1647-1648.
23. Cole, R.B; Harrata, A.K., *Rapid Commun. Mass Spectrom.*, **1992**, 6, 536-539.
24. Harrata, A.K.; Domelsmith, L.N.; and Cole, R.B., *Biol. Mass Spectrom.*, **1993**, 22, 59-67.
25. Cole, R.B.; Harrata, A.K., *J. Am. Soc. Mass Spectrom.*, **1993**, 4, 546-556.
26. Harrata, A.K.; Domelsmith, L.N.; and Cole, R.B., **1993**, Cotton Dust Proceedings - 17th Cotton Dust Res. Conf. (R.R. Jacobs; P.J. Wakelyn; L.N. Domelsmith, Eds.) National Cotton Council, Memphis, TN, pp. 310-313.
27. Chan, S. and Reinhold, V.N., *Anal. Biochem.*, **1994**, 218, 63-73.
28. Honeybourne, D., Wales, D.S., Watson, A., Lee, W.R., and Sager, B.F., **1982**, Byssinosis-causative Agent and Clinical Aspects (A Critical Literature Research Review), Shirley Institute Publication S.43, England.
29. Ellakkani, M.A.; Alarie, Y.C.; Weyel, D.A.; Mazumdar, S.; and Karol, M.H., *Toxicol. Appl. Pharmacol.*, **1984**, 74, 267-284.
30. Ryan, L.K. and Karol, M.H., *Am. Rev. Respir. Dis.*, **1989**, 140, 1429-1435.

31. Griffiths-Johnson, D.; Ryan, L.K.; Spear, K.L.; and Karol, M.H., **1989**, Cotton Dust Proceedings - 13th Cotton Dust Res. Conf. (R.R. Jacobs; P.J. Wakelyn, Eds.) National Cotton Council, Memphis, TN, pp. 101-104.
32. Galanos, C.; Luderitz, O.; and Westphal, O., *European J. Biochem.*, **1969**, 9, 245-249.
33. Kato, N.; Ohta, M.; Kido, N.; Ito, H.; Naito, S.; and Kuno, T., *J. Bacteriol.*, **1985**, 162, 1142.
34. Mayer, H.; Bock, E.; and Weckesser, J. *FEMS Microbiol. Lett.* **1983**, 17, 93-96.
35. Domelsmith, L. N.; DeLucca, A. J.; and Fischer, J. J., **1990**, Cotton Dust Proceedings - 14th Cotton Dust Res. Conf. (R.R. Jacobs; P.J. Wakelyn; L.N. Domelsmith, Eds.) National Cotton Council, Memphis, TN, pp. 13-19.
36. Krauss, J. H.; Seydel, U.; Weckesser, J.; Mayer, H. *European J. Biochem.* **1989**, 180, 519.
37. Dodonov, A.F.; I.V. Chernushevich; V.V. Laiko, **1994**, "Electrospray Ionization on a Reflecting Time-of-Flight Mass Spectrometer" In: R.J. Cotter, Ed. Time-of-Flight Mass Spectrometry, American Chemical Society Symposium Series 549, Washington, D.C., pp. 108-123.
38. Blais, J.-C.; Viari, A.; Cole, R.B.; and Tabet, J.-C., *Int. J. Mass Spectrom. Ion Proc.*, **1990**, 98, 155-166.
39. Rietschel, E.T.; Wollenweber, H.-W.; Brade, H.; Zahringer, U.; Lindner, B., Seydel, U.; Bradaczek, H.; Barnickel, G.; Labischinski, H.; and Giesbrecht, P., **1984**, Rietschel, E.T., Ed., Handbook of Endoxin, Vol. 1, Chemistry of Endotoxin, Elsevier Science Publishers, NY, pp. 187-220.
40. Qureshi, N.; Mascagni, P.; Ribi, E.; Takayama, K. *J. Biol. Chem.*, **1985**, 260, 5271-5278.
41. Wang, Y. and Cole, R.B., submitted for publication.
42. Domelsmith, L.N. and Fischer, J.J., **1989**, Cotton Dust Proceedings - 13th Cotton Dust Res. Conf. (R.R. Jacobs; P.J. Wakelyn, Eds.) National Cotton Council, Memphis, TN, pp. 97-100.
43. Domelsmith, L.N. and Rousselle, M.A., **1987**, Cotton Dust Proceedings - 11th Cotton Dust Res. Conf. (R.R. Jacobs; P.J. Wakelyn, Eds.) National Cotton Council, Memphis, TN, pp. 105-108.
44. Domelsmith, L.N. and Rousselle. M.A., **1988**, Cotton Dust Proceedings - 12th Cotton Dust Res. Conf. (R.R. Jacobs; P.J. Wakelyn, Eds.) National Cotton Council, Memphis, TN, pp. 76-79.
45. Tang, L. and Kebarle, P., *Anal. Chem.*, **1991**, 63, 2709-2715.
46. Wang, G. and Cole, R.B., *Anal. Chem.*, **1994**, 66, 3702-3708.
47. Karol, M.H.; Ogundiran, A.; Gatty, C.; Millner, P.; Domelsmith, L.N.; Rousselle, M.A., **1987**, Cotton Dust Proceedings - 11th Cotton Dust Res. Conf. (R.R. Jacobs; P.J. Wakelyn, Eds.) National Cotton Council, Memphis, TN, pp. 116-118.
48. Takayama, K.; Qureshi, N.; Mascagni, P.; Nashed, M.A.; Anderson, L.; and Raetz, C.R.H., *J. Biol. Chem.*, **1983**, 258, 7379-7385.

RECEIVED November 13, 1995

Chapter 10

Microcolumn Liquid Chromatography–Electrospray Ionization Tandem Mass Spectrometry

Analysis of Immunological Samples

Ashley L. McCormack[1], Jimmy K. Eng[1], Paul C. DeRoos[2], Alexander Y. Rudensky[2], and John R. Yates, III[1]

[1]Department of Molecular Biotechnology and [2]Department of Immunology, University of Washington, Seattle, WA 98195

Coupling separation by micro-column reverse-phase liquid chromatography with direct structural analysis by tandem mass spectrometry is a powerful strategy for the determination of the amino acid sequences of peptides in mixtures of peptides associated with class I and II major histocompatibility complexes (MHC). Methods to sequence peptides presented by two MHC class II alleles associated with rheumatoid arthritis and by a MHC class II presentation mutant are described. Results of the analysis of peptide components in subcellular fractions are presented. Automated acquisition of tandem mass spectra combined with amino acid and nucleotide database searching, using uninterpreted spectra, is described.

In the past several years tandem quadrupole mass spectrometry has made an impact in many areas of biology, especially immunology and the study of antigen presentation and processing. Micro-column reverse-phase liquid chromatography electrospray ionization tandem mass spectrometry (μLC-MS/MS) is well suited to the analysis of highly complex mixtures, including mixtures of peptides associated with class I and II major histocompatibility complexes (MHC) (*1-15*).

The study of the presentation of peptide antigens by MHC molecules is complicated by the vast number of proteins and peptides processed and presented on the surface of cells. MHC class I molecules present peptides synthesized in the cell and stimulate cytotoxic T cell responses to kill the infected cell. Estimates of more than 2,000 different peptides associated with each of the human class I alleles, HLA-A2.1 and HLA-B7, and the murine class II allele, I-A^{d}, have been given (*16*). Further complicating the analysis is the fact that peptides presented by class I MHC molecules are usually 8-10 residues in length. Isolation and fractionation of the peptides involve immunoaffinity purification of the MHC molecules, acid extraction of the bound peptides and reverse-phase high performance liquid chromatography (HPLC) of the complex mixture. After several stages of HPLC, each fraction often contains many peptides. Pool sequencing of mixtures by Edman degradation has been successfully used to identify conserved residues in peptides bound to specific

0097–6156/95/0619–0207$12.00/0

MHC class I alleles (*17*). Hunt and coworkers were the first to use μLC-MS/MS to sequence individual antigens and to identify allele specific binding motifs (*1,2*). Antigens naturally presented by several class I alleles, HLA-A1, HLA-A2.1, HLA-A3, HLA-A11, HLA-A24, and HLA-B7, have been sequenced using μLC-MS/MS (*1, 5, 7, 11*).

To identify class I antigens recognized by specific T cells, Hunt and coworkers coupled cytotoxic T lymphocyte (CTL) based assays and detection to μLC-MS/MS (*6, 9*). By splitting the flow eluting from the μLC to collect small amounts of material in the wells of a microtiter plate, they detected peptides using immunoreactivity (*9*). The identification of antigens with immunoreactivity to specific CTL's requires associating activity with peptide m/z values and then obtaining sequence information for that peptide. The sequences of an antigen recognized by specific CTL's with reactivity against melanoma tumor cells was identified in this manner (*9*).

In contrast to class I proteins, MHC class II molecules present peptides from proteins found in vesicles of the endosomal pathway and stimulate cytokine secretion by helper T cells. Class II MHC molecules present peptides that are 10-30 residues in length with heterogeneity frequently present in the processing of the N-terminus and C-terminus of the peptides (*16*). This is a result of the relaxed binding specificity allowed by the open ends of the binding groove in class II molecules (*16*). Consequently, pool sequencing of HPLC fractionated mixtures by Edman degradation is much more complicated. However, using a combination of mass spectrometry and Edman degradation has been successful in obtaining the amino acid sequence of individual peptides (14, 18). Individual class II antigens have been sequenced using μLC-MS/MS, including bound peptides isolated from murine I-A^d and from human DR4 molecules associated with rheumatoid arthritis (*3, 14*).

The processing of MHC-peptide complexes has been studied extensively; however the exact mechanisms are unknown. Class I subunits, heavy chain and β2-microglobulin, complex and bind peptides in the endoplasmic reticulum (ER) (*19*). Peptides are generated in the cytosol by proteases and are then transported to the endoplasmic reticulum where class I molecules select the peptides with the proper binding motif (*19*). In contrast, the alpha and beta subunits of class II molecules complex in the endoplasmic reticulum with invariant chain and move through the golgi complex into the endosomal pathway, before binding peptides (*19*), Figure 1. Invariant chain is believed to prevent binding of peptides in the endoplasmic reticulum. Within the endocytic vesicles, invariant chain is enzymatically cleaved which allows the class II molecules to complex antigenic peptides. Finally, the MHC-peptide complex moves through the endosomal pathway to the surface.

Studies of processing pathways require less specific information in relation to bioactivity. These studies involve obtaining information about the size, nature and sequence of random antigens present at various stages in the pathway. In addition, processing pathways can be perturbed by the intentional creation of mutations in molecules involved in processing and the resulting phenotypes identified. Studies to characterize the consequences of these phenotypes at the antigen level lead to a more complete understanding of the processing pathway. These studies benefit from rapid methods to survey the population of antigens at various stages of the processing pathway as well as from methods for the rapid interpretation of the data. By using μLC-MS/MS, bound peptides isolated from the class I allele, A2.1, of a mutant cell line were sequenced and the results suggest an additional pathway for processing MHC class I antigens (2). In addition, peptides isolated from class II DR3 molecules of two different presentation mutant cell lines have been sequenced (*4, 15*).

In this paper we present methods to study several problems in immunology, including sequencing peptides complexed to class II alleles associated with rheumatoid arthritis (RA), characterizing peptide components of subcellular fractions, and sequencing peptides associated with a class II presentation mutant.

Experimental Section

Isolation of peptides bound to class II alleles associated with RA. Peptide mixtures isolated from MHC alleles associated with rheumatoid arthritis were generously provided by Professor Michael P. Davey of the Oregon Health Sciences University and Department of Veteran Affairs Medical Center, Portland OR. Materials for the generation of MHC peptides were obtained from the following sources. N-[2-Hydroxyethyl]piperazine-N'-[2-ethanesulfonic acid] (HEPES) buffer, L-glutamine, fetal calf serum, RPMI 1640 media, and protein A-agarose were purchased from Gibco BRL (Gaithersburg, MD). Centricon 10 membrane filters were obtained from Amicon (Danvers, MA). CNBr activated Sepharose was obtained from Pharmacia LKB (Piscataway, NJ). Dulbecco's phosphate buffered saline (PBS), Nonidet P-40 (NP-40), and n-octyl-D-glucopyranoside were purchased from Sigma Chemical Co. (St. Louis, MO). Phenylmethylsulfonyl fluoride (PMSF), 1,10-phenanthroline, pepstatin A, and diethylamine were obtained from ICN (Irvine, CA). The antibodies, L243 (anti-DR, IgG2a) and OKT3 (anti-CD3, IgG2a), are prepared from hybridomas obtained from American Type Tissue Culture.

The cell culture and isolation procedures have been described (*13*). Briefly, Epstein-Barr virus (EBV) transformed cells homozygous for HLA-DRB1*0401 and HLA-DRB1*0404 were cultured in RPMI 1640 with 10-15 mM HEPES, 2 mM L-glutamine, and 5% fetal bovine serum (*20*). Cells were harvested by centrifugation and re-suspended in lysis buffer, PBS, 1% NP-40, 1 mM PMSF, 0.26 mg/mL 1,10-phenathroline, 50 μg/mL pepstatin A, 2 mg/mL ethylenediaminetetraacetic acid (EDTA). Lysate was immediately frozen at -20°C.

Frozen lysate from approximately 1 x 10^{10} cells was placed in a 37 °C water bath until 50% thawed. Large debris was removed by centrifugation at 2000 x g for 10 minutes; lysate was further clarified by ultra-centrifugation at 150,000 x g for 30 minutes (*17*). DR/peptide complexes were purified as described by Buus et al. with modifications (*21*). The supernatant was directly loaded onto the antibody columns equilibrated with PBS, 1% NP-40. The columns were configured in series in the following order: Tris (general non-specific), OKT3 (isotype matched, nonspecific) and L243 (DR specific). The columns were washed with a minimum of 10 volumes PBS, 1% NP-40, followed by at least 5 volumes of PBS, 0.5% NP-40, 0.1% sodium dodecyl sulfate (SDS). The L243 column was removed from the series and was washed with 5 or more volumes of PBS, 1% n-octyl β-D-glucopyranoside. The DR/peptide complexes were eluted with 0.5 M NaCl, 0.15 M diethylamine, and n-octyl β-D-glucopyranoside at pH 10.5 into 2 mL fractions. The fractions containing DR/peptide were detected by BCA assay performed in a microtiter plate. A total of 2.1 mg of DR molecules were isolated from 1 x 10^{10} cells.

All fractions containing DR/peptide were pooled and concentrated in a Centricon 10 to a final volume of 200-300 μL. Acetic acid (2.5 M) was added to the concentrated DR/peptide mixture in the Centricon 10 and mixed thoroughly with a pipette (*22*). The acid-eluted peptides were separated from the α and β chains of the DR molecule by filtration through the original Centricon 10 (*3*).

The peptide filtrate was fractionated by reverse-phase HPLC using a Vydac 2.1 x 250 mm C18 column. Fractions were concentrated to approximately 25 µl and were frozen at -20 °C. Fractions from HLA DRB1*0401 were analyzed by automated Edman degradation, using standard protocol, and mass spectometry. Fractions from HLA DRB1*0404 were analyzed by mass spectrometry. Approximately 1-5 µl aliquots from each fraction were used for analysis by mass spectrometry.

Isolation of peptides from subcellular fractions. Cell culture materials, including RPMI 1640, L-glutamine, HEPES, antibiotics, and fetal calf serum were purchased from Gibco. Lysis buffer materials, including Tris, NP-40, EDTA, iodoacetamide, PMSF, pepstatin, TLCK, aprotinin, and leupeptin were purchased from Sigma.

Cell culture and subcellular fractionation procedures have been described previously (*23, 24*). Briefly, murine cells, LB27.4, were maintained in RPMI 1640, supplemented with 200 mM L-glutamine, 10 mM HEPES, antibiotics, and 5% fetal calf serum. Cells were harvested at $1x10^6$ cells/mL. Cells were washed extensively with PBS. ^{125}I-transferrin was added and the cells were allowed to sit at 37°C for ten minutes. The cells were put on ice to slow endocytosis and washed several times. The cells were lysed in 10 mM Tris buffer (pH 7.4), containing 150 mM NaCL, 1% NP-40, 5 mM EDTA, 50 mM iodoacetamide, 1 mM PMSF, 2µg/ml pepstatin, 2µg/ml Nα-p-tosyl-L-lysine chloromethyl ketone (TLCK), 3U/ml aprotinin, and 20µg/ml leupeptin, using a homogenizer. The cellular material was centrifuged for 10 minutes at 2000 rpm and for 15 minutes at 3000 rpm to remove nuclei and mitochondria. The lysed material was added to the top of a sucrose column(0.4 M to 1.7 M) which was made with a gradient maker. The column was spun overnight at 25,000 rpm. The material was then fractionated. The fractions were assayed for known proteins and the fractions were pooled. After several washes the vesicles were lysed with acetic acid and the mixtures were spun through a Centricon membrane filter with a 10,000 dalton cut-off. The pooled fractions were also fractionated using a Centricon membrane filter, with a 3000 dalton cut-off, to separate high molecular weight material from low molecular weight material. Fractions were concentrated. Aliquots were loaded onto the µLC column and were washed for as much as 30 minutes with 100% A(0.5% (v/v) acetic acid), prior to analysis by automated µLC-MS/MS.

Isolation of peptides bound to class II molecules from mutant cells. Peptide mixtures were generously provided by Professor Donald Pious of the University of Washington. Cell culture materials, including RPMI 1640, L-glutamine, HEPES, and fetal calf serum were purchased from Gibco. Protein A-sepharose CL-4B and sepharose CL-4B were purchased from Pharmacia LKB.

The procedures have been described (*15*). Briefly, cell lines 8.1.6 and 9.5.3 were derived from EBV transformed cells by ethyl methane sulfate mutagenesis and immunoselection. Cells were grown in RPMI 1640 supplemented with 2 mM L-glutamine, 20 mM HEPES, and 10% fetal calf serum. $6x10^8$ cells were incubated in 0.15 M citrate/phosphate buffer pH 4 or Dulbecco phosphate buffered saline pH 7 at 37°C for 16 hr in the absence of exogenous peptide. After neutralization, cells were lysed in 10 mM Tris buffer (pH 6.8), containing 150 mM NaCl, 0.5% NP-40, 2 mM PMSF, 0.1 mM iodoacetimide, 1 µg/mlpepstatin, 10 µg/ml leupeptin, 0.5 µM EDTA, and 0.2% NaN_3. Lysates were centrifuged at 100,000 x g for 1 hr at 4 °C. Filtered supernatant was loaded onto a set of immunoaffinity columns in series: Sepharose CL-4B, Protein A Sepharose CL-4B conjugated with mouse IgG, and Sepharose CL-4B conjugated with VI-15 (DR-specific). The anti-DR column was removed and the MHC molecules were eluted with 50 mM glycine buffer, pH 11.5. Yields were approximately 60 μg of DR molecules from $6x10^8$ cells. Bound peptides were removed by treatment with 10% (v/v) acetic acid, pH 2. The peptides were separated from DR molecules by centrifugation through prewashed a Centricon membrane filter with a 10,000 dalton cutoff. Peptide mixtures were concentrated to 150 μl. 1-10 μl aliquots were loaded onto the μLC column and washed for 5-10 minutes with 95% A prior to analysis by μLC-MS.

Preparation of micro-capillary columns. Micro-columns were made by using the method of Kennedy and Jorgenson employing 98 μm i.d. fused-silica capillary tubing obtained from Polymicro Technologies (Tucson, AZ) (*25*). The columns were packed with PerSeptive Biosystems (Boston, MA) POROS 10 R2, a 10 μm reverse-phase packing material, to a length of 15-20 cm. A frit was constructed at the end of a 30 cm piece of fused silica (*25*). A frit was created by tapping the end of the column into a vial of underivatized silica, 5 μm in diameter, and sintering the silica in the end of the capillary with an open flame of an ethanol burner. A polypropylene eppendorf centrifuge tube (1.5 mL) was filled with ~100 μg of packing material and 1 mL of methanol. The tube was placed in a homemade high pressure packing device, the column was inserted, and the end of the column placed in the suspension. Helium gas at a pressure of ~400 psi was used to drive the packing material into the column while the progress of the packing was followed using a microscope. Packing was continued until the material filled a length of the capillary corresponding to 15-20 cm. The pressure was slowly dropped to zero. The column was conditioned by rinsing with 100% solvent A (0.5% (v/v) acetic acid) and with a linear gradient of 0 to 80% B (80% acetonitrile in 0.5 % acetic acid). The columns and chromatography conditions were not designed for maximal separation of peptides. The columns were tested by performing μLC-MS or automated μLC-MS/MS using 0.5-2 pmol human angiotensin I, obtained from Sigma. The width of the elution peak and the intensity of the triply charged ion were used to determine if the column was functional.

Micro-column High Performance Liquid Chromatography. HPLC grade solvents, methanol, acetonitrile, acetic acid, were purchased from Fisher Scientific. Samples were injected onto the column using the high pressure packing device, as previously described (*13*). During injection the effluent from the end of the column was collected with a 1-5 μL graduated glass capillary to measure the amount of liquid displaced from the column. Once a sufficient volume had been injected, the column was connected to the HPLC pumps. Micro-HPLC was performed using Applied Biosystems (Foster City, CA) 140B dual syringe pumps. The flow rate from the pumps was 100μL/min. The solvent stream was split, ~50/1, precolumn, to produce a final flow rate of 1-2 μL/min. The mobile phase used for gradient elution is described above. The gradient was linear from 0-60% B or 0-90% B over 30 minutes. The fritted end of the column was inserted directly into the electrospray needle.

Electrospray ionization mass spectrometry. Mass spectra were recorded on a Finnigan MAT (San Jose, CA) TSQ700 equipped with an electrospray ionization source as previously described (*13*). Electrospray ionization was performed using the following conditions. The needle voltage was set at 4.6 kV. The sheath and auxiliary gases consisted of nitrogen gas (99.999%) and were set at 20-25 psi and 5-10 units, respectively. The heated capillary temperature was set at 150°C. A sheath liquid flowed around the end of the column at a flow rate of 1.5 µL/min and was a methanol/water (70:30) mixture containing 0.1% acetic acid.

The mass spectrometer was tuned and calibrated by infusing a standard protein, such as horse heart apo-myoglobin (Sigma) in 50/50 methanol/water containing 0.5% acetic acid. The electrospray conditions were optimized to give maximum signal while the signal was most stable. The instrument was calibrated using the resolution described above for each of the quadrupoles and using average mass-to-charge ratios. The mass spectrometer was also tuned by infusing a peptide mixture diluted in 50/50 acetonitrile/water in 0.5% acetic acid with the addition of a sheath liquid (as described above), to approximate µLC-MS conditions. The spray conditions were optimized to give the most stable and intense signal.

A 1-5 µL aliquot of sample was injected onto the column to record the molecular weight of the peptides in the sample. It was estimated that the quantities of material ranged from 0.1 - 1 pmol as compared to the ion current produced by a standard peptide, such as angiotensin. The quantity of peptide contained in the mixture was not measured since sufficient material was present to perform the experiments. Spectra were acquired as peptides eluted from the µLC by scanning Q3 at a rate of 500 u/sec over the range 400-1500 m/z. Peak widths ranged from 1.5-2.0 u. Signal was detected with a conversion dynode/electron multiplier.

Sequence analysis of peptides was performed during a second HPLC analysis by selecting the precursor ion with a 2-3 u (full width at half height) wide window in the first mass analyzer and passing the ions into a collision cell filled with argon to a pressure of 3-4 mtorr. Collision energies were on the order of 10 to 50 eV (E_{lab}). The fragment ions produced in Q2 were transmitted to Q3, which was scanned at 500 u/sec over a mass range of 50 u to the molecular weight of the precursor ion to record the fragment ions. Peak widths in the second mass analyzer ranged from 1.5-2.0 u. The electron multiplier setting was 200-400 V higher than that used to record the molecular weight.

Automated tandem mass spectrometry. Tandem mass spectra were acquired using Instrument Control Language (ICL) as described (*26*). The ICL program acquires two mass spectra, scanning Q3 over 400-1400 m/z in 1 sec. If an ion is present in the scan and the calculated ion abundance is above a specified value, then five product ion spectra are acquired using the m/z value of the base peak in the last mass spectrum. Using the m/z value and the assumption that the parent is doubly-charged, the program calculates the scan range and the collision offset. Product ion spectra were generated using argon as the collision gas (3-4 mtorr) and collision energies (laboratory frame) on the order of 10 to 50 eV. The spectra were acquired by scanning Q3 over the specified range in 2 sec.

Database searching. Amino acid sequence databases were searched directly with tandem mass spectra by a computer algorithm previously described (*26-28*). All computer algorithms were written on a DEC station 5000/200 computer by using the C programming language under the Ultrix operating system. The OWL database, version 24.0, and genpept database, version 87, were obtained as ASCII text files in the FASTA format from the National Center for Biotechnology Information (NCBI) by anonymous ftp. Species specific databases were created by removing the protein sequences derived from *Homo sapiens* and *Mus musculus* to create subsets of the OWL or genpept database.

The computer algorithm is described by Eng et al. in detail (*27*). Briefly, a tandem mass spectrum file is preprocessed to reduce noise by calculating the square root of all ion intensity values and normalizing to an intensity of 100. The experimental mass spectrum is further modified by removing the precursor ion and all but the most abundant 200 ions. This file is used to prescreen amino acid sequences from the database. To perform the search of the database, protein sequences and accession numbers are retrieved and analyzed in batches of 250 until the entire database has been searched. Amino acid sequences from each protein are generated by summing the mass of the linear amino acid sequence from the amino terminus (n). If the mass of the linear sequence exceeds the mass of the unknown peptide, then the algorithm returns to the amino terminal amino acid and begins summing the amino acid masses from the n+1 position. This process is repeated until every linear amino acid sequence combination has been calculated. When the mass of the amino acid sequence is within ± 3 daltons of the mass of the unknown peptide, the predicted m/z values for the type b- and y-ions are generated and compared to the fragment ions in the spectrum. A preliminary score is generated for each sequence and the top 500 candidate peptide sequences are ranked and stored. A final analysis of these 500 amino acid sequences is performed using a correlation function. Using this function a reconstructed tandem mass spectrum for each candidate amino acid sequence is compared to the modified experimental tandem mass spectrum. Cross-correlation values are calculated, ranked and reported. The final results are ranked on the basis of a normalized correlation coefficient.

Spectra were also used to search nucleotide databases, as described by Yates and coworkers (*28*). The European Molecular Biology Laboratory (EMBL) Nucleotide Sequence Database (Release 41, December 1994) was obtained by anonymous ftp from the European Bioinformatics Institute (ftp.ebi.ac.uk). Nucleotide sequence databases were divided into taxonomic classification and used without further subdivision. Spectra of peptides isolated from DR4 molecules were searched through a primate subset of the nucleotide database.

Results and Discussion

Class II Alleles Associated with Rheumatoid Arthritis. We are interested in the sequence motifs of peptides derived from Class II alleles associated with rheumatoid arthritis (RA), a chronic inflammatory disease resulting from a malfunctioning immune system (*29*). One current model of RA involves the exposure of a genetically susceptible individual to an environmental stimulus, which results in an unregulated auto-immune response to tissue in joints. Persons with RA have inherited certain MHC class II DR4 alleles, including HLA DRB1*0401 and HLA DRB1*0404 (*29*). The two proteins differ in sequence in the binding cleft with a substitution of K to R at position 71. While antigen binding motifs have not been fully defined, each MHC class II type appears to have a specific core structural motif (*18*). Our focus is to determine if the peptides isolated from the two alleles contain unique motifs which control allele-specific binding.

Peptides isolated from cells homozygous for the MHC Class II allele HLA DRB1*0401 were fractionated by HPLC and sequenced using a combination of μLC-MS/MS and database searching, Table I. In the first step, each fraction was analyzed by μLC-MS to determine the molecular weight of the peptides in the mixture. In the next step, another aliquot was loaded onto the column and as the peptides eluted, the most abundant ions were selected manually or automatically for collision-induced fragmentation and generation of product ion spectra. Each product ion mass spectrum was used to search a protein or nucleotide database (*27, 28*). For example, the spectrum generated from an $(M+2H)^{+2}$ present in fraction 46, shown in Figure 2, was used to search the OWL database (*27*). The results of the search did not yield a peptide sequence. The spectrum was also used to search the primate subset of the nucleotide database (*28*). The sequence of the peptide was found to be from a human mRNA for an open reading frame (ORF). A similarity search of the sequence to others in the database was performed using the program BLAST (*30*). Sequence similarity was identified to an immunoglobulin heavy chain from *Xenopus laevis*. Automated μLC-MS/MS analysis of fraction 46 yielded tandem mass spectra of two other peptides from the same region of the Human mRNA ORF, Table I.

Peptides isolated from DRB1*0404 were fractionated by HPLC and sequenced using Edman degradation and μLC-MS, Table I. Initially, automated Edman degradation was performed; however, each fraction contained many peptides. For example, in one fraction, one peptide was observed by Edman and five different m/z values were observed by μLC-MS. Preliminary sequences corresponding to the major signals in automated Edman degradation were confirmed using molecular weight analysis by mass spectrometry.

The majority of peptides identified from HLA DRB1*0401 and HLA DRB1*0404 are nested sets of similar peptides originating from many class I HLA A and B alleles. By screening a large number of synthetic peptides using MHC-binding assays, Sette and et al. proposed a preliminary core motif for HLA DRB1*0401 molecules, summarized in Table II (*31*). The sequences of the peptides, as shown in Table I, isolated from both HLA DRB1*0404 and HLA DRB1*0401, are consistent with the core motif proposed by Sette an coworkers. This result suggests that to determine unique motifs which control allele-specific binding a very large number of peptides will have to be sequenced. Further studies will include sequencing peptides isolated from HLA DRB1*0402, a class II allele not associated with RA, which has a residue change to E at position 71.

Peptides Isolated From Subcellular Fractions. We have developed methodology to study the major peptide components isolated from subcellular fractions of murine B lymphoma cells (*23*). Previous research has suggested that MHC class II molecules move from the golgi complex to late endosomes where proteolytic dissociation of invariant chain takes place (*23*). Then class II molecules accumulate in a dense endosomal compartment, positioned between late endosomes and dense lysozomes, termed MIIC. Endogenous peptides bind to class II molecules in MIIC compartments (*23*). Finally the complex moves through the endosomal pathway to the cell surface. Exogenous peptides have also been found to accumulate in MIIC vesicles, where binding to class II molecules may occur (*23*). Our preliminary study focused on characterizing the peptide components in subcellular fractions, including identifying protein origins and identifying the cleavage sites in the protein. Lysed cellular material was fractionated by density using sucrose gradients. Fractions were pooled according to the type of endocytic vesicle using known marker protein assays and vesicles were lysed under acid conditions. Mixtures were fractionated using a 10,000 dalton cut-off membrane filter and analyzed directly with automated μLC-MS/MS and database searching.

Complex mixtures of peptides were found in all fractions, except those enriched in plasma membrane. In the other fractions, large peptides, with molecular weight distributions of approximately 3500 to 11,700 daltons were observed, as well as a complicated mixture of numerous smaller peptides. To facilitate sequencing, large molecular weight peptides were separated from smaller peptides using a 3000 dalton cut-off membrane filter. The low molecular weight peptide mixtures were analyzed using automated μLC-MS/MS and the uninterpreted spectra were used to search the *Mus musculus* subset of the OWL database. Preliminary sequence data was obtained on peptides in fractions containing late endosomes and endoplasmic reticulum, Table III, and in fractions containing MIIC, late endosomes and ER, Table IV.

In most of the peptides, the cleavage sites are consistent with an acid protease, such as cathepsin D or E. It has been suggested that acid proteases, such as cathepsin D and E, are important in the processing of MHC class II-peptide complexes (*24, 32*). Peptides from human keratin and calreticulin were also identified; these proteins are common contaminants in protein samples handled by humans. (Land, A., FinniganMAT, personal communication, 1994) Future studies will include pulsing the cells with hen egg lysozyme, followed by subcellular fractionation, and identification of lysozyme peptides in each pooled fraction by μLC-MS/MS.

Class II Presentation Mutants. We are studying the peptides bound to class II DR3 molecules isolated from a mutant cell line, which exhibits reduced binding to exogenously provided peptides and complete dissociation of dimers in SDS polyacrylamide gel electrophoresis (SDS-PAGE) (*15, 33*). This phenotype is derived from a single point mutation in the HLA-DMB gene, which maps within the MHC class II region. The function of the DMB-gene product is unknown and research suggests that it plays a key role in the formation of class II-peptide complexes, Figure 1.

Previous research has shown DR3 isolated from cells with a deletion mutation are complexed primarily with peptides from invariant chain rather than a complex mixture of peptides derived from many proteins (*4*). The instability of these molecules were corrected by incubating them at low pH with exogenously provided peptide. By comparing cells with perturbations of the normal processing mechanism, information relative to the steps and requirements for processing can be obtained.

Peptide mixtures were analyzed directly by μLC-MS without prior fractionation. Mixtures isolated from the DR3 molecules of progenitor cells after incubation at pH 7.4 and pH 4 were analyzed, Figure 3. The pattern of peptides eluted from these molecules at pH 7.4 was consistent with a complex mixture of low abundance peptides. The spectra obtained from the peptides from DR molecules of cells treated at pH 4 were similar to those eluted from pH 7.4 treated cells, indicating that most of the peptides remained bound to the DR3 molecules at low pH.

Peptide mixtures isolated from the DR3 molecules of mutant cells, 9.5.3, after incubation at pH 7.4 and at pH 4 were also analyzed, Figure 3. The peptide mixture eluted from the DR3 molecules of pH 7.4 treated mutant cells consists of several prominent species. Amino acid sequencing by μLC-MS/MS of the three major peptides confirmed that they are a nested set of peptides derived from residues 82-104 of the invariant chain, Table V. Analysis of the peptide mixture isolated from the molecules of cells treated at pH 4 shows that these DR molecules are virtually empty; only trace amounts of invariant chain peptides were bound after pH 4 treatment. In addition, after treatment at pH 4, the DR3 molecules of the mutant cell line, 9.5.3, bound cognate peptide and were stable in SDS-PAGE.

Table I. Peptides Identified by μLC-MS/MS Isolated from Human Class II DR4 molecules

Allele	Sequence[a]	Protein
DRB1*0401	DTQ**F**VRFD**S**DAASQRMEP	MHC Class I (HLA-A)
	DTQ**F**VRFD**S**DAASQRM*EP	
	DDTQ**F**VRFD**S**DAASQRM*EPR	
	VDDTQ**F**VRFD**S**DAASQRMEPR	
	VDDTQ**F**VRFD**S**DAASQRM*EPR	
	DLRS**W**TAAD**T**AAQITQ	MHC Class I (HLA-B)
	DLRS**W**TAAD**T**AAQITQR	
	DLRS**W**TAAD**T**AAQISQ	MHC Class I (HLA-B)
	DLRS**W**TAAD**T**AAQISQR	
	LPS**Y**EEAL**S**LPSKTP	Human mRNA for ORF
	LPS**Y**EEAL**S**LPSKTPE	
	VLPS**Y**EEAL**S**LPSKTPE	
DRB1*0404	SHS**M**RYFH**T**AMSRP	MHC Class I (HLA-B)
	SHS**M**RYFH**T**AMSRPG	
	GSHS**M**RYFH**T**AMSRPG	
	GSHS**M**RYFH**T**AMSRPGRG	
	SHS**M**RYFY**T**AVSRP	MHC Class I (HLA-B)
	SHS**M**RYFY**T**AVSRPG	
	SHS**M**RYFY**T**AVSRPGRG	
	GSHS**M**RYFY**T**AVSRPG	
	SHS**M**RYFY**T**AVSRPGRG	

[a] Amino acid sequences are shown using the single letter codes for the amino acids. The isomers leucine and isoleucine and the isobaric amino acids glutamine and lysine cannot be differentiated using low-energy collision-induced dissociation and were assigned using the sequences retrieved from the database. An asterisk (*) next to a methionine residue indicates that the sulfoxide form was observed. The nine underlined residues corresponds to the proposed binding core, Table II (31). The amino acids displayed in boldface are positions 1 and 6 of the proposed binding core, Table II (31).

TABLE II. Preliminary Motif of Class II HLA DRB1*0401 Peptides

1[a]	2	3	4	5	6	7	8	9
YFW MVLI	X	X	no pos. charge	X	TSV LIM	no pos. charge	X	no charge

[a] Nine residues form the proposed binding core. Residues at positions 1 and 6 are limited to those displayed. Residues at position 4 and 7 cannot be positively charged (K or R) and residues at postion 9 cannot be charged (K, R, D, or E).
SOURCE: Adapted from ref. 31.

Table III. Peptides Identified by μLC-MS/MS of Murine Subcellular Fractions Containing Late Endosomes and ER

Sequence [a]	Protein
WQVKSGTIFDNF	Calreticulin precursor
LGLLPHTFTPTTQL	ATP synthase a chain
FDITADDEPLGRVSFEL	Peptidyl-prolyl cis-trans isomerase
AGFGGGFAGGDGLL	Human keratin

[a] Amino acid sequences are shown using the single letter codes for the amino acids. The isomers leucine and isoleucine and the isobaric amino acids glutamine and lysine cannot be differentiated using low-energy collision-induced dissociation and were assigned using the sequences retrieved from the database.

Table IV. Peptides Identified by μLC-MS/MS of the Murine Subcellular Fractions Containing MIIC Vesicles, Late Endosomes, and ER

Sequence [a]	Protein
TLDDTWAKAHFAIMF	Cytochrome C oxidase polypeptide I
MTFFPQHFLGL	Cytochrome C oxidase polypeptide I
WQVKSGTIFDNF	Calreticulin precursor
AVLGLDL	Calreticulin precursor
LGLLPHTFTPTTQL	ATP synthase A chain
FDITADDEPLGRVSFEL	Peptidyl-prolyl cis-trans isomerase
VVENLQDDFDFN	MMU06922 NCBI gi: 458706

[a] Amino acid sequences are shown using the single letter codes for the amino acids. The isomers leucine and isoleucine and the isobaric amino acids glutamine and lysine cannot be differentiated using low-energy collision-induced dissociation and were assigned using the sequences retrieved from the database.

Table V. Invariant Chain Peptides Identified by μLC-MS/MS isolated from Human Class II DR3 Molecules of a Mutant Cell Line

Residue Number[a]	Sequence[b]
83-103	KPPKPVSKMRMATPLLMQALP
82-103	PKPPKPVSKMRMATPLLMQALP
82-104	PKPPKPVSKMRMATPLLMQALPM

[a] Residues are numbered by counting from the amino terminus of the invariant chain p33 form (34).
[b] Amino acid sequences are shown using the single letter codes for the amino acids. The isomers leucine and isoleucine and the isobaric amino acids glutamine and lysine cannot be differentiated using low-energy collision-induced dissociation and were assigned using the known sequence of invariant chain p33 form.
SOURCE: Adapted from ref. 15.

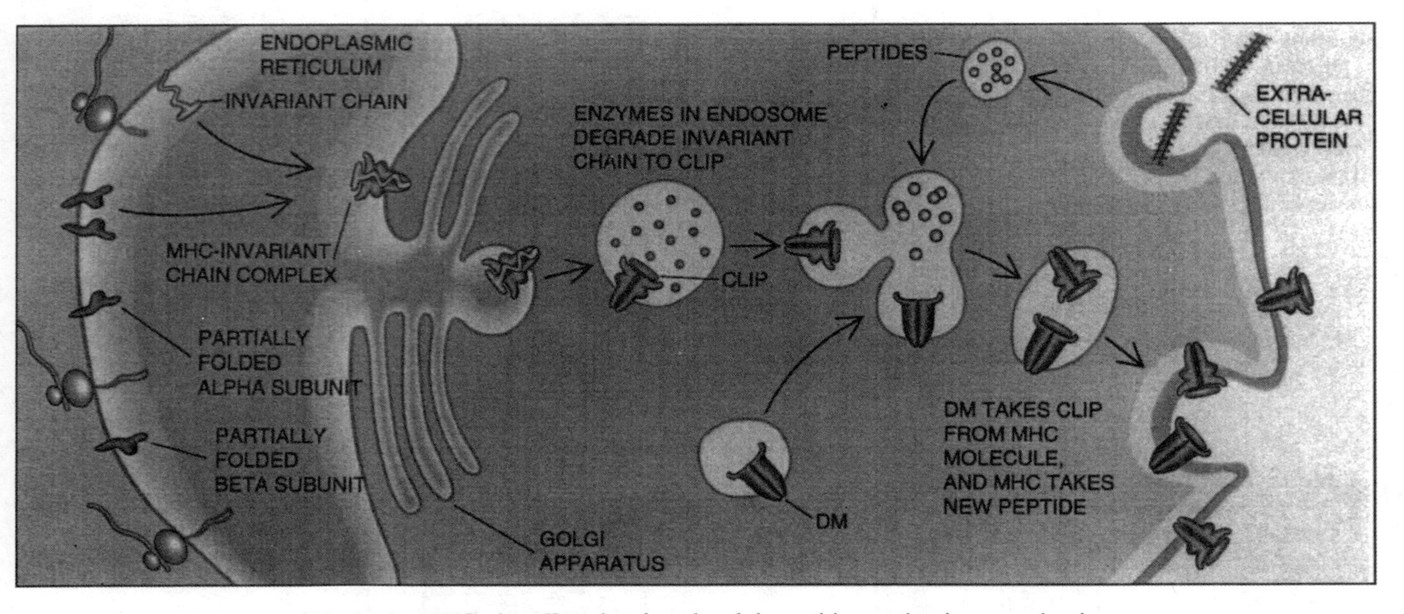

Figure 1. MHC class II molecules, the alpha and beta subunits, complex in the endoplasmic reticulum with invariant chain and move through the golgi complex into the endosomal pathway, before binding peptides. Within the endocytic vesicles, invariant chain is enzymatically cleaved which allows the class II molecules to complex antigenic peptides. Reproduced with permission from How Cells Process Antigens, Victor H Engelhard (August 1994 , p.61). Copyright © 1994 by Scientific American, Inc. All rights reserved.

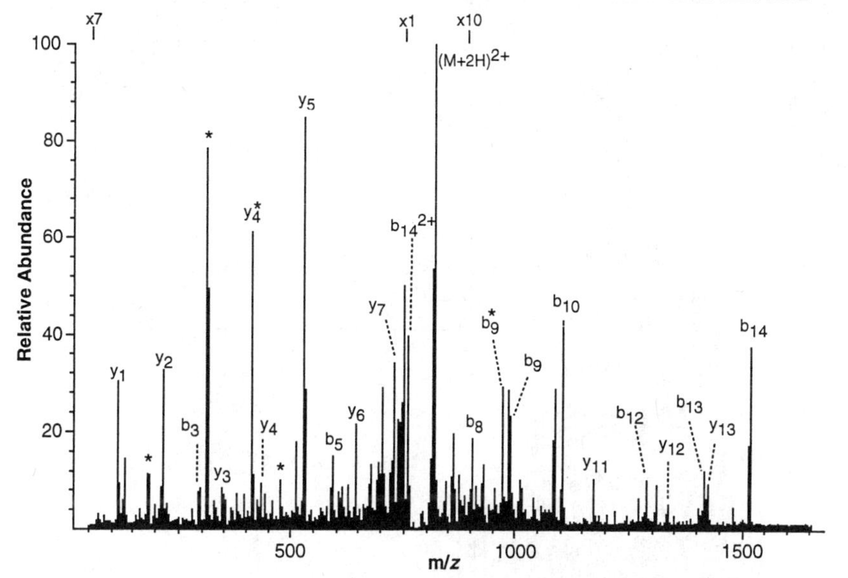

Figure 2. Collision-induced dissociation mass spectrum recorded on the $(M+2H)^{+2}$ ions at m/z 817 of a peptide presented by class II major histocompatibility molecules on the surface of EBV cells homozygous for HLA-DRB*0401. Fragments of type b- and y-ions having the general formulas $H(NHCHRCO)_n^+$ and $H_2(NHCHRCO)_nOH^+$, respectively, are shown above and below the amino acid sequence at the top of the figure. Fragment ions of the type b_ny_n are labeled with an asterisk. Ions observed in the spectrum are underlined. Leu, Ile, Gln, and Lys were assigned by correspondence to the sequences derived from the database. μLC-MS/MS was performed using a 4.5 μl aliquot; collision energy (E_{lab}) was 48 eV and collision gas pressure was 3.5 mtorr. The peptide was eluted using a linear gradient from 0-60% B over 30 minutes, as described in the experimental section.

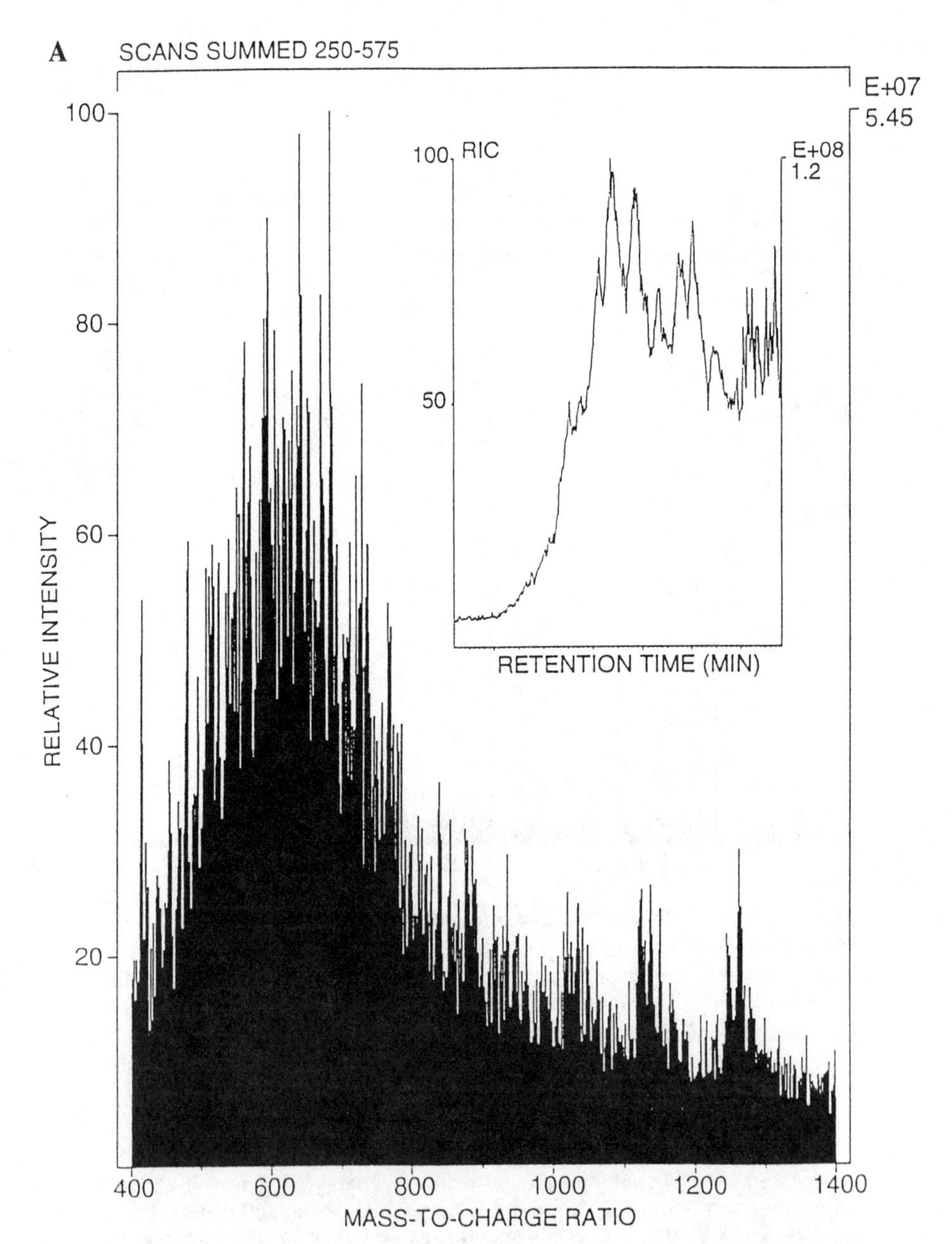

Figure 3a. Micro-column liquid chromatography electrospray ionization mass spectrometry of peptides isolated from DR3 molecules. μLC-MS spectrum (scans 250 to 575 summed) and reconstructed ion chromatogram (insert) of peptide mixture eluted from DR molecules of pH 7.4 treated 8.1.6 cells. (Reproduced with permission from ref. 15. Copyright © 1994 The American Association of Immunologists. All rights reserved.)

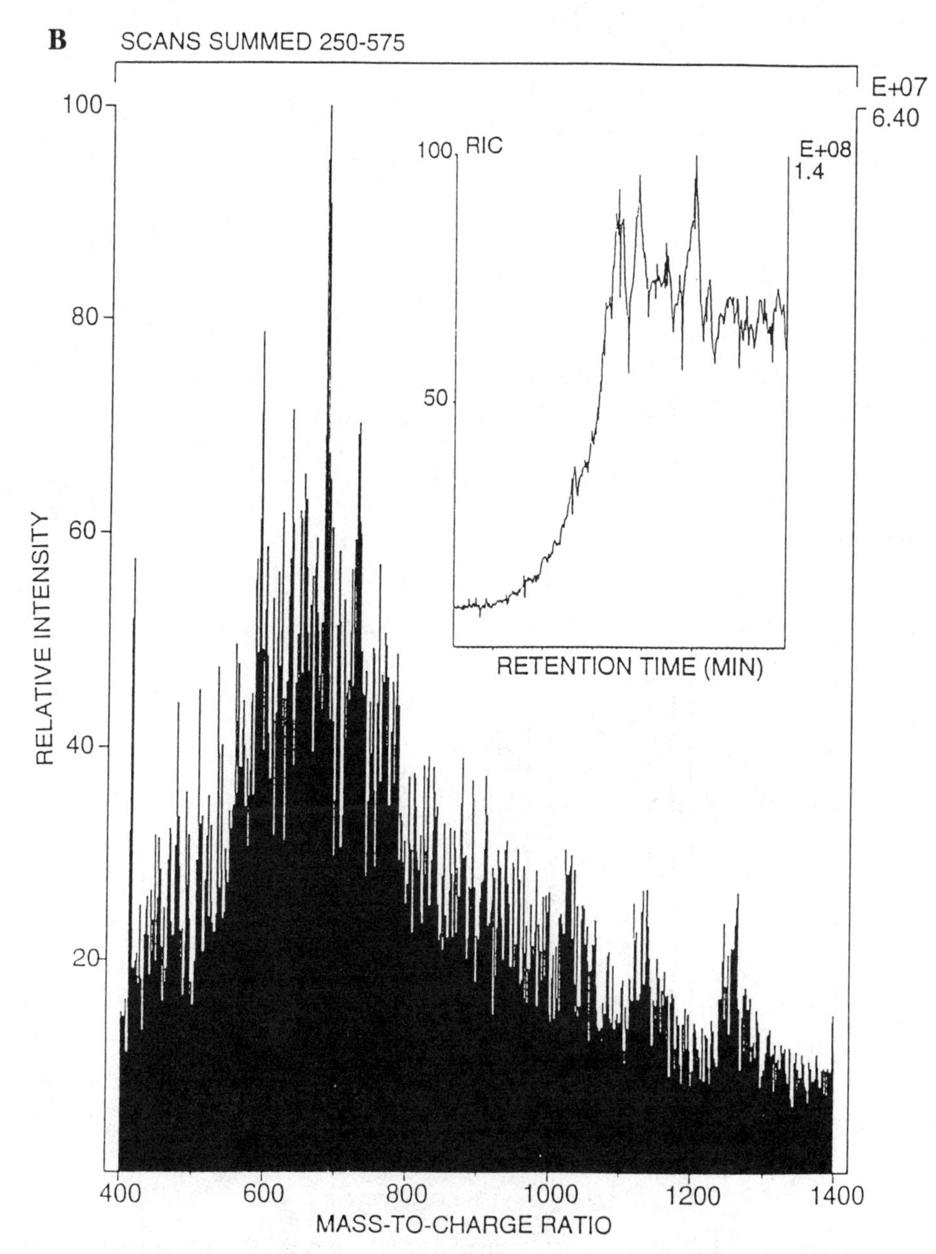

Figure 3b. μLC-MS spectrum (scans 250 to 575 summed) and reconstructed ion chromatogram (insert) of peptide mixture eluted from DR molecules of pH 4 treated 8.1.6 cells. (Reproduced with permission from ref. 15. Copyright © 1994 The American Association of Immunologists. All rights reserved.)

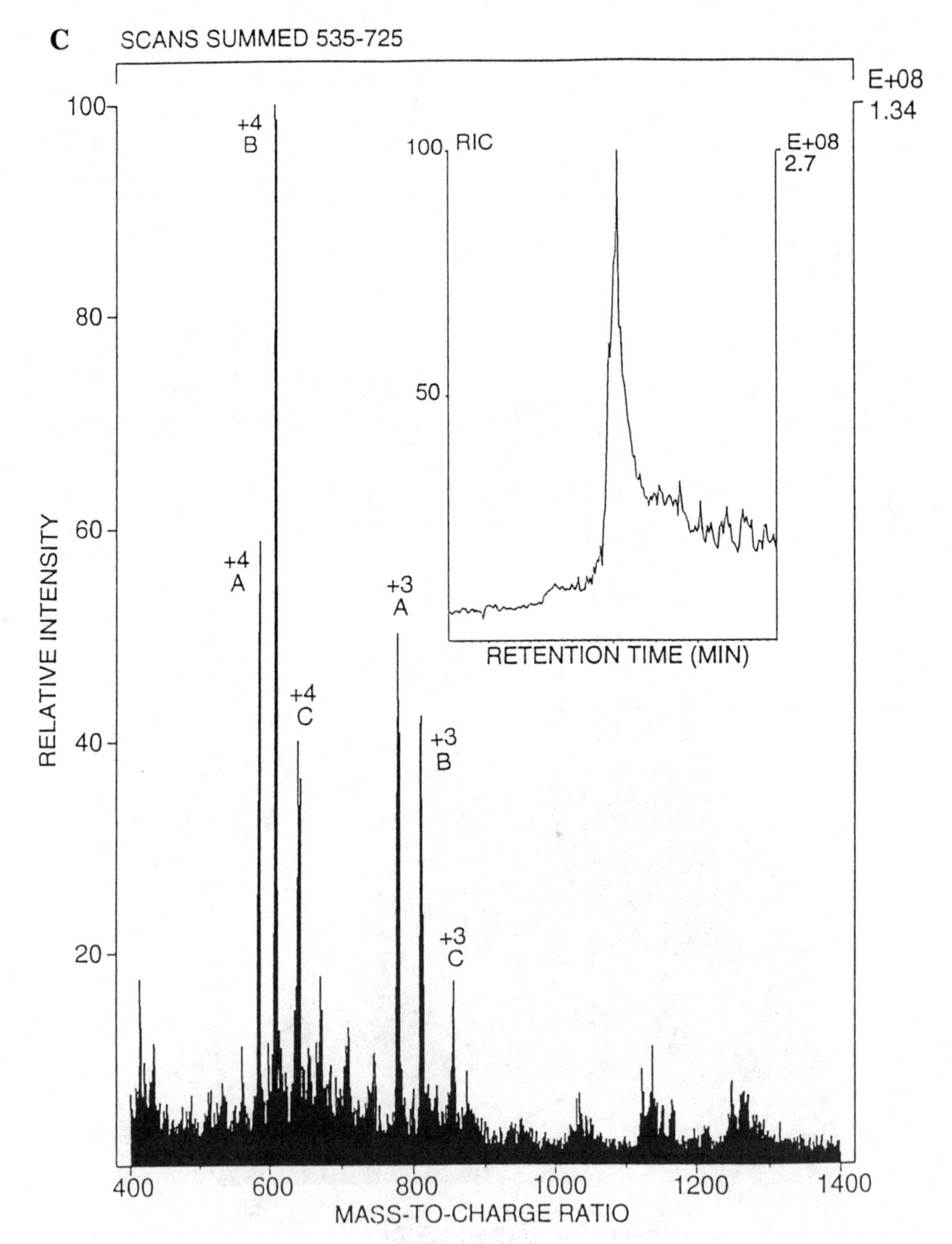

Figure 3c. μLC-MS spectrum (scans 535 to 725 summed) and reconstructed ion chromatogram (insert) of peptide mixture eluted from DR molecules of pH 7.4 treated 9.5.3 cells. Ions correspond to +3 and +4 charge states of the following invariant chain fragments, A, KPPKPVSKMRMATPLLMQALP B, PKPPKPVSKMRMATPLLMQALP C, PKPPKPVSKMRMATPLLMQALPM. (Reproduced with permission from ref. 15. Copyright © 1994 The American Association of Immunologists. All rights reserved.)

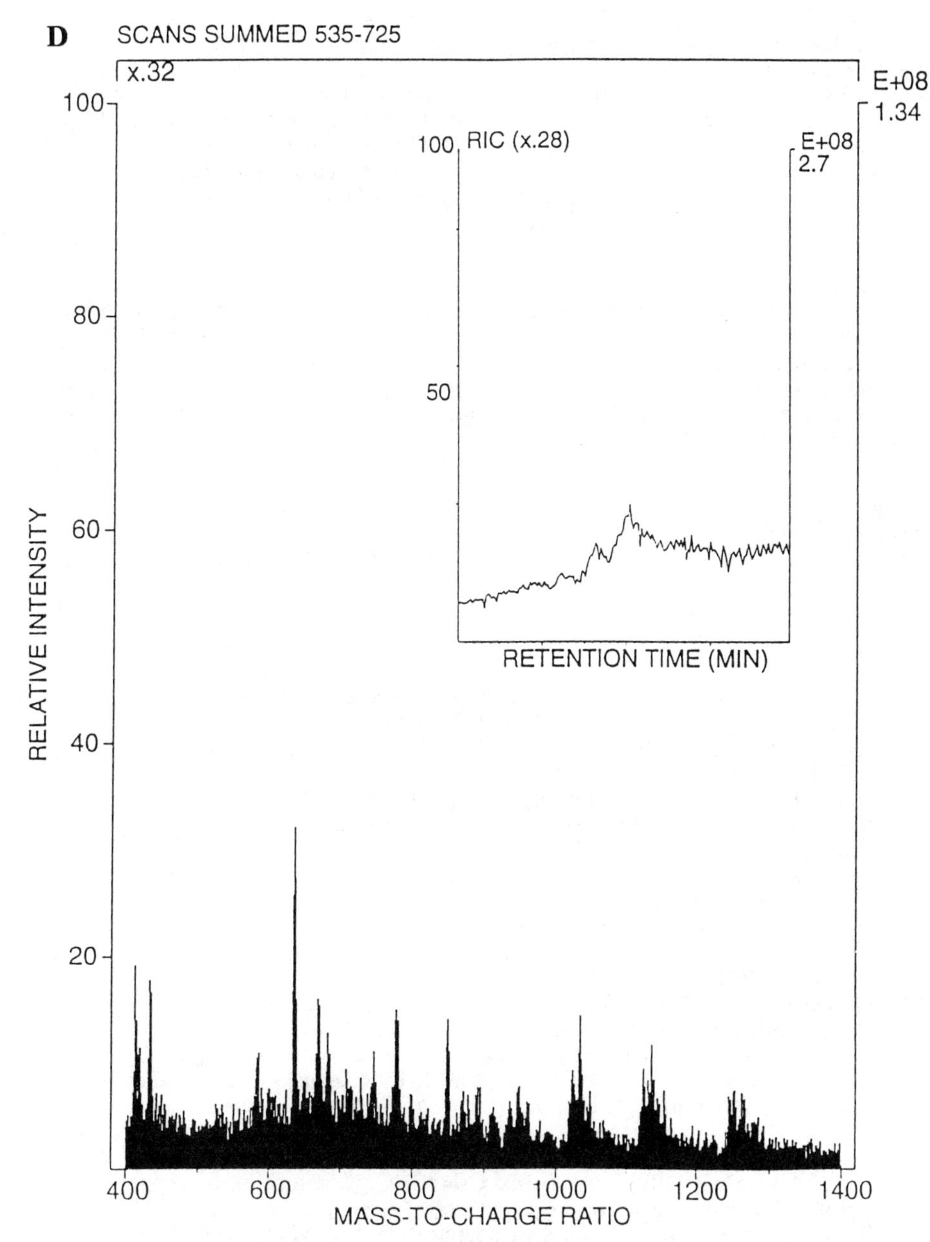

Figure 3d. μLC-MS spectrum (scans 535 to 725 summed) and reconstructed ion chromatogram (insert) of peptide mixture eluted from DR molecules of pH 4 treated 9.5.3 cells. μLC- MS was performed using a 10 μl and a 1.5 μl aliquot of material from 8.1.6 cells and 9.5.3 cells, respectively. The peptides were eluted using a linear gradient from 0-90% B over 30 minutes, as described in the experimental section. (Reproduced with permission from ref. 15. Copyright © 1994 The American Association of Immunologists. All rights reserved.)

It has been suggested that the role of invariant chain is to prevent binding of endogenous peptides to class II molecules in the endoplasmic reticulum and to guide the MHC class II molecules to the endocytic pathway for complexation with peptides. One explanation for the excessive abundance of invariant chain peptides complexed to these class II DR3 molecules from mutant cells is that these invariant chain peptides represent the residual fragments that are normally removed and replaced with peptide. This suggests that the function of the DMB-gene product may be to remove invariant chain peptides, Figure 1.

Conclusions

Micro-column electrospray ionization tandem mass spectrometry has proven to be a powerful technique for the study of antigens presented by MHC molecules. Extending this approach to study antigen processing pathways will provide new insights into the mechanisms of antigen processing, selection, and presentation. The application of mass spectrometry to problems in immunology and other biological areas will continue to expand.

Acknowledgments

This research was supported by the University of Washington's Royalty Research Fund and the National Science Foundation, Science and Technology Center, DIR 8809710.

Literature Cited

1. Hunt,D.F.; Henderson, R.A.; Shabanowitz, J.; Sakaguchi, K.; Michel, H.; Sevilir, N.; Cox, A.L.; Appella, E.; Engelhard, V.H. *Science* **1992**, *255*, 1261.
2. Henderson, R.A.; Michel, H.; Sakaguchi, K.; Shabanowitz, J.; Appella, E.; Hunt, D.F.; Engelhard, V.H. *Science* **1992**, *255*, 1264.
3. Hunt, D.F.; Michel, H.; Dickinson, T.A.; Shabanowitz, J.; Cox, A.L.; Sakaguchi, K.; Appella, E.; Grey, H.M.; Sette, A. *Science* **1992**, *256*, 1817.
4. Sette, A.; Ceman, S.; Kubo, R. T.; Sakaguchi, K.; Appella, E.; Hunt, D. F.; Davis, T. A.; Michel, H.; Shabanowitz, J.; Rudersdorf, R.; Grey, H. M.; Demars, R. *Science* **1992**, *258*, 1801.
5. Huczko, E.L.; Bodnar, W.M.; Benjamin, D.; Sakaguchi, K.; Zhu, N.-Z.; Shabanowitz, J.; Henderson, R.A.; Appella, E.; Hunt, D.F.; Engelhard, V.H. *J. Immunol.*, **1993**, *151*, 2572.
6. Henderson, R.A.; Cox, A.L.; Sakaguchi, K.; Appella, E.; Shabanowitz, J.; Hunt, D.F.; Engelhard, V.H. *Proc. Natl. Acad. Sci., USA* **1993**, *90*, 10275.
7. Engelhard, V.H.; Appella, E.; Benjamin, D.C.; Bodnar, W.M.; Cox, A.L.; Chen, Y.; Henderson, R.A.; Huczko, E.L.; Michel, H.; Sakaguchi, K.; Shabanowitz, J.; Sevilir, N.; Slingluff, C.L.; Hunt, D.F. In *Naturally Processed Peptides*; Sette, A., Ed.; Chemical Immunology; S. Karger: Basel, Switzerland, 1993, Vol. 57; pp 39-62.
8. Sette, A.; DeMars, R.; Grey, H.M.; Oseroff, C.; Southwood, S.; Appella, E.; Kubo, R.T.; Hunt, D.F. In *Naturally Processed Peptides*; Sette, A., Ed.; Chemical Immunology; S. Karger: Basel, Switzerland, 1993, Vol. 57; pp 152-165.
9. Cox, A.L.; Skipper, J.; Chen, Y.; Henderson, R.A.; Darrow, T.L.; Shabanowitz, J.; Engelhard, V.H.; Hunt, D.F. *Science* **1994**, *264*, 716.
10. Chen, Y,; Sidney, J.; Southwood, S.; Cox, A.L.; Sakaguchi, K.; Henderson, R.A.; Appella, E.; Hunt, D.F.;Sette, A.; Engelhard V.H. *J. Immunol.* **1994**, *152*, 2874.
11. Kubo, T.R.; Sette, A.; Grey, H.M.; Appella, E.; Sakaguchi, K.; Zhu, N.-Z.; Arnott, D.; Shabanowitz, J.; Michel, H.; Bodnar, W.M.; Davis, T.A.; Hunt, D.F. *J. Immunol.* **1994**, *152*, 3913.

12. Slingluff, C.L.; Hunt, D.F.; Engelhard, V.H. *Curr. Opin. Immunol.* **1994**, *6*, 733.
13. Yates, J. R. III; McCormack, A. L.; Hayden, J. B.; Davey, M. P. *Cell Biology: A Laboratory Handbook*,. Ed., Celis, J. E., Academic Press: San Diego, 1994, pp. 380-388.
14. McCormack, A.L.; Hayden, J. B.; Davey, M.; Yates, J.R. III *Proceedings of the 42nd ASMS Conference on Mass Spectrometry and Allied Topics*; Chicago, 1994, pp 646.
15. Monji. T.; McCormack, A.L.; Yates, J.R. III; Pious, D. *J. Immunol.* **1994**, *153*, 4468.
16. Engelhard, V.H. *Ann. Rev. Immunol.* **1994**, *12*, 181.
17. Falk, K.; Rötzschke, O.; Stevanovic, S.; Gunther, J.; Rammensee, H.-G. *Nature* **1991**, *351*, 290.
18. Chicz, R.M.; Urban, R.G.; Gorga, J.C.; Vignali, D.A.A.; Lane, W.S.; Strominger, J.L. *J. Exp. Med.* **1993**, *178*, 27.
19. Cresswell, P. *Ann. Rev. Immunol.* **1994**, 12, 259.
20. Coligan, J. E.; Kruisbeck, A. M.; Margulies, D. H.; Shevach, E. M.; Strober, W.,eds. (1992). Current Protocols in Immunology, *Greene Publishing and Wiley Interscience*, New York.
21. Buus, S.; Sette, A.; Colon, S. M.; Jenis, D. M.; Grey, H. *Cell* **1986,** *47*, 1071.
22. Demotz, S.; Grey, H.; Appella, E; Sette, A. *Nature* **1989**, *342*, 682.
23. Rudensky, A.Y.; Maric, M.; Eastman, S.; Shoemaker, L.; DeRoos, P.C.; Blum, J.S. *Immunity* **1994**, *1*, 585.
24. Maric, M. A.; Taylor, M.D.; Blum, J.S. *Proc. Natl. Acad. Sci. USA*, **1994**, *91*, 2171.
25. Kennedy, R. T.; Jorgenson, J. W. *Anal. Chem.* **1989**, *56*, 1128.
26. Yates, J.R. III; Eng, J.; McCormack, A.L.; Schieltz, D. *Anal. Chem.* **1995**, in press.
27. Eng, J.; McCormack, A.L.; Yates, J. R. III *J. Am. Soc. Mass Spectrom.* **1994**, 5, 976.
28. Yates, J.R. III; Eng, J.; McCormack, A.L *Anal. Chem.* **1995**, in press.
29. Wordsworth, B.P.; Lanchbury, J.S.S.; Sakkas, L.L.; Welsh, K.I.; Panayi, G.S.; Bell, J.I. *Proc. Natl. Acad. Sci. USA*, **1989**, *86*, 10049.
30. Altschul, S. F.; Gish, W.; Miller, W.; Myers, E.W.; Lipman, D. J. *J. Mol. Biol.* **1990**, *215*, 403.
31. Sette, A.; Sidney, J.; Oseroff, C.; del Guercio, M.-F.; Southwood, S.; Arrhenius, T.; Powell, M.F.; Colon, S.M.; Gaeta, F.C.A.; Grey, H.M. *J. Immunol.* **1993**, *151*, 3163.
32. Bennett, K.; Levine, T.; Ellis, J.S.; Peanasky, R.J.; Samloff, I.M.; Kay, J.; Chain, B.M. *Eur. J. Immunol.* **1992**, *22*, 1519.
33. Mellins, E.; Smith, L.; Arp, B.; Cotner, T.; Celis, E.; Pious, D. *Nature* **1990**, *343*, 71.
34. Claesson, L.; Larhammer, L.; Peterson, P.A. *Proc. Natl. Acad. Sci., USA* **1983**, *80*, 7395.

RECEIVED September 6, 1995

Chapter 11

Identification of Proteins from Two-Dimensional Electrophoresis Gels by Peptide Mass Fingerprinting

David P. Arnott, William J. Henzel, and John T. Stults

Protein Chemistry Department, Genentech, Inc., 460 Point San Bruno Boulevard, South San Francisco, CA 94080

As part of a project to identify factors involved in congestive heart failure, differences in protein expression levels between normal and enlarged (hypertrophic) heart cells were identified. In initial experiments, two-dimensional gel electrophoresis was used to separate the proteins from normal neonatal rat cardiac myocytes. The proteins were electroblotted to a membrane and identified by staining. Proteins of interest were cleaved into peptides with an *in situ* enzymatic digestion method. The masses of the peptides were determined by capillary high performance liquid chromatography electrospray ionization mass spectrometry and, when possible, partial sequences were obtained by subsequent liquid chromatography tandem mass spectrometry (LC-MS/MS) experiments. These data were used to search a protein sequence database with the program FRAGFIT. The program theoretically cleaves each protein in the database. By comparison of the experimentally-determined peptide masses and the theoretical masses, the program identifies the protein if it exists in the database. A partial sequence from the LC-MS/MS experiment was used to increase the specificity of the search. With this approach, eight proteins present in low picomole quantities on 2-D gels from cardiac myocytes have been identified, with 100 fmol or less of each peptide component required for the mass spectral experiments.

Methods for the determination of a protein's amino acid sequence have evolved considerably during the past forty years. Amino acid sequencing by Edman degradation was the first generation method (*1*). This method, now extensively automated and refined, is still widely used today, though rarely for the sequence determination of an entire protein. Methods for gene cloning and nucleotide sequencing, developed in the 1970's (*2*), made it generally much easier to infer the primary structure from the cDNA sequence. This second generation method is now the one most commonly used. Nonetheless, the determination of partial amino acid sequence by automated Edman degradation, from which oligonucleotide probes are designed, is typically a prerequisite for the cloning process.

The need to determine novel protein sequences is rapidly being supplanted by the need to identify proteins from a database. The explosive growth of protein sequence databases has significantly raised the likelihood that the sequence of a

0097–6156/95/0619–0226$12.00/0

protein of interest is already known. Furthermore, recent strategies for generating partial sequences of cDNA's (expressed sequence tags – EST's) (*3, 4*) should lead to databases that contain partial nucleotide sequences for virtually all human proteins within the next 1-2 years. Finally, the human genome project will, upon its completion, produce the entire sequence for each human gene. Recent proposals suggest that completion of the human genome sequence (*5*) could be completed as early as 2001 (*6*). Sequencing of other genomes is also progressing rapidly.

Although database searching with Edman-generated data is commonplace, a new method, now in its infancy, could potentially become the common method for protein identification. This method no longer involves Edman degradation, but is a liquid chromatography-mass spectrometry-based method (LC-MS/MS) for database searching. This next-generation method has become known as peptide mass fingerprinting, and it is shown conceptually in Figure 1. Peptides are generated enzymatically or chemically from a protein of interest. The peptide masses are determined by mass spectrometry. These masses are used as input to a computer program that theoretically digests each protein in a sequence database according to the specificity of the cleavage reaction used. The experimental peptide set is matched with the peptide set for each protein in the database to determine the identity of the protein, if it exists in the database. As few as 3-4 peptide masses may be sufficient to determine a protein identity, with more masses serving to increase the confidence in the match. This method was originally described by Henzel *et al.* in 1989 (*7*), and its utility subsequently demonstrated by our group (*8-10*) and others (*11-16*). The method has been rapidly adopted (see reviews (*17, 18*)) and used successfully by a growing number of other research groups (19-23). Many data systems sold with commercial mass spectrometers now include software for database matching. Enhancements to the technique include the use of partial peptide sequence (*24*) and high resolution mass measurements (*25*). A particularly innovative approach uses a protein sequence database to generate theoretical MS/MS fragment ion spectra, based on a matching precursor mass, to match with the authentic MS/MS spectrum (*26, 27*).

An important application of peptide mass fingerprinting is the identification of proteins separated by two-dimensional (2-D) electrophoresis. 2-D electrophoresis is an extremely high resolution method of protein separation (*28*), capable of resolving more than 2000 proteins in a single gel (*29*). Comparisons of 2-D gel images have shown differences in protein expression levels between normal and diseased tissue, and altered expression levels in transformed cells and cells treated with a variety of factors (*30*). Identification of proteins on the gels is a difficult task, due to the large numbers of proteins, and the vanishingly small amounts of protein found in a typical spot on a single 2-D gel (approx. 10 pmol for the most abundant proteins on analytical gels). The number of proteins of interest that must be identified can be reduced substantially by comparing 2-D gel images to identify those spots of interest (*31*). Furthermore, the availability of a number of 2-D gel databases (*32-35*), some accessible over the World Wide Web (*36, 37*), can help to identify some of the spots. The use of these databases, however, requires that the same electrophoresis protocol be duplicated precisely and that the same tissue or cell line be used. These constraints limit the widespread usage of 2-D electrophoresis databases based solely upon image matching.

Due to its speed and sensitivity, peptide mass fingerprinting has been used successfully to identify proteins from 2-D gels (*8, 19-23*). One of the keys to this approach is the efficient generation of peptides from the protein spots. Two methods have been used successfully. A protein spot in a gel, following staining and excision of the spot, can be digested in the gel (*38-40*). The peptides are extracted from the gel matrix for subsequent measurement. Alternatively, the proteins can be electroblotted to a membrane where they are subsequently stained. A spot of interest

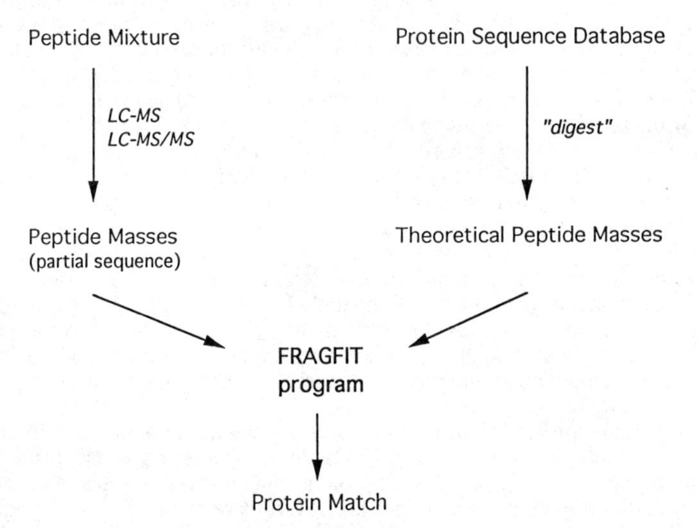

Figure 1. Schematic diagram of peptide mass fingerprinting method. The peptides are generated by *in situ* reduction/alkylation/digestions of electroblotted proteins. Masses for observed peptides are matched against theoretically digested proteins from a database. If the protein is present in the database, a match will be found if the conditions for matching are satisfied (see text).

is cut from the membrane and an *in situ* digest is performed directly on the membrane (*41-45*). The peptides are analyzed after extraction from the membrane. *In situ* digestion on a membrane is favored because fewer auto-proteolytic peptides are observed (*20*).

In our earlier results we showed that proteins from a single 2-D gel of an *E. coli* lysate could be identified by mass alone (*8*). Mass measurement was by matrix-assisted laser desorption/ionization-time-of-flight mass spectrometry (MALDI-TOF). As few as four peptide masses were sufficient to uniquely identify a protein, with as little as 20 fmol of a protein digest needed for the mass measurement. The results were confirmed by conventional automated Edman degradation.

We demonstrate the use of peptide mass fingerprinting for the identification of proteins from 2-D gels of lysates from neonatal rat cardiac myocytes. This work is the first step in a project to identify proteins that are involved in cardiac hypertrophy, one of the major contributors to congestive heart failure. Peptide mass measurement and partial peptide sequence were obtained by LC-MS and LC-MS/MS, respectively, utilizing electrospray ionization (ESI).

Experimental

Materials. All reagents were of the highest quality available. All organic solvents were HPLC grade. Water was purified with a Milli-Q system (Millipore).

Two-dimensional Gel Electrophoresis. The cell pellet derived from a single plate of cultured neonatal rat cardiomyocytes was dissolved in 10 µL lysis buffer (8 M urea, 2% Pharmalyte 3-10 ampholytes, 2% Triton X-100, 2% 2-mercaptoethanol, 0.1 mM 4-(2-aminoethyl)-benzenesulfonylfluoride (AEBSF) protease inhibitor). Insoluble material was removed by centrifugation at 4°C for 45 min in an Eppendorf centrifuge. Two or three such samples were pooled and diluted 2:1 with sample solution (6M urea, 2% Pharmalyte 3-10 ampholytes, 0.5% Triton X-100, 2% 2-mercaptoethanol) yielding a total protein content of approximately 400 µg.

Two-dimensional electrophoresis was carried out using a Pharmacia Multiphor II system. First dimension isoelectric focusing (IEF) was performed with immobilized pH gradient strips (pH 3-10, length 11 cm) rehydrated in a solution containing 6M urea, 0.5% Triton X-100, 10 mM DTT, 2 mM acetic acid. The immobilized pH gradient (IPG) strips were placed in the immobiline strip tray according to the manufacturer's instructions and samples were loaded at the acidic end of each strip. Isoelectric focusing was carried out at 15°C over 16 hours: 300 V for 3 h; 300-2000 V over 5 h; 2000 V for 8 h. The strips were equilibrated for 20 min in SDS buffer (50 mM Tris HCl pH 6.8, 6M urea, 30% glycerol, 1% w/v SDS, 0.005% Bromophenol Blue) and applied (two can be run simultaneously) to Pharmacia SDS-PAGE gels (245 x 180 x 0.5 mm, 12-14 % polyacrylamide gradient, pre-cast on a plastic support film). Electrophoresis proceeded at 20 mA constant current until the dye front migrated 5 mm into the gel, at which point the IPG strip was removed. Electrophoresis then continued at 40 mA constant current until the dye front reached the end of the gel.

Proteins were electroblotted onto poly(vinylidene)difluoride (PVDF) membranes (Immobilon-P^{SQ}, Millipore) at 250 mA constant current in 10 mM CAPS buffer, pH 11, for one hour. The blots were stained with Coomassie Brilliant Blue R-250 (0.1% w/v in 50% methanol) for 1 min, and destained with a solution of 50% methanol, 10% acetic acid until protein spots became visible. The membranes were rinsed thoroughly with deionized water, and allowed to air dry.

In Situ Digestion. Protein spots of interest were cut from the blot, destained, reduced, alkylated, and digested with endoproteinase Lys-C (8). The excised spots

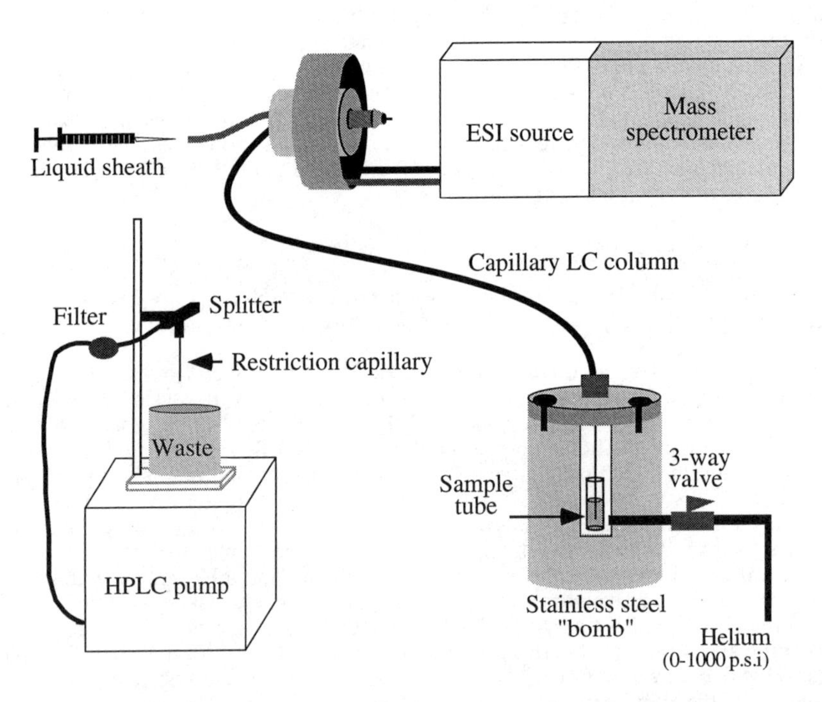

Figure 2. Capillary HPLC-ESI interface. A capillary HPLC column (75-100 μm) is inserted directly into the tip of the ESI needle assembly. The sample is loaded by pressure in a stainless steel "bomb," as shown, with the end of the column placed in the Eppendorf tube that contains the protein digest. Following loading, the capillary column is removed from the bomb and connected via a splitter to the HPLC pump. In this way, an injection loop is avoided, and low flow rates (approx. 0.2 μl/min) are achieved with a conventional HPLC pump. The electrospray stability is enhanced by the use of a 70% methanol/0.1 M acetic acid liquid sheath flowing at 1.2 μl/min.

were transferred to siliconized 0.65 mL microcentrifuge tubes and wetted with 2 µL methanol. Residual stain was removed by the addition of 50 µL deionized water and 200 µL methanol, with vortexing for one minute, followed by addition of 50 µL chloroform and further vortexing. The solvent was then removed by pipette. Destaining, although not essential, was found to reduce interferences in the subsequent mass spectra. Reduction of disulfide bonds was accomplished by addition of 100 µL reduction buffer (10 mM DTT, 5 mM EDTA, 500 mM Tris HCl) to the wet membrane, with incubation at 45°C for one hour. Cysteine residues were then carboxymethylated by addition of 10 µL 200 mM iodoacetic acid in 0.5 M NaOH, followed by incubation in the dark for 20 min. The membranes were washed three times with 10% acetonitrile, and vortexed gently for 20 min in a solution of 0.5% w/v PVP-360, 0.1% acetic acid. The solution was removed and the membranes washed twice with 10% acetonitrile, once with 20% acetonitrile, and once with digestion buffer (50 mM ammonium bicarbonate, 10% acetonitrile). Proteins were digested overnight at 37 °C with 0.05 µg endoproteinase Lys-C (Wako) in 15 µL digestion buffer. The membrane was removed from the digestion buffer, washed once with 5 µL of fresh digestion buffer, which was then pooled with the remaining buffer, acidified with 1 µL glacial acetic acid, and reduced in volume to <5 µL by vacuum centrifugation. The samples were diluted to 10 µL with 0.1 M acetic acid, and stored at -20 °C.

HPLC and Mass Spectrometry. Peptide mixtures were introduced to the ESI source of the triple quadrupole mass spectrometer by microcapillary high performance liquid chromatography (HPLC). Reverse phase microcolumns were constructed according to a procedure derived from Kennedy and Jorgenson (*46*). A frit was created at one end of a 40 cm length of fused silica capillary (75 µm i.d., 180 µm o.d., Polymicro Technology) by sintering a plug of 5 µm silica particles with a microtorch. The resulting frit was pressure tested at 500 psi with 2-propanol driven from a vial in a stainless steel pressure vessel (bomb) pressurized with helium.

The capillary was packed with POROS II/RH (PerSeptive BioSystems) reverse phase beads. A slurry of packing material and 2-propanol in a half-dram vial was sonicated for 30 seconds to disrupt aggregates of beads. A magnetic stir bar was added to the slurry and the vial was placed in the bomb which rested on a magnetic stir plate. Pressurization of the bomb with helium (<100 psi) forced a steady flow of packing into the capillary. Columns were packed to a bed length of 10-15 cm, and washed with 2-propanol and 5% acetic acid at 500 psi. Columns were conditioned with 1 pmol angiotensin and one HPLC gradient prior to first use, and stored with the ends immersed in deionized water.

Samples were loaded onto the column hydrostatically, by the same means used to pack the column (see Figure 2). The amount of sample loaded was determined by measuring the volume of solvent displaced from the column with a 1-5 µL disposable pipet. The HPLC conditions are given in Table I. A pre-column split of the mobile phase (flowing at ~200 µL/min) was used to obtain column flow rates of 0.2 - 0.5 µL/min. A 50 µm i.d. restriction capillary was used to control the back pressure of the system.

Mass spectra were recorded on a Finnigan TSQ-7000 triple quadrupole mass spectrometer with a Finnigan electrospray ionization (ESI) ion source. Analytes were desolvated and transmitted to the first quadrupole by passage through a heated capillary and differentially pumped octapole lens region. Typical operating parameters are listed in Table I. Peptide quantities were estimated from the absolute detector signal with respect to known amounts of previously analyzed synthetic peptides (*47*).

Table I. LC-MS and MS/MS Conditions

Microcapillary HPLC	
Column:	POROS R2 10 μm beads packed 15 cm in fused silica capillary 75 μm i.d. x 180 μm o.d. x 30 cm. run directly to ESI needle.
HPLC pump:	ABI model 140A dual syringe pumps.
Gradient:	Linear; 0-80% acetonitrile in 0.1M acetic acid over 15 min.
Flow rates:	200 μL/min. from the pump, split pre-column to achieve 250 nL/min through the column.
MS Conditions	
ESI voltage:	4.5 kV from needle to capillary.
ESI gases:	38 p.s.i. sheath gas; minimum positive flow of auxiliary gas.
ESI sheath:	70% methanol/ 0.1M acetic acid liquid sheath at 1.5 μL/min.
MS scan:	Quadrupole 1 scanned from 250-1500 u every 1.5 sec.
Resolution:	FWHM = 1.5 u
Detection:	15 KV conversion dynode with electron multiplier set to 1250 V.
MS/MS Conditions	(same as above except as follows)
HPLC flow:	slowed by 50% approximately one minute before peptides elute.
Resolution:	3-4 u window around parent in Q1; FWHM =1.5 u in Q3.
Collison gas:	2.5 mTorr Ar in Q2.
Q3 scan:	Quadrupole 3 scanned from m/z 50 to $[M+H]^+$ mass at 500 u/sec.
Detection:	Electron multiplier voltage raised to 1700 V.

Database Searching. The program FRAGFIT, developed at Genentech, was used for database searching (*8*). For each protein spot, the masses of the peptides were entered along with a partial sequence from the MS/MS spectrum (if available) that corresponded to each peptide. Parameters that can be altered are cleavage specificity, mass tolerance, protein molecular weight range, cysteine alkylation reagent, number of unmatched masses allowed, allowance for methionine oxidation, and allowance for incomplete protease cleavage. All proteins that match according to the input masses and chosen parameters are listed, with those proteins having the most peptides matching listed first. No other scoring procedures are used. The database consists of the merged SWISSPROT, PIR, and GenBank, and currently includes over 166,000 protein sequences. The program was written in C and runs on a DEC Alpha computer that uses the UNIX operating system.

Results

The Coomassie-stained blot from a rat neonatal cardiac myocyte lysate 2-D gel is shown in Figure 3. Spots that were selected (at random) for identification are numbered 1-8. The 2-D gel system used for the separation utilized an IPG strip for the first dimension. The IPG strips have been shown to provide a wider pI range and greater gel-to-gel reproducibility than more traditional tube gels that are based on ampholines (*48*). No pI marker proteins were added to this gel. Molecular weight markers were run in the second dimension only. Coomassie staining of the blot revealed approx. 200 proteins. Since 2-D gels of radiolabeled proteins frequently display in excess of 2000 spots, less than 10% of those are abundant enough to be observed by Coomassie staining. Silver staining can provide detection at lower levels (more than 3000 spots have been observed in a single gel by silver staining (*34*)); however, its use precludes direct analysis of the protein spots.

The selected protein spots were cut from the blot and placed in an Eppendorf tube. Each protein was reduced and alkylated on the membrane in order to obtain optimal proteolytic cleavage. The membrane was then blocked with a polymer to

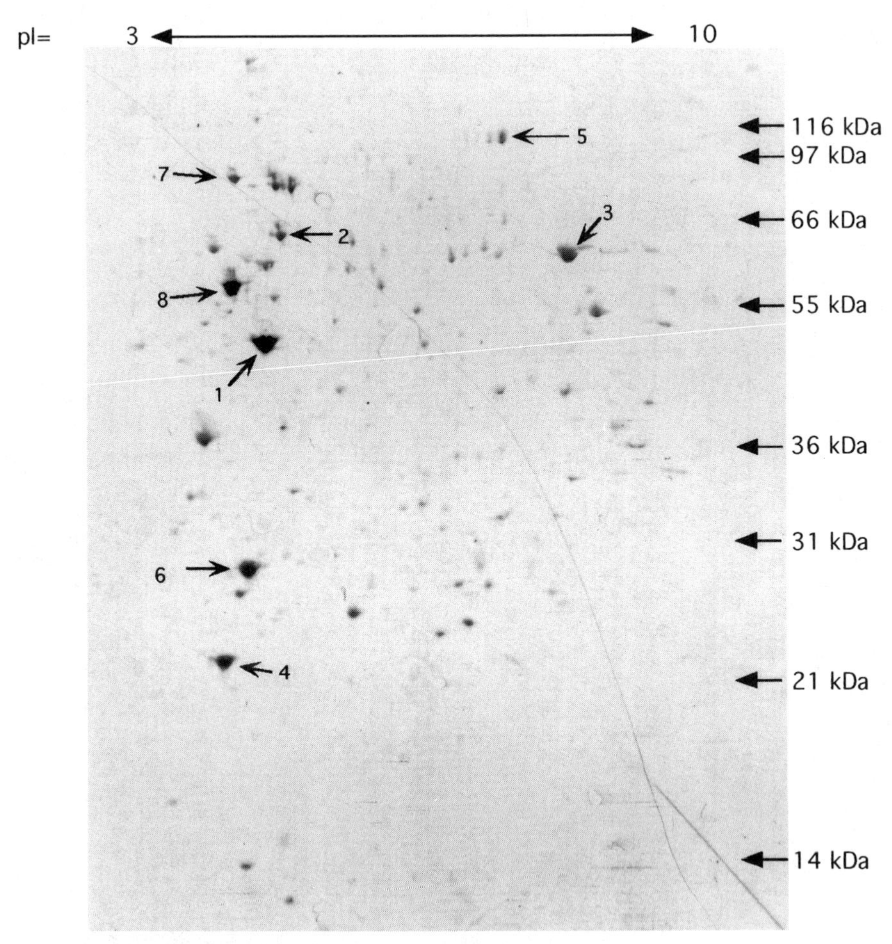

Figure 3. Coomassie-stained blot of neonatal rat cardiac myocyte lysate separated by 2-D gel electrophoresis. The numbered protein spots were selected for identification. Their identities are given in Table II.

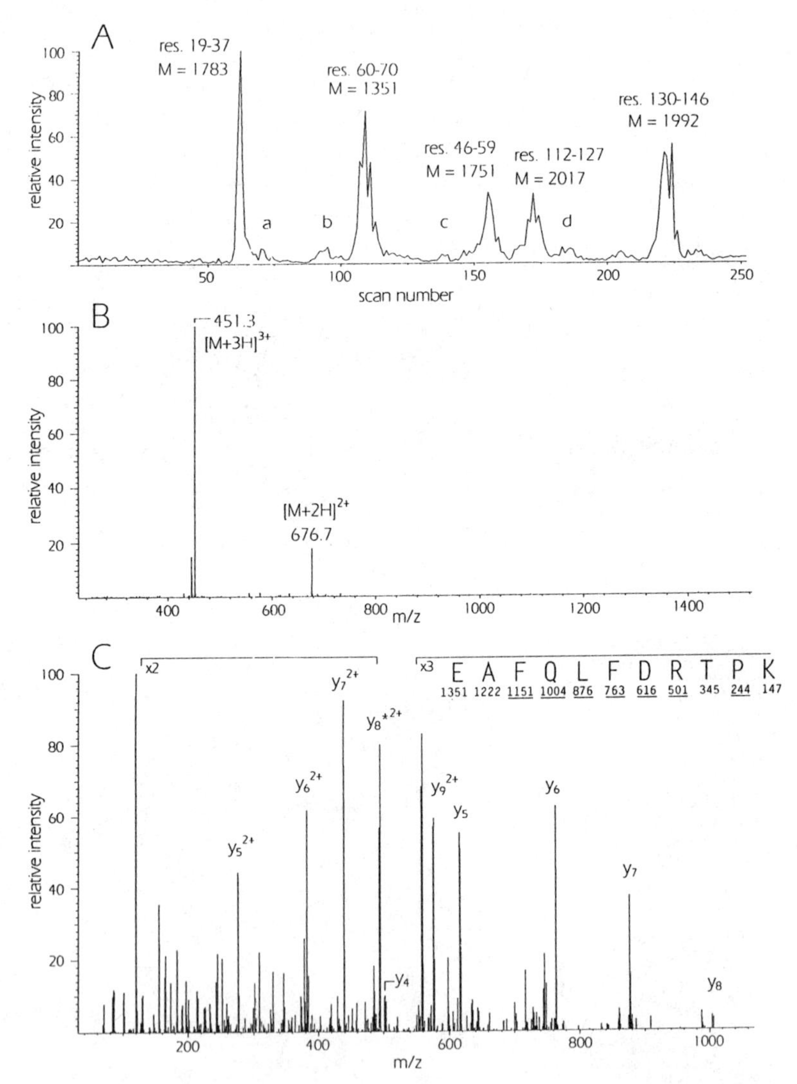

Figure 4. Mass spectral data from peptides derived from spot #6. (A) LC-MS base peak profile. The amino acid residues and nominal molecular mass for each of the main components is labeled. Other minor components are a: res. 176-194, M=2163; b: res. 48-59, M=1509; c: res. 75-104, M=3284; d: res.147-175, M=3170. (B) Mass spectrum for the second major chromatographic peak. (C) MS/MS fragment ion spectrum of precursor m/z 451.3. Only the y-ion series is labeled for simplicity. The inset sequence was taken from the FRAGFIT match. The expected y-ion masses are listed under the sequence; the observed y-ions are underlined. The y_9-ion was observed only in the +2 state. The clearly-evident partial sequence FQ[I,L]F was used as input for FRAGFIT (Ile and Leu cannot be distinguished by mass).

prevent adsorption of the protease to the membrane or re-adsorption of the released peptides. PVP-360 was chosen over the formerly-used PVP-40 due to a lower background observed in UV-detection for capillary HPLC. The protein was digested *in situ* with Lys-C endoprotease. The digestion buffer contained 10% acetonitrile to increase the release of peptides from the membrane.

An aliquot of the released peptide mixture (10%) was used for mass analysis by capillary LC-MS. The capillary HPLC system was developed to minimize sample handling and maximize sensitivity (*49*). The base peak profile is shown in Figure 4A for the peptides from spot #6 in Figure 3. The base peak profile plots the intensity of the most abundant mass spectral peak in each scan as a function of time. The result is similar to the total ion current profile but the contributions from background ion abundances are eliminated, with a resulting enhancement in the signal-to-noise ratio. Each peak is labeled with the molecular mass of the main component, calculated from the observed m/z values. The mass spectrum of the second main component (residue 60-70 peptide) is shown in Figure 4B. Based on the absolute intensities for standards, this spectrum corresponds to approx. 100 fmol for this experiment, which represents 10% of the sample.

The triply-charged ion (m/z 451.3) was chosen as one of the precursor masses for a subsequent LC-MS/MS experiment (*49-51*). The product ion mass spectrum for parent m/z 451.3 is shown in Figure 4C The resolution of MS1 was reduced to allow a 3-4 u window of ions to pass to the center quadrupole in order to obtain maximum signal. The fragment ions are also analyzed at less than unit resolution. Amino acid residues are assigned on the basis of mass differences between peaks. The partial sequence FQ[I,L]F was determined from this spectrum. Leucine and isoleucine cannot be distinguished in the low-energy collisions employed in triple quadrupole mass spectrometers; both possibilities must be used for the search. More complete sequence information can often be obtained with more extensive, time-consuming mass spectral interpretation, but determination of a shorter sequence requires only a few minutes and is sufficient for the database searching (see below).

To compare the amount of data needed to identify the protein in spot #6 uniquely, the output for several FRAGFIT searches is given in Table III. A database search using only the major observed peptide masses returned 4 possible matches when the mass tolerance was set to ±0.5 Da, typical of low-level electrospray data on a quadrupole instrument. When all the data are used (all peptide masses, major and minor, and partial sequence), myosin light chain was the only match. The added specificity of partial sequence is demonstrated more dramatically by a search based on a single mass, with four amino acids determined. Again, only myosin light chain is found. Frequently, methionines are observed to be oxidized. The program includes methionine oxidation (labeled 'MO' in Table III results) as a possibility for each peptide. When partial sequence is included, the program does not require that the positions of the residues within the peptide are known, just that the amino acids are contiguous in the sequence. The current implementation of the algorithm does require that the sequence be given with the most N-terminal residue first. Fortunately, frag-ment ion spectra of most tryptic and Lys-C peptides yield predominantly y-type fragment ions in the higher mass portion of the spectrum, which makes assignment of the sequence directionality straightforward.

Equivalent LC-MS and LC-MS/MS experiments were performed for each of the other spots labeled in Figure 3. The identities of the spots are listed in Table II. In each case, when possible, partial sequence was used in addition to the peptide masses as the FRAGFIT input. Normally, only 2-3 of the peptide masses were selected as precursors for LC-MS/MS experiments. All yielded a minimum of 2-3 amino acids partial sequence.

Table II. Proteins Identified from 2D Gel

Spot	*Protein Name*	Mr_{calc}[a]	pI_{calc}[a]
1	Actin	42 kD	5.3
2	Mitochondrial matrix protein p1 precursor	61 kD	6.1
3	ATP synthase alpha chain, mitochondrial precursor	59 kD	9.4
4	Myosin regulatory light chain 2, cardiac muscle isoform	19 kD	4.9
5	Aconitate hydratase precursor, sequence from pig	86 kD	8.4
6	Myosin light chain 1, ventricular isoform	22 kD	5.1
7	78 kDa glucose regulated protein precursor	73 kD	5.1
8	ATP synthase beta chain, mitochondrial precursor	56 kD	5.3

[a]Molecular mass and pI were calculated from the published sequence of each protein, and do not reflect post-translational processing or modification, or nonlinearity in the scales of IEF or SDS-PAGE.

Discussion

Two-dimensional gel electrophoresis provides extremely high resolution for the separation of very complex protein mixtures, such as whole cell lysates. The use of 2-D gels has grown in recent years due, in part, to the use of new methods that yield improved reproducibility, particularly immobilized pH gradients (*52*). In addition, improved imaging techniques make protein spot quantitation easier and gel comparisons less subjective (*29, 53–55*). Statistical comparisons of spot intensities for multiple gels are used to distinguish real differences in spot intensities from differences due to gel-to-gel variation or cell preparation inconsistencies (*29, 56, 57*). Comparison of 2-D gels has been used to show, among other things, differences in protein expression that arise due to cellular transformation, differences in stages of development, and differences due to treatment with various stimulating factors (*31*).

The utility of 2-D gels relies ultimately on the ability to identify the proteins in the visualized spots. A number of databases have been developed to map the proteins in several cell lines, which should facilitate rapid identification of protein spots of interest. The availability of several of these databases on the World Wide Web makes them useful to researchers around the globe. The proteins from individual 2-D gel spots have been identified in the past by automated Edman degradation (*31, 34, 58*), amino acid analysis (*59*), or immunostaining (*58, 60*). More recently, peptide mass fingerprinting has proven useful and offers advantages over these other methods. The growth of protein sequence databases has also made routine the identification of known proteins from a minimal amount of amino acid sequence. A number of reasons argue in favor of alternatives to Edman degradation. More than 50% of eukaryotic proteins are estimated to be N-terminally blocked and thus resistant to the Edman chemistry (*61*). Methods for generation of internal sequence from fragments at low picomole levels are time- and labor-intensive due to requirements for sample-handling and separation of peptide mixtures. Most modern instruments have cycle times of 20-30 minutes, with setup times per sample of approx. 1 hr. Finally, after significant effort, the sequence found often shows that the protein already exists in a sequence database.

Peptide mass fingerprinting was used to identify proteins in rat neonatal cardiac myocytes. The eight proteins identified (listed in Table II) are known to be expressed at high levels in normal cells . The positions (Mr and pI) of these spots (Figure 3) correspond approximately to the positions of the same proteins in an Internet-accessible rat 2-D gel database, REF52 (*62*). The lack of pI markers in the gel shown here, the difference in the cell lines, and differences in the gel formats preclude precise matching to that database. The REF52 database, like most of the other databases, is based on autoradiograms or phosphorimager scans from 2-D gels of ^{35}S radiolabeled proteins. The images show over 2,000 spots for the most highly resolved gels. In contrast, Coomassie stained 2-D gels or blots show only about 200 spots, reflecting the much higher detection limits. As a result, most of the proteins actually present on the gel or blot are not imaged. This situation is not catastrophic, however, since spots below the detection limit for Coomassie staining are not able to be identified directly by the state-of-the art Edman degradation or peptide mass fingerprinting. Often more protein cannot be loaded on a high resolution gel, because the total protein capacity is limited to <1 mg; overloading a 2-D gel compromises its resolution. This problem can be addressed by the use of thicker, preparative gels, or by fractionating the proteins before loading on the gel.

Two-dimensional gel electrophoresis databases of human cardiac myocyte cytosolic proteins have been established by two separate groups (*33, 34*). Unfortunately, the difference in species (human vs rat) precludes use of these databases for our work. Corbett et al. (*63*) and Jungblut et al. (*34, 64*) have compared normal and hypertrophic cardiac myocytes and found 20 spots that differ between the normal and diseased states.

The detection limit for peptide mass fingerprinting at present is 1-10 pmol for proteins smaller than 100 kD. The limit is due to inefficiencies in blotting, enzymatic digestion, release of peptides from the membrane, and to adsorptive losses of the peptides to tube walls and sample transfer devices. Efforts are continuing to reduce the amount of protein needed for successful identification. Detection limits for the mass measurement have not been limiting for the overall experiment.

Successful protein identification in our laboratory has been limited most often by the mass tolerance of the measurements from linear MALDI-TOF experiments. Results for a FRAGFIT search using the data (five major observed peptide masses in Figure 4A) and parameters listed in Table III, but with a mass accuracy of ±2.0 Da, (typical of linear MALDI-TOF data with external calibration) returned 79 possible matches. The results in Table III show in a single example the improvement provided by greater mass accuracy. The even higher mass accuracy that can be achieved by Fourier transform mass spectrometry (<10 ppm) (*65, 66*) promises to further improve the matching selectivity. While a mass accuracy of ±0.5 Da for the five major peptide masses (Figure 4A) resulted in 4 protein matches from the database (see Table III), a search with a hypothetical mass accuracy of ±0.03 Da (10 ppm for 3,000 Da peptide) for these same five major peaks (theoretical masses used) resulted in a single protein match. Improvement in matching specificity is also produced by use of more peptide masses. Finally, the addition of a minimal partial amino acid sequence provides a dramatic improvement in selectivity for the search (*24*).

In our experience with protein identification on 1-D and 2-D gels by peptide mass fingerprinting, we have not yet encountered a single false positive match. When a single putative match is found, the protein sequence is checked to determine whether other observed masses in the spectrum (e.g., low abundance peaks) also match; most normally do match for a correct identification. The majority of matches are checked subsequently by automated Edman degradation. False negative results (no match even though the sequence is contained in the database) are a more likely result. We typically use liberal parameter settings (e.g., large mass tolerance, many unmatched masses allowed, large protein mass range) for the initial FRAGFIT search in order to minimize missing a match; a large list of matching proteins is typically found at this point, viz. Table III. More conservative parameters are subsequently used to reduce the number of matches and improve the confidence in the search. If a match is only found with a "loose" fit, other confirming data are necessary, e.g., sequence, digestion with a second protease, immunoreactivity, etc.

Table III. FRAGFIT Results for Spot #6

Data Summary

5 major peaks with $[M+H]^+$ 1783.8, 1352.1, 1751.8 2017.6, 1992.6
4 smaller peaks with $[M+H]^+$ 2164.1, 1510.1, 3285.1, 3171.1
MS/MS data yields partial sequences for 1352.1 : FQ[I,L]F and 1783.8 : APE

FRAGFIT Parameters (chosen for all searches)

Lys-C specificity
Oxidized Met allowed
Partial digestion allowed
One unmatched mass allowed
Carboxymethyl Cys assumed

A. *Input:* Masses of the 5 major peaks and mass tolerance of ± 0.5 Da.,
Output: 4 database hits

```
Myosin light chain 1, slow-twitch muscle
     b/ventricular isoform - (22025.01 Da)
Reticulocyte-binding protein 1, PvRBP-1
     =transmembrane-anchored (325851.41 Da)
Reticulocyte binding protein 1 precursor - plasmodium
     vivax (330217.57 Da)
Titin - rabbit (fragment) (751113.32 Da)
[4 specified molwts matched.  not found: 2017.60]
```

B. *Input:* Masses of all peaks, mass tolerance of ± 0.5 Da, and partial sequences,
Output: 1 database hit:

```
Myosin light chain 1, slow-twitch muscle
     b/ventricular isoform - (22025.01 Da)
    1784.03  0.23   19: AAPAPAAAPAAAPEPERPK
    1510.68  0.58   48: IEFTPEQIEEFK
    1752.02  0.22   46: IKIEFTPEQIEEFK
    1352.53  0.43   60: EAFQLFDRTPK
    3284.80  0.30   75:
     ITYGQCGDVLRALGQNPTQAEVLRVLGKPK
3MO 2017.38  0.22  112: MMDFETFLPMLQHISK
    1992.15  0.45  130: DTGTYEDFVEGLRVFDK
1MO 3171.51  0.41  147: EGNGTVMGAELRHVLATLGERLTEDEVEK
1MO 2164.37  0.27  176: LMAGQEDSNGCINYEAFVK
```

C. *Input:* A single mass plus partial sequence: 1352.1 FQ[I,L]F,
Output: 1 database hit:

```
Myosin light chain 1, slow-twitch muscle
     b/ventricular isoform - (22025.01 Da)
```

The increase in matching specificity provided by partial sequence and high mass accuracy can be used to relax the requirement that most of the peptide masses match, yet still maintain a high degree of confidence in the match. The allowance for many unmatched peptides gives the method a new dimension of robustness for several reasons. Since a small number of peptides from a protein are frequently sufficient for matching, the absence of a match for a particular peptide (e.g., due to post-translational modification, protein processing, isoforms, alternative splicing, partial cleavage, database error) is not crucial. Furthermore, a protein can be identified in the presence of peptides from other proteins since only those peptides that match are crucial for the identification. The specificity of the match can be improved by requirements for higher mass accuracy, fewer unmatched peptide masses, and same-species origin. Higher mass accuracy, in particular (see above), can substantially reduce the peptide degeneracy (number of peptides with the "same" mass) for a given observed mass, since amino acid masses are known to better than four decimal places. Post-translationally modified peptides (which are most often not completely characterized, and even then rarely annotated in sequence databases) will not preclude a match. Multiple proteins in a single gel spot can be identified. Database inaccuracies can be tolerated as can isoforms and alternatively spliced proteins. Incomplete sequence entries in the database (see below) can be tolerated.

The present work utilizes capillary HPLC and electrospray ionization mass spectrometry. Although we earlier demonstrated the use of capillary LC-MS for peptide mass fingerprinting (*9, 42*) most of the work in our lab has utilized data acquired by MALDI mass spectrometry. What are the advantages of each technique for peptide mass fingerprinting? MALDI mass spectrometry can provide peptide masses for a mixture, at low levels (<50 fmol), in the presence of buffers, in a short period of time (<10 min). Throughput can be considerable and high sensitivity operation is routine. However, mass accuracy for linear time-of-flight instruments with external calibration is rather poor, typically ±2 Da for peptides. Although the accuracy can be improved to approx. ±0.4 Da with internal mass standards, measurement of sub-picomole samples may yield low S/N and often precludes addition of internal standards which suppress the ionization of the analyte. For the same reason, many peptides in complex mixtures are not observed due to ion suppression which seems to be exacerbated at low sample levels.

The use of small diameter packed capillary HPLC columns (<100 μm) with sample loading that is adapted to small quantities of peptides has opened up the use of LC-MS for peptide mass fingerprinting. Detection limits <50 fmol are possible routinely and mass accuracy is typically better than ±0.5 Da. The HPLC separation often allows more components to be observed since ion suppression is substantially reduced. The ability to determine partial sequence by LC-MS/MS gives much greater specificity to the database searches. Disadvantages include a greater amount of time for the experiment (approx. 30 min per HPLC run), the time required for sequence interpretation of the MS/MS spectra (30 min or more per spectrum), and the challenge of maintaining low femtomole detection limits on a multi-user instrument.

Future Prospects

The preliminary results presented here show that peptide mass fingerprinting is useful for the identification of proteins from normal cardiac myocytes. Experiments are in progress to use an image analysis system to develop a 2-D gel database in order (1) to demonstrate the reproducibility of replicate gels, and (2) to identify proteins that differ between normal and hypertrophic cells.

Continuing improvements in the techniques and methodology for peptide mass fingerprinting suggest that its usefulness should grow further. For example, recent advances in MALDI instrumentation promise to overcome many of the problems discussed earlier. Reflectron TOF instruments (*67*) routinely provide significantly better mass accuracy in the absence of internal standards (±0.02%) than the linear mode of operation. New methods for sample preparation hold promise for even greater mass accuracy and sensitivity (*68, 69*). Fragment ions that arise from metastable decomposition (post-source decay) can be observed with reflector TOF instruments and thus provide partial sequence data (*70, 71*). Peptide ladder sequencing (*72*) with volatile reagents (*73*) also holds promise for providing sequence data at low levels. Finally, significantly greater mass accuracy (±10 ppm) is obtainable by MALDI-FTMS and detection limits similar to TOF appear possible (*65, 66*).

Advances in electrospray instrumentation should also extend the capabilities for peptide mass fingerprinting. In particular, quadrupole ion traps have the potential to extend detection limits below those of conventional quadrupole instruments (*74*). Ion traps are also capable of very high sensitivity tandem mass spectrometry. One of their advantages is the capability to produce MS^n spectra, so that further fragmentation can be achieved to yield potentially more sequence information. ESI-FTMS also has enormous prospects. Chief among its features are extremely high resolution, which allows unambiguous identification of the charge on any ion from the isotope spacing (*75*). At the same time, extremely high mass accuracy for ESI-FTMS can be obtained. Ultrahigh resolution tandem mass spectrometry has also been shown to be capable of generating partial sequences from intact proteins (*76-78*).

One of the most promising aspects of peptide mass fingerprinting is the explosive growth of sequences in protein sequence databases. The number of sequences in our rather eclectic database has nearly doubled in the last two years, and the rate of growth continues to increase. More impressive, and potentially of greater importance has been the phenomenal and until recently unexpected growth of partial cDNA sequence entries in EST (expressed sequence tag) databases (*3, 4*). These sequences come from random sequencing of cDNAs. It is expected that ESTs will be found for virtually all human genes within the next 1-2 years. These EST databases will have significant impact on the peptide mass fingerprinting method. First, the number of unknown sequences will drop to a small number. The advantages of peptide mass fingerprinting with respect to Edman degradation will increase the importance of the technique. Second, the specificity of the matching programs will be more important. The lack of complete sequence for most of the genes means that matching programs must require only a few peptides, and that they be tolerant of many peptides that correspond to the non-sequenced portion of the gene.

The ultimate success of the human genome project in generating a sequence map should further increase the usefulness of the peptide mass fingerprinting method. Complete sequences for all DNA, and by inference all proteins, will further enhance the usefulness of peptide mass fingerprinting.

Acknowledgments

The authors acknowledge the enormous contribution by Colin Watanabe for design, implementation, and continual refinement of the FRAGFIT program. We also acknowledge Kathy King for preparation of the cardiac myocytes, and Susan Wong for assistance with 2-D electrophoresis.

Literature Cited

1. Edman, P. *Acta Chem. Scand.* **1950,** *4*, 238-293.
2. Sanger; Nicklen; Coulson *Proc. Natl. Acad. Sci., USA* **1977,** *74*, 5463-5467.
3. Adams, M. D.; Kelley, J. M.; Gocayne, J. D.; Dubnick, M.; Polymeropoulos, M. H.; Xiao, H.; Merril, C. R.; Wu, A.; Olde, B.; Moreno, R. F.; Kerlavage, A. R.; McCombie, W. R.; Venter, J. C. *Science* **1991,** *252*, 1651-1656.
4. Adams, M. D.; Kerlavage, A. R.; Fields, C.; Venter, J. C. *Nature Genet.* **1993,** *4*, 256-267.
5. Yager, T. D.; Zewert, T. E.; Hood, L. E. *Acc. Chem. Res.* **1994,** *27*, 94-100.
6. Marshall, E. *Science* **1995,** *267*, 783-784.
7. Henzel, W. J.; Stults, J. T.; Watanabe, C. *Third Symposium of the Protein Society*, Seattle, WA 1989.
8. Henzel, W. J.; Billeci, T. M.; Stults, J. T.; Wong, S. C.; Grimley, C. G.; Watanabe, C. W. *Proc. Natl. Acad. Sci., USA* **1993,** *90*, 5011-5015.
9. Henzel, W. J.; Grimley, C.; Bourell, J. H.; Billeci, T. M.; Wong, S. C.; Stults, J. T. *Methods* **1994,** *6*, 239-247.
10. Henzel, W. J.; Billeci, T. M.; Stults, J. T.; Wong, S. C.; Grimley, C.; Watanabe, C. In *Techniques in Protein Chemistry V*; Crabb, J. W., Ed.; Academic Press: San Diego, 1994, pp 3-9.
11. Mann, M.; Hojrup, P.; Roepstorff, P. *Biol. Mass Spectrom.* **1993,** *22*, 338-345.
12. Pappin, D. J. C.; Hojrup, P.; Bleasby, A. J. *Curr. Biol.* **1993,** *3*, 327-332.
13. James, P.; Quadroni, M.; Carafoli, E.; Gonnet, G. *Biochem. Biophys. Res. Commun.* **1993,** *195*, 58-64.
14. Mortz, E.; Vorm, O.; Mann, M.; Roepstorff, P. *Biol. Mass Spectrom.* **1994,** *23*, 249-261.
15. James, P.; Quadroni, M.; Carafoli, E.; Gonnet, G. *Protein Sci.* **1994,** *3*, 1347-1350.
16. Yates, J. R.; Speicher, S.; Griffin, P. R.; Hunkapiller, T. *Anal. Biochem.* **1993,** *214*, 397-408.
17. Cottrell, J. S. *Peptide Res.* **1994,** *7*, 115-124.
18. Patterson, S. D. *Anal. Biochem.* **1994,** *221*, 1-15.
19. Ji, J.; Whitehead, R. H.; Reid, G. E.; Moritz, R. L.; Ward, L. D.; Simpson, R. J. *Electrophoresis* **1994,** *15*, 391-405.
20. Rasmussen, H. H.; Mortz, E.; Mann, M.; Roepstorff, P.; Celis, J. E. *Electrophoresis* **1994,** *15*, 406-416.
21. Gold, M.; Yungwirth, T.; Sutherland, C. L.; Ingham, R. J.; Vianzon, D.; Chiu, R.; vanOostveen, I.; Morrison, H. D.; Aebersold, R. *Electrophoresis* **1994,** *15*, 441-453.
22. Clauser, K. R.; Hall, S. C.; Smith, D. M.; Webb, J. W.; Andrews, L. E.; Tran, H. M.; Epstein, L. B.; Burlingame, A. L. *Proc. Natl. Acad. Sci., USA* **1995,** *92*, 5072-5076.
23. Sutton, C. W.; Pemberton, K. S.; Cottrell, J. S.; Corbett, J. M.; Wheeler, C. H.; Dunn, M. J.; Pappin, D. J. *Electrophoresis* **1995,** *16*, 308-316.
24. Mann, M.; Wilm, M. *Anal. Chem.* **1994,** *66*, 4390-4399.
25. Mann, M.; Laukien, F. H.; Kruppa, G. H.; Watson, C. H.; Wronka, J.; Knobeler, M. *41st ASMS Conference on Mass Spectrometry*, San Francisco 1993; 899a-899b.
26. Eng, J. K.; McCormack, A. L.; Yates, J. R., III *J. Am. Soc. Mass Spectrom.* **1994,** *5*, 976-989.
27. Yates, J. R., III; Eng, J. K.; McCormack, A. L.; Schieltz, D. *Anal.Chem.* **1995,** *67*, 1426-1436.
28. O'Farrell, P. H. *J. Biol. Chem.* **1975,** *250*, 4007-4021.
29. Garrels, J. I. *J. Biol. Chem.* **1989,** *264*, 5269-5282.

30. Celis, J. E.; Olsen, E. *Electrophoresis* **1994,** *15*, 309-344.
31. Celis, J. E.; Rasmussen, H. H.; Leffers, H.; Madsen, P.; Honore, B.; Gesser, B.; Dejgaard, K.; Vandekerckhove, J. *FASEB J.* **1991,** *5*, 2200-2208.
32. Celis, J. E.; Rasmussen, H. H.; Olsen, E.; Madsen, P.; Leffers, H.; Honore, B.; Dejgaard, K.; Gromov, P.; Vorum, H.; Vassilev, A.; Baskin, Y.; Liu, X.; Celis, A.; Basse, B.; Lauridsen, J. B.; Ratz, G. P.; Andersen, A. H.; Walbum, E.; Kjaergaard, I.; Andersen, I.; Puype, M.; Damme, J. V.; Vanderkerckhove, J. *Electrophoresis* **1994,** *15*, 1349-1458.
33. Corbett, J. M.; Wheeler, C. H.; Baker, C. S.; Yacoub, M. H.; Dunn, M. J. *Electrophoresis* **1994,** *15*, 1459-1465.
34. Jungblut, P.; Otto, A.; Zeindl-Eberhart, E.; Pleissner, K.-P.; Knecht, M.; Regitz-Zagrosek, V.; Fleck, E.; Wittmann-Liebold, B. *Electrophoresis* **1994,** *15*, 685-707.
35. Garrels, J. I.; Futcher, B.; Kobayashi, R.; Latter, G. L.; Schwender, B.; Volpe, T.; Warner, J. R.; McLaughlin, C. S. *Electrophoresis* **1994,** *15*, 1466-1486.
36. Appel, R. D.; Sanchez, J. C.; Bairoch, A.; Golaz, O.; Miu, M.; Vargas, J. R.; Hochstrasser, D. F. *Electrophoresis* **1993,** *14*, 1232-1238.
37. Appel, R. D.; Bairoch, A.; Hochstrasser, D. F. *Trends Biochem. Sci.* **1994,** *19*, 258-260.
38. Hess, D.; Aebersold, R. *Methods* **1994,** *6*, 227-238.
39. Jenö, P.; Mini, T.; Moes, S.; Hintermann, E.; Horst, M. *Anal. Biochem.* **1995,** *224*, 75-82.
40. Rosenfeld, J.; Capdevielle, J.; Guillemot, J. C.; Ferrara, P. *Anal. Biochem.* **1992,** *203*, 173-179.
41. Fernandez, J.; DeMott, M.; Atherton, D.; Mische, S. M. *Anal. Biochem.* **1992,** *201*, 255-264.
42. Wong, S. C.; Grimley, C.; Padua, A.; Bourell, J. H.; Henzel, W. J. In *Techniques in Protein Chemistry IV*; Angeletti, R. H., Ed.; Academic Press: San Diego, 1993, pp 371-378.
43. Bauw, G. W.; VanDamme, S.; Puype, M.; Vanderkerchove, B. G.; Ratz, G. P.; Lauridsen, J. B.; Celis, J. E. *Proc. Natl. Acad. Sci., USA* **1989,** *86*, 7701-7705.
44. Iwamatsu, A. *Electrophoresis* **1992,** *13*, 142-147.
45. Fernandez, J.; Andrews, L.; Mische, S. M. *Anal. Biochem.* **1994,** *218*, 112-117.
46. Kennedy, R. T.; Jorgenson, J. W. *Anal. Chem.* **1989,** *61*, 1128-1135.
47. Hunt, D. F.; Henderson, R. A.; Shabanowitz, J.; Sakaguchi, K.; Michel, H.; Sevilir, N.; Cox, A. L.; Appella, E.; Engelhard, V. H. *Science* **1992,** *255*, 1261-1263.
48. Corbett, J. M.; Dunn, M. J.; Posch, A.; Gorg, A. *Electrophoresis* **1994,** *15*, 1205-1211.
49. Hunt, D. F.; Alexander, J. E.; McCormack, A. L.; Martino, P. A.; Michel, H.; Shabanowitz, J.; Sherman, N.; Moseley, M. A.; Jorgenson, J. W.; Tomer, K. B. In *Techniques in Protein Chemistry II*; Villafranca, J. J., Ed.; Academic Press: San Diego, 1991, pp 441-454.
50. Griffin, P. R.; Coffman, J. A.; Hood, L. E.; J. R. Yates, *Int. J. Mass Spectrom. Ion Processes* **1991**, 131-149.
51. McCormack, A. L.; Eng, J. K.; Yates, J. R. *Methods* **1994,** *6*, 274-283.
52. Goerg, A.; Postel, W.; Guenther, S. *Electrophoresis* **1988,** *9*, 531-546.
53. Dunn, M. J. In *Microcomputers in Biochemistry*; Bryce, C. F. A., Ed.; IRL Press: Oxford, 1992, pp 215-242.
54. Patton, W. F. *J. Chromatogr. A* **1995**, 698, 55-87.
55. Collins, P. J.; Juhl, C.; Lognonne, J. L. *Cell. Mol. Biol.* **1994,** *40*, 77-83.
56. Krauss, M. R.; Collins, P. J.; Blose, S. H. *BioTechniques* **1990,** *8*, 218-223.
57. Shi, C. Z.; Collins, H. W.; Garside, W. T.; Buettger, C. W.; Matschinsky, F. M.; Heyner, S. *Mol. Reprod. Devel.* **1994,** *37*, 34-47.

58. Baker, C. S.; Corbett, J. M.; May, A. J.; Yacoub, M. H.; Dunn, M. J. *Electrophoresis* **1992,** *13*, 723-726.
59. Jungblut, P.; Dzionara, M.; Klose, J.; Wittmann-Liebold, B. *J. Prot. Chem.* **1992,** *11*, 603-612.
60. Celis, J. E.; *et al. Electrophoresis* **1991,** *12*, 802-872.
61. Brown, J. L. *J. Biol. Chem.* **1979,** *254*, 1447-1449.
62. Boutell, T.; Garrells, J. I.; Franza, B. R.; Monardo, P. J.; Latter, G. I. *Electrophoresis* **1994,** *15*, 1487-1490.
63. Corbett, J. M.; Dunn, M. J.; Yacoub, M. *2-D PAGE '91: Proceedings of the International Metting on Two-Dimensional Electrophoresis*, London 1991; National Heart and Lung Institute; 202-206.
64. Jungblut, P.; Otto, A.; Regitz, V.; Fleck, E.; Wittmann-Liebold, B. *Electrophoresis* **1992,** *13*, 739-741.
65. Castoro, J. A.; Wilkins, C. L. *Anal. Chem.* **1993,** *65*, 2621-2627.
66. McIver, R. T.; Li, Y.; Hunter, R. L. *Proc. Natl. Acad. Sci., USA* **1994,** *91*, 4801-4805.
67. Cotter, R. J. *Anal. Chem.* **1992,** *64*, 1027A-1039A.
68. Vorm, O.; Mann, M. *J. Am. Soc. Mass Spectrom.* **1994,** *5*, 955-958.
69. Vorm, O.; Roepstorff, P.; Mann, M. *Anal. Chem.* **1994,** *66*, 3281-3287.
70. Kaufmann, R.; Spengler, B.; Luetzenkirchen, F. *Rapid Commun. Mass Spectrom.* **1993,** *7*, 902-910.
71. Kaufmann, R.; Kirsch, D.; Spengler, B. *Int. J. Mass Spectrom. Ion Processes* **1994,** *131*, 355-385.
72. Chait, B. T.; Wang, R.; Beavis, R. C.; Kent, S. B. H. *Science* **1993,** *262*, 89-92.
73. Bartlet-Jones, M.; Jeffery, W. A.; Hansen, H. F.; Pappin, D. J. C. *Rapid Commun. Mass Spectrom.* **1994,** *8*, 737-742.
74. McLuckey, S. A.; VanBerkel, G. J.; Flish, G. L.; Huang, E. C.; Henion, J. D. *Anal. Chem.* **1991,** *63*, 375-383.
75. Winger, B. E.; Hofstadler, S. A.; Bruce, J. E.; Udseth, H. R.; Smith, R. D. *J. Am. Soc. Mass Spectrom.* **1993,** *4*, 566-577.
76. McLafferty, F. W. *Acc. Chem. Res.* **1994,** *27*, 379-386.
77. O'Connor, P. B.; Speir, J. P.; Senko, M. W.; Little, D. P.; McLafferty, F. W. *J. Mass Spectrom.* **1995,** *30*, 88-93.
78. Speir, J. P.; Senko, M. W.; Little, D. P.; Loo, J. A.; McLafferty, F. W. *J. Mass Spectrom.* **1995,** *30*, 39-42.

RECEIVED July 21, 1995

Chapter 12

Lipid Metabolism of the Mosquito Pathogen *Lagenidium giganteum* and Its Hosts

Glycerophospholipids and Hydroxy Fatty Acids

James L. Kerwin

Botany Department 351330, University of Washington, Seattle, WA 98195

Lagenidium giganteum is a facultative parasite of mosquito larvae that is being used for operational control of these medically important arthropods. This water mold cannot synthesize sterols, but needs these lipids to enter its reproductive cycle. In order to infect mosquitoes, this Oomycete must produce motile zoospores by either a sexual or asexual process. The mosquito host, which also is a sterol auxotroph, provides suitable sterols and other nutrients for *L. giganteum* development, and in turn must obtain these lipids from its larval diet. Lipid-mediated morphogenesis plays a pivotal role in this host-parasite relationship, and determines in part whether the parasite will recycle and persist in mosquito breeding areas after an inital field application. The transition from mycelium (or pre-sporangium) to motile infective zoospore to infected mosquito larva involves complex developmental changes. We have used electrospray mass spectrometry and tandem mass spectrometry to examine changes in glycerophospholipid molecular species and oxygenated fatty acids in order to identify factors promoting reproduction *in vivo* and *in vitro* by this microbial pest control agent.

The interaction of parasitic and pathogenic microorganisms with other microorganisms, plants, insects, mammals or other animals involves a complex series of events: recognition of a suitable host, entry into the organism without triggering an overwhelming host defense response, and subsequent development using nutrients provided by the host. Morphogenesis and reproduction of pathogenic microorganisms is completely dependent upon utilization of host nutrients, with or without altering host metabolism. Strategies range from fairly passive adsorption of membrane fatty acids by viruses (1), relatively benign coexistence which implies minimal disruption of host metabolism by the male stage of some species of intracellular microsporidians (2,3), or complete destruction of insect tissue by bacteria transmitted by entomopathogenic nematodes (4).

Lagenidium giganteum (Oomycetes: Lagenidiales) is a facultative parasite of mosquito larvae that was recently registered by the U.S. Environmental Protection Agency for operational mosquito control (5). Oomycetes are a unique group of

0097–6156/95/0619–0244$12.00/0

organisms long considered members of the fungal kingdom. Recently, however, based on a number of ultrastructural and molecular studies, it has become evident that this group is more closely related to chromophyte algae and other heterokont protists than to the true fungi (6-9). This mosquito parasite is unique in that it cannot synthesize sterols (10), a characteristic shared only in this group by two related plant parasitic genera, *Pythium* and *Phytophthora* (11,12). It is interesting to note that mosquitoes, like other insects, are also sterol auxotrophs, and must obtain these lipids from their diet in order to complete their development. Unlike its larval hosts, *L. giganteum* can develop vegetatively in the absence of an exogenous sterol source, but requires these lipids to enter either asexual or sexual reproduction (13, 14). Despite numerous investigations over the last 30 years, the bases for sterol-mediated morphogenesis of sterol auxotrophic Oomycetes are not known.

In order to infect mosquitoes, motile zoospores must be produced through either a sexual or an asexual reproductive process. As part of our efforts to develop *L. giganteum* as an operational mosquito control agent, we have examined various aspects of this organism's lipid metabolism. We have focussed on this area not only because of the sterol requirement, but because the lipid composition of this parasite can be readily altered by changing the growth medium, which in turn regulates morphogenesis of the Oomycete (15-17). Regulation of morphogenesis noted in small scale liquid culture can be mimicked in pilot scale (up to 1000 L) fermentation (18), which has allowed multi-hectare aerial applications to assess the field efficacy of this microbial pest control agent (19, 20).

The strategy used by *L. giganteum* to kill its larval hosts and ensure its recycling and persistence following introduction into a mosquito breeding site is a combination of those listed above for parasites and pathogens. It does not produce any toxins, and host death occurs only after starvation. In a successful infection, the Oomycete proliferates throughout its host, completely filling the larval body cavity. An apparently healthy mosquito, upon microscopic examination, can be filled with vegetative mycelium, including over 75% of its head capsule, without obvious effect on larval behaviour. It is only after complete utilization of host nutrients that the mosquito dies, and the fungus enters its reproductive stage where it either releases asexual zoospores within ca. 24 hours which seek out new hosts, or in producing dormant oospores that ensure survival of the organism should the breeding site dry out or mosquito larvae not be present.

Commercial interest has been focussed on optimizing production of the sexual stage of *L. giganteum* because this dormant propagule is resistant to abrasion and desiccation, with storage for several years without special handling. We have shown that, in addition to sterols, eicosanoids and other hydroxy fatty acids (21) and cell membrane and nuclear cell membrane composition (22) regulate parasite entry into its sexual stage. The main advantage of using this microbial pest control agent is its ability to recycle for months or even years following a single application (23, 24). Recycling will depend in large part on larval diet, which in turn will control growth and morphogenesis of the parasite. In order to understand the role of the mosquito larval host in regulating *L. giganteum* development, we have initiated a study of comparative lipid metabolism as the parasite proceeds from the mycelial stage, through the formation of infective zoospores, and the effect of parasite infection on larval lipid composition. Many of these changes may be subtle, and we have made extensive use of electrospray mass spectrometry (ESI-MS) and tandem mass spectrometry (MS/MS) to characterize glycerophospholipids, and more recently sphingolipids, eicosanoids and other oxygenated fatty acids. This chapter gives examples of the utility of ESI-MS and MS/MS in examing comparative lipid metabolism in a host-parasite interaction.

Materials and Methods

The California strain (ATCC No. 52675) of *L. giganteum* was maintained in 1 L (2500 mL fernbach flasks) of PB2 medium (g/L distilled water): 2.0 Proflo cottonseed extract; 1.5 glucose; 0.5 hydrolyzed lactalbumin; 1.0 egg yolk; 0.1 fish peptone; 0.15 $CaCl_2 \cdot 2H_2O$; 0.15 $MgCl_2 \cdot 6H_2O$; 0.005 $Fe(NO_3)_2 \cdot 9H_2O$; pH 7.5. Defined culture medium, DM2, was a variation of Gleason's medium as previously described (22). Mycelial cultures were grown for 6 - 8 days to insure the Oomycete had reached a stationary growth phase. Zoospores were induced by filtering mycelium from the growth media and diluting the cells in distilled water. The Bakersfield colony of *Culex tarsalis* was obtained from W.R. Reisen, Arbovirus Field Station, University of California, Berkeley. Larvae were reared on guinea pig chow/Tetramin® fish food/yeast extract (1/1/0.01, w/w/w) using a 16:8 light-dark photoperiod. Japanese quail were used as a blood source. Late III and early IV instar larvae were infected with the Oomycete by exposing them to dense zoospore suspensions for ca. 20 hrs. Exposed larvae were removed and placed in fresh water. After 2 to 4 days, moribund larvae were removed and examined microscopically. Larvae completely filled with *L. giganteum* were used for subsequent lipid analyses.

Mycelia and uninfected (control) larvae were collected on filter paper and washed with distilled water. Zoospores were collected by siphoning dense suspensions onto dry ice, followed by centrifugation at 6000 x g as described (25). Infected larvae, because of their relatively fragile nature, were collected individually with pipettes, followed by centrifugation (1000 x g) to obtain a pellet of infected larvae. Glycerophospholipids were obtained by Folch extraction (26) and used without further purification. Free fatty acids and hydroxy fatty acids were obtained by base hydrolysis of lipid extracts followed by partition into ethyl acetate. Following centrifugation, the organic phase was collected, and the aqueous phase was extracted two more times with ethyl acetate (27). The combined organic phases were concentrated under nitrogen, with 25 µL pyridine added just before the ethyl acetate was completely evaporated. Removal of the ethyl acetate was completed, and the residual pyridine was vortexed with 4 mL water and applied to a 6 mL C_{18} Sep-Pak cartridge pre-equilibrated with water. The cartridge was washed with 6 mL water, and the oxygenated fatty acids eluted with 6 mL acetonitrile (28). The acetonitrile was removed under nitrogen, and the sample stored at -70° C in ethanol until mass spectrometric analysis.

Lipid samples were analyzed by ESI as described (29) using a triple-quadrupole Sciex API III instrument (PE/SCIEX) Thornhill, Ontario, Canada). Samples were infused into the source using a 50 µM i.d. fused silica transfer line using a Harvard Apparatus pump at 2-3 µL/min. Positive ion MS and MS/MS were run with an orifice voltage of 70 to 90 volts and negative ion spectra at -65 to -90 volts. The interface temperature was 52° C. For tandem mass spectrometry precursor ions were selected with the first quadrupole (Q1) for collision-induced dissociation with argon in the second quardupole (Q2). The third quadrupole (Q3) was scanned with a mass step of 0.2 Da and 1 ms/step. Precursor ion transmission was maximized by reducing the resolution of Q1 to transmit a 2 to 3 *m/z* window about the selected parent ion, and Q3 resolution was adjusted to approximately 50% valley between peaks 3 Da apart. Samples were dissolved at ca. 50 to 1000 pg/mL of 5 mM ammonium acetate in methanol to facilitate ionization.

All peaks are reported without the mass defect (rounded down to the nearest whole mass unit).

Results

Glycerophospholipids and hydroxy fatty acids. It has been shown that the combination of positive and negative ion ESI-MS and MS/MS can provide rapid and unparalleled resolution of nanogram quantities of glycerophospholipid (GPL) molecular species with minimal sample preparation (29, 30). This includes diacyl, alkyl/acyl and alkenyl/acyl species. Positive ion ESI-MS can be used to obtain a profile of whole cell (organelle or chromatographically purified classes or subclasses) GPL using less than a nanogram of total lipid. Comparison of the positive ion spectra for *L. giganteum* mycelium and zoospores (Figure 1A) shows that significant differences in GPL molecular species are generated during the transition from mycelium (or presporangia) to zoospores. This is not unexpected since these two stages in the life cycle involve the evolution of a relatively inert (stationary phase) cell with a well-defined cell wall to a small motile propagule delimited only by a cell membrane. Similar differences are found in the GPL portion of the positive ion spectra of control and infected mosquito larvae (Figure 1B).

Headgroup composition can then be identified using positive ion MS/MS, due to formation of an ion corresponding to either the headgroup itself or to the loss of the headgroup from the $[M + H]^+$ precursor ion. For instance, positive ion MS/MS of the precursor at *m/z* 716 from uninfected *C. tarsalis* shows a well-defined product ion at *m/z* 575, corresponding to loss of phosphoethanolamine from the precursor ion (Figure 2). There are a number of combinations of fatty acyl moieties that can generate a glycerophosphoethanolamine (GPE) species with this molecular weight. Negative ion MS/MS was then used to resolve individual molecular species, with the larval peak at *m/z* 714 (the negative ion peak corresponding to *m/z* 716 in the positive ion spectrum) shown to consist of two species, 16:1/18:2 and 16:0/18:3 GPE (Figure 2). Note that fatty acyl positional information is not generated by ESI-MS/MS. In biological systems the less saturated fatty acid is usually found in the *sn*-1 position of glycerophospholipids, and this is how molecular species are reported here as a matter of convention.

The resolution and sensitivity of ESI-MS/MS is demonstrated by examination of the tandem mass spectrum of a minor component of uninfected *Culex* mosquito larvae, the $[M-H]^-$ peak at *m/z* 762, which shows the presence of 16:1/22:5, 16:0/22:6, 18:2/20:4 and 18:1/20:5 GPE species (Figure 3). Another example shows the presence of several unique GPE species produced by *L. giganteum.* Positive and negative ion MS/MS of GPL extracts showed the presence of GPE species that contained highly unsaturated 20-carbon fatty acids in both *sn*-1 and *sn*-2 positions (Figure 4, Table I). To my knowledge, this has not been found in mammals (31, 32), other animals (33) or higher plants (34), where fatty acids with three or more sites of unsaturation in the *sn*-1 position are usually found with fatty acids having 0-2 unsaturated bonds in the *sn*-2 position. Only in diatoms and related chromists have comparable unsaturated molecular species been documented (35), further supporting the relationship of Oomycetes with this group of organisms.

There were significant changes in GPE molecular species found in the developmental sequence of mycelium → zoospore → uninfected larva → infected larva (Table I). As further confirmation of the effect of growth medium on *L. giganteum* membrane composition, the GPE molecular species composition was also examined where the Oomycete was grown on defined medium in the presence of isoprenoids that do (cholesterol, ergosterol) or do not (squalene, cholestane) promote reproduction (Table II). As was found in previous studies documenting changes in whole cell fatty acid composition (15 - 17), diet, whether *in vivo* or *in vitro*, has a significant effect on the lipid metabolism of this parasite. One common

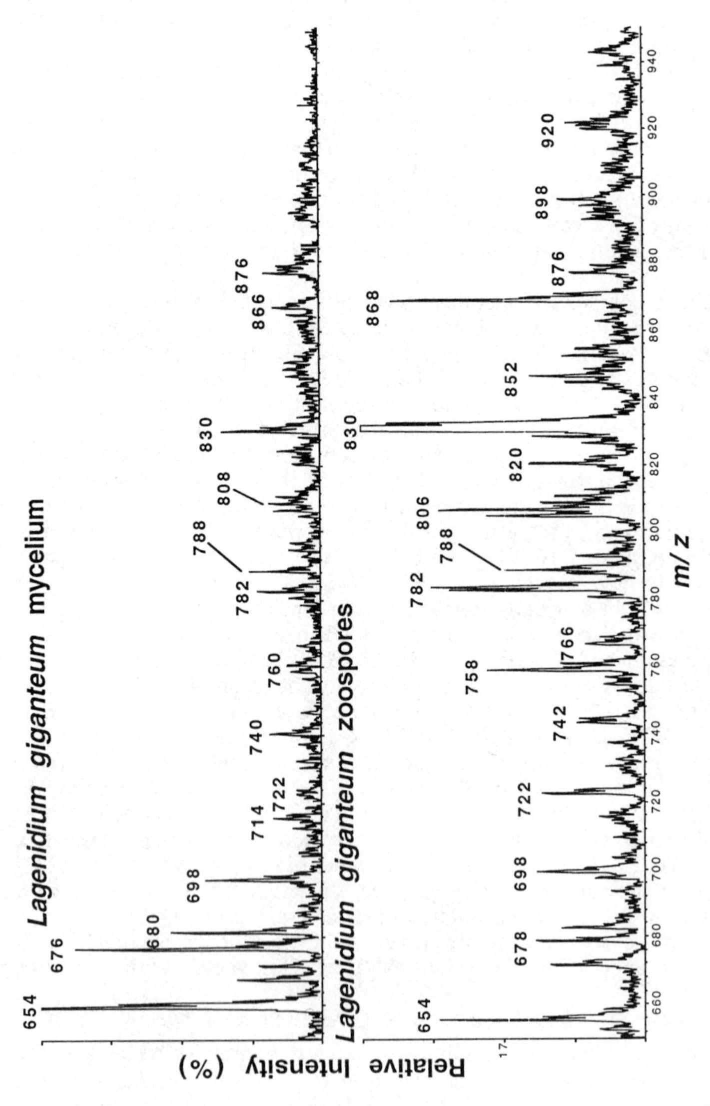

Figure 1A. Comparison of the glycerophospholipid region of the positive ion mass spectra of *Lagenidium giganteum* mycelium and zoospores.

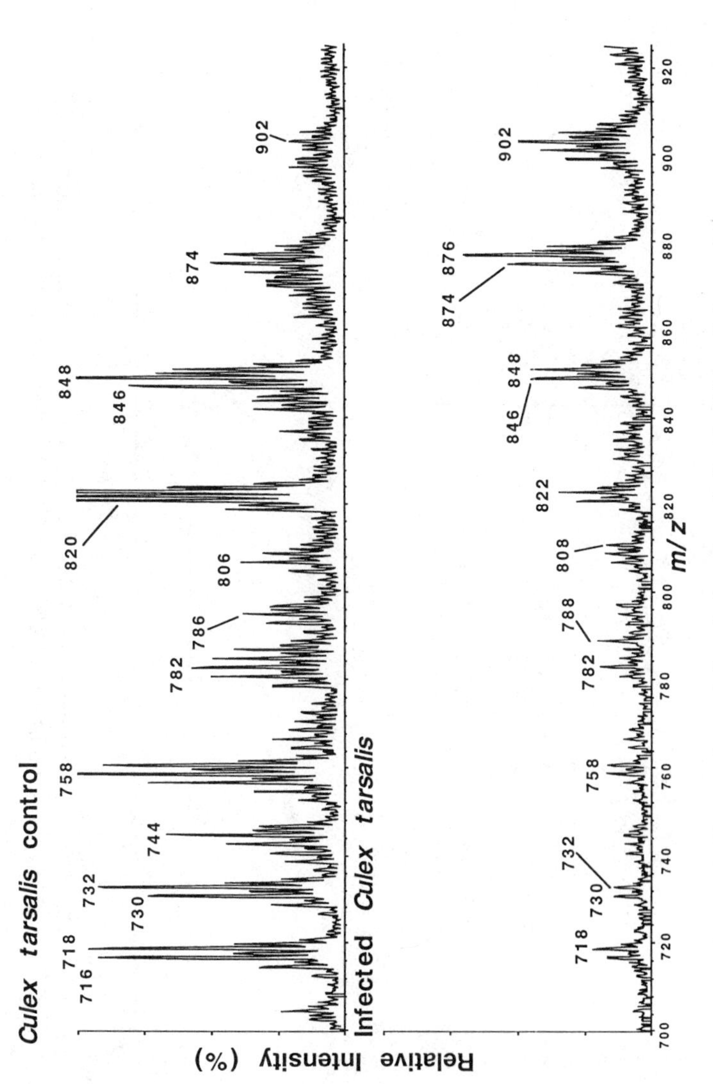

Figure 1B. Comparison of the glycerophospholipid region of the positive ion mass spectra of uninfected and infected *Culex tarsalis* larvae.

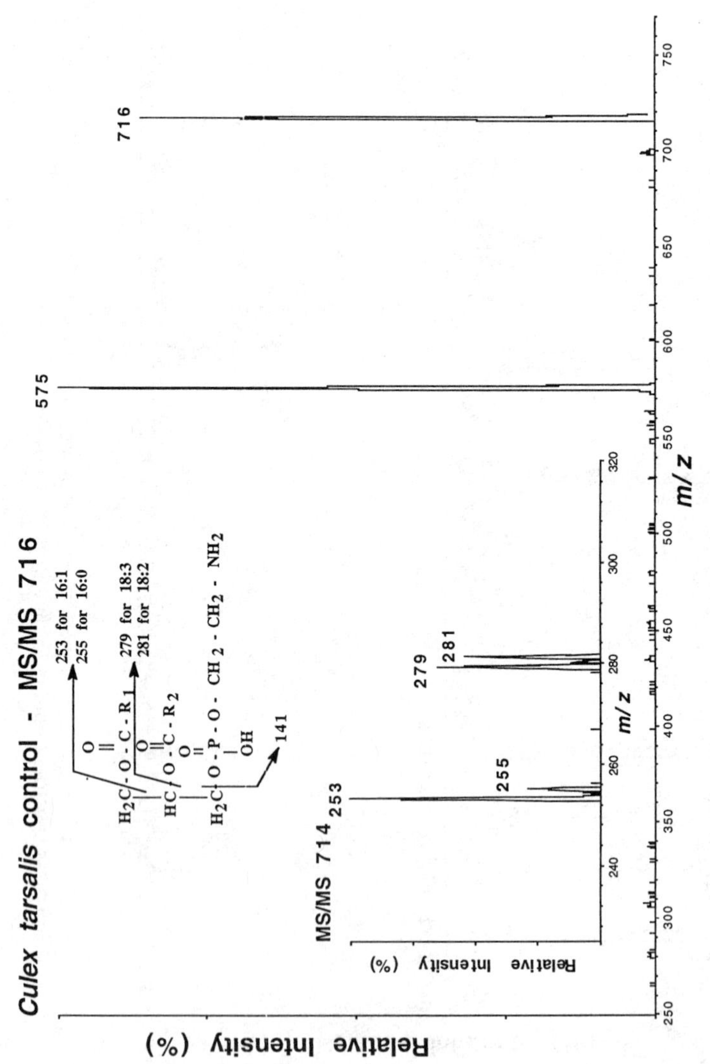

Figure 2. Positive ion MS/MS spectrum of *m/z* 716 and negative ion MS/MS spectrum of *m/z* 714 from uninfected *Culex tarsalis* larval polar lipid extracts.

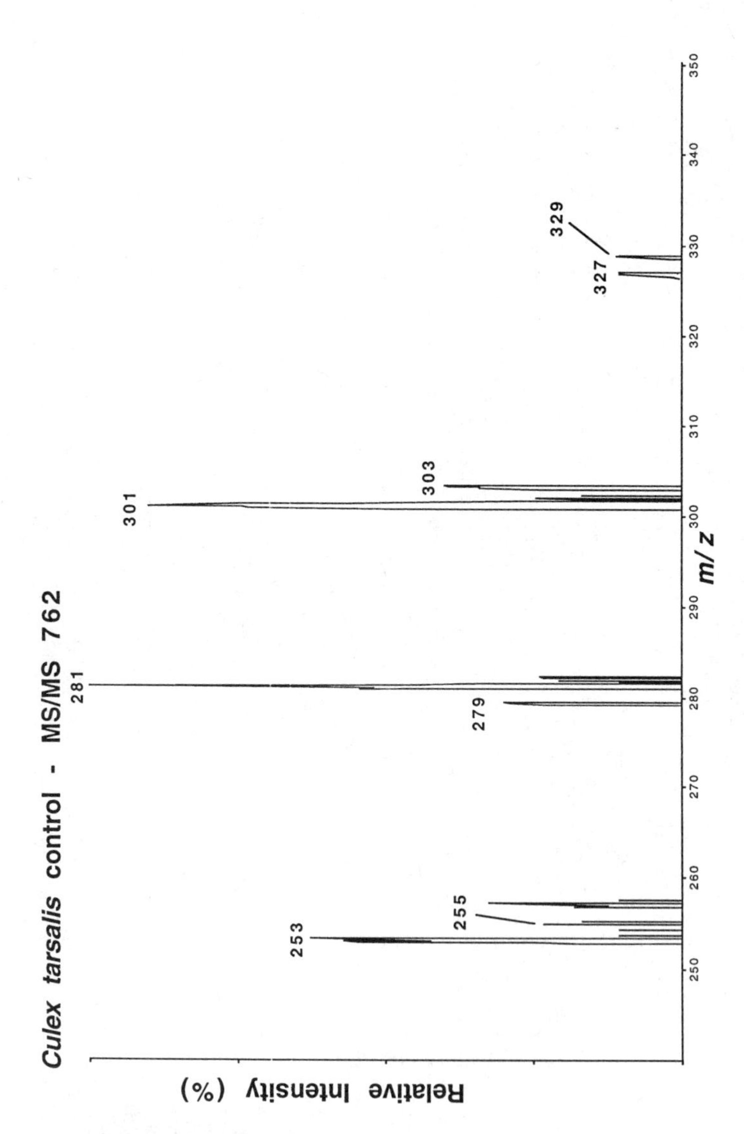

Figure 3. Partial negative ion MS/MS spectrum of *m/z* 762 from uninfected *Culex tarsalis* larval polar lipid extracts.

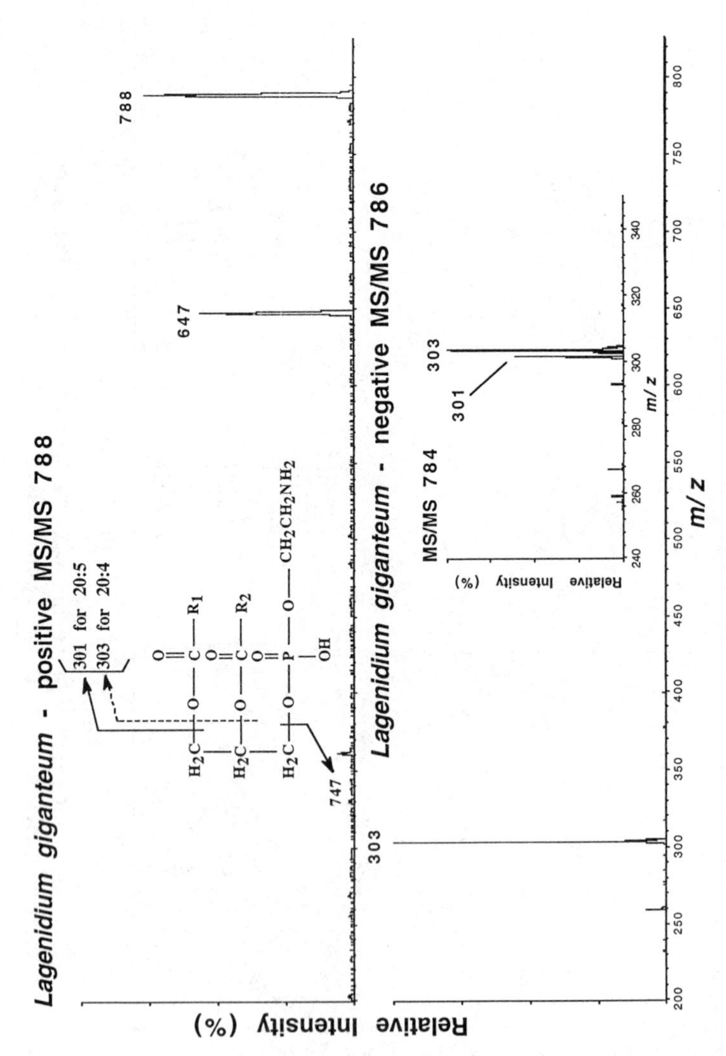

Figure 4. Positive ion MS/MS spectrum of *m/z* 788 and negative ion MS/MS spectra of *m/z* 784 and 786 from *Lagenidium giganteum* mycelium.

Table I. Glycerophosphoethanolamine Molecular Species

Molecular	Species	mol% composition[1]		
	Mycelium	Zoospores	*Cx.* control	Infected *Cx.*
14:2/16:1 e	2.9±0.3	0.5±0.3	0	0
14:1/16:0 e	1.8±0.3	0.3±0.07	0	0
14:0/16:0 a	0.2±0.01	0.7±0.4	0	0
16:0/14:0 a	1.2±0.04	0	0	0
14:1/16:1	0	0	0.7±0.02	0
14:1/16:0	0	0	1.2±0.3	0
14:0/16:0	1.0±0.2	0	0	0
15:0/15:0*	1.5±0.3	0	0.2±0.1	0
12:1/20:4	2.2±0.4	0.7±0.3	0	0
14:0/18:1 e	1.5±0.3	1.6±0.07	0	0
16:1/16:1 e	0	0	0.4±0.02	0
16:1/16:0 e	0 .1±0.4	0	0.4±0.03	0
16:0/16:0 e	8.3±0.7	0	0.7±0.09	0
16:0/16:1 a	1.2±0.8	0	0.9±0.1	0
17:0/15:0* e	10.2±0.8	0	0	0
14:0/18:0 a	0.6±0.2	0	0	0
17:0/15:0* a	2.1±0.8	0	0	0
16:0/16:0 a	2.4±0.9	0	0	0
18:0/14:0 a	1.3±0.5	0	0	0
16:1/16:1	0	0	2.4±0.07	0
16:0/16:1	0	1.6±0.7	12.5±0.5	0
16:0/16:0	0.8±0.02	1.1±0.3	0	5.2±0.8
16:0/17:0*	2.3±0.08	0	0	0
14:0/20:4 e	0.6±0.06	6.4±3.9	0	0
14:0/20:4 a	1.3±0.06	16.7±1.4	0	0
14:2/20:4	0.6±0.06	0	0	0
14:0/20:3	0.3±0.03	0	0	0
16:1/18:3 e	0.07±0.01	0	0	0
18:3/16:0 a	1.4±0.4	0.9±0.6	0	0
18:2/16:0 a	1.6±0.8	3.6±2.6	0	0
16:0/18:1 e	0	0	0.6±0.2	0
16:0/18:1 a	0	0	1.3±0.3	0
18:1/16:0 a	0	0	0	3.2±0.2
18:0/16:0 a	0	0	0	1.0±0.04
16:1/18:3	0	0	1.2±0.2	2.8±0.1
16:0/18:3	1.5± 0.2	0	0.8±0.02	0
16:1/18:2	0.8±0.1	2.2±0.8	2.9±0.08	7.5±1.6
16:0/18:2	0	1.1±0.3	4.1±0.5	0
16:1/18:1	0	0	9.2±1.2	0
16:1/18:0	0	1.1±0.6	5.9±0.3	0
16:0/18:1	2.4±0.8	0	7.4±0.4	16.3±1.1
16:0/18:0	2.0±0.9	0	0	4.0±1.3
16:1/20:4 e	0.04±0.01	1.3±0.4	0	0
18:0/18:2 e	0	0	1.1±0.8	0
18:2/18:1 a	0	0	0	2.4±0.2
18:0/18:2 a	0	0	3.2±0.3	0
18:0/18:1 e	2.7±1.1	0	0	0

Continued on next page

Table I. *Continued*

Molecular	Species	mol% composition		
	Mycelium	Zoospores	*Cx.* control	Infected *Cx.*
18:0/18:1 a	0.7±0.1	0.9±0.1	6.8±1.3	3.5±0.1
18:0/18:1 a	0.7±0.1	0.9±0.1	6.8±1.3	3.5±0.1
17:0/19:0*	1.7±0.3	0	0	0
20:0/16:0 a	1.7±1.0	0	0	0
16:1/20:5	0	0	0.4±0.06	2.4±0.6
16:1/20:4	0	0	0.4±0.01	1.9±0.1
16:0/20:5	0	0	1.2±0.04	0
16:0/20:4	5.1±0.8	1.8±0.7	0	0.6±0.06
16:0/20:3	1.3±0.3	0	0	0
18:1/18:4	0	0	1.2±0.04	0
18:2/18:3	0	0	1.4±0.05	1.0±0.07
18:2/18:2	0	0	0	0.6±0.06
18:1/18:3	0	0	0	1.6±0.2
18:0/18:3	0	0	1.8±0.07	0
18:1/18:2	0	0	4.9±0.2	4.9±1.9
18:1/18:1	0.3±0.02	7.6±1.6	6.0±0.6	2.7±0.4
18:0/18:2	0.8±0.06	0	3.0±0.3	2.3±0.3
18:0/18:1	0	0	3.4±0.07	2.7±1.2
18:2/20:4 a	0.4±0.1	1.4±0.2	1.3±0.2	0
18:0/20:5 a	0	0	2.9±0.6	1.0±0.3
18:1/20:4 a	0	9.5±3.3	0	0
18:0/20:4 e	0	0	0	1.0±0.3
18:0/20:4 a	1.0±0.4	6.5±2.3	0	1.9±0.3
16:1/22:5	0	0	0.6±0.07	0
16:0/22:6	0	0	0.4±0.04	1.0±0.3
18:3/20:4	0	0	0	1.5±0.2
18:2/20:5	0	0	5.0±2.6	1.1±0.2
18:2/20:4	3.3±0.5	1.6±1.3	0.4±0.04	1.2±0.3
18:1/20:5	0	0	1.3±0.2	1.8±0.5
18:1/20:4	2.5±0.07	4.6±1.4	0.8±0.04	1.3±0.05
18:2/20:3	0.4±0.01	0	0.3±0.02	0
18:0/20:5	0	0	0	1.6±0.06
18:0/20:4	0.03±0.01	4.7±0.7	0	1.8±0.08
20:4/20:5	2.5±1.2	0	0	0
20:4/20:4	9.4±1.0	11.6±4.6	0	10.1±0.8
20:3/20:5	0	0	0	2.5±0.2
20:3/20:4	1.2±0.1	9.4±3.4	0	4.6±0.2

[1] mean ± std. dev. (n=3).

Table II. Glycerophosphoethanolamine Molecular Species - *Lagenidium giganteum* Grown on Defined Media

Molecular Species	mol % composition[1]				
	Culture Medium[2]				
	DM2	DM2CHO	DM2ERG	DM2CH	DM2SQN
16:1/16:1	0.81±0.10	0.56±0.14	0.85±0.30	0.62±0.06	0.68±0.08
14:0/18:2	1.51±0.19	1.04±0.26	1.57±0.56	1.10±0.16	1.25±0.14
14:0/18:1	2.58±0.14	2.17±0.35	2.26±0.24	2.61±0.55	2.43±0.34
16:0/16:1	5.01±0.28	4.21±0.67	4.39±0.45	5.07±1.06	4.93±0.33
16:0/16:0	9.19±2.05	6.52±0.94	6.67±0.58	8.68±0.63	7.34±1.30
18:3/16:0 e	6.26±1.37	10.35±3.84	8.56±2.15	5.49±0.61	7.77±0.87
16:1/18:2 e	3.52±0.77	5.82±2.16	4.81±1.20	3.09±0.34	4.37±0.49
14:0/20:4	0.82±0.19	0.74±0.04	0.95±0.23	0.91±0.16	0.95±0.12
16:1/18:3	0.44±0.10	0.40±0.02	0.51±0.12	0.49±0.08	0.51±0.07
16:0/18:2	2.04±0.45	1.60±0.35	1.97±0.30	1.69±0.24	2.04±0.07
16:0/18:1	13.81±5.91	11.17±2.48	10.33±2.39	11.69±3.49	10.82±0.82
16:0/18:0	1.35±0.38	1.57±0.07	1.38±0.34	2.14±0.88	1.59±0.20
16:0/20:4	8.58±1.47	12.99±2.87	10.34±2.61	7.88±1.42	8.85±0.79
16:0/20:3	2.68±0.12	2.86±0.87	2.68±0.40	2.09±0.05	2.36±0.21
18:0/18:2	2.17±0.74	2.30±1.13	2.75±1.19	2.58±0.61	2.09±0.38
18:3/20:4 e	0.75±0.20	0.56±0.19	0.97±0.23	0.95±0.43	0.69±0.04
18:2/20:4 e	0.31±0.04	0.30±0.07	0.57±0.13	0.34±0.05	0.30±0.03
18:2/20:4 a	2.75±0.93	2.54±0.62	2.81±1.20	2.99±0.17	2.96±0.31
18:3/20:4	3.46±0.40	3.14±1.02	3.94±0.77	2.84±0.08	3.55±0.96
18:2/20:4	7.01±0.24	6.61±0.86	8.68±1.26	6.52±1.51	8.20±0.16
18:1/20:4	10.51±1.99	9.91±5.48	11.43±3.42	15.68±2.4	11.00±0.55
18:2/20:3	3.50±0.66	3.30±1.83	3.81±1.14	5.23±0.81	3.67±0.19
18:0/20:4	2.71±0.26	1.84±0.27	2.50±0.06	2.37±0.27	2.55±0.81
18:1/20:3	3.10±0.29	2.10±0.30	2.85±0.07	2.71±0.31	2.92±0.93
20:3/20:3	5.13±1.14	4.90±1.17	2.43±0.32	4.18±2.05	6.08±3.28

[1]mean ± std. dev. (n=3); "a" denotes alkyl-acyl molecular species; "e" denotes alkenyl-acyl species; all others are diacyl species. Complementary data for purified nuclei has been previously published (22).
[2]DM2 - defined medium; DM2CHO - defined medium + cholesterol; DM2ERG - defined medium + ergosterol; DM2CHN - DM2 + cholestane; DM2SQN - DM2 + squalene.

feature of cultures grown on either a defined culture medium, a complex medium, or in mosquito larvae is a significant contribution to the glycerophospholipids of *L. giganteum* from GPE species with 20-carbon polyunsaturated fatty acids in both *sn*-1 and *sn*-2 positions - 20:3/20:3 for defined medium, and 20:4/20:4 in the latter two instances. This is interesting since the Oomycete can synthesize arachidonic acid (20:4ω6), but the 20:4/20:4 molecular species was not synthesized by cultures grown on the defined medium DM2, which was designed to support only vegetative growth. The fungus cannot synthesize 20:5ω3 or any 22-carbon fatty acids, and although these are present in the PB2 (complex) growth medium (from the fish meal), these (poly)unsaturated species were not incorporated to any significant extent in GPE molecular species (Table I); however, these (poly)unsaturated species were found in examination of whole cell fatty acid methyl ester (FAME) profiles (data not shown).

Mosquitoes cannot synthesize arachidonic acid or other 20- or 22-carbon polyunsaturated fatty acids and require an exogenous source of 20:4ω6 or structurally related lipids for growth and morphogenesis (36, 37). The larvae used in this investigation obtained these essential lipids from the fish food included as a part of their diet. The larvae incorporated these fatty acids into GPE species, e.g. the series 16:1/20:5 - 16:0/20:5 and 16:1/22:6 - 18:1/20:4 (Table I). Some of these fatty acids were either assimilated unchanged or remetabolized into similar species by the parasite, e.g. 18:0/20:5 a, 16:0/22:6, 18:2/20:5 and 18:1/20:5 GPE. None of these GPE molecular species species were found in mycelia cultured on PB2 or in zoospores (Table I).

Perhaps the most interesting observation is the occurrence of GPE molecular species marked by an asterisk that had peaks in their negative ion MS/MS spectra corresponding to the presence of odd-carbon length molecular species: 15:0, 17:0 and 19:0 (Table I). Product ions corresponding to the monounsaturated species of these fatty acids were also present in glycerophospholipid classes other than GPE (data not shown). These fatty acids are not synthesized by the fungus but are present in the undefined PB2 culture medium. They are also present in whole cell FAME profiles of the parasite as monitored by gas chromatography but make up less than 0.1 mol % of total fatty acid present. A more intriguing possibility is that these species correspond to the presence of hydroxy fatty acids incorporated into GPE and other GPL species. This possibility was examined by sep-pak purification of a lipid fraction containing free fatty acids and their oxygenated metabolites.

We have completed an extensive study on the use of ESI to identify FFA, hydroxy fatty acids, prostaglandins and leukotrienes (Kerwin *et al.*, unpublished observations). Whole cell profiles of free fatty acids and oxygenated fatty acids can be obtained using negative ion ESI-MS (Figure 5). Saturated and monounsaturated fatty acids produce negligible fragmentation under the conditions we employ for ESI-MS/MS, with the precursor $[M - H]^-$ dominating the spectra of these lipids as shown for oleic acid (*m/z* 281) from the *L. giganteum* MS spectrum (Figure 6). This can be contrasted with the MS/MS spectra of the precursor at *m/z* 239 and 267, which could correspond to 15:1 and 17:1 and show extensive fragmentation (Figure 7A). These species were tentatively identified as 14- and 16-carbon oxygenated metabolites (Figure 7B).

Analysis of a second peak from the mycelium oxygenated/free fatty acid extract, *m/z* 295, further supports the probable incorporation of hydroxy fatty acids into GPE molecular species. This peak could correspond to the presence of 19:1 (less than 0.05% of total cell FAME), but more likely it is due to mono-OH-18:2 moieties (Figure 8). When compared to products formed by the well-characterized reaction of soybean lipoxygenase with linoleic acid, it appears that this mosquito parasite is producing a mixture of 9- and 13-hydroxy octadecadienoic acids (Figure 8).

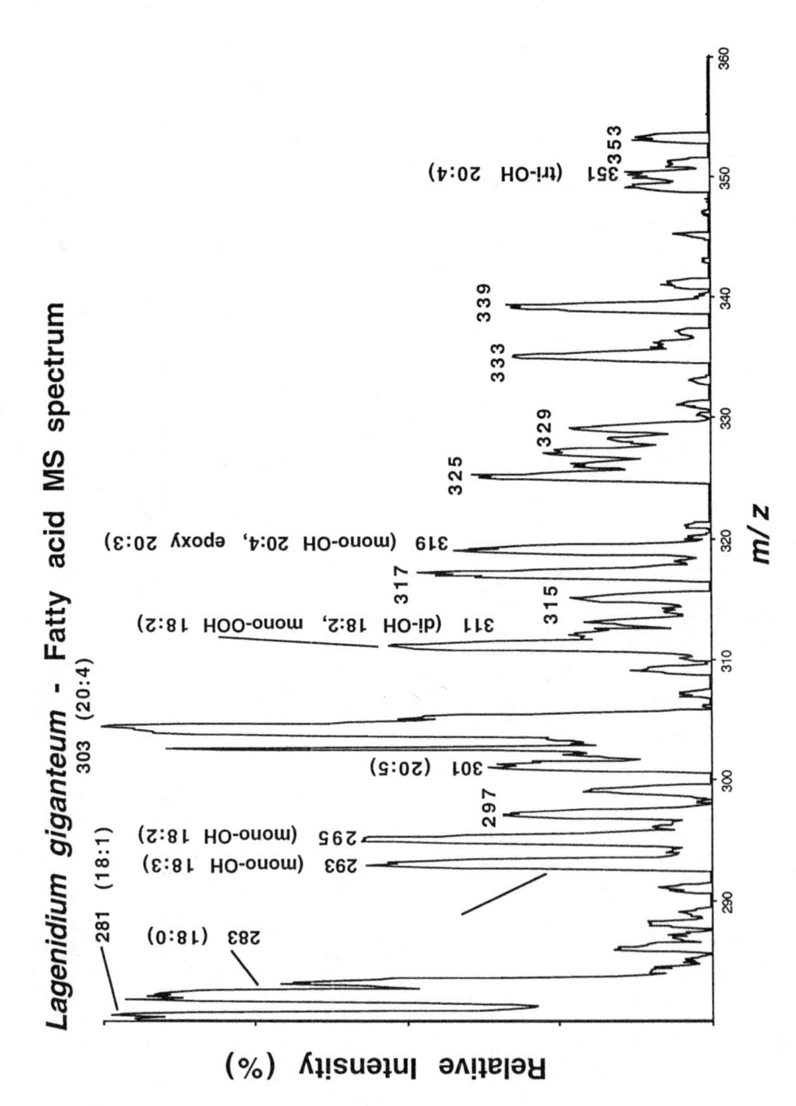

Figure 5. Negative ion MS spectrum from *m/z* 280 to *m/z* 360 of fatty acids and oxygenated fatty acids from base-hydrolyzed *Lagenidium giganteum* mycelial lipid extract.

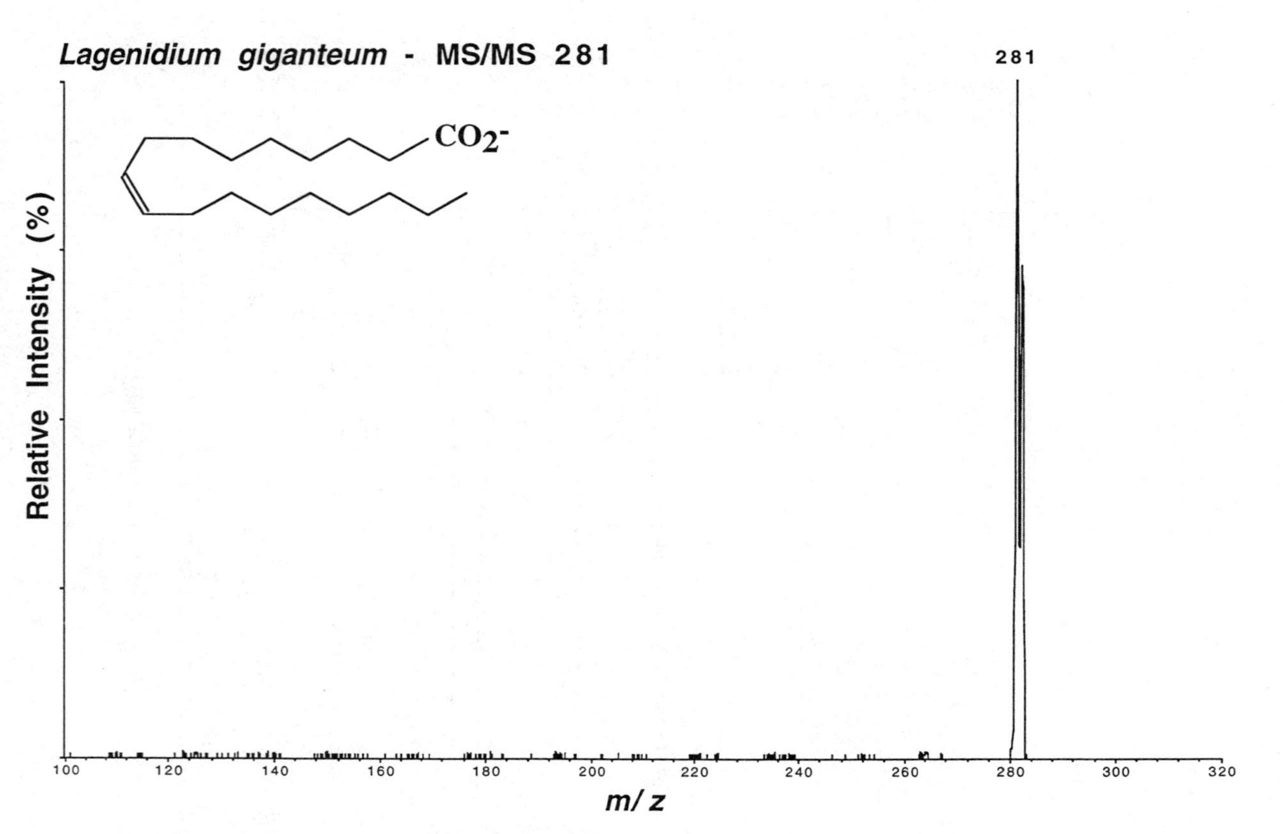

Figure 6. Negative ion MS/MS spectrum of *m/z* 281 from *Lagenidium giganteum* mycelial lipid extract.

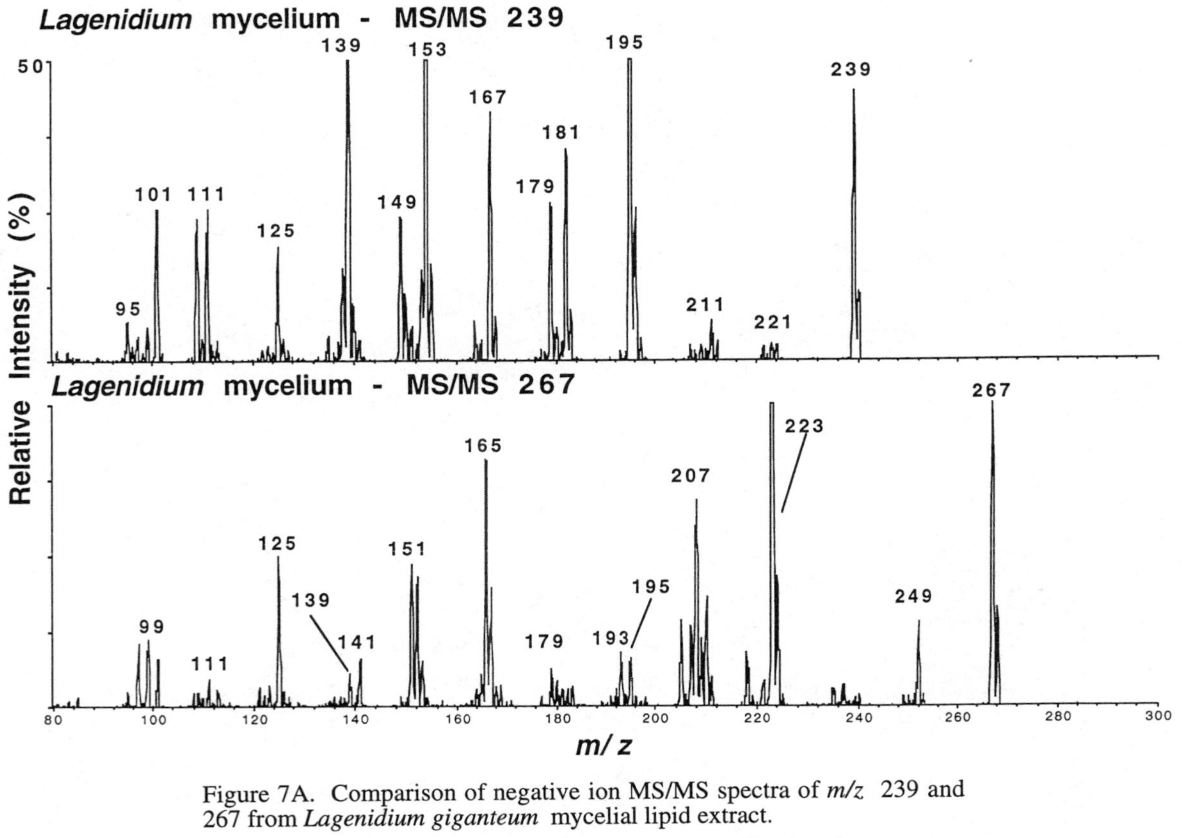

Figure 7A. Comparison of negative ion MS/MS spectra of *m/z* 239 and 267 from *Lagenidium giganteum* mycelial lipid extract.

Continued on next page

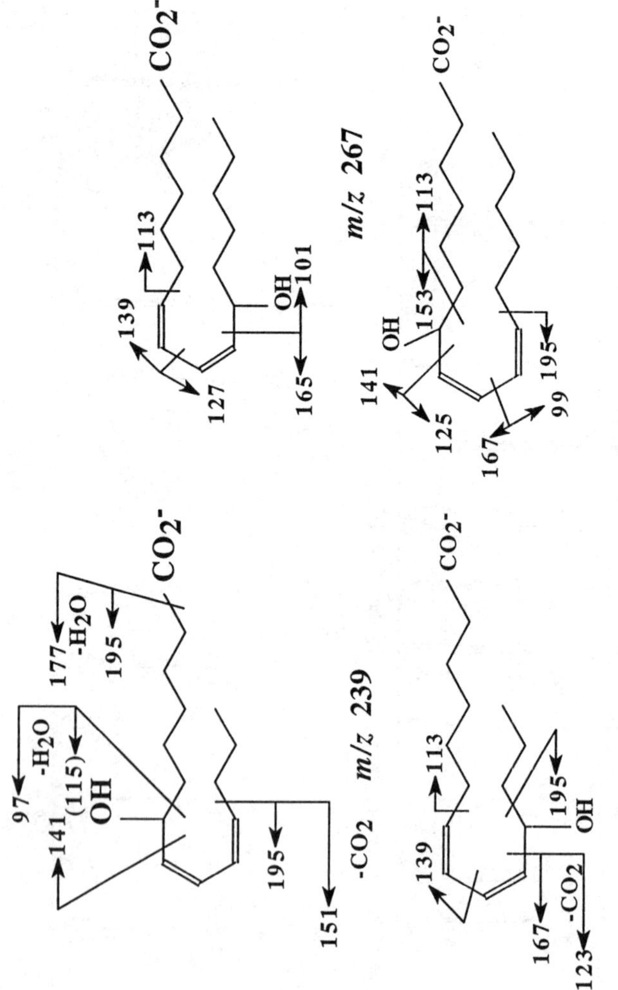

Figure 7B. Structures of compounds consistent with spectra in Figure 7A.

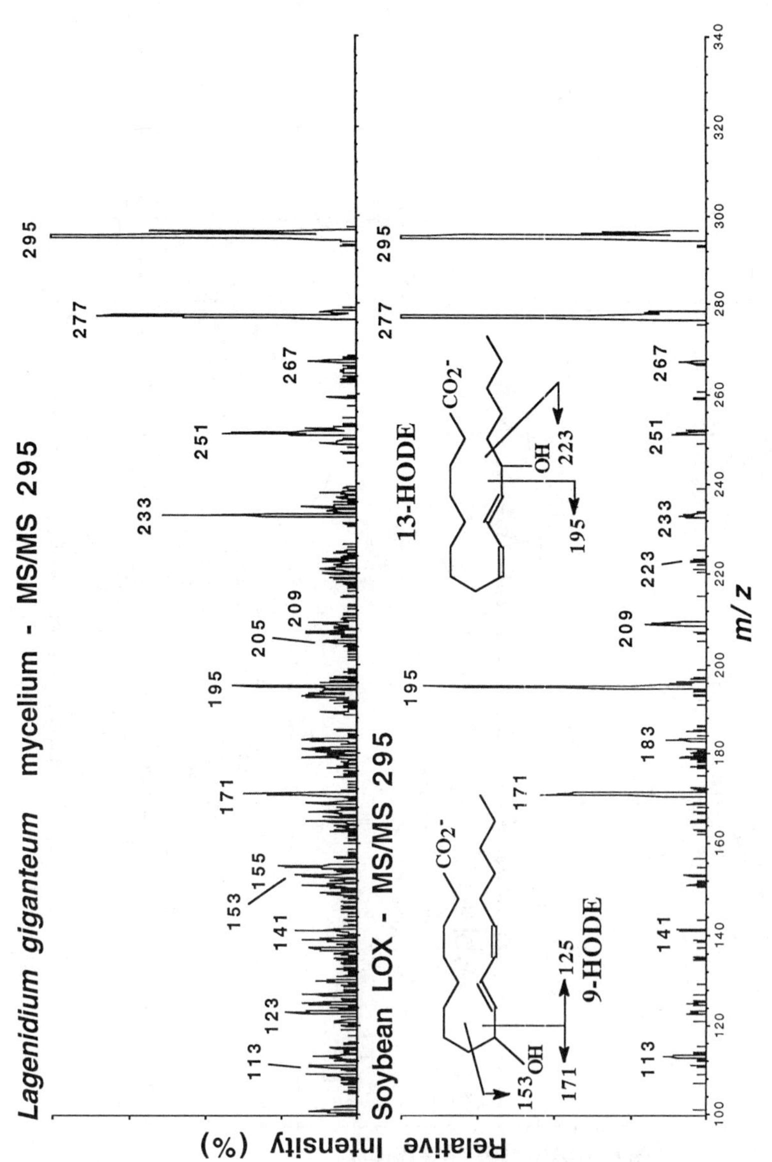

Figure 8. Comparison of negative ion spectra of *m/z* 295 from *Lagenidium giganteum* mycelium and the products generated by incubation of linoleic acid (18:2ω6) with soybean lipoxygenase (LOX).

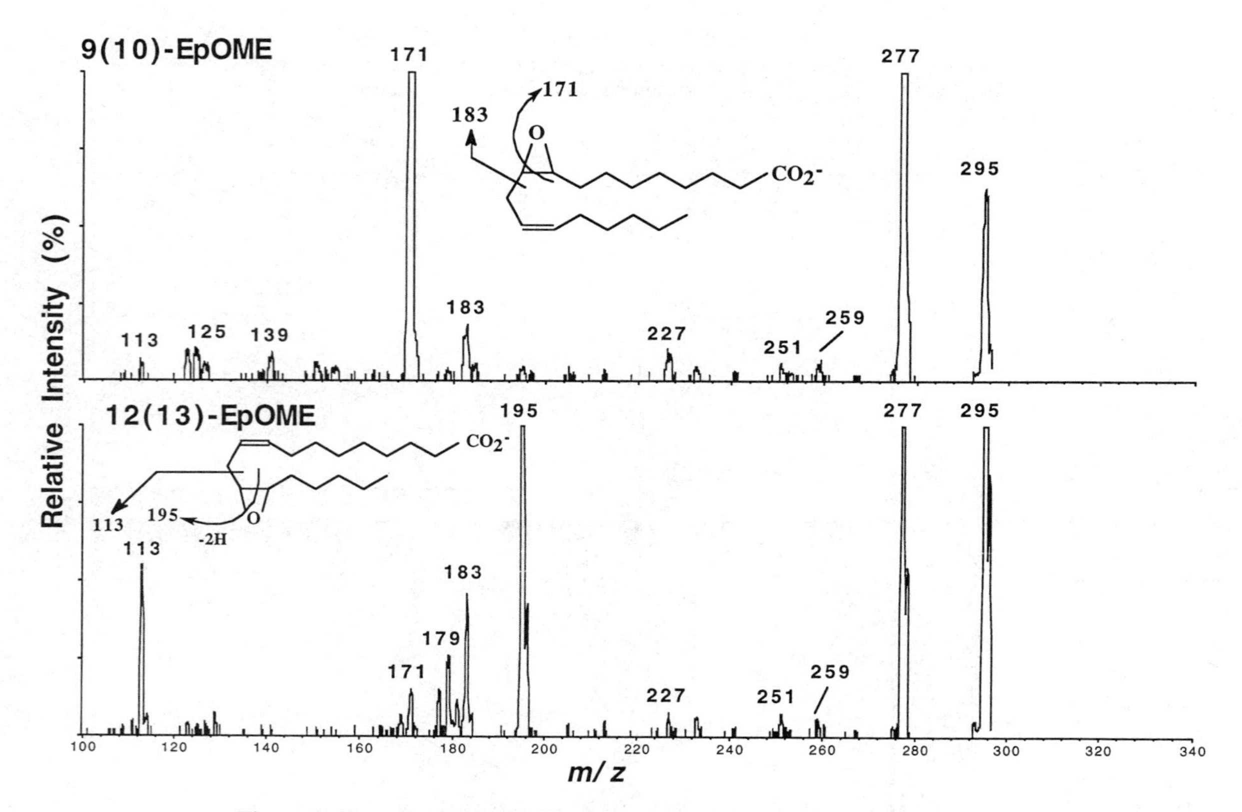

Figure 9. Negative ion MS/MS spectra of 9(10)- and 12(13)-epoxy - 18:1 standards.

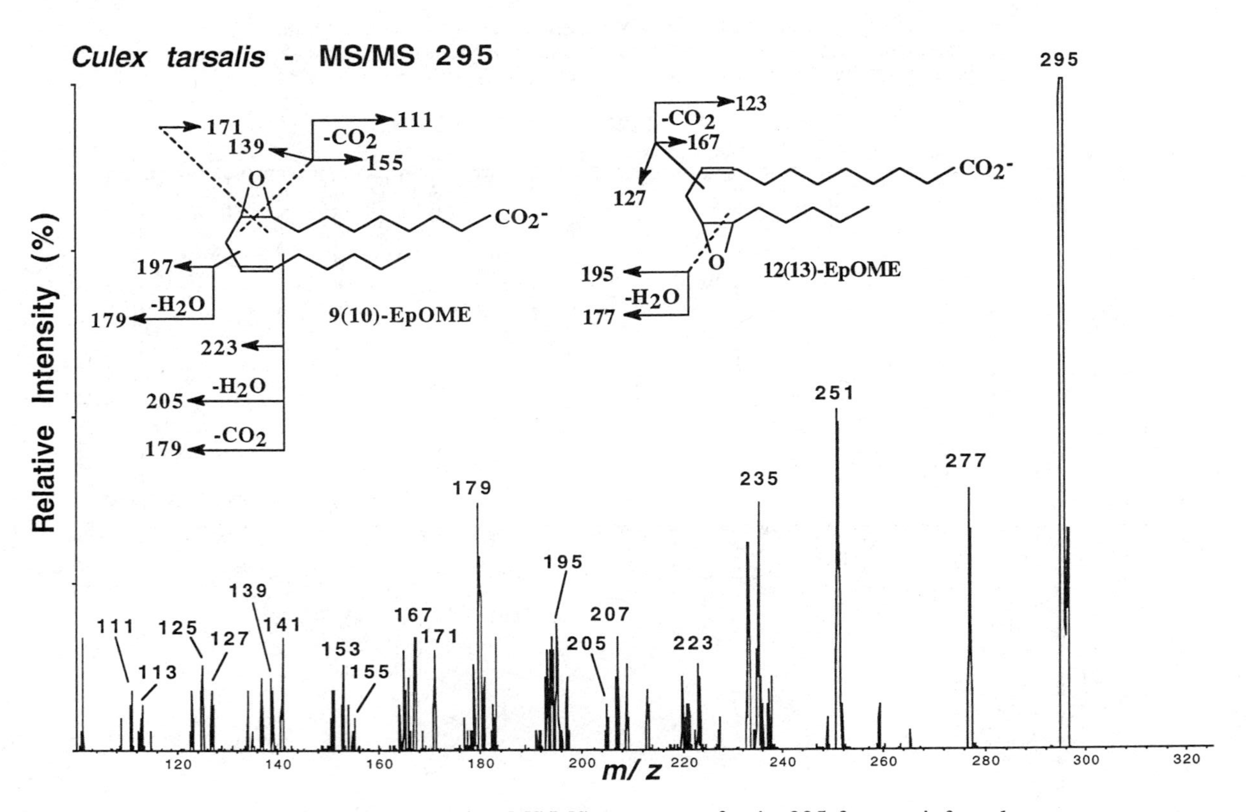

Figure 10. Negative ion MS/MS spectrum of *m/z* 295 from uninfected *Culex tarsalis* lipid extracts.

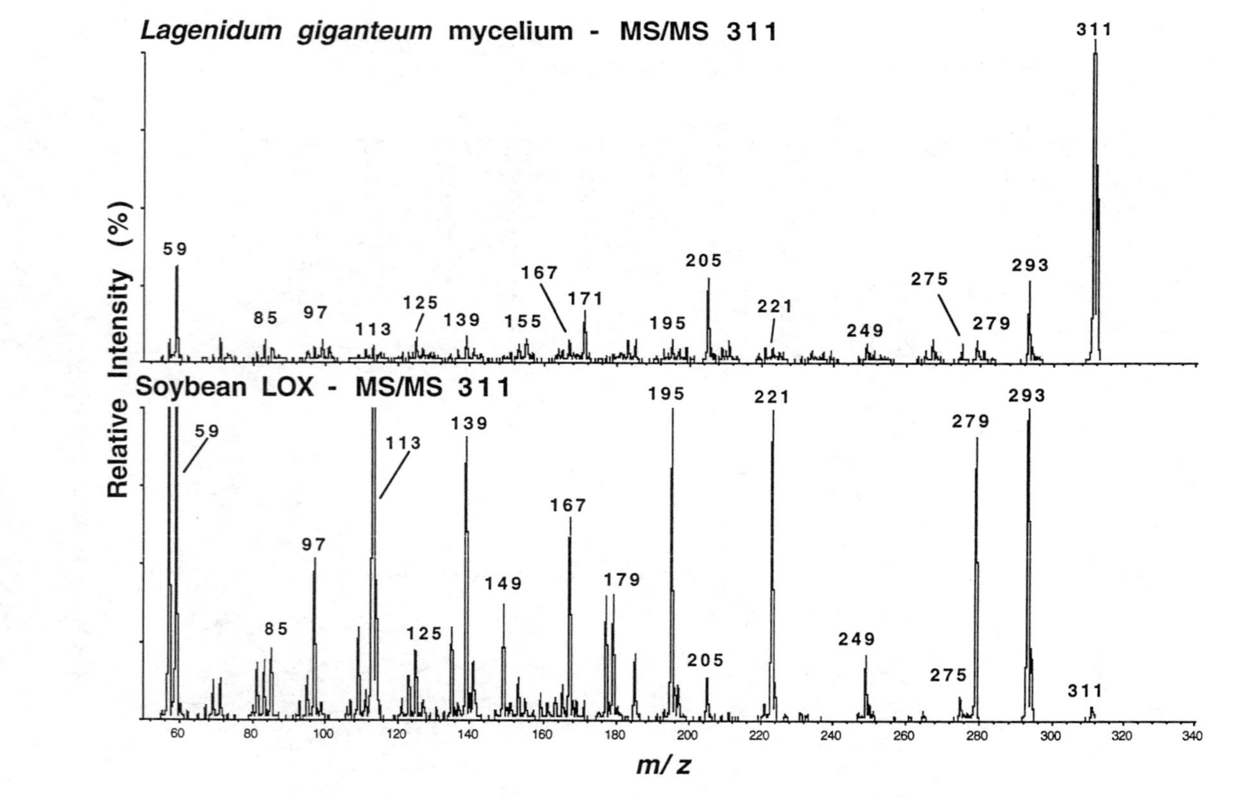

Figure 11. Comparison of negative ion spectra of m/z 311 from *Lagenidium giganteum* mycelium and the products generated by incubation of linoleic acid with soybean lipoxygenase (LOX).

There are other oxygenated compounds, however, that can generate comparable spectra. Negative ion MS/MS spectra of two common epoxy-18:1 standards (Figure 9) shows that fragmentation is again centered around the oxygen-containing moiety, and several major fragments in the mono-OH-18:2 compounds (Figure 8) are also present in the 18:1 epoxides. Comparison with the MS/MS spectrum of the precursor at *m/z* 295 from *C. tarsalis* larvae (Figure 10), which was generated using an ionizing voltage 20 V greater than that used to generate the spectra in Figure 9, also supports the possible synthesis by mosquito larvae of epoxides, but again this is also consistent with mono-OH-18:2 metabolites (Figure 8).

MS/MS spectra of the precursor at *m/z* 319 from extracts of *L. giganteum* and mosquito larvae are even more complicated (spectra not shown), due to the synthesis of a complicated suite of oxygenated metabolites with this mass. There are at least 6 common mono-OH-20:4 metabolites, and several epoxy-20:3 compounds synthesized by eukaryotic organisms. This can be further complicated by enzymatic and nonenzymatic processes that can produce cyclic products that produce a peak at *m/z* 319. Although it is possible to differentiate between two or three discrete mono-OH-20:4 species in one sample (unpublished observations), the presence of epoxy and other oxygenated derivatives makes definitive identification impossible without chromatographic separation. However, since ESI mass spectrometer systems are readily interfaced with high pressure liquid chromatographs, it should be possible to resolve eicosanoids and other oxygenated fatty acids, including racemic mixtures, with minimal loss of sensitivity.

A final example of the sensitivity of ESI involves the direct detection of very labile hydroperoxy species. Comparison of the MS/MS spectrum of the precursor at *m/z* 311 from linoleic acid (18:2) incubated with soybean lipoxygenase shows a spectrum (Figure 11) that corresponds to the -OOH products from which 9-HODE and 13-HODE (Figure 8) are derived by dehydration. Although the peak intensities differ, it is obvious that *L. giganteum* is synthesizing the same or very similar compounds (Figure 11). The precursor ion from the mycelium extract is much stronger than that from the lipoxygenase extract, suggesting there are additional (more stable) species present, possibly di-OH-18:2 moieties.

Conclusions

Results reported here demonstrate that ESI-MS and MS/MS are powerful tools in examining several lipid classes in biological samples. Because of its sensitivity, resolution, and lack of matrix interference, ESI-MS is the only option currently available to examine endogenous levels of many compounds, such as eicosanoids that are often found in picogram quantities. With the increasing availability of these mass spectrometer systems, ESI-MS and MS/MS should be considered as alternatives to or used in combination with more standard methods of mass spectral analysis.

Acknowledgment

This research was supported in part by grants from the National Institutes of Health (AI 22993 and AI34339).

References

1. Barnes, J.A.; Pehowich, D.J.; Allen, T.M. *J. Lipid Res.* **1987,** *28*, 130.
2. Becnel, J.J. *Dis. Aquat. Org.* **1992,** *13*, 17.
3. Sprague, V.; Becnel, J.J.; Hazard, E.I. *Crit. Rev. Microbiol.* **1992,** *18*, 285.
4. Gaugler, R.; Kaya, H.K. *Entomopathogenic nematodes in biological control;* CRC Press: Boca Raton, Florida, 1990; pp. 1-365.

5. Anon. *Federal Register* **1992,** *57(219)*, 53570.
6. Beakes, G.W. *The Chromophyte algae: Problems and Perspectives*; Clarendon Press: Oxford, 1989; pp 323-340.
7. Gunderson, J.H.; Elwood, H.; Ingold, A.; Kindle, K.; Sogin, M.L. *Proc. Natl. Acad. Sci. USA* **1987**, *84*, 5823.
8. Cavalier-Smith, T. *The Chromophyte Algae: Problems and Perspectives;* Clarendon Press: Oxford, 1989; pp. 381-407.
9. Barr, D.J.S. *Mycologia* **1992,** *84,* 1.
10. Warner, S.A.; Domnas, A.J. *Exp. Mycol.* **1981,** *5,* 184.
11. Elliot, C.G. *Adv. Microbial Physiol.* **1977,** *15,* 121.
12. Hendrix, J.W. *Science* **1964,** *144,* 1028.
13. Domnas, A.J.; Srebro, J.P.; Hicks, B.F. *Mycologia* **1977,** *69,* 875.
14. Kerwin, J.L. Washino, R.K. *Exp. Mycol.* **1983,** *7,* 109.
15. Kerwin, J.L.; Washino, R.K. *Can. J. Microbiol.* **1986,** *32,* 294.
16. Kerwin, J.L.; Simmons, C.A.; Washino, R.K. *J. Invertebr. Pathol.* **1986,** *47,* 258.
17. Kerwin, J.L.; Duddles, N.D.; Washino, R.K. *J. Invertebr. Pathol.* **1991,** *58,* 408.
18. Kerwin, J.L.; Dritz, D.A.; Washino, R.K. *J. Am. Mosq. Control. Assoc.* **1994,** *10,* 451.
19. Kerwin, J.L.; Washino, R.K. *J. Am. Mosq. Control Assoc.* **1986,** *2,* 182.
20. Kerwin, J.L.; Washino, R.K. *J. Am. Mosq. Control Assoc.* **1987,** *3,* 59.
21. Kerwin, J.L.; Simmons, C.A.; Washino, R.K. *Prostagland. Leuko. Med.* **1986,** *23,* 173.
22. Kerwin, J.L.; Tuininga, A.R.; Wiens, A.M.; Wang, J.C.; Torvik, J.J.; Conrath, M.L.; MacKichan, J.K. *Microbiol.* **1995,** *141,* 399.
23. Fetter-Lasko, J.L.; Washino, R.K. *Environ. Entomol.* **1983,** *12,* 635.
24. Jaronski, S.; Axtell, R.C. *Mosq. News* **1983,** *43,* 332.
25. Kerwin, J.L.; Grant, D.F.; Berbee, M.L. *Protoplasma* **1991,** *161,* 43.
26. Folch, J.; Lees, M.; Sloane-Stanley, G.H. *J. Biol. Chem.* **1957,** *226,* 497.
27. Wakayama, M.; Dillwith, J.W.; Blomquist, G.J. *Insect Biochem.* **1986,** *16,* 895
28. Yamane, M.; Abe, A.; Yamane, S. *J. Chromatogr.* **1994,** *652,* 123.
29. Kerwin, J.L.; Tuininga, A.R.; Ericsson, L.H. *J. Lipid Res.* **1994**, *35*, 1102.
30. Han X.; Gross, R.W. *Proc. Natl. Acad. Sci USA* **1994**, *91*, 10635.
31. Chapkin, R.S.; Akoh, C.C.; Miller, C.C. *J. Lipid Res.* **1991,** *32,* 1205.
32. Murphy, R.C.; Harrison, K.A. *Mass Spectrom. Rev.* **1994,** *13,* 57.
33. Bell, M.V.; Dick, J.R. *Lipids* **1993,** *28*, 19.
34. Justin, A.-M.; Mazliak, P. *Biochim. Biophys. Acta* **1992,** *1165,* 141.
35. Arao, T.; Sakaki, T.; Yamada, M. *Phytochem.* **1994,** *36,* 629.
36. Dadd, R.H. *J. Nutr.* **1980,** *110,* 1152.
37. Stanley-Samuelsson, D.W.; Dadd, R.H. *Insect Biochem.* **1983,** *13,* 549.

RECEIVED July 21, 1995

Chapter 13

Characterization of Polyunsaturated Phospholipid Remodeling in Mammalian Cells by High-Performance Liquid Chromatography–Electrospray Ionization Mass Spectrometry

Hee-Yong Kim[1], Tao-Chin Lin Wang[2], and Yee-Chung Ma[1]

[1]National Institute of Alcohol Abuse and Alcoholism, National Institutes of Health, Building 10, Room 3C–103, Bethesda, MD 20892
[2]National Institute of Mental Health, National Institutes of Health, Building 10, Room 3D–40, Bethesda, MD 20892

Characteristics of cell membranes are greatly affected by the composition of phospholipid bilayers. The cellular membrane phospholipid composition is maintained by complex biosynthetic and remodeling processes, including phospholipase, base exchange and acyltransferase reactions. In order to understand these complex biochemical mechanisms, analysis of the phospholipid profile in biomembranes is necessary. Using electrospray mass spectrometry coupled to reversed phase HPLC, phospholipid molecular species of different head groups and fatty acyl compositions could be separated and detected in a single run. Sensitivity at the picomole level was attained with a linear response range over 2 orders of magnitude. Application of this technique to remodeling of polyunsaturated phospholipids in C-6 glia cells and the rodent brain is demonstrated.

Phospholipid bilayers constitute the major part of cell membranes. Since the characteristics of cell membranes are influenced by the phospholipid composition, the maintenance of proper phospholipid composition in membranes is of great importance. Distinctive lipid composition is maintained in specific cell membranes, probably through complex biosynthetic and remodeling processes including phospholipase, base exchange and acyltransferase reactions (1, 2). Neuronal membranes are highly enriched in polyunsaturated fatty acids (3). These polyunsaturates can modulate various neuronal functions as membrane components (4) or as the free fatty acid form after mobilization upon stimulation (5). Polyunsaturated fatty acids, released from distinctive lipid pools in response to stimuli, are mostly reincorporated into membrane lipids while a certain portion escapes to be metabolized to biologically active compounds through various oxygenative pathways. In order to understand these complex biochemical processes, analysis of the phospholipid profile in cell membranes is necessary.

Analysis of phospholipids from a biological matrix is a difficult task since most of the biomembranes contain a mixture of a variety of molecular species with different phospho-head groups. Conventional analysis techniques including column and argentation thin-layer chromatography (6,7) are not acceptable due to limited resolving power as well as limited sensitivity. High performance chromatographic analysis after derivatization with a chromophore (8) improves the sensitivity and resolving power; however, the added analysis step is laborious and also time consuming. In addition, confirmation of each peak identity must be established using other techniques such as gas chromatography (GC), GC-mass spectrometry (MS) or fast atom bombardment MS (9) after collecting and derivatizing each fraction. Therefore, an alternative technique which can provide reasonable separation, sensitivity and structural information in an efficient manner is desirable. Such techniques have been developed in our laboratory using thermospray liquid chromatography / mass spectrometry (LC/MS) which allows the direct detection of samples eluting from an LC column without any derivatization (10, 11). However, difficulties in quantification imposed by non-linear response curves (12) together with insufficient sensitivity to monitor the remodeling of labeled lipid molecules prompted us to develop another LC/MS technique using electrospray LC/MS for phospholipid analysis (13).

Experimental

A Hewlett-Packard 5989 mass spectrometer with HP electrospray source was used for collecting most of the data presented here. In this system, a liquid flow at typically 1-5 μL/min is introduced through a needle with the aid of a nebulizing gas, nitrogen, to a high electric field which generates the charged droplets. Typically the column effluent was split 1:100 using a commercial splitter. In some cases, a Finnigan TSQ-700 mass spectrometer with a Finnigan electrospray/atmospheric pressure chemical ionization (APCI) source was employed, and a flow rate of 0.1 mL/min was introduced into the ion source after the flow was split by 1:5 or 1:4. The heated nitrogen gas was supplied to aid desolvation and orientation of produced ion clusters through a capillary. Ions exiting from the capillary were introduced into the mass analyzer through focusing lenses. Spectra generated in this way usually contain molecular ion species with minimal fragmentation. However, changing the electric voltage at the end of the capillary can induce fragmentation for certain compounds.

Electrospray Mass Spectra of Phospholipids

Various phospholipids were detected as their protonated or natriated molecules as shown in figure 1. The spectra were obtained using a capillary exit voltage set at 200 V. Sodium attachment was more noticeable for acidic phospholipids, phosphatidylserine (PS) and phosphatidylinositol (PI). Under this condition only PI produced significant diacylglyceryl (DG) fragments. Diacylglyceryl fragment ions contain molecular species information and were common to all phospholipid classes with a given fatty acyl composition. Therefore, producing diacylglyceryl fragments was beneficial, since less ions needed to be monitored in comparison to the case with molecular ions. Formation of diacylglyceryl fragments was induced by raising the capillary exit voltage as shown for 18:0 22:6-PE in figure 2. Generally, the molecular ion intensity was maximum at an exit voltage of 200 V and decreased at higher exit voltages. At around 300 V, the intensity of the diacylglyceryl fragment was at a maximum. Similar fragmentation could be induced for phosphatidic acid (PA), PI and PS, although the extent of fragmentation was molecular species dependent. Phosphatidylcholine (PC) was resistant to fragmentation to DG even with the highest capillary exit potential applicable under the present configuration.

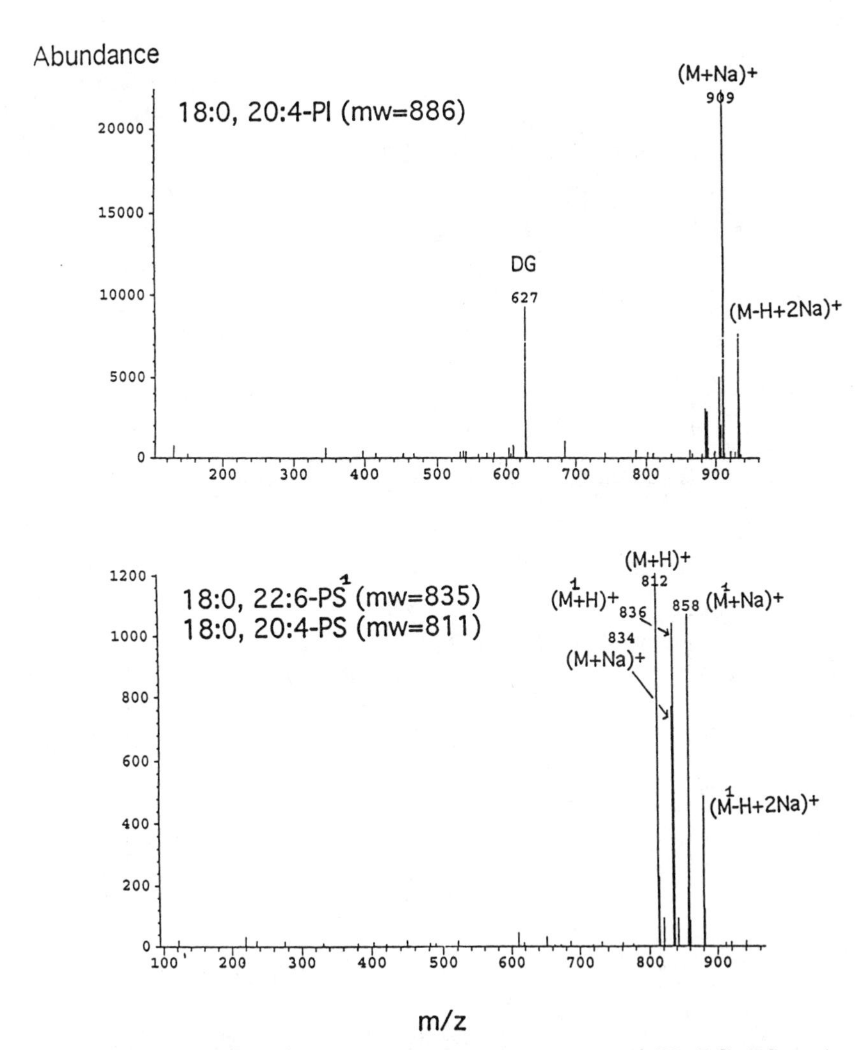

Figure 1. Positive ion electrospray mass spectra of PI, PC, PS and sphingomyelin (from ref. 13). Approximately 40-80 pmoles (0.4 to 0.8 femtomole after split) of standards were directly introduced by the flow injection technique using methanol:hexane:water 96:3:1.

Continued on next page

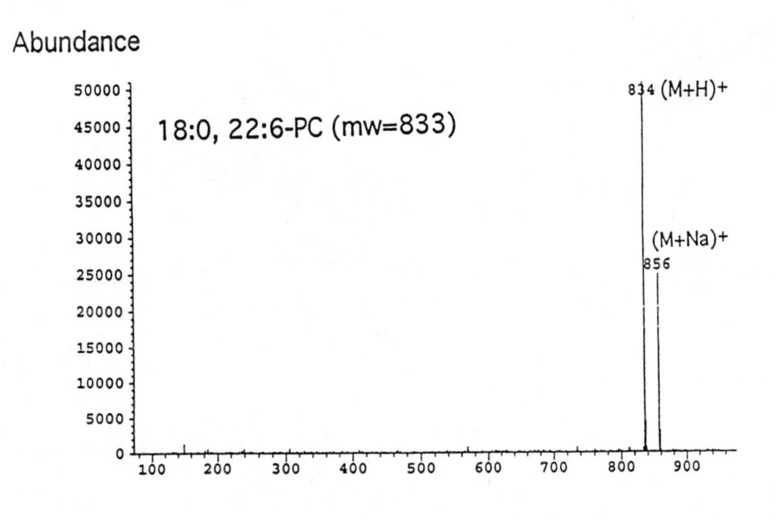

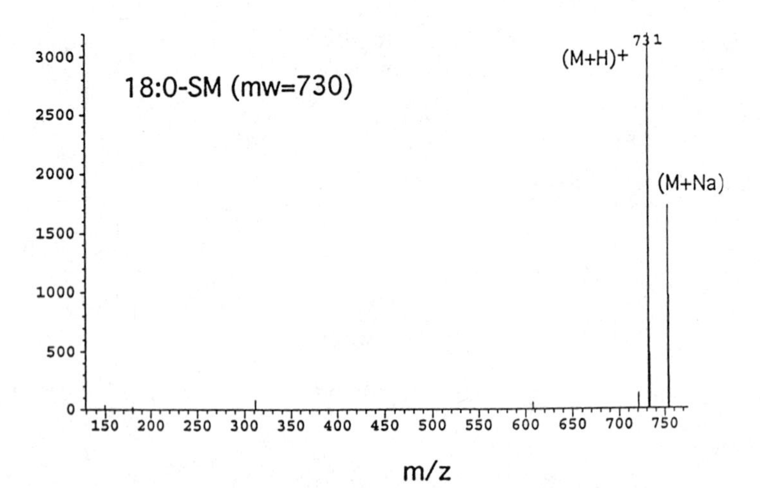

Figure 1. *Continued*

Sensitivity of the technique

The sensitivity of the technique was assessed using the protonated molecule of 18:0 20:4-PC (Figure 3). The ion trace was generated from the injection of 1.35 pmoles of 18:0 20:4-PC by direct flow injection at a flow rate of 5 μL/min after a 100:1 flow split. This split allows only 1 % of the sample to be introduced into the mass spectrometer. Under this condition the present detection limit assessed for various PC molecular species was in the 0.5-1 pmole range, which represents approximately a 20 fold greater sensitivity in comparison to the limit (20 pmoles) obtained by the thermospray technique (11), even after the split. Response was reproducible with no significant peak broadening, which indicates that direct coupling of LC for chromatographic peak detection is possible with this technique.

High Performance Liquid Chromatography (HPLC) / Electrospray MS of Standard Phospholipids

Chromatographic integrity of the HPLC separation was well maintained during transit through the capillary line to the mass spectrometer. An example is shown in figure 4 for 18:0 20:4 containing phospholipids. Each molecular species can be monitored by using either the DG fragment or the $(M+H)^+$. Due to the difficulty to induce DG fragments from PC, monitoring protonated molecules was required for PC. Since diacylglyceryl ion resulted from the loss of the head group from the intact molecule, all phospholipids with the same fatty acyl composition of 18:0 20:4 can be detected as DG fragments at the same mass to charge ratio of 627 Da regardless of their head group identity. Therefore, with a proper chromatographic system which can separate a given molecular species of PS, PI and PE, only one diacylglyceryl fragment ion needed to be monitored. We have developed such a rapid chromatographic system using reversed phase HPLC (2.1 mm x 15 cm) with the mobile phase containing a mixture of 0.5% ammonium hydroxide in water : methanol : hexane changing from 12:88:0 to 0:88:12 after holding at the initial solvent composition for 3 min (12). As shown in this figure, PS eluted first, followed by PE and then PC. Under the same condition, PI and PA eluted between PS and PE (data not shown). With ESI the intensity of DG fragment ions as well as protonated molecular ions appeared to be affected by the nature of the phospho-head group and was less dependent on fatty acyl composition within a given phospholipid class. This phenomenon is opposite to the results obtained by the thermospray technique (12) where the response was more dependent on the fatty acyl composition than on the head group characteristics. Therefore, this technique might be more suited for studies involving comparison of molecular species within a phospholipid class while the thermospray technique may be more useful for phospholipid remodeling between phospholipid classes.

An example is shown for brain cytosolic phospholipids containing 20:4n6 or 22:6n3 in figure 5. The cytosolic fraction was obtained after centrifugation of brain homogenate at 100,000xg for 1 hour. All phospholipid classes except PC could be analyzed using common DG fragment ions as 18:0 22:6, 16:0 22:6, 18:0 20:4 and 16:0 20:4 species were monitored by the ion traces of 651, 624, 628 and 599 Da, respectively in this figure. Unexpectedly, after centrifugation the cytosolic fraction still contained considerable levels of phospholipids, indicating that phospholipids exist in soluble forms in the cytosol, possibly associated with protein. The cytosolic phospholipids, however, were enriched in phosphatidylinositol-containing 20:4n6 in comparison to synaptosomal or whole brain phospholipids (data not shown), suggesting that preferential processes may exist in association of phospholipids with proteins such as ones involved in phospholipid transport (14).

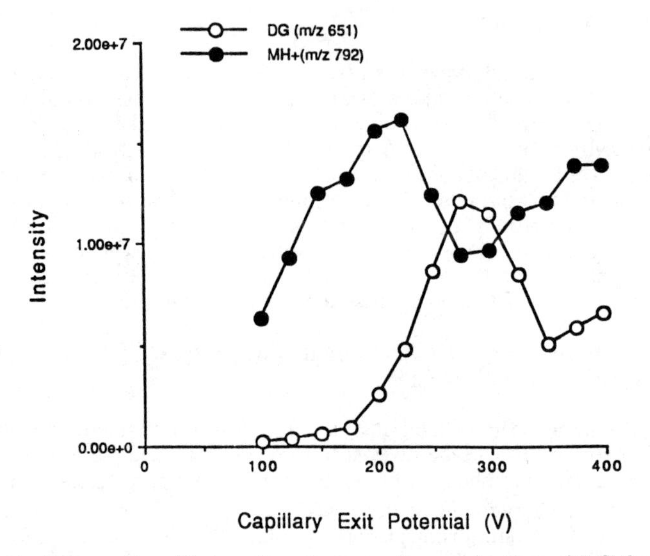

Figure 2. Effect of capillary exit voltage on the production of DG fragment ions for 125 pmoles of 18:0 22:6-PE (from ref. 13). The standard was injected by the flow injection technique as in Figure 1.

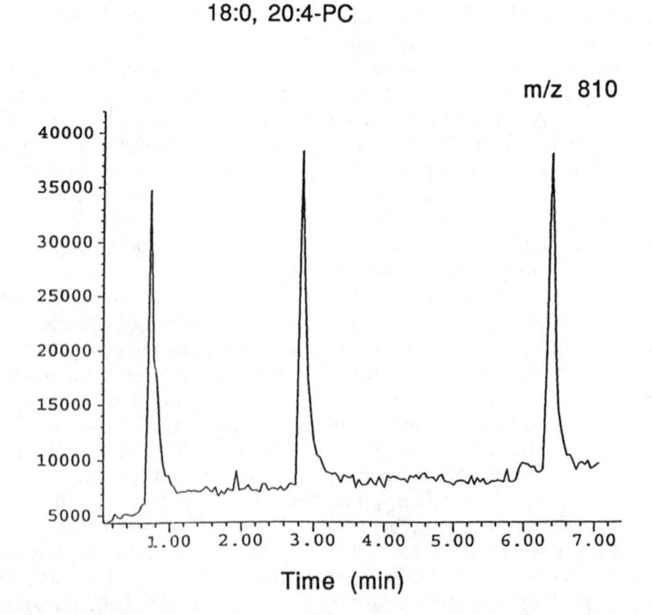

Figure 3. Selected ion trace obtained from 1.35 pmoles (before split) of 18:0,20:4-PC. The standard was repeatedly injected by the flow injection technique as in Figure 1.

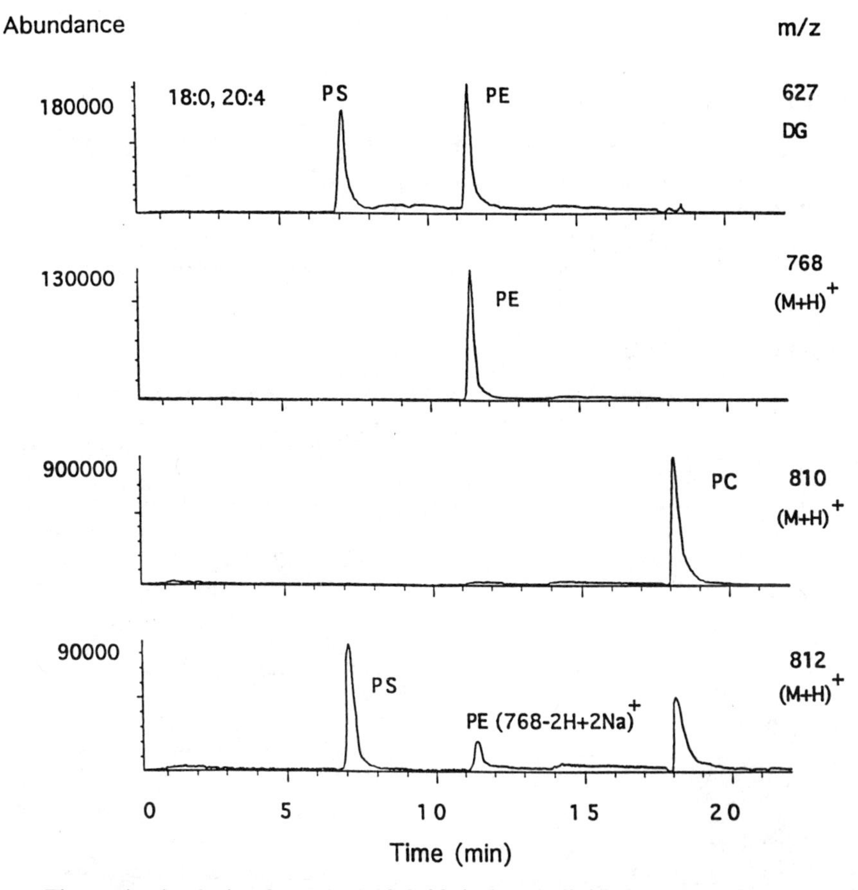

Figure 4. Analysis of standard 18:0 20:4-phospholipids by reversed phase HPLC/ESP-MS. Approximately 40 pmoles of each phospholipid were injected and $(M+H)^+$ and DG ions were selectively monitored to obtain mass chromatograms.

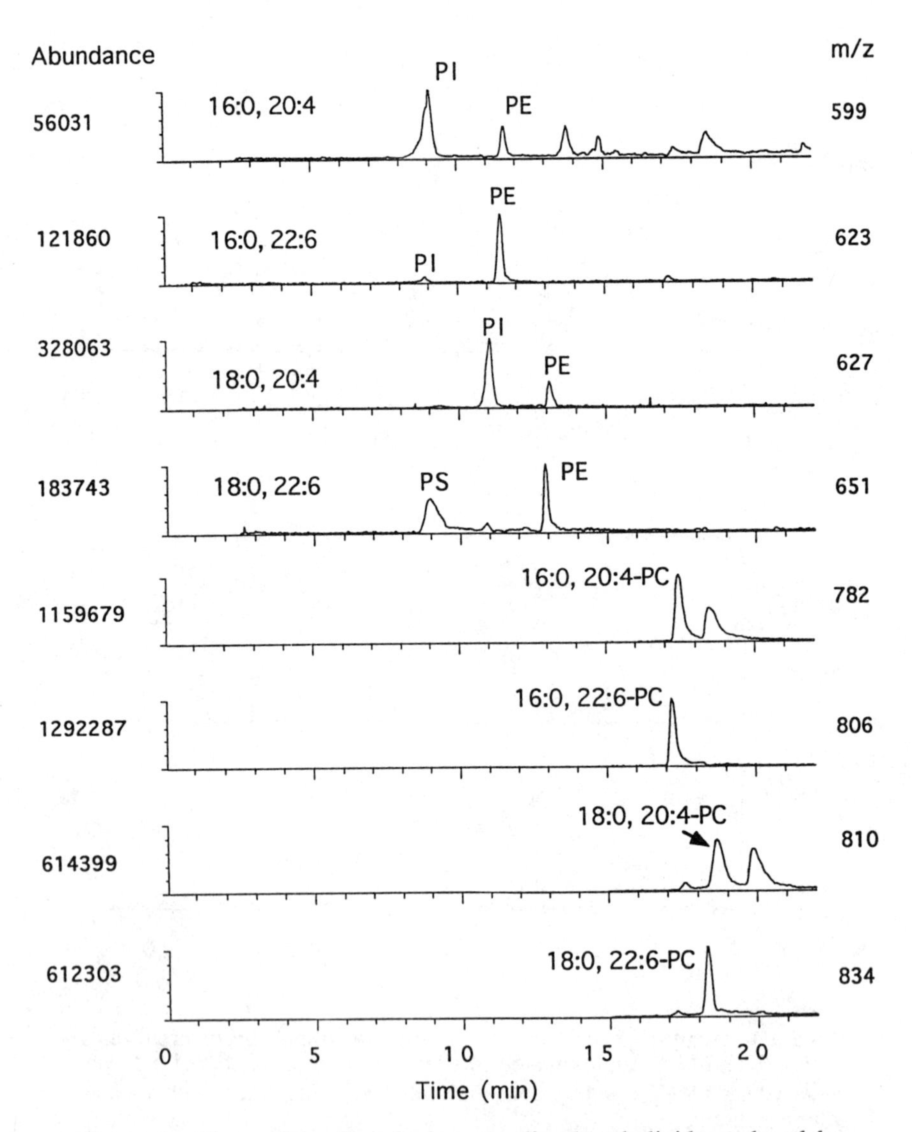

Figure 5. The profile of rat brain cytosolic phospholipids analyzed by reversed phase HPLC/ESP-MS. Approximately 10 nmoles of total phospholipids were injected. In the upper four pannels, DG fragment ions were monitore while MH^+ ions were monitored in the bottom 4 pannels for PC molecular species.

Quantitative Aspects

The response is linear within the two orders of magnitude range (2-200 pmoles). In a higher concentration range, the suppression of signal occured. By using stable isotope labeled standards, however, linear response was extended to the low nmole range, indicating that accurate quantitative analysis can be done over a wide range of sample amounts.

Incorporation of 22:6n3 to C-6 Glioma Cells

It has been indicated that enrichment of docosahexaenoic acid (22:6n3) in neuronal membranes may be important to maintain proper neuronal function. However, the mechanism involved in enrichment of this fatty acid is not clearly understood. In an attempt to characterize the process of 22:6n3 enrichment in neuronal membranes, we investigated the incorporation and remodeling of 22:6n3 using C-6 glioma cells as a model system. C-6 glioma is a cell line of neuronal origin and has been shown to preserve astrocytic characteristics (15). Cells were plated and raised with culture medium containing 100 µM 22:6n3 and incorporation of this fatty acid into various phospholipids was monitored for 72 hours. Without supplementation, the 22:6n3 content in cell membranes was less than 2% at any time point. After 24 hours of incubation, incorporation of 22:6n3 reached near the plateau based upon the fatty acid analysis by GC. In order to evaluate the changes, sample amounts assessed by phosphorous assay (16) were kept comparable. At 24 hours, 22:6n3 appeared in PA, PI and PC as is shown in the HPLC/mass chromatogram of m/z 651 in figure 6a. For PI, PA and PS, 22:6n3 was incorporated into phospholipids containing 18:0, presumably at the sn-1 position, while incorporation into PC preferred species containing 16:0 at the sn-1 position. After 48 hours of incubation, 22:6n3 containing PE increased while PC, PI and PA decreased as shown in figure 6b. Further decrease in these species was observed after 72 hours of incubation (data not shown) with a continued increase of 22:6n3 in PE. Plasmalogen species also increased at 48 hours and further accumulated at 72 hours while the level of 22:6-PS did not change considerably. Although the molecular species monitored did not represent the entire 22:6n3 species in C-6 cells, it appeared that glioma cells initially incorporated 22:6 into PC, PA and PI and remodeled to PE and PE plasmalogen species. The incorporation of PS may occur at the initial stage and turnover of this species may not be as active as PC, PA or PI. Radiolabeling studies performed in parallel under this condition confirmed the HPLC/ESI-MS results.

Polyunsaturated Phospholipid Turnover in Mouse Brain

As stated above, neuronal membranes are rich in long chain polyunsaturated fatty acids, especially 22:6n3, mainly esterified in phospholipids (3). A high level of 22:6n3 is maintained and in fact it is very difficult to deplete this fatty acid from brain tissues. The underlying mechanism for enriching 22:6n3 in brain phospholipids is not clearly known. Selective incorporation, active on-site biosynthesis and slow turnover may be included in underlying mechanisms. Turnover of polyunsaturates involves various biochemical processes including hydrolysis of fatty acyl moiety, head group exchange and reincorporation.

By labeling phospholipid molecular species with stable isotopes and monitoring by mass spectrometry, turnover of each molecular species in complex mixtures can be evaluated as has been demonstrated in mast cells (17). In order to monitor the turnover of brain polyunsaturates, mouse brain phospholipids were first labeled with d_5-20:4n6 and d_5-22:6n3 by feeding d_5-18:2n6 and d_5-18:3n6 during pregnancy and lactation period. Through in vivo elongation and desaturation processes 18:2n6 and 18:3n3 are

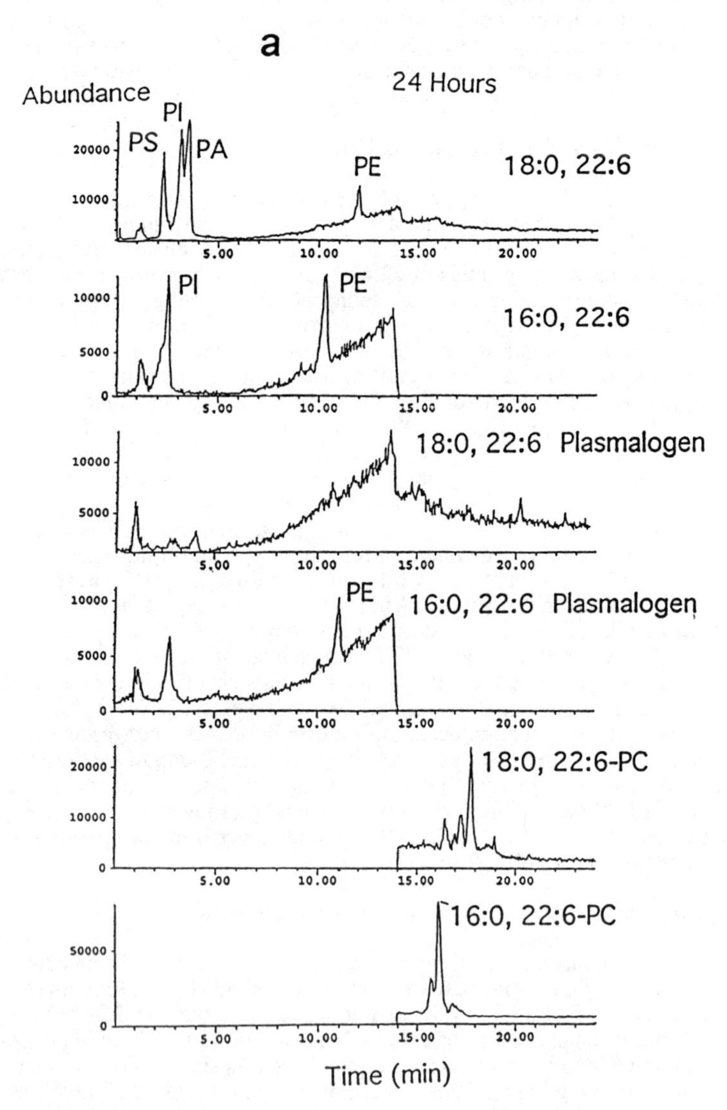

Figure 6. Mass chromatogram of 22:6n3 containing phospholipids obtained from C-6 glia cells after incubation with 100 μM 22:6n3 for 24 (a) and 48 (b) hours (from ref. 13).

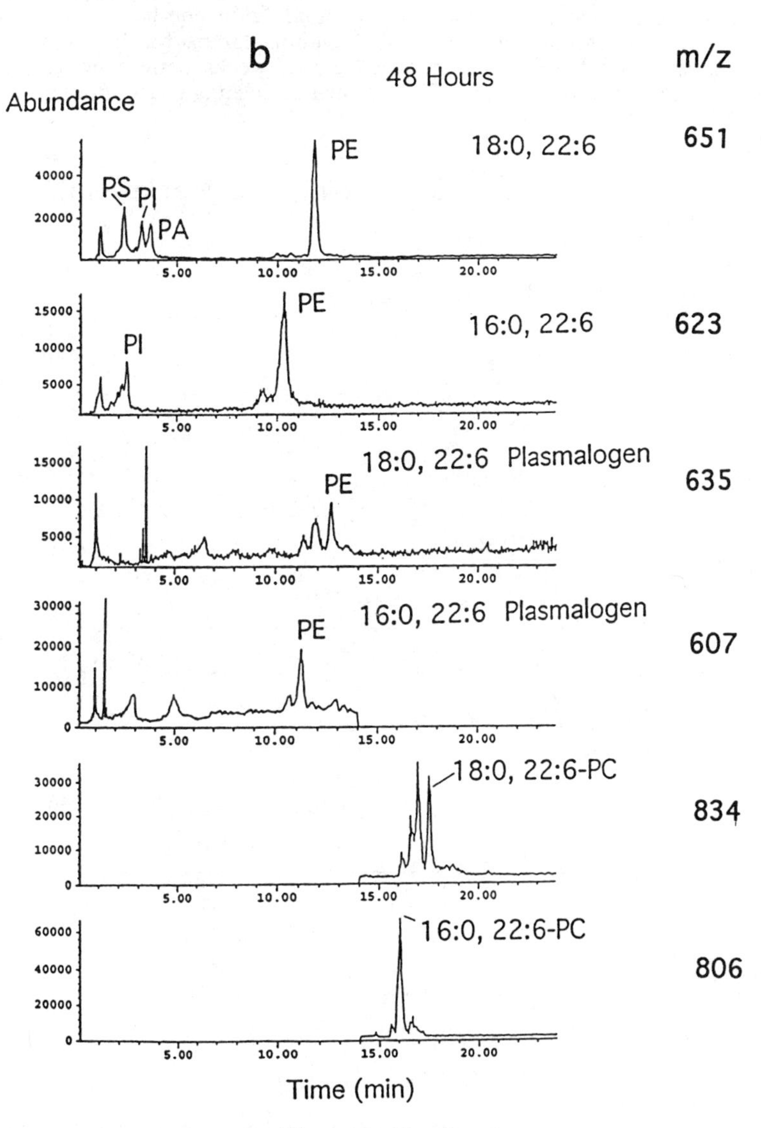

Figure 6. *Continued*

metabolized to 20:4n6 and 22:6n3 (17,18). After weaning, pups were fed non-labeled 18:2n6 and 18:3n3 in place of labeled compounds. At the weaning stage, 20:4 and 22:6-containing phospholipid species were labeled at the comparable extent with d0/d5 ranging 1.6-2.0. Turnover of polyunsaturated phospholipids was followed by monitoring replacement of labeled phospholipids with non-labeled species using electrospray mass spectrometry (19). Mass chromatograms for three representative molecular species in brain lipid extracts at age of 12 weeks, corresponding to 9 weeks after weaning, are shown in figure 7. Although calculating exact turnover rates may

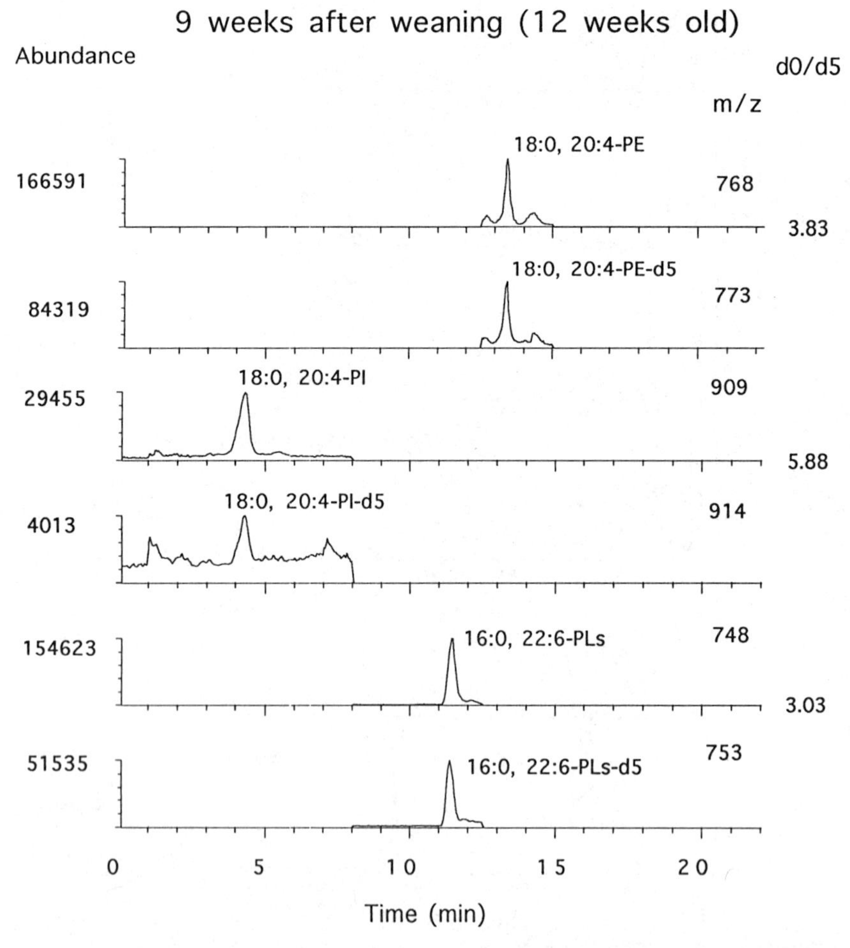

Figure 7. Representative mass chromatograms of brain phospholipids containing 20:4n6 or 22:6n3, obtained from mice at 9 weeks after weaning. Brain lipids were labeled by feeding dams with d_5-18:2n6 and d_5-18:3n3 during the pregnancy and lactation period. After weaning (at 3 weeks), pups were fed non-labeled 18:2n6 and 18:3n3 instead of labeled compounds. PLs; plasmalogen.

involve the determination of pool size at each time point, turnover under a given circumstance could be readily compared for each phospholipid molecular species by monitoring changes in the ratio of d0/d5. Since 18:2n3 was supplied 4 times higher than 18:3n3 with this experimental protocol, the turnover rate was faster for 20:4-species in comparison to 22:6-species. However, within 20:4 containing species, it was easily determined that turnover of 18:0 20:4-PI was more active than 18:0 20:4-PE. Interference of other molecular species containing the same number of carbons and double bonds was minimal at least for 20:4 and 22:6 containg species. The fatty acid analysis indicated that the content of 16:1n7, 18:2n6, 18:3n3, 20:5n3 and 22:5n3 were minimal in the normal brain (less than 1%). Separately, samples were run by thermospray LC/MS using identical conditions. From the analysis of monoglyceride ions it was obvious that the contribution from 18:1 22:5, 18:2 22:4, 16:1 20:3, 18:1 20:5, 18:2 20:4-species was either absent or negligible in comparison to the major molecular species monitored. In addition, the sn-1 fatty acyl chain appeared to have more influence on the retention behaviour under our HPLC condition; therefore, some of these molecular species could be separated. Additional confirmation of these results may be achieved by electrospray mass spectral analysis of fatty acid carboxylate anions using product ion scans as reported with the fast atom bombardment (FAB) technique (20) .

As demonstrated in these examples, the electrospray HPLC/MS technique can be easily applied to studies involving phospholipid biosynthesis or remodeling. The technique is sensitive enough to trace labeled phospholipids at low picomole levels with excellent reproducibility. In the presence of internal standards, accurate quantification of each molecular species can be achieved with a reasonable dynamic range. We believe that this technique will facilitate the investigation of phospholipid metabolism under physiological conditions where trace metabolic changes may need to be elucidated.

References

1. Van den Bosch, H. Ann. Review Biochem. 1974, 43, 243-277.
2. MacDonald, J. I. S.; Sprecher, H. Biochim. Biophys. Acta 1991, 1084, 105-121.
3. Salem, N., Jr.; Kim, H. Y.; Yergy, J. A. Health Effect of Polyunsaturated Fatty Acids in Seafoods (Simopoulos A. P. and Kifer R. R., eds.); Academic Press, New York, New York, 1986, pp. 263-317.
4. Slater, S. J.; Kelly, M. B.; Taddeo, F. J.; Ho, C.; Rubin, E.; Stubbs, C. D. J. Biol. Chem. 1994, 269, 4866-4871.
5. Graber, R.; Sumida, C.; Nunez, E. A. J. Lipid Mediators Cell Signalling 1994, 9, 91-116.
6. Arvidson, G. A. E. J. Chromatogr. 1975, 103, 201-204.
7. Arvidson, G. A. E. J. Lipid Res. 1965, 6, 574-577.
8. Blank, M. L.; Robinson, M.; Fitzgeral, V. and Snyder, F. J. Chromatogr. 1984, 298, 473-482.
9. Kayganichi, K.A.; Murphy, R. C. Anal. Chem. 1992, 64, 2965-2971.
10. Kim, H. Y.; Salem, N., Jr. Anal. Chem. 1986, 58, 9-14.
11. Kim, H. Y.; Salem, N., Jr. Prog. Lipid Res. 1993, 32, 221-245.
12. Ma, Y. C.; Kim, H. Y. Anal. Biochem. 1995, 226, 293-301.
13. Kim, H. Y.; Wang, T. C. L.; Ma, Y. C. Anal. Chem. 1994, 3977-3982.
14. Wirtz, K. W. A.; Gadella, T. W. J. Experientia 1990, 46, 592-599.
15. Pfeiffer, S. E.; Betschart, B.; Cook, J.; Mancini, P.; Morris, R. Cell, Tissue and Organ Cultures in Neurobiology (Fedoroff, H. ed.); Academic Press, New York, New York, 1977, pp. 287-346.

16. Nelson, G. J. Blood Lipids and Lipoproteins: Quantitation, Composition and Metabolism (Nelson, G.J., ed.); Wiley-Interscience, New York, 1972.
17. Kayganich-Harrison, K. A.;Murphy, R. C. Biol Mass Spectrom 1994, 23, 562-571
18. Pawlosky, R.; Sprecher, H; Salem, N. Jr. J. Lipid Res. 1992, 33, 1711-1717.
19. Kim, H. Y.; Ward, G.; Ma. Y. C.; Wang, T. C. L.; Salem, N., Jr. manuscript in preparation.
20. Kayganich, K. A.; Murphy, R. C. Anal. Chem. 1992, 64, 2965-2971.

RECEIVED October 5, 1995

Chapter 14

Application of Electrospray Ionization Mass Spectrometry to the Analysis of Oligodeoxynucleotides

Charles R. Iden[1], Robert A. Rieger[1], M. Cecilia Torres[1], and LeRoy B. Martin[2]

[1]Department of Pharmacological Sciences, Health Sciences Center, State University of New York at Stony Brook, Stony Brook, NY 11794–3400
[2]Fisons Instruments, 55 Cherry Hill Drive, Beverly, MA 01915

Electrospray mass spectrometry is a powerful technique for the analysis of oligodeoxynucleotides providing accurate molecular masses from the multiply charged negative ions formed by the electrospray volatilization/ionization process. Fragment ions are generally absent unless formed by secondary collisions near the exit of the source where pressures are relatively high. Use of the electrospray technique for the analysis of oligodeoxynucleotides is superior to other ionization techniques for verification of site-specific incorporation of modified deoxynucleotides, especially for oligomers containing more than ten deoxynucleotides where FAB/MS is not effective. ESI spectra may be generated on 10-100 picomoles of an oligomer on a single quadrupole mass spectrometer system, and mass accuracy of ±0.01% is attainable with careful calibration of the instrument. Spectra are presented for oligomers containing modified nucleobases and sugar moieties, as well as a hybrid DNA/RNA oligomer in which an O-methyl moiety is positioned at the 2'-position of adenosine. The technique is also effective for the elucidation of mechanisms by which unstable oligomers may disintegrate into unknown products, and the known sensitivity of 8-oxo-2'-deoxyguanosine-containing oligomers to the postsynthetic ammonia deprotection procedure has been examined. ESI/MS was used to characterize the major products of oligomer decomposition. Deoxynucleotide sequence of an oligomer cannot be determined from the electrospray mass spectrum. However, MS/MS experiments were conducted on multiply charged oligodeoxynucleotide molecular ions generated by ESI on a triple quadrupole instrument. Product ion spectra from collision induced dissociation are complex, and product ions may differ in charge state from the precursor ion. However, families of ions were identified from which the sequence of the oligomer was verified.

0097–6156/95/0619–0281$12.00/0

With the recent introduction of effective ionization methods for labile biopolymers, mass spectrometry is emerging as an indispensable technique for biochemical and biomedical research (*1*). Electrospray ionization (ESI) (*2-4*) has revolutionized the analysis of DNA, RNA, and oligonucleotides because it presents significant advantages over other methods used for analysis of these compounds (*5*). Excellent sensitivity and the formation of multiply charged molecular ions permit accurate mass determinations and facilitate the use of low mass range instruments for the analysis of high molecular weight compounds. Prior to the development of the electrospray technique, fast atom bombardment (FAB) was useful for the analysis of small oligomers (*6-8*); however, no ionization technique was available routinely for mass analysis of large oligomers. In the negative ion mode, FAB mass spectra contain an intense molecular ion $(M-H)^-$ and two series of fragment ions formed by the cleavage of the carbon-oxygen bond in the sugar-phosphate backbone at the C3' or C5' positions on the ribose moiety. These ions are sufficient to permit determination of the nucleotide sequence of the oligomer. However, the sensitivity of the FAB technique is not outstanding and oligomers containing more than ten deoxynucleotides are difficult to analyze.

In recent years matrix assisted laser desorption ionization (MALDI) has been used for the analysis of oligodeoxynucleotides (*9-15*). Picolinic acid, 3-hydroxypicolinic acid, 2,5-dihydroxybenzoic acid, and a host of other compounds have been used successfully as matrix materials, and abundant negative ions have been analyzed in both time-of-flight (TOF) and Fourier transform mass spectrometers (FT/MS). Sensitivity is excellent, and only femtomoles of sample are required. Linear MALDI-TOF mass spectra show a dominant molecular ion $(M-H)^-$ with little or no fragmentation of the oligomer (*9*). In some cases, doubly charged molecular ions $(M-2H)^{2-}$ are present in the spectrum. Mass accuracies of 0.01% can be achieved, especially when internal mass standards are used. Oligodeoxynucleotide sequence can be obtained from the spectra of samples that have been partially digested enzymatically prior to analysis (*10*) or directly from infrared matrix assisted laser desorption ionization (IR-MALDI) at 2.94 μm with a reflectron TOF instrument (*15*). In contrast to TOF mass analysis, considerable fragmentation of an oligomer is found with the MALDI desorption/ ionization process when FT/MS mass analysis is performed, and it is strongly dependent on the matrix material in use (*13,14*).

Electrospray volatilization/ionization of oligodeoxynucleotides forms a series of negatively charged molecular ions of the form $(M-nH)^{n-}$ (*16-18*), and similar ions have been reported in the analysis of small ribonucleic acids (*19*). Fragment ions are generally absent from the mass spectrum; however, adducts with adventitious metal cations (Na^+, K^+) occur frequently and lead to broad mass peaks, since individual adducts in high charge states are not always resolved with a low resolution mass analyzer. Desalting techniques (*20*) and procedures for replacing metal ions with ammonium using cation-exchange resins have been successful in improving peak shape (*21*); ammonium adducts are not seen in the mass spectrum. Moreover, addition of small amounts (1-5%) of organic bases such as triethylamine to the solvent have been used to adjust pH, dramatically increasing signal:noise and peak resolution (*22*). Mass accuracy of 0.01% is attainable on a single quadrupole mass analyzer as an average of the molecular mass values obtained from each of the multiply charged peaks in the mass spectrum. In many cases this level of accuracy is sufficient to obtain the deoxynucleotide composition of an oligomer with mass less than 5000 Da (*23*). Recently, high resolution FT/MS analysis of large oligomers provided mass measurement accuracy of a few ppm and was used to verify the molecular mass of 50- to 100-mer DNA and RNA sequences (*24*).

The mechanisms of spray formation, solvent evaporation, and ion formation were reviewed by Kebarle and Tang (*25*), and are presently undergoing intense examination (*26-30*). An initial spray of highly charged, small droplets occurs through formation of a Taylor cone at the end of the metal capillary when placed in a large electric field. At a sufficiently high fields, the cone tip becomes unstable, and a thin filament of liquid is drawn from the tip, breaking up into small droplets as it moves away. Solvent evaporation occurs promptly, shrinking the diameter of the droplets. Two proposals for ion formation have garnered significant attention. The Charged Residue Model (*2*) proposes a series of coulombic explosions reducing the droplet size until a single solute molecule containing some of the droplet charge remains after the evaporation of all solvent molecules. The Ion Evaporation Model (*31*) proposes that multiply charged ions are transferred to the gas phase prior to total evaporation of the solvent in the charged droplet. Factors which control the relative abundance of the charge states are not well characterized (*30,32,33*); however, overall sensitivity of the technique is excellent, and only 10-100 pmoles of an oligonucleotide will generate a full-scan mass spectrum with a good signal:noise ratio.

Since little energy is transferred to the solute molecule during the volatilization/ionization process, few fragment ions are formed, and deoxynucleotide sequence information is not available directly from the mass spectrum. However, collisions in the nozzle skimmer region of the source, an area of relatively high pressure, cause molecular ion fragmentation for some species, and this effect can be accentuated by optimizing lens voltages. While lack of fragmentation is a distinct disadvantage for those seeking information on deoxynucleotide sequence, others have taken advantage of the characteristic stability of the molecular ion to analyze noncovalently bound complexes which remain intact throughout the electrospray process. ESI mass spectra of duplex DNA have been obtained in which intense molecular ions of the duplex form are found distributed among the molecular ions of each single strand (*34*). Conditions have also been determined for the analysis of higher order complexes, such as the quadruplex structure formed by the decamer, 5′-CGCGGGGGCG-3′ (*35*), and drug-DNA noncovalent complexes (*36*).

Deoxynucleotide sequence can be determined through tandem mass spectrometry MS/MS or $(MS)^n$ experiments in which fragmentation is induced through energetic collisions with argon gas atoms. Initial work on the collisional induced dissociation of oligonucleotides was reported by McLuckey *et al.*, who performed experiments in a quadupole ion trap at relatively low collision energies from which a set of rules for the decomposition of multiply charged oligonucleotides was devised (*37,38*). By coupling an ESI source to a Fourier Transform mass spectrometer (FT/MS), Little *et al.* were able to use the high resolving power of the FT/MS instrument to obtain accurate masses of structurally significant fragment ions formed in the nozzle skimmer region of the source or by collision assisted dissociation (*39*). These data were used to determine the complete sequence of a 14-mer and to verify the sequence of oligomers as large as 25-mers. In addition, sequencing of modified oligomers (methylphosphonates) was shown in the positive ion mode using MS/MS (*40*).

We report here the use of ESI mass spectrometry on a single quadrupole instrument to verify the synthesis of oligodeoxynucleotides which contain modified components in both the base and sugar portions of the molecule and to investigate oligomer degradation. Examination of the collision induced dissociation of oligonucleotide multiply charged molecular ions was performed on a triple stage quadrupole instrument, and ion types which provide sequence information are identified.

Experimental

Oligodeoxynucleotides were synthesized on an Applied Biosystems Model 394 DNA Synthesizer using the phosphoramidite chemistry. Resins and standard reagents were purchased from Applied Biosystems (Foster City, CA); phosphoramidites for modified deoxynucleotides were purchased from Glen Research (Sterling, VA) or provided by Dr. Francis Johnson, Department of Pharmacological Sciences, SUNY-Stony Brook. After completion of a synthesis, oligomers were released from the resin by treatment with 28% ammonia at 55° C for sixteen hours, evaporated to dryness in a Savant Speed Vac SC100 (Farmingdale, NY), reconstituted in distilled water, and filtered through 0.45 micron PVDF centrifuge filter (Whatman, Inc., Clifton, NJ). A two step high pressure liquid chromatography (HPLC) purification procedure was performed using a Hamilton PRP-1 column (4.1 mm x 250 mm) on a Waters HPLC system, which included two 6000A pumps, a U6K injector, a 440 UV absorbance detector, and a Hewlett Packard 3394 integrator. For dimethoxytrityl-protected (DMT) oligomers, a linear gradient of 16 - 33% acetonitrile in 0.1 M triethylammoniun acetate (TEAA) buffer, pH 6.8, in 15 min was used at a flow rate of 2 mL/min. A fraction containing approximately 10 O.D. (260 nm) of the DMT-protected oligomer was collected and evaporated to dryness. The DMT-protecting group was removed with 250 μL of 80% acetic acid for 30 min at room temperature. The oligomer was dried, reconstituted in 250 μL of distilled water, and rechromatographed using a gradient of 9-15% acetonitrile in 0.1 M TEAA buffer, pH 6.8, over a period of 15 minutes. The oligomer was collected as a pure fraction, dried, and stored at -20° C.

Electrospray mass spectra were acquired on a Fisons Instruments Trio-2000 mass spectrometer (Danvers, MA). A continuous infusion of 60:40 acetonitrile/water (8 μL/min) was provided by a syringe pump (Harvard Apparatus, South Natick, MA). Oligomers were dissolved in 60:40 acetonitrile/water containing 1% triethylamine (200 pmoles/μL). Approximately 10 μL were introduced through a Rheodyne injector with a 10 μL sample loop. The ion source was operated with 3.7 kV on the source capillary which was positioned approximately 4 mm in front of the counter electrode. Nitrogen was supplied as the nebulizer gas (15 L/h) and as the drying gas (300 L/h). Lens voltages were adjusted to give the maximum signal with 2′-deoxycytidine-5′-monophosphate. The instrument was scanned over the *m/z* range 200-1400 in 8 sec. Data were acquired in the multichannel analyzer mode (MCA or MCB). Raw spectra consisted of ten scans which were averaged, digitally filtered, and baseline adjusted.

Decomposition products of an 8-oxo-2′-deoxyguanosine-containing oligomer in basic solution were examined by placing 2-3 O.D. (260 nm) of the oligomer, 5′-GTTCAXTTGC-3′ (X=8-oxodG), in an Eppendorf tube, adding 28% ammonia, sealing the tube tightly, and heating at 55° C. After a predetermined time, the ammonia was evaporated on a Savant SpeedVac SC100, and the products of the reaction were chromatographed by HPLC as described above using conditions for deprotected oligomers. Products were collected as pure fractions and dried under vacuum. Prior to ESI/MS analysis, the fraction was reconstituted in 60:40 acetonitrile/water containing 1% triethylamine at a concentration of 50-100 pmoles/μL and injected into the mass spectrometer.

MS/MS data were acquired on a Fisons Instruments VG Quattro triple stage quadrupole mass spectrometer (Danvers, MA). An oligonucleotide was dissolved in 50:50 isopropanol/water at a concentration of 20 pmole/μL and infused into the ESI source at 5 μL/min. Nitrogen was used for the nebulizing gas (15 L/h) and drying gas (200 L/h), and the ion source was maintained at 60° C. The collision gas was argon (9 mtorr) and the collision energy was 23 eV. Product ion mass spectra were acquired over the *m/z* range 50-1650, and spectra were scanned repetitively in the MCA mode to increase the signal:noise ratio.

Results and Discussion

ESI Mass Spectra of Oligodeoxynucleotides. Electrospray ionization mass spectrometry can be used routinely to determine the molecular mass of synthetic oligodeoxynucleotides and verify the product of the automated synthesis. While not necessary for standard oligomers where the chemistry is well-defined, mass measurement of oligomers containing modified deoxynucleotides, site-specifically included in a sequence, is essential to ensure that the modified deoxynucleotide and the oligomer have remained intact throughout the automated chemistry in the DNA synthesizer, during deprotection, and during the purification procedures.

Modifications to the oligonucleotide structure can be situated in the base moiety, the sugar, and in the phosphate linkage between deoxynucleosides. Figure 1 shows an ESI mass spectrum of an oligomer in which the base moiety of one of the deoxynucleotides was altered. A 17-mer, 5′-TCGAGGAXGATCCATTC-3′, was synthesized with 7-deaza-2′-deoxyadenosine site-specifically located at the X-position. The ESI mass spectrum contains eight molecular ions with charge states that range from -4 to -11. The average molecular mass determined from the data is 5193.3 Da, and the calculated value is 5193.4 Da. Thus, the change at the 7-position in deoxyadenosine from N to CH was verified by the mass spectrum. Figures 2 and 3 show the spectra of oligomers in which a single ribose moiety was modified. The first spectrum (Figure 2) is that of a DNA/RNA hybrid in which 2′-O-methyladenosine was incorporated into a 22-mer with the nucleotide sequence, 5′-CATGCTGATCCTXCTACTCTTC-3′. This ESI mass spectrum contains ten molecular ions with charges that range from -5 to -14. The molecular mass obtained from the data is 6626.3 Da and that calculated from the molecular formula is 6626.4 Da. The next spectrum (Figure 3) is that of a 23-mer in which the X-location is deoxyaristeromycin, a carbasugar derivative of deoxyadenosine in which the cyclic oxygen atom in deoxyribose was replaced by a methylene moiety. The base sequence is 5′-CTCTCCCTTCXCTCCTTTCCTCT-3′, having a molecular mass of 6761.4 Da; the mass determined from the spectrum is 6760.8 Da.

In the calculations of molecular mass, M_r, the average mass of each of the atoms present was used (averaged over the natural isotopic abundance) rather than the

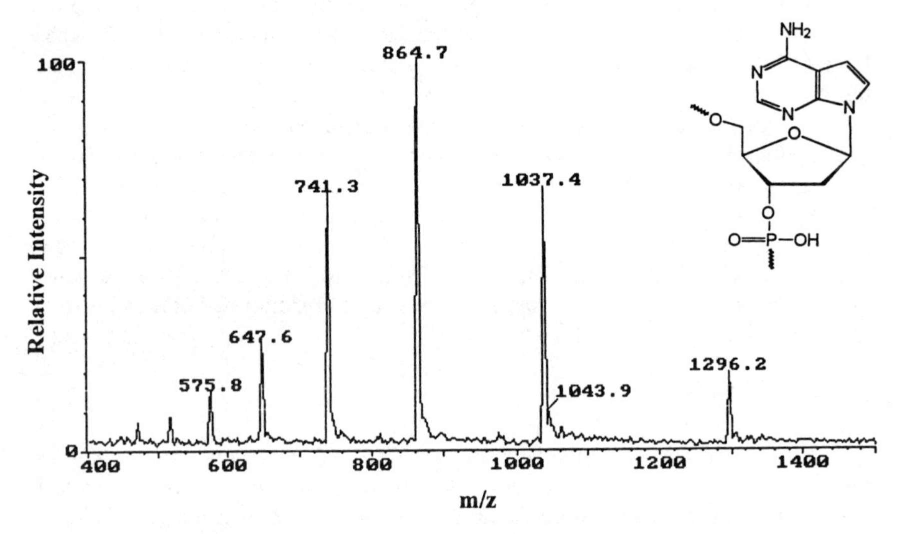

Fig. 1. Electrospray mass spectrum of an oligomer, 5′-TCGAGGAXGATCCA-TTC-3′, containing 7-deaza-2′-deoxyadenosine at the X-position.

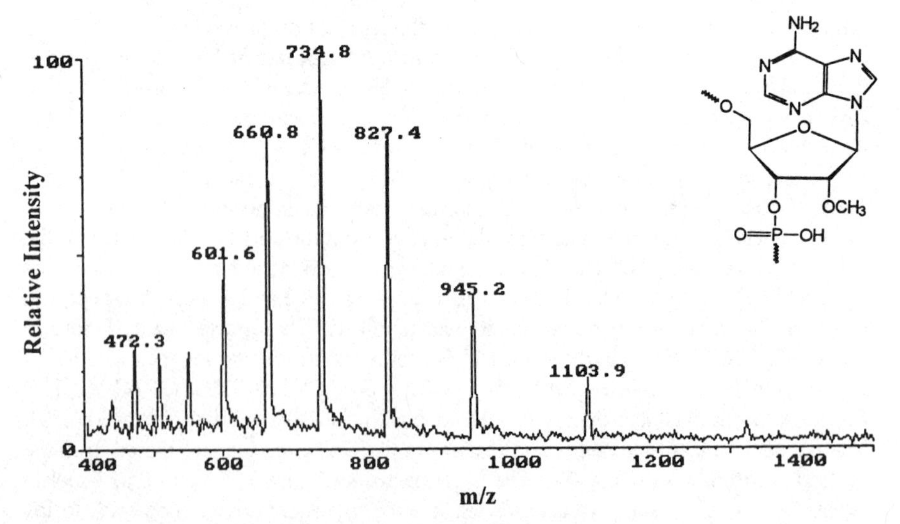

Fig. 2. ESI mass spectrum of a DNA/RNA hybrid, 5′-CATGCTGATCCTXC-TACTCTTC-3′, in which the X-position is the protected ribonucleotide, 2′-O-methyladenosine.

mass of a specific isotope, since the data do not display unit mass resolution in the region of the molecular ion and individual isotopic peaks cannot be resolved. While the average mass value obtained from the data is generally quite good when compared to the calculated value (< 0.01%), the statistical error in the determination of the average mass value is approximately 0.06% for the data we have presented. None of the spectra has significant peaks due to cation (Na^+ or K^+) adduction, although small natriated peaks may appear on the high mass side of the larger molecular ions. Although samples were not exposed to Na^+ during synthesis or purification, adventitious cations will bind to the oligomer tightly and appear in the mass spectrum. When used in large concentration during ion exchange purification, Na^+ or K^+ will bind to one or more of the phosphate linkages, producing many adducts in the mass spectrum. Even after thoroughly desalting the oligomer, natriated peaks may continue to appear in the ESI mass spectrum. For large molecular masses and high charge states, these adducts may not be resolved from either the molecular ion or each other and contribute to a broad, unresolved peak.

Degradation of Oligomers Containing 8-Oxo-2′-deoxyguanosine. In many cases during oligodeoxynucleotide synthesis or deprotection, a modified base or sugar moiety will not be stable to all of the conditions and reagents with which it is in contact. In these cases there may be significant degradation of the deoxynucleotide or the oligomer which can be identified during HPLC purification if, indeed, this technique is used. Oligomers containing 8-oxo-2′-deoxyguanosine have been reported to be unstable to the conditions of release from the resin and deprotection (28% ammonia at 55° C for 16 h) (*41,42*), and ESI mass spectra have been crucial for experiments to determine how an oligomer degrades under these conditions.

An oligonucleotide, 5′-GTTCAXTTGC-3′ with X = 8-oxo-2′-deoxyguanosine, was synthesized using standard methods; however, during the deprotection step, the material was treated with an antioxidant, 2-mercaptoethanol, to prevent degradation (*41*). After HPLC purification, the oligomer was subjected to 28% ammonia at 55° C for up to 48 h without the antioxidant present. Degradation products were separated by HPLC (Figure 4). Two apparent products of the reaction appeared as intense peaks in the chromatogram (Peak 1 and Peak 2) and were collected as pure fractions and submitted for ESI/MS analysis.

The mass spectrum of Peak 1 showed it to be a single substance with a molecular mass of 1244.7 Da as determined from the data (Figure 5a). This mass corresponds to the 3′-terminal portion of the original oligomer, *i.e.*, the four deoxynucleotides, TTGC, and a phosphate moiety at the 5′-end, namely 5′-p-TTGC-3′ whose calculated mass is 1244.8 Da. This product is apparently cleaved from the 8-oxo-2′-deoxyguanosine sugar moiety during a β-elimination process. The ESI mass spectrum of Peak 2 (Figure 5b) actually contained two substances that were unresolved in the HPLC chromatogram. The first substance, with a measured mass of 1558.8 Da, produced the major ions in the mass spectrum at *m/z* 778.3, 518.3, and 388.9. The molecular mass corresponds to that of the five deoxynucleotides at the 5′-end of the oligomer with a terminal phosphate, namely 5′-GTTCA-p-3′ and indicates there is cleavage of the C5′-O bond on the sugar moiety of 8-oxo-2′-deoxyguanosine. The second substance contained in this fraction generated three ions at *m/z* 835.6, 556.7, and

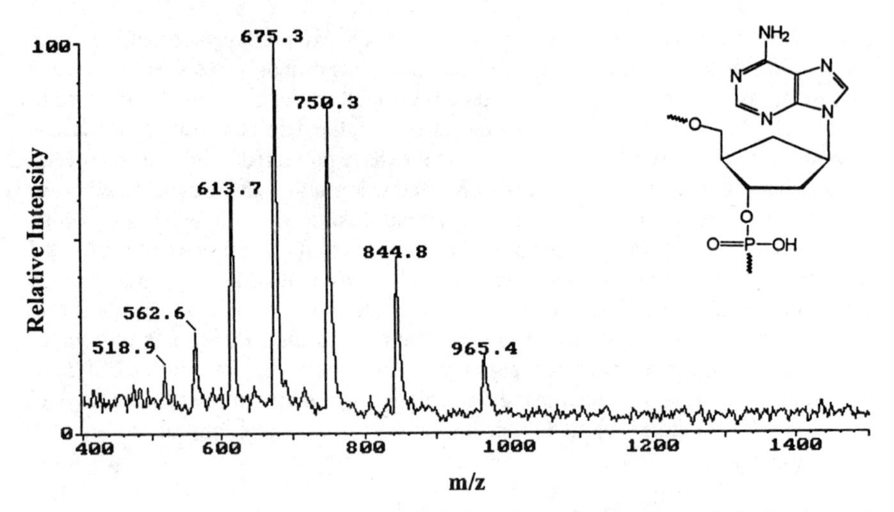

Fig. 3. ESI mass spectrum of a 23-mer, 5′-CTCTCCCTTCXCTCCTTTCCTCT-3′, containing the carbasugar (see text) analogue of 2′-deoxyadenosine, *i.e.*, 2′-deoxyaristeromycin, at the X-position.

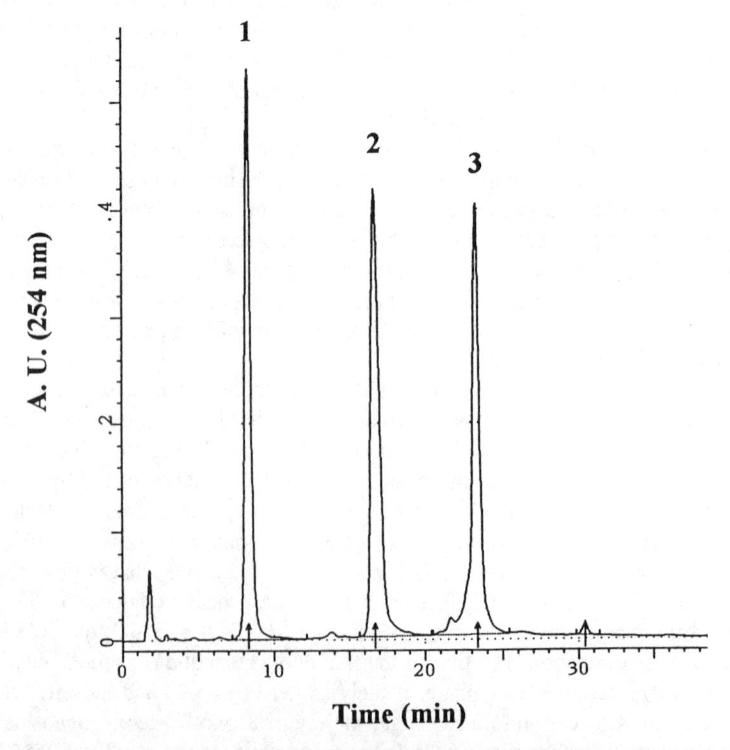

Fig. 4. HPLC chromatogram of a six hour digest of 5′-GTTCAXTTGC-3′, where X is 8-oxo-2′-deoxyguanosine, in 28% ammonia at 55° C. Reaction products were found in fractions under Peak 1 and 2; the original oligomer is found in Peak 3.

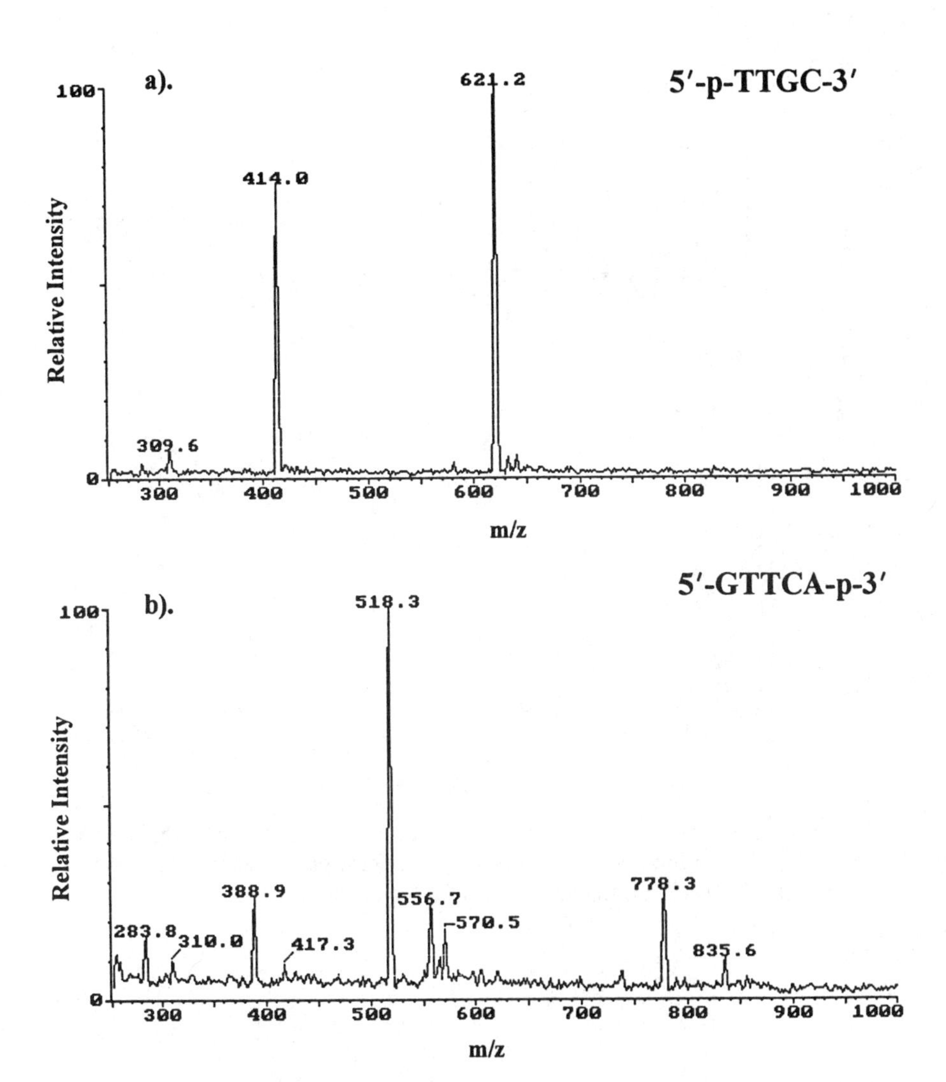

Fig. 5. ESI mass spectra of fractions of the reaction products found under (a) Peak 1 and (b) Peak 2 of the chromatogram in Figure 4.

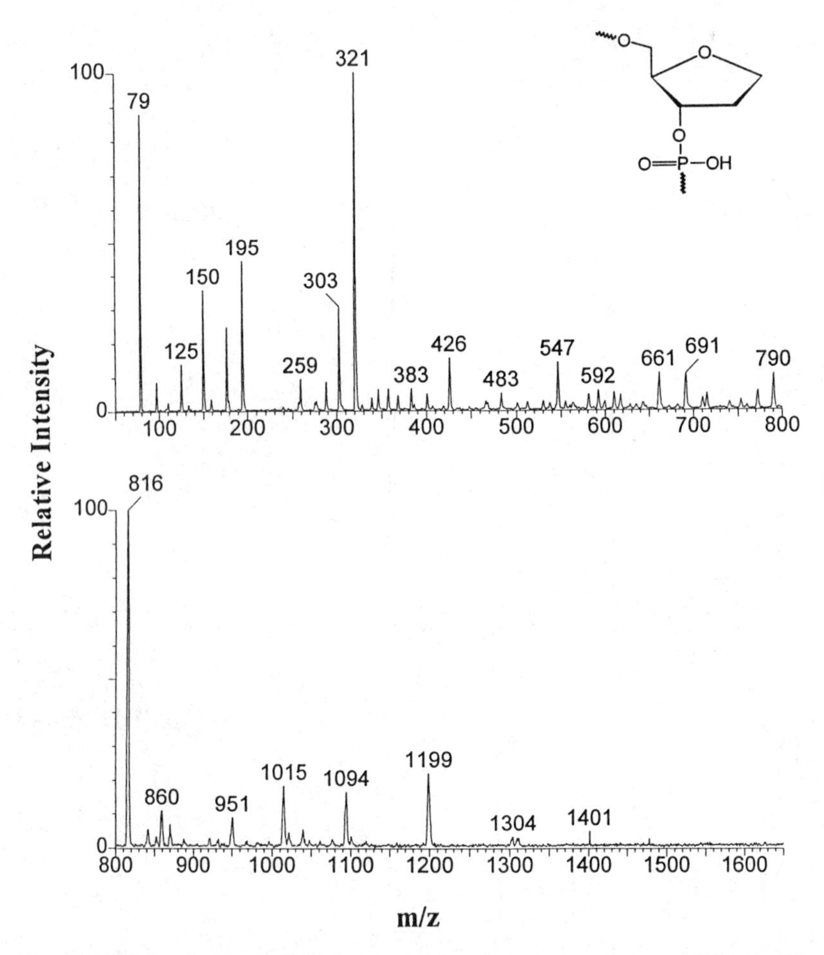

Fig. 6. Collision induced dissociation product ion mass spectrum of the doubly charged molecular ion (*m/z* 816) formed in the electrospray mass spectrum of 5′-GCTXCT-3′, where X is the furan model for the abasic site.

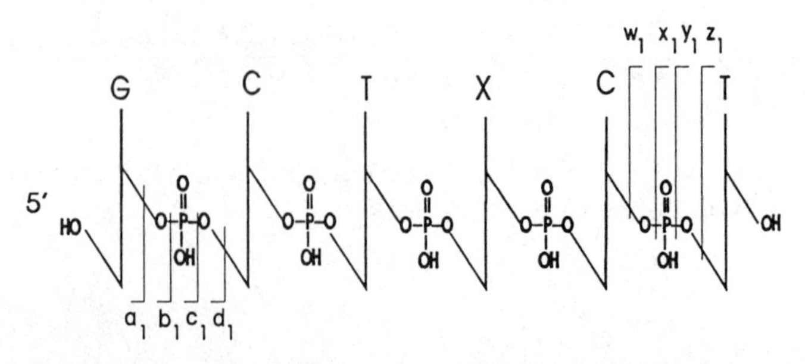

Fig. 7. Product ion nomenclature for the ions formed by collision induced dissociation of the oligomer, 5′-GCTXCT-3′, where X is the furan model for the abasic site.

417.3 and has a molecular mass of 1673.3 Da. This mass indicates a structure similar to the previous one with additional elements of a deoxyribose; however the precise structure is not yet known and cannot be inferred from mass spectrometry alone.

Deoxynucleotide Sequence Verification by Collision Induced Dissociation of Multiply Charged Molecular Ions in a Triple Quadrupole Instrument. In the absence of significant fragmentation, collision induced dissociation of ESI-generated oligonucleotide molecular ions on a tandem quadrupole instrument is an alternative technique to obtain structurally important ions. However, the product ion spectrum is complicated by the fact that precursor ions formed in the electrospray ionization technique are multiply charged. In most cases the singly charged oligonucleotide molecular ion is very weak or absent from the spectrum, although instrumental tuning and modification to the solvent may enhance the lower charged states. Consequently, when multiply charged molecular ions are used in the analysis, product ions may exist in one or more charge states and differ from that of the precursor ion, making the interpretation of the product ion spectrum complex.

Collision induced dissociation was performed on the molecular ions of the oligomer 5′-GCTXCT-3′, where X represents the furan model for the abasic site (*43*) in which the base moiety has been replaced by a hydrogen atom. The product ion spectrum obtained from the doubly charged molecular ion (*m/z* 816) of this oligomer is shown in Figure 6. The spectrum is replete with intense fragment ions, both singly and doubly charged, and several ion series can be discerned. For the purposes of this presentation, the ion nomenclature introduced by McLuckey *et al.*(*37*), which is similar to that used for describing fragmentation of peptides, has been adopted (Figure 7). The ion type is given a lower case letter designation with a subscript denoting the number of deoxynucleotides present; the charge state appears as a superscript. The first series which is readily discernable consists of doubly charged ions appearing at *m/z* 691 and 547 and form a series, w_n^{2-}, which contain the 3′-terminus of the oligomer. These ions are w_5^{2-} and w_4^{2-}, respectively, and are formed by the cleavage of the 3′-carbon-oxygen bond in the sugar-phosphate backbone of the doubly charged molecular ion. The series appears to be truncated at n = 4, since no ions with n = 1-3 are found. A similar, but extended series of singly charged ions with the same structure, w_n^{1-}, can be found at *m/z* 1094, w_4^{1-}; 790, w_3^{1-}; 610, w_2^{1-}; and 321, w_1^{1-}. In addition, loss of water produces a related ion series of the general form $(w_n\text{-}H_2O)^{1-}$ which are present at *m/z* 1076, 772, 592, and 303.

A fourth series of ions of the type y_n^{1-} is also present in the spectrum in which the 5′-terminal phosphate is missing. This series consists of *m/z* 1304, y_5^{1-}; 1015, y_4^{1-}; 711, y_3^{1-}; and 530, y_2^{1-}. A final series of ions can be discerned in the spectrum and are formed by loss of the nucleobase followed by β-elimination from the sugar moiety, that is $(a_n\text{-}B_n)^{1-}$. Finally, the low *m/z* region of the spectrum contains ions for each of the negatively charged nucleobases in the oligomer (*m/z* 110, 125, and 150), phosphate ions (*m/z* 79 and 97), and deoxyribose-phosphate species (*m/z* 177 and 195).

Collision-induced dissociation of multiply charged molecular ions created from oligodeoxynucleotides by electrospray ionization generates complex product ion spectra each of which contains several ion series valuable for verification of sequence and the molecular weight of modified nucleotides in the sequence. Unfortunately, none of these series is complete; however, the w-series does contain all possible ions in the sequence if both the singly and doubly charged product ions are included. The data differ considerably from that obtained on an ion trap mass spectrometer, where w_n and (a_n-B_n) type ions predominated (*37,38*).

Acknowledgments

The authors would like to thank Professor Francis Johnson for supplying reagents for the synthesis of modified oligodeoxynucleotides and for many helpful discussions. This research was supported by the National Institutes of Health through grants ES04068 and CA47995.

Literature Cited

1. Siuzdak, G. *Proc.Natl. Acad. Sci. USA* **1994**, *91*, 11290-11297.
2. Dole, M.; Mack, L. L.; Hines, R. L.; Mobley, R. C.; Ferguson, L. D.; Alice, M. B. *J. Chem. Phys.* **1968**, *49*, 2240-2249.
3. Yamashita, M.; Fenn, J. B. *J. Chem. Phys.* **1984**, *88*, 4451-4459.
4. Whitehouse, C. M.; Dreyer, R. N.; Yamashita, M.; Fenn, J. B. *Anal. Chem.* **1985**, *57*, 675-679.
5. Smith, R. D.; Loo, J. A.; Edmonds, C. G.; Barinaga, C. J.; Udseth, H. R. *Anal. Chem.* **1990**, *62*, 882-899.
6. Grotjahn, L.; Frank, R.; Blocker, H. *Nucleic Acids Res.* **1982**, *10*, 4671-4678.
7. Grotjahn, L.; Frank, R.; Blocker, H. *Biomed. Mass Spectrom.* **1985**, *12*, 514-524.
8. Iden, C. R.; Rieger, R. A. *Biomed. Environ. Mass Spectrom.* **1989**, *18*, 617-619.
9. Wang, B. H.; Biemann, K. *Anal. Chem.* **1994**, *66*, 1918-1924.
10. Pieles, U.; Zürcher, W.; Schär, M.; Moser, H. E. *Nucleic Acids Res.* **1993**, *21*, 3191-3196.
11. Wu, K. J.; Shaler, T. A.; Becker, C. H. *Anal. Chem.* **1994**, *66*, 1637-1645.
12. Hathaway, G. M. *Biotechniques* **1994**, *17*, 150-155.
13. Hettich, R.; Buchanan, M. *J. Am. Soc. Mass Spectrom.* **1991**, *2*, 402-412.
14. Stemmler, E. A.; Hettich, R. L.; Hurst, G. B.; Buchanan, M. V. *Rapid Commun. Mass Spectrom.* **1993**, *7*, 828-836.
15. Nordhoff, E.; Karas, M.; Cramer, R.; Hahner, S.; Hillenkamp, F.; Kirpekar, F.; Lezius, A.; Muth, J.; Meier, C.; Engles, J. W. *J. Mass Spectrom.* **1995**, *30*, 99-112.
16. Covey, T. R.; Bonner, R. F.; Shushan, B. I.; Henion, J. *Rapid Commun. Mass Spectrom.* **1988**, *2*, 249-256.
17. Reddy, D. M.; Rieger, R. A.; Torres, M. C.; Iden, C. R. *Anal. Biochem.* **1994**, *220*, 200-207.
18. Potier, N.; van Dorsselaer, A.; Cordier, Y.; Roch, O.; Bischoff, R. *Nucleic Acids Res.* **1994**, *22*, 3895-3903.
19. Limbach, P. A.; Crain, P. F.; McCloskey, J. A. *J. Am. Soc. Mass Spectrom.* **1995**, *6*, 27-39.
20. Stults, J. T.; Marsters, J. C. *Rapid Commun. Mass Spectrom.* **1991**, *5*, 359-363.
21. Vollmer, D. L.; Gross, M. L. *J. Mass Spectrom.* **1995**, *30*, 113-118.

22. Greig, M.; Griffey, R. H. *Rapid Commun. Mass Spectrom.* **1995**, *9*, 97-102.
23. Pomerantz, S. C.; Kowalak, J. A.; McCloskey, J. A. *J. Am. Soc. Mass Spectrom.* **1993**, *4*, 204-209.
24. Little, D. P.; Thannhauser, T. W.; McLafferty, F. W. *Proc. Natl. Acad. Sci. USA* **1995**, *92*, 2318-2322.
25. Kebarle, P.; Tang, L. *Anal. Chem.* **1993**, *65*, 972A-986A.
26. Smith, D. P. H. *IEEE Trans. Ind. Appl.* **1986**, *IA-22*, 527-535.
27. Wilm, M. S.; Mann, M. *Int. J. Mass Spectrom. Ion Processes* **1994**, *136*, 167-180.
28. Fernández de la Mora, J.; Loscertales, I. G. *J. Fluid Mech.* **1994**, *260*, 155-184.
29. Wang, G.; Cole, R. B. *Anal. Chem.* **1994**, *66*, 3702-3708.
30. Tang, L.; Kebarle, P. *Anal. Chem.* **1993**, *65*, 3654-3668.
31. Iribarne, J. V.;Thomson, B. A. *J. Chem. Phys.* **1976**, *64*, 2287-2294.
32. Cheng, X.; Gale, D. C.; Udseth, H. R.; Smith, R. D. *Anal. Chem.* **1995**, *67*, 586-593.
33. Bleicher, K.; Bayer, E. *Biol. Mass Spectrom.* **1994**, *23*, 320-322.
34. Bayer, E.; Bauer, T.; Schmeer, K.; Bleicher, K.; Maier, M.; Gaus, H.-J. *Anal. Chem.* **1994**, *66*, 3858-3863.
35. Goodlett, D. R.; Camp II, D. G.; Hardin, C. C.; Corregan, M; Smith, R. D. *Biol. Mass Spectrom.* **1993**, *22*, 181-183.
36. Gale, D. C.; Goodlett, D. R.; Light-Wahl, K. J.; Smith, R. D. *J. Am. Chem. Soc.* **1994**, *116*, 6027-6028.
37. McLuckey, S. A.; Van Berkel, G. J.; Glish, G. L. *J. Am. Soc. Mass Spectrom.* **1992**, *3*, 60-70.
38. McLuckey, S. A.; Habibi-Goudarzi, S. *J. Am. Chem. Soc.* **1993**, *115*, 12085-12095
39. Little, D. P.; Chorush, R. A.; Speir, J. P.; Senko, M. W.; Kelleher, N. L.; McLafferty, F. W. *J. Am. Chem. Soc.* **1994**, *116*, 4893-4897.
40. Baker, T. R.; Keough, T.; Dobson, R. L.; Riley, T. A.; Hasselfield, J. A.; Hesselberth, P. E. *Rapid Commun. Mass Spectrom.* **1993**, *7*, 190-194.
41. Bodepudi, V.; Iden, C. R.; Johnson, F. *Nucleosides Nucleotides* **1991**, *10*, 755-761.
42. Shibutani, S.; Gentles, R. G.; Iden, C. R.; Johnson, F. *J. Am. Chem. Soc.* **1990**, *112*, 5667-5668.
43. Takeshita, M.; Chang, C.-N.; Johnson, F.; Will, S.; Grollman, A. *J. Biol. Chem.* **1987**, *262*, 10171-10179.

RECEIVED July 21, 1995

Chapter 15

Noncovalent Complexes of Nucleic Acids and Proteins Studied by Electrospray Ionization Mass Spectrometry

Richard D. Smith, Xueheng Cheng, Brenda L. Schwartz, Ruidan Chen, and Steven A. Hofstadler

Environmental Molecular Sciences Laboratory, Pacific Northwest Laboratory, 902 Battelle Boulevard, P.O. Box 999, Richland, WA 99352

The use of electrospray ionization-mass spectrometry (ESI-MS) for the characterization of noncovalent complexes of biomacromolecules from solution is based upon the gentle nature of the electrospray process. The important aspects of ESI source and interface conditions are described, with particular emphasis placed upon methods for distinguishing, and conditions for avoiding, artifacts due to gas phase complexes that may arise from non-specific associations and aggregation due to ESI processes. Studies conducted under suitably gentle interface conditions and low concentrations are generally required, and it has now been shown that a wide range of specific interactions and associations in solution can be readily detected. Specific examples cited include tetrameric proteins (i.e., intact quaternary structure), oligonucleotide duplexes, duplex-drug complexes, and protein-inhibitor complexes. The potential for rapid affinity screening of combinatorial libraries and conducting competitive binding studies are suggested by recent results.

Molecules which form structurally specific noncovalent associations in solution are of fundamental biological importance. Both the number and strength of the (generally) relatively weak bonds which comprise these liquid-phase noncovalent complexes is reflected by their respective dissociation constants under specific solution conditions (K_D). The gentleness of electrospray ionization (ESI) coupled with an appropriate mass spectrometer and atmosphere to vacuum interface permits the analysis of noncovalent complexes. There are now numerous reports showing that ESI-mass spectrometry (MS) can be used to detect intact structurally specific noncovalent associations from solution (*1-39*). Several types of noncovalent

0097–6156/95/0619–0294$12.25/0

associations in solution, including enzyme-substrate, receptor-ligand, host-guest, protein-nucleic acid, multimeric proteins (*11,20,25,28,33,35*), enzyme complexes (*19*), and DNA duplex-drug complexes (*2,8*), have been demonstrated to survive transfer into the gas phase.

The observation of either specific or nonspecific complexes by ESI-MS requires that the complex survive the electrospray droplet dispersal, droplet break-up and subsequent desolvation processes (by heating in the higher pressure regions of the source, or collisional activation at lower pressures). The observation of nonspecific complexes can be considered within the context of the model outlined previously (*18,38*), and is summarized in Figure 1. Although a number of details of the ESI process remain a subject of speculation and debate, and there is only indirect evidence of the 3-D structural conformation of large ions in the gas phase, it is now incontestable that many weak associations can survive transfer to the gas phase and are stable for periods of at least seconds. The present article does not explicitly discuss stronger complexes that are largely due to electrostatic associations (such as protein-metal ion complexes). We also focus on complexes involving larger biopolymers, where the stability of the complexes in the gas phase may be an effect arising from contributions due to multiple noncovalent bonds. Thus, the success for larger molecules evident in this article may not be similarly reflected by studies of small molecule complexes (particularly where electrostatic contributions may also be weak).

Our aim in this article is to summarize the current understanding of the conditions most appropriate for the study of weak noncovalent associations, and to highlight recent results from our laboratory that illustrate the range of associations subject to study and potential applications.

Interface Conditions for Observation of Noncovalent Associations

There are several important instrumental variables in the interface region which can be manipulated to either preserve noncovalent complexes or cause their dissociation into individual components. The interface conditions used to preserve such complexes are inherently more gentle than those conventionally employed for ESI-MS analysis. Figure 2 shows the various regions in the electrospray process for ion desolvation/activation, and some of the variables which can be adjusted to affect observation of noncovalent complexes. In instruments using a glass or metal ion inlet capillary, adjusting the temperature of the ion inlet capillary can cause thermal desolvation of electrosprayed ions and, if not excessive, can allow noncovalent complexes to be detected intact. (It should be noted that some instruments utilize a sampling aperture or cone that effectively eliminates this heating region. Although at least some noncovalent complexes have been shown to be detected with this arrangement, it remains to be determined whether this difference imposes any fundamental limitations upon the detection of noncovalent complexes with such instruments.) Another variable that can be manipulated to induce collisional activation in the interface region is the capillary-skimmer voltage difference (ΔCS)

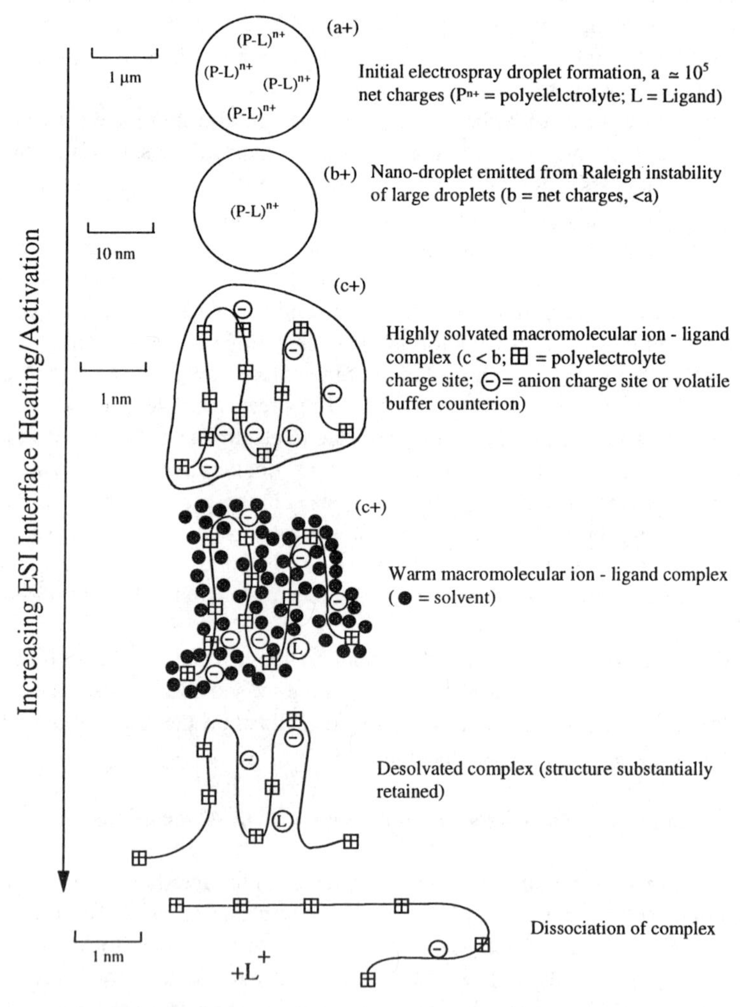

Figure 1. Model for formation of macromolecular ions during electrospray ionization (*18,38*), involving formation of small nanometer diameter droplets formed from the initial micron size droplets. Significant numbers of counter ions may remain at this stage, with most residing close to charges of opposite polarity. In subsequent steps, solvent and some solvent "residue" and charge are shed, including more volatile buffer components. The loss of higher structure and noncovalent associations is primarily driven by the extent of heating and repulsive Coulombic forces after removal of the solvent. Coulombic considerations also make it likely that in dissociation of a complex (e.g., loss of ligand, L) each species will be charged, and that the smaller product will carry a disproportionate fraction of the total charge (*35*).

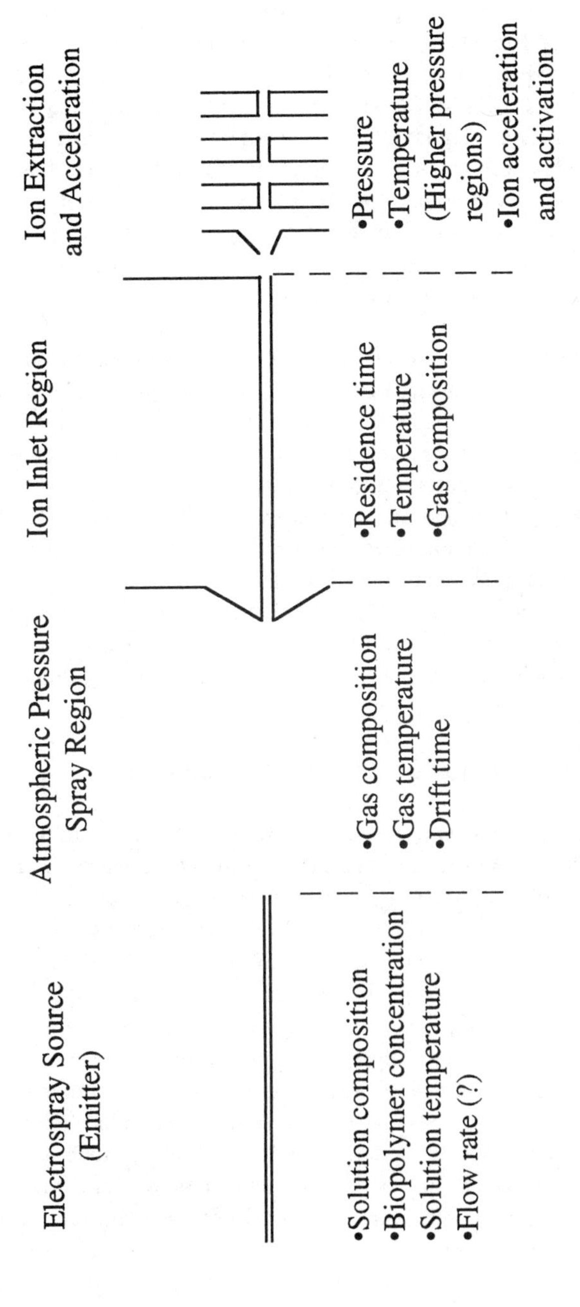

Figure 2. Four regions of an ESI-MS interface and variables relevant to the preservation of weak noncovalent complexes.

(Figure 2), or orifice-skimmer voltage bias for those interfaces incorporating a sampling orifice rather than a capillary inlet. Relatively low capillary temperatures and low ΔCS voltages are typically employed to provide the gentle interface conditions necessary for intact noncovalent associations to be observed. These source variables can be adjusted to provide harsher interface conditions, which will generally decrease adduction/solvation of the molecular ions, but may also result in dissociation of the noncovalent complex. It is not unreasonable that the kinetics of the macroion complex desolvation process may be of significance in this regard; the rate of heating (or desolvation) and the ion internal energy vs. time profile may be important. Also, use of a heated countercurrent drying gas at atmospheric pressure in the atmospheric pressure electrospray region has shown to provide additional ion desolvation for high *m/z* ions, reducing the amount of heating necessary in further steps.

Generally, selection of the instrumental variables in the electrospray interface region in order to allow preservation of noncovalent associations involves a compromise between providing adequate heating/activation for ion desolvation while minimizing complex dissociation. The extent to which the complex can be heated is expected to depend on the nature of its binding, the rate of heating, residence time in this region, and the relative stabilities of the nonspecific complexes (with solvent, buffer components, etc.) compared to the specific complex. At present, there is insufficient knowledge of the relevant kinetics and thermodynamics for even simple model systems to predict whether such ESI-MS methods will be successful for any specific application. However, it must be noted that success has now been demonstrated for a wide range of systems, and many of the general criteria for successful studies are increasingly evident.

Solution Conditions for Retention of Noncovalent Associations

In addition to instrumental considerations, proper solution conditions must be utilized that are compatible with ESI and preservation of the intact complexes. For conventional ESI-MS experiments, solution conditions are generally acidic (for detection of positive ions), where more extensive gas-phase charging of biomolecules allows detection at relatively low *m/z* (<2000). However, many noncovalent associations of physiological relevance are unstable under such solution conditions. For example, the physiologically active forms of many proteins are multimeric, and the destruction of the higher order quaternary structures generally occurs in acidic solution. In these instances, as with many other noncovalent interactions of biological interest, it is desirable to maintain aqueous solution conditions near neutral pH. The use of "volatile" buffers, such as ammonium acetate, is generally preferred for this purpose, since the use of nonvolatile salts can lead to excessive adduction and decreasing sensitivity, making precise molecular weight measurements problematic.

In the selection of buffers for ESI it is also important to keep in mind the need for appropriate control experiments, or that an independent analysis be performed comparing, for example, ammonium acetate vs. established physiological buffers. In many cases, however, a restriction to volatile buffers is of little importance, since solution pH, temperature and ionic strength are often the primary factors of concern.

Utilizing buffers at physiological pH generally results in a reduced amount of gas-phase charging of the biomolecule, presumably due to the retention of a relatively compact higher order structure, and which results in Coulombic constraints upon maximum charging (*28*). Therefore, for certain large proteins, mass spectrometers capable of extended m/z ranges (>3000) are generally required for the detection of ions corresponding to the multimeric form of the protein.

Contributions Due to Nonspecific Solution Associations

The fact that mass spectra illustrating the intact quaternary structure of proteins, such as the hemoglobin tetramer (*28*), and non self-complementary oligonucleotide duplexes can be produced (*2,8*) without contributions from nonspecific or random aggregation, constitute convincing evidence for the feasibility of studying such interactions by ESI-MS. Of greatest biological interest are noncovalent associations which are highly structurally specific in nature and whose strength of interactions is reflected by the dissociation constant (K_D) which is typically determined for a specific set of solution conditions. Since association rates in solution are generally diffusion limited, dissociation constants provide a measure of relative complex lifetime in solution. However, in addition to such specific associations in solution, a broad range of nonspecific associations, having a range of interaction strengths, also occur.

The most common type of both nonspecific and specific noncovalent interactions in solution are generally those involving the solvent and buffer species (see Figure 1). Water is thought to be an essential component of the structure of most proteins. There are often a number of well localized (i.e., "specific" or internal) water molecules that are considered essential to protein conformation. Water corresponding to several percent of the protein mass is generally quite specific in its association with hydrophilic protein functional groups on the protein surface, and may organize into clathrate-like structures around hydrophobic patches. The failure to observe the "internal" water molecules in typical ESI-MS spectra would appear to support the view that the protein has "unfolded" and lost its native tertiary structure. However, in some cases it has been shown that both nonspecific and specific noncovalent associations can be maintained in the gas phase, indicating preservation of higher order structure even after the complete removal of any residual solvent molecules. How large biopolymer ions are "vacuum dried" and can apparently retain substantial "memory" of their solution structures remains to be determined. The interpretation of charge state distributions to draw inferences regarding gas phase structure is suspect since, as we have previously demonstrated, the charge state distribution may, in some cases, be "locked in" earlier in the ESI process (*11,28*). Other solution components can have interactions of variable specificity and affinity. In particular, charge carrying components (e.g., salts, organic ions, ionic surfactants) can have a broad range of interactions that are primarily electrostatic in nature. Interactions with such species are readily observed by ESI-MS, and are generally mitigated by the incorporation of volatile solution or buffer constituents that are readily shed by mild heating/activation in the ESI interface.

Any attractive solution interactions could be reflected by the enhanced production of gas-phase complexes, and contributions due to aggregation generally become significant at higher concentrations (the precise concentration for this is likely to depend upon both the species involved and the interface design). "False positives" from the use of high concentrations is likely the most common artifact encountered in studies aimed at detection of noncovalent complexes. For example, peaks due to presumably nonspecific aggregates (or clusters) of biopolymers can be observed in many ESI-MS spectra, as discussed previously (*3*). They can arise from either nonspecific associations in solution or, possibly, due to aggregation occurring in the charged droplet after electrospray dispersal. Ding and Anderegg (*41*) have recently demonstrated the role of high concentrations, showing that homodimers of non complementary oligonucleotides to be significant for high concentrations. Winger et al. have previously observed that such nonspecific aggregation apparently plays a role in the production of higher m/z (lower charge states) ions (*38*). Interestingly, different degrees of aggregation were found from the same solution for the negative and positive ESI-mass spectra, indicating that the ESI process can affect the extent of aggregation observed. It appears likely that dimers and larger aggregates observed in ESI spectra can reflect **both** solution associations as well as contributions induced during the electrospray process. The issue raised by these observations is clear: if one wishes to conduct studies under conditions that exploit the detection of specific associations, one must first establish that contributions from nonspecific or ESI derived associations are not significant under the selected ESI conditions. To do this with high confidence, one must concurrently demonstrate that the associations detected are specific (as opposed to more random aggregation) for the system of interest, or a suitable model system.

Distinguishing Specific and Nonspecific Noncovalent Associations in ESI-MS

The simple observation of peaks indicative of a complex by ESI-MS constitutes insufficient evidence upon which to infer a structurally specific interaction. Distinguishing between structurally specific noncovalent interactions arising from solution in preference to nonspecific aggregation possibly resulting from the electrospray process, appears to be generally feasible with careful selection of experimental conditions. However, the necessary conditions (e.g., low analyte concentrations) may not be achievable with certain instrumentation (due to sensitivity limitations in performance, for example) for specific systems (e.g., due to low binding affinity), and thus experiments may include contributions from nonspecific associations. There are several methods which can be used to provide evidence that the observed gas-phase interactions are structurally specific and reflect those associations existing in solution, as we outlined earlier (*18*).

Relative Abundance. If adjustment of interface conditions allows the acquisition of mass spectra in which the complex dominates over the dissociated species, this constitutes good supporting evidence of a specific association in solution. The majority of literature reports of putative structurally specific associations meet this test. Cases particularly suspect might include spectra showing "evidence" for a claimed "homodimer" in the presence of a greater abundance of monomer. Artifacts arising due to aggregation during the ESI process are quite sensitive to solution concentration, and it is generally desirable to conduct studies at the lowest feasible concentrations.

Stoichiometry. Generally, complexes (such as A·B) should also be observed with the preferred or reasonable stoichiometry, i.e., without comparable contributions which might arise from random aggregation (i.e., A·A, B·B, $A{\cdot}B_2$, $A_2{\cdot}B$, $A{\cdot}B_3$, etc.) in order to be attributed to solution association. Convincing examples include non-self complementary oligonucleotide duplexes (*8,15,30*) in the absence of random dimers and tetrameric proteins in the absence of trimers and pentamers (*11,20,25,28,33,35*).

Gas Phase Lability. Complexes due to weak binding forces (e.g., H-bonding, hydrophobic interactions, etc.) will often be quite labile and readily dissociated under more severe interface conditions or tandem mass spectrometry (MS/MS) experiments. Such studies can help to distinguish noncovalent complexes from possible covalent associations inadvertently introduced during sample handling. However, it should be noted that gas phase complexes due to electrostatic interactions may be quite strong in the gas phase, in the absence of the solvent, and that the cumulative contributions of many weak bonds should make complexes having larger binding domains more stable (*41*).

Dissociation due to Modification of Solution Conditions. Changes in solution temperature, pH, the presence or absence of buffer components, the addition of organic solvents, etc., can destroy many specific associations in solution and should produce a corresponding change in the ESI mass spectra. Such studies are readily conducted and can provide strong evidence that associations do not arise from aggregation processes late in the electrospray process, but by themselves cannot be considered definitive.

Sensitivity to Structural Modifications. Perhaps the most convincing demonstration of a specific interaction is obtained when a variant of one of the complex components produces a substantial change in the relative intensity of the complex in the mass spectrum. For example, it has been shown that minor changes in the sequence of the S-peptide component of the Ribonuclease S complex (S-peptide·S-protein), known to strongly affect association behavior in solution, also produced large differences in complex stability by ESI-MS (*14*). Another example is the relative absence of iminobiotin association observed with streptavidin by ESI-

MS (see below) compared with the stoichiometric and relatively stable association observed with biotin, a structurally similar compound having much higher binding affinity. Such experiments demonstrate in the most unambiguous manner that the interaction is specific in nature, and that its observation in the gas phase depends upon its presence in solution.

While it may be impossible to prove beyond doubt that a specific association exists in solution solely on the basis of ESI-MS results, combinations of the above criteria provide the basis for a high level of confidence in such studies. Ideally, the ESI-MS data will also be supported by data from independent studies, although when the necessary conditions become well defined the extension to less well characterized systems is increasingly justified. Most of the criteria noted above can be applied to "unknown" complexes; however, the extrapolation to such systems is only justified when well characterized model systems (such as those discussed below) routinely yield acceptable results. It is important to recognize that constraints arising from the design of particular instrumentation (including insufficient sensitivity) may serve to prevent the study of specific noncovalent associations. When the appropriate experimental technique is practiced with amenable instrumentation, the sensitivity and speed of ESI-MS methods suggest its use in resolving questions of relative binding, stoichiometry, and as a precursor to more costly and extensive biochemical or biological studies.

Model Systems and Specific Applications

Oligonucleotide-Drug Associations. Gale et al. have reported ESI-MS studies of the noncovalent complexes formed between a minor groove binding molecule and a 12 base pair self-complementary oligonucleotide (*2*). The association constant for the binding of distamycin (Dm) to a 16 base pair oligonucleotide duplex is 1.3×10^7 M^{-1}. By varying the ratio of distamycin A to oligonucleotide, oligonucleotide duplex as well as 1:1 and 2:1 distamycin A/oligonucleotide duplex noncovalent complexes were observed. These observations were consistent with NMR results for distamycin A to oligonucleotide duplex concentration ratios for the same oligonucleotide sequence. For the two 14-base pair sequence specific oligonucleotides, Dm was observed specifically associated with the sequence specific oligonucleotide duplex, and nonspecific duplexes complexed with Dm were not observed. These noncovalent complexes were shown to be readily dissociated in the gas phase into their component parts. In subsequent work, it was observed that specific noncovalent complexes could form between both self-complementary and non self-complementary oligonucleotides and other minor groove binding molecules (i.e., pentamidine and Hoechst 33258) (*8*). Variation of several ESI interface parameters as well as collision induced dissociation methods were utilized to characterize the nature and stability of the noncovalent complexes. Upon collisional activation the noncovalent complexes dissociated into single-stranded oligonucleotides and single-stranded oligonucleotides associated with a minor groove binding molecule. Other results of Gale et al. have shown that ESI-Fourier transform ion cyclotron resonance (FTICR) mass spectrometry could be effective for the detection of specific drug-oligonucleotide duplex non-covalent complexes, as illustrated in Figure 3 (*30*). Figure 3 shows a high resolution ESI-FTICR spectrum for an annealed 50 μM distamycin - 100 μM oligonucleotide duplex in 30 mM NH_4OAc solution. A charge state distribution for both the oligonucleotide duplex and 1:1 distamycin/oligonucleotide duplex is observed. The peak centered ca. *m/z* 1214 shows contributions due to both monomer ions and duplex ions, as evident from the isotopic envelope. The peak at *m/z* 1457 is an odd charge state of the oligonucleotide duplex (where the monomer cannot contribute). After annealing, the concentrations of 1:1 Dm/oligonucleotide duplex and oligonucleotide duplex will be equivalent if all of the Dm binds to an oligonucleotide duplex. Peaks representative of both the intact oligonucleotide duplex and the 1:1 Dm/oligonucleotide duplex are observed having approximately equal intensities, consistent with this expectation. These results also show that the specific noncovalent complexes were stable in the FTICR ion trap for indefinite periods (*30*).

Multimeric Proteins. Electrospraying noncovalent multimeric proteins from near-neutral pH aqueous solutions has been shown to preserve higher order structural associations, with the consequence of more limited charging for the molecular ion in the gas phase than observed under more conventional ESI conditions. The more compact solution structures imposes Coulombic constraints that apparently result in the formation of multimeric protein ions at high *m/z* with relatively narrow charge state distributions. It has been shown, at least in cases studied to date, that the observation of ions corresponding to the intact multimeric associations requires the use of mass spectrometers capable of extended *m/z* range (>3000) detection. As an example, Figure 4 shows the ESI-FTICR mass spectrum for the tetrameric protein streptavidin (*35*), whose charge state distribution is centered at ~*m/z* 3800.

Given the proper solution conditions and gentle ESI interface conditions, the intact specific complexes for this and other multimeric proteins (*28*) were detected. These observations were consistent with predicted solution behavior and were in preference to other nonspecific forms of aggregation (i.e. trimers, pentamers, etc.) that might be formed under some interface conditions during the ESI process. Either thermal and/or collisional activation in the interface region was found to readily dissociate these tetrameric species, confirming their noncovalent nature.

As an example, ESI-MS spectra for concanavalin A (Con A) or other multimeric proteins may used to demonstrate that complexes observed by ESI-MS are both noncovalent and specific. Con A is a lectin composed of four identical polypeptide subunits (25.5 kDa each) which has a pH dependent solution equilibrium between the dimeric and tetrameric forms. The ESI mass spectra of Con A in aqueous 10 mM NH_4OAc (pH 6.7) obtained with an extended *m/z* range quadrupole mass spectrometer shows peaks indicative of the intact tetramer as well as the dimer (the only significant species observed (*11*)), both of which should be present under the solution conditions employed. If these ions arose from nonspecific aggregation, then peaks indicative of trimeric and possibly pentameric species should be observed

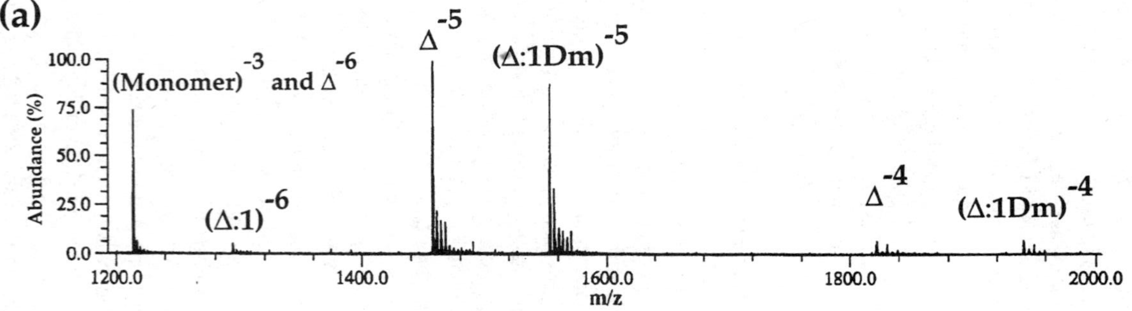

Figure 3. ESI-FTICR mass spectrum (a) for an annealed 50 μM distamycin A-100 μM oligonucleotide duplex in a 30 mM NH_4OAc buffer solution (pH 8.3), showing high resolution detection of the specific noncovalent complex (*30*). The spectrum is dominated by contributions of the 5- charge states of the intact duplex (Δ) and duplex-drug complex (Δ:1). Sodium adduction is particularly evident for the duplex and duplex-drug species. (b) An expanded view of the molecular ion regions for these species, showing the high resolution of FTICR allows resolution of the 1-Da isotopic spacings.

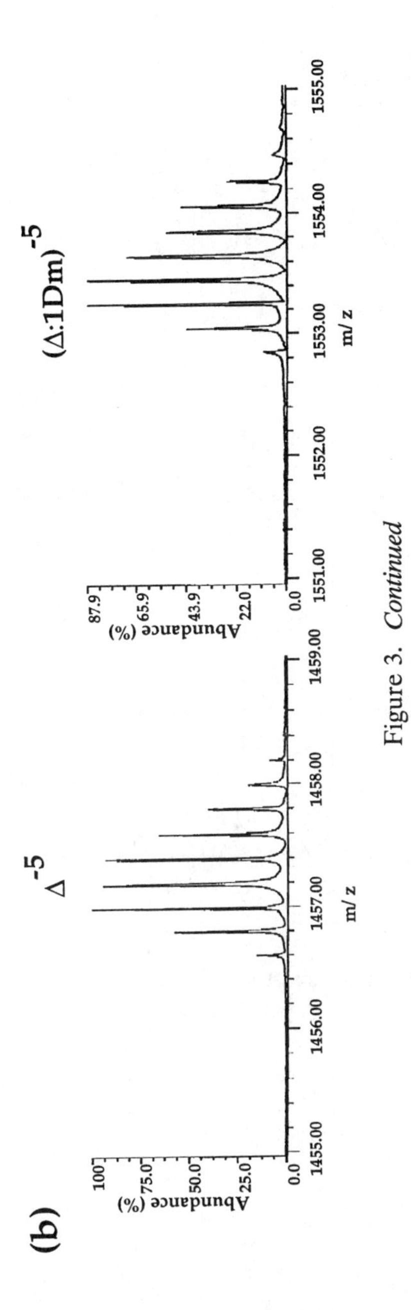

Figure 3. *Continued*

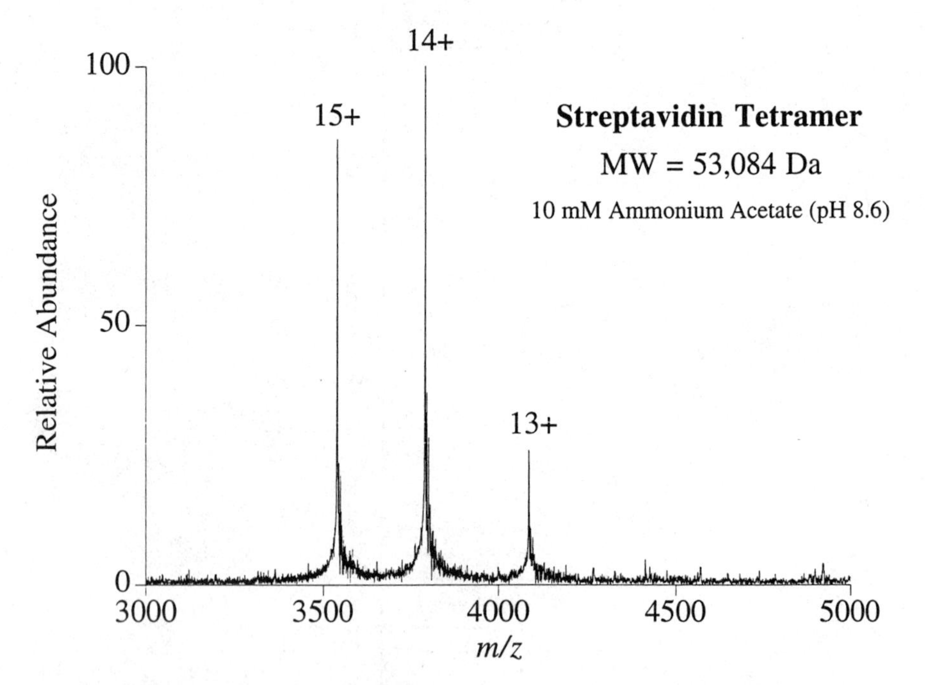

Figure 4. ESI-FTICR mass spectrum of the intact streptavidin tetramer (*35*).

(even though they are not known to form in solution). The absence of these peaks strongly suggests that the dimeric and tetrameric species evident in the mass spectra arise from specific solution associations. At higher capillary inlet temperatures, the dimeric and tetrameric complexes of Con A readily dissociate due to the more severe interface conditions to form primarily the monomeric subunit with a charge state distribution centered at *m/z* 1,600. The much greater average charge for the monomer species indicates that the quaternary structure is lost relatively early in the electrospray process, presumably due to rapid heating of the electrosprayed droplets, before the charge state is determined. The loss of at least some higher order structure provides greater spatial separation of charge sites in the gas phase, allowing higher charge states to persist for the monomer ions. Since the Con A complex is pH sensitive, altering the solution conditions (and thus the species present in solution) should affect the species observed with ESI-MS. The ESI-MS spectrum of Con A in a lower pH solution (pH 5.7) using gentle interface conditions shows peaks primarily attributed to the dimeric species, as expected from known solution behavior.

Additional studies for heating/activation in the ESI atmosphere-vacuum interface region for the intact noncovalent multimeric proteins avidin, Con A, and adult human hemoglobin have also been reported (*28*). For harsher interface conditions, obtained by increased heating of the metal capillary inlet, the tetramers dissociated to produce a series of charge states for the monomer and trimer species. The relatively low charge state, high *m/z* trimers were correlated with the appearance of relatively high charge state monomers at much lower *m/z*. A mechanism was proposed accounting for these observations in which two different pathways for monomer ion formation were possible. The highest charge state monomer ions were attributed to the earlier break-up of the tetramer before its charge state was "locked in", i.e., before desolvation was complete. It was also suggested that the monomer might lose portions of its higher order structure, be "elongated" through a Coulombically driven mechanism, and then be ejected from the aggregate. Gas-phase dissociation of the tetramer (after initial ion formation/charge state determination) could account for the observed lower charge state monomer and trimer species having approximately complementary charge state distributions. Schwartz et al. have recently confirmed this hypothesis by MS/MS studies using FTICR (*35*), showing that the gas-phase dissociation of the tetramer results in higher *m/z* trimers and lower *m/z* monomers.

Avidin and Streptavidin Complexes with Biotin and Iminobiotin. One of the most well-known and strongest protein-ligand interactions found in nature is the avidin-biotin complex. This highly selective noncovalent complexation has had wide applicability in biochemistry, immunochemistry, and affinity chemistry by virtue of its low dissociation constant ($K_D \sim 10^{-15}$ M). Avidin ($M_r \sim 64{,}000$) is a glycoprotein found in egg white, and its quaternary structure is a noncovalent tetramer composed of four identical subunits which associate into the active form. In addition, each subunit can accommodate one biotin molecule (M_r = 244 Da), consequently, the tetrameric form of avidin is capable of noncovalently binding four biotins. Schwartz et al. have shown that comparison of the *m/z* values for the ions of the avidin tetramer to those observed for the avidin-biotin tetramer complexes are consistent with the known solution behavior for one biotin noncovalently attached to each of the four subunits of avidin (*5*). The non-glycosylated tetrameric protein, streptavidin ($M_r \sim 54$ kDa), displays an analogous complex with biotin in solution. Schwartz et al. have also shown that the stability by ESI-MS of the streptavidin complex with iminobiotin (a biotin analog with $K_D \sim 10^{-7}$ M) was much weaker than the biotin complex, qualitatively consistent with solution behavior (*33*). For example, at relatively low protein:ligand molar ratios, immediate and complete biotin binding to streptavidin was readily observed, indicating that the stoichiometrically specific 1:4 tetrameric noncovalent complex was preserved. Under identical solution and gentle interface conditions, iminobiotin binding was not observed to occur, reflecting the known solution protein-ligand dissociation constants, and indicating the ability of ESI-MS to probe the relative binding affinities of these small molecule ligands to the intact streptavidin tetramer. Only under extremely gentle interface conditions was the partial binding of iminobiotin to streptavidin observed. These results again emphasize that specific solution binding is the origin of the noncovalent complexes detected in the gas phase, since the very similar nature of biotin and iminobiotin would lead one to expect very similar behavior for any process that does not depend on their subtle structural differences.

Protein-Inhibitor Complexes. Recent studies by Cheng et al. at our laboratory, conducted in collaboration with Professor George Whitesides and co-workers (Harvard University), have indicated the potential for measuring relative binding affinities simultaneously with small libraries of inhibitors, and have suggested the potential for the extension to substantially larger "combinatorial" libraries (*31*). These studies exploited the high resolving power and multi-stage MS capabilities of FTICR. Cheng et al. initially studied two inhibitor mixtures with bovine carbonic anhydrase (BCA) where inhibitor dissociation constants (K_D's) spanned 2-3 orders of magnitude (approximately 10^{-6} M to 10^{-9} M) (*31*). Systems that include species with such low binding affinities are of interest for defining the range of applicability of ESI-MS. One inhibitor system studied consisted of 16 to 18 compounds having hydrocarbon/ fluorohydrocarbon linker groups of variable length (*31*). The ESI-MS measurements also showed the complexes contained one Zn cation. Zn is known to bind to three histidine residues in the active site and is involved in the catalytic process. Figure 5 shows an example, illustrating the relatively high *m/z* of the complexes and limited charge state distribution observed, consistent with the tightly folded structure of the enzyme-inhibitor complexes (*31*).

Two informative control experiments were performed to further ascertain the specific nature of the noncovalent complexes observed using ESI-MS (*31*). In the first experiment, the pH was adjusted to below pH 4 using acetic acid; the resultant mass spectra showed no contributions due to the noncovalent complexes. The Zn metal was lost from the higher charge states of the BCA ions formed from acidic solution, consistent with an acid-catalyzed denaturation of the enzyme. In the

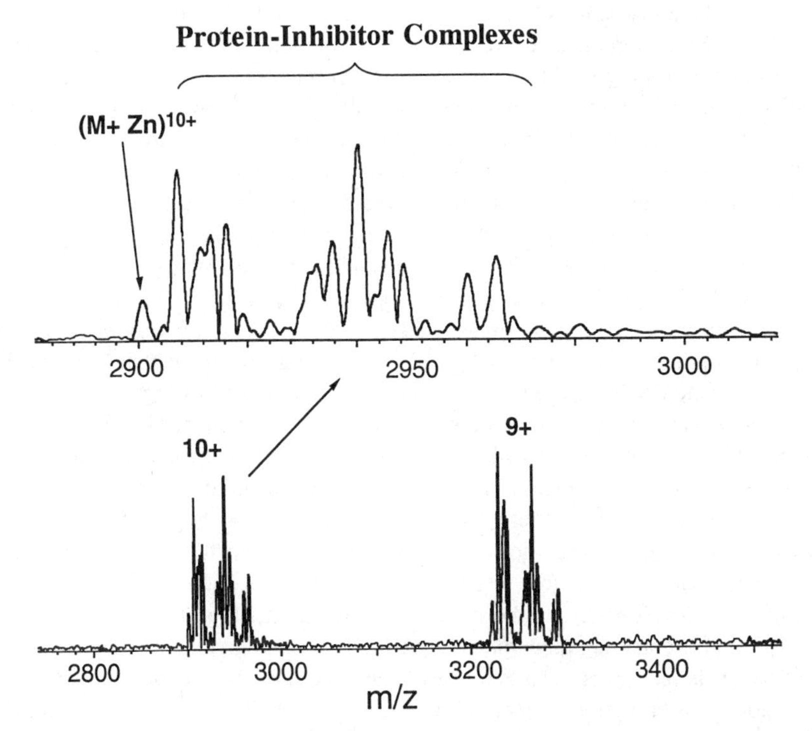

Figure 5. Positive-ion ESI-FTICR spectrum obtained for a set of 17 inhibitors (0.05 μM each) with bovine carbonic anhydrase II (BCA) (1.0 μM in 10 mM ammonium acetate at pH 7) (*31*). The bottom shows the narrow charge state distribution observed, typical of cases where noncovalent associations and (presumably) a compact three-dimensional structure is maintained using gentle ESI conditions. The top spectrum shows details of the 10+ charge state region, and indicates a small contribution of intact BCA (due to incorporation of one zinc atom). A series of peaks corresponding to the enzyme-inhibitor complexes were observed and have abundances approximately proportional to their binding constants (*31*).

second control experiment a mixture of BCA and apoBCA (i.e., BCA without Zn in the binding pocket) at pH 7 in 10 mM NH_4OAc was examined. Inhibitor complexes were observed with BCA, but not with apoBCA (*31*). These results also showed that neither ESI nor the FTICR detection methods cause any significant dissociation of the complexes. It is also noteworthy that the structural information for the noncovalently bound inhibitors could be obtained in one FTICR experiment (*31*).

The most important aspect of this work was that the individual inhibitors from a 7 component library could be isolated from the dissociated complexes, and their relative abundances were shown to correlate well with the known binding constants. Beyond this, the isolated inhibitors could be dissociated and the products measured with high resolution to aid in their identification, as shown in Figure 6. Thus, these methods appear to offer the potential to rapidly evaluate the binding of sets of compounds to a specific target molecule in one experiment that consumes only very limited amounts of both the target compound and the ligand "library".

The above results provide further support for the use of ESI in the study of noncovalent associations and suggest that even relatively weak associations are amenable to study (K_D's of > 10^{-6} to 10^{-7}). It should be noted that even weaker complexes should be amenable to study if based upon stronger electrostatic interactions. In contrast, the components of the 18 component library studied by Cheng et al. have substantial hydrophobic components to their interactions. These results also support the potential for obtaining semi-quantitative information about relative binding constants. These results are in sharp contrast to previous efforts that did not provide good correlation with relative binding, but the studies may not be directly comparable. The fact that no dissociation of the complexes was observed during the ESI-MS process in the studies of Cheng et al. suggests that this may be a prerequisite for obtaining more quantitative information regarding relative binding in solution. It is reasonable that gas phase abundances of the complexes under such conditions might better reflect relative abundances in solution, and provide conditions under which relative binding affinities can be derived from appropriately designed studies. This approach is attractive compared to dissociation studies of the relative stabilities of complexes in the gas phase, since there is no firm basis for deriving solution binding data from the gas phase behavior of these complex systems reported to date. Clearly, there is a need for further study to define the range of solution conditions and complexes amenable to relative binding studies.

Summary and Concluding Remarks

In summary, we have attempted in this Chapter to convey some of the considerations relevant to the detection of structurally specific noncovalent associations from solution. The observation of intact noncovalent complexes by ESI-MS requires a balance between providing sufficient ion desolvation, while ensuring that the interface conditions are mild enough to preserve the complex. Generally, some experimentation is required to determine the optimum solution and instrumental conditions (i.e., settings in the atmosphere-vacuum interface region, which also vary

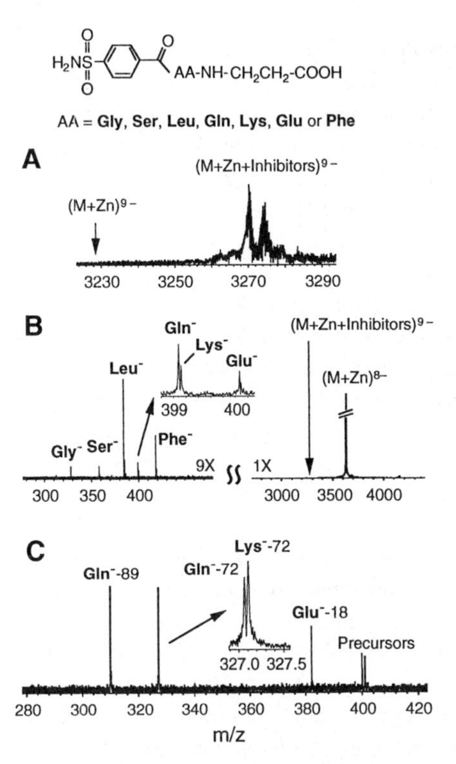

Figure 6. Negative ESI-FTICR mass spectra of BCA (7.0 μM) with a set of seven inhibitors (7.0 μM each) in 10 mM ammonium acetate at pH 7 (*31*) and their study in multi-step dissociation experiments. (A) The mass spectrum of complex ions of BCA with inhibitors (9- charge state) before ion isolation; and (B) isolation of 9- complex ion species using selective ion accumulation (*34*) and the subsequent dissociation of the complexes. (The relative intensities of the dissociated inhibitors showed reasonable correlation with their known binding affinities.) The complexes were completely dissociated, producing BCA ions with one less charge (8-) and singly charged inhibitor ions. (The arrow points to the *m/z* of the precursor species that has been completely dissociated.) (C) Shows the spectrum obtained after isolation of three inhibitor ions (from B) at *m/z* 399-401 followed by their dissociation.

between instruments of different design) required to detect intact noncovalent associations due to differences in noncovalent complex systems (i.e., nature of the binding, K_D values, M_r, etc.) and mass spectrometric instrumentation (i.e., types of electrospray interfaces). It appears that not all instrumentation is equally well suited for such applications. However, once appropriate conditions are determined they appear to be generally applicable for study of complexes having similar composition.

Some larger noncovalent complex systems inherently may be more difficult to observe by ESI-MS at lower *m/z* (<3000) due to the lesser extent of gas phase charging arising from the more compact solution structures, Coulombic effects, and near neutral pH conditions. This is particularly evident for multimeric proteins, but of less significance for less compact structures (e.g., oligonucleotide duplexes).

The ability to detect noncovalent complexes from solution is, of course, of only limited interest unless it can be effectively applied to "unknown" complexes, i.e. where stoichiometry and/or relative binding affinity is not established. At the present time "false negatives", in which a suspected or expected complex is not observed, must be viewed with caution. Unless a complex of similar size and nature can be readily detected under the same experimental conditions it is difficult at this time to be confident that some aspect of either the interface conditions or instrument design prevents detection. (This assumes that the solution conditions are suitable; e.g., that oligonucleotide duplexes, for example, have been prepared in an appropriate buffer and been properly annealed.) On the other hand "false positives" should be more readily avoidable by applying proper experimental conditions (e.g., low analyte concentrations, etc.) and testing against the criteria noted earlier. In this regard, it should also be remembered that noncovalent associations that are relatively strong and nonspecific in nature (such as those arising primarily from electrostatic associations) will likely also be detected by ESI-MS. Thus, if proper experimental procedures are followed and the noted criteria applied, one must proceed with caution before dismissing a complex detected by ESI-MS as a "false positive". The **absence** of a literature report of such a complex, particularly for less well studied systems, cannot be taken as the sole reason for dismissing the mass spectrometric results. In such cases it would appear prudent to apply independent methods, such as affinity capillary electrophoresis. The fact that such independent experiments can often be difficult to conduct, and ambiguous in interpretation, highlights the potential role of mass spectrometry in such applications.

The application of ESI-MS for the study of noncovalent associations in solution constitutes a new capability for mass spectrometry with potentially broad applications in biological research. A better understanding of both the ESI process and the role of instrumental parameters is clearly desirable. Recent results showing the basis for the initial screening of compounds for their binding to specific macromolecular targets suggests important future applications are likely in this area. The detection of noncovalent complexes also provides a basis for more conventional approaches, such as the variation of ligand concentration and the use of Scatchard plots for determining binding constants (*40*). Establishing relative binding affinities in competitive binding experiments for biopolymers based upon relative abundances of the complexes in the gas phase also appears feasible, at least in some cases. Taken in the context of the range of mass spectrometric capabilities for accurate mass measurements, tandem methods for obtaining structural information and high sensitivity, significant contributions to biochemical and biological research appear likely.

Acknowledgments

The authors wish to thank Drs. David C. Gale (Finnigan MAT), James E. Bruce and Karen J. Light-Wahl (PNL) for helpful discussions and the U.S. Department of Energy, and Laboratory Directed Research and Development of Pacific Northwest Laboratory for support of this research. Pacific Northwest Laboratory is operated by Battelle Memorial Institute for the U.S. Department of Energy, through Contract No. DE-AC06-76RLO 1830.

Literature Cited

1. Ganem, B.; Li, Y.-T.; Henion, J. D., *J. Am. Chem. Soc.* **1991**, *113*, 6294-6296.
2. Gale, D. C.; Smith, R. D. *J. Am. Chem. Soc.* **1994**, *116*, 6027-6028.
3. Smith, R. D.; Light-Wahl, K. J.; Winger, B. E.; Loo, J. A. *Org. Mass Spectrom.* **1992**, *27*, 811-821.
4. Haskins, N. J.; Ashcroft, A. E.; Phillips, A.; Harrison, M. *Rapid Comm. Mass Spectrom.* **1994**, *8*, 120-125.
5. Schwartz, B. L.; Light-Wahl, K. J.; Smith, R. D. *J. Amer. Soc. Mass Spectrom.* **1994**, *5*, 201-204.
6. Ganem, B.; Li, Y. T.; Henion, J. D. *J. Amer. Chem. Soc.* **1991**, *113*, 7818-7819.
7. Hsieh, Y.-L.; Li, Y.-T.; Henion, J. D.; Ganem, B. *Biol. Mass Spectrom.* **1992**, *23*, 272-276.
8. Gale, D. C.; Smith, R. D., submitted for publication.
9. Li, Y.-T.; Hsieh, Y.-L.; Henion, J. D.; Ocain, T. D.; Schiehser, G. A.; Ganem, B. *J. Am. Chem. Soc.* **1994**, *116*, 7487-7493.
10. Li, Y.-T.; Hsieh, Y.-L.; Henion, J. D.; Senko, M. W.; McLafferty, F. W.; Ganem, B. *J. Am. Chem. Soc.* **1993**, *115*, 8409-8413.
11. Light-Wahl, K. J.; Winger, B. E.; Smith, R. D. *J. Am. Chem. Soc.* **1993**, *115*, 5869-5870.
12. Ganem, B.; Henion, J. D. *Chemtracts* **1993**, *6*, 1-22.
13. Ganem, B.; Li, Y.-T.; Henion, J. D. *J. Am. Chem. Soc.* **1991**, *113*, 7818-7819.
14. Ogorzalek Loo, R. R.; Goodlett, D. R.; Smith, R. D.; Loo, J. A. *J. Am. Chem. Soc.* **1993**, *115*, 4391-4392.
15. Light-Wahl, K. J.; Springer, D. L.; Winger, B. E.; Edmonds, C. G.; Camp, D. G. II; Thrall, B. D.; Smith, R. D. *J. Am. Chem. Soc.* **1993**, *115*, 803-804.
16. Ganem, B.; Li, Y.-T.; Henion, J. D. *Tetrahedron Lett.* **1993**, *34*, 1445-1448.
17. Bayer, E.; Bauer, T.; Schmeer, K.; Bleicher, K.; Maier, M.; Gaus, H.-J. *Anal. Chem.* **1994**, 66, 3858-3863.
18. Smith, R. D.; Light-Wahl, K. J. *Biol. Mass Spectrom.* **1993**, *22*, 493-501.
19. Ganem, B.; Li, Y-T.; Hsieh, Y-L.; Henion, J. D.; Kaboord, B. F.; Frey, M. W.; Benkovic, S. J. *J. Am. Chem. Soc.* **1994**, *116*, 1352-1358.
20. Loo, J. A.; Ogorzalek Loo, R. R.; Andrews, P. C. *Org. Mass Spectrom..* **1993**, *28*, 1640-1649.

21. Baczynskyj, L.; Bronson, G. E.; Kubiak, T. M., *Rapid Commun. Mass Spectrom.* **1994**, *8*, 280-286.
22. Jaquinold, M.; Leize, E.; Potler, N.; Albrecht, A-M.; Shanzer, A., Van Dorsselaer, A.,*Tetrahedron Letters* **1993**, *34*, 2771-2774.
23. Hopfgartner, G.; Piguet, C.; Henion, J. D.; Williams, A. F. *Helvetica Clinica Acta* **1993**, *76*, 1759-1766.
24. Goodlett, D. R.; Camp II, D. G.; Hardin, C. C.; Corregan, M.; Smith, R. D. *Biol. Mass Spectrom.* **1993**, *22*, 181-183.
25. Tang, X-J.; Brewer, C. F.; Saha, S.; Chernushevich, I.; Ens, W.; Standing, K. G. *Rapid Commun. Mass Spectrom.* **1994**, *8*, 750-754.
26. Ganguly, A. K.; Pramanik, B. N.; Huang, E. C.; Tsarbopoulos, A.; Girjavallabhan, V. M.; Liberles, S. *Tetrahedron* **1993**, *49*, 7985-7996.
27. Aplin, R. T.; Robinson, C. V.; Schofield, C. J.; Westwood, N. J. *J. Chem. Soc., Chem. Commun.* **1994**, 2415-2417.
28. Light-Wahl, K. J.; Schwartz, B. L.; Smith, R.D. *J. Amer. Chem. Soc.* **1994**, *116,* 5271-5278.
29. Camilleri, P.; Haskins, N. J. *Rapid Commun. Mass Spectrom.* **1993**, *7*, 603-604.
30. Gale D. C.; Smith R. D., unpublished work.
31. Cheng, X; Chen, R.; Bruce, J. E.; Schwartz, B. L.; Anderson, G. A.; Hofstadler, S. A.; Gale, D. C.; Smith, R. D.; Gao, J.; Sigal, G. B.; Mammen, M.; Whitesides, G. M. *J. Amer. Chem. Soc.*, submitted.
32. Cheng, X.; Gao, Q.; Smith, R. D.; Simanek, E. E.; Mammen, M.; Whitesides, G. M. *Rapid Commun. Mass Spectrom.*, **1995**, *9*.
33. Schwartz, B. L.; Gale, D. C.; Smith, R. D.; Chilkoti, A.; Stayton, P. S. *J. Mass Spectrom.*, in press.
34. Bruce, J. E.; Van Orden, S.L.; Anderson, G. A.; Hofstadler, S. A.; Sherman, M. G.; Rockwood, A. L.; Smith, R. D. *J. Mass Spectrom.*, **1995**, *30*, 124-133.
35. Schwartz, B. L.; Bruce, J. E.; Anderson, G. A.;Hofstadler, S. A.; Rockwood, A. L.; Smith, R. D.; Chilkoti, A.; Stayton, P. S. *J. Amer. Soc. Mass Spectrom.*, **1995**, *6*, 459-465.
36. Goodlett, D. R.; Ogorzalek Loo, R. R.; Loo, J. A.; Wahl, J. H.; Udseth, H. R.; Smith, R. D. *J. Am. Soc. Mass Spectrom.*, **1994**, *5,* 614-622.
37. Loo, J. A.; Ogorzalek Loo, R. R.; Goodlett, D. R.; Smith, R. D.; Fuciarelli, A. F.; Springer, D. L.; Thrall, B. D.; Edmonds, C. G. *Techniques in Protein Chemistry IV*, **1993**, Angeletti, R.H., Ed., Academic Press, 23-31.
38. Winger, B. E.; Light-Wahl, K. J.; Ogorzalek Loo, R. R.; Udseth, H. R.; Smith, R. D. *J. Am. Soc. Mass Spectrom.*, **1993**, *4,* 536-545.
39. Katta, V.; Chait, B. T. *J. Amer. Chem. Soc.* **1994**, *113,* 8534-8535.
40. Lim, H.-K.; Hsieh, Y.-L.; Ganem, B.; Henion, J. D. *J. Mass Spectrom.*, **1995**, *30*, 708-714.
41. Ding, J.; Anderegg, R. J. *J. Am. Soc. Mass Spectrom.*, **1995**, *6,* 159-164.

RECEIVED July 21, 1995

Chapter 16

Analysis of Xanomeline, a Potential Drug for Alzheimer's Disease, by Electrospray Ionization Tandem Mass Spectrometry

Todd A. Gillespie, Thomas J. Lindsay, J. David Cornpropst, Peter L. Bonate, Theresa G. Skaggs, Allyn F. DeLong, and Lisa A. Shipley

Department of Drug Disposition, Lilly Research Laboratories, Indianapolis, IN 46285

Xanomeline, (3-[4-(hexyloxy)-1, 2, 5-thiadiazol-3-yl]-1, 2, 5, 6-tetrahydro-1-methylpyridine) is a selective M_1 muscarinic agonist currently under investigation for the potential symptomatic treatment of Alzheimer's disease. The characterization of the metabolites of xanomeline has been accomplished utilizing electrospray tandem mass spectrometry (ES-MS/MS). The use of ES-MS/MS has been employed in the identification of numerous oxidative metabolites of xanomeline resulting from extensive first pass metabolism. In addition, on-line liquid chromatography (LC) coupled with ES-MS/MS has been utilized to further characterize the metabolites of xanomeline after subcutaneous dosing in rats and transdermal dosing in humans. The advantages and limitations of on-line LC/ES-MS/MS for biotransformation studies are demonstrated in this application.

Electrospray (ionspray) ionization coupled with tandem mass spectrometry (ES-MS/MS) has proven to be an invaluable tool in the characterization of trace level mammalian metabolites of pharmaceutical candidates (*1, 2*). Furthermore, the combination of on-line LC with ES-MS/MS has become the method of choice for analysis of polar, thermolabile drug candidates and their respective metabolites in complex biological matrices (*3, 4*). The utilization of this methodology from the initial discovery phase of a potential drug candidate through human clinical trials and beyond can enhance a drug's rapid development (*5, 6*). Electrospray typically generates abundant $[M+H]^+$ species and avoids the use of high temperatures which may cause thermal degradation of metabolites (particularly polar conjugates). In addition, tandem mass spectrometry provides both the selectivity and sensitivity required for detection of trace level (ng/mL) metabolites in the presence of large concentrations of endogenous components.

Identification of the metabolites from a new drug substance typically follows three steps in our laboratories. Initially animal urine, plasma or microsomal incubates are examined to tentatively identify major metabolites. Second, *in vivo* metabolism and balance studies are performed in different animal species utilizing radiolabeled drug substance and third, human studies are performed with

0097–6156/95/0619–0315$12.00/0

radiolabeled material when possible. The applications discussed in this work will focus on the second and third steps of this process using specific examples to illustrate these steps.

Xanomeline, (3-[4-(hexyloxy)-1, 2, 5-thiadiazol-3-yl]-1, 2, 5, 6-tetrahydro-1-methylpyridine) is a selective M_1 muscarinic agonist currently under investigation for the potential symptomatic treatment of Alzheimer's disease.

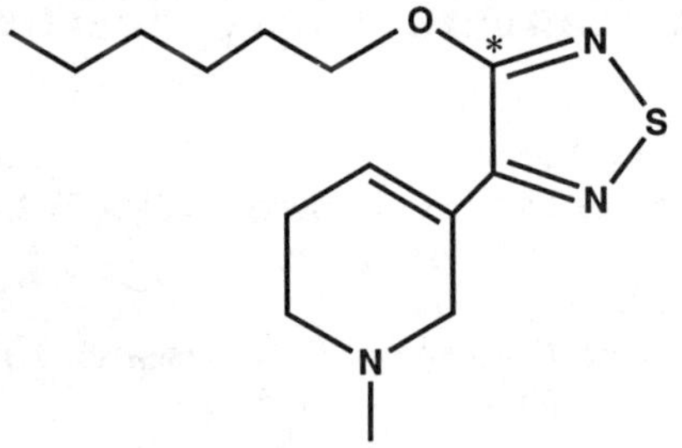

Xanomeline (* denotes ^{14}C-radiolabel)

Previous pharmacokinetic/balance studies conducted in rats and monkeys with radiolabeled drug suggested extensive metabolism of the parent compound (*7*). ES/MS/MS has been utilized to characterize the urine and plasma metabolites of xanomeline in humans (*8-10*). Furthermore, on-line LC/ES-MS/MS has been used to identify the urinary metabolites in rat after a subcutaneous dose of xanomeline in preparation for metabolite identification in humans after transdermal dosing of xanomeline. This work shows the advantages and limitations of using on-line LC/ES-MS/MS for characterization of numerous oxidative metabolites of xanomeline.

Experimental

A biotransformation study was conducted in human volunteers using ^{14}C-xanomeline at an oral dose of 75 mg (50μCi). Both urine and plasma samples were collected for analysis, however, only the urine was utilized for metabolite identification due to the low dose of xanomeline administered. Therefore, pooled human plasma for metabolite identification was obtained from a pharmackinetic study in human volunteers administered an oral dose of 150 mg (free base) of xanomeline. In addition, urine and plasma from rats was collected after subcutaneous administration of 10 mg/kg (free base) of ^{14}C-xanomeline to F344 rats.

Sample Preparation. Initial work with xanomeline utilized LC with fraction collection. Twenty mL of urine from each subject's 0-6 hour urine collection was pooled (80 mL), freeze-dried and the residue extracted with 1-2 mL of methanol (MeOH)/water (50/50;v/v), then centrifuged. A 200 μL aliquot of the supernatent was injected onto a Zorbax RX-C8 column (25cm x 0.46mm id) and separated by using a linear gradient consisting of 80/20 50 mM ammonium acetate (NH_4OAc)/MeOH for 30 minutes, then ramped to 100% MeOH over 75 minutes. Fractions were then collected in 1.25 minute increments and based on the radiochromatogram were then dried for analysis.

Human plasma collected at 1, 1.5, 2, and 4 hour time points was pooled (30 mL) for analysis. The pooled plasma was diluted with an equal volume of 50 mM NH_4OAc (pH 5) split into approximately 5 mL aliquots and applied to CN/C18 (1cc, Bond Elut) combination SPE ("piggy-backed") cartridges. The "piggybacked" cartridges were washed with 50 mM NH_4OAc (pH 5) then separated and the C18 cartridges eluted with 2 mL of MeOH. The CN cartridges were eluted with 2 mL of 200 mM NH_4OAc/MeOH (13.8 mL of 56.6% concentrated NH_4OH brought to 1 L with MeOH). The eluents in both cases were collected in silanized glass tubes and evaporated to dryness using nitrogen (N_2). The dryed eluents were reconstituted with MeOH, combined and centrifuged. The supernatent was then transferred to a silanized glass tube and taken to dryness with N_2. (Recovery data obtained for this procedure indicated a 66% recovery from the elution step and a 31% loss in the load and wash steps.) This residue was then reconstituted with 2.5 mL of 50 mM NH_4OAc (pH 5), then injected (200 µL injections until complete) onto a Zorbax RX-C8 column (25cm x 0.46mm id). Separation was obtained using a linear gradient consisting of 100% 50 mM NH_4OAc ramped to 100% MeOH over 60 minutes and held for an additional 5 minutes at a flow rate of 1 mL/min. Fractions were then collected and dried for analysis.

All other sample analyses were performed utilizing on-line LC/MS and LC/MS/MS. The rat urine and human urine collected from voluteers administered xanomeline via a patch were centrifuged, transferred to glass test tubes and then diluted 1:1 with 100 mM NH_4OAc.

LC/MS and MS/MS. All experiments were performed on a Finnigan MAT TSQ700 triple stage quadrupole (MS/MS). Initial work with xanomeline in human urine and plasma utilized an Analytica of Branford electrospray (ES) interface. In most cases, samples (collected fractions) were directly infused with a Harvard Apparatus syringe pump at a rate of 2 µL/min after reconstitution with 100 - 300 µL of MeOH/1% acetic acid (50/50;v/v). Positive ion detection was achieved with a voltage between -3500 to -3800 V applied to the Analytica ES interface and with 20 mL/min of nitrogen (N_2) drying gas @ 250°C. All direct infusion analyses were performed in the profile mode from m/z 100 - m/z 500 at 1 sec/scan and averaged for 1 min.

Select human urine samples (fractions) were analyzed by LC-ES/MS and LC/ES-MS/MS. Liquid chromatographic analyses were performed using a Waters 600-MS liquid chromatographic pump. A Basic column (5cm x 0.40mm id) from YMC, Inc. was coupled to the Analytica ES interface. An isocratic system of acetonitrile/0.2M acetic acid (50/50;v/v) was used at a flow rate of 1 mL/min. A flow of 2.5 µL/min was introduced into the Analytica ES interface by splitting the LC flow of 1 mL/min with a low dead volume tee. A voltage of -4200 V (applied to the canister) was used with 30 mL/min of nitrogen (N_2) drying gas @ 250°C to obtain positive ion detection. Analyses were performed in centroid mode at 1 sec/scan. The collisionally activated dissociation (CAD) analyses were obtained using a collision offset energy of -20 eV and a collision pressure of 2.0 mTorr of argon (Ar).

On-line LC/MS and LC/MS/MS were performed using a Finnigan atmospheric pressure ionization (API) source with a "high flow" ES interface. The diluted xanomeline urine samples were analyzed on a Zorbax RX-C8 column (25cm x 4.6mm id) using 100 µL injections. A linear gradient was employed consisting of 100% 50 mM NH_4OAc ramped to 100% MeOH over 60 minutes and held for an additional 5 minutes at a flow rate of 1 mL/min. The LC eluant was split 1:4 to allow 250 µL/min into the Finnigan ES interface and 750 µL/min to

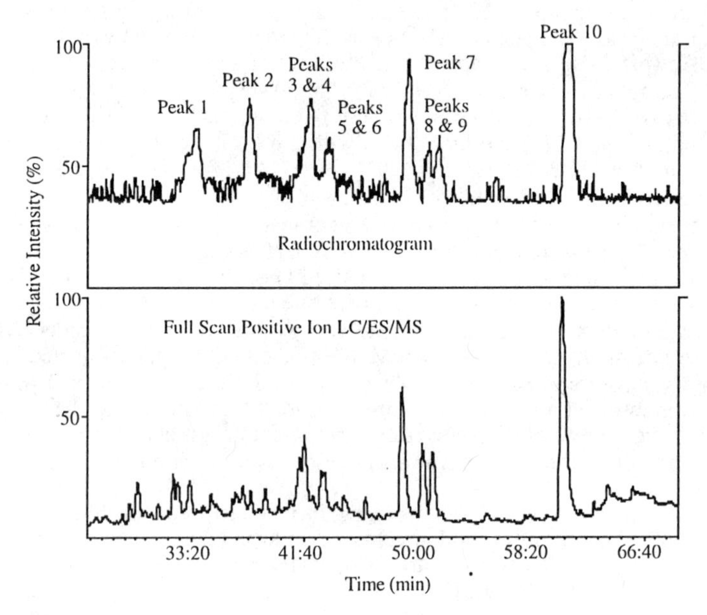

Figure 1. Full scan positive ion LC/ES-MS trace and the radiochromatogram for 24 hour rat urine after subcutaneous dosing.

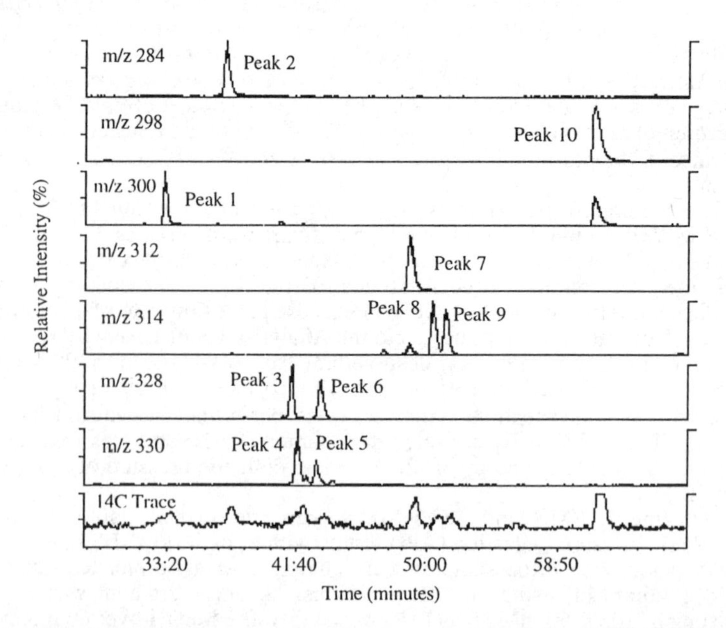

Figure 2. LC/ES-MS mass chromatograms of selected ions for 24 hour rat urine after subcutaneous dosing.

either a Raytest Ramona 5-LS radio chemical detector with a solid flow cell or a Waters 490-MS UV multiwavelength detector. Positive ion detection was achieved with the Finnigan ES using a spray voltage of +4500 V (applied to the needle), capillary heater temperature of 275°C, sheath gas of 70 psi (N_2), and an auxiliary gas flow of 15-20 mL/min (N_2). Analyses were performed in centroid mode at 1 sec/scan. The CAD analyses utilized a collision offset energy of -30 eV and a collision gas pressure of 1.6 mTorr (argon).

Results and Discussion

Rat Urine. At present, all ADME (absorption, distribution, metabolism, and elimination) work performed within our laboratories involves metabolism and balance studies which utilize radiolabeled compound. Figure 1 shows the full scan electrospray positive ion RIC (reconstructed ion chromatogram) trace and radiochromatogram from a 24 hour urine sample (~1 mL) collected from a rat after subcutaneous dosing of radiolabeled xanomeline (^{14}C-xanomeline). A gradient consisting of 50 mM NH_4OAc and MeOH was used over a period of 60 minutes to separate potential metabolites. In our laboratory we typically use shorter (<30 min), more rapid gradients, however, it was determined from previous studies that such a gradient was not appropriate for this analysis. The radiochromatogram was obtained by splitting the effluent leaving the LC column to the electrospray interface and a radiochemical detector. This type of rapid on-line analysis coupled with an appropriate selective detector can provide a number of advantages. For instance, the use of electrospray allows the rat urine to be centrifuged, transferred to a sample vial, diluted with buffer and directly injected on the column without further sample clean-up (such as fraction collection). This allows for more rapid feedback of information to the pharmacologist and metabolism scientist. In addition, many peaks are present in the ES-RIC trace, however, the radiochromatogram shows the presence of only 8 potential metabolite peaks. Coupling the electrospray interface with the radiochemical detector allows us to eliminate endogenous material which may produce erroneous ions corresponding to potential metabolites (*11-12*). Furthermore, the radiochromatogram allows one to make accurate quantitative assessments of the metabolite peaks present. However, in general, radiolabeling of the parent drug must be designed to avoid formation of metabolites not containing the label. This is important in that reliable quantitative data cannot be obtained based on ion intensity, due to the relative differences in the molecular ion yields for each metabolite. Figure 2 shows the ions which were selected from the peaks in the RIC which corresponded to the peaks identified in the radiochromatogram.

The use of MS/MS in metabolite identification is invaluable both to determine if a peak is produced by a potential metabolite or endogenous material and to elucidate the possible structure by comparing the product ion mass spectrum of the unknown peak to the product ion mass spectra of synthetic standards. To elucidate the structure of each xanomeline metabolite, the same diluted rat sample was analyzed further by LC/ES-MS/MS to simultaneously obtain the product ion mass spectra of each $[M+H]^+$ ion identified in Figure 2. The most abundant peak in the radiochromatogram, Peak 10, produced an ion at m/z 298. The product ion mass spectrum for this ion along with the proposed fragmentation scheme is shown in Figure 3. The loss of 17 amu from the $[M+H]^+$ ion produces the fragment ion at m/z 281 which is a characteristiclos of -OH. In addition, the fragment ion at m/z 60 indicates the presence of an N-oxide. Furthermore, this ion allows for the differentiation of N-oxidation on the tetrahydropyridine ring from hydroxylation on the side chain, which produces the same molecular ion. Comparison of this

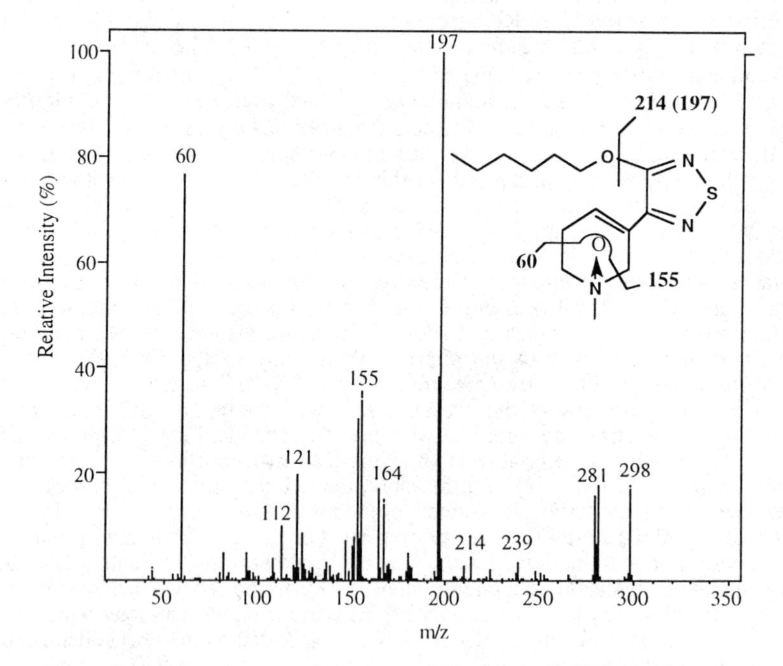

Figure 3. Product ion mass spectrum of Peak 8 (m/z 298) in 24 hour rat urine after subcutaneous dosing.

metabolite with that of a synthetic standard confirms that the major urinary metabolite present in rat urine after subcutaneous dosing is N-oxide xanomeline. In addition, the radiochromatogram showed one peak for Peaks 3 and 4. However, upon examination of the product ion mass spectra for the ions at m/z 328 and 330, two peaks are found to coelute for both Peaks 3 and 4. Figure 4 shows the product ion spectra for these two ions. At least 10 putative metabolites have been examined and identified by MS/MS and are listed in Table I. In addition, Table II shows identical information for all of the synthetic standards examined.

Table I. Characteristic Product Ions of Proposed Rat Urine Metabolites

Peak #[a]	$[M+H]^+$	Characteristic Product Ions (m/z)	Proposed Metabolite Identification
1	300	60, 155, 164, 197, 214	N-oxide, butyl carboxylic acid
2	284	44, 87, 155, 170, 198	butyl carboxylic acid
3	328	60, 155, 164, 197, 214, 252, 270	N-oxide, hexyl carboxylic acid
4	330	60, 99, 155, 197, 214	N-oxide, dihydroxyl
5	330	60, 99, 155, 197, 214	N-oxide, dihydroxyl
6	328	60, 155, 164, 197, 214	N-oxide, ketone, hydroxyl
7	312	60, 99, 155, 164, 197, 214, 294	N-oxide, ketone
8	314	60, 83, 121, 153, 155, 164, 197, 214, 296 (297)	N-oxide, hydroxyl
9	314	60, 83, 121, 153, 155, 164, 197, 214, 296 (297)	N-oxide, hydroxyl
10	298	60, 121, 155, 164, 197, 214, 281	N-oxide

[a] Peaks are identified in Figure 2.

Table II. Characteristic Product Ions of Synthesized Standards

Standard (#)	$[M+H]^+$	Characteristic Product Ions (m/z)
xanomeline (LY246708)	282	44, 112, 155, 170, 198
N-oxide (LY301908)	298	60, 121, 155, 164, 197, 214, 281
ω–1 hydroxyl (LY287546)	298	44, 114, 155, 198, 280
ω–1 ketone (LY287781)	296	44, 99, 155, 170, 198
butyl carboxylic acid (LY297542)	284	44, 87, 121, 155, 170, 198, 223, 241
hexyl carboxylic acid (LY296482)	312	44, 97, 155, 170, 198, 251, 269

The hexyl side chain and tetrahydropyridine ring appear to be the major sites of metabolism of xanomeline with the thiadiazole ring of xanomeline appearing to be metabolically inert in the rat. No intact parent was detected in these metabolism studies, however, xanomeline has been identified and quantitated (at levels < 1 ng/mL) in plasma utilizing an LC/MS/MS assay optimized specifically for xanomeline *(13)*. These data correspond favorably with the previous metabolite

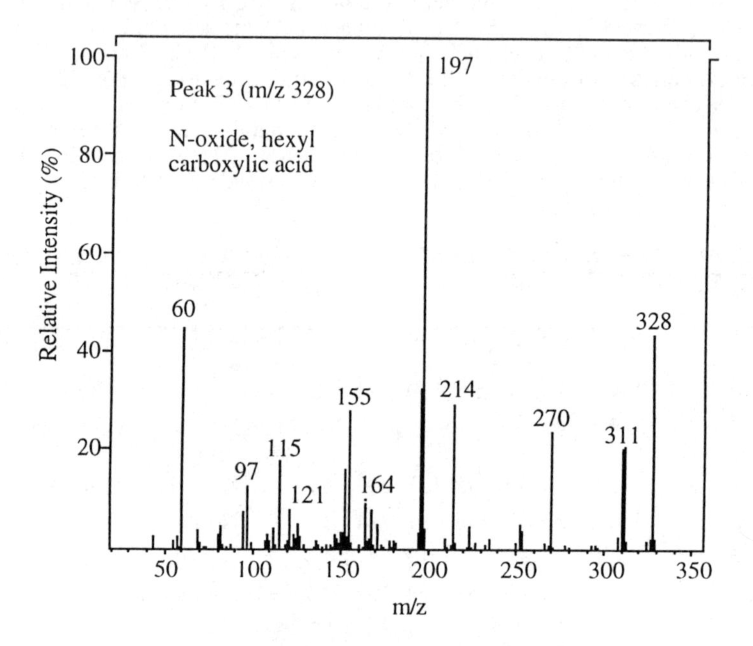

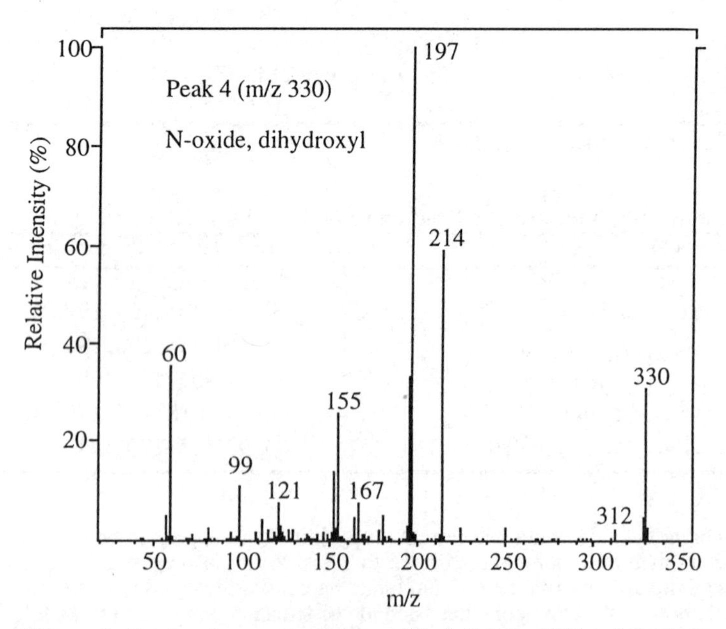

Figure 4. Product ion mass spectra of Peak 3 (m/z 328) and Peak 4 (m/z 330) in 24 hour rat urine after subcutaneous dosing.

identification performed after rats received an oral dose of ^{14}C-xanomeline tartrate (60 mg/kg, free base). However, N-oxidation appears to be the major route of metabolism in rats dosed by the subcutaneous route. In this study, urinary metabolites with side chain oxidations also undergo N-oxidation while in the oral study not all the metabolites with side chain oxidations were oxidized on the nitrogen. These differences in urinary metabolic profiles of xanomeline following different routes of dose administration may be explained by extrahepatic metabolism of xanomeline following subcutaneous administration.

Human Urine and Plasma. Initial work to determine the metabolic profile of xanomeline in humans was performed with an "older" electrospray interface. This interface would not allow the consistent use of an on-line gradient system, therefore, fractions were collected off-line utilizing an LC gradient system to separate the potential metabolites. The collected fractions were then analyzed either by LC/ES using an isocratic system with a short LC column (5 cm) to provide additional clean-up of urine fractions or by direct infusion (DI) for the collected plasma fractions.

Figure 5 shows the radiochromatogram generated from the 24 hour urine collection of four human volunteers after receiving an oral dose of ^{14}C-xanomeline tartrate (75 mg/kg, free base) and the UV chromatogram from 30 mL of pooled plasma from human volunteers after receiving an oral dose of xanomeline tartrate (150 mg, free base). Eight potential metabolite peaks in urine and in plasma were isolated, purified and characterized by ES-MS and ES-MS/MS. Due to differences in the chromatographic systems utilized the retention times for several of the identified metabolites in urine and plasma do not correspond to each other . The most prominent peak observed in both human urine and plasma produced an ion at m/z 284 (mass spectrum not shown). The product ion mass spectra of m/z 284 from the major metabolite peak in both urine and plasma along with a proposed fragmentation scheme are shown in Figure 6. Comparison of this metabolite with that of a synthetic standard (Table II) confirms that the major metabolite present in human urine and plasma after oral dosing is the butyl carboxylic acid of xanomeline. A variety of other oxidative metabolites were identified in both human urine and plasma. Table III shows a summary of the human urine and plasma metabolites identified at present by utilizing either ES-MS/MS or LC/ES-MS/MS with fraction collection. Figure 7 shows the proposed metabolic pathway for xanomeline in humans. The hydroxyl and ketone metabolites have been identified in other studies.

The hexyl side chain and tetrahydropyridine ring appear to be the major sites of metabolism of xanomeline with the thiadiazole ring of xanomeline appearing to be metabolically inert in man. No parent was identified utilizing this methodology. However, as was mentioned earlier with rat plasma, xanomeline has been detected in human plasma at levels < 1 ng/mL by utilizing a specific assay. The identified metabolites correspond favorably with previous identification performed in rat, dog, and monkey urine after dosing orally with xanomeline tartrate. Comparison of these human metabolites with those identified in the urine of the rat after subcutaneously dosing with xanomeline shows side chain oxidation to be more prevalent than N-oxidation.

On-line LC/ES-MS/MS was attempted with human urine employing the electrospray interface used for the subcutaneous rat urine work described earlier. Ten to twenty milliliters of 0 - 24 hour urine from human volunteers administered a dose of xanomeline via a patch was concentrated, reconstituted and injected on the same chromatography system used for the rat urine sample. Figure 8 shows the full scan electrospray positive ion RIC trace for an injection of the concentrated human

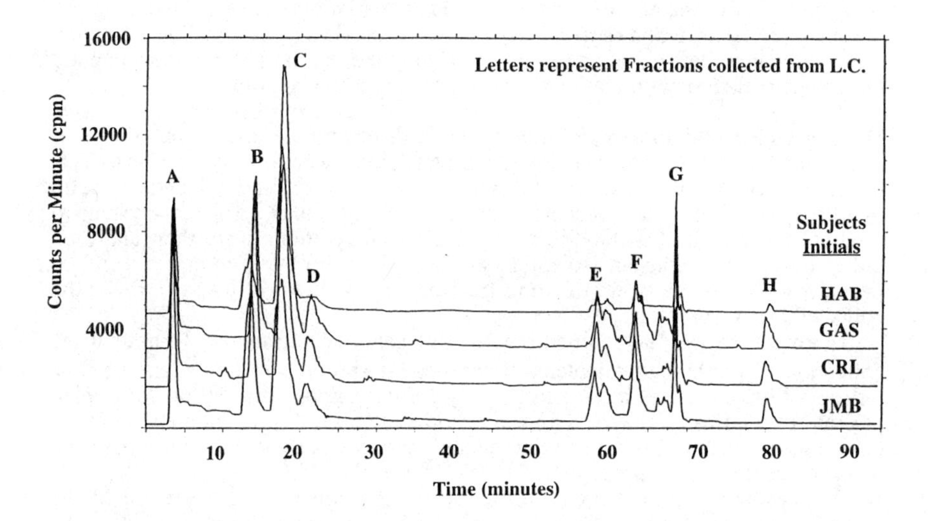

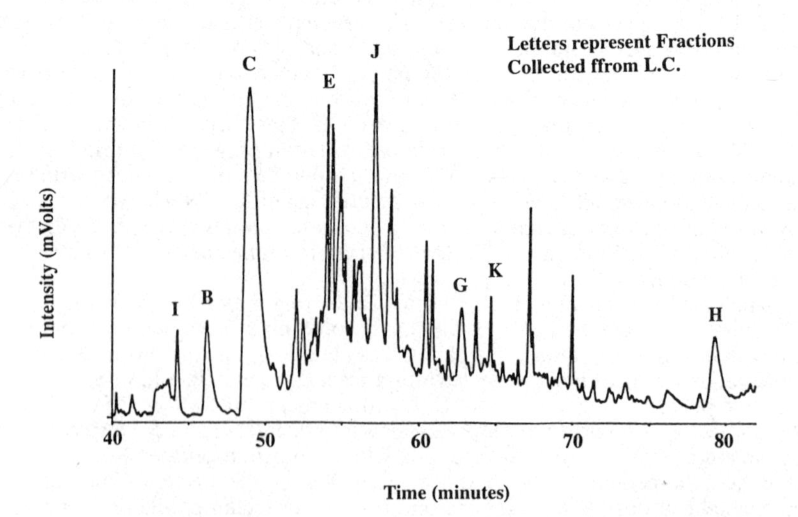

Figure 5. Radiochromatogram for urine (top) and a UV chromatogram for plasma (bottom) for human volunteers after oral dosing.

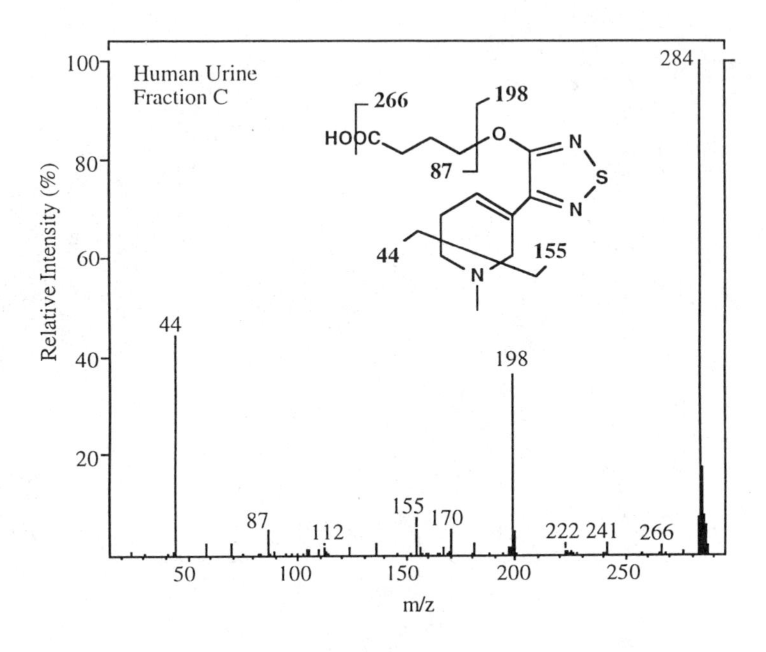

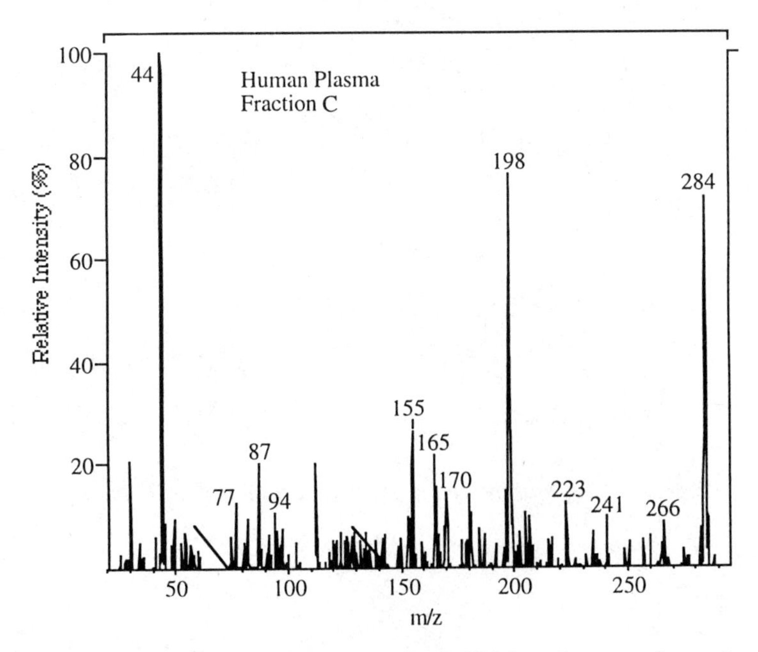

Figure 6. Product ion mass spectra of m/z 284 from human urine and human plasma (Fraction C).

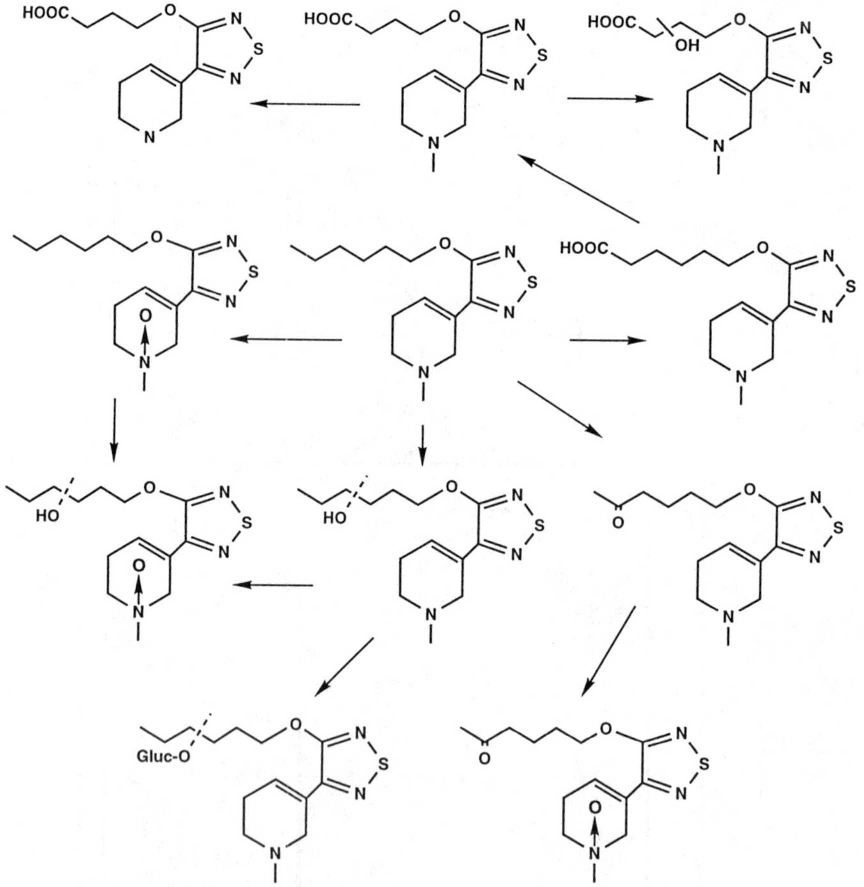

Figure 7. Proposed metabolic pathway for xanomeline in humans.

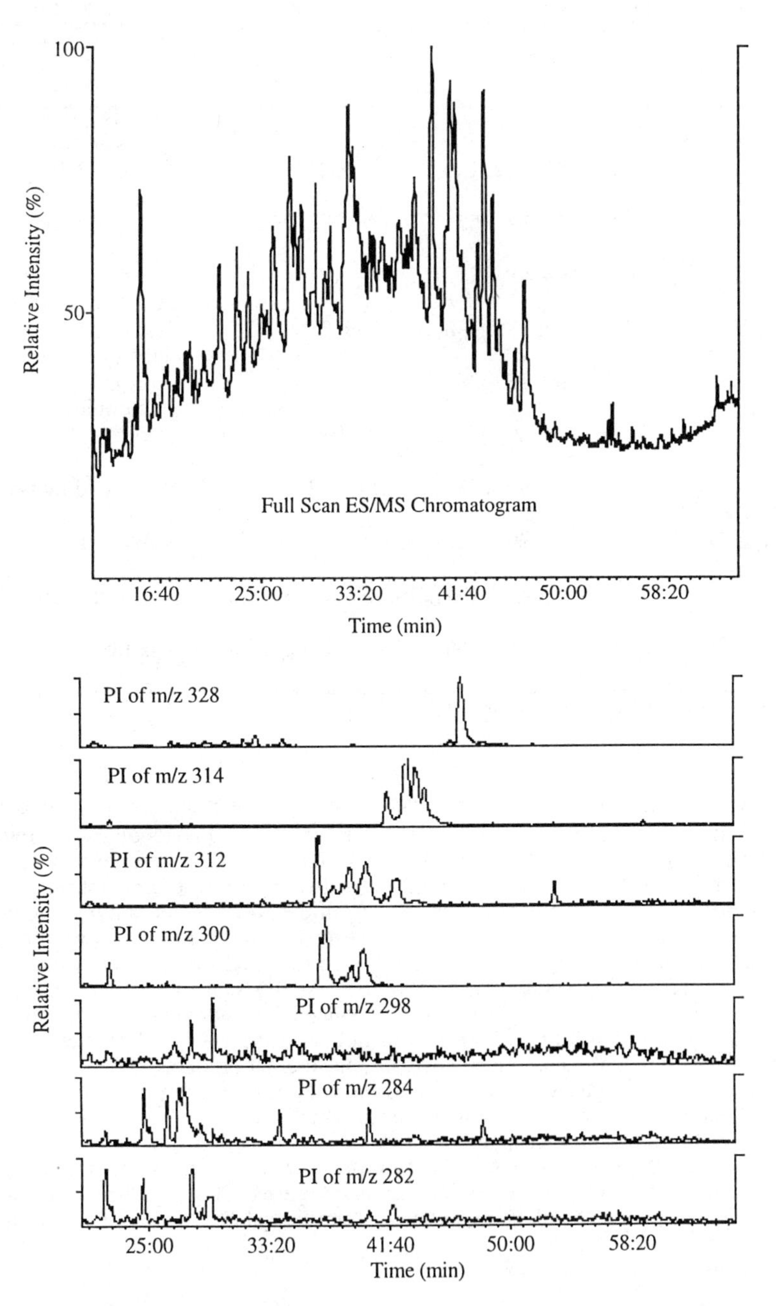

Figure 8. Full scan positive ion LC/ES-MS trace and product ion (PI) chromatograms of selected ions for human urine after transdermal dosing.

Table III. Characteristic Product Ions of Proposed Human Metabolites

Fraction[a] (Urine or Plasma)	$[M+H]^+$	Characteristic Product Ions (m/z)	Proposed Metabolite Identification
I	300	44, 89, 105, 133, 282	hydroxyl, butyl carboxylic acid
B	270	87, 112, 155, 198, 223, 241	desmethyl, butyl carboxylic acid
C	284	44, 87, 155, 170, 198, 223, 241, 266	butyl carboxylic acid
D	300	60, 99, 121, 155, 164, 197, 214, 282	N-oxide, butyl carboxylic acid
E	298	97, 155, 198, 251, 269, 280	desmethyl, hexyl carboxylic acid
E	474	89, 133, 280, 298	hydroxyl glucuronide
J	312	44, 69, 97, 155, 170, 198, 251, 294	hexyl carboxylic acid
F	330	60, 99, 155, 197, 214, 312	N-oxide, dihydroxyl
G	312	60, 99, 155, 197, 214	N-oxide, ketone
K	314	60, 155, 197, 214, 296 (297)	N-oxide, hydroxyl
H	298	60, 112, 155, 164, 197, 214, 281	N-oxide

[a] Fractions are identified in Figure 5.

urine sample along with selected product ion mass chromatograms. The selected product ions shown represent the $[M+H]^+$ ions of potential metabolites previously identified in humans and/or rats. Numerous peaks are present at relative retention times similar to the metabolites identified in the rat after administration of the subcutaneous dose of xanomeline tartrate. Both the N-oxide and butyl carboxylic acid metabolites of xanomeline were confirmed to be present. However, the product ion mass spectra of many of these ions revealed that the peaks were not related to xanomeline and/or its metabolites. Analysis of the product ion mass spectra indicate that these peaks were produced by endogenous urine related material. A larger volume (~200 mL) of human urine was used to increase the signal of the potential metabolites, however this attempt produced the same results. The data suggests that utilization of a larger volume of human urine results in a larger concentration of endogenous material which interferes (or masks) with the detection of the xanomeline metabolites. Therefore, to obtain an adequate concentration of potential metabolites and proper clean-up of the biological sample, fraction collection may be necessary in many cases for drug candidates with extensive first pass metabolism or following transdermal administration of a drug.

Conclusions

On-line LC/ES-MS/MS has proven to be an invaluable tool in the characterization of the metabolites of xanomeline in a variety of mammalian species. The analysis of urine and plasma from humans and animals has established the presence of numerous oxidative metabolites. The hexyl side chain and tetrahydropyridine ring appear to be the major sites of metabolism of xanomeline with the thiadiazole ring appearing to be metabolically inert.

This data along with other applications performed in our laboratory point out the advantages and limitations of on-line LC/ES-MS/MS in biotransformation studies. It's ability to handle gradient solvent systems containing a large percentage of aqueous mobile phase and wide dynamic range of compound response has proven invaluable for metabolism studies compared with previous LC/MS methodology. In addition, ES typically generates abundant $[M+H]^+$ species and avoids the use of high temperatures which may cause thermal degradation of metabolites. The utility and speed of this methodology for performing animal metabolism studies is obvious. However, due to lower concentrations of metabolites, extensive first pass metabolism, and matrix interferences, biotransformation studies of "potent" potential drug candidates in humans pose difficult problems for on-line LC/ES-MS/MS techniques. Future work with column switching methodology for preconcentration and selective cleanup of the samples (*14*) may help to eliminate many of the above mentioned limitations.

Literature Cited

1. Bruins, A.P. *J. Chromatogr.* **1991**, *554*, 39-46.
2. Baillie, T.A. *Int. J. Mass Spectrom.. Ion Proc.* **1992**, *118/119*, 289-314.
3. Baillie, T.A. *Proc. 42nd ASMS Conf. Mass Spectrom. Allied Topics*, **1994**, *Chicago, IL*, 862.
4. Bordas-Nagy, J.; Murphy, D.M.; Anderson, K.M.; Blake, T.J.; Rhodes, G.R.; and Dulik, D.M. *Proc. 42nd ASMS Conf. Mass Spectrom. Allied Topics*, **1994**, *Chicago, IL*, 865.
5. Weidolf, L.; and Covey, T.R. *Rap. Comm. in Mass Spectrom.*, **1992**, *6*, 92-196.
6. Fouda, H.G.; Avery, M.J.; Cole, M.J.; and Schneider, R.P. *Proc. 42nd ASMS Conf. Mass Spectrom. and Allied Topics*, **1994**, *Chicago, IL*, 871.
7. Cornpropst, J.D.; Occolowitz, J.D.; Gillespie, T.A.; and Shipley, L.A. *Int. Soc. for Study Xenobiotics*, **1993**, *Tucson, AZ* (abstract).
8. Gillespie, T.A.; Bonate, P.L.; Cornpropst, J.D.; DeLong, A.F.; and Shipley, L.A. *Proc. 42nd ASMS Conf. Mass Spectrom. and Allied Topics*, **1993**, *San Francisco, CA*, 36.
9. Shipley, L.A.; Cornpropst, J.D.; Brown, T.J.;Satterwhite, J.H.; Gillespie, T.A.; Lachno, D.R.; Carter, G.V.; and Lucas, R.A. *Int. Soc. for Study Xenobiotics*, **1993**, *Tucson, AZ* (abstract).
10. DeLong, A.F.; Henry, D.P.; Bonate, P.L.; Gillespie, T.A.; and Satterwhite, J.H. *3rd Int. Conf. Alzheimer's & Parkinsons Diseases*, **1993**, *Chicago, IL* (abstract).
11. Spreen, R.C.; Bui, K.H.; Savidge, R.D.; Raybuck, D.L.; and Birmingham, B.K. *Proc. 42nd ASMS Conf. Mass Spectrom. and Allied Topics*, **1994**, *Chicago, IL*, 360.
12. Scheider, R.P.; Davis, K.M.; Inskeep, P.B.; and Fouda, H.G. *Proc. 42nd ASMS Conf. Mass Spectrom. and Allied Topics*, **1994**, *Chicago, IL*, 345.
13. Murphy, A.T.;Bonate, P.L.;Kasper, S.C.;Gillespie, T.A.; and DeLong, A.F. *Biological Mass Spectrom*, **1994**, *23*, 621-625.
14. Van der Graf, J.;Niessen, W.M.A.; and Tjaden, U.R. *Proc. 42nd ASMS Conf. Mass Spectrom. and Allied Topics*, 1993, *San Francisco, CA*, 286.

RECEIVED July 21, 1995

Chapter 17

High-Performance Liquid Chromatography with Atmospheric Pressure Ionization Tandem Mass Spectrometry as a Tool in Quantitative Bioanalytical Chemistry

John D. Gilbert, Timothy V. Olah, and Debra A. McLoughlin

Department of Drug Metabolism, Merck Research Laboratories, West Point, PA 19486

Quantitative bioanalytical chemistry has been revolutionized in the past five years by the development of Atmospheric Pressure Ionization Tandem Mass Spectrometry (API-MS/MS) as a detection technique for liquid chromatography. The power of the technique stems firstly from the high specificity of the MS/MS detector and secondly, from the ion source's compatibility with conventional liquid chromatography (LC) conditions. The freedom of interferences from endogenous biological substances enables assays for drugs in biological fluids to be developed in days rather than weeks. The ruggedness of LC with API-MS/MS permits batch processing with rapid sample throughput and sensitivities previously attainable only by radioimmunoassay or negative ion gas chromatography-mass spectrometry. The availability of such powerful assays is having a profound effect on the pharmaceutical industry's drug discovery and development activities.

Quantitative bioanalytical chemistry is a cornerstone of pharmaceutical research. The challenges of bioanalysis stem from the need to accurately and reproducibly measure part per million to part per trillion quantities of analyte in complex biological matrices, full of potentially interfering endogenous or drug-related substances. The innovativeness of modern pharmaceutical discovery groups continually challenges bioanalysts since the new generations of drugs are much more potent and require increasingly sensitive assays. The most common bioanalytical methods used for the measurement of drugs in biological samples are high performance liquid chromatography (HPLC), gas-liquid chromatography (GLC), gas chromatography-mass spectrometry (GC/MS), immunoassay and, most recently, liquid chromatography-mass spectrometry (LC/MS).

Immunoassay is very sensitive, normally requires no sample preparation, and can efficiently process samples in large batches with relatively inexpensive equipment. However, the development of an immunoassay for a small non-immunogenic molecule

0097–6156/95/0619–0330$12.25/0

requires the preparation of one or more immunogens and a delay of several months after immunization for animals to produce antisera suitable for testing. Immunoassay may also lack specificity with respect to metabolites. The technique is generally used as a contingency for substances which cannot be detected at sufficiently low concentrations using chromatographic methods.

Capillary gas chromatography affords excellent chromatographic resolution and, when combined with mass spectrometry, provides assays of excellent specificity and sensitivity. For many years GC/MS was the ultimate weapon of the bioanalyst, and it spawned the sub-science of microderivatization in which new reagents and efficient microchemistries were developed both to improve the volatility and thermal stability of analytes and to enhance their detectability by mass spectrometry. Halogen-containing derivatives used in conjunction with electron capture negative chemical ionization GC/MS can yield assays of particularly high sensitivities and good precision. Two recent examples are those described for clenbuterol *(1)* with a lower quantifiable limit (LQL) of 5 pg/ml and a coefficient of variation (CV) at the LQL of 12.8%, and the serotonin 5-HT, receptor antagonist, zacopride *(2)* (LQL 10 pg/ml; CV = 3.2%). The high resolution of the capillary column generally renders the use of gas chromatography tandem mass spectrometry (GC/MS/MS) for quantitation unnecessary.

Sensitivities of GC/MS/MS methods are generally not as good as those using GC/MS since, in contrast to LC/MS/MS-based procedures (below), the loss of signal due to the collision process is not offset by a significant reduction in the background noise. However, GC/MS/MS has been used to increase the specificity of assays for tebufelone in crude plasma extracts *(3)* and for prostaglandins E-M and F-M in urine *(4)*.

The development of HPLC greatly advanced the science of quantitation because of its simplicity, speed and applicability to polar and thermally labile analytes. In contrast with GLC, derivatization was generally unnecessary to ensure good chromatography although it could be used to enhance sensitivity. HPLC with ultraviolet detection has, however, been likened to GLC with flame ionization detection in that it affords little specificity. GLC had evolved with the development of element-specific detectors (N, P, S and halogen) but it was GC/MS with selected ion monitoring that ultimately provided the specificity needed for sensitive quantitative analysis.

In the case of HPLC, the use of fluorometric and electrochemical detectors could, in some cases provide additional specificity. However, the HPLC analysts would cast envious glances at their GC/MS counterparts who were using a technique, intrinsically much more restrictive from a chromatographic viewpoint, but which was compatible with a highly specific mass spectrometric detector. Their frustration increased with increasing demands for specific assays of polar substances at low to sub-ng/ml sensitivities, but LC/MS technology for many years did not advance past the pragmatic but water-intolerant moving belt interface *(5)*, which was largely incompatible with the increasing power of reverse-phase HPLC. In direct liquid introduction (DLI) - LC/MS *(6,7,8)* the HPLC effluent passed through a small (25 micron) pin-hole aperture into the ion source where the solvent (up to 70% water) served as a reagent gas for chemical ionization. A splitter was needed with conventional HPLC columns, although the interface could accommodate flows of 5-50 µl/min from microbore columns. Plugging of the orifice was common and great care was required to avoid particulates *(9)*.

The development of thermospray *(10)* provided a liquid introduction interface able to cope with LC flows of 1-2 ml/min without splitting. The ion source still operated at low pressure, with the solvent being removed by a single-stage mechanical pump. Control of the capillary temperature was critical since this determined the droplet size. Excessively large droplets would not produce ions while inordinately high temperatures would result in a "dry vapor" with analyte decomposition. Total ion currents tended to be unstable, especially with gradient elution, and reliable quantitation required the use of co-eluting, stable isotope-labeled internal standards rather than chromatographically-resolved homologs. Additionally, since no electrical field was applied, both positive and negative ions were formed together, thus reducing sensitivity. Despite these limitations, sensitive assays were established for several drugs including budesonide *(11)*, mirfentanil *(12)*, pyridostigmine *(13)* and sumatriptan *(14)*.

Around 1985, one commentator voiced the frustration of the HPLC community as follows:- " ...despite some advances LC/MS is still a Cinderella technique and will remain that way until a manufacturer has the courage to introduce a specifically-designed LC/MS system and not, as is currently the case, advertise LC/MS as an interesting but problematic bolt-on to GC/MS *(15)*." About the same time, at Cornell University, the potency of atmospheric pressure LC/MS/MS as a quantitative bioanalytical tool was being unequivocally demonstrated on just such an instrument *(16)*.

The fundamental problem of LC/MS with conventional HPLC flow rates was the inability of the mass spectrometer to pump away all of the solvent vapor from the eluent. This hurdle, partly surmounted in thermospray by the provision of additional pumping to the source, is completely obviated by use of a no-vacuum, atmospheric pressure ion source *(17)*. The ionized effluent is simply sprayed close to a pin-hole orifice (larger than that used with DLI) which separates the high vacuum analyzer from atmospheric pressure. Gas-phase ions are formed in the source region either by electrospray or atmospheric pressure chemical ionization (APCI) using a corona discharge, and these ions are sampled through the orifice into the mass analyzer. Little solvent penetrates into the analyzer and ion clusters are removed by passage through a curtain of dry nitrogen.

The API III mass spectrometer developed by Sciex was the first to provide a reliable means to combine the power of reverse phase HPLC with the exquisite specificity and sensitivity of the mass spectrometer. Total ion current (TIC) signals are very stable and largely independent of mobile phase composition and interface temperatures. Conventional HPLC conditions are eminently accommodated. The instrument is remarkably easy to use and automate, and performs equally well in the positive and negative ionization modes. The efficiency of ionization provides excellent sensitivity, and the tandem MS/MS detector gives an extra dimension of specificity, efficiently filtering out chemical noise from the HPLC background. The orifice does not clog, and since there are no lenses, insulation or electrical connections, the ion source can get dirty with no adverse effect on instrument sensitivity.

Two interfaces are commonly used. Both are designed to convert the HPLC effluent into droplets, but ionization is effected by different mechanisms which have been described in great detail elsewhere *(16-20)*. The heated nebulizer interface combines heat and pneumatics to volatilize the effluent. The droplets are ionized by use of a corona discharge from the tip of a needle held at high voltage (± 4-5 kV) relative to the sampling orifice. The analyte is ionized by charge transfer chemical ionization *via* the ionized mobile phase (water, solvents and/or additives). The application of heat enables handling of reverse-phase HPLC effluent flows from 0.2 - 2 ml/min. Since the vapor temperature is only *ca.* 125°C, thermal degradation is normally not observed except for the most labile analytes. The principal advantage of the heated nebulizer is its compatibility with conventional 4.6 mm HPLC columns, flow rates and mobile phases without splitting of the effluent prior to passage into the mass spectrometer.

The ionspray interface is ideally suited for polar, thermally labile analytes, since pre-formed ions in solution are emitted into the gas phase without any application of heat. The HPLC effluent is pumped through a sprayer maintained at a high voltage (up to ± 6Kv) so that a mist of highly charged droplets is formed. As the droplets evaporate, the analyte ions are ejected into the gas phase. The use of this pneumatic spray accommodates higher flows than electrospray alone. With conventional HPLC columns, effluent splitting is still required; a flow of 20-50 µl/min into the ion source is optimal. The interface is, however, ideal for use with 1 mm i.d. microbore HPLC columns. These are less rugged than conventional columns but perform satisfactorily provided the cleanliness of sample extracts is correctly addressed. With both interfaces, separate generation of positive and negative ions is accomplished simply by changing the polarity of the applied voltage. It is important to stress that these interfaces generate ions by different processes. The heated nebulizer uses atmospheric pressure chemical ionization (APCI) while ionspray is a pneumatically assisted electrospray. Most of the assays developed in our laboratories have used the heated nebulizer but the recent introduction of fast (turbo) ionspray enables column flow rates of between 0.5 and 0.8 ml/minute to be used which will now improve its popularity. However, high capacity ionspray should be regarded as a complement to rather than substitute for APCI since the choice depends critically of the chemical properties of the analyte.

The development of the atmospheric pressure ionization interface finally provided bioanalysts with a remarkable tool which was readily embraced by the pharmaceutical industry and is now used routinely for the quantitation of drugs in biological fluids. The power of these assays is having a profound effect on the way pharmaceutical companies discover and develop drugs. API LC/MS/MS has not only changed the way in which quantitation of analytes in biological fluids is performed but also has influenced our approach to basic research. The remarkable specificity and sensitivity of the technique have armed us with an additional tool for simultaneous determination of metabolites, for the use of isotopes in pharmacokinetic research and for the evaluation of potentially less specific methods. The following sections summarize some of the important applications described both in the literature and from our own experiences. The use of LC/MS/MS as an aid in drug discovery and to assist in our compliance with the Good Laboratory Practice (GLP) regulations are also briefly described.

LC/MS *vs.* LC/MS/MS

Before illustrating the various applications of LC/MS with atmospheric pressure ionization, some discussion of the relative merits of LC/MS *vs.* LC/MS/MS is warranted. This is an important issue, since there is a considerable cost difference between single stage and tandem instruments.

Despite the numerous advantages that HPLC has over GLC, the latter is superior in two important respects. First, the resolution of capillary GLC columns is much beyond the capabilities of any liquid chromatographic technique other than capillary electrophoresis and second, the chemical noise associated with HPLC is much greater than that found with GLC. These two factors frequently necessitate the use of HPLC with tandem MS/MS detection if very sensitive and specific assays for drugs in biological fluids are required.

In LC/MS, selected ion monitoring (SIM) is used as the operating mode in which the mass spectrometer monitors a single ion specific for the target substance. There are several examples of satisfactory bioanalytical methods being developed using detection by SIM. Fouda *(22)* developed an assay for the determination of the renin inhibitor CP-80,794 using the heated nebulizer with negative-ion SIM. The lower limit of quantification was 50 pg/ml. LC/MS methods for the determination of the HMG CoA reductase inhibitors, SQ 33,600 *(23)* and pravastatin *(24)*, using the ion spray interface and negative-ion detection have also been reported. Both drugs were isolated from plasma using elegant solid phase extraction procedures. The biggest problem encountered with the development of these methods was finding a chromatographic system with suitable selectivity to resolve the analytes from endogenous interferences. Gradient elution had to be used for the determination of pravastatin which increased analysis times to approximately 9 minutes.

When MS/MS detection is used, the parent pseudomolecular ion is selected by the first mass filter. Fragmentation of the parent ion is then induced by collision with atoms of argon and an intense and/or characteristic product ion is selected by the second mass analyzer. This procedure is described as selected reaction monitoring (SRM) or, when two substances are being examined simultaneously as in the case of drug accompanied by an internal standard, multiple reaction monitoring (MRM).

Weidolf and co-workers *(25)* noted that although the limit of detection by LC/MS with SIM was 10-12 pg for standards of sulfoconjugated metabolites of the synthetic anabolic steroid boldenone, as compared to 110 - 370 pg using MRM, MS/MS showed improved selectivity for structure confirmation and quantitation in biological extracts. Unless rigorous sample clean-up strategies are employed or fortuitous chromatographic conditions can be found, ions originating in the mobile phase or from endogenous substances present in biological extracts can result in considerable background noise when SIM is used. The effect is most pronounced at lower m/z values. In such circumstances, sensitivity can be significantly compromised by background noise. The ability of the tandem mass spectrometer, operating in the MRM mode, to detect specific pseudomolecular ions in combination with specified product ions while the analytes of interest are eluting, affords an extra dimension of specificity over single stage SIM, which discriminates simply on the basis of an analyte's molecular weight. The limitations of SIM detection are generally not apparent simply by the examination of a clean standard solution of the drug.

Shown in Figure 1 *(21)* is a comparison of chromatograms of a standard solution containing simvastatin and lovastatin as their pentafluorobenzyl (PFB) esters analyzed by both LC/MS, using SIM, and LC/MS/MS using MRM. The signal intensities, in counts per second, are shown in the right hand corner of each chromatogram. The absolute signals for both derivatives are approximately 30-fold greater when LC/MS is used. Although tandem LC/MS/MS is less sensitive in terms of absolute signal, because of the transmission losses through the collision cell and second mass analyzer, the use of MRM to monitor a specific parent/product-ion combination reduces the background noise much more than it reduces the absolute ion signal. This greatly improves the signal-to-noise ratio and thus provides better overall detectability than single stage LC/MS. Comparison of mass chromatograms from a derivatized extract of plasma containing 0.5 ng/ml of simvastatin and 5 ng/ml of the internal standard, lovastatin, is shown in Figure 2. The background noise in the SIM trace for simvastatin at the injection point is around 13 thousand counts per second (c.p.s.) while the noise due to endogenous substances immediately prior to elution of the analyte peak increases to 23 thousand c.p.s. The signal from the simvastatin derivative is 29 thousand c.p.s. The corresponding chromatograms obtained using LC/MS/MS in the MRM mode are strikingly different, with a background noise essentially independent of retention time (*ca.* 11 c.p.s.), and a signal for the pentafluorobenzoate ester of simvastatin of 260 c.p.s. Thus, the signal-to-noise ratio for the determination of the analyte, simvastatin, is much improved by the use of LC using MS/MS detection (signal-to-noise *ca.* 24:1) as opposed to LC/MS (signal-to-noise of around 2:1), because the background is reduced by approximately 3 orders of magnitude (from 23,000 c.p.s. to 11 c.p.s.) when LC/MS/MS is used. The net result is a 10 to 20-fold gain in practical assay sensitivity when tandem MS/MS detection is used. The superiority of LC/MS/MS over LC/MS for the determination of the 5α-reductase inhibitor, L-654,066 in plasma, has been similarly demonstrated *(26)*.

A number of examples of very sensitive assays, utilizing HPLC with tandem MS/MS detection are appearing in the literature. Kaye and co-workers *(27)* developed a robust assay with high sensitivity for the $alpha_1$-adrenoceptor antagonist, abanoquil, in plasma after "it became clear early on in our investigation that, even with a lengthy clean-up procedure, it would be necessary to use the mass discriminatory power of the triple quadrupole to achieve high selectivity and the maximum signal-to-background ratio." By using MRM, they were able to achieve a lower quantifiable limit of 10 pg/ml. Gilbert *et al.* (*J. Pharm. Biomed. Anal.* in press) have been able to measure plasma levels of the antiarrhythmic agent MK-499 to 13 pg/ml with a CV of 6% using an LC/MS/MS-based assay which proved four times more sensitive than the established radioimmunoassay.

Despite the mounting evidence of the advantages of utilizing HPLC with tandem MS/MS, rather than single stage MS, for the measurement of drugs in biological fluids, there are some who continue to advocate LC/MS detection. As long as a chromatographic separation is employed, they argue, LC/MS can be used in place of LC/MS/MS and with much superior sensitivity *(28)*. There are areas in pharmaceutical development where the sample matrix is relatively simple and LC/MS can be correctly utilized (e.g. the determination of impurities in drugs). However,

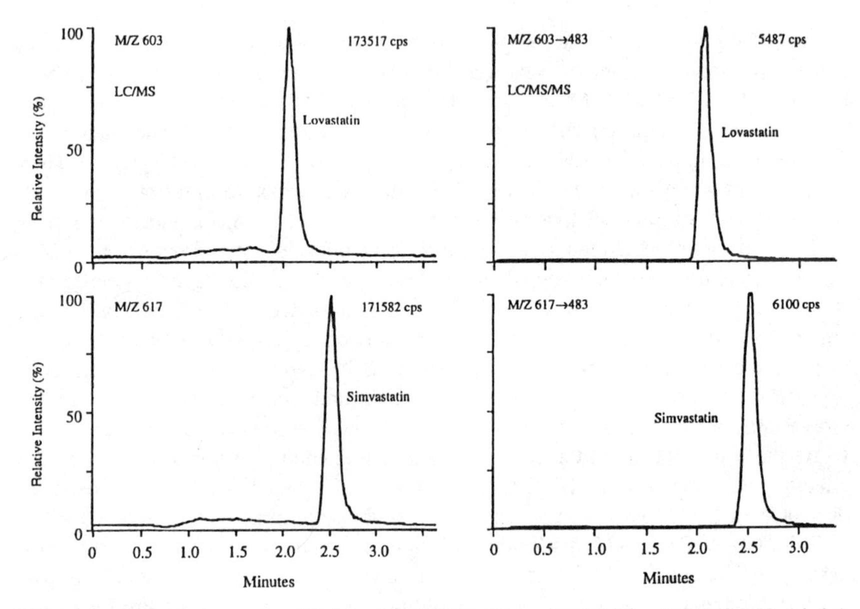

Figure 1. Chromatograms, SIM by LC/MS (left) and MRM by LC/MS/MS (right) of an aqueous standard containing lovastatin and simvastatin β-hydroxyacids as their PFB esters (10 ng of each on-column). Adapted from reference 21 by permission of the Royal Society of Chemistry.

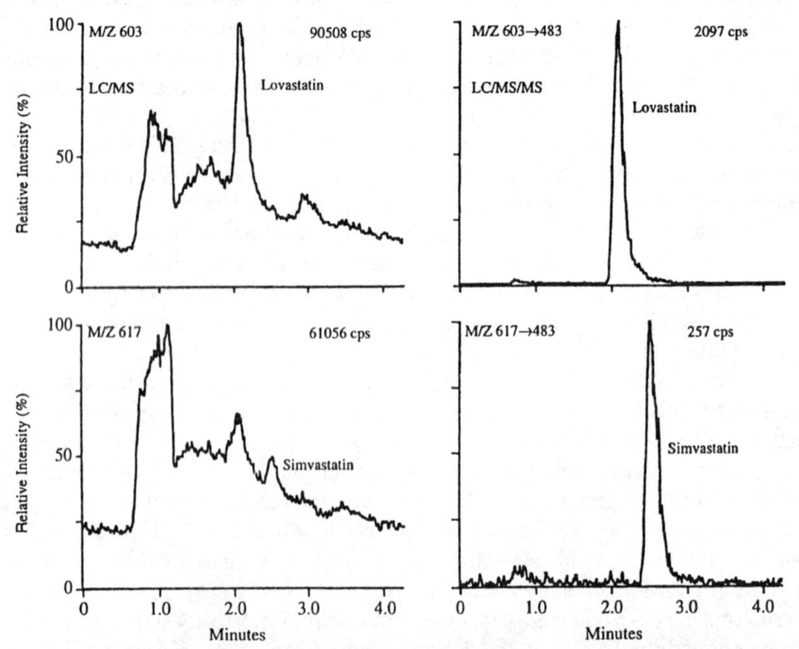

Figure 2. SIM (left) and MRM (right) chromatograms of a derivatized extract of plasma containing simvastatin at 0.5 ng/ml. Adapted from reference 21 by permission of the Royal Society of Chemistry.

the challenge facing the bioanalyst is the routine, fast and accurate measurement of minute quantities of analyte in extracts of complex biological samples, and the sensitivity of such assays is much more dependent on signal-to-noise ratio than is the case with pristine standard solutions. Experienced bioanalysts who are fortunate to possess triple quadrupole mass spectrometers will rarely make their jobs more difficult by turning off the second mass analyzer.

Assay Comparisons and Cross-Validations. The adoption of methods based on LC/MS is frequently prompted by the need to improve assay sensitivity, specificity or efficiency. In several cases, direct comparisons have been made on the performance of the older assay with that of the LC/MS procedure which supercedes it. The extent to which assays are compared varies considerably. Constanzer (29) compared the accuracy and precision of HPLC-based assays for cyclobenzaprine in plasma using ultraviolet (UV) and MS/MS detection. Both procedures performed satisfactorily to 0.5 ng/ml, but the LC/MS/MS method improved the sensitivity to 0.1 ng/ml and shortened the chromatographic run time from twelve to three minutes thus tripling the assay throughput. Similarly, the use of HPLC with MS/MS rather than UV detection greatly improved the sensitivity and efficiency of the assays for finasteride in plasma and semen *(30)*. Fraser *(31)* compared assays based on GC/MS and LC/MS/MS for the quantitation of a new acetylCoA-cholesterol acyltransferase (ACAT) inhibitor in human plasma. Both assays showed satisfactory precision and accuracy to 0.5 ng/ml but sample preparation for the LC/MS/MS procedure was much more efficient since derivatization and tedious multiple extraction steps were avoided. Additionally the chromatographic run times of the GC/MS- and LC/MS/MS-based assays were 30 and 2 minutes, respectively. Such improvements in assay efficiency expedite the reporting of analytical data and are much sought after by all levels of management within the pharmaceutical industry.

A deeper level of assay comparison involves not only the examination of assay acceptance criteria but also some degree of statistical analysis of the results generated by both the established method and the new LC/MS procedure. Wang-Iversen *(23)* compared the assay based on negative ion spray LC/MS for SQ-33600 in human serum with an existing procedure using HPLC with ultraviolet detection. Pools of control plasma spiked with the test substance were assayed "blind" by both methods. Good agreement (94-103%) was observed indicating that the assays were equivalent over the concentration range examined, which was high (0.14 to 7.6 μg/ml) to accommodate the much less sensitive LC-UV assay.

The development of a drug from discovery to regulatory approval may span many years. During this period, bioanalytical methods evolve, first in response to the need for greater sensitivity (e.g. preclinical studies in animals to phase I studies in man) and second, with advances in analytical technology. There can, however, be a reluctance to adopt new methods for fear that the results obtained by the new, more efficient (and perhaps more specific) method might be different from those obtained by the established procedure. In such instances a cross-validation of the two methods is advisable in which real test samples rather than spiked controls are examined to highlight any potential interference by metabolites. A method based on negative chemical ionization GC-MS was developed in the authors' laboratories for the determination of the HMG CoA reductase inhibitors lovastatin and simvastatin in plasma *(32)*. This assay measured both the inactive pro-drug lactones along with the pharmacologically active acids resulting from *in vivo* hydrolysis. To improve sample throughput and reliability, the assay for simvastatin was superceded *(21)* by a procedure based on LC/MS/MS. The sensitivity of the latter (0.5 ng/ml) was slightly poorer than that obtained by GC/MS (0.2 ng/ml) but the chromatographic run time was five times shorter and a simpler sample preparation procedure was employed. Since both drugs are extensively metabolized, it was important to definitively cross-validate the assays. Plasma obtained from three individuals (30 samples) were accordingly assayed by both methods. The results, compared by a paired t-test and ratios analysis, showed very good agreement and no significant differences between the methods. The more efficient LC/MS/MS-based assay is now preferred in our laboratories.

If an LC/MS assay is being developed as an alternative to radioimmunoassay (RIA) or as a means to confirm its specificity, the use of clinical test samples for cross validation is mandatory. Several of the LC/MS/MS methods developed in our laboratories were in response to RIAs which were suspected of being non-specific. Comparing assay results from spiked control plasma is pointless in this situation.

Several such comparisons *(26,33,34)* showed obviously (1.5 to 2-fold) greater drug concentrations obtained by RIA than by LC/MS/MS. This clearly indicated that the RIA was not specific because of cross-reactivity by structurally related metabolites. We were fortunate that LC/MS/MS was available as an alternative procedure to accurately measure these drugs in plasma to the required concentrations of *ca.* 0.5 ng/ml.

Prior to the availability of LC/MS/MS, cross-validation of RIAs was rarely possible and reliance had to be made on a series of never-quite-definitive tests to assess specificity. These include parallelism experiments to detect potential cross-reacting metabolites with non-parallel displacement curves, accuracy by the method of standard addition and fractionation of plasma or urine preparations by HPLC and subsequent analysis of the fractions by RIA *(35-37)*. The latter procedure is generally useful but can prove troublesome to execute when dealing with very low concentrations because of manipulative losses during extraction and reconstitution of fractions.

RIA has been our preferred method for the analysis of the class III antiarrhythmic agents L-691,121 *(37)* and MK-499 (Gilbert, J.D. *et al. J. Pharm. Biomed. Anal.*, in press). These very potent drugs are sometimes administered at sub-milligram doses and plasma concentrations range from 50 pg/ml to 10 ng/ml. The RIA for MK-499 had been used to support its development program from preclinical animal studies through phase III human clinical studies, and had been extensively validated using plasma and urine from a variety of species. All the tests conducted indicated that the RIA was specific. However, it was not until an LC/MS/MS-based method was developed with a sensitivity (13 pg/ml) comparable to, and in fact exceeding, that of the immunoassay (50 pg/ml), that a cross-validation could be conducted. Seventy plasma samples obtained from two separate clinical studies were assayed and the analytical results obtained by the two methods were in excellent agreement as determined by using paired t-test, linear regression and assay ratio analysis. This provided definitive evidence of the specificity of the RIA. The more detailed statistical analysis including linear regression is needed to show the identity of the two assays rather than just differences (t-test) between methods (38). The regression analysis is illustrated in Figure 3. Plasma concentrations (expressed as ng/ml) obtained by RIA (the test method) are plotted against data obtained by LC/MS/MS (the reference method). The regression equation and the index of determination (r^2) are shown. A log scale is used only to improve the clarity of the illustration and the external lines are the reproducibility limits (±R) calculated from the combined variances of the individual assays. The availability of LC/MS/MS-based methods which enable cross-validation of immunoassays has greatly increased our confidence in the latter's specificity.

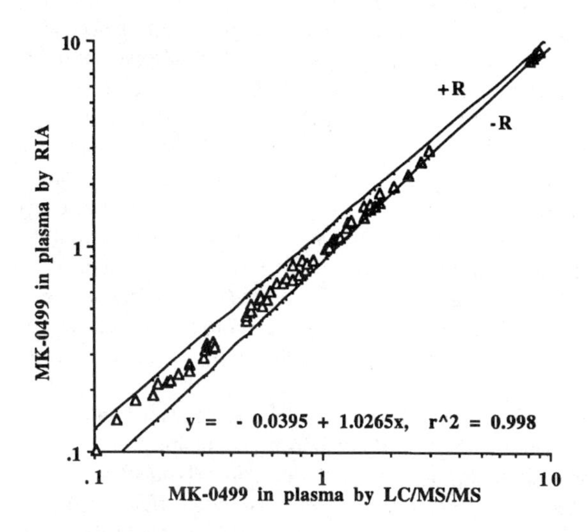

Figure 3. Regression analysis of cross-validation data from assays for the determination of MK-0499 in the plasma of patients by both RIA and LC/MS/MS. The external lines (±R) are the reproducibility limits defined from the combined variances of the individual assays. Adapted from Gilbert *et al. J. Pharm. Biomed. Analysis*, in press with kind permission for Elsevier Science Ltd, The Boulevard, Langford Lane, Kidlington, OX5 1GB, UK.

Isotopes

One of the advantages of mass spectrometric detection is its capacity to resolve isotopes, and for many years GC/MS assays used in conjunction with stable isotopes have found application in bioanalytical and pharmacokinetic research *(39)*. Stable isotope-labeled drugs have been widely regarded as excellent internal standards in quantitative methods. More elegantly, stable isotopes were used in pharmacokinetic studies by enabling simultaneous administration of drugs by different routes and monitoring of the labeled and unlabeled drug to determine absolute bioavailability in single, rather than crossover, experiments *(40-42)*. Stable isotopes have also found use as biological "internal standards" in comparative bioavailability experiments where different solid formulations are administered orally along with a solution of the labeled drug. The comparison of relative rather than absolute pharmacokinetic parameters enhances the statistical power of such studies by minimizing intra-subject variations and significantly reduces the number of participating subjects *(43,44)*. The decline in popularity of GC/MS in preference to HPLC without mass spectrometric detection diminished studies of this type for many years. However, with the recent development of powerful LC/MS technology, there is renewed interest in the application of stable isotopes.

As previously mentioned, the use of LC/MS with thermospray almost necessitated the use of stable isotope-labeled internal standards if adequate assay precision was to be obtained *(14)*. Possibly as a carryover from these thermospray experiences, many assays using API LC/MS have also used stable isotopes, although because of the latter's high total ion current stability, there is no real need to do so. In our laboratories, we are very happy to use stable isotopes as internal standards when they are available *(33,45)* and are of sufficient isotopic purity. At least 99 atoms percent is preferred if analyte concentrations are to be measured over several orders of magnitude. The molecular weight of the internal standard should ideally be at least 3 amu different from that of the analyte. In general, homologs are perfectly satisfactory internal standards and these have been used for the majority of our assays, since they are generally readily available free from any traces of analyte. The normal criteria for selection of internal standards apply. Recovery using the chosen extraction procedure should be similar to that of the analyte (preferably as high as possible if best precision is to be obtained), with a similar response factor and complete resolution from the analyte, either chromatographically or by its unique parent/product ion combination.

Only a few pharmacokinetic uses of stable isotopes in conjunction with LC/MS have so far been reported. Avery *(46)* demonstrated identical pharmacokinetic profiles after intravenous administration of unlabeled and $^{2}H_{3}$-labeled tenidap to rats, and thus the absence of a biological kinetic isotope effect. We have recently simultaneously measured the ophthalmic and oral bioavailability of timolol in dogs, in a single, rather than a 3-way crossover experiment, using stable isotopes. The unlabeled drug was administered as commercial Timoptic eye-drops (dose 0.4 mg) while the oral and i.v. formulations consisted of the $^{13}C_{3}$- and $^{2}H_{9}$-labeled isotopes, respectively (both doses *ca.* 3 mg). LC/MS/MS was used as the detection technique, and the N-isopropyl homolog of timolol served as the internal standard. Timolol was derivatized by conversion to its oxazolidinone (using phosgene), not to enhance sensitivity but to improve chromatography (45). The chromatographic run time and the lower quantifiable limit were 3 minutes and 0.1 ng/ml, respectively. Plasma concentration-time curves of unlabeled timolol given ophthalmically (OP), $^{13}C_{3}$-labeled timolol given orally (PO) and $^{2}H_{9}$-labeled timolol given intravenously (IV) are shown in Figure 4. The ease and simplicity of such experiments may help stimulate interest in the use of stable isotopes in conjunction with LC/MS for pharmacokinetic research.

Because of their high potencies, new drug candidates are sometimes being administered labeled with ^{14}C of very high specific activities. The content of the labeled species can be 10-50% of the dose. If mass spectrometry is used for quantitation, the labeled drug will usually not be detected. For example, during the cross-validation of RIA and LC/MS/MS methods for the antiarrhythmic agent MK-0499 plasma taken from patients dosed with the ^{14}C-labeled drug were assayed by both methods. When the analytical data were compared, we were surprised that the plasma concentrations obtained by LC/MS/MS were, throughout the plasma concentration

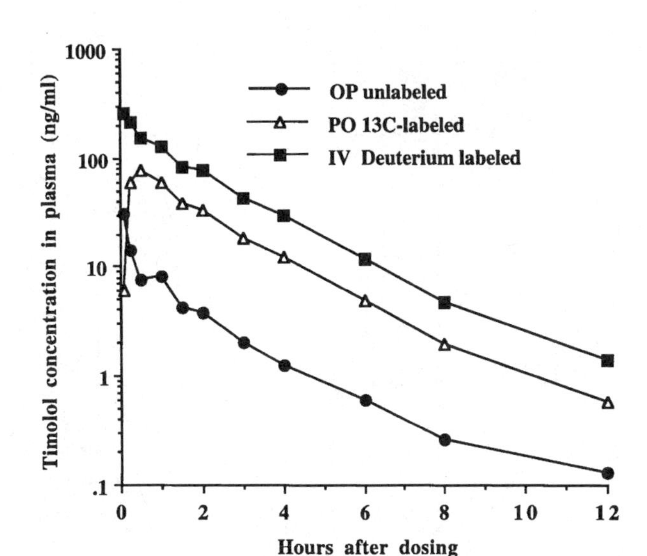

Figure 4. Plasma concentration-time curves of timolol in the plasma of a dog simultaneously dosed ophthalmically with Timoptic solution, orally with $^{13}C_3$-labeled timolol and intravenously with $^2H_9$-labeled timolol.

time profiles, consistently 35% lower than the results obtained by RIA. This was of concern since it suggested that the immunoassay was overestimating the concentrations of the drug in plasma possibly due to cross-reactivity by a metabolite. However, further investigation disclosed that patients had been dosed with ^{14}C-labeled MK-0499 with a specific activity of 47 μCi/mg. Since the maximum specific activity (assuming incorporation of one ^{14}C-atom per molecule) was 131.3 μCi/mg, 35.7% of the molecules of the drug given to the patients (and subsequently measured in plasma) contained the ^{14}C-atom. The labeled molecule, with a pseudomolecular ion at m/z 470, was invisible to the LC/MS/MS assay which was designed to monitor the product ion of the unlabeled parent mass at m/z 468. The product ion mass spectra of unlabeled and ^{14}C-labeled MK-0499 are shown in Figure 5. The unlabeled drug was monitored using the parent/product ion combination m/z 468 → 253 while the ^{14}C-labeled drug was detected using the m/z 470 → 138 combination.

Plasma concentration time curves of labeled and unlabeled MK-0499 after administration of the ^{14}C-labeled drug (700 μg total by intravenous infusion over one hour) are shown in Figure 6. The sums of the concentrations of the two species separately measured by LC/MS/MS are in good agreement with the values obtained by RIA.

We had a similar experience with MK-0462, a $5HT_{1D}$ agonist under development for the treatment of migraine. Analytical support for this substance was by LC/MS/MS, and when a radiolabeled study was conducted in volunteers, the assay was modified to measure both the unlabeled and ^{14}C-labeled species. The plasma concentration-time curves of the labeled and unlabeled drug are compared to the total radioactivity profile in Figure 7.

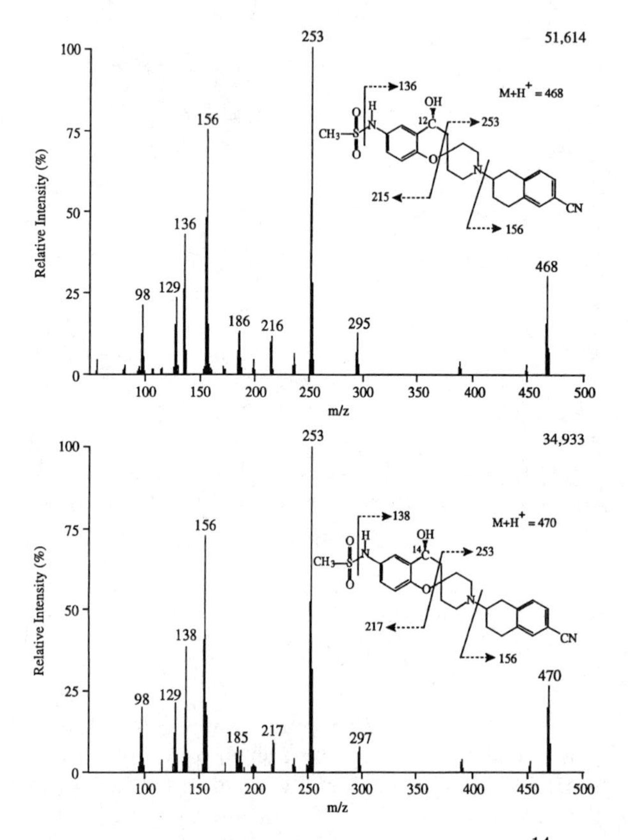

Figure 5. Product-ion mass spectra of unlabeled (upper) and ^{14}C-labeled (lower) MK-0499. Adapted from Gilbert *et al. J. Pharm. Biomed. Analysis*, in press with kind permission for Elsevier Science Ltd, The Boulevard, Langford Lane, Kidlington, OX5 1GB, UK.

Of course, if one assumes that the specific activity of the drug remains constant after administration, i.e. there is no biological kinetic isotope effect, such heroics are not strictly necessary but they serve as elegant illustrations of the power of LC/MS/MS as a tool for the quantitation of labeled drugs. In such circumstances LC/MS/MS is much superior to LC/MS alone, since the use of a parent/product mass combination can frequently select a product ion for the unmodified substance excluding the labeled atom while a complementary product ion carrying the label is used for monitoring the labeled drug. This is demonstrated in the product ion mass spectrum of ^{14}C-labeled MK-0499 in Figure 5. The amount of signal generated by the mass combination m/z 468 → 253 used for the unlabeled drug on two potentially useful ion combinations for the labeled drug was investigated. The cross reactivity on the m/z 470 → 253 channel was 6.1%; i.e. 10 ng of unlabeled drug produced an apparent quantity of 0.61 ng of ^{14}C-labeled drug. When the alternative m/z 470 → 138

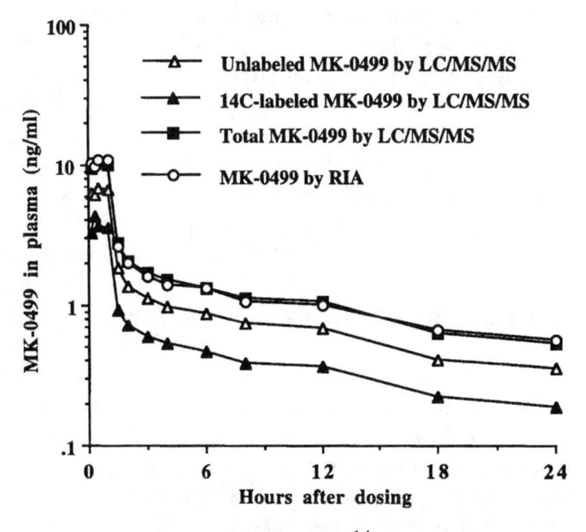

Figure 6. Plasma concentration-time curves for ^{14}C-labeled and unlabeled MK-0499, determined by both RIA and LC/MS/MS, after intravenous administration to a volunteer.

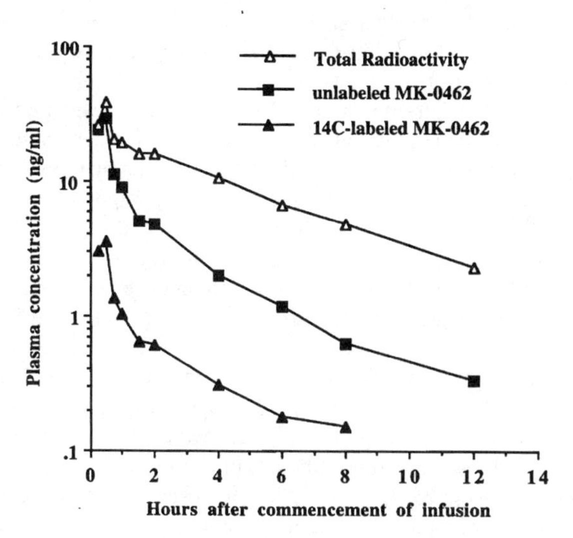

Figure 7. Mean plasma concentration-time curves of unlabeled and ^{14}C-labeled MK-0462 administered intravenously to a volunteer. The total radioactivity profile is included for comparison.

combination was used the cross reactivity was 1.0%. The lower interference with this combination of ions is predictable from the drug's fragmentation pattern. Generally only one carbon-14 atom is incorporated into the test drug so the difference in mass is only 2 daltons and there is a high risk of cross-reactivity. Under these circumstances, it is important that this interference is carefully monitored, since a cross-reactivity of, say, 3% would result in an overestimation of 27% for the labeled drug present at only 10% of the dose.

As more potent drugs are developed, the specific activities of ^{14}C-labeled species administered to animals and man will more commonly exceed what previously could be regarded as just traces of radioactivity. Caution needs to be exercised if the principal analytical method for such substances is based on mass spectrometry.

Simultaneous Determination of Drugs and Metabolites

The specificity of LC/MS/MS also enables the simultaneous quantitation of drugs and their metabolites. Covey, Lee and Henion *(16)* first demonstrated the great power of APCI LC/MS/MS with the development of analytical methods for the determination of illegal drugs and their metabolites in horse urine. The detection of the phenothiazine tranquilizers, promazine and propiopromazine, in biological extracts is particularly difficult due to their extensive metabolism. However, by utilizing MS/MS, these researchers rapidly screened equine urine extracts for metabolites of the parent drugs based on common structural features derived from the product ion spectra. When applied to biological extracts, LC/MS/MS detection can quantitate co-eluting metabolites based upon their unique parent/product ion combinations. Pharmacokinetic profiles of the analgesic, phenylbutazone and its two major metabolites were also presented, illustrating the fast sample throughput that is possible with this mode of detection.

The assay for the determination of the 5-alpha-reductase inhibitor, MK-0434, and its two principal metabolites in plasma provides another example *(34)*. The metabolites are a pair of diastereomeric alcohols with identical product spectra and, whose molecular weights differ from that of the parent compound by only two atomic mass units. An LC/MS/MS-based method was developed which could measure all three substances in plasma at concentrations in the range 0.5 - 50 ng/ml. MRM chromatograms of extracts of plasma from a volunteer receiving MK-0434 (25 mg p.o.) are shown in Figure 8. Retention times are shown in minutes. The parent/product ion combinations used to monitor the species of interest are m/z 378 → 310, MK-0434; m/z 358 → 290, L-654,066 (internal standard); and m/z 380 → 312, the metabolites L-694,579 and L-691,919. The upper chromatograms are of plasma collected prior to drug administration while the lower chromatograms are of plasma collected 6 hours postdose. The concentrations of MK-0434, L-694,579 and L-691,919 in the plasma sample were 25, 3.7 and 15.8 ng/ml, respectively. Due to their identical mass spectra, chromatographic resolution of the metabolites was required. Even so, the analysis time was less than four minutes.

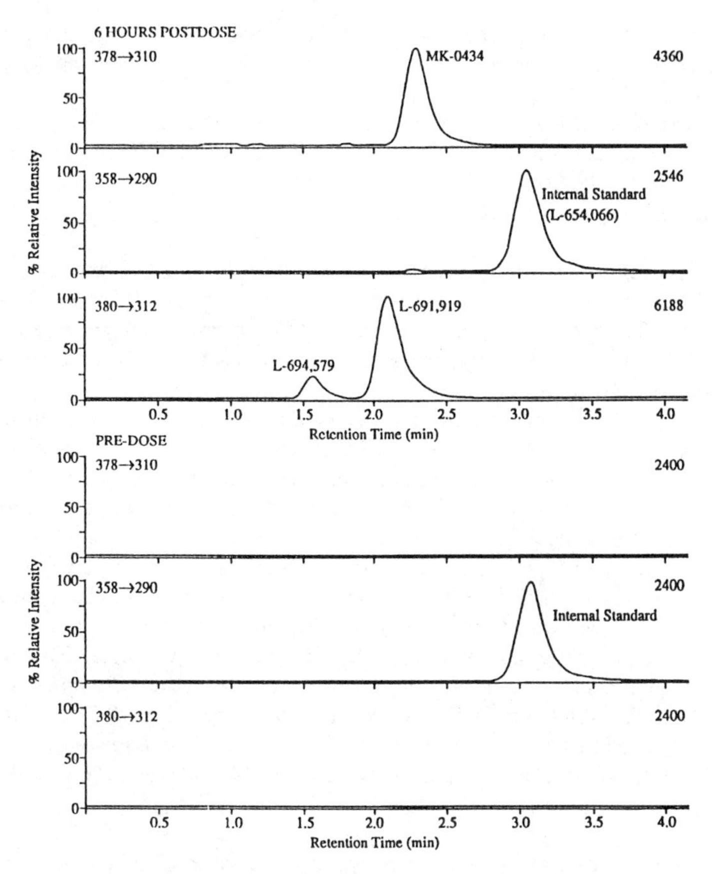

Figure 8. MRM chromatograms of extracts of plasma from a volunteer receiving MK-0434. This LC/MS/MS assay separately quantitated the parent drug and its diasteromeric metabolites. L-654,066 was used as the internal standard. Adapted from reference 34 with kind permission for Elsevier Science Ltd, The Boulevard, Langford Lane, Kidlington, OX5 1GB, UK.

Despite the intrinsic specificity of the tandem mass spectrometer, careful assessment of assay specificity in post-dose biological fluids is critical when short chromatographic run times are employed. Matuszewski (Merck Research Laboratories, personal communication) observed that although an assay with a two-minute analysis run-time for the oxytocin antagonist L-368,899 in spiked control urine appeared satisfactory in terms of accuracy and precision, analysis of clinical samples was not possible. Several oxygenated metabolites [M+16] which were present in substantial quantities in post-dose urine samples interfered with the detection of the internal standard, which was a methylated analog [M+14] of the parent drug. The assay was satisfactorily revalidated using improved chromatographic conditions which resolved the metabolites from both the analyte and its internal standard. Adequate chromatographic resolution is important. Although the intrinsic specificity of MS/MS might suggest that assays can be performed without chromatographic separation, even in the absence of potentially interfering metabolites analysis of biological extracts with little or no chromatographic resolution frequently results in signal suppression and inferior precision. Good chromatographic practices will almost always ensure quality results.

Assays have also been published for the simultaneous determination of plasma levels of the angiotensin converting enzyme inhibitor, fosinopril (24) and the HMG CoA reductase inhibitors, simvastatin *(21)* and dalvastatin *(47)*, along with their pharmacologically-active metabolites.

Automation of Sample Preparation. An autosampler, used in conjunction with the LC/MS/MS instrument, allows batches of several hundred samples to be run unattended. The remarkable reliability of the APCI mass spectrometer is such that the rate-limiting step is no longer the analysis time but rather the sample preparation and data handling steps.

Simple solid phase extraction (SPE) methods are generally developed in our laboratories. Typically SPE cartridges are activated and the sample is loaded, washed and eluted. The eluents are evaporated, and the residue is reconstituted in mobile phase and injected. In order to liberate the analyst from such routine and repetitive work we have used the Gilson ASPEC XL automated solid-phase sample extraction system, in our daily operations. Due to the simplicity of solid phase extraction methods, transfer from manual to automated methodology generally requires few modifications. The sample preparation time is somewhat longer than ideal, with a typical method requiring approximately 10 to 15 minutes per sample. However, the ASPEC is extremely reliable and runs unattended overnight. Samples are then evaporated, reconstituted in mobile phase and assayed by LC/MS/MS the following day. Having used this instrument for the analysis of several thousand plasma samples over 18 months we conclude that it is robust, reproducible and requires minimal routine maintenance.

Automation of Data Processing. The management of data can become overwhelming when such an efficient analytical technique as LC/MS/MS is used. A data management system is almost a necessity when the analysis of large numbers of biological samples are performed routinely.

We were reluctant to support the development and implementation of an in-house data management program, and after evaluating a variety of LIMS packages, we purchased Drug Metabolism Laboratory Information Management System Plus (DMLIMS+, PennComp, Wayne, PA). DMLIMS+ is compatible with a variety of data collection systems, and allows the user to pre-establish a database based on study design information. Analytical data are then summarized in terms of dose group, demographics or subject number. Results are automatically entered into the database as they are obtained. The user may, at any time, generate summaries of data including plasma concentration-time profiles, tables, pharmacokinetic parameters and assay performance reports. Interim comparisons between dose groups or sexes are readily obtained. Data which are required for bioanalytical reports can be generated with a few key strokes and readily inserted into a variety of word processing documents. Additionally DMLIMS+ greatly facilitates the use of Sciex's RAD (routine acquisition and display) software. Incorporation of automated sample preparation and information management into our assays has enabled us to increase our monthly sample through-put by a factor of two. Nonetheless, it is the reliability of the mass spectrometer, such as the Sciex API III, which enables us to take advantage of automated sample preparation and information management systems to enhance our capabilities, not only in developing and validating new methods, but also for rapidly processing samples and reporting analytical data of high quality in a consistently timely fashion.

Support of Basic Research

The need for quantitative bioanalysis transcends many of the activities of pharmaceutical research. The examples we have cited are taken largely from published methods for drugs which became marketed products or were developed to a significant extent before being abandoned.

Assays in support of drug discovery are rarely published but are of profound importance in the selection of drug candidates. Having established that a series of newly synthesized substances show the required pharmacological activity *in vitro* the research chemist next needs to know whether the substances are systemically absorbed, since if they are not, they are unlikely to be effective *in vivo*. Because these substances are potent, assays with reasonably high sensitivities (1 - 10 ng/ml) are required. These assays are only initially needed as screens, but it is important that some reasonable degree of validation is conducted, since erroneous data from an unproven or unreliable assay could seriously misdirect the research effort. The time taken to develop sensitive assays has frustrated our colleagues in basic research for many years. However, when LC/MS/MS is used, sensitive and reliable assays can frequently be established and validated, at least in terms of accuracy and precision, in a few days, and results can be reported within a day or two of sample receipt.

In a recent example from our laboratories, a team of two analysts developed and validated assays for eight potential β_3-agonists in plasma and supported sixteen bioavailability studies in rats and dogs all in the space of 2 months. The assays had lower quantifiable limits ranging from 1 to 5 ng/ml with CVs at the LQLs of 4-15%. The information provided was extremely important in helping to focus the synthetic efforts of the drug discovery group.

Support of drug discovery is an increasingly important part of our bioanalytical activities. Frequently, the sensitivities of the assays developed are not particularly challenging and could be achieved, albeit more slowly, by HPLC with conventional detectors. However, it is difficult to envisage a more important application of our new LC/MS/MS technology than rapidly providing reliable bioanalytical data to assist in the selection of the new drug candidates which are the life-blood of our industry.

Good Laboratory Practice

Much of the bioanalytical work undertaken by pharmaceutical companies is conducted in compliance with the FDA's Good Laboratory Practice (GLP) regulations *(48)*. The use of LC/MS/MS for drug quantitation in biological fluids has improved our GLP compliance in two important respects. First, the sensitivity and lack of interference from endogenous substances enable the establishment of conservative but consistently attainable lower quantifiable limits. Second, the speed and reliability of such assays indirectly improves the quality of data, since the addition of extra standards and quality controls provides no appreciable burden to the analyst. These factors minimize the need for repeat analysis and improve our efficiency. A useful review of mass spectrometry, both quantitative and qualitative, used in a GLP environment is in press (Boyd, R.K., *J. Am. Soc. Mass Spectrom.*, in press).

Conclusions

GC/MS has over the years made a considerable contribution to quantitative bioanalytical science. The development of API mass spectrometers harnessing the power of reverse phase HPLC heralds a quantum leap in bioanalytical technology which was difficult to envisage only a few years ago.

Analytical methods can now be established for analytes at low to sub ng/ml concentrations in a few days. Hundreds of biological extracts can be assayed per day, unattended. A dirty ion source has no significant impact on assay performance and the mass spectrometer is so reliable and easy to use that multiple operators can usefully share the equipment with minimum down-time (49). The best operators are generally analysts experienced in chromatography who have been properly trained in the fundamentals of mass spectrometry, such as mass calibration and resolution.

API LC/MS/MS is still in its infancy as a quantitative bioanalytical tool, but already it has had a profound impact on the speed with which drugs can be discovered and developed. From our perspective, the technology is remarkable (though long overdue) and, in the hands of the dedicated analysts who have so enthusiastically adopted it, is revolutionizing the way we conduct our bioanalytical business. Our next principal challenge will likely be devising procedures for the pharmacokinetic screening of the vast numbers of potential drug candidates which are being synthesized using combinatorial methods. The magnitude of this task is just becoming apparent, how it will be accomplished is unclear, but it is certain that LC/MS/MS will play an invaluable role.

Acknowledgments

The authors gratefully acknowledgment the assistance of Mrs. Maureen Hetzel and Mr. Jeff Campbell in preparing this manuscript.

Literature Cited

1. Girault, J.; Gobin, P. and Fourtillan, J.B. *Biomed. Environ. Mass Spectrom.* **1990**, *19,* 80-89.
2. Girault, J.; Longueville, D.; Ntzanis, L.; Coufin, S. and Fourtillan, J.B. *Biol. Mass Spectrom.* **1994**, *23*, 572-580.
3. Strife, R.J. and Simms, J.R. *J. Am. Soc. Mass Spectrom.* **1992**, *3*, 372-377.
4. Schweer, H.; Meese, C.O. and Seyberth, H.W. *Anal. Biochem.* **1990**, *189*, 54-58.
5. McFadden, W.H.; Schwartz, H.L. and Evans, S. *J. Chromatogr.* **1976**, *122*, 389-396.
6. Baldwin, M.A. and McLafferty, F.W. *Org. Mass Spectrom.* **1973**, *7*, 1111-1112.
7. Arpino, P.J.; Dawkins, B.G. and McLafferty, F.W. *J. Chromatogr. Sci.* **1974** *12*, 574-578.

8. Arpino, P.J.; Krien, P.; Vajta, S. and Devant, G. *J. Chromatogr.* **1981**, *203*, 117-130.
9. Mauchamp, B. and Krien, P. *J. Chromatogr.* **1982**, *236*, 17-24.
10. Blakely, C.R.; Carmody, J.J. and Vestal, M.L. *Anal. Chem.* **1980**, *52*, 1636-1641.
11. Lindberg, C.; Blomquist, A. and Paulson, J. *Biol. Mass Spectrom.* **1992**, *21*, 525-533.
12. Buggé, C.J.L.; Tucker, M.D.; Garcia, D.B., Kvalo, L.T. and Wilhelm, J.A. *J. Pharm. Biomed. Anal.* **1993**, *11*, 809-815.
13. Malcolm, S.L.; Madigan, M.J. and Taylor, N.L. *J. Pharm. Biomed. Anal.* **1990**, *8*, 771-776.
14. Oxford, J.; and Lant, M.S. *J. Chromatogr.* **1989**, *496*, 137-146.
15. Gilbert, M.T. In *High Performance Liquid Chromatography*; Wright, Bristol, 1987, 52.
16. Covey, T.R.; Lee, E.D. and Henion, J.D. *Anal. Chem.* **1986**, *58*, 2453-2460.
17. Bruins, A.P.; Covey, T.R. and Henion, J.D. *Anal. Chem.* **1987**, *59*, 2642-2646.
18. Huang, E.C.; Wachs, T.; Conboy, J.J. and Henion, J.D. *Anal. Chem.* **1990**, *62*, 713A-725A.
19. Wachs, T.; Conboy, J.C.; Garcia, F. and Henion, J.D. *J. Chromatogr. Sci.* **1991**, *29*, 357-366.
20. Bruins, A.P. *Mass Spectrom. Reviews* **1991**, *10*, 53-77.
21. Gilbert, J.D.; Olah, T.V.; Morris, M.J.; Schwartz, M.S. and McLoughlin, D.A. In *Methodological Surveys in Bioanalysis of Drugs (Reid, E.; Hill, H.M. and Wilson, I.D. Eds.)*, Royal Society of Chemistry: Cambridge, UK, 1994, Vol. 23, pp 157-167.
22. Fouda, H.; Nocerini, M.; Schneider, R. and Gedutis, C. *J. Am. Soc. Mass Spectrom.* **1991**, *2*, 164-167.
23. Wang-Iverson, D.; Arnold, M.E.; Jemal, M. and Cohen, A.I. *Biol. Mass Spectrom.* **1992**, *21*, 189-194.
24. Wang-Iverson, D. and Cohen, A.I. In *Methodological Surveys in Bioanalysis of Drugs* (Reid, E.; Hill, H.M. and Wilson, I.D. Eds.), Royal Society of Chemistry: Cambridge, UK, 1994, Vol. 23, pp 177-190.
25. Weidolf, L.O.G.; Lee, E.D. and Henion, J.D. *Biomed. Environ. Mass Spectrom.* **1988**, *15*, 283-289.
26. Gilbert, J.D.; Olah, T.V.; Barrish, A. and Greber, T.F. *Biol. Mass Spectrom.* **1992**, *21*, 341-346.
27. Kaye, B.; Clark, M.W.H.; Cussans, N.J.; Macrae, P.V. and Stopher, D.A. *Biol. Mass Spectrom.* **1992**, *21*, 585-589.
28. Goodley, P.C.; Gartiez, D. and McManus, K.T. Hewlett-Packard MS Application Note, 93-5, July 1993.
29. Constanzer, M.; Chavaz, C. and Matuszewski, B. *J. Chromatogr.* **1995**, *666*, 117-126.
30. Constanzer, M.L.; Chavaz, C.M. and Matuszewski, B.K. *J. Chromatogr.* **1994**, *658*, 281-287.
31. Fraser, I.J.; Clare, R.A. and Pleasance, S. In *Methodological Surveys in Bioanalysis of Drugs* (Reid, E.; Hill, H.M. and Wilson, I.D. Eds.), Royal Society of Chemistry: Cambridge, UK, 1994, Vol. 23, pp 113-120.

32. Morris, M.J.; Gilbert, J.D.; Hsieh, J.Y.-K.; Matuszewski, B.K.; Ramjit, H.G. and Bayne, W.F. *Biol. Mass Spectrum.* **1993**, *22*, 1-8.
33. Gilbert, J.D.; Hand, E.L.; Yuan, A.S.; Olah, T.V. and Covey, T.R. *Biol. Mass Spectrom.* **1992**, *21*, 63-68.
34. Olah, T.V.; Gilbert, J.D.; Barrish, A.; Greber, T.F. and McLoughlin, D.A. *J. Pharm. Biomed. Anal.* **1994**, *12*, 705-712.
35. Yuan, A.S.; Hand, E.L.; Olah, T.V.; Barrish, A.; Fernández-Metzler, C. and Gilbert, J.D. *J. Pharm. Biomed. Anal.* **1993**, *11*, 427-434.
36. Hand, E.L.; Gilbert, J.D.; Yuan, A.S., Olah, T.V. and Hichens, M. *J. Pharm. Biomed. Anal.* **1994**, *12*, 1047-1053.
37. Greber, T.F.; Olah, T.V.; Gilbert, J.D.; Porras, A.G. and Hichens, M. *J. Pharm. Biomed. Anal.* **1994**, *12*, 483-492.
38. Gilbert, M.T.; Barinov-Colligon, I. and Miksic, J.R. *J. Pharm. Biomed. Anal.* **1995**, *13*, 385-394.
39. Wolen, R.L. and Garland, W.A. In *Synthesis and Applications of Isotopically Labeled Compounds*; Baillie, T.A. and Jones, J.R., Eds., Elsevier: Amsterdam, The Netherlands, 1988, 147-156.
40. Mikus, G.; Fischer, C.; Heuer, B.; Langen, C. and Eichelbaum, M. *Br. J. Clin. Pharmacol.* **1987**, *24*, 561-569.
41. Kasuya, Y.; Mamada, K.; Baba, S. and Matsukura, M. *J. Pharm. Sci.*, **1985**, *74*, 502-507.
42. Vandenheuvel, W.J.A.; Carlin, J.R. and Walker, R.W. *J. Chromatog. Sci.* **1983**, *21*, 119-124.
43. Heck, H.A.; Buttrill, S.E.; Flynn, N.W.; Dyer, R.L.; Anbar, M.; Cairns, T.; Dighe, S. and Cabana, B.E. *J. Pharmacokinet. Biopharm.* **1979**, *7*, 233-248.
44. Gammans, R.E.; MacKenthun, A.V. and Russell, J.W. *Br. J. Clin. Pharmacol.* **1984**, *18*, 431-437.
45. Olah, T.V.; Gilbert, J.D. and Barrish, A. *J. Pharm. Biomed. Anal.* **1993**, *11*, 157-163.
46. Avery, M.J.; Mitchell, D.Y.; Falkner, F.A. and Fouda, H.G. *Biol. Mass Spectrom.* **1992**, *21*, 353-357.
47. Hsu, S.-H.; Schlater, T. and Rich, L. In *Methodological Surveys in Bioanalysis of Drugs* (Reid, E.; Hill, H.M. and Wilson, I.D. Eds.), Royal Society of Chemistry: Cambridge, UK, 1994, Vol. 23, pp 169-176.
48. *Code of Federal Regulations, Title 21 "Food and Drugs",* **1 April 1994**, Part 58: *Good Laboratory Practice for Nonclinical Laboratory Studies.* Published by the Office of the Federal Register, National Archives and Records Administration, Washington, DC.
49. Busch, K.L. *Spectroscopy*, **1995**, *10*, 24-25.

RECEIVED July 21, 1995

Chapter 18

Analysis of Diarrhetic Shellfish Poisoning Toxins and Metabolites in Plankton and Shellfish by Ion-Spray Liquid Chromatography–Mass Spectrometry

M. A. Quilliam and N. W. Ross

Institute for Marine Biosciences, National Research Council of Canada, 1411 Oxford Street, Halifax, Nova Scotia B3H 3Z1, Canada

The analysis of diarrhetic shellfish poisoning toxins at the ng/g level in shellfish and plankton was achieved using ionspray liquid chromatography-mass spectrometry (LC-MS). Complex mixtures of naturally-occurring derivatives of okadaic acid and dinophysistoxins-1 and -2 were also detected and identified in mussel digestive glands and in *Prorocentrum lima* cultures. LC-MS facilitated the discovery of enzyme-catalyzed hydrolysis and transesterification reactions of some of these derivatives during extraction. Two improved extraction procedures were then developed: one that inactivates the enzyme before extraction, and another that takes advantage of the enzymatic action to convert all ester derivatives to parent toxins for quantitative assessment of toxicity.

Diarrhetic shellfish poisoning (DSP) is a severe gastrointestinal illness caused by consumption of shellfish contaminated with toxigenic dinoflagellates such as certain *Dinophysis* and *Prorocentrum* species (*1-3*). It has been recognized as a significant threat to public health and the economy in Japan and Europe since its discovery in the late 1970s. Recently, incidents of DSP have been confirmed in both North America (*4*) and South America (*5*). The main toxins responsible for DSP are okadaic acid (OA) (*6*) and the dinophysistoxins, DTX1 (*1*) and DTX2 (*7*) (see Table I). These toxins have been shown to be potent phosphatase inhibitors (*8*), a property which can cause inflammation of the intestinal tract and diarrhea (*9*). OA and DTX1 have also been shown to have tumour promoting activity (*10*).

A number of naturally-occurring derivatives of OA and the DTX toxins have been observed. The term DTX3 was first used to describe a group of compounds in which saturated or unsaturated fatty acyl groups are attached through the 7-OH group of DTX1 (see Table I) (*1*). These acyl derivatives also possess toxic activity (*11*).

0097–6156/95/0619–0351$12.00/0

Since DTX3 has only been observed in shellfish tissue, it has been suggested that they are probably metabolic products (*12*) and not *de novo* products of toxin-producing microalgae. More recently, using liquid chromatography-mass spectrometry (LC-MS) analysis to directly examine chromatographic fractions obtained from digestive gland extracts of toxic mussels, it has been shown that any of the parent toxins, OA, DTX1 or DTX2, can be acylated with a range of saturated and unsaturated fatty acids from C_{14} to C_{18} (*13*).

Table I. Structures of okadaic acid and its known naturally-occurring derivatives

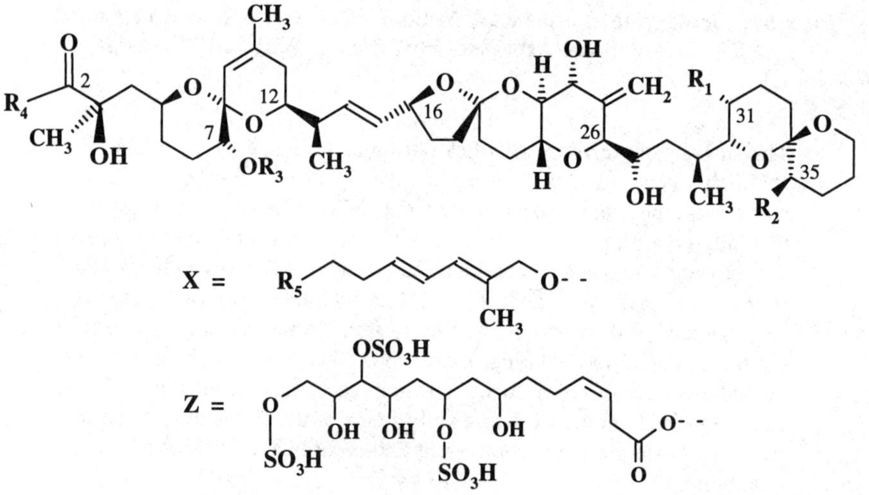

	R_1	R_2	R_3	R_4	R_5	*Mol. Wt.*	
1	CH_3	H	H	OH	—	804.5	Okadaic acid (OA)
2	CH_3	CH_3	H	OH	—	818.5	DTX1
3	H	CH_3	H	OH	—	804.5	DTX2
4	(H or CH_3)		Acyl	OH	—	1014-1082	DTX3
5	CH_3	H	H	X	OH	928.5	OA diol ester
6	CH_3	H	H	X	Z	1472.6	DTX4

Two naturally occurring ester derivatives of OA, named diol esters, were originally isolated from a strain of *P. lima* (*12*). Later, new diol esters were isolated from a Caribbean strain of *P. maculosum* (previously *P. concavum*), and one of these (compound 5, Table I) was also found in an eastern Canadian strain of *P. lima* (*14*).

Although such diol esters are not phosphatase inhibitors, they have the potential to be readily hydrolyzed in the digestive tract to yield the active parent DSP toxin. Recently, a water-soluble DSP toxin has been isolated from the eastern Canadian strain of *P. lima* and named DTX4 (*15*) (see Table I). This compound is an even more complicated derivative of OA, in which the primary hydroxyl of diol ester 5 is esterified with a trisulfated end group.

Mass spectrometry is a valuable tool for the detection and identification of marine toxins. Besides providing high sensitivity and selectivity, mass spectrometry yields useful structural information. Fast atom bombardment ionization has proven valuable for generating mass spectra of DSP toxins (*16,17*). Electron ionization and chemical ionization mass spectra of OA, and its synthetic methyl, pentafluorobenzyl and trimethylsilyl ester and ether derivatives, have also been shown to be useful (*17*). However, the most sensitive method reported to date for DSP toxins is ionspray mass spectrometry (*18*), which is also very well-suited to their analysis by combined LC-MS (*13,16,19-21*). This paper will report on the application of ionspray LC-MS to the analysis of DSP toxins in shellfish and plankton. A number of new DSP toxin derivatives have been detected, and methods for their extraction and analysis are described.

Methods

Materials. Batch cultures of *Prorocentrum lima* (200 mL; 70 days old; *ca.* 7 x 10^4 cells/mL) were kindly provided by Dr. J. L. McLachlan. This particular strain was recovered from Mahone Bay, Nova Scotia and has been shown to produce both OA and DTX1 (*22,23*).

Extraction of shellfish tissues. A portion of homogenized tissue sample (2.0 g) was weighed accurately into a 50-mL plastic centrifuge tube. Aqueous 80% methanol (8.0 mL) was added and the mixture was homogenized using a Polytron mixer for 3 min at 6K-10K rpm. The sample was then centrifuged at 4000xg or higher for 10 min. An aliquot (5.0 mL) of the supernatant was transferred to a 15-mL glass centrifuge tube and extracted twice with 5-mL aliquots of n-hexane by vortex mixing for 0.5 min. After discarding the hexane layers, 1 mL of water and 6 mL chloroform were added to the tube, and the mixture was vortex mixed for 0.5 min. The lower chloroform layer was transferred to a 50-mL glass concentrator tube. The extraction of the aqueous layer was repeated with another 6 mL of chloroform. The combined chloroform layers were evaporated to dryness under a stream of nitrogen. The residues were then dissolved in 0.5 mL methanol by vortex mixing and transferred to an LC autosampler vial. This extract contained 2 g tissue equivalent per mL of solution. An overall extraction efficiency of better than 95% has been demonstrated (*21*). Additional cleanup and pre-concentration can be performed by solid phase extraction (*21*).

Extraction of plankton. Several 30 mL aliquots of a well-mixed batch culture of *P. lima* were transferred to 50-mL plastic centrifuge tubes and centrifuged for 10 min at 6600xg. The supernatants were decanted carefully to avoid disturbing the cell pellets.

The cell pellets were extracted using several different methods, the principal ones being as follows. *Method 1 (80% methanol):* Each cell pellet was resuspended in 2 mL of methanol/water (8:2) and sonicated for 1 min in pulse mode (50% duty cycle, 375 watts; model VC375 Vibracell ultrasonic processor with tapered microtip; Sonics & Materials Inc., Danbury, CT) while cooling in an ice bath. After centrifugation for 10 min at 6600xg, the supernatant was decanted. The pellet was twice rinsed (vortex mixing, centrifugation) with 1 mL methanol/water (8:2). Supernatants were combined and made to 5.0 mL. Extracts were passed through a 0.45 μm filter prior to analysis. *Method 2 (French press):* Four cell pellets were each resuspended in 0.2 mL 50 mM TrisHCl pH 7.4, combined and passed through a French press (chilled in a cold room) at pressures >10 Kpsi. A 1-mL aliquot of buffer was used to wash remaining residues through the press and the sample was then brought to 2.0 mL with buffer. Aliquots (0.5 mL) were mixed with 2.0 mL methanol, vortex mixed and centrifuged. Pellets were twice rinsed (vortex mixing, centrifugation) with 1 mL methanol/water (8:2). Supernatants were combined and made to 5.0 mL. *Method 3 (freeze/thaw):* Each cell pellet was resuspended in 0.5 mL of TrisHCl buffer and immersed in liquid nitrogen. The sample was then allowed to thaw at room temperature and left in the dark for 24 hr. Then 2 mL of methanol was added, and the sample was sonicated and extracted as in method 1. *Method 4 (boiling):* Each cell pellet was resuspended in 0.5 mL of TrisHCl buffer and immersed in boiling water for 3 min. After this, 2 mL of methanol was added followed by sonication and extraction as in method 1.

LC-MS analysis. Experiments were performed using a model HP1090M LC system equipped with a ternary DR5 pumping system and variable volume injector (Hewlett-Packard, Palo Alto, CA) coupled with an API-III+ triple-quadrupole mass spectrometer (Perkin-Elmer/Sciex, Thornhill, Ontario, Canada) through an ionspray LC-MS interface. High purity air was used as nebulizing gas (approximately 0.5 L/min) and a potential of 5000 volts was applied to the interface needle. Effluent from the LC column was split with a low dead volume coaxial splitter to provide a 30 μL/min flow to the MS. Shellfish and plankton extracts were analyzed for OA, DTX1 and DTX2 using the positive ion mode, a 2.1 mm i.d. x 25 cm column packed with 5 μm Vydac 201TP octadecylsilica (Separations Group, Hesperia, CA) at ambient temperature, isocratic elution with a mobile phase of aqueous 70-80% methanol containing 0.1% trifluoroacetic acid (TFA), a flow rate of 0.2-0.3 mL/min, and an injection volume of 5 μL. DTX3 toxins in the same extracts were analyzed using a 2.1 mm i.d. x 25 cm column packed with 5 μm Vydac 214TP C4-silica and 0.2 mL/min aqueous 80% methanol containing 0.1% TFA. Plankton extracts were analyzed for DTX4 toxins using the negative ion mode, a 2 x 150 mm column packed with 5 μm Zorbax Rx-C8 (Chromatographic Specialties, Brockville, Ont.), 0.2 mL/min flow rate, and gradient elution with an aqueous acetonitrile-ammonium acetate (1 mM, pH 7) mobile phase programmed from 20% to 50% acetonitrile over 15 min.

Results and Discussion

Ionspray Mass Spectra. It has been shown previously (*16*) that OA can be ionized using ionspray. The mass spectrum (see Figure 1a) exhibits an abundant $[M+H]^+$ ion at m/z 805.5 and occasionally shows $[M+Na]^+$ and $[M+NH_4]^+$ ions in the flow injection mode or during LC-MS if the mobile phase is contaminated with salts. The best sensitivity is achieved with an aqueous methanol (70-80%) mobile phase acidified with 0.1% TFA (*21*), and under such conditions, about 10 ng is required to produce a recognizable spectrum. This sensitivity is at first surprising because DSP toxins are not basic compounds, but it should be noted that OA and other polyether compounds are known to be quite ionophoric in nature.

Although ionspray generally produces a spectrum showing almost no fragment ions, it is possible to induce fragmentation in the high pressure expansion region just after the sampling orifice. The Sciex technology uses collision-induced dissociation (CID) in this region to remove solvent molecules clustered to analyte ions. By further increasing the potential difference between the sampling orifice and the RF-only quadrupole which funnels ions from the interface region to the first mass filter, the collision energy may be increased to the point where CID of ions is induced. The spectrum shown in Figure 1a, acquired with a moderate potential difference of 25 V, shows fragment ions due to consecutive losses of water molecules. At higher potentials (70-100 V), extensive fragmentation can be induced while still maintaining high sensitivity (*21*). This fine control over fragmentation is a useful feature for deriving structural information and for generating multiple ions for selected ion monitoring. The only disadvantage over a tandem mass spectrometry (MS/MS) is that precursor ions (e.g., $[M+H]^+$) are not mass selected prior to CID. Therefore, complete chromatographic separation of analytes must be achieved to acquire pure spectra.

The positive ionspray spectrum of the OA diol ester 5 (Table I) is shown in Figure 1c. At a moderate orifice potential (25 V), it exhibits extensive fragmentation by elimination of $C_8H_{12}O$ to give an ion at m/z 805.5, the same as the $[M+H]^+$ ion for OA (Figure 1a). This is very useful for structure elucidation. At low orifice potentials, this fragmentation can be greatly reduced.

The acidic solvent used for positive ionspray is well-suited to the chromatographic separation of DSP toxins, since the TFA suppresses ionization of the carboxyl function. A Vydac 201TP stationary phase provides excellent peak shape and separation selectivity (see below for examples). The detection limit for LC-MS of OA in the selected ion monitoring mode is 40 ng/mL in a sample extract (*21*). With a 1 mm i.d. column, 50 µL/min flow and a 1 µL injection volume, a mass detection limit of 40 pg has been achieved. However, we usually perform LC-MS with 2 mm i.d. columns using a 5 µL injection volume and a flow rate of 200 µL/min with a post-column split that directs 20-30 µL/min to the MS. We prefer such columns because they are more readily available and usually give higher column efficiency. Although the ionspray interface can tolerate a 200 µL/min flow rate, the best flow for optimum signal-to-noise ratio is 20-50 µL/min. Since ionspray MS behaves as a concentration detector (*24,25*), there is no sacrifice in concentration detection limit. Some sample is wasted via the split but this is rarely a problem for the analysis of shellfish and plankton samples.

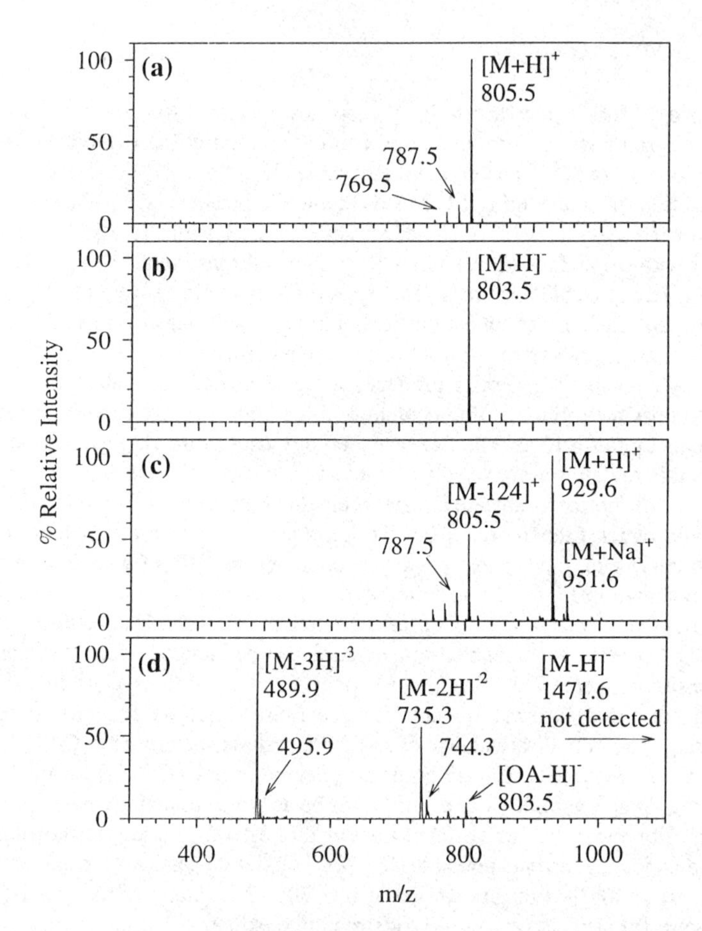

Figure 1. Ionspray mass spectra of OA (a,b), OA diol ester (c) and DTX4 (d) in the positive (a,c) and negative ion modes (b,d).

The acidic DSP toxins can also be ionized in the negative ionspray mode. Excellent sensitivity is possible using an aqueous acetonitrile mobile phase with 1 mM ammonium acetate (pH 7). Figure 1b shows the negative ionspray spectrum of OA. Only the $[M-H]^-$ ion at m/z 803.5 is observed. Not surprisingly, non-acidic esters such as compound 5 cannot be detected in the negative ion mode. The detection limit for OA using negative ion LC-MS is similar to that in the positive ion mode.

The only mass spectra reported to date for DTX4 were obtained using fast atom bombardment ionization (*15*). Positive ionspray gave no signal in the molecular ion region, but a weak erratic fragment ion at m/z 805.5, the same as $[M+H]^+$ of OA, was observed (data not shown). We do not yet understand the mechanism for production of this ion. The negative ionspray spectrum is shown in Figure 1d. This spectrum was acquired from an LC-MS analysis using aqueous acetonitrile with 1 mM ammonium acetate (pH 7). An $[M-H]^-$ ion could not be detected, but abundant signals were observed at m/z 735.3 and 489.9, corresponding to $[M-2H]^{-2}$ and $[M-3H]^{-3}$ ions, respectively. A complementary fragment ion, due to the anion of OA, is observed at m/z 803.5. Some water adduct ions are also observed at m/z 495.9 and 744.3. The LC-MS analysis of DTX4 is discussed later.

Analysis of Shellfish Tissues. The analysis of DSP toxins by LC is complicated by the lack of a strong chromophore in the structures. However, the principal toxins, OA, DTX1 and DTX2, contain a carboxyl group that allows them to be converted to fluorescent ester derivatives for analysis by LC with fluorescence detection (LC-FLD). Early work by Lee *et al.* (*26*) demonstrated the suitability of 9-anthryldiazomethane (ADAM) for this purpose. While this method has been useful, it has also proved to be difficult to use. LC-MS eliminates the need to perform derivatization and therefore minimizes labor-intensive sample preparation. A detection limit of 8 ng/g in shellfish digestive glands (corresponding to 1 ng/g whole edible tissue) has been achieved with a high degree of selectivity and a precision of 2-3% (*21*). This detection limit is well below the legal tolerance levels in Japan and Europe which range from 200 to 600 ng OA equivalent per gram of edible tissue. Figure 2 illustrates the analysis of two mussel tissue samples contaminated with OA and DTX1.

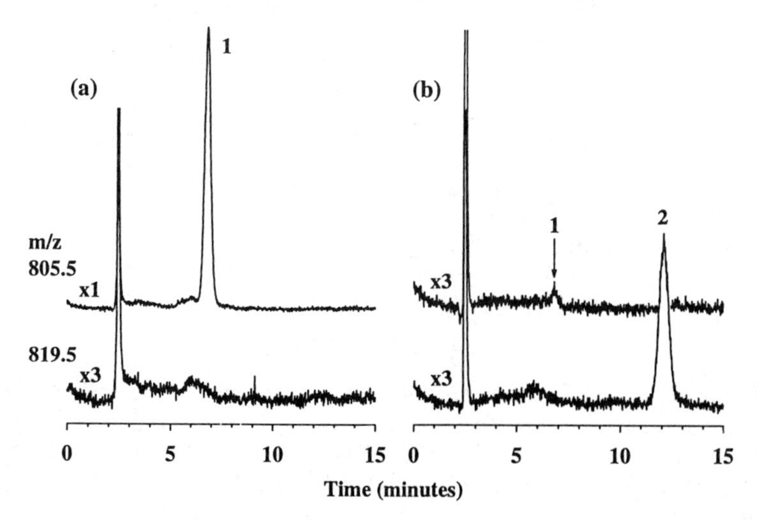

Figure 2. LC-MS mass chromatograms for extracts of mussel digestive glands from (a) Denmark (OA at 930 ng/g); (b) Nova Scotia, Canada (OA and DTX1 at 8 and 230 ng/g, respectively). Scale expansions are shown above the traces. Conditions: as in Methods, with 0.3 mL/min aqueous 70% methanol with 0.1% TFA and selected ion monitoring: m/z 805.5, OA (peak 1); m/z 819.5, DTX1 (peak 2). The peaks at 2.2 min are due to salts eluting at the solvent fronts.

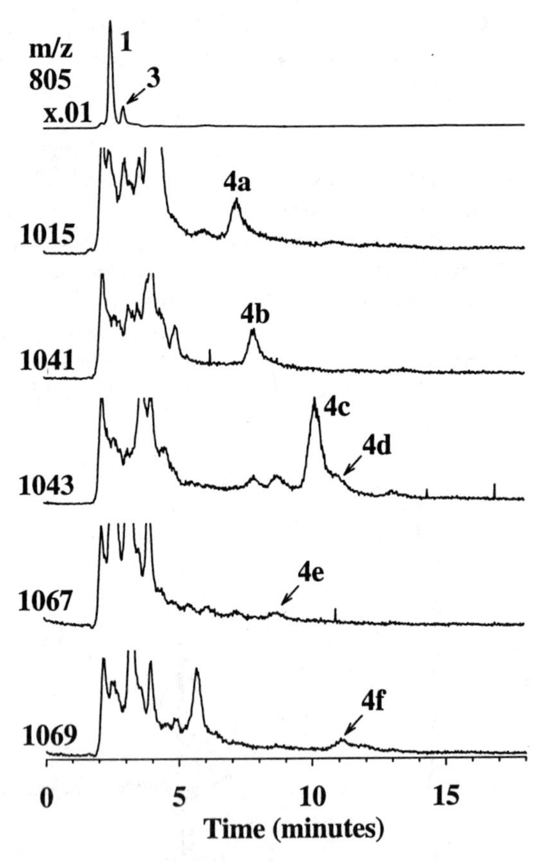

Figure 3. LC-MS mass chromatograms for an extract of Irish mussel digestive glands. This sample contained OA and DTX2 at 10 and 2 µg/g, respectively, and several of their 7-O-acyl esters ranging from 10 to 300 ng/g. The m/z 805 trace is scaled down 100-fold. Conditions: see Methods. Peak identities: 1 = OA; 3 = DTX2; 4a = OA-C14:0; 4b = OA-C16:1; 4c = OA-C16:0; 4d = DTX2-C16:0; 4e = OA-C18:1; 4f = OA-C18:0.

With LC-MS it is also possible to directly measure naturally-occurring derivatives of the DSP toxins. Figure 3 shows the analysis of an extract of Irish mussel digestive glands contaminated with OA, DTX2 and several DTX3 toxins (7-O-acyl esters of OA and DTX2). Identities of the peaks were based on a comparison with some synthetic standards and on fragment ions observed with MS/MS. Direct analysis of DTX3 toxins by the ADAM/LC-FLD method has not been possible due to their lipophilicity; they must be hydrolyzed to the parent toxins first before derivatization.

Analysis of Plankton Samples. A variety of plankton species has been associated with DSP incidents and, in a few cases, chemical analysis of plankton samples taken from affected shellfish production waters have revealed the presence of OA or DTX1 (*27*). Only a few of these plankton species have been grown in culture and shown to produce the toxins. A possible source for DTX2 has not yet been discovered.

Very little method development has been reported on the analysis of DSP toxins in plankton. Several extraction methods have been reported, and solvents such as methanol, acetone and chloroform have been used. In preparation for what we initially considered a "routine" experiment to measure production of toxin by *P. lima* in culture, we investigated some simple extraction procedures. Isolation of cells from cultures by centrifugation and sonication in aqueous 80% methanol, followed by LC-MS analysis, seemed to give excellent reproducibility with replicate subsamples of a single culture in early growth phase. However, erratic results began appearing as the experiment on toxin production proceeded, particularly in the cellular concentration of OA (but not DTX1). It is important to note that these experiments were started before the discovery of DTX4. It was evident from examination of the literature that others had also experienced erratic quantitative results in such experiments (*23*), although it was assumed that there were difficulties with the LC method used. This section reports on what has turned out to be a very interesting biochemical puzzle that was solved with the assistance of LC-MS analyses.

Figure 4a shows the positive ionspray LC-MS analysis of a 70-day old culture sample. OA (peak 1) and DTX1 (peak 2) were detected, as was the diol ester of OA (peak 5). It is interesting that significant levels of diol ester of DTX1 could not be observed in this particular isolate (we have observed it in other *P. lima* isolates). A peak at 5.3 min in the m/z 819 mass chromatogram (peak 1m, Figure 4a) was shown to be the methyl ester of OA by matching its spectrum and retention time with those of a standard. At first this compound was thought to be a natural product of the plankton because both OA and OA diol ester are quite stable in methanolic extracts and were not expected to react with methanol. However, the level of methyl ester was highly variable even between replicate subsamples of culture, sometimes rising higher than that of OA in older cultures. It was found that the percentage of methanol used in the extraction step had a significant effect. Lower percentages increased the level of methyl ester, opposite to what would be expected if this was a problem with extraction yields due to lipophilicity, thus suggesting that there might be a reaction between methanol and the analytes. During the preparation of a sample by centrifugation it is difficult to remove all of the medium from the cell pellet, so the final percentage of methanol was difficult to control. This partially explains the erratic variation in the methyl ester level. Substitution of deuterated methanol in the extraction protocol and subsequent LC-MS analysis showed a shift of the m/z 819.5 ion to m/z 822.5, proving that the methyl ester was entirely an artifact of the extraction procedure. This conclusion was supported by substitution of acetonitrile or tetrahydofuran for methanol. The methyl ester was eliminated entirely. Unfortunately, extraction yields of OA and its diol ester appeared to be much lower overall with these solvents.

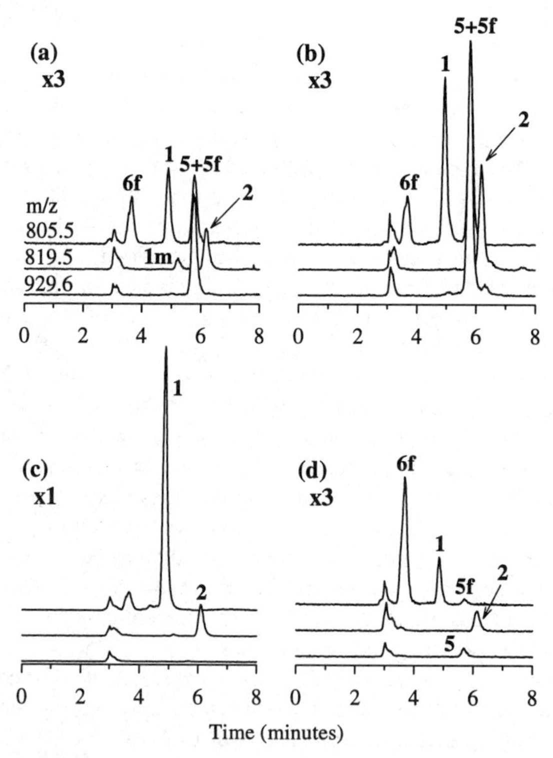

Figure 4. LC-MS analyses of replicate subsamples of *P. lima* culture extracted according to Methods 1 to 4 (a to d, respectively) as described in the Methods section. Peak identities are provided in Table II; 1m = methyl ester of OA; 5f and 6f are fragment ions of compounds 5 and 6. Conditions: as in Methods, with SIM and 0.2 mL/min aqueous 80% methanol with 0.1% TFA.

A number of other observations made when testing different cell disruption methods added to the initial confusion. It was found that cells could be rapidly and completely disrupted by passing them through a French press (see Figure 4b). Subsequent extracts of such material showed increased levels of OA, DTX1 and diol ester. If the disrupted cells were allowed to sit at room temperature for several hours, the diol ester concentration decreased while that of OA increased. Similar results were observed if a cell pellet was resuspended in TrisHCl buffer, frozen in liquid nitrogen and allowed to thaw. Figure 4c shows the results after freezing and subsequently incubating at room temperature for 24 hr. All of the diol ester had disappeared, leaving only OA and DTX1. The above observations suggested enzymatic reactions were at work. To test this hypothesis, the cell pellet was immersed in boiling water for 3 min prior to extraction. This did eliminate the methyl ester formation but the diol ester concentration was found to be very low (see Figure 4d).

Information on the existence of DTX4 then became available (*15*). An additional peak close to the solvent front in the m/z 805.5 chromatogram (peak 6f, see Figure 4) could now be explained. It is now known that this signal is due to a weak positive fragment ion from DTX4. Usually this compound appears as a broad tailing peak since it does not chromatograph very well with an acidic mobile phase and aged columns. In these chromatograms, acquired using a new column, the peak is sharp and is most intense in the extract of the boiled plankton.

A specific LC-MS method for the analysis of DTX4 was then developed. Using negative ionspray, a Zorbax RxC8 column and an aqueous acetonitrile gradient with 1 mM ammonium acetate buffer (pH 7), it was possible to directly measure DTX4 in plankton extracts. Figure 5 shows the full scan LC-MS analysis of the same extract of heated plankton analyzed in Figure 4d. DTX4 (peak 6) was easily detected by the $[M-3H]^{-3}$ and $[M-2H]^{-2}$ ions (see Figure 1d). Interestingly, a number of related new compounds were also detected. The signals from these compounds are evident in the other reconstructed mass chromatograms in Figure 5. Molecular weights, calculated from the multiply charged ions, and tentative identities, based on relative retention times and fragment ions, are given in Table II. Most of the structural variations appear to be associated with the sulfated end group as only one diol ester of OA is observed in this *P. lima* isolate. Some derivatives of DTX1 are also observed (supported by presence of a fragment ion at m/z 817.5) but at lower concentrations than those based on OA. Work is continuing on the detailed structure elucidation of these new compounds.

LC-MS analysis of various extracts revealed that DTX4 was at very low concentrations in fresh French press extracts, and in extracts of samples that had been allowed to sit for several hours after a freeze/thaw cycle. This suggested that enzymatic conversion of DTX4 to diol ester was occurring very rapidly. The conversion of diol ester to OA seems to proceed at a much lower rate and could be caused by another enzyme. A very interesting further observation is that the esterase(s) responsible for these transformations seem to be able to function in aqueous methanol solutions yielding a mixture of OA and OA methyl ester, the ratio of which was dependent on the methanol concentration. The transesterification reaction was maximized at a methanol concentration of 40%.

We are currently investigating the fate of the diol ester and DTX4 derivatives in shellfish. We anticipate that they will be hydrolyzed to some extent by esterases in the shellfish digestive gland as well as those from the plankton itself. Although the diol ester is not toxic, hydrolysis yields OA. Therefore, comprehensive analysis of all the toxins is necessary to properly assess the toxicity of any new plankton isolate. Clearly LC-MS will be an important tool for this task. The fingerprinting of the suite of toxins is also of interest for chemotaxonomic studies. In preliminary experiments, we have seen tremendous variations in toxin profiles between different isolates of *P. lima* and between different *Prorocentrum* species. For those laboratories without access to LC-MS equipment, the freeze/thaw/hydrolyze method should be very useful for assessing the toxic potential of plankton isolates using the ADAM/LC-FLD method.

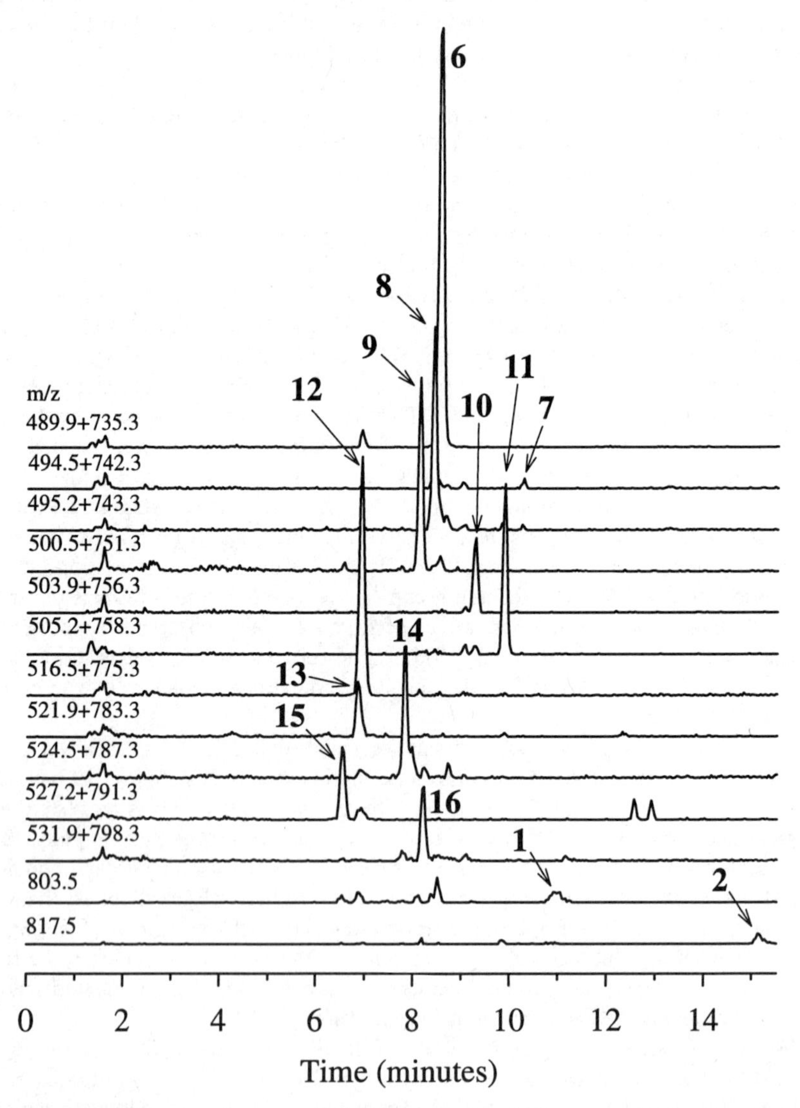

Figure 5. Negative ionspray LC-MS analysis of the same *P. lima* extract analyzed in Figure 4d. Peak identities are provided in Table II. Conditions: see Methods.

Table II. DSP toxins observed in *Prorocentrum lima* extracts

Cmpd. #	*Mol. Wt.*	*Ions Observed (m/z)* [a]	*Toxin Identity* [b]
1	804.5	805.5 (+); 803.5 (-)	Okadaic acid (OA)
2	818.5	819.5 (+); 817.5 (-)	DTX1
5	928.6	929.6, 805.5 (+)	OA diol ester
6	1472.6	489.9, 735.3, 803.5 (-)	DTX4
7[c]	1486.6	494.5, 742.3, 817.5 (-)	DTX4 + CH_2
8	1488.6	495.2, 743.3, 803.5 (-)	DTX4 + O
9	1504.6	500.5, 751.3, 803.5 (-)	DTX4 + 2O
10	1514.6	503.9, 756.3, 803.5 (-)	DTX4 + 42
11[c]	1518.6	505.2, 758.3, 817.5 (-)	DTX4 + CH_2 + 2O
12	1552.6	387.2, 516.5, 775.3, 803.5 (-)	DTX4 + SO_3
13	1568.6	391.2, 521.9, 783.3, 803.5 (-)	DTX4 + SO_3 + O
14	1576.6	524.5, 787.3, 803.5 (-)	DTX4 + 104
15	1584.6	395.2, 527.2, 791.3, 803.5 (-)	DTX4 + SO_3 + 2O
16[c]	1598.6	398.7, 531.9, 798.3, 817.5 (-)	DTX4 + CH_2 + SO_3 + 2O

[a] Positive or negative ionization indicated by (+) or (-) following ions.
[b] Structures of compounds 1, 2, 5 and 6 are given in Table I.
[c] Compounds 7, 11 and 16 are derivatives of DTX1 (methyl group at C_{35}).

In summary, LC-MS has proven itself valuable for the detection and identification of trace levels of DSP toxins in shellfish and plankton. It is clear that the toxin composition in plankton can be very complicated, and care must be taken to avoid enzymatic and/or hydrolytic reactions during sample preparation.

Acknowledgments

The authors wish to thank Dr. J. L. McLachlan for *P. lima* cultures, Mr. W. R. Hardstaff for technical assistance, Dr. J. L. C. Wright for a gift of DTX4 standard, and Dr. R. K. Boyd for review of the manuscript. NRCC No. 38093

Literature Cited

1. Yasumoto, T.; Murata, M.; Oshima, Y.; Sano, M.; Matsumoto, G. K.; Clardy, J. *Tetrahedron* **1985**, *41*, 1019-25.
2. Yasumoto, T.; Murata, M. *Marine Toxins*; Hall, S.; Strichartz, G. R., Eds; American Chemical Society: Washington, DC, 1990; pp. 120-32.
3. Aune, T.; Yndestad, M. In *Algal Toxins in Seafood and Drinking Water*; Falconer, I. R., Ed; Academic Press: London, 1993; pp. 87- 104.

4. Quilliam, M. A.; Gilgan, M. W.; Pleasance, S.; de Freitas, A. S. W.; Douglas, D.; Fritz, L.; Hu, T.; Marr, J. C.; Smyth, C.; Wright, J. L. C. In *Toxic Phytoplankton Blooms in the Sea*; Smayda, T. J.; Shimizu, Y., Eds; Elsevier: Amsterdam, 1993; pp. 547-552.
5. Zhao, J.; Lembeye, G.; Cenci, G.; Wall, B.; Yasumoto, T. In *Toxic Phytoplankton Blooms in the Sea*; Smayda, T. J.; Shimizu, Y., Eds; Elsevier: Amsterdam, 1993; pp. 587- 92.
6. Tachibana, K.; Scheuer, P. J.; Tsukitani, Y.; Kikuchi, H.; Van Engen, D.; Clardy, J.; Gopichand, Y.; Schmitz, F. J. *J. Am. Chem. Soc* **1981**, *103*, 2469-71.
7. Hu, T.; Doyle, J.; Jackson, D.; Marr, J.; Nixon, E.; Pleasance, S.; Quilliam, M. A.; Walter, J. A.; Wright, J. L. C. *Chem. Commun.* **1992**, 39-41.
8. Bialojan, C.; Takai, A. *Biochem. J.* **1988**, *256*, 283-90.
9. Terao, K.; Ito, E.; Yanagi, T.; Yasumoto, T. *Toxicon* **1986**, *24*, 1141-51.
10. Suganuma, M.; Fujiki, H.; Suguri, H.; Yoshizawa, S.; Hirota, M.; Nakayasu, M.; Ojika, M.; Wakamatsu, K.; Yamada, K.; Sugimura, T. *Proc. Natl. Acad. Sci. USA* **1988**, *85*, 1768-71.
11. Yanagi, T.; Murata, M.; Torigoe, K.; Yasumoto, T. *Agric. Biol. Chem.* **1989**, *53*, 525-9.
12. Yasumoto, T.; Murata, M.; Lee, J. S.; Torigoe, K. In *Mycotoxins and Phycotoxins*; Natori, S.; Hashimoto, K.; Ueno, Y., Eds; Elsevier: Amsterdam, 1989; pp. 375-382.
13. Marr, J. C.; Hu, T.; Pleasance, S.; Quilliam, M. A.; Wright, J. L. C. *Toxicon* **1992**, *30*, 1621-30.
14. Hu, T.; Marr, J.; de Freitas, A. S. W.; Quilliam, M. A.; Walter, J. A.; Wright, J. L. C.; Pleasance, S. *J. Nat. Prod.* **1992**, *55*, 1631-7.
15. Hu, T.; Curtis, J.M.; Walter, J.A.; Wright, J.L.C. *J. Chem. Soc., Chem. Comm.* **1995**, 597-599.
16. Pleasance, S.; Quilliam, M. A.; de Freitas, A. S. W.; Marr, J. C.; Cembella, A. D. *Rapid Commun. Mass Spectrom.* **1990**, *4*, 206-13.
17. Bencsath, F. A.; Dickey, R. W. *Rapid Commun. Mass Spectrom.* **1991**, *5*, 283-290.
18. Bruins, A.P.; Covey, T.R.; Henion, J.D. *Anal. Chem.* **1987**, *59*, 2642-2646.
19. Quilliam, M. A.; Thomson, B. A.; Scott, G. J.; Siu, K. W. M. *Rapid Commun. Mass Spectrom.* **1989**, *3*, 145-150.
20. Pleasance, S.; Quilliam, M. A.; Marr, J. C. *Rapid Commun. Mass Spectrom.* **1992**, *6*, 121-127.
21. Quilliam, M.A., *J. AOAC Int.* **1995**, *78*, 555-570.
22. Marr, J. C.; Jackson, A. E.; McLachlan, J. L. *J. Applied Phycology* **1992**, *4*, 17-24.
23. Jackson, A. E.; Marr, J. C.; McLachlan, J. L. In *Toxic Phytoplankton Blooms in the Sea*; Smayda, T. J.; Shimizu, Y., Eds; Elsevier: Amsterdam, 1993; pp. 513-18.
24. Pleasance, S.; Quilliam, M.A.; Boyd, R.K.; Thibault, P.; Covey, T.R. In *Proceedings of 38th Annual Conference on Mass Spectrometry and Allied Topics*; American Society for Mass Spectrometry, 1990, pp. 381-382.
25. Hopfgartner, G.; Bean, K.; Henion, J.; Henry, R. *J. Chromatogr.* **1993**, *647*, 51-61.
26. Lee, J.S., Yanagi, T., Kenma, R., Yasumoto, T. *Agric. Biol. Chem.* **1987**, *51*, 877-881.
27. Lee, J. S.; Igarashi, T.; Fraga, S.; Dahl, E.; Hovgaard, P.; Yasumoto, T. *J. Applied Phycology* **1989**, *1*, 147-52.

RECEIVED September 15, 1995

Chapter 19

Identification of Active-Site Residues in Glycosidases

Stephen G. Withers

Department of Chemistry, University of British Columbia, Vancouver, British Columbia V6T 1Z1, Canada

A strategy has been developed for the reliable labeling of the active site nucleophile in glycosidases which hydrolyse with net retention of anomeric configuration. Electrospray mass spectrometric analysis of the intact protein before and after inactivation allows facile determination of the stoichiometry of labeling. High performance liquid chromatography-electrospray ionization tandem mass spectrometry (HPLC-ESI-MS/MS) analysis of proteolytic digests of these samples with monitoring in the neutral loss mode allows identification of the peptide within the mixture which is labeled, without the need for radioactive tagging. Inspection of the sequence of the protein for candidate peptides of this mass allows the generation of a short list of possible peptides. The true identity of the labeled peptide can be determined after further ESI-MS/MS analysis of the peptide of interest in the product ion mass spectrum. Labeled peptides which do not undergo the necessary facile cleavage in the collision cell of the mass spectrometer can generally be identified within a mixture by comparison of HPLC-ESI-MS data from both unlabeled and labeled samples. A search is made for a peptide which is present in the labeled digest and absent in the unlabeled digest, and is heavier by the expected amount than a peptide which is missing from the labeled digest but present in the unlabeled digest. Examples of these approaches are described with the xylanase from *Bacillus subtilis,* human β-glucocerebrosidase and the exo-xylanase/glucanase from *Cellulomonas fimi*.

Glycosidases are a class of enzymes responsible for the hydrolysis of glycosidic linkages (1-4). They play important roles in a wide variety of medically important biological processes such as digestion, where such enzymes as α-amylase, maltase and lactase are responsible for the degradation of dietary polysaccharides and oligosaccharides to simple monosaccharides which are then absorbed into the bloodstream. Other important roles are in catabolism such as in the lysosomal degradation of glycolipids. Indeed, deficiencies in such enzymes result in serious genetic diseases such as Gaucher and Tay Sachs diseases. A third important role is in glycoprotein processing, where linkage-specific glycosidases are reponsible for the

0097–6156/95/0619–0365$12.00/0

conversion of fully glycosylated precursor glycoproteins into the correct glycoform for their specific function. They are also important in a number of non-medical applications, particularly in biotechnology, where the efficient enzymatic degradation of cellulosic biomass has considerable potential for the generation of alternative energy sources. Another application currently gaining favour is the use of xylanases to assist in the bleaching of wood pulp by degradation of hemicellulose whose presence hinders the decolorisation of lignin contaminants.

Glycosidases fall into two major mechanistic categories: those which hydrolyse the glycosidic bond with net inversion of configuration (inverting enzymes) and those which do so with net retention of anomeric configuration (retaining enzymes) (3). Likely mechanisms for these two enzyme classes were proposed by Koshland (5) over 40 years ago, and have largely stood the test of time. Although the two mechanisms are distinctly different, they do retain a number of features in common, as illustrated in Scheme 1. Inverting glycosidases are believed to function by a single step mechanism in which a water molecule effects a direct displacement at the anomeric centre. This displacement process is general-acid/base catalysed, with one active site amino acid acting as the general base, helping to deprotonate the nucleophilic water molecule, and the other amino acid acting as a general acid, protonating the departing oxygen atom in a concerted fashion as the bond cleaves.

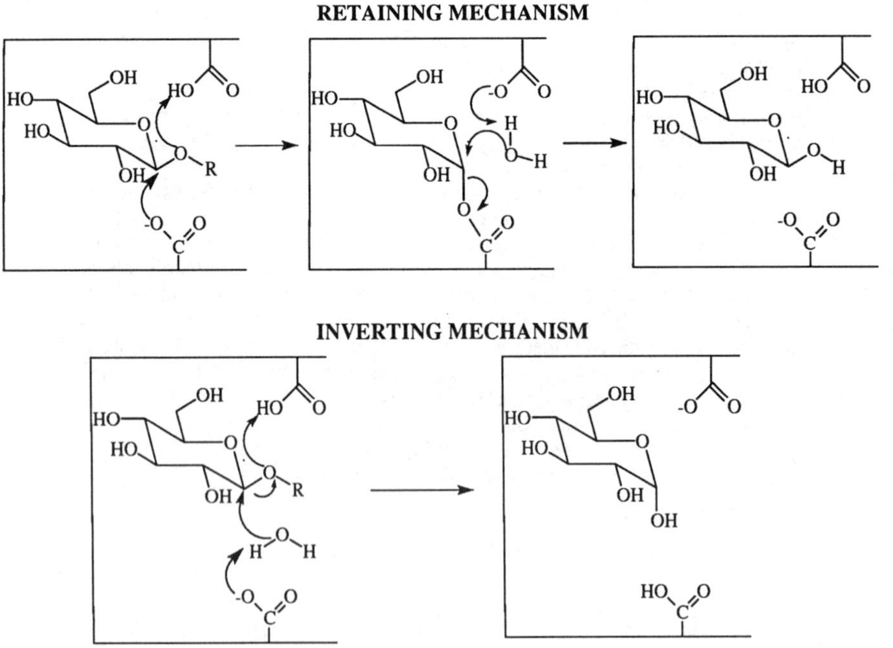

RETAINING MECHANISM

INVERTING MECHANISM

Scheme 1

Retaining glycosidases are generally believed to function through a double displacement mechanism in which a glycosyl enzyme intermediate is formed and hydrolyzed *via* oxocarbenium ion-like transition states as shown in Scheme 1. Again the reaction is facilitated by acid/base catalysis, but in this case it is probable that the same group plays both roles. Transition states for these chemical steps in both classes of enzymes have considerable oxocarbenium ion character.

Despite the fact that 3-dimensional structures of quite a number of these enzymes have been determined recently by X-ray crystallography (see Reference 6 for a recent compilation) the identities of the active site residues in the vast majority of such enzymes are still unknown. Such information is important for an understanding of their catalytic mechanisms and design of effective inhibitors, and is essential for mutagenic studies of the active site region. An alternative means of gaining this information is through the design of suitable, specific mechanism-based inactivators or affinity labels that can be used to tag active site residues. The identities of these residues can then be determined by proteolysis of the labeled protein, followed by purification and subsequent sequencing of the labeled peptide. Perhapsthe most interest lies in the identities of the two key residues, as indicated in Scheme 1, and in all cases reliably studied to date, it is apparent that the two residues are the carboxylic side chains of aspartic or glutamic acids (6). This report will focus upon strategies developed in the author's laboratory for the identification of these two residues, the nucleophile, and the acid/base catalyst, in retaining glycosidases.

Experimental

Purification of the enzymes employed, synthesis of the inactivators and kinetic studies were carried out as described in the following references: *B. subtilis* xylanase (21), human glucocerebrosidase (23), *A. faecalis* β-glucosidase (25), and *C. fimi* exoglycanase (25). Mass spectrometry was carried out on a PE-Sciex API *III* triple quadrupole mass spectrometer (Sciex, Thornhill, Ontario, Canada) equipped with an ion-spray ionisation source. Peptides were separated by reverse phase HPLC on an Ultrafast Microprotein Analyzer (Michrom BioResources Inc., Pleasanton, CA) directly interfaced with the mass spectrometer. In each of the MS experiments, the proteolytic digest was loaded onto a C18 column (Reliasil, 1 X 150 mm), then eluted with a gradient of 0-60% solvent B over 20 minutes followed by 100% B over 2 minutes at a flow rate of 50 μl/minute (solvent A: 0.05% trifluoroacetic acid, 2% acetonitrile in water; solvent B: 0.045% trifluoroacetic acid, 80% acetonitrile in water). A post-column splitter was present in all experiments, splitting off 85% of the sample into a fraction collector and sending 15% into the mass spectrometer.

In the single quadrupole mode (LC/MS) the quadrupole mass analyzer was scanned over a *m/z* range 300-2400 Da with a step size of 0.5 Da and a dwell time of 1 ms per step. The ion source voltage (ISV) was set at 5 kV and the orifice energy (OR) was 80 V.

In the neutral loss scanning mode, MS/MS spectra were obtained by searching for the mass loss corresponding to the loss of the sugar from a peptide ion in the singly charged state. Thus, scan range: m/z 300-2400; step size: 0.5; dwell time: 1ms/step; ion source voltage (ISV): 5 kV; orifice energy (OR): 80; RE1=115; DM1=0.16; R1=0 V; R2=-50 V; RE3=115; DM3=0.16; Collision gas thickness: 3.2-3.6 X 10^{14} molecules/cm^2. (CGT=320-360). To maximize the sensitivity of neutral loss detection, normally the resolution (RE and DM) is compromised without generating artifact neutral loss peaks.

Specific details for each case are provided in the references quoted.

Identification of the active site nucleophile

The mechanism for retaining glycosidases involves the formation of a glycosyl-enzyme intermediate in which the sugar is covalently attached to the protein *via* the carboxylic side chain of a glutamic or aspartic acid. In order to identify the nucleophilic amino acid residue it was necessary to develop techniques for the trapping of this intermediate. This required differential manipulation of the rates of formation (glycosylation) and hydrolysis (deglycosylation) of the intermediate such that the rate constant for the deglycosylation step was slowed enormously from its typical values ($t_{1/2}$ around 1-10

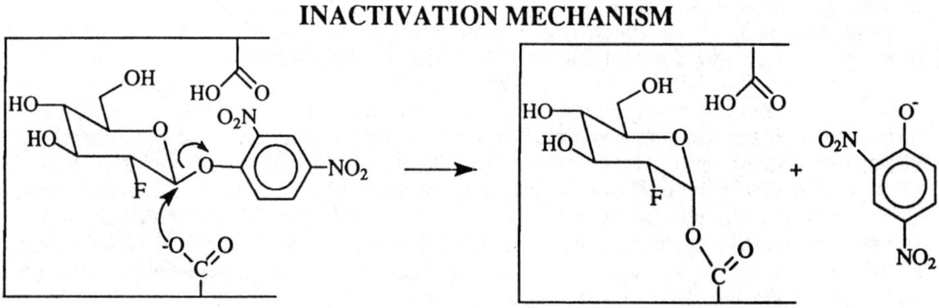

INACTIVATION MECHANISM

Scheme 2

ms). This was achieved through the use of 2-deoxy-2-fluoro-glycosides with good leaving groups such as dinitrophenolate or fluoride (7, 8). The presence of the fluorine substituent at C-2 slows *both* the glycosylation and deglycosylation steps in two ways. One of these is due to the fact that the hydroxyl substituent at C-2 plays a crucial role in transition state stabilisation in glycosidases by making key interactions (worth more than 8 kcal/mol), probably predominantly hydrogen bonding interactions, with the enzyme active site (9). Removal, or diminution, of these interactions by replacement of the hydroxyl with fluorine, a substituent of limited hydrogen bonding potential, destabilises both transition states significantly. The second way in which the fluorine substituent slows the two steps is through inductive destabilisation of the two electron-deficient transition states. Fluorine is much more electronegative than a hydroxyl, thus the positive charge developed at the transition state will be significantly destabilised by the presence of this substituent, thereby slowing both steps. The consequence of these two effects combined is a significant (up to 10^6 - 10^7 fold) reduction in the rates of both steps (10, 11). Incorporation of a good leaving group speeds up the glycosylation step relative to the deglycosylation step with the effect that the intermediate is accumulated. Incubation of the enzyme with its corresponding 2-deoxy-2-fluoroglycoside therefore results in time-dependent inactivation, *via* the accumulation of a relatively stable 2-deoxy-2-fluoroglycosyl-enzyme intermediate as shown in Scheme 2.

The stoichiometry of inactivation can be demonstrated easily by electrospray mass spectrometric analysis of labeled and unlabeled enzyme. Observation of a mass increase of the expected amount for the attachment of a single sugar label provides excellent evidence for the desired 1:1 stoichiometry of labeling, and this has indeed been the case for all enzymes studied using this approach to date, highlighting the specificity of this class of mechanism-based inactivator (12). Identification of the labeled amino acid residue requires proteolysis of the labeled enzyme, high performance liquid chromatographic (HPLC) separation of the resultant peptide mixture, location of the labeled peptide within this mixture, and finally determination of its sequence. The approach traditionally used has involved the synthesis of a radiolabeled version of the inactivator, then use of this radioactive "handle" to find the peptide of interest for subsequent sequencing, typically by the Edman degradation. This approach, while frequently successful, is extremely time-consuming since the synthesis of a radiolabeled version of the inactivator is usually difficult, and in some cases essentially impossible. An efficient strategy, which obviates the need for synthesis of radiolabeled versions of the inactivator, has been developed in the author's laboratory recently, and has proved extremely valuable in the rapid identification of the labeled peptides. This approach builds upon the pioneering work of others both in the characterisation of intact labeled proteins and in the mass spectrometric identification of the sites of labeling (13-20). The approach involves HPLC-ESI-MS/MS analysis of the proteolytically digested labeled enzyme, using the mass spectrometer in the neutral loss mode and monitoring for a fragment species that is specific for the sugar-peptide linkage, as described below.

***Bacillus subtilis* xylanase.** Inactivation of the *B. subtilis* xylanase was accomplished using 2',4'-dinitrophenyl 2-deoxy-2-fluoro-β-xylobioside to trap an intermediate (21). Peptic hydrolysis of the 2-deoxy-2-fluoroxylobiosyl-labeled xylanase resulted in a mixture of peptides which was separated by reversed phase-HPLC. This chromatogram reveals a large number of peaks, which arise from every peptide in the mixture (Fig.1). The labeled peptide was then identified by using neutral loss tandem mass spectrometry. In this method ions are subjected to fragmentation in the second quadrupole by collisions with an inert gas (Ar). Under these conditions, homolytic cleavage of the ester linkage between the sugar inhibitor and the peptide occurs, resulting in the loss of a neutral sugar residue of known mass (267), but leaving the peptide moiety with its original charge. Scanning of the two quadrupoles in a linked

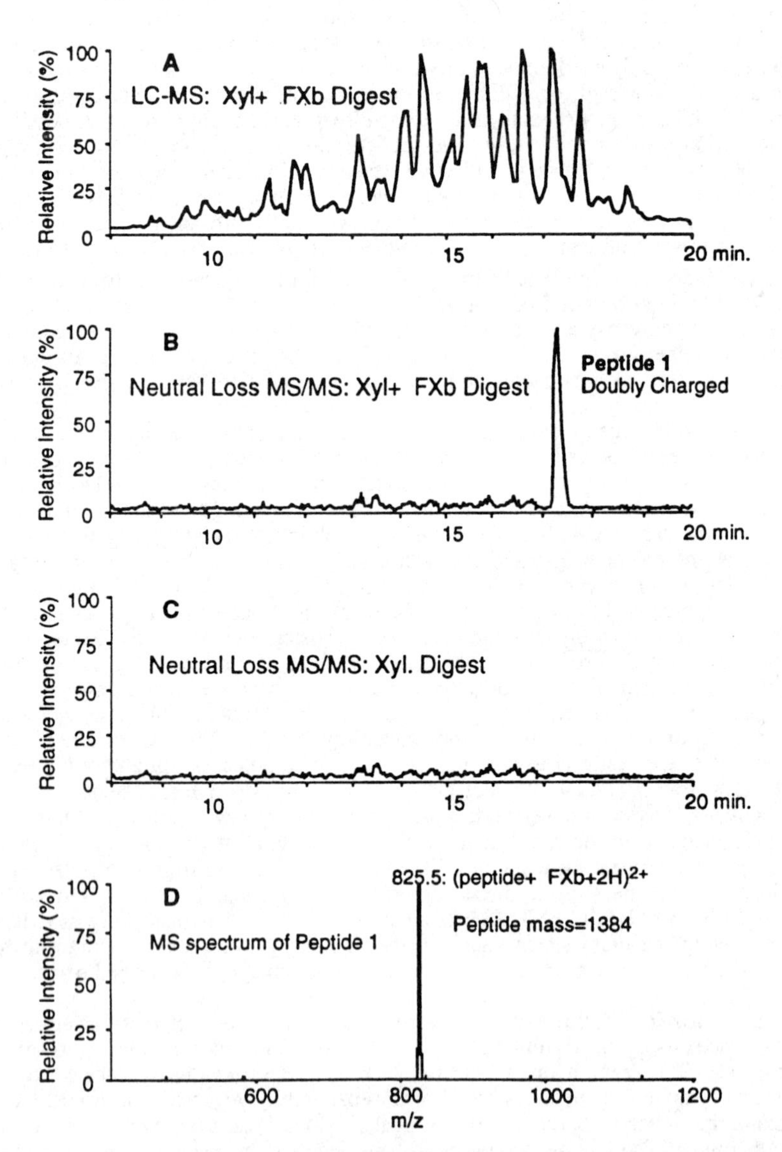

Figure 1. ESI-MS experiments on peptic digests of *Bacillus subtilis* xylanase (Xyl). A) LC-MS total ion chromatogram (TIC) of digest of xylanase labeled with 2-deoxy-2-fluoro-xylobiose (Fxb). Neutral loss mass spectra ($\Delta m = 133.5$) of digests of xylanase: B) labeled with 2-deoxy-2-fluoro-xylobiose, and C) unlabeled. D) ESI-mass spectrum of peptide 1 from B).

mode, in which they are offset by the desired *m/z* ratio corresponding to the mass of the sugar lost, permits only those ions having a mass that is less than that of the lost sugar moiety by a mass of 267 to pass through the third quadrupole and be detected. If the peptide bears more than one charge, it is also necessary to look for *m/z* differences of one half, or one third of the mass of the neutral species, depending upon the charge carried. In this case no signal was detected when the mass spectrometer was scanned in the neutral loss MS/MS mode searching for the mass loss of 267. However, a single peak was observed when scanned for the mass loss 133.5 (Fig.1B), yet no such peak was detected in the neutral loss spectrum of the unlabeled xylanase digest (Fig.1C). Thus a doubly charged peptide corresponding to the active site glycopeptide is being selectively detected. An *m/z* value of 825.5 (Fig.1D) was measured for this peptide, corresponding to a molecular weight of 1649 Da {(825.5 X 2) - 2H}. Since the mass of the lost sugar is 267, the unlabeled peptide must have a molecular weight of 1383 (1649 - 267 + 1H).

Upon searching the amino acid sequence of the enzyme for all possible peptides with this mass, only three with a mass of 1383 +/- 1 could be identified, namely YGWTRSPLIEY (residues 69-79), GWTRSPLIEYY (residues 70-80), and PLIEYYVVDSW (residues 75-85). They all contain the same Glu78 residue and the third peptide also contains an aspartate (Asp83). Absolute identification of the labeled peptide was obtained in a second experiment, without a need for purification, by further fragmentation of the peptide of interest in the product ion scan mode.

After selection of the parent ion (*m/z* 825.5) in the first quadrupole it was subjected to collision induced fragmentation in the second quadrupole and the masses of the daughter ions were detected in the third quadrupole as shown in Figure 2. The peak at *m/z* 1385 arises from the peptide ion of interest in the singly charged state (MH^+) after loss of its sugar in the collision cell. Other peaks represent fragmentation essentially from the C-terminus; N-terminal fragments are not observed since the loss of the charged N-terminal amino acid produces undetected neutral peptides. The peak at m/z 1204 (b10) arises from loss of C-terminal tyrosine (*m/z* = 181) from the parent ion peak at *m/z* = 1385 while the other peaks (b9, b8, b6, b5, and b3) result from the respective losses of EY, IEY, PLIEY, SPLIEY, RSPLIEY, and TRSPLIEY fragments from the C-terminus. This sequence information is sufficient to unambiguously identify the labeled peptide as YGWTRSPLIEY (residues 69-79), thereby indicating that the catalytic nucleophile in *B. subtilis* xylanase is Glu78. This result was confirmed by isolation of the labeled peptide and conventional solid phase sequencing.

Gratifyingly, this amino acid residue is indeed found in the active site of the enzyme in the 3-dimensional structure determined crystallographically (22). Further, mutants in which that residue had been replaced by a glutamine were found to be completely devoid of catalytic activity, as expected for replacement of the catalytic nucleophile (22).

Human lysosomal glucocerebrosidase. Although the results of the neutral loss scan were quite unambiguous in the case of the xylanase, there being no other significant peaks in the scan, this is not always the case. Indeed, if the neutral species being lost has the same mass as one of the amino acids, or as a fragmentation product thereof, the scan can contain a considerable number of peaks. Fortunately this can easily be interpreted by carrying out an equivalent experiment on a digest of the unlabeled enzyme, revealing all those peaks *not* due to the labeled peptide. This is illustrated below in the study carried out on the identification of the active site nucleophile of human glucocerebrosidase (23). Peptic digestion of the 2-deoxy-2-fluoroglucose (FGlc)-inhibited glucocerebrosidase resulted in a mixture of peptides (Fig. 3A). In this case, however, neutral loss mass scans under conditions suitable for detection of the 2-fluoroglucosyl-labeled peptide revealed two significant peaks (peptide 1 and peptide 2,

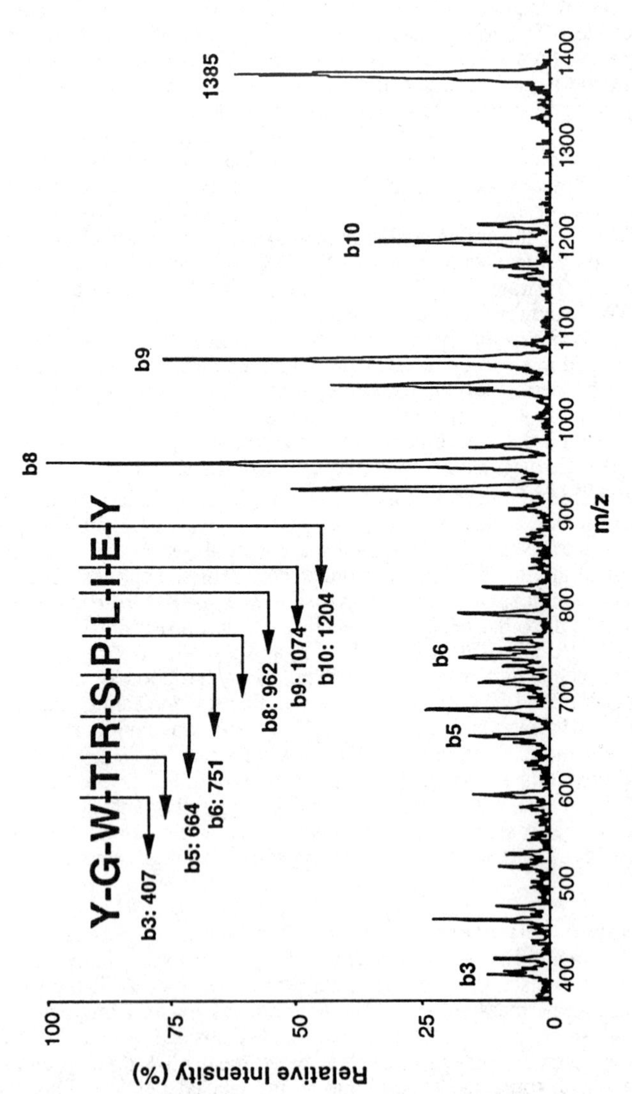

Figure 2. Product ion mass spectrum of labeled xylanase peptide.

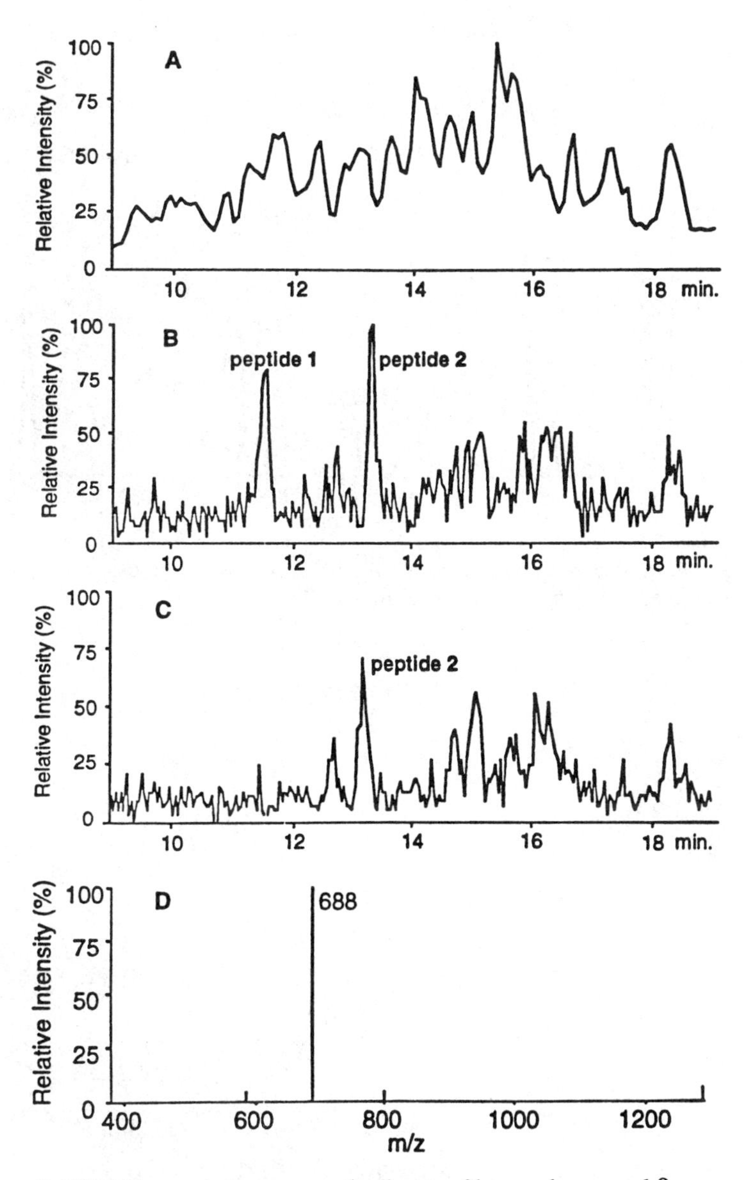

Figure 3. ESI-MS experiments on peptic digests of human lysosomal β-glucocerebrosidase. A) LC-MS total ion chromatogram (TIC) of digest of glucocerebrosidase labeled with 2-deoxy-2-fluoro-glucose, TIC chromatogram. Neutral loss mass spectra ($\Delta m = 165$) of digests of xylanase: B) labeled with 2-deoxy-2-fluoro-glucose (FGlu).; C) unlabeled. D) ESI-mass spectrum of peptide 1 from B). Reproduced with permission from reference 23.

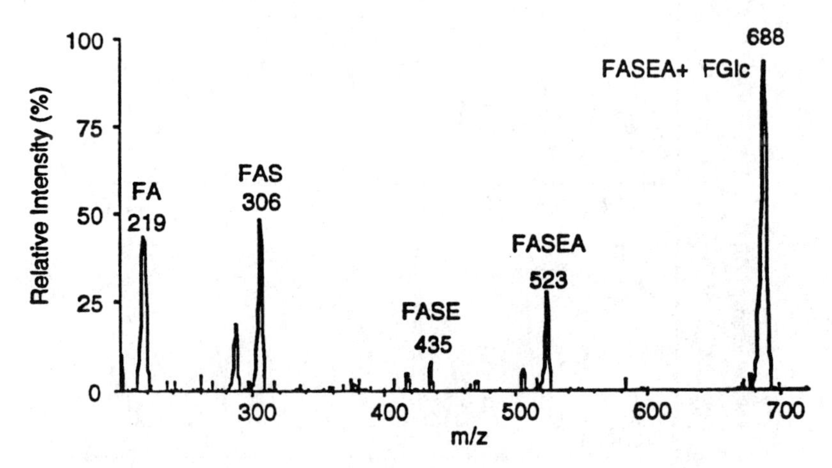

Figure 4. Product ion mass spectrum of labeled β–glucocerebrosidase peptide. Reproduced with permission from reference 23.

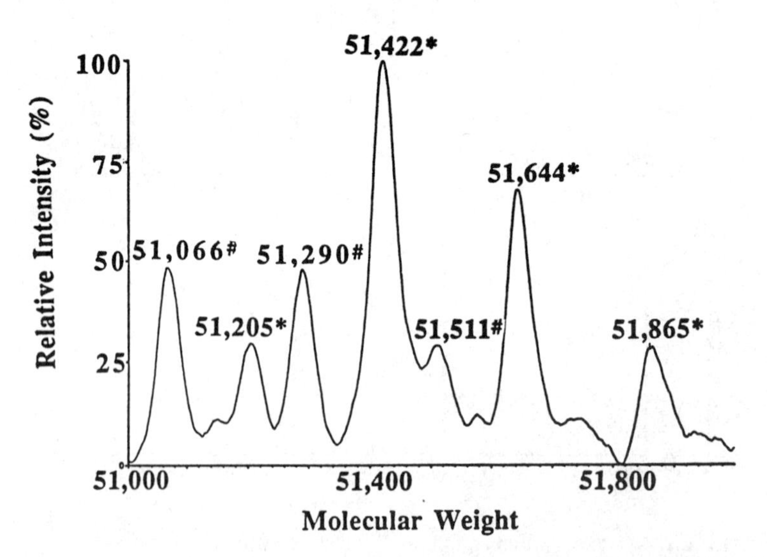

Figure 5. ESI-reconstructed mass spectrum of *Agrobacterium faecalis* β-glucosidase inactivated with N-bromoacetyl β-D-glucopyranosylamine: (*) corresponds to (+) Met β-glucosidase, (#) corresponds to (-) Met β-glucosidase. Reproduced with permission from reference 25.
(Copyright 1993 Elsevier Science).

Fig. 3B), as well as some smaller ones. As can be seen, peptide 2 also was present in the sample derived from the unlabeled enzyme (peptide 2, Fig. 3C), indicating that this was not the peptide of interest. Determination of the mass of the labeled peptide (peptide 1, Fig 3D) again allowed selection of a number of candidates, from which the actual peptide was shown to be FASEA by further analysis in the product ion mode, as shown in Fig. 4. In addition, the peptide of interest was purified and the sequence confirmed by standard Edman degradation. The importance of the residue (Glu340) was verified by kinetic analysis of a purified enzyme which was altered at this residue by site-directed mutagenesis (23). Mutants modified at this position were essentially catalytically inactive. This result corrected the previously published report, based upon affinity labeling with conduritol B epoxide, that Asp443 was the essential catalytic nucleophile (24).

Identification of other residues: the acid/base catalyst.

The other major amino acid residue of interest in the active site of retaining glycosidases is the acid/base catalyst. No methods have yet been developed for the reliable identification of this important residue, though several different classes of affinity label have been tried. In some cases these did label the residue of interest, but unfortunately no one label has proved successful in all cases. Reagents which have labeled the acid/base catalyst include the exocyclic glycosyl epoxides, the conduritol epoxides (1), and the N-bromoacetyl glycosylamines (25, 26). In several of these cases, in fact, it was not clear at the time that it was the acid/base catalyst which had been labeled. However subsequent structural or mechanistic work has clarified this point. In order to illustrate the approaches employed, and to highlight the application of electrospray mass spectrometry to the problem, the labeling of the acid/base catalyst in the exo-xylanase/glucanase from *Cellulomonas fimi* will be described.

Labeling of *C. fimi* exo-xylanase/glucanase. N-Bromoacetyl β-glycosylamines were first used as affinity labels with *Escherichia coli* β-galactosidase (27), the labeled amino acid being identified as Met501. More recently we have described their use as affinity labels for both the *Agrobacterium faecalis* β-glucosidase and the *Cellulomonas fimi* exo-xylanase/glucanase (25). In the former case, mass spectrometric analysis of the enzyme labeled with N-bromoacetyl β-glucosylamine revealed that, even though simple pseudo-first order kinetics of inactivation had been determined, the inactivated enzyme was labeled with at least three equivalents of inactivator (Fig. 5). The picture is further complicated by the presence of two native β-glucosidase species of masses 51,205 and 51,066, corresponding to species with (+-Met) and without (-Met) an N-terminal methionine residue. The presence of this N-terminal methionine residue, due to incomplete processing at the higher expression levels employed, has previously been shown to have no effect upon kinetic parameters. However in the case of the *Cellulomonas fimi* exo-xylanase/glucanase (Cex), inactivation by N-bromoacetyl β-cellobiosylamine resulted in the addition of only a single equivalent of inactivator (m = 382) per enzyme, as shown by the mass increase of ($\Delta m = 378 \pm 8$) in Fig. 6. The multiple labeling seen for the β-glucosidase precluded simple identification of the site of labeling which led to inactivation, thus no further studies were performed on this system. However, the single labeling seen for the *C. fimi* exo-xylanase/glucanase provided an ideal system for study.

The identification of the site of labeling in this case was attempted *via* comparative analysis of HPLC-ESI-MS profiles of peptic digests of the labeled and unlabeled enzymes, using the mass information to "realign" the profiles and thereby correct for any mass offset between HPLC chromatograms. A search was then made for a peptide which was present in the labeled sample, but absent from the unlabeled

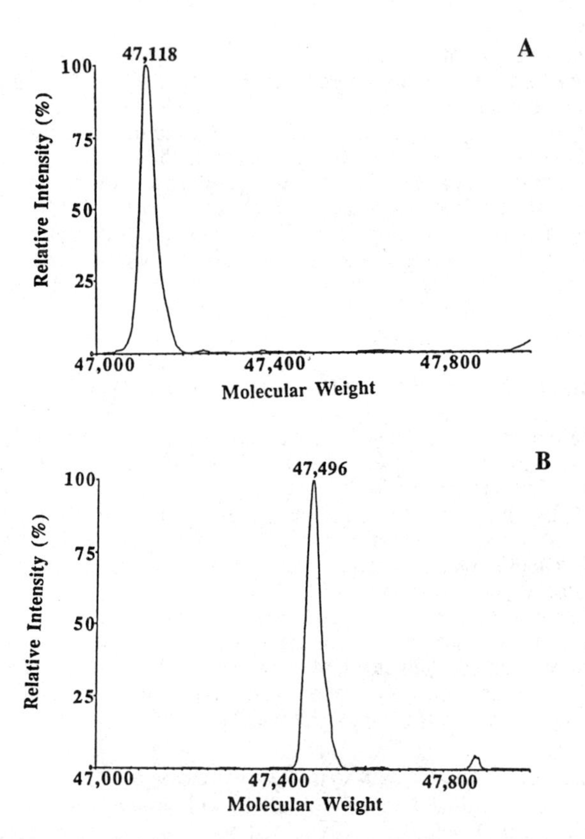

Figure 6 ESI-reconstructed mass spectrum of A) *Cellulomonas fimi* exo-xylanase/glucanase; B) *Cellulomonas fimi* exo-xylanase/glucanase inactivated with N-bromoacetyl β-D-cellobiosylamine.

sample, which was greater in mass by exactly the mass of the label (Δm = 382) than another peak which was present in the unlabeled sample, but (ideally) absent from the labeled. After searching through all the peaks only one peptide was found which met these criteria. Fig. 7 shows the mass spectrum of all the peptides eluting from the HPLC between 24 and 25 minutes, the region in which the peptide of interest eluted. As can be seen, a peptide of m/z = 1028 is present in the digest of the labeled protein (Fig. 7B), but is not found in the equivalent digest of the unlabeled enzyme (Fig. 7A). Correspondingly, a peptide of m/z = 646, smaller by the mass of the label (382) is found in the digest of the unlabeled enzyme (Fig. 7A), but only at lower intensity in the digest from the labeled (Fig. 7B). Ideally this peptide would not be found in the sample of the labeled protein digest, but is presumably present because of incomplete labeling or, more likely, because of partial degradation of the labeled peptide. A search of the amino acid sequence of Cex revealed only nine peptides of this mass which could be generated from this protein. The peptide of *m/z* = 1028 was then further purified by reverse phase HPLC, and sequenced by Edman degradation using a modified Edman reagent (28, 29). The great advantage of this approach, in which a mass spectrometer is

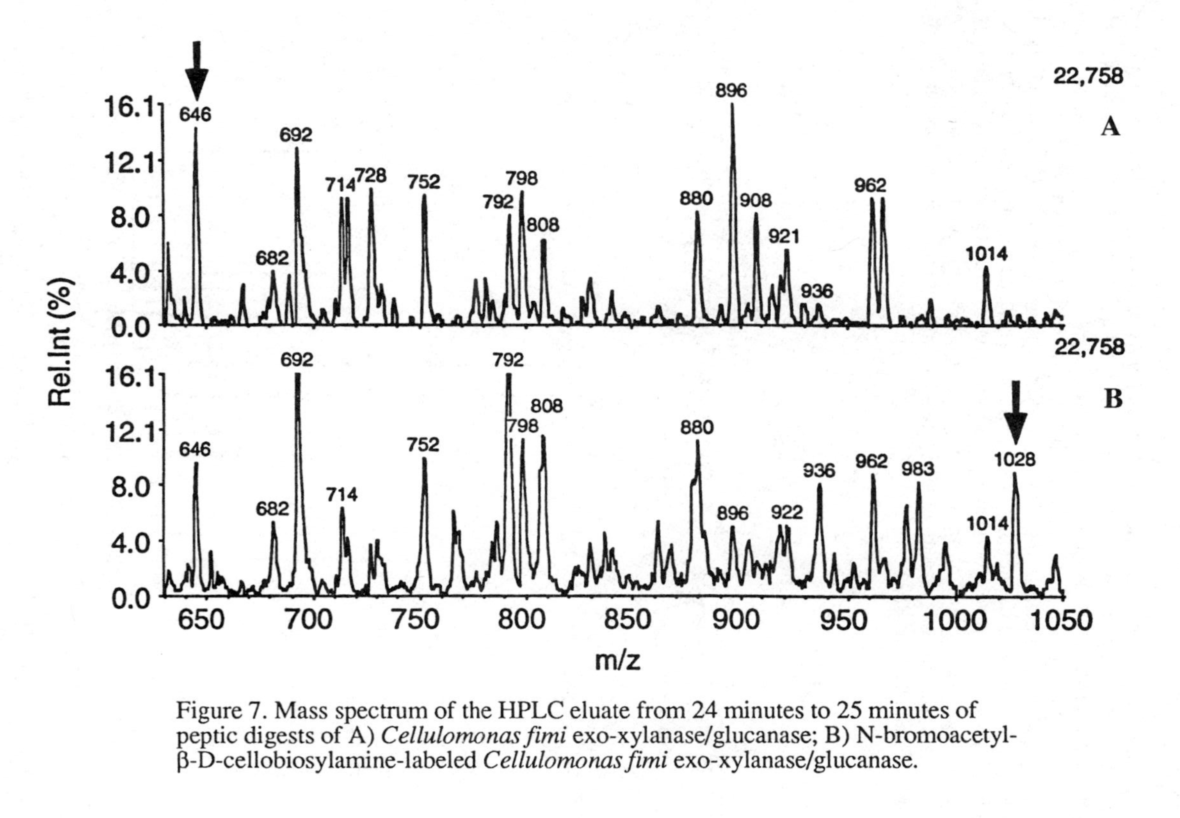

Figure 7. Mass spectrum of the HPLC eluate from 24 minutes to 25 minutes of peptic digests of A) *Cellulomonas fimi* exo-xylanase/glucanase; B) N-bromoacetyl-β-D-cellobiosylamine-labeled *Cellulomonas fimi* exo-xylanase/glucanase.

used to identify the product modified phenylthiohydantoins, is that the mass of the modified amino acid phenylthiohydantoin is directly obtained. Using this approach the sequence DVVNEA was determined for this peptide, where the fifth cycle resulted in the release of a phenylthiohydantoin derivative of m/z = 822. This is exactly the mass predicted for the phenylthiohydantoin of N-bromoacetylcellobiosylamine-modified glutamic acid. Further confirmation of this identity was obtained by "on-line" product ion mass spectral analysis of this derivative. The fragments (Fig. 8) observed are entirely consistent with the expected structure.

Alignment of the sequence of this peptide with the amino acid sequence of the enzyme (30) indicates that the modified residue corresponds to Glu127. Gratifyingly this residue was recently identified as the most likely candidate for the acid-base residue on the basis of a detailed kinetic analysis of mutants modified at conserved glutamic and aspartic acid residues (31). In addition, this residue has recently been located in the active site of the enzyme through X-ray crystallographic determination of the 3-dimensional structure (32).

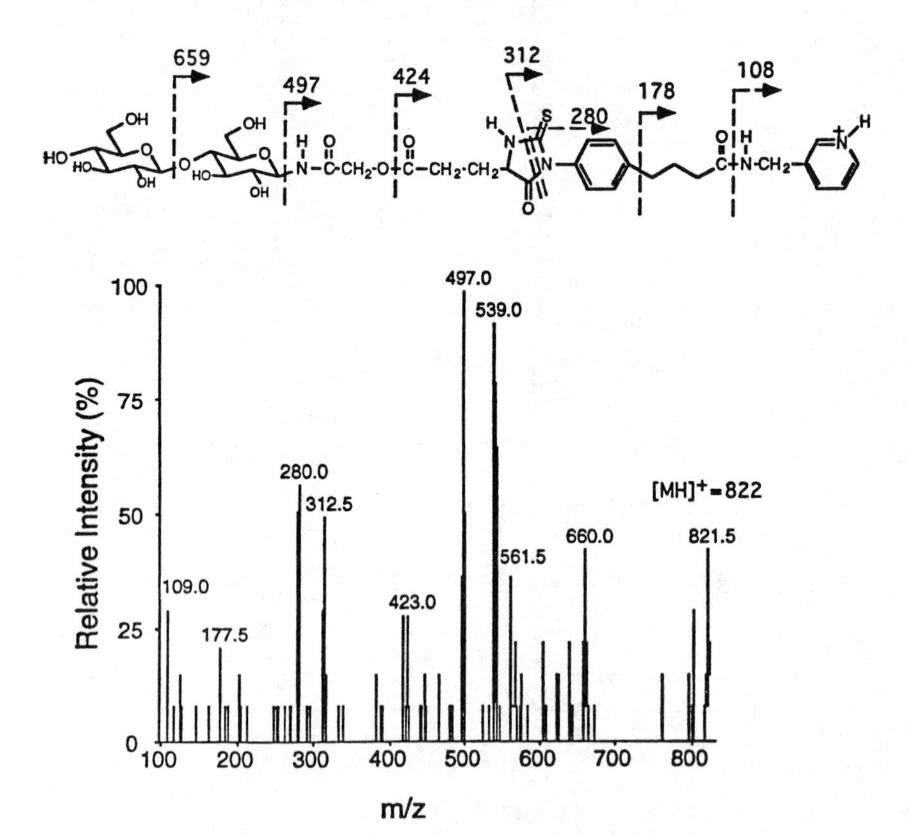

Figure 8. Product ion mass spectrum of the modified phenylthiohydantoin released in cycle 5 of the Edman degradation of the labeled DVVNEA peptide from Cex.

Conclusion

The two key active site residues in retaining glycosidases, the nucleophile and the acid-base catalyst, have been identified by specific chemical labeling strategies in conjunction with electrospray mass spectrometric analysis of proteolytic digests. This approach provides a rapid and reliable method for the identification of these important residues without the need for the synthesis and use of radiolabeled reagents. The validity of the approach has been confirmed in several cases by subsequent X-ray crystallographic determination of the 3-dimensional structures.

Acknowledgments

I thank all the individuals whose names are provided in the appropriate references for their invaluable contributions to this work. In particular I would like to thank Dr Ruedi Aebersold, with whom I have collaborated extensively on this work. In addition I thank the Natural Sciences and Engineering Research Council of Canada and the Protein Engineering Network of Centres of Excellence of Canada for financial support.

References

(1) Legler, G. *Adv. Carb. Chem. Biochem.* **1990**, *48*, 319-385.
(2) Mooser, G. *The Enzymes* **1992**, *20*, 187-233.
(3) Sinnott, M. L. *Chem. Rev.* **1990**, *90*, 1171-1202.
(4) Sinnott, M. L. In *Enzyme Mechanisms*; M. I. Page and A. Williams, Ed.; Royal Society of Chemistry: London, 1987; pp 259-297.
(5) Koshland, D. E. *Biol. Rev.* **1953**, *28*, 416-436.
(6) McCarter & Withers, *Curr. Opin. Struct. Biol.* **1994** *4*, 885-892.
(7) Withers, S. G.; Rupitz, K.; Street, I. P. *J. Biol. Chem.* **1988**, *263*, 7929-7932.
(8) Withers, S. G.; Street, I. P.; Bird, P.; Dolphin, D. H. *J. Amer. Chem. Soc.* **1987**, *109*, 7530-7531.
(9) McCarter, J.; Adam, M.; Withers, S. G. *Biochem. J.* **1992**, *286*, 721-727.
(10) Street, I. P.; Rupitz, K.; Withers, S. G. *Biochemistry* **1989**, *28*, 1581-1587.
(11) Street, I. P.; Kempton, J. B.; Withers, S. G. *Biochemistry* **1992**, *31*, 9970-9978.
(12) Withers, S. G.; Aebersold, R. A. *Protein Science* **1995**, *4* , 361-372.
(13) Carr, S. A.; Roberts, G. D. *Anal. Biochem.* **1986**, *157*, 396-406.
(14) Fenselau, C.; Heller, D. N.; Miller, M. S.; White III, H.B. *Anal. Biochem.* **1985**, *150*, 309-314.
(15) Biemann, K.; Martin, S. A. *Mass Spectrom. Rev.* **1987**, *6*, 1-76.
(16) Biemann, K.; Scoble, H. A. *Science* **1987**, *237*, 992-998.
(17) Fenn, J. B.; Mann, M.; Meng, C. K.; Wong, S. F.; Whitehouse, C. M. *Science* **1989**, *246*, 64-71.
(18) Covey, T. R.; Bonner, R. F.; Shushan, B. I.; Henion, J. D. *Rapid. Commun. Mass Spectrom.* **1988** *2*, 249-256.
(19) Hunt, D. F.; Henderson, R. A.; Shabanowitz, J.; Sakaguchi, K.; Michel, H.; Sevilir, N.; Cox, A. L.; Apella, E.; Engelhard, V. H. *Science* **1992** *255*, 1261-1263.
(20) Henderson, R. A.; Michel, H.; Sakaguchi, K.; Shabanowitz, J.; Apella, E.; Hunt, D. F., Engelhard, V. H. *Science* **1992** *255*, 1264-1266.
(21) Miao, S.; Ziser, L.; Aebersold, R.; Withers, S. G. *Biochemistry* **1994**, *33*, 7027-7032.
(22) Wakarchuk, W. W.; Campbell, R. L.; Sung, W. L.; Davoodi, J.; Yaguchi, M. *Protein Sci.* **1994**, *3*, 467-475.

(23) Miao, S.; McCarter, J. D.; Grace, M.; Grabowski, G.; Aebersold, R.; Withers, S. G. *J. Biol. Chem.* **1994**, *269*, 10975-10978.
(24) Dinur, T.; Osiecki, K. M.; Legler, G.; Gatt, S.; Desnick, R. J.; G., G. *Proc. Natl. Acad. Sci. U.S.A.* **1986**, *83*, 1660-1664.
(25) Black, T. S., Kiss, L., Tull, D. and Withers, S. G. *Carbohydr. Res.* **1993** *250*, 195-202.
(26) Veresztessy, Z.; Kiss, L.; Hughes, M. *Arch. Biochem. Biophys.* **1994***315*, 323-330.
(27) Naider, F.; Bohak, Z.; Yariv, J. *Biochemistry* **1972** 11, 3202-3207.
(28) Bures, E. J.; Nika, H.; Chow, D. T.; Morrison, H.; Hess, D.; Aebersold, R. *Anal. Biochem.* **1995** *224*, 364-372.
(29) Hess, D.; Nika, H.; Chow, D. T.; Bures, E. J.; Morrison, H.; Aebersold, R. *Anal. Biochem.* **1995** *224*, 373-381.
(30) Gilkes, N. R.; Langsford, M. L.; Kilburn, D. G.; Miller, R. C. Jr.; Warren. R. A. J. *J. Biol. Chem.* **1984**, *259*, 10455-10459.
(31) MacLeod, A. M.; Lindhorst, T.; Withers, S. G.; Warren, R. A. J. *Biochemistry* **1994**, *33*, 6571-6376.
(32) White, A.; Gilkes, N. R.; Withers, S. G.; Rose, D. *Biochemistry* **1994** *33*, 12546-12552.

RECEIVED November 7, 1995

Chapter 20

Electrospray Ionization Mass Spectrometric Investigation of Signal Transduction Pathways

Determination of Sites of Inducible Protein Phosphorylation in Activated T Cells

Julian D. Watts[1], Michael Affolter[2], Danielle L. Krebs[3], Ronald L. Wange[4], Lawrence E. Samelson[4], and Ruedi Aebersold[1,5]

[1]Department of Molecular Biotechnology, University of Washington, Box 357730, Seattle, WA 98195–7730
[2]NESTEC Ltd. Research Center, Vers-chez-les-Blanc, P.O. Box 44, CH–1000, Lausanne, Switzerland
[3]Department of Microbiology, University of British Columbia, Vancouver, British Columbia V6T 1Z3, Canada
[4]Cell Biology and Metabolism Branch, National Institute for Child Health and Development, National Institutes of Health, Bethesda, MD 20892

We have developed methodologies for the determination of sites of protein phosphorylation by tandem microbore column chromatography, with on-line detection by electrospray ionization mass spectrometry (ESI-MS), of phosphopeptides recovered from two-dimensional (2D) phosphopeptide maps. We have applied these methods to the identification of sites of tyrosine phosphorylation induced on the lymphocyte-specific protein tyrosine kinase ZAP-70 and the T cell receptor (TCR) ζ subunit *in vivo*, following stimulation of T cells via their TCR. Our approach is generally applicable to the investigation of signal transduction pathways since it involves direct ESI-MS analysis of peptides recovered from 2D phosphopeptide maps, currently the analytical technique of choice for the detailed analysis of protein phosphorylation sites, and is fully adaptable for the analysis of serine- and threonine-phosphorylated peptides.

It has been known for some time that a cell's ability to respond to external stimuli, be they chemical, physical or hormonal, is regulated to a large part by a wide range of cell surface receptor complexes, which in turn are linked to a variety of intracellular protein complexes and enzymes. The biochemical mechanism by which the triggering of a cell surface receptor is interpreted by the cell, resulting in a defined biological response, is often referred to as signal transduction. Of particular

[5]Corresponding author

0097–6156/95/0619–0381$13.75/0

importance in such processes is the role of protein phosphorylation/dephosphorylation of key signal transduction pathway components, events which are carried out by protein kinases and phosphatases, respectively, with many of the critical early phosphorylation events being performed by protein tyrosine kinases (PTK). The main obstacle to unraveling signal transduction pathways is that each cell type expresses many different receptors, with each capable of eliciting a different set of biochemical responses. For example, as is shown in Figure 1, engagement of either the T cell receptor (TCR) or interleukin-2 (IL-2) receptor of cultured T cells leads to the rapid tyrosine phosphorylation of different subsets of cellular proteins.

Figure 2 gives an overview of an activated TCR complex, its subunit structure, and the selection of associated PTKs and co-receptor molecules, known to be involved in TCR-mediated signaling, which are discussed below. Within T cells, a number of PTKs have been shown to play a role in the regulation of a range of

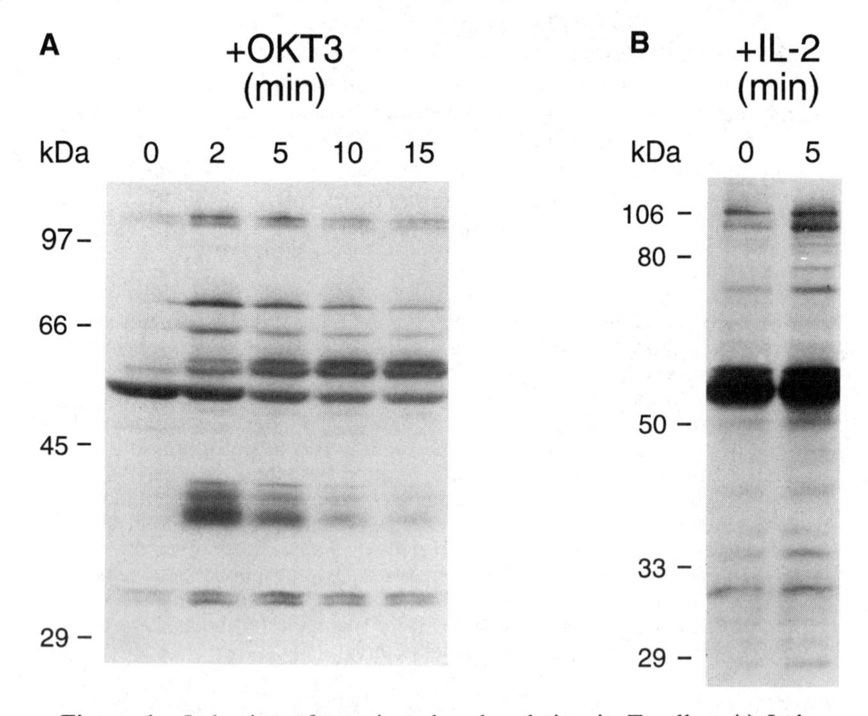

Figure 1: Induction of tyrosine phosphorylation in T cells. **A)** Jurkat (human) T cells (2 x 10^6/lane) stimulated for the indicated times with the anti-(human) CD3ε mAb OKT3 at 10 μg/ml essentially as described elsewhere (41, 42). **B)** IL-2-dependent splenic (murine) T cells (3.3 x 10^6/lane) were prepared, starved of, and then re-stimulated with recombinant IL-2 at 100 units/ml for the indicated times as described (22). Following SDS-PAGE and transfer to nitrocellulose, tyrosine phosphorylated proteins were visualized with the anti-phosphotyrosine mAb 4G10 detected by an enhanced chemiluminescence system. The relative mobility of molecular mass standards (kDa) are also indicated.

biochemical and cellular responses, generated by the engagement of the TCR (reviewed in 1-3). These include members of the *src* family, in particular p56*lck* and p59*fyn*, which interact with the CD4/CD8 co-receptors and the TCR respectively, along with the *syk* family member, ZAP-70. p56*lck* and p59*fyn* have been extensively studied in a number of model and transgenic cellular systems, and at least one or both have been shown to be vital components in multiple TCR-mediated signal transduction pathways, including positive and negative selection during thymocyte development (4-7), cell proliferation (6, 8), IL-2 production (9, 10) and TCR-mediated killing (11, 12). The activation of both p56*lck* and p59*fyn* is regulated via reversible tyrosine phosphorylation at a conserved C-terminal tyrosine residue, most likely involving another PTK, p50*csk* (13-15) and the transmembrane protein tyrosine phosphatase CD45 (16-18). p56*lck* has also been implicated in IL-2-mediated T cell responses through its interaction with the IL-2 receptor (19-22).

Following TCR engagement, a number of its subunits, in particular the ζ and CD3ε chains, become multiply tyrosine phosphorylated on a conserved motif, referred to as an immunoreceptor tyrosine-based activation motif (ITAM) (reviewed in references 1, 3 and 23). The ZAP-70 PTK binds specifically to the doubly tyrosine phosphorylated ITAMs of the TCR ζ and CD3ε subunits (ζ having three such motifs and CD3ε having one) (24-29), subsequently becoming itself both tyrosine phosphorylated and activated (24, 26-28, 30, 31). This ZAP-70/TCR ζ ITAM interaction is known to require both SH2 domains of ZAP-70 (32) and both phosphorylated tyrosines of the ITAM (27). SH2 domains are conserved folded structures common to a number of signal transduction proteins, whose function is to recognize and bind phosphotyrosine-containing polypeptide sequences (reviewed in reference 33).

Recent studies have shown that ZAP-70, like p56*lck* and p59*fyn*, is vital to the induction of a full TCR-mediated response. These include the analysis of immunodeficient individuals lacking ZAP-70 expression (34-36), or in cells in which the binding of ZAP-70 to phospho-ζ is blocked (31). Of great interest has been the recent demonstration that engagement of the same TCR by variant ligands can induce different T cell responses, with an optimal receptor engagement inducing ζ phosphorylation, ZAP-70 recruitment, tyrosine phosphorylation and activation, ultimately leading to cell proliferation and IL-2 secretion. On the other hand, sub-optimal receptor engagement induces a (presumed to be) differentially, or incompletely phosphorylated form of ζ (37, 38), the cells ultimately entering an inactive (anergic) state in one system (37). While this lower apparent molecular mass phospho-ζ isoform does appear to bind ZAP-70, the kinase does not become tyrosine phosphorylated or activated in the other system (38). It is thus clear that the identification of the sites of phosphorylation induced on molecules such as ZAP-70 and ζ are vital to the elucidation and understanding of the signal transduction pathways in which they are involved.

Currently, techniques for the analysis of protein phosphorylation rely largely on two-dimensional (2D) phosphopeptide mapping of isolated proteins which have been phosphorylated either via *in vitro* kinase assays, or *in vivo* following metabolic labeling of cells with radiolabeled phosphate. Due to the small quantities of such material recoverable from typical cell lysates and the often low stoichiometry of phosphorylation observed on proteins of interest, direct biochemical determination of phosphorylation sites inducible *in vivo* is almost impossible.

In order to address these problems, we combined and adapted current techniques in column chromatography and electrospray ionization mass spectrometry (ESI-MS).

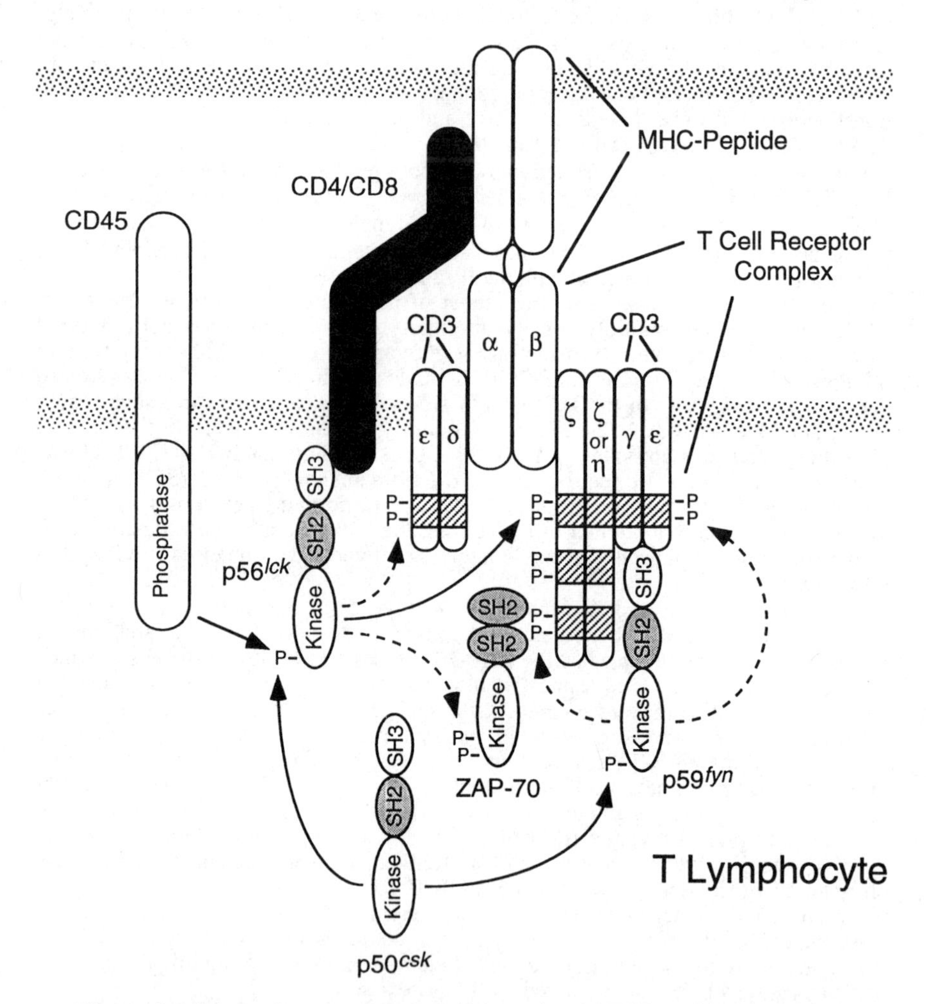

Figure 2: Schematic representation of an activated TCR and its associated PTKs, bound to the MHC-peptide complex on the surface of an antigen presenting cell. The subunits of a normal TCR (α, β and ζ and the CD3 chains γ, δ and ε) are labeled, with their respective ITAMs indicated (hatched boxes). The PTKs $p56^{lck}$, $p59^{fyn}$, $p50^{csk}$ and ZAP-70 are shown, along with their conserved domain structures: the catalytic domain (Kinase), SH2 and SH3 domains, which recognize and bind respectively phosphotyrosine-containing and proline-rich polypeptide sequences (reviewed in references 1-3). Inducibly phosphorylated tyrosine residues are indicated (P). Generally accepted phosphorylation/dephosphorylation events associated with TCR-mediated signaling are shown by solid arrows. Phosphorylation events thought likely to occur (at least to some degree), but lacking in substantial *in vivo* supporting data, are indicated by broken arrows.

Taking advantage of the affinity of phosphopeptides for immobilized ferric ions (39), we connected microbore immobilized metal affinity chromatography (IMAC) and microbore reversed phase high performance liquid chromatography (HPLC) columns in series, with on-line analysis of column eluates by ESI-MS (IMAC-HPLC-ESI-MS). With this system we were able to determine both the sites of ZAP-70 autophosphorylation and those induced on ZAP-70 by p56^{lck} *in vitro*, by using recombinantly expressed and isolated ZAP-70 and p56^{lck} and the recovery of tryptic phosphopeptides from 2D phosphopeptide maps. By comparison of these data with tryptic phosphopeptide maps of *in vivo*-phosphorylated ZAP-70, isolated from stimulated Jurkat T cells following ^{32}P-metabolic labeling, we were able to determine the sites of tyrosine phosphorylation induced on ZAP-70 following TCR engagement. This methodological approach is summarized in Figure 3. Using a simpler, single column, HPLC-ESI-MS system, expressed p56^{lck} and a synthetic peptide corresponding to the entire cytoplasmic fragment of the TCR ζ subunit (residues 52-164), we determined the mobilities of all the major potential tyrosine-phosphorylated tryptic fragments of ζ in a 2D map. We were then similarly able to identify TCR-induced sites of tyrosine phosphorylation on ζ *in vivo* by the comparison of its 2D phosphopeptide map with that obtained from the *in vitro*-derived samples.

Since many of the important phosphoprotein components of cell signaling pathways have now been identified, the availability of expressed forms of these proteins means that comparison of *in vitro* and *in vivo* generated phosphopeptide maps and subsequent analysis by ESI-MS of the *in vitro*-derived phosphopeptides should be a more rapid approach than by mutagenic analysis of all potential phosphorylation sites. This approach thus represents a direct bioassay for the biochemical determination of the exact phosphorylation state of any phosphoprotein of interest, which is, by its very nature, more reliable than indirect analysis via mutagenesis. The availability of ESI-MS-based rapid, sensitive and direct approaches for the determination of inducible phosphorylation sites on ZAP-70 and other proteins of interest involved in signal transduction pathways should thus greatly facilitate investigation of the specific roles played by these modifications via site-directed mutagenesis and subsequent expression of these mutant proteins in appropriate transgenic or cell culture systems.

Materials and Instrumentation.

Chemicals and reagents. Chemicals and solvents required for 2D phosphopeptide mapping were purchased from BDH. All other solvents were from Fisher Scientific. Sequencing grade trifluoroacetic acid and acetonitrile/water (1:4) were obtained from Pierce and Applied Biosystems, respectively. 18 mΩ water for all buffers and solvent mixtures was purified on a Barnstead NANOpure water system. Sequencing grade modified trypsin was from Promega, cellulose TLC plates (20 x 20 cm) were from Kodak. Nitrocellulose membranes were from Schleicher and Schuell, RPMI 1640 culture media from the Terry Fox Laboratories, and all other cell culture reagents from Gibco BRL. Recombinant IL-2 was obtained from Genzyme, reagents and standards for polyacrylamide gel electrophoresis (PAGE) were from BioRad, ^{32}P-labeled reagents from ICN, and ATP, polyvinyl pyrrolidone-40 and protein G-Sepharose from Sigma. Protein A-Sepharose was from Pharmacia, ECL reagents were from Amersham, and HRP-conjugated secondary antibodies for immunoblotting from Calbiochem.

Synthetic peptides were generously provided by Dr. I. Clark-Lewis (University of British Columbia). Baculovirus-expressed p56^{lck} and a synthetic peptide corresponding to the cytoplasmic domain of the TCR ζ subunit (residues 52-164) were prepared as described elsewhere (40). Baculovirus expression vector for ZAP-70 was also prepared as described (41), with expression being performed by Invitrogen (San Diego). Sepharose beads conjugated to a synthetic ζ peptide (residues 52-164) in either a phosphorylated or non-phosphorylated form were

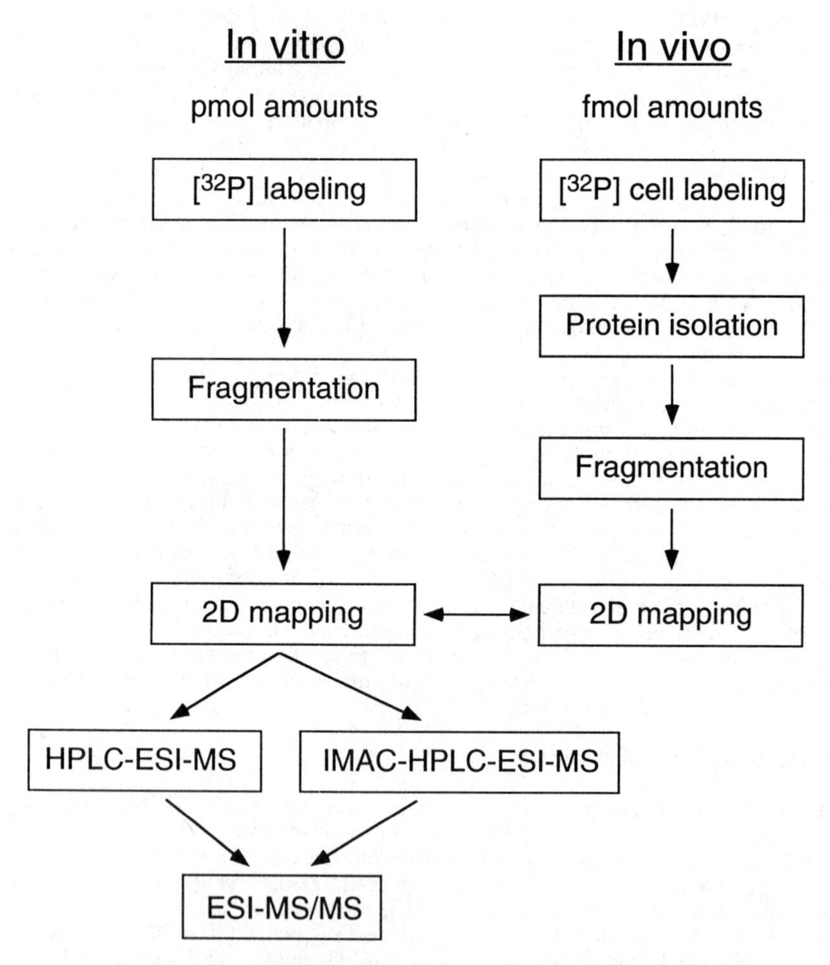

Figure 3: Schematic representation of the experimental approach, utilizing both *in vivo*- and *in vitro*-derived material for the determination of protein phosphorylation sites. Due to the low fmol amounts of phosphoproteins of interest available from *in vivo* sources, the *in vivo*-derived samples are correlated at the level of 2D phosphopeptide mapping, while larger (pmol) amounts of *in vitro*-prepared samples are analysed via ESI-MS.

prepared essentially as described (26) and were generously provided by Dr. P. Orchansky (University of British Columbia). CD11.3 cells were a gift from Dr. H.-S. Teh (University of British Columbia), the monoclonal antibodies (mAb) OKT3 and 2C11 were prepared as described (42), 9E10 mAbs were prepared as ascites fluid, and 4G10 mAbs were a gift from Dr. M. Gold (University of British Columbia). Anti-ZAP-70 polyclonal antiserum was a gift from Dr. J. Bolen (Bristol-Myers Squibb) and C305 mAbs were prepared as described (31). The cytoplasmic fragment of erythrocyte band 3 protein (cfb3) was prepared as described (43) and was known to be a substrate for the ZAP-70 related kinase, $p72^{syk}$ (44).

Column Chromatography and ESI-MS Instrumentation: Column chromatography, plumbing and solvent systems employed were exactly as described previously (41, 45). Fused silica capillaries were from Polymicro Tech. Inc. Teflon tubing was from Mandel Scientific, chelating Sepharose Fast Flow from Pharmacia, and microbore C_{18} reversed phase HPLC columns used were from either Micro-Tech Scientific or LC Packings (Hypersil). C_4 reversed phase columns (4.6 x 250 mm) were purchased from Vydac. The flow to the IMAC column (250 µm x 3 mm) was driven by a Harvard Apparatus via a Rheodyne 8125 injector, the IMAC and HPLC columns being connected by a Rheodyne 7000 valve. The solvent gradient to the C_{18} column (250 µm x 150 mm) was delivered by the pump system of an Ultrafast Microprotein Analyzer (Michrom BioResources Inc.), and the outlet from the column was on-line to the electrospray probe tip of a PE Sciex API*III* triple quadrupole mass spectrometer.

Mass Spectrometric and Data Analyses. ESI-MS analyses were performed on a PE Sciex API*III* triple quadrupole mass spectrometer, equipped with a pneumatically assisted ESI source (ion spray). The mass spectrometer was scanned repetitively over a mass to charge ratio (m/z) range of either 300-2000, using a step of 0.5 units, a dwell time of 1.0 msec (single scan duration = 3.4 sec) (41) over the m/z range of 300-2400, 0.5 Da step, 1.0 msec dwell/step (single scan duration: 4.2 sec) (45), and an 80 V orifice potential. For tandem mass spectrometry (MS/MS) analyses, doubly charged parent ions were selected in quadrupole 1 for collision-induced decomposition (46, 47) with an argon gas (thickness of ~4×10^{14} molecules/cm^2) in a radio frequency only quadrupole 2; scanning quadrupole 3 was set at 1.0 Da step, with a 5.0 msec dwell/step. Data analysis software was also obtained from PE Sciex.

Results and Discussion.

Over the last few years, there has been a significant increase in the development of MS-based approaches for the analysis of protein phosphorylation. These include fast atom bombardment (FAB) MS (48-50), ESI-MS (51-53), liquid secondary ion MS (54) and matrix-assisted laser desorption time-of-flight MS (55). However, in order to identify phosphorylation sites on relevant signal transduction proteins, it is often only possible to characterize *in vivo*-derived samples at the level of a 2D phosphopeptide map, due to the low quantities and stoichiometry of phosphorylation occurring in cells. Thus as is summarized in Figure 3, it would be useful to devise MS-based technology for the analysis of phosphopeptides recovered from the 2D TLC plates used in such analyses. Some attempts have been made to apply both FAB and ESI-MS analyses to samples recovered from phosphopeptide maps, both involving off-line reversed phase HPLC purification steps (46, 56). More recently, we used ESI-MS on-line to an HPLC column for the detection of such recovered phosphopeptides (45). However, in each of these studies, sensitivity was relatively poor (10-20 pmol of injected sample). This was due in part to the high chemical background observed in the samples recovered from the 2D phosphopeptide maps, as well as inefficiencies in the off-line sample recovery and transfer methods.

Analysis of ZAP-70 phosphorylation sites by IMAC-HPLC-ESI-MS. In order to provide a useful tool for protein chemists and cell biologists, a significant increase in sensitivity was required. Unrelated work had revealed that phosphorylated peptides and proteins have an affinity for immobilized Fe^{3+} ions (39). Attempts to couple Fe^{3+} IMAC columns on-line to ESI-MS detection (57) had been made. However, this resulted in the simultaneous detection of all phosphopeptide species eluted from the column, and again with poor detection sensitivity (10-20 pmol). Additionally, this approach leads to the elution of phosphopeptides in a sub-optimal buffer for ESI-MS analysis, necessarily resulting in compromise. To overcome these limitations, we linked a microIMAC column and microbore HPLC column in series (41), the instrumentation set-up being summarized in Figure 4. A key element in the set-up was a switching valve (*) which was used to isolate the IMAC column, driven by a low-pressure syringe drive, from the high pressure HPLC system. Samples eluted from the IMAC column could then be collected in the sample loop (Figure 4B) and then easily applied on-line to the HPLC (Figure 4C). An advantage of this system over previous ones was that it allowed both the removal of some of the background signal resulting from the IMAC elution conditions, as well as providing separation of multiple phosphopeptides if present. It also allowed solvent exchange, which facilitated ultimate elution of the sample in a buffer optimized for efficient ionization.

We compared this IMAC-HPLC-ESI-MS system to the same set-up, but without the IMAC column (HPLC-ESI-MS). Extensive details for the instrumentation and column chromatography conditions for both set-ups have been described (41, 45). Loading known amounts of the same peptide recovered from a 2D phosphopeptide map, the advantages of the two column system over the single column system in both sensitivity and background signal were clear (Figure 5). In the HPLC-ESI-MS system, any peptide eluting ~22 min post-injection would be difficult to find due to background ions resulting from the 2D mapping procedure (Figure 5A). In Figure 5B, the signal peak for a lower loading (5 pmol) is well resolved above the background, with the signal to noise ratio (~15:1) indicating that much smaller sample loadings would be possible, even for peptides eluting late from the HPLC column.

To calibrate the IMAC-HPLC-ESI-MS system, we then used a synthetic phosphopeptide as a standard, which was serially diluted and injected directly onto the two column system. Under these conditions, we obtained a detection sensitivity for the system of at least 250 fmol, if we assumed that the mass of the peptide was unknown, and much lower (50 fmol or less) should the peptide sequence be known (data not shown, see reference 41). As can be seen from the signal to noise ratio in Figure 5B (~15:1), obtained with a 5 pmol load on the IMAC-HPLC-ESI-MS system of a phosphopeptide recovered from a 2D map (spot G in Figure 6B), phosphopeptide samples actually derived from the TLC plate can be meaningfully analyzed in the 0.5 to 1 pmol range. However, there is a high degree of variability inherent in preceding preparation steps: multiple phosphorylation sites and varying stoichiometry, varying efficiency of phosphorylation, digestion efficiency (which will depend also on the endoprotease employed), relative solubility of resulting phosphopeptides when recovering from a gel, membrane or TLC plate. Thus the amount of starting material required to yield the required ~1 pmol for ESI-MS analysis can vary tremendously, and must be evaluated independently on a case-by-case basis.

We then chose to apply the IMAC-LC-ESI-MS technology to the determination of sites of tyrosine phosphorylation inducible on ZAP-70 in order to demonstrate viability as a general method for the investigation of phosphorylation events occurring in signal transduction pathways. To facilitate the *in vitro* studies, we used recombinantly expressed (murine) $p56^{lck}$ and (human) ZAP-70. The human T cell line, Jurkat, which expresses high levels of both PTKs, was used to facilitate the *in vivo* studies. The recombinant human ZAP-70 was isolated from expressing cells via a C-terminal peptide affinity tag that had been cloned into the ZAP-70 gene (41) and is recognized by the mAb 9E10 (58). Recombinant expressed $p56^{lck}$ was partially purified as described (40). The expressed ZAP-70 was immunoprecipitated with

9E10 and incubated with or without the addition of purified recombinant p56*lck* prior to analysis by SDS PAGE and transfer to nitrocellulose (Figure 6A). Tryptic digestion of the ZAP-70 band from Figure 6A, lane 2 was performed, and recovered phosphopeptides were analysed by 2D phosphopeptide mapping (Figure 6B). By carefully removing the cellulose matrix from the TLC plate containing each phosphopeptide (labeled A to G), the peptides could be resolubilized in acetonitrile/water (1:4 v/v) and concentrated for injection onto the IMAC-HPLC-ESI-MS system. The data set for the analysis of spot G is shown in Figure 5B, where the singly charged ion ($M+H^+$ = 1247.0) and doubly charged ion ($[M+2H]^{2+}$ = 624.5) shown in inset correspond to a phosphopeptide mass of M = 1246.0 for the indicated peak. This matched a computer-predicted mass of 1166.6 for a partial tryptic fragment of ZAP-70, residues 176-186 (KLYSGAQTDGK), plus an additional 80 mass units due to a single phosphate ester group. Each phosphopeptide spot (A to G) from Figure 6B was similarly analyzed, each giving mass values corresponding to an expected ZAP-70-derived tryptic phosphopeptide, the results being summarized in Table I below. Spots A and B (derived from a partial digestion around tyrosine 292) and spot C (derived from tyrosine 126) were found to be the autophosphorylation sites of ZAP-70. This was done by analyzing the ZAP-70 band from lane 1, Figure 6A as well as ZAP-70 autophosphorylated *in vitro* following isolation from Jurkat T cells (data not shown, see reference 41). Additional sites of phosphorylation on ZAP-70 were seen in the presence of p56*lck* at tyrosines 69 and 178 (spots C and G, respectively) as well as spots resulting from the ZAP-70 tryptic fragment containing tyrosines 492 and 493 in both its singly and doubly phosphorylated forms (spots E and F, respectively).

We next investigated sites of phosphorylation induced on ZAP-70 in Jurkat T cells following ^{32}P metabolic labeling and stimulation of the TCR with the anti-CD3ε mAb, OKT3. ZAP-70 was isolated from cell lysates using synthetic phospho-ζ (to which ZAP-70 naturally binds) coupled to agarose beads (26, 41). Figure 6C shows an increase in the phosphorylation of a 70 kDa band (p70), presumably ZAP-70, observed following TCR stimulation. A 2D phosphopeptide map (Figure 6D) of the 70 kDa ZAP-70 band from lane 2 of Figure 6C revealed a subset of the peptides (spots A, E and F) previously observed *in vitro* with p56*lck*. Spots A, E and F from both *in vitro* (Figure 6B) and *in vivo* (Figure 6D) derived 2D maps were again eluted from the TLC plates, like spots pooled together, and re-analyzed on new 2D maps. A single spot result in each case confirmed the assignment indicated in Figure 6D (data not shown, see reference 41). Thus the combination of the 2D mapping and ESI-MS data could demonstrate that the tyrosine 292 autophosphorylation site and both tyrosines 492 and 493 were the principle sites on which ZAP-70 became phosphorylated in the activated Jurkat T cells. Since tyrosines 492 and 493 are not sites of ZAP-70 autophosphorylation, at least one other T cell PTK, possibly a *src*-family member such as p56*lck* or p59*fyn* must be involved in ZAP-70 phosphorylation *in vivo*. All the phosphorylation sites thus far identified on ZAP-70 are summarized in Figure 7, along with an indication of the conditions under which each has so far been observed.

Analysis of TCR ζ phosphorylation sites by HPLC-ESI-MS. As indicated in Figure 2, it is known that ZAP-70 binds to the phosphorylated forms of ζ and CD3ε within the TCR complex (24-30). This interaction requires the binding of both the SH2 domains of ZAP-70 to the two phosphorylated tyrosines of a single ITAM (27, 32). ITAMs are conserved protein sequences ($YXX(L/I)X_{(6\text{-}8)}YXX(L/I)$) and are found in a number of lymphocyte receptor subunits. In T cells, ITAMs are found in the TCR ζ subunit, its alternately spliced isoform η, as well as the CD3 chains γ, δ and ε, and have also been identified in B cells in the Igα and Igβ subunits of the B cell receptor (reviewed in 1, 3, 23, 59). Since the recent demonstration that different phospho-ζ isoforms can be induced *in vivo* by differential stimulation of the same TCR (37, 38), it would thus be extremely useful to determine which tyrosines residues in ζ can become phosphorylated following TCR engagement.

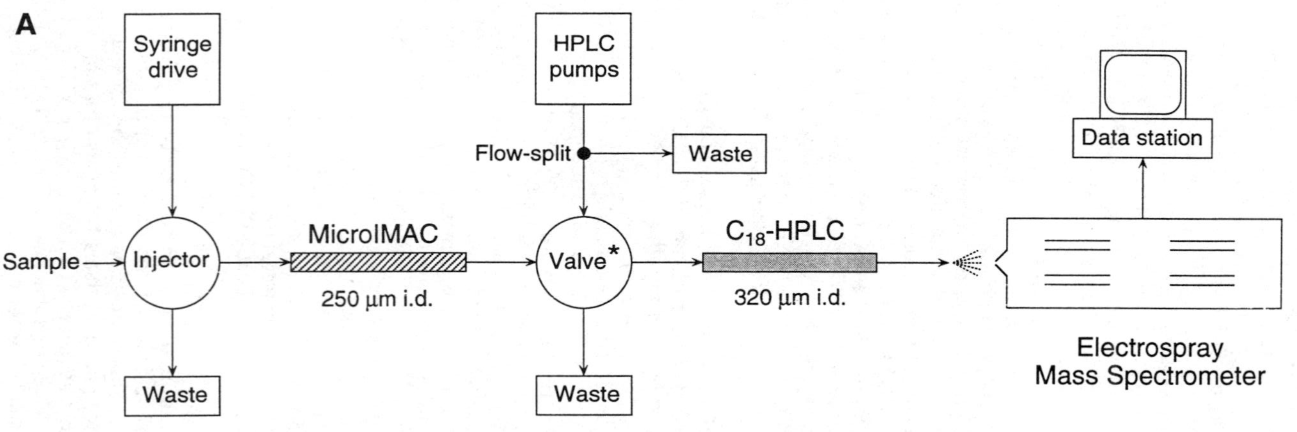

Figure 4: **A)** Schematic of the IMAC-HPLC-ESI-MS instrumentation employed. Microbore IMAC and HPLC columns were connected on-line to a triple quadrupole electrospray mass spectrometer as shown, described in detail elsewhere (41). Expanded views of the switching valve (*) used to link the microIMAC column to the HPLC system are shown. **B)** Valve position for IMAC column loading, washing and subsequent elution of bound phosphopeptides into the sample loop. **C)** Valve position for loading the contents of the sample loop onto the HPLC system. (Reproduced with permission from reference 41. Copyright 1994, The American Society for Biochemistry and Molecular Biology, Inc.).

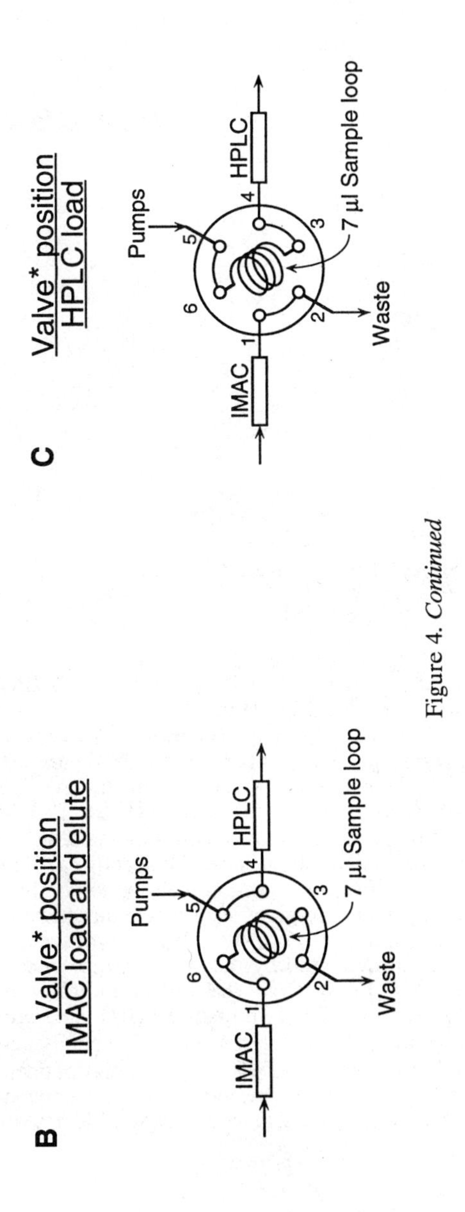

Figure 4. *Continued*

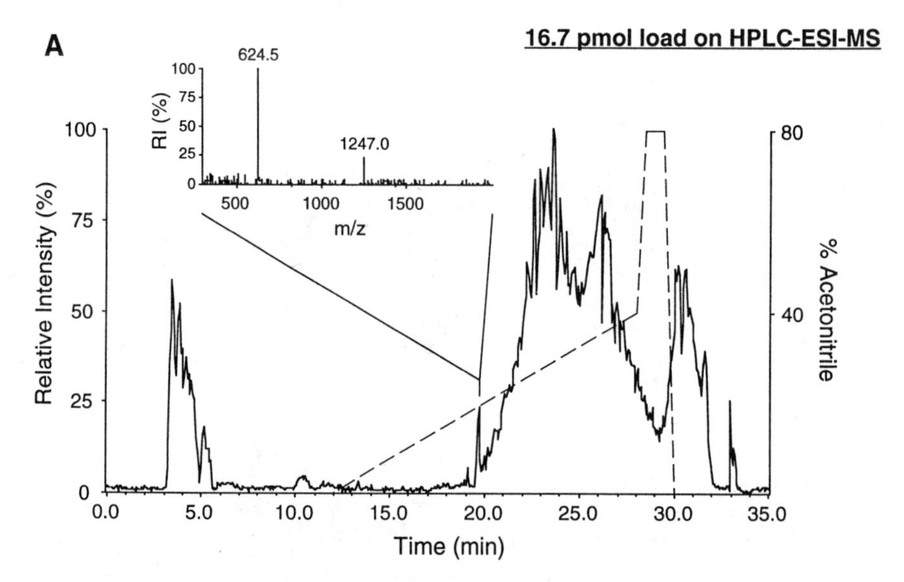

Figure 5: Comparison of HPLC-ESI-MS and IMAC-HPLC-ESI-MS. A single phosphopeptide resulting from the *in vitro* phosphorylation of expressed ZAP-70 by expressed p56*lck* following trypsin digestion and elution from a 2D phosphopeptide map (spot G in Figure 6B) was prepared as described (41). This was analyzed on both the **A)** HPLC-ESI-MS as described elsewhere (45) or the **B)** IMAC-HPLC-ESI-MS system as described (41). Following Cerenkov counting, peptide quantities given (pmol) were estimated from the known ATP specific activities used during the kinase reaction. Solid lines represent the total ion currents obtained as a function of time (min) and are given as a percentage relative to the largest signal peak in the data set. Broken lines represent the acetonitrile concentration gradient delivered by the HPLC pump system. Insets show the complete mass/charge (m/z) spectra for the indicated peaks following background subtraction. Peak intensities (RI) are again given as a percentage, relative to the largest signal peak. The singly charged ($M+H^+$ = 1247.0) and doubly charged ($[M+2H]^{2+}$ = 624.5) phosphopeptide species are both indicated. (Reproduced with permission from reference 41. Copyright 1994, The American Society for Biochemistry and Molecular Biology, Inc.).

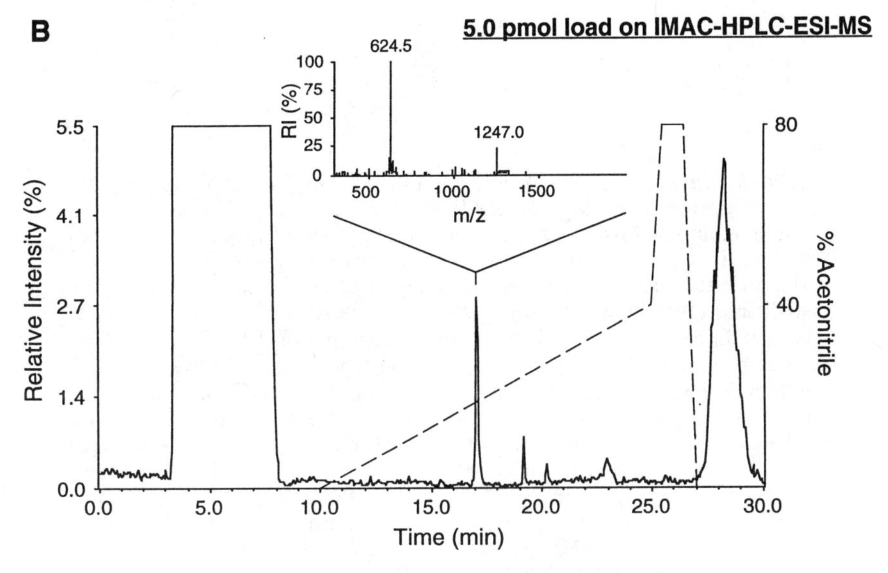

Figure 5. *Continued*

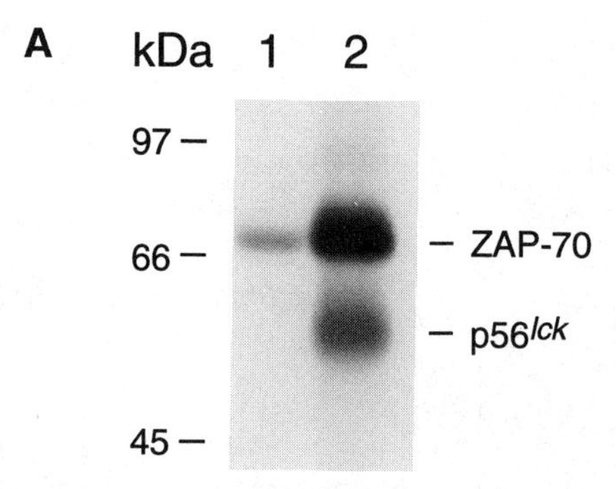

Figure 6: *In vitro* (panels A and B) and *in vivo* (panels C and D) 2D phosphopeptide mapping of ZAP-70. **A)** Expressed ZAP-70 was immunoprecipitated from cell lysate by means of an affinity tag recognized by the 9E10 mAb, and immune complex kinase assays were then performed in the absence (lane 1) or presence (lane 2) of $p56^{lck}$ as described (41). Visualization was by autoradiography. The relative mobility of molecular mass standards (kDa) are indicated, along with that of ZAP-70 and $p56^{lck}$. **B)** 2D tryptic phosphopeptide mapping was performed following SDS-PAGE and *in situ* digestion of ZAP-70 phosphorylated with $p56^{lck}$ (panel A, lane 2) as described (41). The orientation of the positive and negative electrodes during electrophoresis are indicated along with the sample origin (O). Visualization was by autoradiography. **C)** ^{32}P-labeled Jurkat cell lysates (1.6 x 10^7/lane) were precipitated with either phospho-ζ-beads (pζ) or non-phospho-ζ-beads (non-pζ), with or without stimulation with the anti-CD3ε mAb OKT3 (10 μg/ml) as indicated (described in 41). Phosphoproteins were visualized by autoradiography following SDS-PAGE and transfer to nitrocellulose. The relative mobility of molecular mass standards (kDa) are indicated. **D)** 2D tryptic phosphopeptide map of the 70 kDa band indicated in lane 2 of panel C (p70), digested *in situ* on the membrane as described (41, 64). The orientation of the positive and negative electrodes during electrophoresis are indicated along with the sample origin (O). Visualization was again by autoradiography. (Adapted from reference 41).

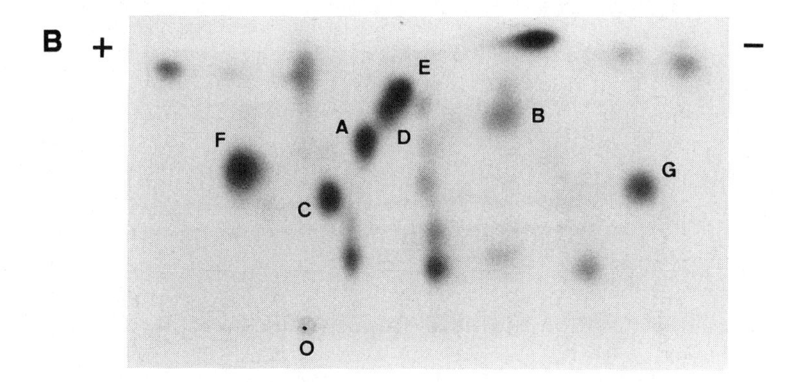

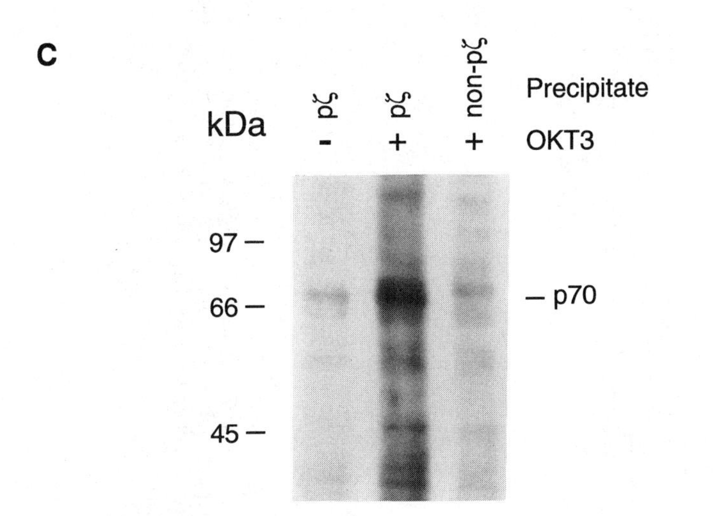

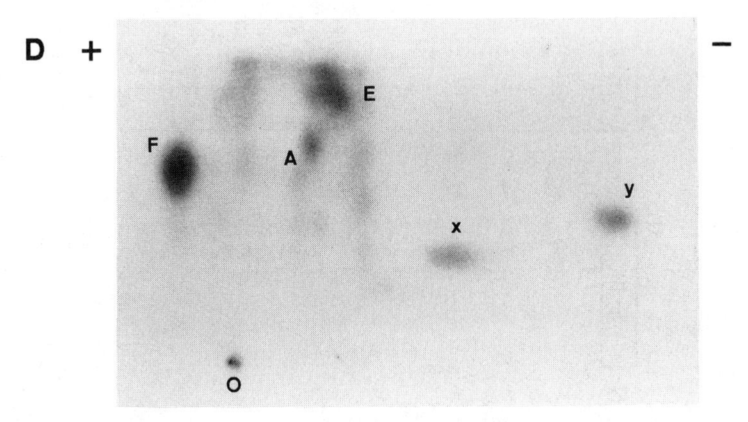

Figure 6. *Continued*

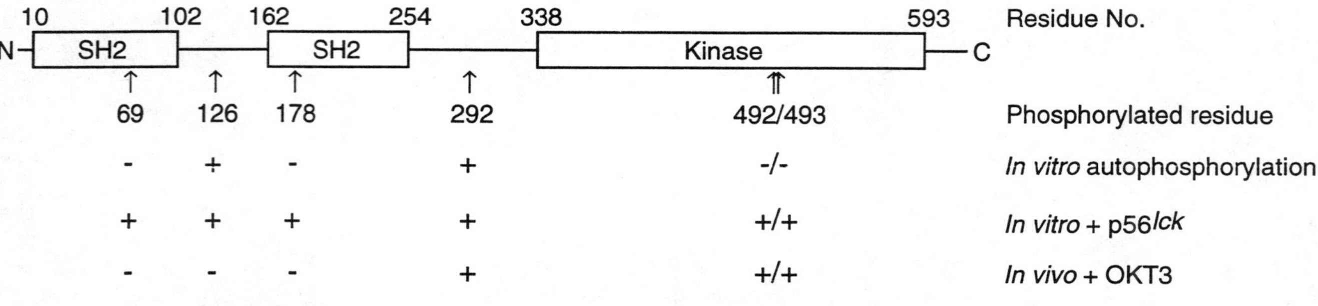

Figure 7: Schematic representation of inducible phosphorylation sites on ZAP-70. Residue numbers and the boundaries of the SH2 and kinase domains are taken from reference 24. The location and residue numbers of the characterized tyrosine phosphorylation sites so far identified on ZAP-70 are also indicated, along with the conditions under which they were been observed, summarized from Table I, Figure 5 and reference 41. (Reproduced with permission from reference 41. Copyright 1994, The American Society for Biochemistry and Molecular Biology, Inc.).

To facilitate these studies, we utilized a synthetic peptide corresponding to residues 52-164 of murine ζ. This peptide includes all six of the cytoplasmic tyrosines of ζ, which are collectively contained within the three potential ITAMs present in ζ. This peptide was phosphorylated *in vitro* with $p56^{lck}$, repurified over a wide-bore C_4 HPLC column, and cleaved with trypsin (45). The advantage of using trypsin for mapping ζ is the presence of at least one potential tryptic cleavage site between each of the tyrosine residues (see Figure 8B). The disadvantage of trypsin is presence of several multiple tryptic cleavage sites in ζ, which result in a more complex 2D phosphopeptide map than had been hoped for (see Figure 8A). Nevertheless, the abundance of material afforded by the use of a synthetic molecule made analysis of the *in vitro*-derived phosphopeptide map by HPLC-ESI-MS simple (45). Figure 8B summarizes the MS data obtained, displaying the peptide fragment contained in each of the spots (labeled A to J in Figure 8A). The assignments indicated in Figure 8B were additionally confirmed by MS/MS analysis of each spot. This was particularly useful to confirm the identity of the partial tryptic fragments (spots F and H) as well as the two anomalous tryptic cleavage sites, both derived from cleavage C-terminal of glycines 135 and 137 (spots G and J) (data not shown, and see reference 45).

We next investigated the phosphorylation of TCR ζ in the murine T cell line CD11.3, stimulated with the mAb 2C11, which recognizes the murine CD3ε subunit. In order to resolve the ζ subunit away from other small, phosphorylated TCR subunits, anti-TCR immunoprecipitates were run on non-reducing/reducing 2D gels (60). As can be seen in Figure 9A, ζ, being a disulfide-linked dimer, drops below the diagonal following reduction, whereas other TCR-associated molecules do not. Following TCR engagement, we see the δ, ε and ζ subunits (labeled d, e and z, respectively) inducibly tyrosine phosphorylated (Figure 9A, left two panels). When the same stimulation and immunoprecipitations are performed on ^{32}P-metabolically labeled cells (Figure 9A, right two panels), we see additional inducible phosphorylation of the TCR γ subunit (labeled g), which occurs on serine residue(s) (61, 62). Figure 9B shows the 2D phosphopeptide map obtained following tryptic cleavage the high molecular mass isoform of ζ (labeled p21). By comparison with the *in vitro*-derived data (Figure 8A), we appeared to have generated phosphopeptides resulting from 5 of the six tyrosine residues (spots A to E), the partial fragment related to spot A (spot F) and what is probably the major (anomalous) tryptic fragment resulting from the sixth tyrosine residue (spot G), which does not produce a perfect tryptic fragment *in vitro* either. As before, elution and co-mapping of the spots observed in Figure 9B with their suspected counterparts, eluted from the *in vitro*-derived map (Figure 8A), confirmed their identity, though spots E and F were hard to assign since they overlapped in the *in vivo* phosphopeptide map (data not shown). Thus it appears that at least five, if not all six, of the cytoplasmic tyrosines of the ζ ITAMs can become phosphorylated *in vivo* following TCR engagement.

Interpretation of 2D phosphopeptide maps: potential pitfalls. Analytical 2D phosphopeptide mapping has been used almost universally for the analysis of protein phosphorylation sites. Frequently, the number of spots obtained in a 2D phosphopeptide map are taken to indicate the number of phosphorylation sites present in the protein under investigation. However, the examples presented and analyzed in detail here (Figures 6 and 8) make it clear that a phosphopeptide map is not necessarily representative of the phosphorylation state of the protein, and highlight a number of potential pitfalls that could easily lead to the misinterpretation of a 2D phosphopeptide map without additional analysis beyond that level.

Since every 2D phosphopeptide map runs slightly differently, it is important to be confident that when comparing the *in vivo* and *in vitro*-derived maps (e.g. Figure 6B and 6D) that identical spots (A, E and F in the case of ZAP-70) are assigned correctly. This can be done by eluting the peptides spots from the cellulose matrix as before and pooling together spots believed to represent the same phosphopeptide. In this way, the co-migration as a single spot on a second 2D phosphopeptide map of the

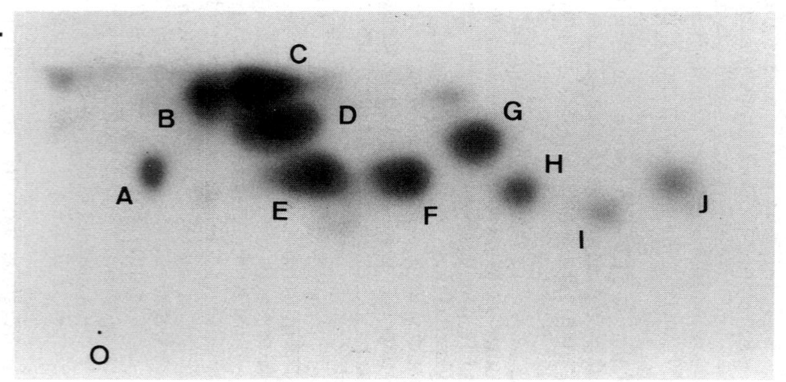

Figure 8: 2D phosphopeptide mapping of *in vitro*-phosphorylated TCR ζ. **A**) Synthetic murine TCR ζ (residues 52-164) was phosphorylated *in vitro* with p56*lck*, digested with trypsin and analyzed on a 2D phosphopeptide map. The orientation of the positive and negative electrodes during electrophoresis are indicated along with the sample origin (O). Visualization was by autoradiography. Phosphopeptide spots (labeled A to J) were eluted and analyzed by HPLC-ESI-MS as described (45). ζ-derived phosphopeptide fragments thus identified are summarized in **B**). The location of potential tryptic cleavage sites (arrows) and the positions and residue numbers of the phosphotyrosine residues (bold) are shown. Peptide fragments A to J identified by HPLC-ESI-MS are also indicated. Amino acid residue numbers referred to relate to those for the entire TCR ζ molecule. (Adapted from reference 45).

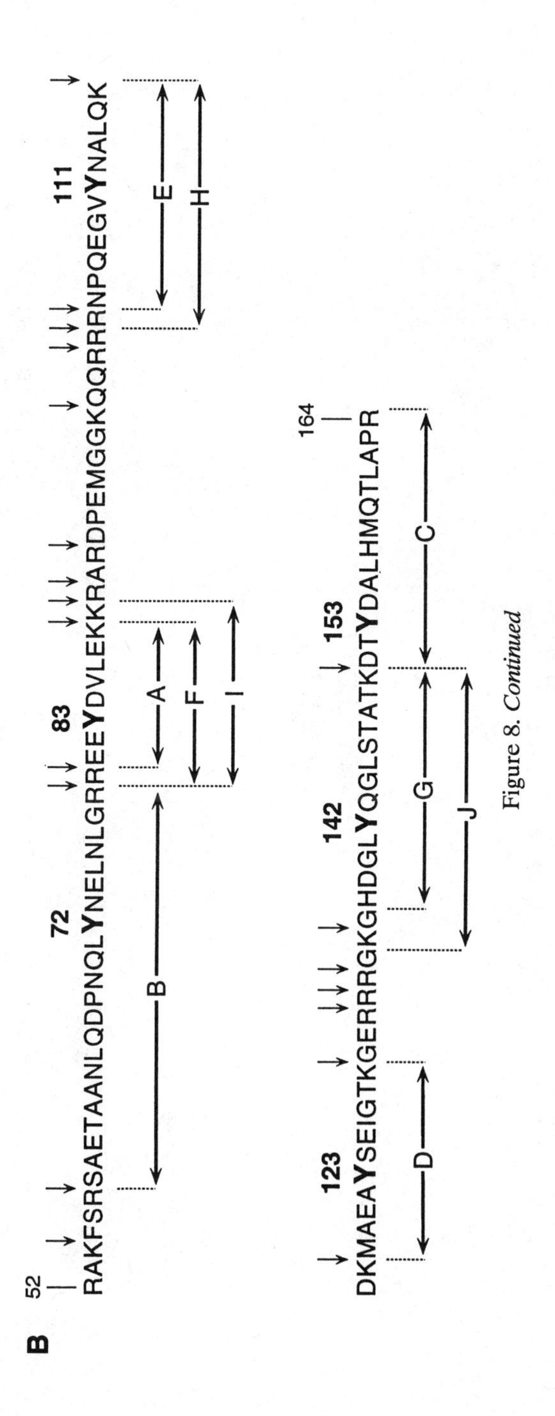

Figure 8. *Continued*

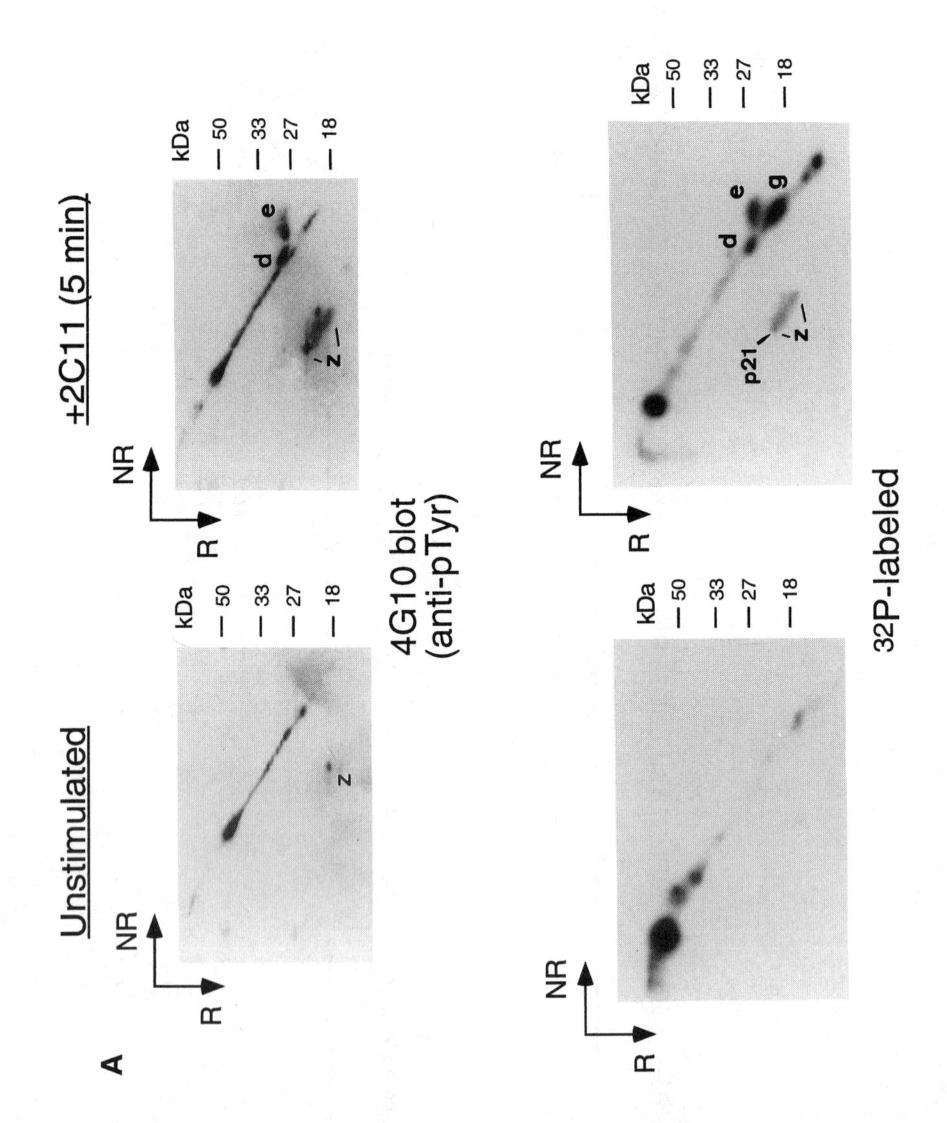
A
Unstimulated
+2C11 (5 min)
NR
R
kDa
50
33
27
18
d
e
g
z
p21
4G10 blot
(anti-pTyr)
32P-labeled

Figure 9: Tyrosine phosphorylation and 2D phosphopeptide mapping of TCR ζ in T cells. **A)** 5 x 107 CD11.3 (murine) T cells were incubated for 5 min in the presence or absence of the anti-(murine) CD3ε mAb 2C11 (10 μg/ml) with or without prior ^{32}P-labeling, as indicated. TCR complexes were prepared by precipitating with 2C11, run on non-reducing/reducing 2D gels and transferred to nitrocellulose essentially as described (60). Non-32P-labeled cell precipitates were analyzed by immunoblotting with 4G10 and visualized by enhanced chemiluminescence. 32P-labeled cell precipitates were visualized directly on the membrane by autoradiography. The relative mobility of molecular mass standards (kDa) are also indicated, along with the directions of the reducing (R) and non-reducing (NR) electrophoresis dimensions. The positions of the δ, ε, γ and ζ subunits are indicated (d, e, g and z, respectively) where appropriate, along with the maximally phosphorylated TCR ζ isoform (p21) subjected to further analysis. **B)** 2D tryptic phosphopeptide map of the ^{32}P-labeled p21 TCR ζ isoform indicated in panel A, digested *in situ* on the membrane as described (41, 64). Phosphopeptide spots A to G matching those shown in Figure 8 are labeled, along with the orientation of the positive and negative electrodes during electrophoresis and the sample origin (O).

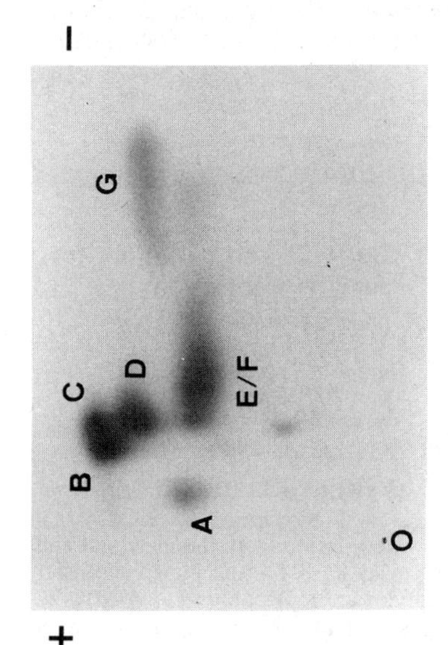

in vivo and *in vitro*-derived ZAP-70 spots A, E and F confirmed as correct the assignments we had made for the *in vivo* sites of ZAP-70 phosphorylation (data not shown, see reference 41). From Figure 6D, we also suspected that one of the spots labeled x and y might represent *in vivo* phosphorylation of ZAP-70 at tyrosine 178 (spot G, Figure 6B). However, neither spot x nor y were found to co-migrate with spot G when run on a second 2D map (data not shown, see reference 41). Thus without the co-mapping experiment, it would have been easy to conclude (probably incorrectly) that tyrosine 178 of ZAP-70 is also phosphorylated *in vivo*. It is also worth bearing in mind that in order to perform such co-mapping of *in vivo* and *in vitro* derived phosphopeptides, sufficient incorporation into each sample must be obtained to permit autoradiographic exposure of the original 2D maps, and still have sufficient radioactive counts remaining in each phosphopeptide spot for mixing together and developing the second 2D co-map.

As a final confirmation of correct assignment, synthetic peptides can be made and phosphorylated *in vitro* prior to re-analysis. For example, peptides corresponding to the ZAP-70 regions containing tyrosine 292 and both 492 and 493 were synthesized and phosphorylated *in vitro* with p56*lck* and repurified by HPLC. In the case of the peptide containing tyrosines 492 and 493, care had to be taken to resolve the singly and doubly phosphorylated forms, which were identified by ESI-MS. Aliquots of these phosphopeptide species were then mixed with the corresponding peptide spots recovered from the ZAP-70 phosphopeptide map shown in Figure 6B and re-run on a second phosphopeptide map. The appearance of a single spot in each case was again used to confirm the assignments indicated in Table I (data not shown).

Another problem is often the appearance of multiple spots patterns derived from a single phosphorylation site. This is apparent in Figure 8 (e.g. spots A, F and I) and in Figure 6 (spots A and B), and is usually due to multiple enzymatic cleavage sites within the protein. For analyzing ζ however, the advantage of trypsin being able to cleave between each tyrosine residue outweighed the complication of the multiple partial digestion fragments obtained. Additionally, despite taking the utmost precautions in reagent quality and sample preparation, anomalous cleavage fragments can also occur. While mapping ζ, we observed tryptic cleavage C-terminal of glycine residues (spots G and J), which could not have been anticipated in advance. These facts should also be considered when interpreting phosphopeptide maps without performing further analyses: the occurrence of, say, three spots does not necessarily

Table I

Spot	Observed mass	Expected mass	Residue nos.[a]	Inferred peptide sequence[a]
A	1727.99 ± 0.01	1647.78	284 - 298	(R) IDTLNSDG**pY**TREPAR (I)
B	1883.24 ± 0.36	1803.88	283 - 298	(R) RIDTLNSDG**pY**TREPAR (I)
C	631[b]	551.28	125 - 128	(R) D**pY**VR (Q)
D	1272.24 ± 1.06	1191.63	64 - 75	(R) QLNGT**pY**AIAGGK (A)
E	1381.49 ± 0.71	1301.6	485 - 496	(K) ALGADDS**pYY**TAR[c] (S)
F	1461.99 ± 0.7	1301.6	485 - 496	(K) ALGADDS**pYpY**TAR (S)
G	1246.49 ± 0.7	1166.6	176 - 186	(R) KL**pY**SGAQTDGK (F)

[a]ZAP-70 residue numbers and relevant peptide sequences were inferred from the observed and expected peptide masses, allowing for the addition of 80 mass units for the phosphate-ester group. Amino acid residues flanking observed phosphopeptides are indicated in parentheses to display tryptic cleavage sites in full. [b]Manual assignment based on the observation of the singly charged species alone. [c]Contains a single (undetermined) phosphotyrosine residue.

mean that there are three phosphorylation sites, nor can it be assumed that each spot represents a predictable fragment for the endoprotease employed. The latter point is particularly important if peptide sequencing is subsequently being employed to determine an *in vivo* phosphorylation site (where radioactive counting of sequencing cycle eluates are taken and compared to the sequence of predicted cleavage fragments) without the prior knowledge that the peptide under analysis actually represents a *bona fide* total enzymatic digestion fragment.

Phosphorylation sites may also occur in regions of proteins where the endoprotease of choice cuts infrequently, resulting in large peptide fragments. Such peptides have lower solubility in aqueous buffers, and may be difficult or impossible to recover, especially from a very hydrophobic membrane such as polyvinyldifluoride or from an *in situ* digest in a high percentage polyacrylamide gel. For example, the principal site of serine phosphorylation on p56*lck* in activated T cells has been found to be on serine 59 (22, 42). Cleavage with trypsin produces a large peptide containing this residue which recovers poorly, if at all, from a membrane digest or from an *in situ* gel digest. The large size of this peptide also results in poor mobility on the 2D mapping system (J.D.W., unpublished observations). Two simple precautions can be taken to ensure that a 2D phosphopeptide map well represents the phosphorylation sites present in the protein of interest. Firstly, the use of more than one endoprotease can reveal additional phosphorylation sites (the serine 59 site in p56*lck* was identified using V8 protease (42)). Secondly, since phosphopeptides are radiolabeled with ^{32}P, Cerenkov counting the gel or membrane pieces after digestion shows how efficient the phosphopeptide recovery was. For the tryptic digests of ZAP-70 above (Figure 6) for example, recovery from the membrane was consistently 80 to 90%, suggesting that the major ZAP-70-derived phosphopeptide species were represented in the 2D maps obtained.

Conclusions. The data presented above all involve the analysis of phosphotyrosine-containing peptides. Since many early signaling events involve the activation of PTKs and the subsequent tyrosine phosphorylation of a number of cellular proteins, the techniques described here already have a wide range of applications. However, less than 1% of the protein phosphorylation events in cells occur on tyrosine, the majority of down-stream signaling and regulation events both in the cytoplasm and in the nucleus requiring key serine and/or threonine phosphorylation events. The techniques described here are readily adaptable to the analysis of serine/threonine phosphorylated peptides. For example, we identified the major TCR-induced phosphorylation site on the P56*lck* T cell PTK as serine 59 in a similar fashion (42).

From the data presented above, an MS-based approach to the analysis and determination of protein phosphorylation sites is clearly viable and can be used as a diagnostic assay to determine the phosphorylation state of any protein of interest. Its application to the determination of signal transduction pathways has successfully been demonstrated by the identification of a number of phosphorylation sites on two proteins instrumental in the response of T cells to the engagement of a specific receptor (the TCR), namely the ZAP-70 PTK, and the ζ subunit of the TCR with which it associates. A number of studies have shown the importance of these molecules in TCR-mediated signaling, as discussed above. For example, if ZAP-70 is prevented from binding ζ *in vivo*, then it does not become phosphorylated or catalytically activated (Figure 10) (31). In this experiment, a phosphatase-resistant phosphotyrosine analog-containing peptide was introduced into Jurkat T cells by pre-treatment of the cells with the permeabilizing agent tetanolysin (63). Subsequent stimulation of the TCR revealed that pre-treatment of the cells with the synthetic ITAM analog (lane 4, T) blocks both activation and tyrosine phosphorylation of ZAP-70, whereas no peptide (lane 3) or treatment with a synthetic ITAM analog control peptide to which ZAP-70 cannot bind (lane 5, C) lead to the normal levels of ZAP-70 activation and tyrosine phosphorylation observed in control stimulated cells (lane 2).

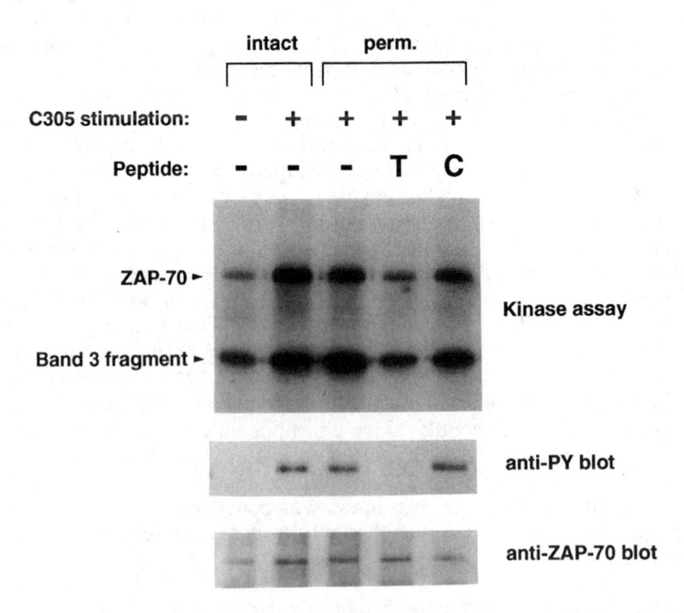

Figure 10: Activation and tyrosine phosphorylation of ZAP-70 can be blocked *in vivo* by a non-hydrolyzable phosphotyrosine analog-containing synthetic ITAM peptide. Tetanolysin-permeabilized Jurkat T cells (perm.) were incubated in the absence (-) or presence (T) of a synthetic ITAM peptide based on the C-terminal ITAM sequence from (human) TCR ζ, in which the phosphatase-resistant phosphotyrosine amino acid analog, difluorophosphonomethyl phenylalanine had been used during synthesis (31). Cells not treated with tetanolysin (intact), or tetanolysin-permeabilized cells treated with an ether-blocked non-ZAP-70-binding control peptide (C) were also analysed. Cells were subsequently stimulated where indicated with the anti-(human) TCR β chain mAb C305 (65) for 2 min. ZAP-70 was immunoprecipitated from cell lysates with an anti-ZAP-70 polyclonal antiserum and subjected to an *in vitro* kinase assay, using cfb3 as a protein substrate (31, 44). The level of tyrosine phosphate on the ZAP-70 was determined by anti-phosphotyrosine (anti-PY) immunoblotting aliquots of each immunoprecipitate (middle panel). The relative amounts of ZAP-70 in each sample was determined by performing an immunoblot on the membrane from the kinase assay (upper panel) with the anti-ZAP-70 antiserum (lower panel). (Reproduced with permission from reference 31. Copyright 1995, The American Society for Biochemistry and Molecular Biology, Inc.).

Since it is known that signaling via the TCR is compromised if ZAP-70 is not able to bind ζ (31), with disruptions in the ZAP-70 gene also being known to cause severe auto-immune disorders in both mice and humans (34-36), the results shown in Figure 10 suggest that targeted *in vivo* intervention in a (defective) signaling pathway could have therapeutic applications in the future. Together with the recent demonstration that sub-optimal TCR engagement leads to the appearance of multiple phospho-ζ isoforms and T cell responses (37, 38), these experiments show the potential value of the MS-based technologies in facilitating the rapid unraveling of the many important biological and biomedical problems linked to signal transduction pathways.

Acknowedgments

The authors would like to thank the following individuals at the University of British Columbia: Dr. Ian Clark-Lewis for peptide syntheses, including TCR ζ, Dr. Patricia Orchansky for providing phospho-ζ-conjugated agarose beads and advice on non-reducing/reducing PAGE, Dr. Hung-Sia Teh for providing the CD11.3 cell line and Dr. M. Gold for 4G10 mAbs. We are also grateful to J. Bolen (Bristol-Myers Squibb) for ZAP-70 anti-serum, P. Low (Purdue University) for cfb3, T. Burke and P. Roller (NCI, NIH) for synthesis of non-hydrolyzable phosphotyrosine analog-containing ITAM peptides, and E Bonvini (FDA) for tetanolysin. R.A. would like to thank the Medical Research Council of Canada and the National Science Foundation of the United States for funding.

References

1. Isakov, N.; Wange, R. L.; Samelson, L. E. *J. Leukocyte Biol.* **1994,** *55,* 265-271.
2. Peri, K. G.; Veillette, A. in *Chem. Immunol.* Samelson, L.E. Ed.: Karger, Basel: 1994, Vol. 59, pp. 19-39.
3. Chan, A. C.; Desai, D. M.; Weiss, A. *Ann. Rev. Immunol.* **1994,** *12,* 555-592.
4. Abraham, K. M.; Levin, S. D.; Marth, J. D.; Forbush, K. A.; Perlmutter, R. M. *J. Exp. Med.* **1991,** 1421-1432.
5. Van Oers, N. S. C.; Garvin, A. M.; Davis, C. B.; Forbush, K. A.; Carlow, D. A.; Littman, D. R.; Perlmutter, R. M.; Teh, H.-S. *Eur. J. Immunol.* **1992,** *22,* 735-743.
6. Molina, T. J.; Kishihara, K.; Siderovski, D. P.; Van Ewijk, W.; Narendran, A.; Timms, E.; Wakeham, A.; Paige, C. J.; Hartmann, K.-U.; Veillette, A.; Davidson, D.; Mak, T. W. *Nature* **1992,** *357,* 161-164.
7. Van Oers, N. S. C.; Garvin, A. M.; Cooke, M. P.; Davis, C. B.; Littman, D. R.; Perlmutter, R. M.; Teh, H.-S. *Adv. Exp. Med. Biol.* **1992,** *323,* 89-99.
8. Cooke, M. P.; Abraham, K. M.; Forbush, K. A.; Perlmutter, R. M. *Cell* **1991,** *65,* 281-291.
9. Davidson, D.; Chow, L. M. L.; Fournel, M.; Veillette, A. *J. Exp. Med.* **1992,** *175,* 1483-1492.
10. Abraham, N.; Miceli, M. C.; Parnes, J. R.; Veillette, A. *Nature* **1991,** *350,* 62-66.
11. Molina, T. J.; Bachmann, M. F.; Kündig, T. M.; Zinkernagel, R. M.; Mak, T. W. *J. Immunol.* **1993,** *151,* 699-706.
12. Karnitz, L.; Sutor, S. L.; Torigoe, T.; Reed, J. C.; Bell, M. P.; McKean, D. J.; Leibson, P. J.; Abraham, R. T. *Mol. Cell. Biol.* **1992,** *12,* 4521-4530.
13. Bergman, M.; Mustelin, T.; Oetken, C.; Partanen, J.; Flint, N. A.; Amrein, K. E.; Autero, M.; Burn, P.; Alitalo, K. *EMBO J.* **1992,** *11,* 2919-2924.
14. Okada, M.; Nada, S.; Yamanashi, Y.; Yamamoto, T.; Nakagawa, H. *J. Biol. Chem.* **1991,** *266,* 24249-24252.

15. Takeuchi, M.; Kuramochi, S.; Fusaki, N.; Nada, S.; Kawamura-Tsuzuku, J.; Matsuda, S.; Semba, K.; Toyoshima, K.; Okada, M.; Yamamoto T. *J. Biol. Chem.* **1993,** *268,* 27413-27419.
16. Ostergaard, H. L.; Trowbridge, I. S. *J. Exp. Med.* **1990,** *172,* 347-350.
17. Biffen, M.; McMichael-Phillips, D.; Larson, T.; Venkitaraman, A.; Alexander, D. *EMBO J.* **1994,** *13,* 1920-1929.
18. Mustelin, T.; Pessa-Morikawa, T.; Autero, M.; Gassmann, M.; Andersson, L. C.; Gahmberg, C. G.; Burn, P. *Eur. J. Immunol.* **1992,** *22,* 1173-1178.
19. Hatakeyama, M.; Kono, T.; Kobayashi, N.; Kawahara, A.; Levin, S. D.; Perlmutter, R. M.; Taniguchi, T. *Science* **1991,** *252,* 1523-1528.
20. Minami, Y.; Kono, T.; Yamada, K.; Kobayashi, N.; Kawahara, A.; Perlmutter, R. M.; Taniguchi, T. *EMBO J.* **1993,** *12,* 759-768.
21. Horak, I. D.; Gress, R. E.; Lucas, P. J.; Horak, E. M.; Waldmann, T. A.; Bolen, J. B. *Proc. Natl. Acad. Sci. (USA)* **1991,** *88,* 1996-2000.
22. Watts, J. D.; Welham, M. J.; Kalt, L.; Schrader, J. W.; Aebersold, R. *J. Immunol.* **1993,** *151,* 6862-6871.
23. Weissman, A. M. in *Chem. Immunol.* Samelson, L.E. Ed.: Karger, Basel: 1994, Vol. 59, pp. 1-18.
24. Chan, A. C.; Iwashima, M.; Turck, C. W.; Weiss, A. *Cell* **1992,** *71,* 649-662.
25. Irving, B. A.; Chan, A. C.; Weiss, A. *J. Exp. Med.* **1993,** *177,* 1093-1103.
26. Van Oers, N. S. C.; Tao, W.; Watts, J. D.; Johnson, P.; Aebersold, R.; Teh, H.-S. *Mol.Cell.Biol.* **1993,** *13,* 5771-5780.
27. Iwashima, M.; Irving, B. A.; Van Oers, N. S. C.; Chan, A. C.; Weiss, A. *Science* **1994,** *263,* 1136-1139.
28. Straus, D. B.; Weiss, A. *J. Exp. Med.* **1993,** *178,* 1523-1530.
29. Gauen, L. K. T.; Zhu, Y.; Letourneur, F.; Hu, Q.; Bolen, J. B.; Matis, L. A.; Klausner, R. D.; Shaw, A. S. *Mol. Cell. Biol.* **1994,** *14,* 3729-3741.
30. Wange, R. L.; Kong, A.-N. T.; Samelson, L. E. *J. Biol. Chem.* **1992,** *267,* 11685-11688.
31. Wange, R. L.; Isakov, N.; Burke Jr., T. R.; Otaka, A.; Roller, P. P.; Watts, J. D.; Aebersold, R.; Samelson, L. E. *J. Biol. Chem.* **1995,** *270,* 944-948.
32. Wange R. L.; Malek, S. N.; Desiderio, S.; Samelson, L. E. *J. Biol. Chem.* **1993,** *268,* 19797-19801.
33. Fry, M. J.; Panayotou, G.; Booker, G. W.; Waterfield, M. D. *Protein Sci.* **1993,** *2,* 1785-97.
34. Elder, M. E.; Lin, D.; Clever, J.; Chan, A. C.; Hope, T. J.; Weiss, A.; Parslow, T. G. *Science* **1994,** *264,* 1596-1599.
35. Chan, A. C.; Kadlecek, T. A.; Elder, M. E.; Filipovich, A. H.; Kuo, W.-L.; Iwashima, M.; Parslow, T. G.; Weiss, A. *Science* **1994,** *264,* 1599-1601.
36. Arpala, E.; Shahar, M.; Dadi, H.; Cohen, A.; Roifman, C. M. *Cell* **1994,** *76,* 947-958.
37. Sloan-Lancaster, J., Shaw, A. S.; Rothbard, J. B.; Allen, P. M. *Cell* **1994,** *79,* 913-922.
38. Madrenas, J.; Wange, R. L.; Wang, J. L.; Isakov, N.; Samelson, L. E.; Germain, R. N. *Science* **1995,** *267,* 515-518.
39. Andersson, L.; Porath, J. *Anal. Biochem.* **1986,** *154,* 250-254.
40. Watts, J. D.; Wilson, G. M.; Ettehadieh, E.; Clark-Lewis, I.; Kubanek, C.-A.; Astell, C. R.; Marth, J. D.; Aebersold, R. *J.Biol.Chem.* **1992,** *267,* 901-907.
41. Watts, J. D.; Affolter, M.; Krebs, D. L.; Wange, R. L.; Samelson, L. E.; Aebersold, R. *J. Biol. Chem.* **1994,** *269,* 29520-29529.
42. Watts, J. D.; Sanghera, J. S.; Pelech, S. L.; Aebersold, R. *J.Biol.Chem.* **1993,** *268,* 23275-23282.
43. Wang, C. C.; Badylak, J. A., Lux, S. E.; Moriyama, R., Dixon, J. E.; Low, P. S. *Protein Sci.* **1992,** *1,* 1206-1214.
44. Harrison, M. L.; Isaacson, C. C.; Burg, D. L.; Geahlen, R. L.; Low, P. S. *J. Biol. Chem.* **1994,** *269,* 955-959.

45. Affolter, M.; Watts, J. D.; Krebs, D. L.; Aebersold, R. *Anal. Biochem* **1994,** *223*, 74-81.
46. Payne, D. M.; Rossomando, A. J.; Martino, P.; Erickson, A. K.; Her, J.-H.; Shabanowitz, J.; Hunt, D. F.; Weber, M. J.; Sturgill, T. W. *EMBO J.* **1991,** *10*, 885-892.
47. Biemann, K. in *Methods in Enzymology* McCloskey, J. A. Ed.: Academic Press, San Diego: 1990, Vol. 193, pp. 455-479.
48. Boyle, W. J.; van der Geer, P.; Hunter, T. in *Methods in Enzymology* Hunter, T.; Sefton, B.M., Eds.; Academic Press, San Diego: 1991, Vol. 201, pp. 110-149.
49. Hettasch, J. M.; Sellers, J. R. *J. Biol. Chem.* **1991,** *266*, 11876-11881.
50. Ruiz-Gomez, A.; Vaello, M. L.; Valdivieso, F.; Mayor, F. *J. Biol. Chem.* **1991,** *266*, 559-566.
51. Michel, H.; Griffin, P. R.; Shabanowitz, J.; Hunt, D. F.; Bennet, J. *J. Biol. Chem.* **1991,** *266*, 17584-17591.
52. Rossomando, A. J.; Wu, J.; Shabanowitz, J.; Hunt, D. F.; Weber, M. J.; Sturgill, T. W. *Proc. Natl. Acad. Sci. (USA)* **1992,** *89*, 5779-5783.
53. Huddleston, M. J.; Annan, R. S.; Bean, M. F.; Carr, S. A. *J. Am. Soc. Mass Spectrom.* **1993,** *4*, 710-717.
54. Greenwood, J. A.; Scott, C. W.; Spreen, R. C.; Caputo, C. B.; Johnson, G. V. *J. Biol. Chem.* **1994,** *269*, 4373-4380.
55. Yip, T. T.; Hutchens, T. W. *FEBS Lett.* **1992,** *308*, 149-153.
56. Erickson, A. K.; Payne, D. M.; Martino, P. A.; Rossomando, A. J.; Shabanowitz, J.; Weber, M. J.; Hunt, D. F.; Sturgill, T. W. *J. Biol. Chem.* **1990,** *265*, 19728-19735.
57. Nuwaysir, L. M.; Stults, J. T. *J. Am. Soc. Mass Spectrom.* **1993,** *4*, 662-669.
58. Evan, G. I.; Lewis, G. K.; Ramsay, G.; Bishop, J. M. *Mol. Cell. Biol.* **1985,** *5*, 3610-3616.
59. Gold, M. R.; Matsuuchi, L. in *International Review of Cytology* Jeon, K. Ed.: Academic Press, San Diego: 1995, Vol. 157, pp. 181-276.
60. Orchansky, P. L.; Teh, H.-S. *J. Immunol.* **1994,** *153*, 615-622.
61. Samelson, L. E.; Harford, J. B.; Schwartz, R. H.; Klausner, R. D. *Proc. Natl. Acad. Sci. (USA)* **1985,** *82*, 1969-1973.
62. Patel, M. D.; Samelson, L. E.; Klausner, R. D. *J. Biol. Chem.* **1985,** *262*, 5831-5838.
63. Sarosi, G. A.; Thomas, P. M.; Egerton, M.; Phillips, A. F.; Kim, K. W.; Bonvini, E.; Samelson, L. E. *Int. Immunol.* **1992,** *4*, 1211-1217.
64. Aebersold, R. H.; Leavitt, J.; Saavedra, R. A.; Hood, L. E.; Kent, S. B. H. *Proc. Natl. Acad. Sci. (USA)* **1987,** *84*, 6970-6974.
65. Weiss, A.; Stobo, J. D. *J. Exp. Med.* **1984,** *160*, 1284-1299.

RECEIVED August 22, 1995

Chapter 21

Characterization of Recombinant Glycoproteins from Chinese Hamster Ovary Cells

Michael F. Rohde, Viswanatham Katta, Patricia Derby, and Robert S. Rush

Amgen, Inc., 1840 DeHavilland Drive, Thousand Oaks, CA 91320

Production of recombinant proteins may be accomplished in a variety of cellular expression systems. Bacterial expression is often preferred because yields are high and manipulation of the genetics of bacteria is relatively easy. However, when the goal is to produce proteins that are as nearly identical as possible to the non-recombinant protein, this may not always be the best choice, since bacterial systems do not usually form the correct disulfide pairs nor add carbohydrate. If the desire is to produce a protein with glycosylation that is similar to the human protein, expression of proteins in tissue culture from Chinese hamster ovary cells is one method of choice, since a great body of knowledge exists for this cell system. Traditional, non-mass spectrometric methods for characterization of carbohydrates are often quite involved and can require large quantities of sample. Many methods have recently evolved which take advantage of mass spectrometry to provide a maximum of information with a much lower requirement for sample. Additionally, information can be obtained on glycosylated and non-glycosylated peptides in the same experiment. This paper will focus on the application of electrospray mass spectrometry for the analysis of recombinant proteins produced in mammalian cell culture systems. Emphasis will be placed on strategies to optimize data obtained while minimizing sample preparation and isolation efforts. Examples will include confirmation of expected carbohydrate structures, as well as the use of ESI-MS in identifying unexpected structures. Also we will present results using a modified, miniaturized electrospray source which enhances quality of the data signal with lower sample amounts.

The complete structural determination of the carbohydrate moieties attached to a given protein is a daunting analytical problem. Extensive volumes of the methodologies are available that cover most of the traditional methods (*1*). The complexity of these structures is far greater than that which is normally found in the amino acid portion of the underlying protein. The monomeric carbohydrate units can

0097–6156/95/0619–0408$12.00/0

be joined in a variety of linkages involving different carbon-oxygen-carbon bonds. Also different anomeric arrangements can exist at each glycosidic bond. Even the number of individual monosaccharides involved in a structure at a given amino acid side chain exhibits considerable variety. The only simplicity seems to lie in which amino acids are used for carbohydrate attachment; only serine, threonine and asparagine are involved in most cases (*2*).

Nonetheless, a great deal of insight can be gleaned from the application of relatively simple common analytical techniques. One of the more powerful of these is the method of electrospray ionization mass spectrometry (ESI-MS). While this approach will not easily answer all questions regarding a given carbohydrate structure, significant information can be obtained from this method. In order to maximize the information content derived from ESI-MS, one must take a careful and rational approach. Given the complexity and potential variety of structures at a given amino acid attachment site, it is preferable to isolate each site for study. It would be ideal to further isolate the individual carbohydrate isoforms associated with a given site, but this is a challenge that is not easily met. Also, in the manipulations undertaken to isolate subspecies, one must be careful not to modify any of the carbohydrate structures by exposure to extremes of pH.

This report is intended to elaborate on some of the approaches taken in the analysis of carbohydrate structures found on the recombinant glycoproteins that are produced in Chinese Hamster Ovary (CHO) cells. This cell line is chosen since it is a mammalian cell line known to have the appropriate cellular machinery to add the carbohydrate structures normally found on glycoproteins produced in human cells; also, CHO cells are easier to grow in culture than are most human cell lines. A great deal is known about the genetics of the CHO cell, and the cell line is genetically stable, which makes the expression of recombinant proteins a relatively easy task (*3*). While by no means an exhaustive compilation of all possible approaches, it is intended that this report will provide some examples of the power of ESI-MS in glycoprotein analysis.

MATERIALS AND METHODS

Protein Isolation. The preparations of recombinant human erythropoietin (rHuEPO) used in this study were purified from *E. coli* and CHO cells producing rHuEPO. Purification procedures have been previously published (*4, 5*).

Proteolytic digestions. Samples of unreduced rHuEPO in 20 mM citrate, pH 7.0 were digested with trypsin (Boehringer Mannheim, sequencing grade) at a substrate to enzyme ratio of 50:1; incubation was carried out at 37° for six hours. Digestions with *Staphylococcus aureus* V8 protease (endoproteinase Glu-C, ProMega) were carried out at 25° for 18 hr with a substrate to enzyme ratio of 10:1. All digests were then immediately subjected to high pressure liquid chromatography (HPLC) separation or stored frozen.

HPLC Chromatography. Two systems were used for different applications. One was a Hewlett Packard (Palo Alto, CA) 1090M with a Vydac C4 reversed phase column, 4.6 x 250 mm; peptides were separated with linear gradients of water and acetonitrile containing 0.06% trifluoroacetic acid (TFA) (*6*). Elution was monitored either by absorption at 214 nm (A_{214}) or from ion current signals in the mass spectrometer (see details below). When desired, fractions were collected manually, either as the total effluent from the UV flow cell, or as a portion of the effluent, with the other portion of the flow stream going directly to the mass spectrometer. The second system used was a microbore HPLC from Michrom BioResources (Auburn, CA); this system allowed use of 1 x 150 mm reversed phase columns operated at 50 µl/min and did not require a flow split for connection to the ESI-MS.

Edman Sequencing. Collected peptides were examined by Edman sequencing using commercial instrumentation from Applied Biosystems (Models 470A, 476A or 477A) or Hewlett Packard (Model G1000A).

Electrospray Ionization Mass Spectrometry. A Sciex API III triple quadrupole ionization mass spectrometer (Thornhill, Ontario) was utilized for the off-line infusion and on-line LC/MS experiments. The instrument was calibrated using a mixture of polypropylene glycols. Two different kinds of electrospray needles were used: one was the standard Sciex ionspray needle (spray voltage 4500-4800 V, nebulization gas pressure 35-45 psig, curtain gas at 0.6 L/min) and the other was a patented direct spray method (*7,8*) that utilized a 50 micron id, 360 micron OD fused silica capillary whose end was etched to a sharp point and gold plated (no nebulization gas or sheath liquid, curtain gas at 0.6 L/min). The second source provided a means to spray liquids at low flow rates, in the range 250-400 nL/min, and was exclusively used for off-line infusion experiments (*9*). Lower infusion rates allowed slower scan rates, hence higher resolution. Typically 20 scans were acquired in multichannel mode to increase the signal to noise ratio and the accuracy of mass measurements. A step size of 0.33 m/z units was employed for data acquisition. When greater mass accuracy was required, i.e., measurements of the masses of the O-acetylation structures, a step size of 0.1 m/z units was used and the instrument was operated at better than unit resolution. For most of the infusion experiments the samples were dissolved in a mixture of water/methanol/acetic acid (50/50/3 v/v/v) and occasionally in water/methanol/formic acid (50/50/0.2 v/v/v) to verify that the observed acetylation was not caused by the acetic acid solvent. The instrument was operated in positive ion mode.

The orifice potential affects both the desolvation of the ions and the extent of collisional activation in the interface region in the Sciex instrument. Huddleston *et al.* have described a stepped orifice method to identify glycopeptides (*10*). Typically an orifice potential of 125 V was used while scanning the low m/z region (150-500) and a normal low orifice potential of 65 V was used in the high m/z region (600-2400). Diagnostic carbohydrate fingerprints are observed in the low m/z region while normal multiply charged molecular ion spectra are observed in the higher m/z region. Data analysis employed SCIEX MacSPEC v3.22 software.

On-line LC/MS. On-line LC/MS analysis employed a Michrom UMA (Auburn, CA) high performance liquid chromatography system with a 1 x 150 mm C4 or C18 Vydac column interfaced to the SCIEX API III. Elution of bound peptides was accomplished with a linear gradient of acetonitrile (2-75%) at a flow rate of 0.05 ml/min in 0.06% trifluoroacetic acid while monitoring both absorbance at 214 nm and the total ion current (TIC) from the MS. Column effluent was directly electrosprayed using a standard Sciex needle at ambient pressure and temperature.

RESULTS AND DISCUSSION

Direct analysis of intact proteins. The most direct application of ESI-MS to the analysis of proteins is that of the intact protein. While this is many times an effective approach for simple proteins, this method suffers greatly in cases of increased complexity due to amino acid side chain modifications. Figure 1a is the mass spectrum of a homogeneous protein, produced in *E. coli* and devoid of carbohydrate; the spectrum contains well defined signals with sufficient resolution to identify the presence of two protein species in the preparation, differing by a single alanine, which has been post-translationally removed from the amino terminus. Several ions are observed which represent the multiple charge states of the protein species present. The determined M_r values correspond to the predicted mass of the full length protein, and the protein lacking one alanine, a mass difference of 71 Da. In this preparation, these two species are observed with roughly equivalent intensities; furthermore, one can even ascribe a low level signal to a third species lacking an additional amino acid, also processed from the amino terminus. These assignments were confirmed by Edman sequencing.

Figure 1b is the same protein produced in CHO cells; the CHO cells have produced a protein with several specific sites of glycosylation. Note that the overall signal is so broad that it is not possible to ascertain any of the individual features of the protein. The diffuse nature of the signal can be ascribed to heterogeneity in the carbohydrate, especially in the number of terminal N-acetylneuraminic acid (NeuAc) residues.

If the glycoprotein in Figure 1b is treated with neuraminidase to enzymatically remove the NeuAc groups attached to the oligosaccharide chains, the desialylated protein produces a simplified spectrum. While still composed of a multiplicity of species (Figure 1c), the removal of the NeuAc heterogeneity allows conversion to a M_r scale; this is shown in Figure 1d. The spectrum is still complex, but we can at least observe individual mass peaks with spacing consistent with known carbohydrate structures associated with this cell line. The biochemical subcellular machinery of the CHO cells will most commonly produce a set of structures on specific asparagines; this can be summarized in Scheme 1 (cf. [*11*]).

Scheme 1:

$$\text{NeuAc}_k - (\text{LacNAc})_{l+m} - \text{Man}_3 - \text{GlcNAc} - \underset{}{\overset{\text{Fuc}_n \atop |}{\text{GlcNAc}}}$$

NeuAc is N-acetylneuraminic acid, $k = 0\text{-}l$; LacNAc is lactosamine or Gal-GlcNAc, $l = 2\text{-}4$, $m \geq 0$; Man is mannose; Fuc is fucose, $n = 0,1$. The fucose branches from the terminal GlcNAc; attachment to the Asn is through the reducing end of the terminal GlcNAc. The most frequent variations are lactosamine additions (either branching or extensions) and NeuAc capping. Other structures have been observed which arise from incomplete cellular processing of a high mannose precursor core and other hybrid structures, but have not been observed in the examples in this discussion. In Figure 1d the predominant spacing is 366 Da, which is the residue mass of LacNAc. This structural feature is the central building block that generates the differences between various branch oligosaccharide structures known as bi- tri-, and tetra-antennary. LacNAc extensions may also occur as a linear extension of one of the bi- tri- or tetra-antennary arms (*12*). In the sample examined in Figure 1d, the spectrum is simplified by the fact that all Asn-linked structures are substituted with fucose (*12*). Other proteins we have studied have had mixtures of oligosaccharide chains on the protein, some with and some without fucose. This considerably increases the complexity of the spectrum, therefore, it is desirable to obtain information on the carbohydrate structures associated with individual amino acid sites of glycosylation. To do this, we must examine less complex portions of the glycoprotein, one at a time.

Peptide mapping. With the complexity seen in Figures 1b-d, limited insight can be obtained into the structure of a given glycoprotein. To elucidate further details of the structure, it is necessary to fragment the protein into smaller parts. Individual details of the whole structure may be better understood by examination of specific, well defined pieces. This is most commonly done by specific proteolytic digestion, followed by reversed phase separation of the resulting peptide mixture.

The upper panel of Figure 2 presents a total ion chromatogram (TIC) from m/z 550-2350, of a tryptic digest of recombinant erythropoietin from CHO cells. The chromatogram contains a series of sharp, well resolved peaks; some are a little

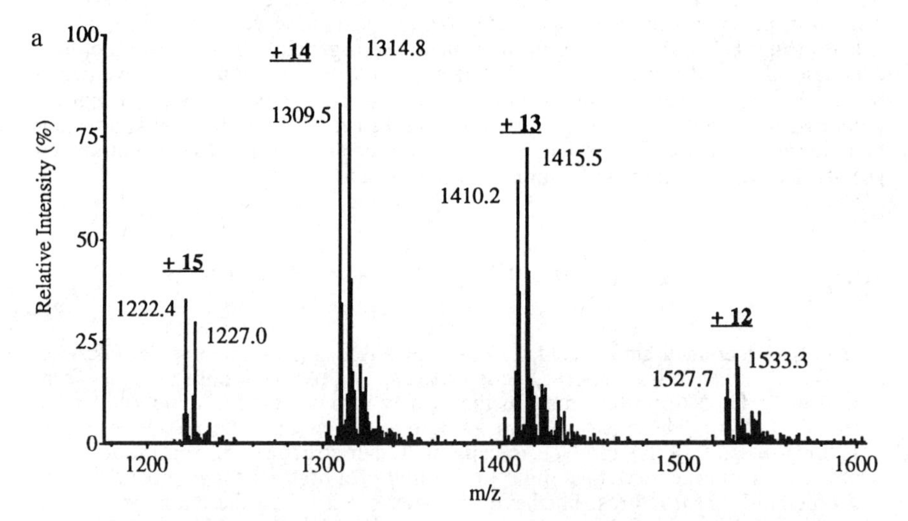

Figure 1. a) ESI-MS spectrum of non-glycosylated rHuEPO produced in *E. coli.* ; the charge state (underlined numbers) and the peak m/z values are given. b) ESI-MS spectrum of purified rHuEPO produced in CHO cells. c) ESI-MS Spectrum of protein from panel b) treated with neuraminidase to remove sialic acids. d) Reconstructed ion spectrum of neuraminidase treated recombinant rHuEPO from CHO cells. Calculated molecular weights are indicated above the major peaks; the mass difference between major peaks is also given. Minor peaks marked with asterisks are on the average 80 Da higher than the major peaks.

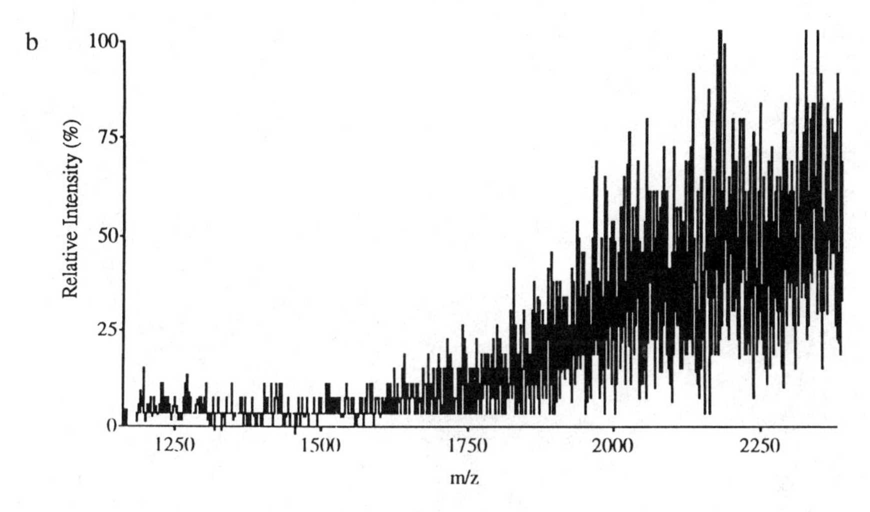

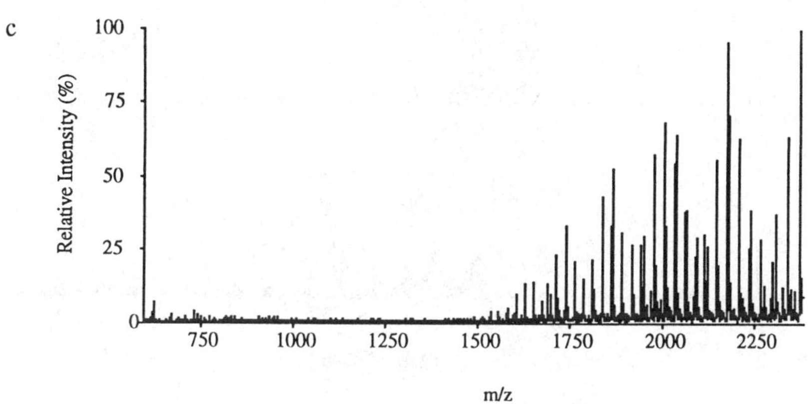

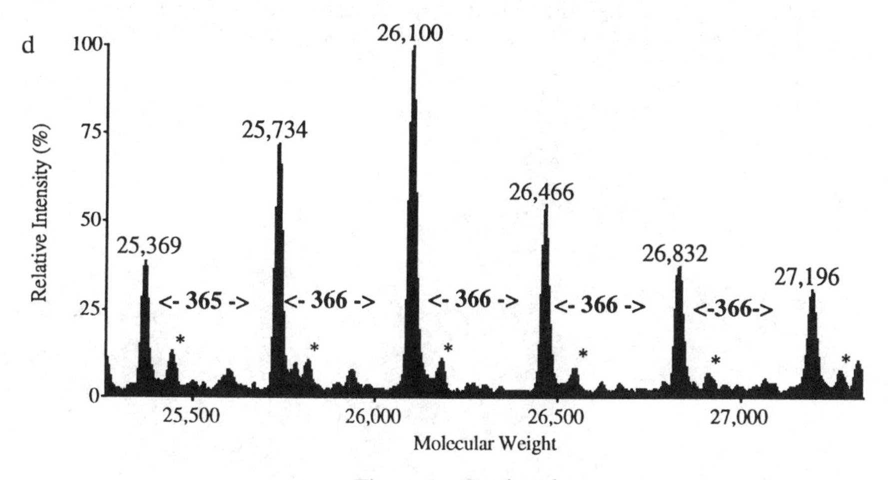

Figure 1. *Continued*

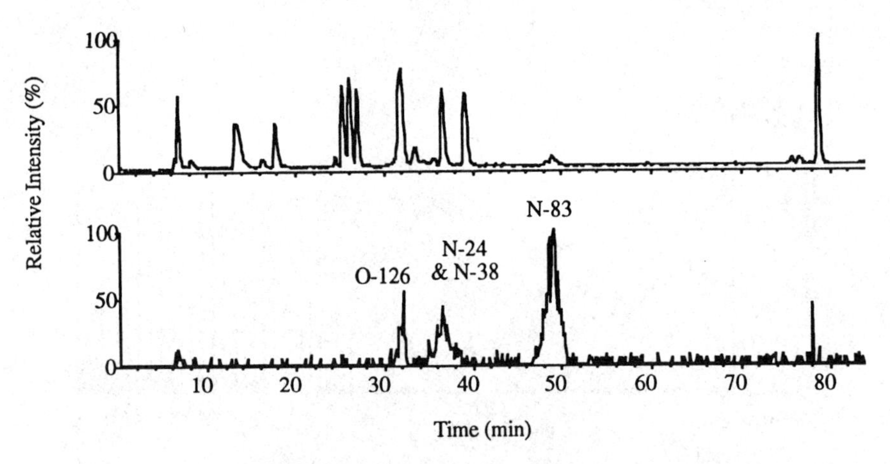

Figure 2. Reversed phase chromatographic separation of a tryptic digest of rHuEPO isolated from CHO cells. The upper trace includes signals from all ions; the lower trace includes only signals in the range m/z 365-367.

broader than others. The broader ones include the peptides with the sites of glycosylation. Using only UV absorbance, peak width is often the first indication of potential glycosylation. Traditionally, all chromatographic peaks would be collected and further analyzed by Edman sequencing, or mass spectrometry. The approach of Edman sequencing allows identification of the amino acid sequence of the isolated peptides. Edman sequencing, while extremely valuable, does not provide much information about modified amino acids, other than a missing assignment at the cycle of modification. The combination of the peptide sequence and the determination of the mass of the peptide can provide valuable information about modified peptides. The known sites of glycosylation in EPO are: serine-126, (O-126); asparagines-24, -38 and -83, (N-24, N-38 and N-83) (13).

For the O-126 peptide (lower panel, Figure 2), Edman sequencing determined the sequence of E-A-I-S-P-P-D-A-A-X-A-A-P-L-R, where no assignment could be made at the position indicated by "X". The gene sequence of this position codes for a serine (*13,14*); the predicted tryptic peptide mass of this peptide would be 1466 Da. In Figure 3a, ESI-MS of this peptide reveals the presence of two primary glycopeptide ions with masses of 2123 and 2415 Da (observed as the doubly charged ions at m/z 1061.5 and 1207.7); the ion at 799.5 is the triply charged ion of the 2415 Da species and the ion at m/z 898.5 is a co-eluting nonglycosylated peptide. The masses at 2123 and 2415 are those expected from the mucin type of O-linked carbohydrate produced in CHO cells: Ser-GalNAc-Gal, with either one or two NeuAc. No signal is observed for a peptide with the combination GalNac-Gal and no NeuAc, but a peak under 5% relative intensity is present with the mass of an unglycosylated peptide at m/z 733.8.

Identification of glycopeptides based on broad HPLC peaks and unassignable cycles in the Edman sequence runs is of limited utility. An exceptionally powerful addition to simple LC-ESI-MS is the use of a stepped orifice scan method to identify which peptides contain signature ions for carbohydrate chains (*10*). This method provides direct information as to which regions of the chromatogram contain the glycopeptides; signature ions at m/z 204 and 366 are reliable indicators of the presence of glycopeptides. Signal at m/z 204 indicates the presence of hexose; m/z 366 is likewise indicative of hexose-hexosamine. Using this approach, we can generate a second display of a given LC-ESI-MS run that localizes regions of the chromatogram with signature ions for carbohydrate. This produces a reconstructed ion chromatogram (RIC); the lower panel of Figure 2 is the RIC for ions from m/z 365-367. In it we were able to identify the broad peak at about 48 min as another glycosylated peptide, N-83.

The analysis of this peptide proved to be more complex. The O-126 site, discussed above, is a relatively simple structure, and much can be inferred from direct inspection; more complex structures are found in Asn-linked carbohydrate chains and obviously require more detailed analyses. The mass spectrum of the 48 min region is given in Figure 3b, and is also presented on a molecular weight scale in 3c, for clarity of the ensuing discussion.

Edman sequence determination of this peptide was G-Q-A-L-L-V-X-S-S-Q-P-W-E-P-L-Q-L-H-V-D-K, where the gene sequence for position X is asparagine-83. The calculated peptide mass is 2361 Da; the determined mass exceeds this by several thousand daltons. In order to begin to make assignments for this increase in mass, we need to examine the literature for the types of asparagine-linked carbohydrate structures that mammalian cells are capable of producing. In the case of this protein, extensive studies have been reported for the carbohydrate structures released from the recombinant and the human urinary proteins; detailed characterization of the variety of structures and glycosidic linkages have been determined employing traditional analytical approaches of methylation and fast atom bombardment mass spectrometry (*12,15,16*). Common Asn-linked carbohydrates have a core oligosaccharide structure of two units of N-acetylglucosamine with the reducing end

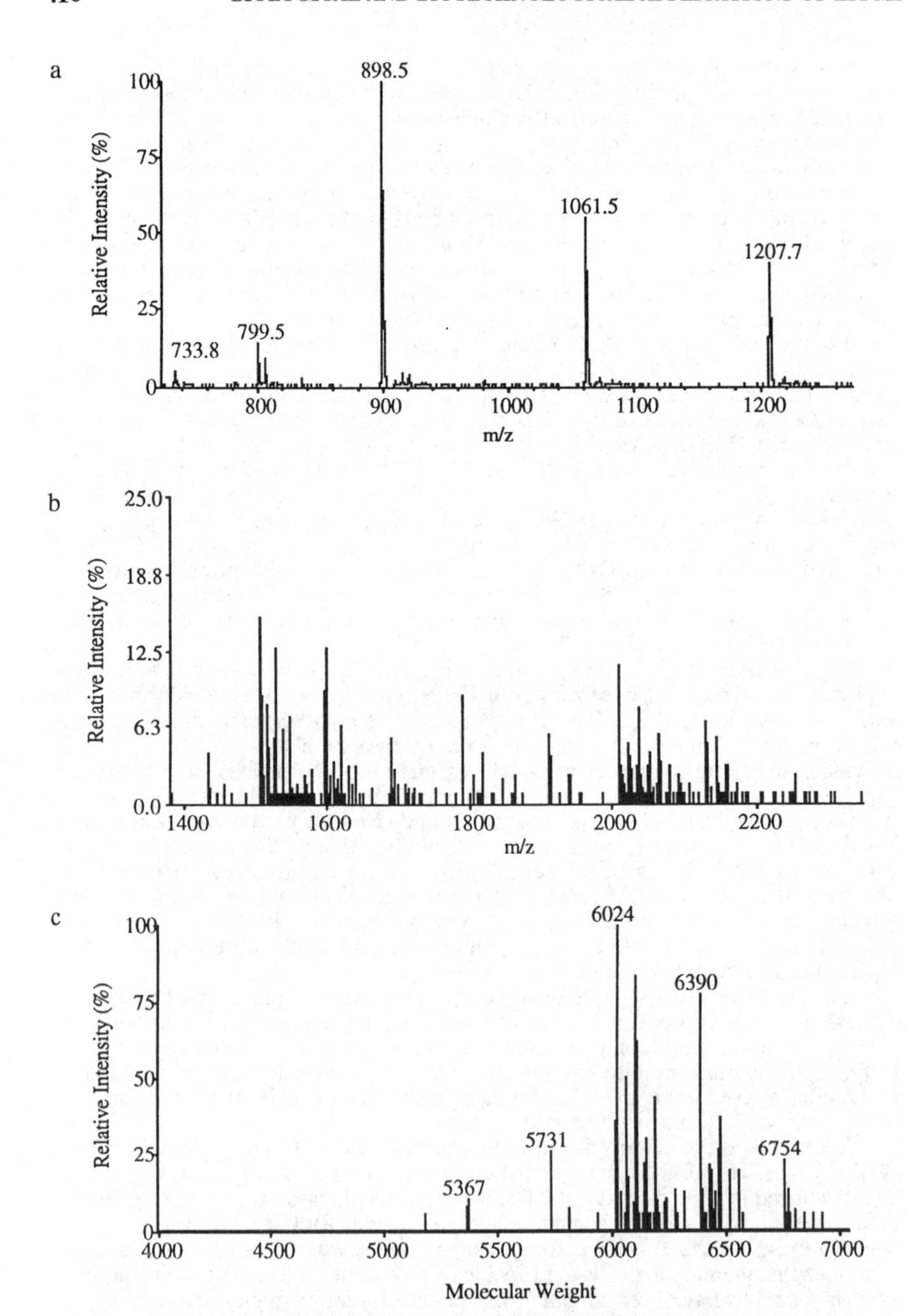

Figure 3. a) Mass spectrum of tryptic peptide O-126 from Figure 2; peak m/z values are indicated. b) Mass spectrum of tryptic peptide N-83 from Figure 2; c) reconstructed mass spectrum of ions from panel 3b; calculated molecular weights are indicated above the principal peaks.

attached to the terminal nitrogen of the asparagine side chain, and a branched tri-mannose with two to four chains extending from two of these mannose residues. These chains have the basic LacNAc repeat; the non-reducing end of the LacNAc can be capped with NeuAc residues; a fucose residue is linked to the initial GlcNAc attached to the Asn. All these potential structures were generalized above in Scheme 1.

Comparisons are made between observed masses and the myriad of predicted structures associated with this peptide. Such a comparison is given in Table I.

Table I. Matches between predicted and observed masses for tryptic N-83

# NeuAc (k)	*# Fuc (n)*	*Number of LacNAc (l+m)* 2	3	4	5	6	7
0	0	3983	4349	4714	5079	5445	5810
0	1	4130	4495	4860	5226	5591	5956
1	0	4275	4640	5005	5371	5736	6101
1	1	4421	4786	5152	5517	5882	6248
2	0	4566	4931	5297	5662	6027	6393
2	1	4712	5077	5443	5808	6173	6539
3	0	4857	5223	5588	5953	6319	6684
3	1	5003	5369	5734	6099	6465	6830
4	0	5149	5514	5879	6244	6610	6975
4	1	5295	5660	6025	6391	6756	7121

Matches between potential structures and those confirmed by ESI-MS are underlined. We find that the predominant species contain three or four NeuAc on a base oligosaccharide structure with predominately four and five LacNAc units. Evidence is present for three NeuAC with three and four LacNAc, as well as for four NeuAc with four, five and six LacNAc residues. This approach identifies the major MS peaks, but obviously does not distinguish isobaric structures. More detailed analysis requires the inclusion of other techniques. These can include the use of structure-specific glycosidic enzymes, chemical modifications of the carbohydrates, or reaction with structure specific lectins (*1*). The current discussion will touch on chromatographic and electrophoretic separations, but will not go into the other mentioned approaches.

For rHuEPO, the two other Asn linked glycosylation sites, N-24 and N-38, are also observed (lower panel in Figure 2). The RIC indicates that glycopeptides elute at 37 min, but the only molecular ions found correspond to nonglycosylated peptides not including N-24 and N-38. Edman sequence analysis indicated that a tryptic peptide with sites N-24 and N-38 was indeed present, but no ions corresponding to a carbohydrate-containing peptide were detected. The reasons for this paradox include a number of factors: the peptide alone is of high mass, the dual carbohydrate substitution sites greatly increase this mass, and there is a lack of sufficient protonation sites on the peptide to provide for a mass to charge ratio in the range of the quadrupole system employed for this analysis. It is possible to obtain mass measurements on this peptide by varying the ionization conditions to produce higher charge states or by use of an instrument with extended mass analysis range, but these approaches still do not provide ready access to structural information on the two individual sites. The power of glycopeptide RIC scans is abundantly evident in this example; strong signals at m/z 204 and 366 appear in a region where the molecular ion cannot be observed.

To obtain information on these two glycosylation sites, it is necessary to reconsider the choice of protease. In this case, there is a glutamic acid residue

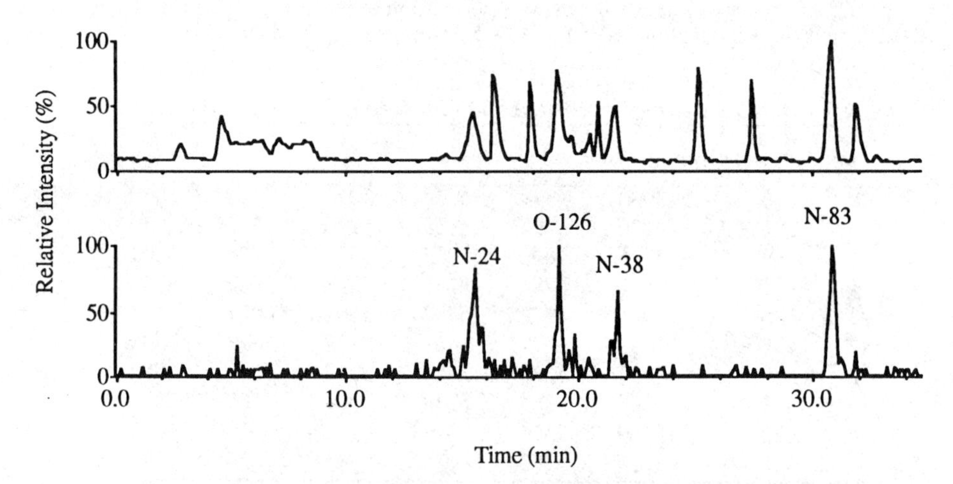

Figure 4. ESI-MS of a reversed phase chromatographic separation of a Glu-C digest of rHuEPO; the upper trace includes signals from all ions; the lower trace includes only signals in the range m/z 203-205.

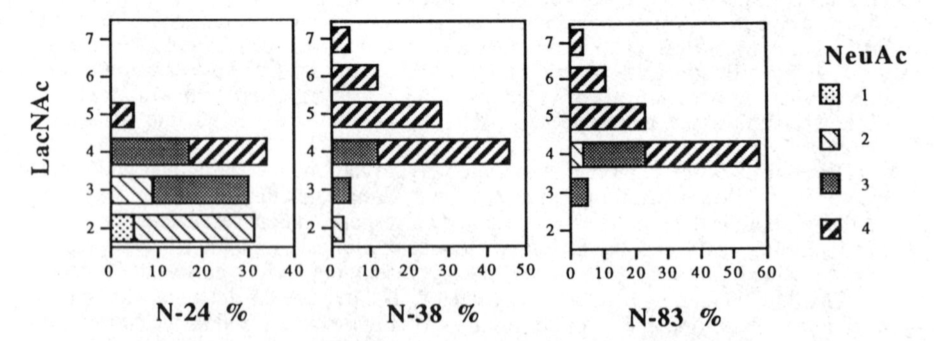

Figure 5 Relative distributions of LacNAc and NeuAc. Plotted for each Asn site are the relative proportion of each combination of LacNAc and NeuAc, determined by ion intensities.

between N-24 and N-38, thus endoproteinase Glu-C was chosen for the alternate peptide map shown in Figure 4.

Applying the same principles to this digest that were discussed for the tryptic N-83 site, we are able to determine the distribution of carbohydrate masses for each of the four sites, and find some remarkable differences in the three Asn linked sites (Figure 5). Absolute quantitation is not to be implied in this analysis, but Asn-24 clearly has less of the highly extended LacNAc structures than do Asn-38 and Asn-83. These site specific differences in degree of LacNAc substitution has been observed by several other techniques (*6,16-19*), many involving a great deal more work than the single peptide digest LC-ESI-MS analysis reported here. This once again demonstrates the exceptional power and utility of this method.

Expanded HPLC. When a reversed phase separation of a glycopeptide mixture is closely examined by either UV or total ion current, the separation seems to be dominated by the peptide backbone, and all glycoforms elute in close proximity. When the separation is examined by LC-ESI-MS, the pattern of ions detected shows the beginnings of separation by variations in glycoforms. This can be observed by displaying either a three variable contour map of elution time, mass to charge ratio and intensity (*20*), or as a comparison plot of the elution of a given ion versus time (Figure 6). For the ions selected in this display, it appears that addition of LacNAc imparts a limited but sufficient decrease in hydrophobicity to effect slightly earlier elution in the reversed phase gradient. This effect has recently been noted by others (*21*).

The observation of partial separation of glycoforms led us to investigate whether an expanded collection of several fractions across a UV peak might allow better separation of the individual glycoforms associated with a given peptide sequence. Collected fractions could then be subjected to more detailed ESI-MS. Figure 7 depicts the result of such an experiment for the N-83 Glu-C peptide of rHuEPO. This separation was performed off-line from the mass spectrometer; hand collected fractions were analyzed by Edman sequence analysis and ESI-MS infusion. Partial separation into two fractions is seen in the A_{214} trace; the peptide sequences of all collected fractions were identical, thus the splitting can only be due to differences in the carbohydrate structure. This is confirmed by examination of the mass spectra; as we step across the HPLC subfractions, we find that the number of LacNAc found is less in the later fractions than in earlier ones (*22*). Careful examination of these trends and of LC-ESI-MS contour plots confirm the observation that increasing LacNAc appears to be a strong determinant for earlier elution; NeuAc content is linked to LacNAc levels, but appears to have less influence on elution position. This is based on comparison of variations of NeuAc in structures with similar levels of LacNAc. Taken all together, these analyses provides clear confirmation that while separation in reversed phase chromatography is dominated by the peptide component, a contribution from the carbohydrate portion also exerts its influence.

Additional substituents. More intriguing results from these narrowly collected fractions are found in the expanded mass spectra. A variety of minor mass peaks are associated with many of the major ions (Figure 8). The ability to observe these minor ions is partly possible due to use of the gold tip electrospray needle (*9*). The reduced size of the needle allows slow infusion rates, allowing the tip to be positioned closer to the inlet orifice of the mass spectrometer; this combination results in increased sensitivity for limited quantity samples. Careful examination of the minor ions found here reveal mass spacings of 42 Da; this difference is consistent with the presence of O-acetyl substitutions on the NeuAc groups. This assignment is further supported by three additional observations: 1) the number of + 42 Da species goes up by approximately two per NeuAc; 2) the number of + 42 Da species increases with retention time, consistent with the expected increase in hydrophobicity due to

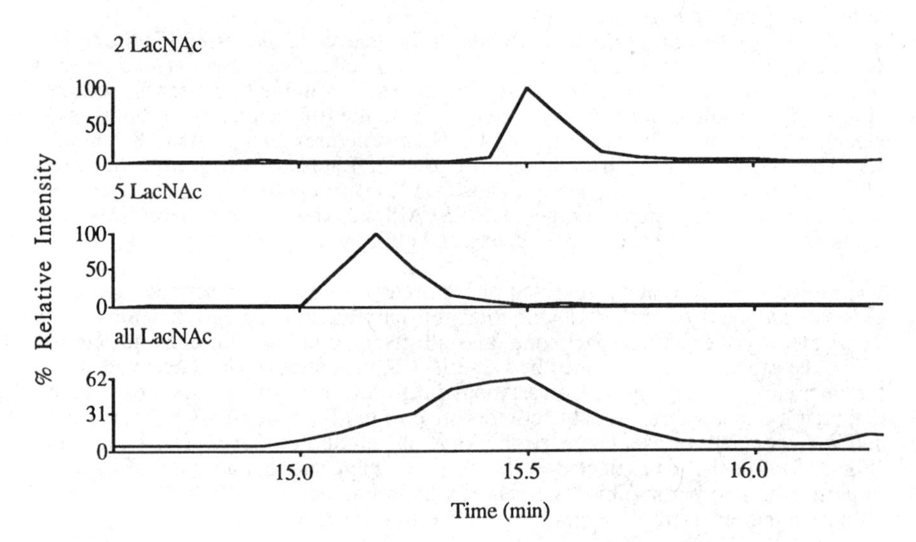

Figure 6. LC-ESI-MS of the N-24 region of Figure 4; the upper panel includes signals derived from m/z 1347-1349 (2 LacNAc), the middle panel shows signals from m/z 1906-1908 (5 LacNAc) and the lower panel shows signals from the range 1347-1908 (all LacNAc).

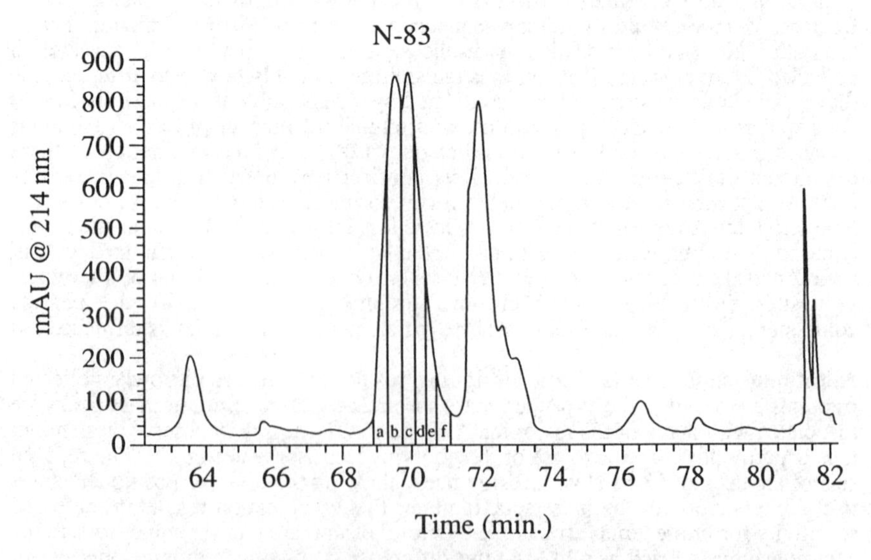

Figure 7. Expanded scale of N-83 Glu-C peptide; fraction collection is indicated by vertical drops from the A_{214} trace.

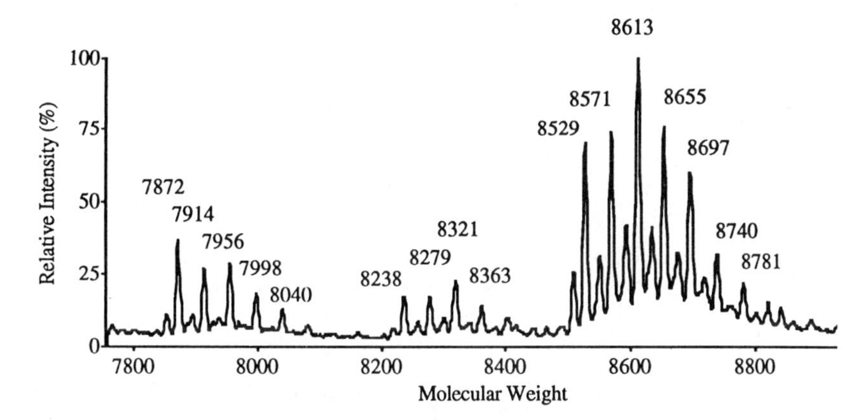

Figure 8. The reconstructed ion spectra of fraction d from Figure 7. Sample was infused with a micro electrospray tip. Major mass peaks are labeled.

addition of acetate groups to a relatively hydrophilic hydroxyl group; 3) stepped orifice potential scans show the presence of ions at m/z 334 and 376 (data not shown), consistent with one and two O-Acetyl NeuAc groups. The presence of O-acetylated NeuAc has been confirmed by observation of the loss of the + 42 Da species upon treatment with base and by NMR analysis of the enzymatically released NeuAcs from the intact protein (*22*); historical precedence and biological significance are also discussed in the same paper.

Additionally, suggestion of sulfate or phosphate on the carbohydrate chains is also indicated by the minor + 80 Da peaks in Figure 1d; each of these + 80 Da peaks is associated with a LacNAc-containing mass peak. This observation is intriguing in light of the reported sulfation of rHuEPO (*23*) and tissue plasminogen activator (*24*). Further research is needed to confirm this assignment of the + 80 Da peaks in Figure 1.

In general, these detailed analyses emphasizes the power of ESI-MS as an analytical tool for difficult problems such as glycoprotein analysis; also underscored is the breadth of information that this technology can uncover.

Future directions. The questions that we ask next are: what structural features govern the limitations in separation, and how can the analytical process benefit by finding better separations to feed into the ESI-MS analytical system? Clearly, the maximum benefit would be realized by further optimizing the reversed phase separation seen in Figure 7. Work is in progress to achieve this goal by employing higher resolution reversed phase chromatography media, mobile phases and different separation formats.

Another approach would be to employ a different basis for separation. One that we have found offers initial promise is capillary electrophoresis (*19*). Under the conditions described, it is possible to resolve multiple glycopeptide peaks with an HPLC purified, single amino acid sequence glycopeptide fraction (Figure 9). In capillary electrophoresis, the separation is on the basis of charge to mass ratios, thus it is not surprising that this method affords higher resolution than reversed phase. Presumably, the multiple peaks in Figure 9 arise from glycosylation differences all on the same peptide backbone. However, further analysis is complicated by new

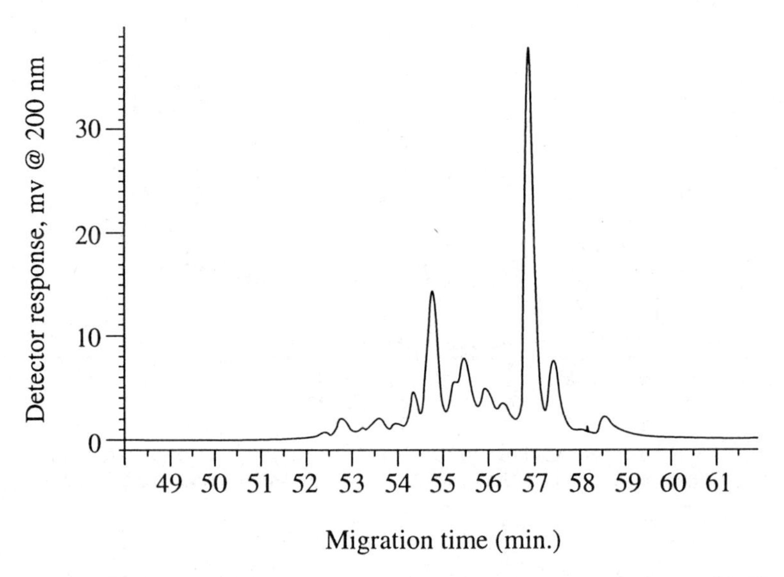

Figure 9. Capillary electrophoretic separation of rHuEPO tryptic peptide N-83.

problems that are introduced by this approach. Sample injection volumes are much more limited, thus methods of detection must be used that provide good sensitivity. Collection of separated peaks is a problem due to the necessity of maintaining constant voltage conditions during separation and collection; some progress has been made in this area, although we have not extended it to complex glycopeptides (*25*). Lastly, the buffer conditions needed to obtain satisfactory resolution (sodium phosphate and the ion pairing agent heptane sulfonate) introduce components that interfere with the ESI-MS detection event. Recently described techniques may provide the solution to these problems. These include the miniaturized electrospray tip (microspray) used in some of the above analyses, also developed and described by Wilm and Mann (*26*). Packing that tip with reversed phase chromatography media may likewise aid in signal enhancement; this would allow desalting of the peptide by binding in low organic solvent, followed by release and introduction to the electrospray source (Mathais Mann, personal communication).

References

1. *Guide to Techniques in Glycobiology*; Academic Press, Inc.: San Diego, 1994.
2. Kobata, A. In *Biology of Carbohydrates*; Robbins, V. G. a. P. W., Ed.; Wiley: New York, 1984; Vol. 2, pp 87.
3. Cullen, B. R. *Methods Enzymol* **1987,** *152*, 684-704.
4. Takeuchi, M.; Inoue, N.; Strickland, T. W.; Kubota, M.; Wada, M.; Shimizu, R.; Hoshi, S.; Kozutsumi, H.; Takasaki, S.; Kobata, A. *Proc Natl Acad Sci USA* **1989,** *86*, 7819-7822.
5. Narhi, L. O.; Arakawa, T.; Aoki, K. H.; Elmore, R.; Rohde, M. F.; Boone, T.; Strickland, T. W. *J Biol Chem* **1991,** *266*, 23022-23026.

6. Derby, P. L.; Strickland, T. W.; Rohde, M. F. In *Techniques in Protein Chemistry*; Angeletti, R. H., Ed.; Academic Press, Inc.: San Diego, 1993; Vol. IV, pp 161-168.
7. Jorgenson, J. W. *Microelectrospray Method and Apparatus. W.O. Patent 92/20437 B01D 59144, H01J49/00, 1992* .
8. Dohmeier, D. M.; Jorgenson, J. W. *Microelectrospray Method and Apparatus. U.S. Patent 5,115,131, 1992* .
9. Lewis, K. C.; Dohmeier, D. M.; Jorgenson, J. W.; Kaufman, S. T.; Zarrin, F.; Dorman, F. D. *Anal Chem* **1994,** *66*, 2285-2292.
10. Huddleston, M. J.; Bean, M. F.; Carr, S. A. *Anal Chem* **1993,** *65*, 877-884.
11. Carr, S. A.; Hemling, M. E.; Folena Wasserman, G.; Sweet, R. W.; Anumula, K.; Barr, J. R.; Huddleston, M. J.; Taylor, P. *J Biol Chem* **1989,** *264*, 21286-21295.
12. Sasaki, H.; Bothner, B.; Dell, A.; Fukuda, M. *J Biol Chem* **1987,** *262*, 12059-12076.
13. Lin, F. K.; Suggs, S.; Lin, C. H.; Browne, J. K.; Smalling, R.; Egrie, J. C.; Chen, K. K.; Fox, G. M.; Martin, F.; Stabinsky, Z.; et al. *Proc Natl Acad Sci U S A* **1985,** *82*, 7580-7584.
14. Jacobs, K.; Shoemaker, C.; Rudersdorf, R.; Neill, S. D.; Kaufman, R. J.; Mufson, A.; Seehra, J.; Jones, S. S.; Hewick, R.; Fritsch, E. F.; et al. *Nature* **1985,** *313*, 806-810.
15. Tsuda, E.; Goto, M.; Murakami, A.; Akai, K.; Ueda, M.; Kawanishi, G.; Takahashi, N.; Sasaki, R.; Chiba, H.; Ishihara, H.; Mori, M.; Tejima, S.; Endo, S.; Arata, Y. *Biochemistry* **1988,** *27*, 5646-5654.
16. Sasaki, H.; Ochi, N.; Dell, A.; Fukuda, M. *Biochemistry* **1988,** *27*, 8618-26.
17. Higuchi, M.; Oh-Eda, M.; Kubonina, H.; Tomonoh, K.; Shimonaka, Y.; Ochi, N. J. *J Biol Chem* **1992,** *267*, 7703-7709.
18. Linsley, K. B.; Chan, S.-Y.; Chan, S.; Reinhold, B. B.; Lisi, P. J.; Reinhold, V. N. *Anal Biochem* **1994,** *219*, 207-217.
19. Rush, R. S.; Derby, P. L.; Strickland, T. W.; Rohde, M. F. *Anal Chem* **1993,** *65*, 1834-1842.
20. Ling, V.; Guzzetta, A. W.; Canova-Davis, E.; Stults, J. T.; Hancock, W. S.; Covey, T. R.; Shushan, B. I. *Anal Chem* **1991,** *63*, 2909-2915.
21. Medzihradszky, K. F.; Maltby, D. A.; Hall, S. C.; Settineri, C. A.; Burlingame, A. L. *J Am Soc Mass Spectrom* **1994,** *5*, 350-358.
22. Rush, R. S., Derby, P. L., Smith, D. M., Merry, C., Rogers, G., Rohde, M. F. and Katta, V., *Anal Chem* **1995,** *67*, 1442-1452.
23. Strickland, T. W.; Adler, B.; Aoki, K.; Asher, S.; Derby, P.; Goldwasser, E.; Rogers, G. *J. Cellular Biochemistry* **1992,** *Supplement 16D*.
24. Pfeiffer, G.; Schmidt, M.; Strube, K. H.; Geyer, R. *Eur J Biochem* **1989,** *186*, 273-86.
25. Boss, H. J.; Rohde, M. F.; Rush, R. S. *Analytical Biochemistry* **1995** (in press).
26. Wilm, M.S.; Mann, M., *Int. J. Mass Spectrom. Ion Processes* **1994,** *136*, 167-180.

RECEIVED November 14, 1995

Chapter 22

Complications in the Determination of Molecular Weights of Proteins and Peptides Using Electrospray Ionization Mass Spectrometry

Catherine Fenselau and Michele Kelly[1]

Structural Biochemistry Center, University of Maryland–Baltimore County, 5401 Wilkens Avenue, Baltimore, MD 21228

Complications are discussed that have been encountered in assigning molecular weights to proteins and peptides from electrospray mass spectra. A variety of causes can contribute molecular ions to spectra that are heavier or lighter than the actual values, or multiple peaks in the region of the molecular ion. Examples are taken from studies of the mutation and processing of *gag* preproteins in lentevirus, and from the determination of the primary structure of the orphan drug bovine adenosine deaminase.

The demonstration of the ability of electrospray mass spectrometry to provide accurate molecular weights of proteins, its easy retrofit on mass spectrometers already in the field, and its compatibility as an interface between mass spectrometry and high pressure liquid chromatography or capillary electrophoresis have led to immediate acceptance and widespread implementation in mass spectrometry laboratories, and also in steadily increasing numbers of biochemistry laboratories. The application of mass spectrometry to biochemical problems has also been catalyzed by matrix assisted laser desorption ionization (MALDI). The reliable determination of molecular weights by electrospray is being exploited in protein folding studies that employ H/D exchange, determination of molar ratios in non-covalent complexes, and characterization of unstable reaction intermediates. Electrospray ionization mass spectrometry has also found widespread application in studies of protein processing and modifications of proteins by enzymic and chemical reactions. In the present report we discuss the reliability of molecular weight determinations made by electrospray with illustrations from studies of protein processing, post-translational modification, and peptide mapping.

[1]Current address: Ciba-Geigy Corporation, 556 Morris Avenue, Summit, NJ 07901

0097–6156/95/0619–0424$12.00/0

As has been pointed out (1), molecular weight determinations by electrospray and MALDI mass spectrometry are significantly more accurate than estimations based on the widely used polyacrylamide gel electrophoretic (PAGE) methods. More significant figures are provided, and the measurement is not altered by hydrophobicity or chemical modification. Mass resolution (the ability to separate proteins with similar molecular weights) is also higher with mass spectrometry. Some examples of comparative determinations are provided in Table I. Nonetheless, gel electrophoresis is a convenient and popular technique for proteins, and developmental efforts are underway to interface gel electrophoresis with MALDI and electrospray mass spectrometry.

Table I. Examples of Protein Molecular Weights Determined by Gel Electrophoresis and Electrospray Mass Spectrometry

Protein	*PAGE*	*ESMS*	*Reference*
Kinase C	7000	10460 ± 1, 10537 ± 1	2
	23000	17288 ± 7, 17362 ± 7	2
Nucleocapsid	13000	7283 ± 5	3
alpha amylase-1	45000	45529 ± 8, 45774 ± 11	4
		45833 ± 9, 46077 ± 11	

The general experience of the community also indicates that protein analysis by mass spectrometry provides information that is highly complementary to that inferred from gene sequences. Post-translational modifications, including processing, may be revealed by the lack of agreement between measured and inferred molecular weights, and occasionally the gene sequence is shown to be incorrect. Several examples are summarized in Table II. Although analysis by mass spectrometry does destroy the sample, its accuracy, sensitivity and speed make it the method of choice for direct examination of proteins.

Table II. Examples of Protein Molecular Weights Determined by Gene Sequencing and Mass Spectrometry

Protein	*Gene Sequence*	*Mass Spectrometry*	*Reference*
Glyoxalase I	18442	19440 ± 16	5
S. aureus V8	29024	29994 ± 3	6

We have recently published collaborative studies of the processing of preproteins transcribed by the *gag* (group antigen) genes in both human immunodeficiency virus (HIV) (7) and bovine immunodeficiency virus (BIV) (3). In both cases the gene sequence had been determined, and molecular weights were expected to reveal proteolytic sites in the preprotein. Not unexpectedly, this straightforward strategy was compromised by the presence of modifications such as myristoylation of the amino terminus in HIV (though not in BIV), errors in the protein sequences deduced from gene sequences, and mutations that occurred during cell culture. Some of the results of these processing studies are summarized in Table III. Assignments in the first and fourth columns are based on mobility in gel electrophoresis, and calculated values are based on gene and protein sequences.

TABLE III. Electrospray Mass Spectrometry Analysis of Lentevirus Proteins Processed from *gag* Preproteins

	MW			MW	
HIV (7)	*Calc'd*	*Observed*	*BIV (3)*	*Calc'd*	*Observed*
p17	15056	15269 ± 4	p16	14627	14628 ± 6
p24	25551	25752 ± 50	p26	24596	24610 ± 14
p7	6451	6451 ± 1	p13	7287	7283 ± 5

Observed values in Table III were determined by electrospray mass spectrometry. Functions in cell growth and replication for the proteins in Table III have been assigned by others. Mutations were found to be associated frequently with p17 from HIV, and to a lesser extent with p24 from HIV. However, no mutations were detected in p7 from HIV cells that had successfully replicated in culture. This is consistent with the requirement of this zinc finger protein for viral genome recognition during budding, genomic RNA packaging, and early events in viral infection.

Several experimental problems were encountered in these molecular weight determinations that led to the detection of multiple molecular ions in some samples. After the cultured cells were lysed under appropriate biosafety conditions (7), proteins were stabilized in solutions of thioethanol and guanidine hydrochloride and fractionated by complex reverse phase high pressure liquid chromatography (HPLC). Even with rechromatography, complete purification was sometimes difficult and samples collected for mass spectrometry sometimes contained more than one protein (3). Considerable heterogeneity or genetic mutation was encountered, and the resulting protein and peptide isoforms often had very similar chromatographic behavior, but different molecular weights.

One protein was converted to a mixture by partial phosphorylation in the cultured virus. The electrospray spectrum of *gag* p24 isolate from cultured HIV cells (8) was much more complex than that of HIV p24 prepared by recombinant techniques (9). Earlier work with capsid proteins from other HIV strains had revealed electrophoretic heterogeneity and partial phosphorylation (7). This is assumed to contribute to the complexity of the electrospray spectrum, along with sequence heterogeneity confirmed by Edman chemistry and collisional activation in tandem mass spectrometry experiments. In contrast, the molecular weight of the major form of p26 (>90%) isolated from BIV and determined by electrospray ionization, agreed to within 2 Da with that predicted by the gene sequence and was judged not to be phosphorylated.

Multiple proteins can also derive from imprecise processing that leaves ragged protein ends. In this case the mass differences within a series of mass spectral peaks will correspond to additional amino acids at the amino or carboxyl termini. Ragged ends and thus mixtures can also be formed artifactually by chemical cleavages that occur during protein purification and preparation of samples for mass spectrometry. Mass spectrometry is poorly equipped to distinguish these histories. In theory, the protein could be purified in $H_2{}^{18}O$, under which conditions isotopically labelled oxygen would be incorporated into any artifactual hydrolysis products for recognition by mass spectrometry (10).

Artifactual chemical modifications sometimes occur during protein purification and preparation of peptide maps. For example, thioethanol, frequently used in purification, can form disulfide bonds with unprotected cysteines, and sulfur in methionine is readily oxidized to sulfoxides and sulfones.

Another source of multiple ions in mass spectra of proteins is cleavage in the electrospray ionization source of small pieces from either the amino or carboxyl terminus. This phenomena is generally attributed to collisions occurring in the electrospray source. It is usually manifest as a series of peaks whose mass differences correspond to amino acid residues, and it has been termed microsequencing by Tom Covey (11). A selection of examples is presented in Table IV, and includes cleavages at proline, glutamate and other residues. Primary structures and actual molecular weights of the proteins were established independently in the appropriate references. In at least one case (15) the denatured protein has been found to fragment more readily than the folded protein.

Table IV. Examples of Protein Truncation in the Electrospray Source

Mol. Wt. Observed	*Neutral fragment*	*Ionized protein*	*Reference*
25554	acetyl	S-K•••	12
25469	acetyl-S	K •••	12
21978	M-F	P •••	13
15416	M	A •••	13
7122	A-S	Q •••	3
5953	acetyl-M	D-P•••	15
5824	acetyl-M-D	P •••	15
3722	acetyl-A	Q •••	14

The amino acid and acetyl groups listed as neutral fragments were shown by independent means to be bonded to the amino acids shown in the next column as amino termini of the ionized proteins actually detected. The truncated proteins listed in Table IV were observed as fully protonated y class ions (9,16) formed by cleavages between the carbonyl and amino moieties of amide bonds near the amino terminus. It is difficult to make the distinction

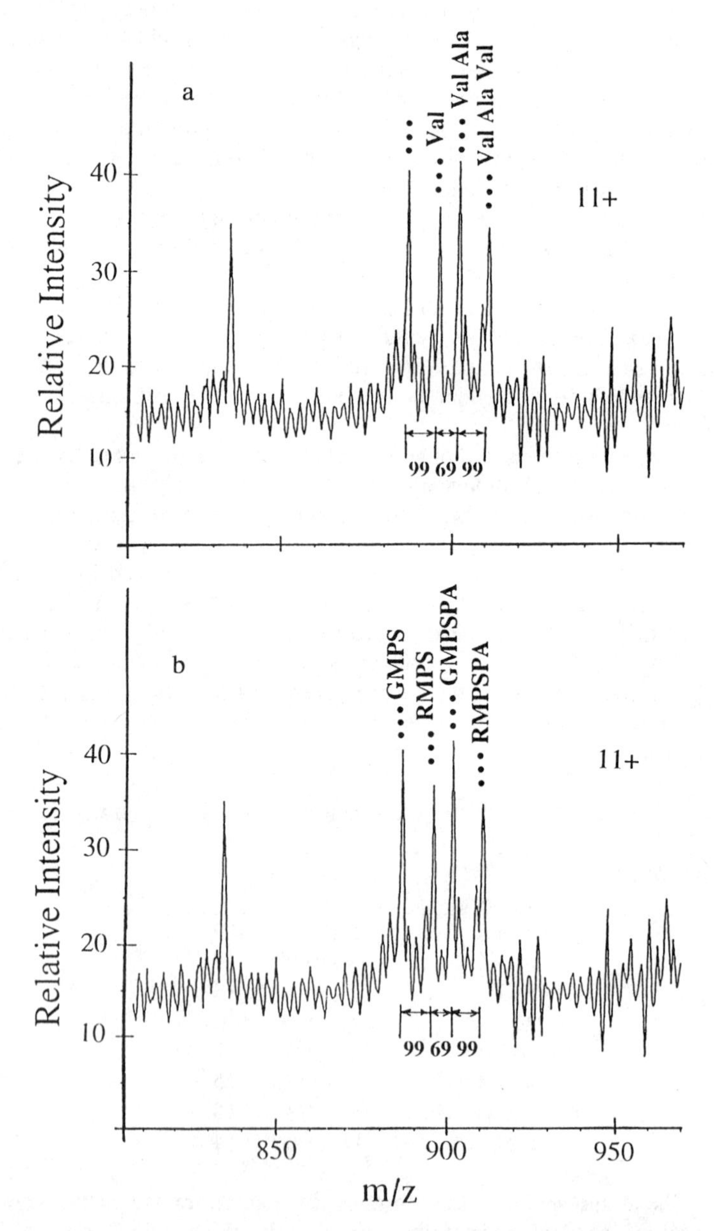

Figure 1. The +11 charge state of a polypeptide comprising amino acid residues [218-356] from bovine adenosine deaminase with two structural interpretations that accomodate the mass differences between the four peaks.

between gas and solution phase processes leading to losses of residues at the amino terminus, since fully protonated new amino termini are formed in either process. Experimental approaches to confirm that multiple protein species are the source of multiple peaks in electrospray mass spectra, rather than cleavages in the ionization source, include varying the voltages that control collision energies in the electrospray source (17) in order to reduce or increase gas phase fragmentation. Inspection of relative abundances of putative fragment ions in different charge states in a single spectrum may reveal enhancement in higher charge states due to higher collision energies. Parallel analysis by MALDI may reveal multiple components, as may examination by a high resolution separation technique such as isoelectric focussing (18) or capillary electrophoresis (5).

Fragmentation in the electrospray source of a carbonyl-amine bond near a protein's carboxyl terminus may be distinguished from solvolytic processes by the loss, in the former case, of the terminal hydroxyl group ($\Delta M = 17$) (9,16).

Part of an interesting electrospray spectrum is shown in Figure 1, in which four peaks are recorded in the +11 charge state of a polypeptide derived by cleavage of bovine adenosine deaminase with NBPS-skatole (14). Mass increments are close to those that would be expected from in-source fragmentation (microsequencing) of a Val-Ala-Val sequence at the amino terminus, as indicated in Figure 1a. However, further examination of the enzyme's primary structure revealed that this pattern reflects both glycine/arginine heterogeneity in the protein ($\Delta M = 99$) and hydrolysis of the last two residues, proline and alanine, ($\Delta M = 168$), as annotated on the spectrum in Figure 1b.

Electrospray mass spectrometry is also a very powerful technique for mapping the molecular weights of peptides released from proteins by cleavage with chemical reagents or proteolytic enzymes. This is particularly effective when an electrospray source is used in an interfaced HPLC electrospry mass spectrometry instrumental system. Table V contains a mass map of the peptides released by incubation of the orphan drug bovine adenosine deaminase (whose sulfhydryl groups have been stabilized as acetamide derivatives) with the Arg-specific protease clostripain (14). The experimental measurements were made at the beginning of the determination of the primary structure of this protein and the calculated values are based on the sequence that was eventually determined by collisional activation in tandem mass spectrometry experiments. Most match within 1 Da, and all match within 2 Da. It can be seen that the amino terminus peptide [2-33] was truncated by cleavage of acetyl-alanine resulting from ion molecule collisions in the electrospray source. The acetylated first residue was identified in peptides produced by other proteolytic agents, and is consistent with the molecular weight of the intact protein. The carboxyl terminus was found in other studies not to be ragged, as Table V suggests, but to comprise arginine/glycine heterogeneity in the sixth position from the end. Those molecules containing arginine were cleaved at residue 351, while the carboxyl terminus (residue 356) is retained in proteolytic products containing glycine as the sixth residue from the carboxy terminus.

Table V. Mass Map of Peptides From The Digestion of Bovine Adenosine Deaminase with Clostripain

Peptide	*Calculated*	*Electrospray Mass Spectrometry*
2 - 33	3721.3	3722.1
34 - 80	5251.1	5249.7
81 -100	2345.7	2345.3
101 - 148	5332.9	5333.0
149 - 210	7018.9	7018.8
211 - 234	2509.8	2511.1
235 - 250	1890.0	1890.0
251 - 281	3773.2	3771.4
282 - 323	5054.6	5054.2
324 - 351(R)	3282.8	3282.6
324 - 356(G)	3667.2	3666.4

(R) Arginine at position 351
(G) Glycine at position 351

Despite the cautionary words presented here, this chapter closes with the same theme on which it opened, that the power of electrospray mass spectrometry to provide molecular weight determinations can be of great value to biochemists. In our laboratory, for example, in addition to determining molecular weights of unknown proteins and their proteolytic peptide products, we have used it to assess the complexed metal content of native and reconstituted metallothioneins (15), to define the stoichiometry of products of covalent reactions of metallothioneins with anticancer drugs (19) and to characterize oxidation products from reactions of native and apo HIV p7 with potential antiviral drugs (20).

Acknowledgements

We thank NIH and NSF for financial support and we acknowledge our excellent collaborators in the AIDS Vaccine Program of the National Cancer Institute and the Octamer Research Foundation.

Literature Cited

1. Jardine, I. *Methods in Enzymology* **1990**, *193*, 441-455.
2. Dianoux, A.C.; Stasia, M.J.; Garin, J.; Gagnon, J.; Vignais, P.V. *Biochemistry* **1992**, *31*, 5898-5905.
3. Tobin, G.J.; Sowder II, R.C.; Fabris, D.; Hu, M.Y.; Battles, J.K.; Fenselau, C.; Henderson, L.E.; Gonda, M.A. *J. of Virology* **1994**, *68*, 7620-7627.

4. Sogaard, M.; Andersen, J.S.; Roepstorff, P.; Svensson, B. *Bio/Technology* **1993**, *11*, 1162- .
5. Hua, S.; Vestling, M.M.; Murphy, C.M.; Bryant, D.K.; Height, J.J.; Fenselau, C.; Theibert, J.; Collins, J.H. *Int. J. Peptide Protein Res.* **1992**, *40*, 546-550.
6. Lu, T.; Creighton, D.J.; Antoine, M.; Fenselau, C.; Lovett, P.S. *Gene* **1994**, *150*, 93-96.
7. Henderson, L.E.; Bower, M.A.; Sowder II, R.C.; Serabyn, S.A.; Johnson, D.G.; Bess, Jr., J.W.; Arthur, L.O.; Bryant, D.K.; Fenselau, C. *J. of Virology* **1992**, *66*, 1856-1865.
8. Fenselau, C.; Yu, X.; Bryant, D.; Bowers, M.A.; Sowder II, R.C.; Henderson, L.E. In: *Mass Spectrometry for the Characterization of Microorganisms*; C. Fenselau, ed.; American Chemical Society: Washington, DC, **1994**; *541*, 159-172.
9. Carr, S.A.; Hemling, M.E.; Bean, M.F.; Roberts, G.D. *Anal. Chem.* **1991**, *63*, 2802-2824.
10. Rose, K.; Savoy, L.A.; Simona, M.G.; Offord, R.E.; Wingfield, P. *Biochem. J.* **1988**, *250*, 253-259.
11. Covey, T. **1995** private communication.
12. Murphy, C.M. Ph.D. Dissertation, University of Maryland Baltimore County, Baltimore, MD **1994**.
13. Loo, J.A.; Edmonds, C.G.; Smith, R.D. *Anal. Chem.* **1993**, *65*, 425-438.
14. Kelly, M.A. Ph.D. Dissertation, University of Maryland Baltimore County, Baltimore, MD **1994**.
15. Yu, X.; Wojciechowski, M.; Fenselau, C. *Anal. Chem.* **1993**, *65*, 1355-1359.
16. Biemann, K. *Biomedical and Environmental Mass Spectrometry* **1988**, *16*, 99-111.
17. Loo, J.A.; Udseth, H.R.; Smith, R.D. *Rapid Commun. Mass Spectrom.* **1988**, *2*, 207-210.
18. Payne, D.J.; Skett, P.W.; Aplin, R.T.; Robinson, C.V.; Knowles, D.J.C. *Biological Mass Spectrometry* **1994**, *23*, 159-164.
19. Yu, X.; Wu, Z.; Fenselau, C. *Biochemistry* **1995**, *34*, 3377-3385.
20. Yu, X.; Hathout, Y.; Fenselau, C.; Sowder II, R.C.; Henderson, L.E.; Rice, W.G.; Mendeleyev, J.; Kun, E. *Chem. Res. Toxicology* **1995,** *8*, 586-590.

RECEIVED October 31, 1995

Chapter 23

Use of Hyphenated Liquid-Phase Analyses–Mass Spectrometric Approaches for the Characterization of Glycoproteins Derived from Recombinant DNA

A. Apffel[1], J. Chakel[1], S. Udiavar[1], W. S. Hancock[1,3], C. Souders[2], and E. Pungor, Jr.[2]

[1]Hewlett-Packard Laboratories, 3500 Deer Creek Road, MS26U–R6, Palo Alto, CA 94304
[2]Berlex Biosciences, 430 Valley Drive, Brisbane, CA 94005

The analysis of recombinant Desmodus Salivary Plasminogen Activator (DSPAα1), a heterogeneous glycoprotein, is demonstrated through the use of High Performance Liquid Chromatography (HPLC), High Performance Capillary Electrophoresis (HPCE), Electrospray Liquid Chromatography Mass Spectrometry. (ES-LC/MS) and Matrix Assisted Laser Desorption Ionization - Time of Flight Mass Spectrometry (MALDI-TOF MS). The protein is analyzed at three specific levels of detail: the intact protein, proteolytic digests of the protein and fractions from the proteolytic digest. A method for "on-column" collection of HPLC fractions for subsequent transfer and analysis by HPCE and MALDI-TOF is shown.

For many "real world" applications in bioscience no single technique, however powerful, is sufficient to provide the total picture. In these cases, the challenge is to be able to exploit the strengths of various complimentary techniques to develop a complete, reliable and robust solution to a given problem. A further challenge is to effectively integrate the data produced by these analytical techniques into purity and product consistency information suitable for the characterization of a protein pharmaceutical. The transformation of the data from a mass of unrelated facts derived from both the intact molecules as well as series of enzymatically produced fragments to a coherent understanding of the nature of the pharmaceutical product is not a trivial one, rather an essential part of the development of improved analytical methodology.

[3]Corresponding author

0097–6156/95/0619–0432$17.00/0

It is interesting to note the evolution of the concept of multidimensional or hyphenated techniques. Ten years ago, gas chromatography/mass spectrometry (GC/MS) was considered a "multidimensional" analysis; five years ago, liquid chromatography/mass spectrometry (LC/MS) was "hyphenated"; now these are at best a single part of a higher dimensional approach to analyses. Similarly, a few years ago, the application of Electrospray LC/MS to peptide mapping was a revolutionary approach supplying a great deal of information in an astonishingly short time when compared to the earlier approaches of sequencing each and every peak collected from an High Performance Liquid Chromatography (HPLC) peptide map. Now, such an analysis is routine, and the more challenging problems are the characterization of post-translational modifications of proteins which introduce significant heterogeneity and therefore cannot be solved in a single step.

While such maps have become the primary quality control (QC) method for the characterization of small proteins, it has recently been realized that the mapping approach has limitations. For example, the analysis of a much larger protein, e.g. fibrinogen (MW of 350 kDa), heterogeneous glycoproteins or antibodies (MW of 150 kDa) is hindered by the complexity of the range of peptides generated by an enzymatic digestion. The limitations of the map are particularly evident with the analysis of a glycoprotein in which many hundreds of glycoforms have the potential of adding an enormous number of peaks in the map. Often, however, such peptides are not observed because of limitations of the detection system or due to the inability of the separation method to discriminate between closely related species. Table 1 illustrates the enormous heterogeneity that can be present in a typical glycoprotein. This table is derived from Tissue Plasminogen Activator (rTPA) which has three sites of glycosylation. The first site contains mainly high mannose forms while the second and third sites contain complex structures. With the additional factor of optional glycosylation at the second site, one can calculate that this glycoprotein will contain over 11,000 major forms. With such substantial microheterogeneity, it is often observed that the abundance of glycopeptides is close to the detection limit, and the small glycopeptide peaks are obscured by the non-glycosylated fragments. Figure 1 shows a typical situation where the intensity of a major set of glycopeptides is found to be 1/5th to 1/40th of the level of a non-glycosylated sample. Such complexity makes a single reversed phase HPLC separation combined with on-line UV detection of limited utility.

The advent of commercially available combined HPLC and Electrospray Ionization Mass Spectrometry (ESI LC/MS) systems compatible with conventional HPLC methodology has increased the power of peptide mapping considerably[1,2]. ESI LC/MS in combination with in-source collisionally induced dissociation (CID) has been used effectively to identify sites of N- and O-linked glycosylation[3,4,5]. However, even this technique is limited by insufficient resolution resulting from the large number of very similar peptides caused by variable protein glycosylation and enzymatic digests of moderately sized glycoproteins. It is therefore necessary to employ a range of techniques with orthogonal selectivity in order to characterize such samples. A

Table 1: Possible glycoforms for rTPA[a]

Site I	High Mannose: 6	Hybrid: 4		Total: 10
Site II	Complex Tetraantennary : 10	Complex Triantennary: 8	Complex Biantennary : 6	Total: 24
Site III	Complex Tetraantennary : 10	Complex Triantennary: 8	Complex Biantennary : 6	Total 24
	rTPA	(24*24*10)*2[b]		Total: 11520

a. Derived from data presented by A. Guzzetta[2].

b. The factor of two is due to partial glycosylation of the second site.

significant issue is to efficiently utilize such multidimensional data in a way that multiple samples can be characterized for a product development program.

We have, therefore, investigated the use of combinations of HPCE, HPLC, ESI LC/MS and MALDI-TOF MS to allow for characterization of enzymatic digests of underivatized glycoprotein samples. The separation selectivity of HPLC and CE is sufficiently orthogonal to yield a great deal of comparative information. Although ESI LC/MS is extremely powerful for peptide mapping, when dealing with intact proteins the mass analyis is limited to relatively homogeneous proteins. Due to the complexity of the ion envelope generated by ESI-LC/MS, which is observed for a protein with substantial heterogeniety, spectra become difficult, if not impossible, to interpret. In addition, as the mass range is limited to <80 kDa, well below the limits obtained with MALDI-TOF MS (up to 800kDa). Thus in combining data from the two mass spectrometric techniques, again highly complementary information is obtained.

As an example of a heterogeneous glycoprotein, DSPAα1, a single chain plasminogen activator derived from Desmodus Rotundus (vampire bat) salivary glands[6] was chosen. DSPAα1 is a serine protease that plays a role in clot lysis. DSPAα1 displays approximately 70% sequence homology with the double chain serine protease, Tissue Plasminogen Activator (tPA), and may have a fibrin specificity with a strict dependence on polymeric fibrin as a cofactor[7]. DSPA is known to be heterogeneous when expressed in CHO cells[8]. It is a large (441 amino acids) complex molecule of calculated average molecular weight 49,508 Da (non-glycosylated) with six sites for potential glycosylation, four O-linked and two N-linked based *on minimum consensus sequences*. It has been proposed, based on homology with other serine proteases that there are 28 cysteines forming 14 disulfide bridges.

Experimental

HPLC

The HPLC separation was performed on a Hewlett Packard 1090 Liquid Chromatography system equipped with a DR5 ternary solvent delivery system, diode array (DAD) UV/VIS detector, autosampler and heated column compartment (Hewlett-Packard Co., Wilmington, DE). All HPLC separations were done using a Vydac C_{18} (Catalog # 218TP54, Hesperia, CA, U.S.A.) 5 µm particle, 300A pore size reverse phase column. A standard solvent system of H_2O (solvent A) and acetonitrile (solvent B), both with 0.1% TFA was used with a flowrate of 0.2 mL/min. The gradient for the separation was constructed as 0-60% B in 90 minutes. The column temperature was maintained at 45°C throughout the separation.

A schematic for "on-column" fraction collection is shown in Figure 2. The effluent from the HPLC (at 200 µL/min) is mixed with a non-elutropic solvent, (0.1% TFA) at 1000 µL/min delivered with an Eldex Model A-30-S Metering Pump (Eldex Laboratories, Inc. Napa, CA). The resultant mixture was selectively directed to one of four possible collection columns, or a bypass, using a Valco Model CST6UW 6 position 14 port electrically actuated valve (Valco Instrument Co. Inc., Houston, TX).

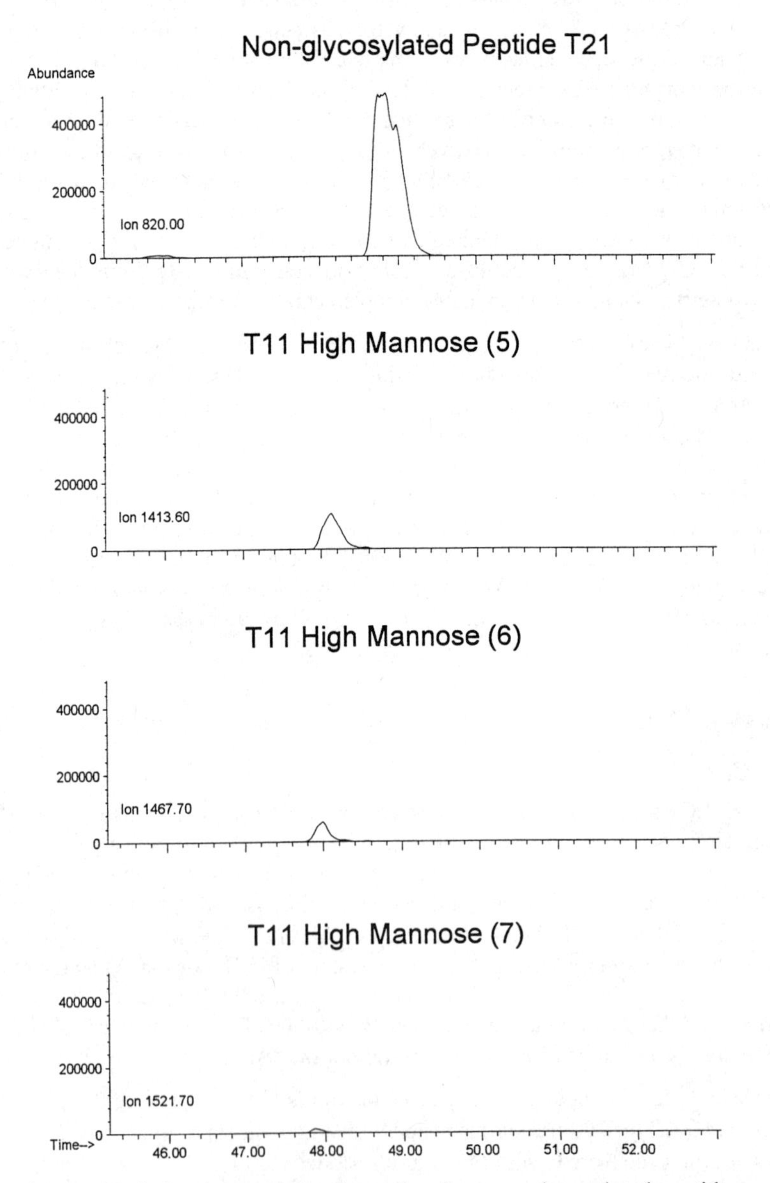

Figure 1: Relative response for glycosylated vs non-glycosylated peptides.

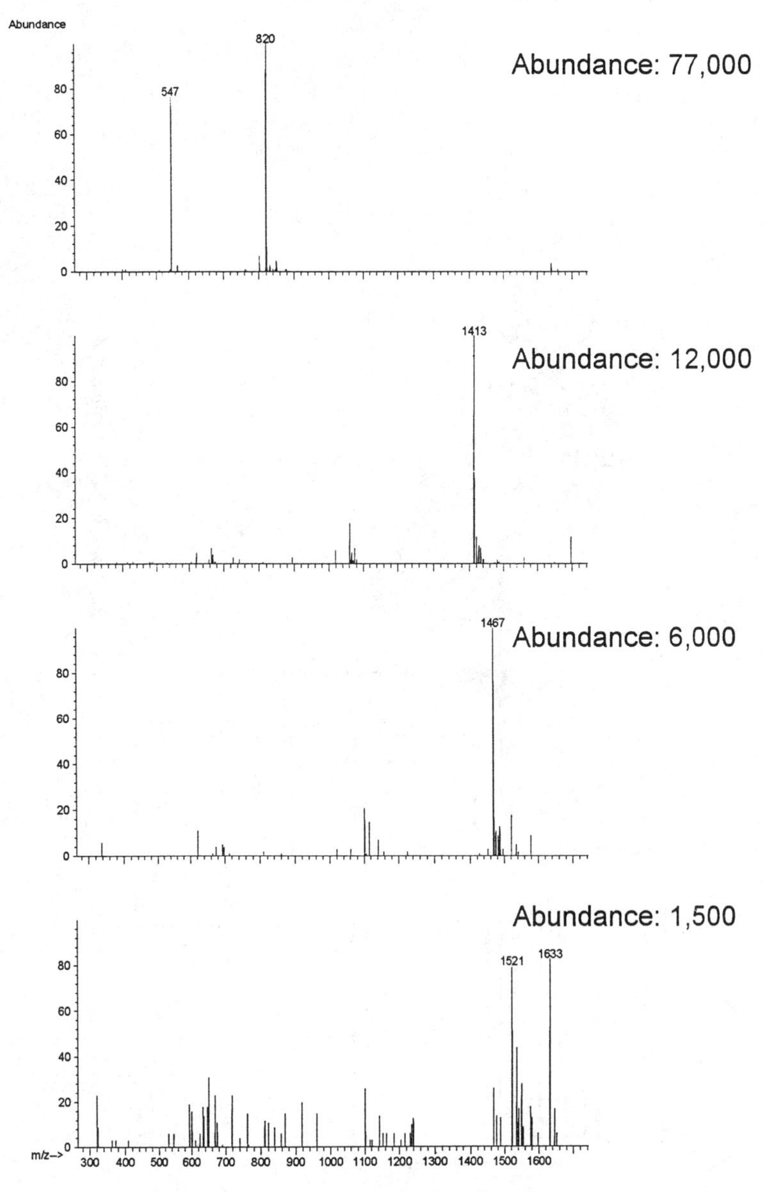

Figure 1. *Continued*

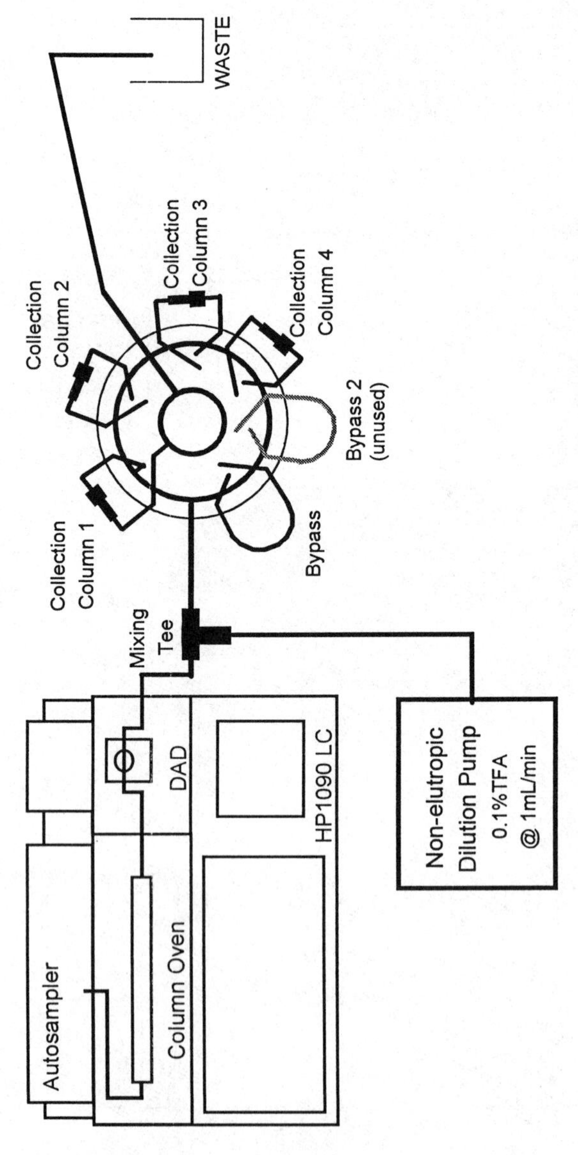

Figure 2: Schematic of On-Column Fraction Collection.

The collection columns consisted of HP Hydrophobic Sequencing Cartridge Columns (P.N. G1073A) in a Column Adapter (P.N. G1007A). Fractions were eluted from the collection columns using a gradient from 0-90%B/5 minutes at a flow rate of 0.2 mL/min and collected in eppendorf vials. The fraction collection process was repeated four times with 250 μL of sample at a concentration of 1.2 nmol/mL. Thus approximately 5 nmoles of each homogeneous peptide was collected. A lower amount is present for the heterogeneous glycoforms. The samples were then eluted in approximately 200 μL and concentrated using a speedvac to a final volume of 100 μL resulting in an approximately 10 fold increase in concentration.

HPCE

HPCE was done on a Hewlett-Packard HP^{3D}CE system with a built in UV/VIS Diode Array Detector and MS Windows Chemstation 3D Software. A phosphate deactivated fused silica capillary of internal diameter 50 μm and 56 cm effective length with a 150 μm extended light path (bubble cell) (HP P.N. G1600-62232). The background electrolyte consisted of 100 mM sodium phosphate buffer at pH 2.4 and 100 mM NaCl. Injections were made by pressure for 10-30 sec at 50 mbar. For peptide separations, the injection was followed by a trailing electrolyte consisting of 1 mM phosphoric acid at 5 kV for 2 minutes. The use of this discontinuous buffer system served to sharpen the peaks from the injection plug. The separation was monitored at 214 nm and 280 nm.

Mass Spectrometry

Electrospray LC/MS

Mass spectrometry was done on a Hewlett-Packard 5989B Quadrupole Mass Spectrometer equipped with extended mass range, high energy dynode detector (HED) and a Hewlett-Packard 59987A API-Electrospray source with high flow nebulizer option.

To counteract the signal suppressing effects on Electrospray LC/MS of trifluoroacetic acid, a previously reported method[9], referred to as the "TFA Fix" was employed. The "TFA Fix" consisted of post column addition of a 75% propionic acid; 25% isopropanol solvent at a flow rate of 100 μL/min. The TFA Fix was delivered using an HP 1050 HPLC Pump and was teed into the column effluent after the DAD detector and after the column switching valve. Column effluent was diverted from the MS for the first 5 minutes of the chromatogram, during which time, excess reagents and unretained components eluted.

For the in-source Collisionally Induced Dissociation (CID) method for detecting fragments indicative of glycopeptides[3], the capillary exit (CapEx) voltage was set to 200V instead of the standard 100V. Data acquisition was done in Selected Ion Monitoring (SIM) mode, monitoring ions at m/z 204, 292, and 366, each with a 1 Da window and a dwell time of 150 msec resulting in an acquisition rate of 1.5 Hz. Data was filtered in the time domain with a 0.05 min gaussian time filter.

MALDI-TOF

Mass spectra were generated with a Hewlett-Packard 1700XP (predecessor to the Hewlett-Packard G2025A) MALDI-TOF system. This system utilizes a N_2 laser (337 nm) for desorption/ionization, a two stage ion source 29.5 KeV acceleration, a linear 1.7 meter flight tube and a dual MCP detector.

Spectra were acquired at laser powers slightly above the ionization threshold using a matrix consisting of either 3,5 -dimethoxy-4-hydroxycinnamic acid (sinapinic acid, HP P.N. G2005A) or a multicomponent matrix consisting of 2-5-dihydroxybenzoic acid and L-fucose (both from Aldrich, Milwaukee, WI).

Intact Protein: DSPAα1 (0.55 mg/mL) was mixed 1:1 with a 100 mM sinapinic acid matrix solution and two 1 μL aliquots (~10 pmol total) were sequentially deposited onto the probe tip and vacuum dried in the HP G2024A sample prep accessory. Data were collected at a 100 MHz sampling rate and a total of 100 laser shots were summed. The mass scale was calibrated using a protein standard mixture (HP P.N. G2053A) as an external calibrant.

Protein Digest Mixture and HPLC Pooled Fractions: The entire digest mixture sample was prepared with two different matrices. For each preparation, data were collected at a 400 MHz sampling rate and a total of 30-40 laser shots were summed. The first sample preparation comprised of diluting the digest mixture (0.35 mg/mL) 1:9 with 100 mM sinapinic acid. Two 0.5 μL aliquots (~600 fmol total) were sequentially deposited onto the probe tip and vacuum dried. The alternate sample preparation involved diluting the digest mixtures 1:9 with the multicomponent matrix consisting of 20 mg/mL of both 2,5-dihydroxybenzoic acid and fucose with 1 mg/mL 5-methoxysalicylic acid in 9:1 TFA/ethanol. Two 0.5 μL aliquots (approx. 600 fmole total) were sequentially deposited onto the probe tip and vacuum dried. This multicomponent matrix has been shown to give significant enhancement in spectral reproducibility, signal intensity and mass resolution.[10]

The HPLC fractions were mixed 1:1 with a 100 mM sinapinic acid, and two 0.5 μL aliquots were sequentially deposited onto the probe tip and vacuum dried. Data were collected at a 400 MHz sampling rate and a total of 75 laser shots were summed for each fraction.

The mass scale was calibrated for the peptide mass range using a peptide standard mixture (HP P.N. G2052A) as an external calibrant.

Sample preparation

DSPA α1 was reduced, alkylated and digested with Endoproteinase Arg C in a standard manner, as has been reported elsewhere.[11]

Results and Discussion

Analytical Strategy

In describing the case study for the characterization of the DSPAα1 glycoprotein using hyphenated liquid phase analysis/mass spectrometry, we hope to accomplish to goals; to demonstrate that the analytical approach can be used to characterize real world samples, and secondly to demonstrate a range of techniques that can be used in an integrated approach.

Our analytical approach we have developed for the characterization of complex protein products is to apply a series of different analytical techniques at increasing levels of detail starting with the intact protein and stepwise focusing on a more specific and detailed scope of information. It is important to provide structure and heterogeneity information on the parent molecule as it represents the actual drug substance. With the molecular complexity of a biopolymer, however, it is not possible to attribute a chemical difference to any observed heterogeneity and thus the next step in the analysis is to use a highly specific process to generate smaller fragments for subsequent analysis. The process is outlined in Figure 3. At each level, there are different strengths of the techniques in our analytical "tool box". For example, in general, MALDI-TOF tends to be a more powerful tool for examining intact proteins that exhibit microheterogeneity, while ESI-LC/MS provides better selectivity and specificity for the separation and analysis of the mixtures produced by enzymatic digests of proteins.

It should be noted that the strategy outlined in Figure 3 is, in practice, not quite as linear as pictured. For example, we have on several occasions been able to identify the sites and character of a specific glycosylation and then been able to reanalyze the intact protein in such a way as to verify the more detailed analysis. Furthermore, there is another dimension added into the strategy through the selective use of chemical and enzymatic modification of the proteins. For example, sialic acid heterogeneity can be examined with the enzyme neuraminidase which allows the production of the asialo product.

The Intact Glycoprotein

At the level of detail of the intact glycoprotein, information can be gathered concerning the gross chemical and structural characteristics of the protein as well as the purity of the sample preparation. A related issue is to define the degree of heterogeneity of variants of the target sequence, i.e. non-product related impurities. With such a protein, there exist multiple opportunities for heterogeneity including glycosylation, varying N-terminal sequences, amino acid modifications and proteolytic clipping. The goal of the production process in a recombinant DNA derived pharmaceutical such as DSPAα1 is to produce a high purity product in a consistent manner. Other variants can be produced by degradative processes that can occur during the purification process or upon storage. Examples of such processes include deamidation (protein component) or loss of sialic acid at lower pH values (carbohydrate component). With a glycoprotein, the most substantial heterogeneity is introduced by the post-translational modification of glycosylation. Therefore, the analytical sample may consist of a mixture of not only potential contaminants, but the enormous number of glycoforms as well, which are an intrinsic part of the target

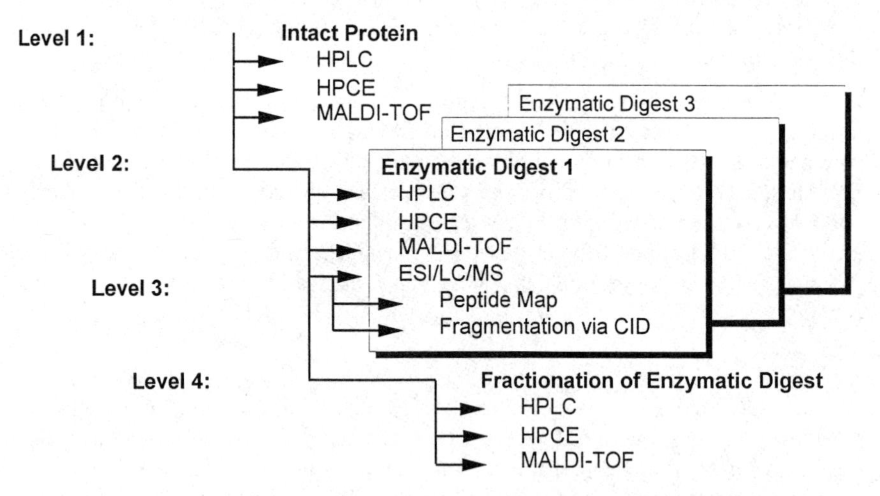

Figure 3: Analytical Strategy

molecule. It is the challenge of the analytical chemist to devise a suitable set of methods that can be applied to such complex samples. Of course, in addition to separation methods discussed below, a wide range of tools including N and C-terminal sequence analysis, x-ray crystallography and biological activity tests can be applied. An appropriate bioassay is of particular importance as such an assay can be used to determine the significance of a particular variant. Key issues with product variants are usually related to loss of activity and potential antigenicity

The initial examination of the intact protein sample was performed by reversed phase HPLC as shown in Figure 4a. From the narrow symetrical peak in the separation, it can be seen that the sample is quite pure, and at least in terms of hydrophobicity, relatively homogeneous. It is clear, however that this separation approach is only suitable for product variants which are separated under the applied conditions[12].

The capillary electrophoresis separation (see Figure 4b) is also used to analyze the intact protein. In addition, the orthogonality of the separation mechanism, being based on charge rather than hydrophobicity, is useful in the further characterization of purity and homogeneity. Such an analysis can be used to monitor charge heterogeneity introduced from the variable sialic acid content in the carbohydrate moiety or in deamidation of the polypeptide backbone. Due to the glycosylation this analysis results in a relatively broad peak, but the absence of other well separated components is consistent with the expectation of low levels of contaminating proteins. These results are consistent with the expectation that this is a highly purified protein which nonethelss exhibits substantial microheterogeniety. Contrast this with, for example, a crude cell extract which would include many unrelated proteins and contaminants.

The analysis of the intact glycoprotein by MALDI-TOF MS (shown in Figure 4c) yields mass based information of product heterogeneity. The two major peaks in the spectrum represent singly and doubly charged ions from the protein. The mass of the protein, determined from the singly charged ion, is 54,111.6 $\pm$ 54 Da compared with the average molecular weight of 49,508.15 Da calculated from the amino acid sequence only. The difference of 4603 Da or approximately 9.3% indicates a mass increase due to glycosylation. The peak width of approximately 5,000 Da relates the degree of microheterogeneity of glycosylation. Based on the expected resolution of the MALDI-TOF MS experiment, if DSPAα1 were homogeneous, it would have a peak width at half height on the order of 500 Da due, in part, to the isotopic distribution.

The results described above show that a combination of the three analytical procedures can give orthogonal 'snap shots' of purity of the drug substance based primarily on differences in hydrophobicity (LC), charge (CE) and mass (MALDI-TOF). In this context, electrospray is less useful due to the complexity of the overlapping ion envelopes resulting from such a heterogeneous sample. One could argue that a combination of the three methods is particularly useful for Quality Control as a significant impurity is less likely to escape detection by at least one of the methods.

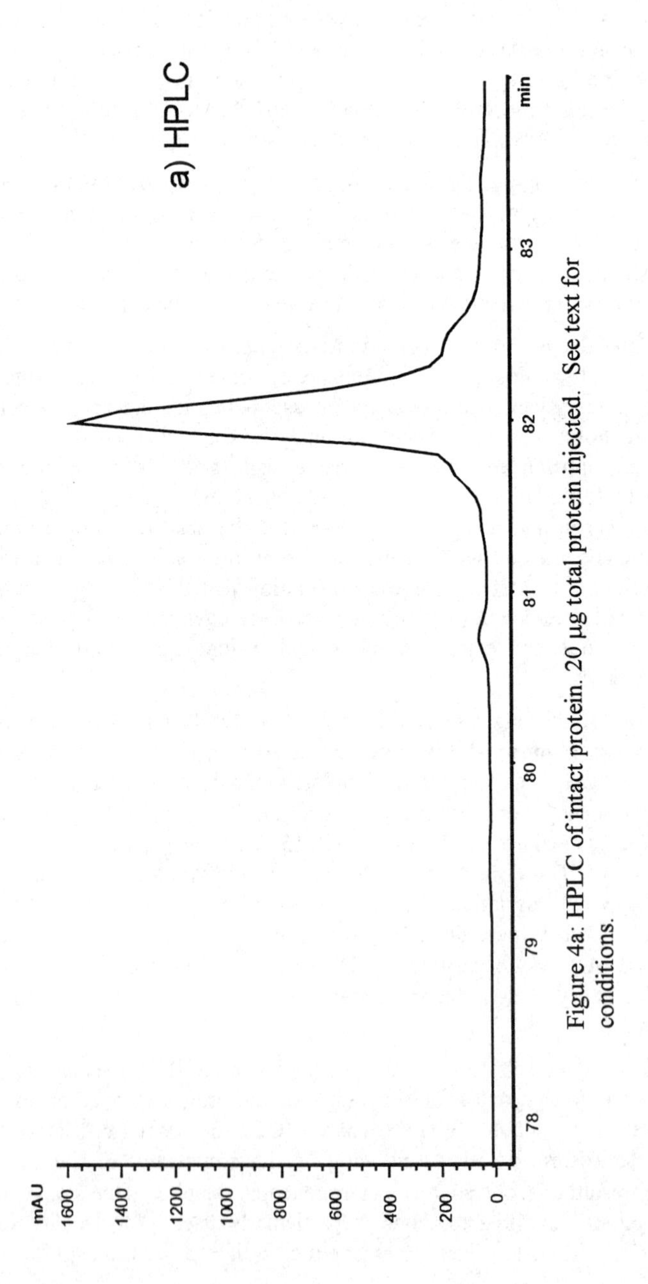

Figure 4a: HPLC of intact protein. 20 µg total protein injected. See text for conditions.

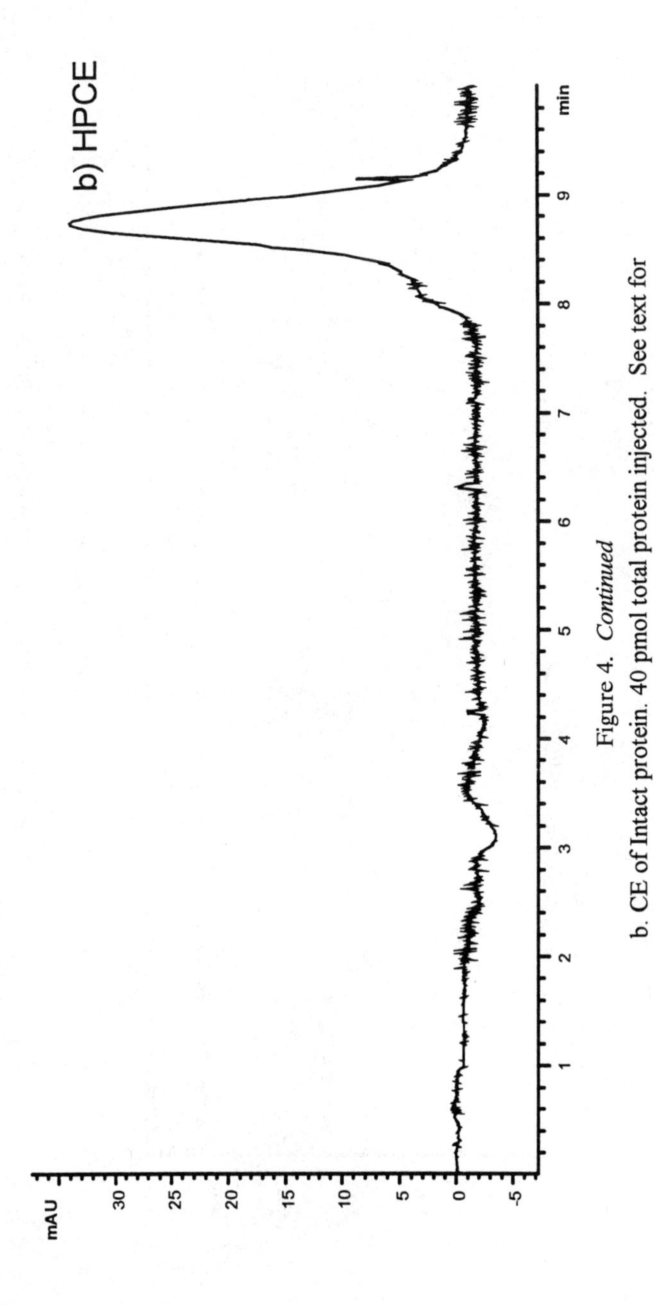

Figure 4. *Continued*

b. CE of Intact protein. 40 pmol total protein injected. See text for conditions

Continued on next page

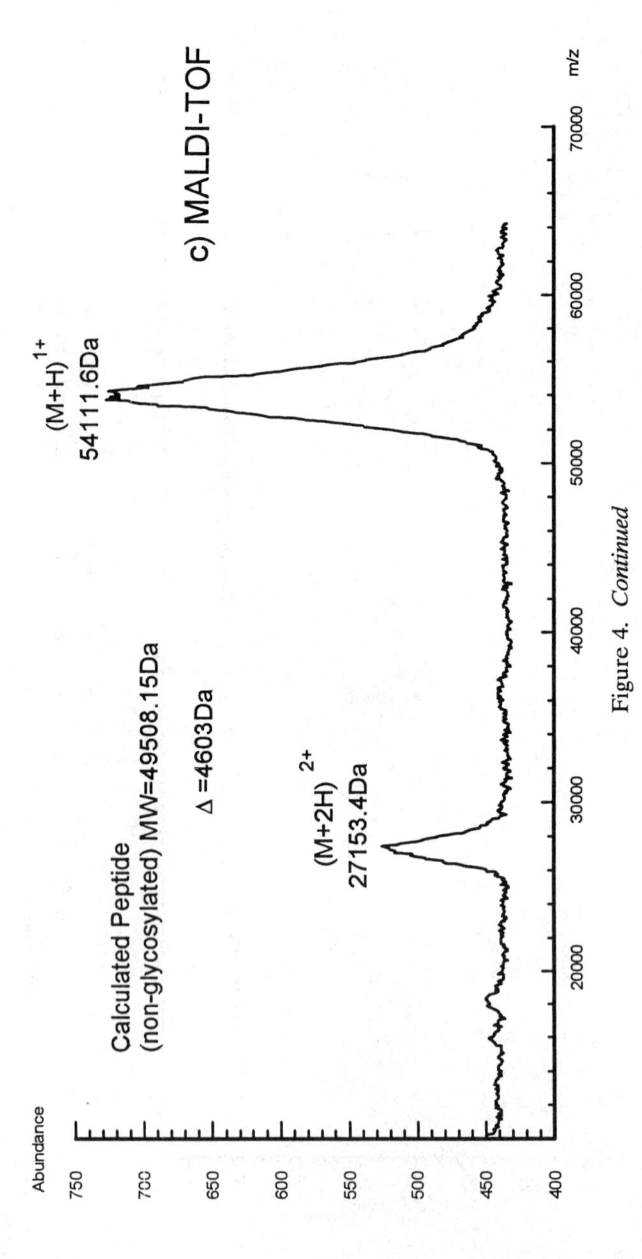

Figure 4. *Continued*

c. MALDI-TOF of Intact protein. 10pmol total protein applied. Matrix: Sinapinic Acid

The Proteolytic Digest

At the next level of detail, the intact glycoprotein is digested through the use of a proteolytic enzyme, resulting in a mixture of not only fragments due to the selectivity of cleavage, but as well different forms of glycosylated fragments. If, as is the case of DSPAα1, the cDNA sequence of the protein is known, the peptide map generated by the separation of the digest can be compared with the predicted sequence.

The generation of peptide maps through the chromatographic or electrophoretic separation of the proteolytic digest (as shown in Figure 5a) is extremely powerful in generating a "fingerprint" of the protein. However, when the separation is monitored with a UV/VIS detector, complete identification of the individual fragments is problematic. Although it has been shown that the individual fragments can be identified by comparing the UV spectra with those of standards, fragment identification usually requires fraction collection and subsequent analysis by mass spectrometry or N-terminal peptide sequencing.

The orthogonal separation mechanisms of reversed phase HPLC and HPCE are evident in comparing the elution patterns of the two separations shown in Figure 5a and b. This can be exploited to resolve areas in one mode which co-elute in the other. The orthogonality of the separation mechanism can also be used in a Quality Control environment as a fingerprint which can detect subtle changes in the protein product. The analysis of the fragments is particularly important as a product variant can escape detection in analysis of the intact molecule because of the broad peaks due to the microheterogeneity of the sample.

MALDI-TOF MS can also be used to characterize the unfractionated enzymatic digest as shown in Figure 5c. In this complex spectrum, a number of the predicted fragments can readily be identified. However, the lack of chromatographic or electrophoretic separation hinders the ability to detect all peptides in such mixtures because the complexity of the sample introduces selectivity in the ionization process which may suppress specific fragments. In principle, fractions can be collected from an HPLC separation and analyzed by MALDI-TOF (see later section), although this results in a loss of chromatographic resolution. Nonetheless, this approach is especially useful for laboratories that lack access to ESI-LC/MS. Often, 60-80% of the expected digest fragments are observed. This approach should be used as a fingerprint in which a manufactured lot is compared with a reference material and lot release could be based on the absence of new 'peaks'. The amount of information can be optimized by judicious choice of matrices. In the data shown in Figure 5c, more detail is obtained by the use of two matrices, namely the multicomponent matrix system and sinapinic acid. In general, lower mass peptides ($<$ 4 kDa) have higher desorption/ionization yields with the multicomponent matrix than with sinapinic acid. More extensive coverage of a digest mixture can be realized via MALDI-TOF analysis of prefractionated HPLC pools of the mixture (see later).

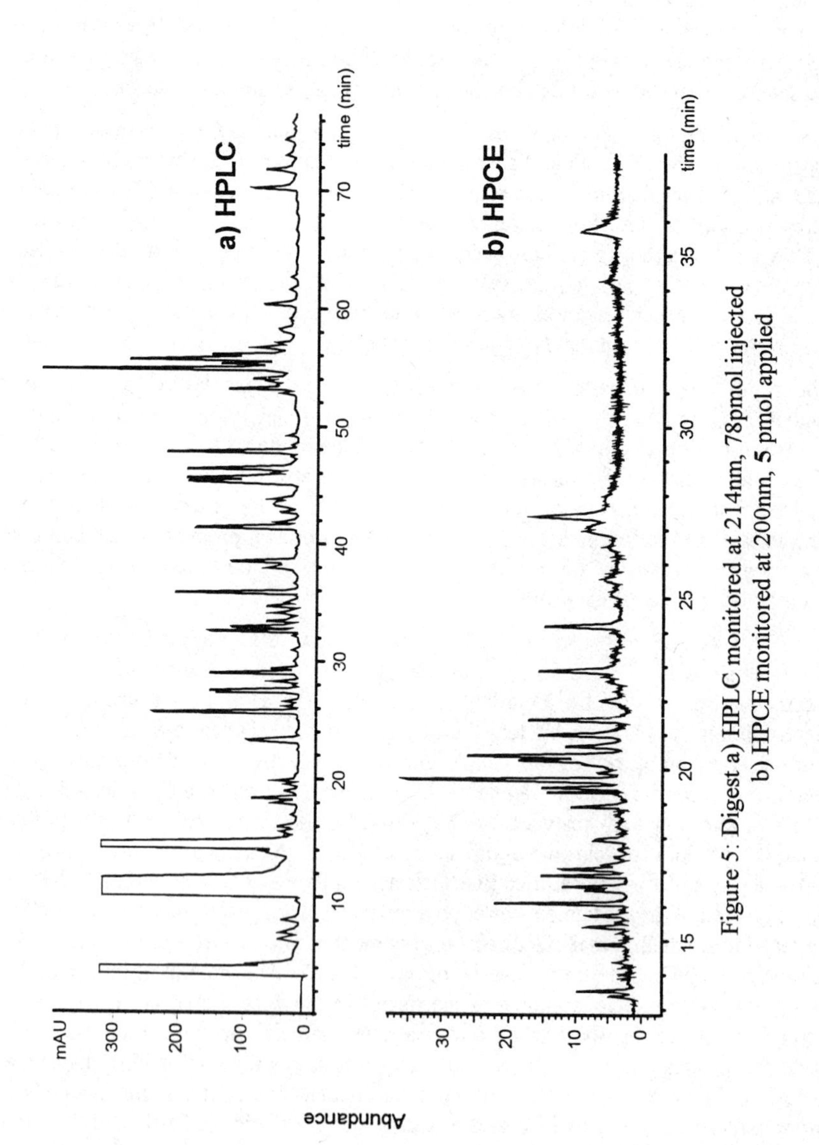

Figure 5: Digest a) HPLC monitored at 214nm, 78pmol injected
b) HPCE monitored at 200nm, 5 pmol applied

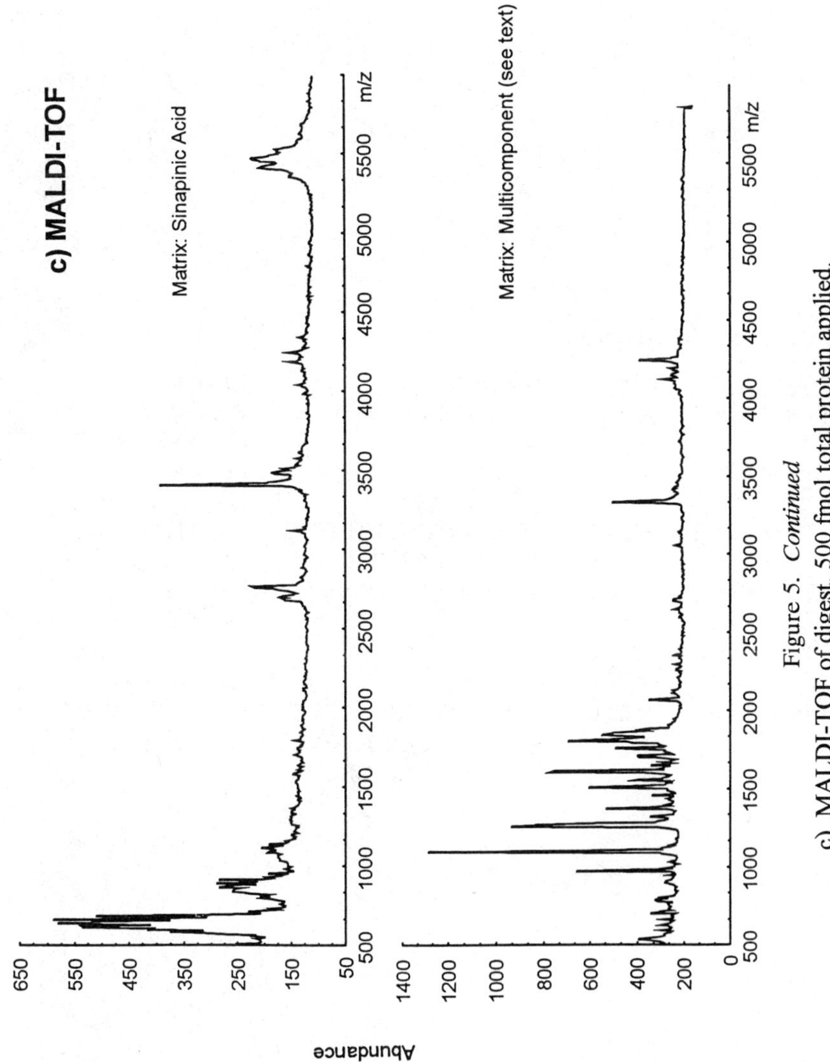

Figure 5. *Continued*

c) MALDI-TOF of digest. 500 fmol total protein applied.

Table 2: Sites of Glycosylation

Fragment	MW	Sequence	Reason
T15[111-123]	1591.8	VECIN WNSSL LTR	Possible N-linked Glycosylation
T43[349-363]	667.4	FLFNK	Possible N-linked Glycosylation
T25[185-212]	3123	EPQLHSTGGLFTDITSH PWQAAIFAQNR	Possible O-linked Glycosylation
T36[293-306]	1556.6	SDSPQCAQESDSVR	Possible O-linked Glycosylation
T39[333-343]	1271.6	SSSPFYSEQLK	Possible O-linked Glycosylation
T45[377-419]	4649.0	SGEIY PNVHD ACQGD SGGPL VCMND NHMTL LGIIS WGVGC GEK	Possible O-linked Glycosylation

A recent method for the characterization of glycoproteins is peptide mapping by the on-line combination of HPLC and electrospray ionization mass spectrometry. In the past, the analyst had to cleaved the carbohydrate with chemical or enzymatic procedures and analyzed the carbohydrates separately. Such an approach is tedious and requires large glycoprotein samples. The ESI LC/MS analysis of the ArgC digest of DSPAα1 is shown in Figure 6. The UV signal at 214 nm is shown in the lower half of Figure 6 for comparison. Note that the sensitivities of the two detectors (UV and MS) are similar and that most peaks show up in both traces although the individual intensities may differ. In this approach, the MS is allowing detection of the glycopeptides even in the absence of significant chromophores in the glycan moiety. The UV detector is also useful because of the presence of the peptide backbone which has significant absorption at 214-220 nm. The combination of UV/MS detection of the separated peptide maps avoids many of the disadvantages of traditional carbohydrate analysis such as derivatization steps or the difficulties of pulsed amperometric detection.

Electrospray ionization is particularly useful in the identification of relatively high molecular weight peptides because of its ability to generate a set of multiply charged ions[13] which can be used to determine the molecular weight of the peptide even when the molecular weight is far in excess of the range available to the mass analyzer for singly charged molecules. For peptides found in enzymatic digests, masses are typically below 5000 Da and the charge states are typically less than +5 (positive ion mode). The mass accuracy of the determination is typically better than 0.02%. In these experiments, mass spectral resolution was maintained at unit resolution across the scanned mass range and the sample was based on 78 pmol total protein. This is an important aid in assigning charge states to ions from peptides. In general, a comfortable working range is 10-100 pmol and allows detection of glycopeptides which are present at low levels, down to approximately 1% of the total.

As reported elsewhere[11] the application of a software interpretation utility allowed detection of essentially all of the peptides predicted from the cDNA sequences. Interestingly, those digest fragments which are expected to be glycosylated (O-linked; T25, T36, T39, T45; N-linked; T15, T43, see Table 2) are either missing or are found at very low abundance relative to other fragments and can be used as a rough predictor of glycosylation sites. This pattern was consistent with carbohydrate characterization studies[14] performed by others. There are also a few lower abundance fragments which were not identified at this stage, but based on previous studies it would be expected that secondary cleavages induced by the protease are the cause of many such fragments.

The location and characterization of the glycopeptides present in enzymatic digests of glycoproteins by electrospray LC/MS can be approached in three ways[15]. If the mass spectral data is presented as a contour map of m/z vs time vs intensity, series of unresolved glycopeptides will appear as diagonal bands since for a given digest fragment, glycoforms with a greater degree of glycosylation will appear as slightly heavier and elute somewhat earlier[1]. This approach is less successful, however, for more heterogeneous samples because the complexity of the 2D plot hinders the

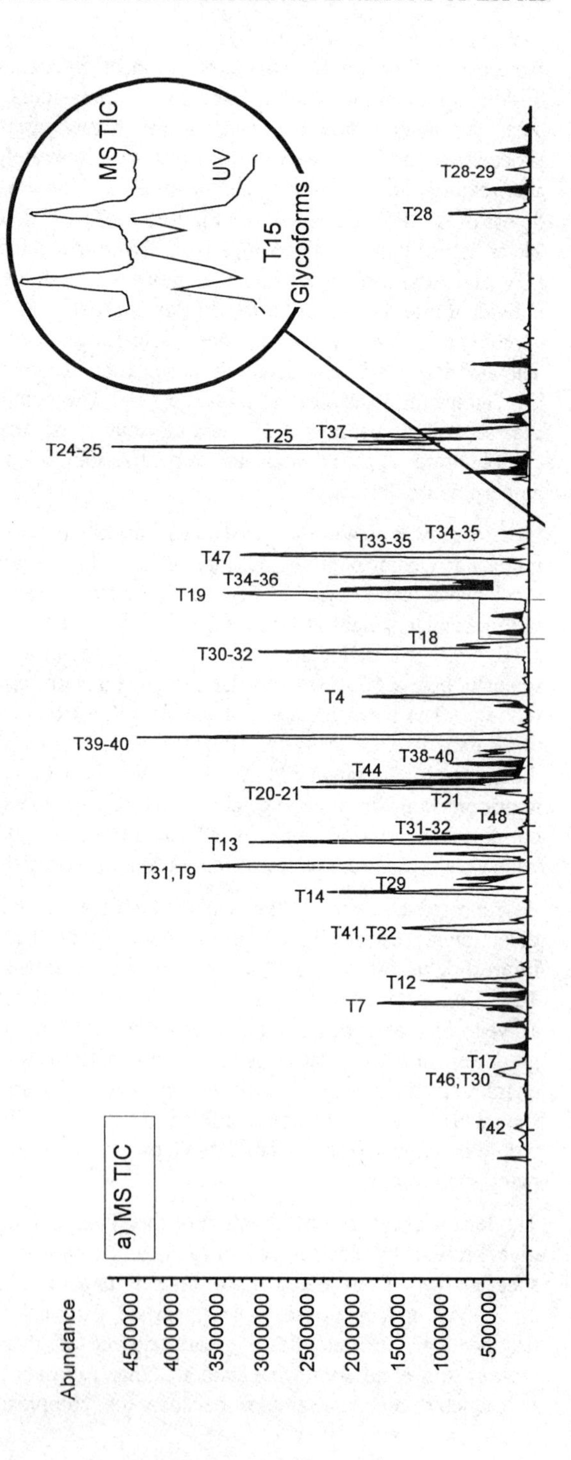

a) MS TIC
Abundance
4500000
4000000
3500000
3000000
2500000
2000000
1500000
1000000
500000
MS TIC
UV
T15
Glycoforms
T28-29
T28
T24-25
T25
T37
T34-35
T33-35
T47
T34-36
T19
T18
T30-32
T4
T39-40
T38-40
T44
T20-21
T21
T48
T31-32
T13
T31,T9
T29
T14
T41,T22
T12
T7
T17
T46,T30
T42

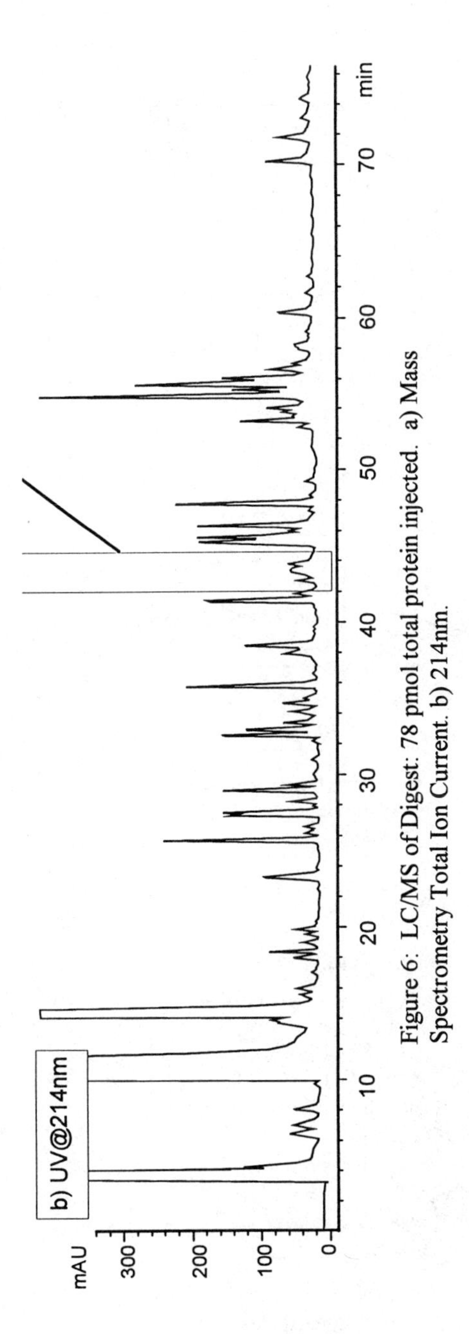

Figure 6: LC/MS of Digest: 78 pmol total protein injected. a) Mass Spectrometry Total Ion Current. b) 214nm.

Complex Biantennary, 0 Sialic Acid

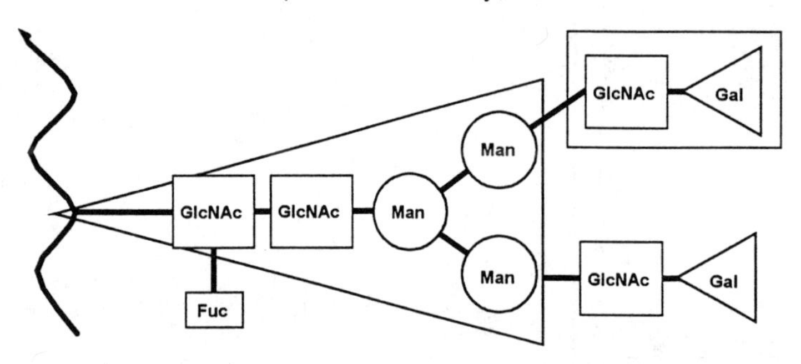

Complex Biantennary, 1 Sialic Acid

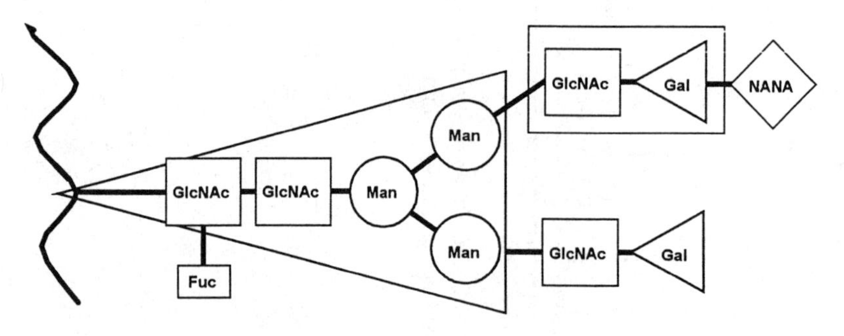

Complex Biantennary, 2 Sialic Acid

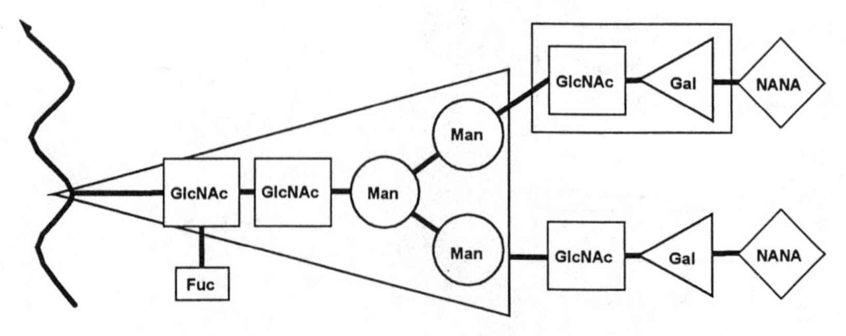

Figure 7: Schematic of N-linked glycoforms

location of the patterns. A second method consists of using in-source collisionally induced dissociation to fragment glycopeptides, producing marker ions indicative of specific types of glycosylation[3]. Finally, if the sequence of the protein is known, fragments and their glycoforms can be predicted from knowledge of the structures and hence the mass of expected peptides. In the absence of other information, the individual sites of glycosylation can be predict based on the minimum consensus sequence required for glycosylation; Asn-X-(Ser/Thr) for the N-linked[16] and (Ser/Thr)-X-X-Pro for the O-linked[17] *structures.* One can then hunt for the specific glycoforms through the use of extracted ion chromatograms[18].

As an example of the use of monitoring carbohydrate structures with this LC/MS approach, we will examine possible oligosaccharides N-linked at asparagine at residue 117, in fragment T15[111-123]. N-linked glycopeptides generated in mammalian cell lines often fall into three general classes; high mannose, complex and hybrid. A common structure is a biantennary form with variable sialic acid content (See Figure 7). If an ion extraction is performed based on the masses of these structures and the T15 peptide , the results shown in Figure 11are obtained. This preliminary identification of these structures can then be confirmed by neuraminidase digestion to remove the sialic acid and repeating the LC/MS analysis[2] . The abundance of these forms is low as can be seen in the insert in Figure 6 where the TIC is essentially flat in this region of the map. In the same region, a single broad peak is observed by UV detection presumably because each of the glycoforms has similar spectra but the significant mass degeneracy defeats the TIC monitoring. This example illustrates the power of combined UV and MS detection for peptide mapping separations.

Two Dimensional Maps

The use of 2D mapping is popular for the analysis of complex biological samples because of the increased peak capacity afforded by the second dimension. It is, therefore, important that the two dimensions should be orthogonal. A popular combination is to perform 2D Gel electrophoresis using isoelectric focusing followed by SDS-PAGE (charge vs molecular weight). We are in the process of exploring the use of contour maps from these experiments as alternatives to the earlier approaches. The contour plot is based on the differences in hydrophobicity (x axis) and mass (y axis). Although in LC separations a weak correlation between the two is observed, they are still essentially orthogonal (See Figure 9 and Figure 10). A feature of this approach is that all of the data from the different charge states is plotted and because of this feature contour maps have been shown to be useful in locating sites of glycosylation. For example, Figure 8 shows the comparison of contour maps obtained from the contour mapping of two complex glycoproteins; a tryptic digest of recombinant Tissue Plasminogen Activator (rTPA) (three sites of glycosylation) and the Arg C digest of DSPAα1 (six sites of glycosylation).

Although the conventional chromatogram view of the peptide map shown in the upper part of Figure 9 would suggest that the rTPA map has more components than the

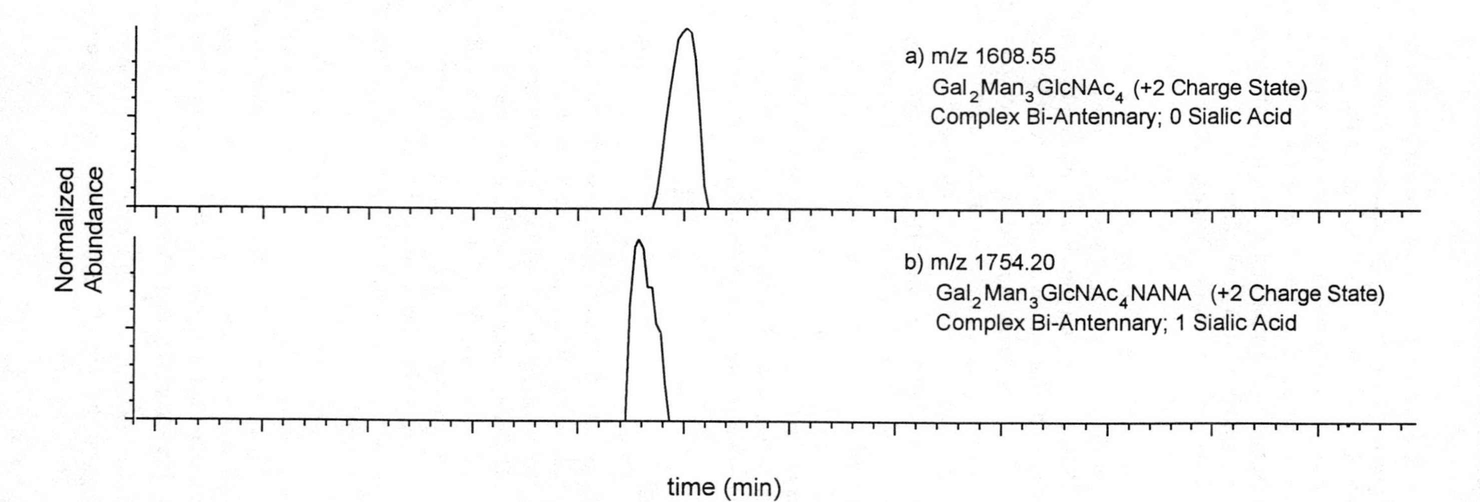

Figure 8: Extracted Ion Detection of Glycopeptides. Scan acquistition (CapEx=100 V). a) Extracted m/z 1608.55 Da. $Gal_2Man_3GlcNAc_4NANA_0$. (+2 Charge state) b) Extracted m/z 1754.2 Da. $Gal_2Man_3GlcNAc_4NANA_1$. (+2 Charge state) c) Extracted m/z 1266.9 Da. $Gal_2Man_3GlcNAc_4NANA_2$ (+3 Charge state). d) Total Ion

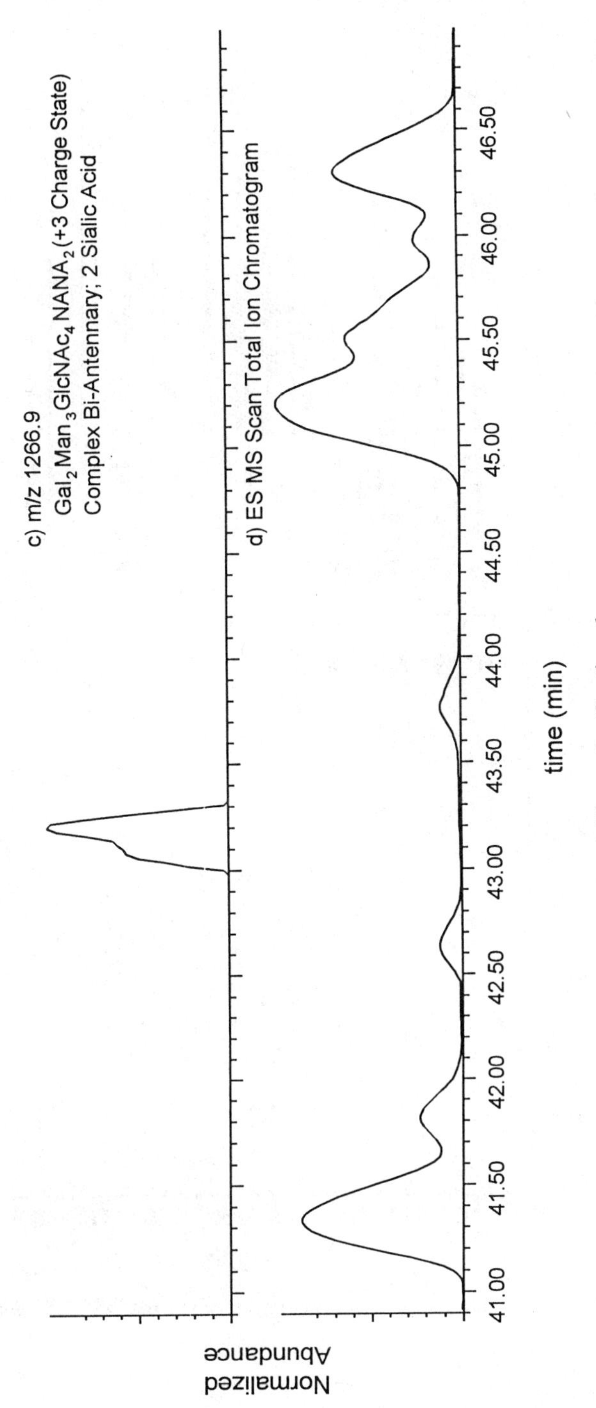

Figure 8. *Continued*

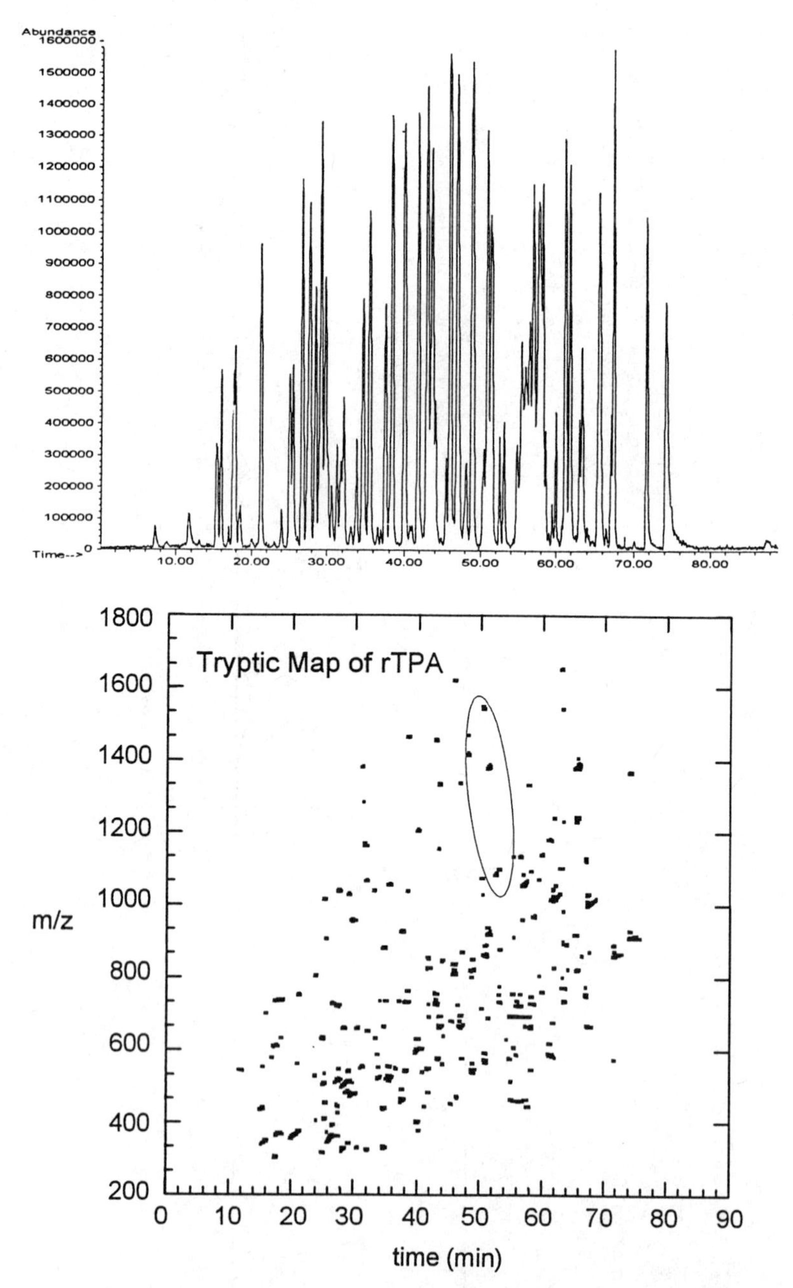

Figure 9 ChromatogramComparison of Contour Maps for DSPAα1 and rTPA.

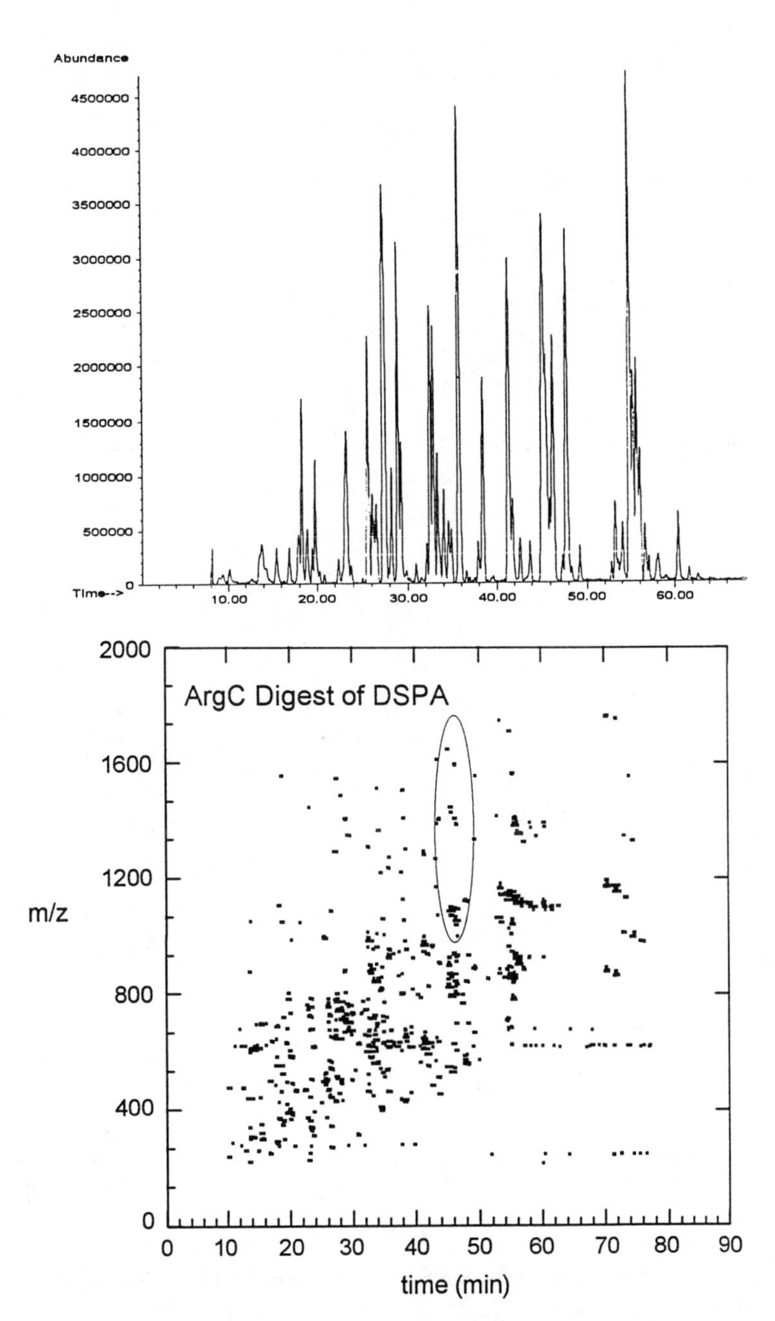

Figure 9. *Continued*

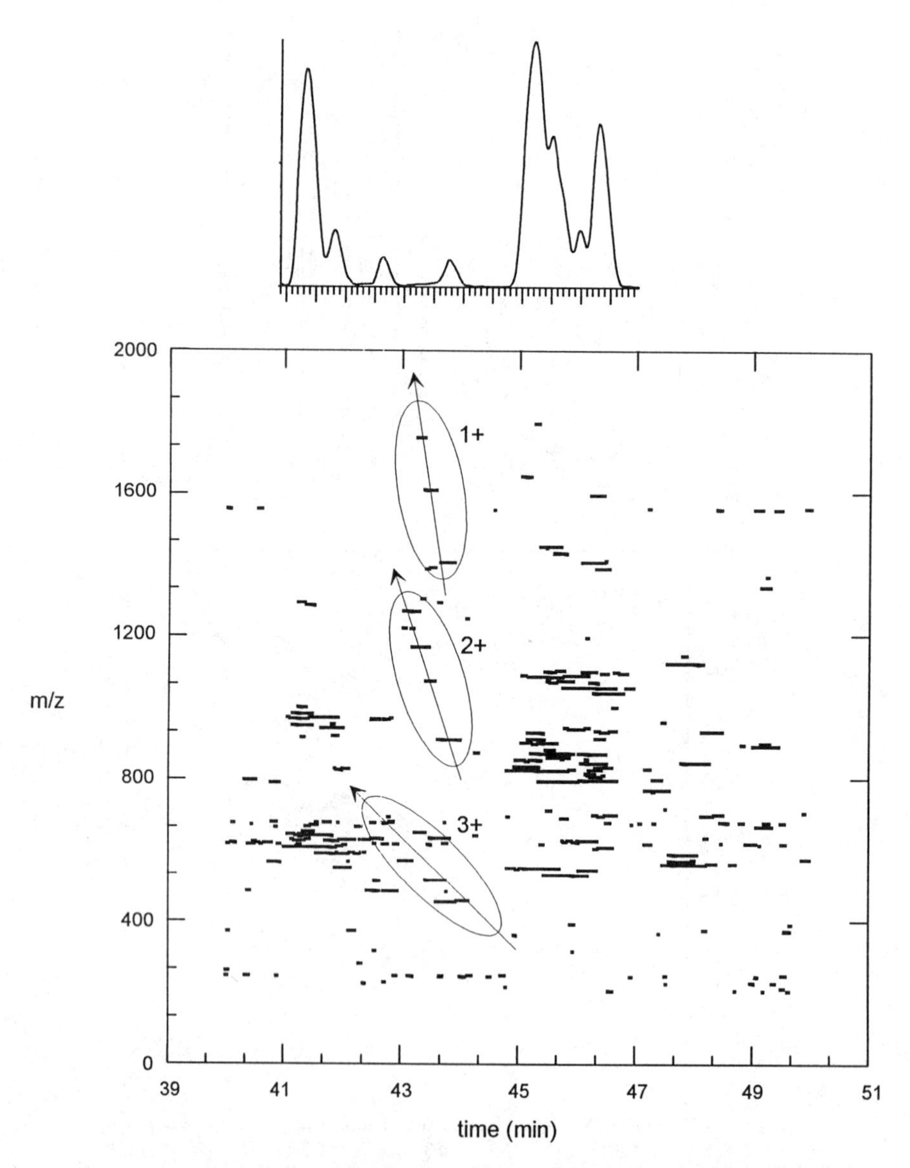

Figure 10: Detail of Contour Map for DSPAα1 ArgC digest. 78 pmol total protein injected. CapEx Voltage=100V. Data for Signal >2000 counts shown

DSPAα1 map, the 2D map shows that the latter digest is much more complex as would be expected from the greater number of glycosylation sites in DSPAα1. The more heterogeneous the sample is (as in the case of DSPAα1), the more glycoforms over which the ion current will be distributed. Unlike the UV chromatogram in which the different unresolved glycoforms contribute to a moderately intense, relatively broad chromatographic peak, the contour map has a single point for each mass state present for the glycoform.

Figure 10 shows a contour for a section of the ArgC map of DSPAα1 from 40-50 minutes. The circled area shows signal due to fragment T15[111-123] which contains complex sialylated biantennary structures with three different charge states (+1, +2 and +3) for variable degrees of sialylation. The visual pattern is striking because the higher the charge state, the closer the pattern and the greater the slope of the diagonal (due to the effective compression of the mass range by multiple charging). We are investigating the application of contour plots and multivariate statistical techniques to the examination of different lots of the drug substance as we believe such an approach may have value in demonstrating consistency in glycosylation and for the detection of significant levels of host cell contaminants.

The use of in-source collisionally induced dissociation in the analysis of DSPAα1 is illustrated in Figure 11. The CID mass spectral acquisition was carried out in Selected Ion Monitoring mode to increase the sensitivity. The marker ions which were monitored are shown in Table 3. The reproducibility of the HPLC separation is sufficiently good that areas identified in the SIM analysis can be accurately compared with separate full spectral acquisitions conducted at lower fragmentation energy. This allows the parent glycopeptide to be identified. The success of this technique can be compared to the identification of glycopeptides from the standard ESI chromatogram. At low collision energies (no CID), extensive microheterogeneity in the glycoforms results in a low abundance of all of the individual ions. Often, the intensities may be difficult to distinguish from background. Note, however, that the sensitivity from the SIM CID data is sufficient since all of the different chromatographically unresolved glycoforms will contribute fragment ions to the electrospray signal of the glycomarkers. The summation of the signals due to the marker ions results in relatively broad looking chromatographic peaks. Based on the CID data, it is possible to infer some glycostructure differentiation. For example, all N-linked structures will generate a m/z 204 ion corresponding to the HexNAc in the backbone structure. All complex N-linked glycopeptides will generate a m/z 366 ion corresponding to the HexNAc+Hex structure indicative of the branching in multi-antennary structures. Only sialyted glycopeptides will generate a m/z 292 ion. Figure 11 relates the major structure of the glycopeptide T15 to the fragments that could be detected in the CID experiment. It can be seen that the CID data in the T15 region (around 45 minutes) would predict structures that would be low in fucose, but contain a substantial amount of sialic acid.

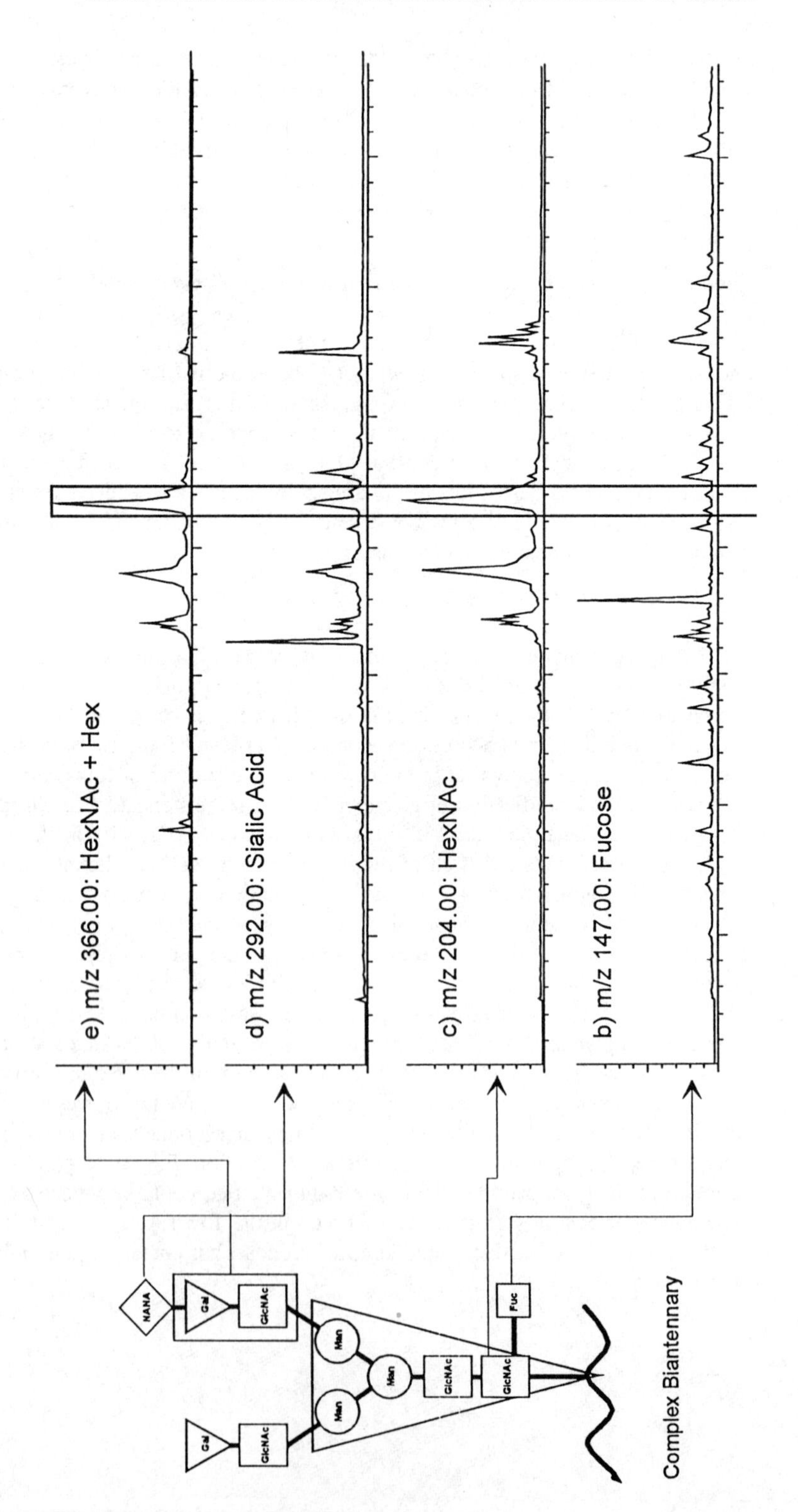
e) m/z 366.00: HexNAc + Hex
d) m/z 292.00: Sialic Acid
c) m/z 204.00: HexNAc
b) m/z 147.00: Fucose
NANA
Gal
GlcNAc
Man
Fuc
Complex Biantennary

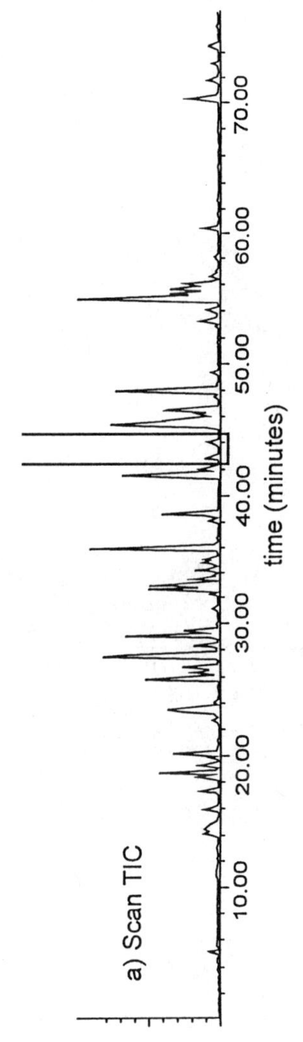

Figure 11: CID detection of Glycopeptides. 78 pmol total protein injected. a) Total ion chromatogram from scan acquistion (CapEx=100 V). b) through e) from selected ion monitoring acquisition (CapEx=200 V). b) m/z 147 Da. c) m/z 204 Da. d) m/z 292 Da. e) m/z 366 Da

Table 3: Glycomarkers

Sugar	Symbol	Chemical Formula	MW	Ion Monitored $(M+1)^{1+}$
Fucose	Fuc	$C_6O_5H_{13}$	146.1	147.1
Hexose	Hex	$C_6O_6H_{12}$	162.1	163.1
N-Acetylhexosamine	HexNAc	$C_8O_6NH_{15}$	203.2	204.2
N-Acetylneuraminic Acid	NANA	$C_{11}O_9NH_{19}$	291.3	292.1
Complex Branch Unit	HexNAc + Hex		365.3	366.3

HPLC Fractionation of the Digest

In many cases, the complexity of the mixture generated by an enzymatic digest of a glycoprotein is too high for all components to be completely resolved by a single dimensional analytical technique. Recently, Jorgenson[19] introduced two dimensional separations based on the combination of HPLC and HPCE. Only a small fraction of the LC effluent is analyzed in the second dimension and thus a pre-separation fluorescent derivatization was required. While this approach produces excellent 2D displays, the analysis is not able to chemically characterize any differences that may be observed in lot-to-lot comparisons. For this reason, we chose to transfer samples from an LC separation to subsequent analysis in a semi-off-line technique. Fractions eluting from the reversed-phase HPLC separation were selectively transferred onto disposable hydrophobic collection columns after the effluent was diluted with a non-elutropic solvent. Because of the highly reproducible retention times of the LC separation, this process could be repeated, loading fractions from multiple LC runs on single collection columns. The hardware setup employed used a 14 port, 6 position valve which would allow 4 collection columns to be randomly accessed in addition to a bypass path. Following fraction collection, the fractions were eluted from the collection column with a rapid gradient. Since the fraction elution solvent doesn't have to be the same as the solvent system used in the peptide map, buffer transfer or desalting can also be accomplished in this step. This had the advantage of concentrating the samples for subsequent analysis and of great flexibility in transferring the sample fraction to subsequent analytical techniques.

This technique is particularly useful in transferring samples to subsequent HPCE separation. As mentioned above, HPCE suffers from rather poor concentration sensitivity due to relatively small injection volumes and short path length UV detection. While on-line sample focusing and preconcentration techniques such as sample stacking[20] and capillary isotachophoresis (CITP)[21,22] are extremely useful in this respect, the low concentration of individual glycopeptides requires additional concentration to fully characterize the sample.

The technique is also useful for transferring samples to inherently static techniques such as MALDI-TOF, in which an on-line flow based approach would be difficult. The fractionation step itself is very important in reducing the complexity of the mixture to a point at which most of the components can be identified mass spectrally by MALDI-TOF. The desalting characteristics of the fraction elution step can be very useful in increasing signal yields and the fraction of components observed.

For the purpose of illustration, we will focus on a single fraction collected from the HPLC peptide map as illustrated in Figure 12. This fraction was chosen because of the high degree of glycosylation present in these areas as shown by the CID studies on glycomarkers. The benefits of the approach are not only the simplification of the analytical sample via fractionation, but a concentration step of approximately 10 fold as well. The fraction was analyzed by HPCE using the same method as had been used for the total digest. The electropherogram of the fraction is shown in Figure 13a. The real strength of using CE to reanalyze HPLC fractions lies in the orthogonality of the separation mechanism. To fully exploit this separation power, a method is needed to

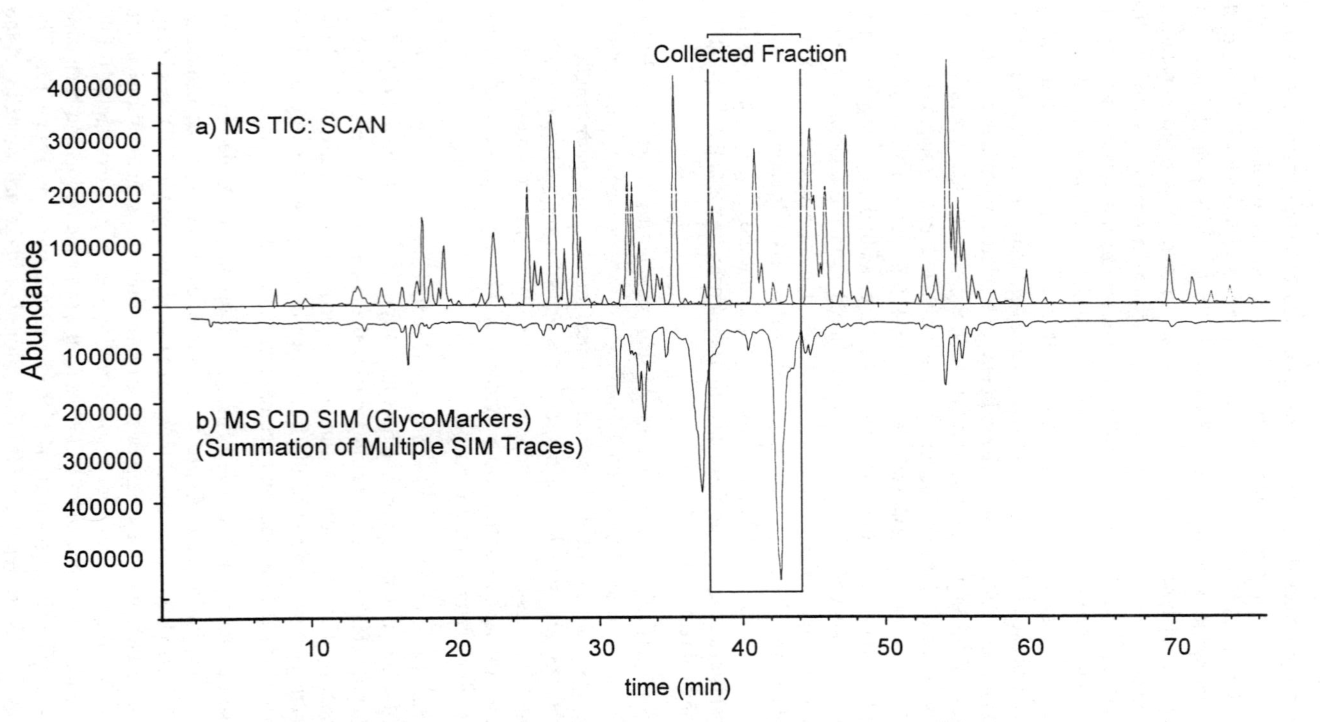

Figure 12: Area of HPLC Fraction Collection. Identification of regions from analytical separation (76 pmol). a) Scan Acquisition Total Ion Current (CapEx= 100V). b) CID SIM Acquisition Total Ion Current (CapEx = 200 V). c) UV at 214nm. The actual fraction collections were done with 312 pmol injected

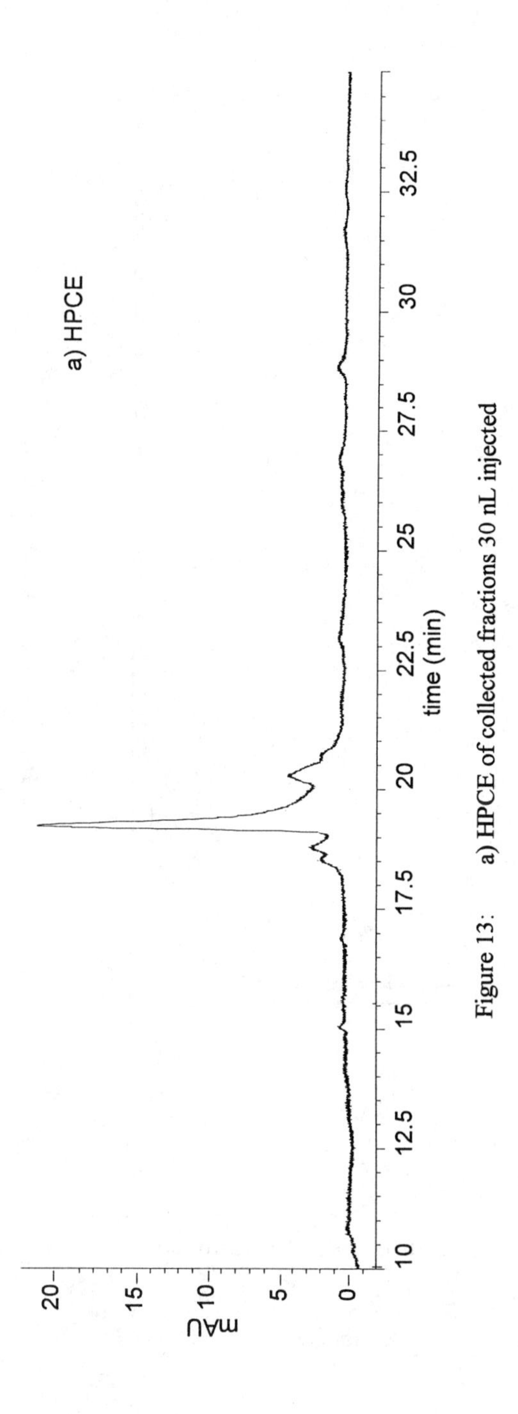

Figure 13: a) HPCE of collected fractions 30 nL injected

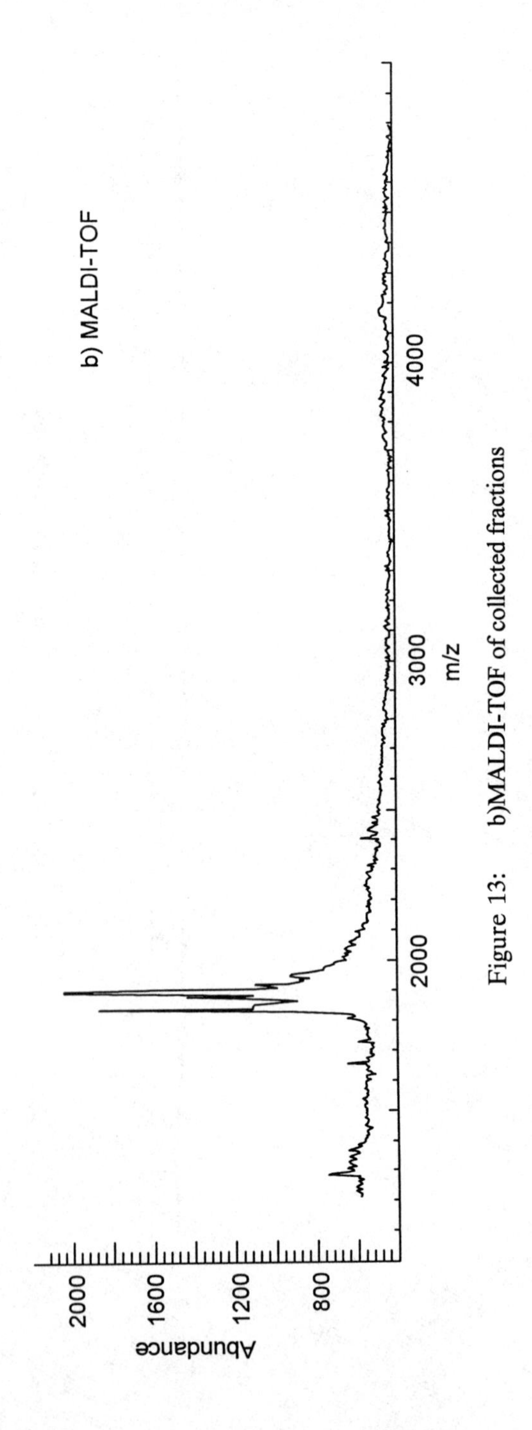

Figure 13: b)MALDI-TOF of collected fractions

couple the CE separation with mass spectrometry, either directly via electrospray CE/MS or in an off-line manner running collected CE fractions by MALDI-TOF. This approach has been used to characterize growth hormone tryptic peptides in a CE separation[23] and other laboratories for other protein samples. These approaches are currently under investigation, and will be reported in a future publication. Alternately, UV spectra could be used to track the elution order of a CE separation transferred from an HPLC fraction, analogously to peak tracking in HPLC peptide mapping[24]. This approach requires that the HPLC separation was already sufficiently resolved to obtain pure spectra or that pure peptide standards exist.

The collected fraction was also analyzed by MALDI-TOF as shown in Figure 13b. This is particularly useful combination because the sensitivity of MALDI-TOF allows clear unequivocal spectra which actually exhibit a selectivity closer to HPCE than HPLC, being based on mass/charge (as can be seen in the similarity of profiles in Figure 13a and Figure 13b). The orthogonality of MALDI-TOF relative to HPLC generates extremely useful information (Figure 13b relative to Figure 12). In analyzing fractions collected from the total digest, the sample complexity has been reduced sufficiently so that the MALDI-TOF spectrum is a good representation of the components present in the fraction. Contrast this with the data obtained for the total digest, in which only a fraction of the components are represented by strong signals in the spectrum. Thus a more complete and accurate picture is obtained by a combination of the two approaches.

Conclusions

This study has shown that there is considerable potential in the analysis of complex glycoprotein samples by hyphenated liquid phase separations and mass spectrometry. Such information should prove invaluable in reducing the approval barriers for the pharmaceutical use of glycosylated proteins produced by mammalian fermentation systems. For example, a high capacity two dimensional analytical method can be used to demonstrate consistency of carbohydrate structures produced under different scale fermentations.

The use of electrospray mass spectrometry on-line with reversed phase HPLC has greatly expanded the power of peptide mapping to identify carbohydrate structures that are attached to asparagine, serine or threonine residues. One can use in-source CID to scan the map for regions with a high concentration of glycopeptides. The presence of certain glycoforms can be achieved by extracted ion monitoring and used to monitor distribution of sialic acid in complex carbohydrate structures.

We have started to explore the use of contour maps (m/z vs retention time) as a facile approach to rapid 2D mapping of complex samples. Such maps are readily available from the data generated by an LC/MS analysis and can give valuable information about glycosylation patterns and product consistency. A problem with current 2D methods is that it is difficult to explain differences that may be observed in comparison of samples. An approach to this problem was described in this report where an automated fraction collection system was developed for the RP-HPLC as first dimension separation. The collection system was designed to concentrate the samples in a buffer suitable for the second dimension CE analysis. The third dimension of this analysis was illustrated by the application of off-line MALDI-TOF to the CE fractions. This analytical approach has great potential as it allows the combination of chemical characterization with a multidimensional analysis approach.

Acknowledgments

The authors would like to acknowledge Sally Swedberg and Bob Holloway at Hewlett-Packard Laboratories for help with capillary electrophoresis, Julie Sahakian and James Kenny at Hewlett-Packard Protein Chemistry Systems for N-terminal sequencing, Steve Fischer at Hewlett-Packard Bay Analytical Operation for aid in mass spectral interpretation, Thabiso M'Timkulu and Joanne Johnson at Berlex Biosciences for carbohydrate analysis, Ray-Jen Chang and Maria Johnson at Berlex Biosciences for N-terminal sequencing, Peter Murakami and Peter Sandel at Berlex Biosciences for amino acid analysis and BaiWei Lin and Joe Traina at Berlex Biosciences for mass spectrometry.

Literature Cited

[1] Ling, V., Guzzetta, A.W., Canova-Davis, E., Stults, J.T., Hancock, W.S., Covey, T.R. and Shushan, B.I., *Anal.Chem.*, 63 (1991) 2909-2915.

[2] Guzzetta, A.W., Basa, L.J., Hancock, W.S., Keyt, B.A. and Bennet, W.F., *Anal.Chem.*, 65 (1993) 2953-2962.

[3] Carr, S.A., Huddleston, M.J. and Bean, M.F., *Protein Science*, 2 (1993) 183-196.

[4] Huddleston, M.J., Bean, M.F. and Carr, S.A.,*Anal.Chem.*,65 (1993) 877-884.

[5] Conboy, J.J. and Henion, J.D., *J.Am.Soc.Mass Spectrom.* 3 (1992) 804-814.

[6] Kraetschmar, J., Haendler, B., Langer, G., Boidol, W., Bringman, P., Alagon, A., Donner, P. and Scheuning, W.D., *Gene* 105 (1991) 229-237.

[7] Witt, W., Maass, B., Baldus, B., Hildebrand, M., Donner, P. and Scheuning, W.D., *Circulation*, 90 (1994) 421-426.

[8] Witt, W., Baldus, B., Bringmann, P., Cashion, L., Donner, P.and Scheuning, W.D., *Blood*, 79 (1992) 1213-1217.

[9] Apffel, A., Fischer, S., Goldberg, G., Goodley, P.C. and Kuhlmann, F.E., *J.Chromatogr*,, Submitted for publication, 1994.

[10] Gusev, A.I., Wilkinson, W.R., Proctor, A. and Hercules, D.M., *Anal.Chem.*, 67 (1995) 1034-1041.

[11] Apffel, A. Chakel, J. Udiavar, S., Hancock, W.S., Souder, C and Pungor, E., *J.Chromatogr.* Submitted for publication (1995)

[12] Oroszlan, P., Wicar, S., Teshima, G., Wu, S.-L., Hancock, W.S. and Karger, B.L., *Anal.Chem.*, 64 (1992) 1623-1631.

[13] Fenn, J.B., Mann, M., Meng, C.K., Wong, S.F. and Whitehouse, C.M., *Science* 246 (1989) 64-71.

[14] M'Timkulu, T. (unpublished results).

[15] Guzzetta, A.W. and Hancock, W.S., *Recent Advances in Tryptic Mapping, CRC Series in Biotechnology*, CRC Press, Boca Raton, FL, in press (1994).

[16] Marshall, R.D., Ann. Rev. Biochem. 41 (1972) 673-702.

[17] Aubert, J.P., Biserte, G., and Loucheux-Lefebre, M-H., Arch. Biochem. Biophys. 175 (1976) 410-418.

[18] Spellman, M.W. Basa, L.J., Leonard, C.K., Chakel, J.A., O'Connor, J.V., Wilson, S.W., and van Halbeek, H., *J.Biol.Chem.* 264 (1989) 14100-14111.

[19] Bushey, M.M. and Jorgenson, J.W., Anal. Chem. 62 (1990) 978-984.

[20] Chein, R.L. and Burgi, D.S., *J.Chromatogr.* 559 (1991) 141-152.

[21] Mikkers, F.E.P., *Thesis* Eindhoven University, Eindhoven, 1980.

[22] Foret, F., Szoko, E. and Karger, B.L., *J.Chromatogr.* 608 (1992) 3-12

[23]Herold, M. and Wu, S.-L., *LC-GC* 12 (1994) 531-533.

[24] Sievert, H.-J. P., Wu, S.-L., Chloupek, R., and Hancock, W.S., *J.Chromatogr.*, 499 (1990) 221-234.

RECEIVED September 13, 1995

Chapter 24

From Protein Primary Sequence to the Gamut of Covalent Modifications Using Mass Spectrometry

A. L. Burlingame, K. F. Medzihradszky, K. R. Clauser, S. C. Hall, D. A. Maltby, and F. C. Walls

Department of Pharmaceutical Chemistry, Mass Spectrometry Facility and Liver Center, University of California, San Francisco, CA 94143–0446

Taking advantage of the remarkable technical advances in mass spectrometry over the last few years, previously difficult or intractable problems in protein biology can now be solved effectively. Currently three soft ionization techniques [liquid matrix secondary ion (LSI), electrospray (ESI) and matrix-assisted laser desorption (MALDI)] may be employed to elicit information on the molecular nature of a sample at several levels of structural detail. Each ionization method suffers from selectivity (or discrimination or suppression) toward any given sample and the type and quality of information which may be obtained varies depending upon the parameters of the ion optical configurations chosen. Thus, in many cases, more than one technique must be explored and employed to obtain eventually the information required. In cases where mass spectral information is obtained from several techniques on the same sample, it is many times complementary. In addition, one can anticipate the introduction of new ion optical strategies periodically. This chapter deals with problems ranging from protein primary sequence determination through identification of a variety of covalent modifications. Examples are presented which illustrate application of all the ionization methods currently available.

Discovery of the ease with which surface active, thermally labile, zwitterionic substances could be sputtered as intact ions from the surface layers of viscous organic substances such as glycerol and thioglycerol by both neutral (*1*) and ionic (*2*) atomic beams initiated a revolution in structural characterization of biological macromolecules. The fact that a preponderance of intact ionic species were observed is a consequence of their formation with relatively low internal energy content. These ionic species (pseudomolecular ions) are formed from a given molecule by protonation/deprotonation in solution or the desorption plume. These early successes led quickly to peptide mass mapping (*3*). The need for structural information quickly paved the way for the development (*4*) and optimization (*5*) of tandem instrumentation for primary sequence determination utilizing collision-induced dissociation (CID) analysis. The introduction of multichannel array detectors led to orders of magnitude increase in system sensitivities in subpicomole

0097–6156/95/0619–0472$17.00/0

range based on simultaneous detection of a percentage of the mass range observed as well as a major decrease in complete spectral recording time (*4, 6*). Further, the advent of electrospray (ESIMS) (*7*) has provided the technical basis for the virtual ideal coupling of high performance liquid chromatography with mass spectrometry (HPLC)-MS. By applying a voltage to the end of a microbore HPLC colum the effluent is continuously atomized into a bath gas at atmospheric pressure thus permitting the direct sampling of protonated analyte species through an orifice followed by their acceleration by an ion gun into the mass spectrometer for mass analysis. Finally, matrix-assisted laser desorption ionization (MALDI) (*8, 9*) and ESIMS have made available two ionization techniques which have considerably higher ionization efficiences than sputtering methods, thus providing for peptide analyses in the low femtomole range. Ions formed in elecrospray appear to have lower internal energies than those formed in sputtering methods (*10*), while MALDI-generated ions readily undergo unimolecular decomposition (*11*). The former has led to the exploitation of CID analysis of doubly protonated molecular species (*12*), while the latter has led to development of instrumental strategies which permit the recording of mass spectra of these metastable fragments [the so-called post source decay (PSD) experiments (*11*)]. Such mass spectra may be recorded by stepping the voltage applied to an electrostatic ion mirror (reflectron), or by use of a curved field electrostatic mirror (*13*).

For characterization of biological macromolecules, high sensitivity and the generation of reliable structural information are necessities (*6, 13*). While application of these methods in a complementary fashion to many problems in biomedical research and biotechnology has been an overwhelming success over the past 5 years (*13, 14*), many of the most important unsolved problems present considerable challenges. These problems could not be tackled without the power and versatility of the most recent instrumental developments in mass spectrometry. This chapter will outline the application of this suite of techniques to a variety of problems encountered in an academic biomedical research environment. Several proteins will be used to illustrate various points, including the cloning of unknown proteins which functionally comprise the signal recognition particle (SRP) in yeast and the identification of proteins from melanoma cell lysates separated by two dimensional sodium dodecylsulphate polyacrylamide gel electrophoresis (2-D SDS PAGE). In addition, the detection and identification of natural and xenobiotic covalent modifications to bovine fetuin and albumin using microbore HPLC-ESI-selected ion monitoring (SIM) are described. Finally, the performance of microbore HPLC coupled with electrospray ionization on a magnetic sector-electro-optical multichannel array instrument is described.

Analysis of complex protein mixtures.

Proteins from enzyme complexes or from cell or tissue lysates may be separated using 1- or 2-D gel electrophoresis. In this laboratory considerable emphasis has been placed on the development of the mass spectrometric-based analyses of proteins isolated from SDS PAGE systems both for cloning of unknown proteins as well as for the identification of complex mixtures of proteins in tumor cell lysates. This effort has brought us to a routine and reliable protocol for working successfully with amounts of protein in the low picomole range. The cornerstone of our strategy is to take advantage of the ability to obtain de novo primary sequences rapidly by high energy collision-induced dissociation (CID) using four-sector tandem mass spectrometry. This capability serves us well in many ways by providing the information required to: 1) carry out gel spot identification unambiguously; 2) search the protein data bases for homology; 3) clone unknown proteins; 4) study the nature of posttranslational modifications directly; and 5) characterize adventitious modifications or artifacts. These concepts are outlined below.

Protein primary sequence determination (signal recognition particle). The signal recognition particle is a conserved ribonucleoprotein complex required for protein targeting to the endoplasmic reticulum membrane. Only two of the proteins (Srp 54p and Sec65p) and the RNA subunit, scR1, in this complex were known previously. Immunoaffinity purification of this particle from yeast revealed five additional proteins. RNA was removed using chelating agents and sucrose gradient sedimentation and the proteins in the complex were separated by SDS-PAGE.

Sufficient primary amino acid sequences were obtained for cloning of these unknown proteins using two mass spectrometric techniques. As described elsewhere (*15*), the proteins were digested directly in the gel using trypsin and the resulting peptides were extracted using 60% acetonitrile with 0.1% TFA. Microbore LC/ESIMS was employed to determine the molecular weight composition of each HPLC peak subjecting 5% of the chromatographic effluent to mass analysis during collection of the remainder of each fraction. The remaining sample of the appropriate HPLC fractions was then analyzed by liquid secondary ionization mass spectrometry (LSIMS) and CID analysis employing a rapid scanning charge coupled device (CCD) multichannel array detector (*4, 6, 16*). For example, two of the unresolved bands (SRP-66 and SRP-68) were excised from a 1-D SDS-PAGE and analyzed together. The chromatographic trace based on UV detection during this LC-MS run is shown in Figure 1a together with the ESIMS base peak ion chromatogram (see Figure 1b). In this case two components were observed in the electrospray spectrum obtained at a retention time of about 24 min (shown in Figure 2). The peptide with the measured average molecular mass of 1003.09 Da by electrospray yielded the high energy CID spectrum shown in Figure 3 (MH^+ (^{12}C) at m/z 1003.6) [for a discussion of the definitions of chemical average vs. monoisotopic mass, see Appendix XI in reference *13*]. Interpretation of this CID spectrum established the amino acid sequence of the peptide as VTTNINWR. The strategies employed for interpretation of high energy CID mass spectra have been reported in detail elsewhere (*6*). It is of interest to note that residue 5 is assigned Ile of the isomeric pair (Leu/Ile). This distinction is possible due to the satellite ions $\mathbf{w_4}$ and $\mathbf{w_4}$' (fragment ions with charge retention at the C-terminus are numbered from the C-terminal amino acid) present at m/z 543 and 557, respectively. These satellite ions form *via* fragmentation between the amino group and the α-carbon *plus* a bond cleavage between the β- and γ-carbon of the "new" N-terminal amino acid of the ion thus formed. Since the amino acid Ile has two different substituents on its β-carbon, it can produce two satellite ions with different mass values due to the fragmentation processes described above. These ions appear 14 and 28 Da higher than the single fragment possible from Leu residues. Oligonucleotide probes were constructed based on this and other such primary amino acid sequences enabling polymerase chain reaction (PCR) to be employed to clone the genes encoding these proteins (*6, 15*).

Two-dimensional SDS-PAGE of melanoma proteins. The strategy employed in this laboratory for the characterization of a selection of proteins separated from a cell culture lysate of human melanoma A375 cells (*16*) is presented in Figure 4. The spots chosen for protein identification from the 2-D gel separation of this lysate are labeled in Figure 4.

The easiest approach for rapid identification of a known protein from a gel slice or spot involves digestion in the gel, extraction of peptides and measurement of the molecular weights of the mixture using MALDI time-of-flight (TOF) mass spectrometry. Then the set of mass values obtained are used to search the protein databases for identical entries. As an example of the nature of the MALDI spectra

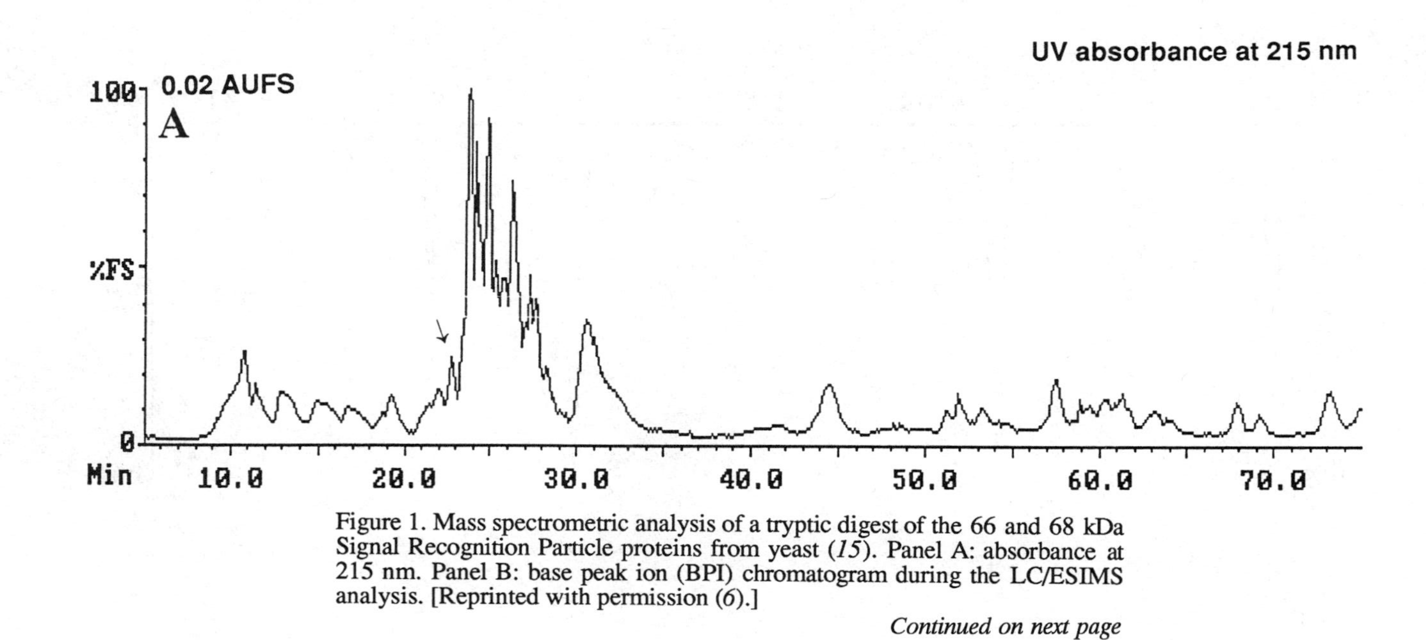

Figure 1. Mass spectrometric analysis of a tryptic digest of the 66 and 68 kDa Signal Recognition Particle proteins from yeast (*15*). Panel A: absorbance at 215 nm. Panel B: base peak ion (BPI) chromatogram during the LC/ESIMS analysis. [Reprinted with permission (*6*).]

Continued on next page

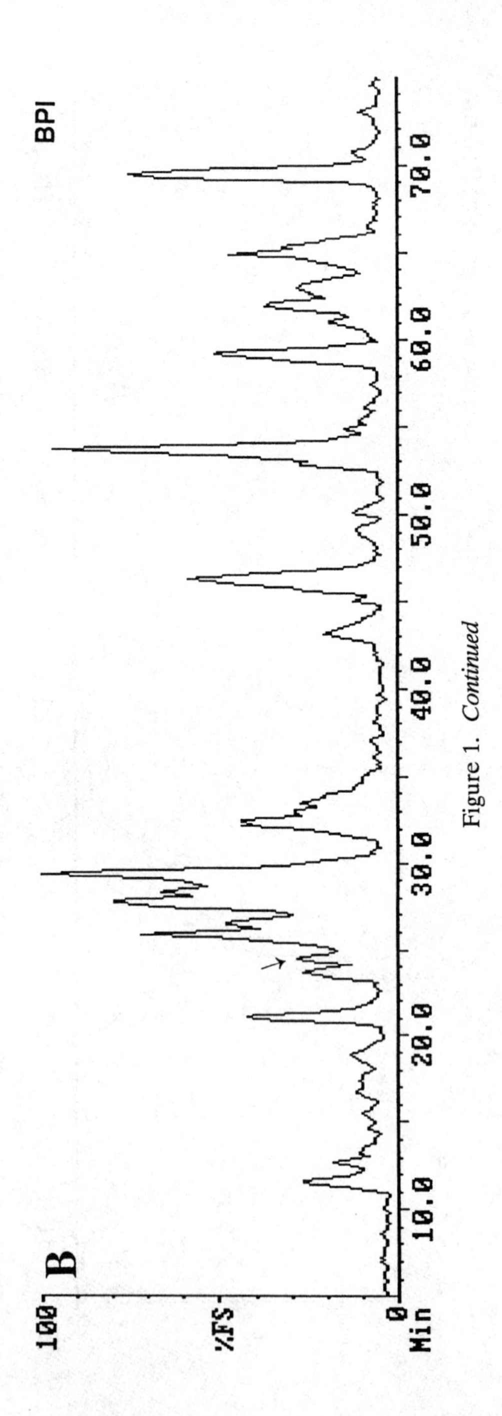

Figure 1. *Continued*

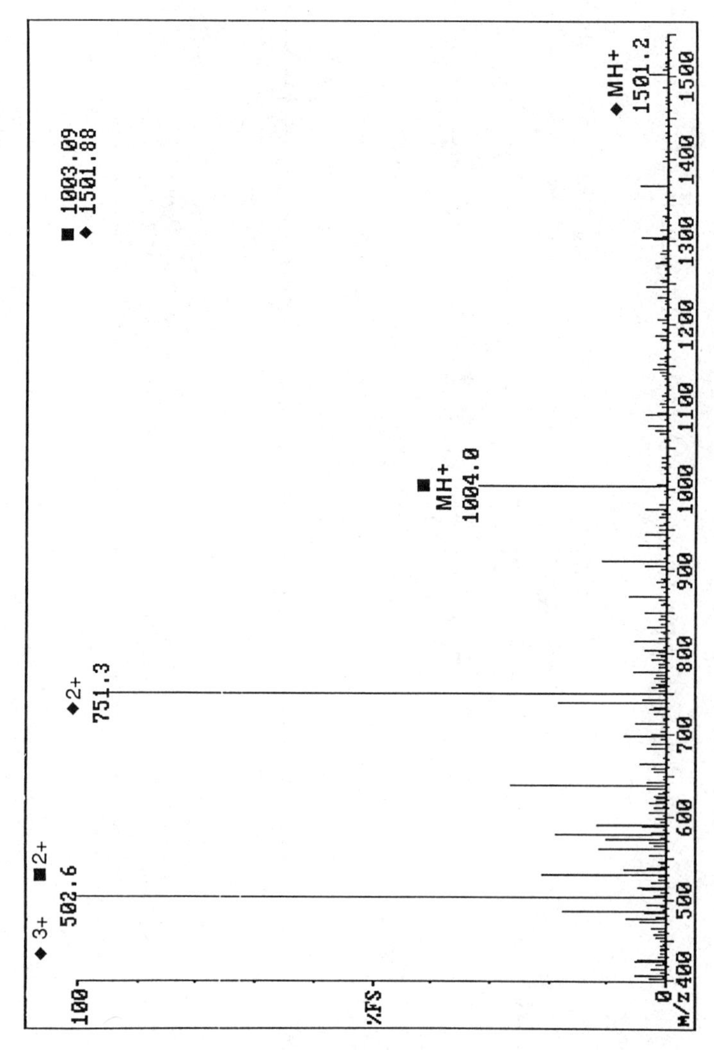

Figure 2. ESIMS spectrum of the fraction labeled in the chromatogram in Figure 1. In this experiment the average molecular masses of the components were determined and are presented. [Reprinted with permission (*6*).]

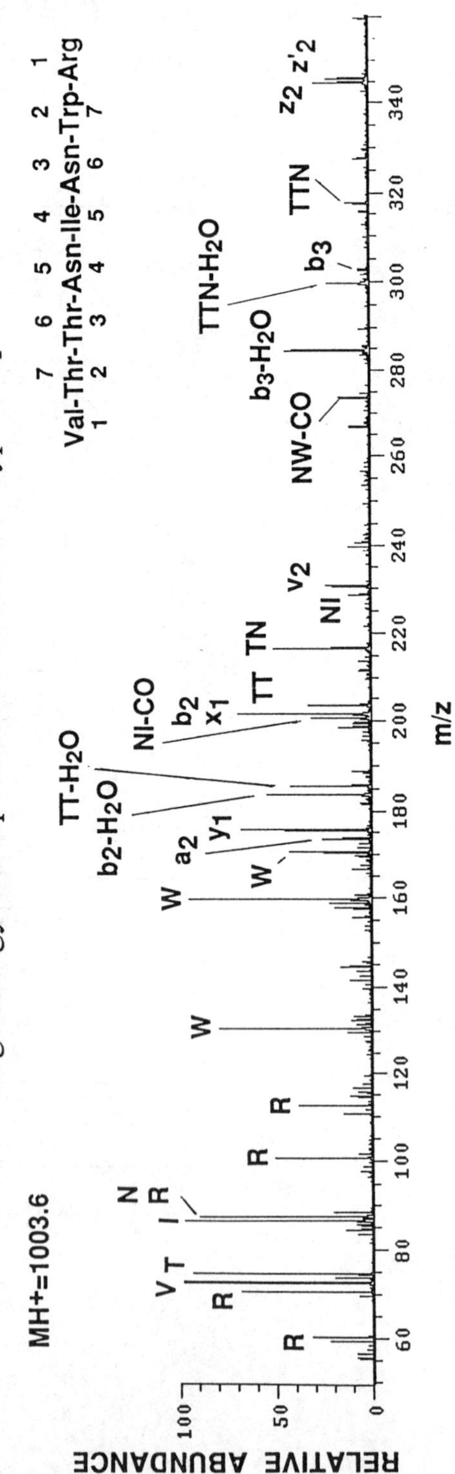

Figure 3. High energy CID spectrum of the smaller tryptic peptide (Figure 2) eluted in the fraction labeled in Figure 1. The C^{12} monoisotopic MH^+ ion (at m/z 1003.6) was chosen as the precursor ion for this analysis. [Reprinted with permission (*6*).]

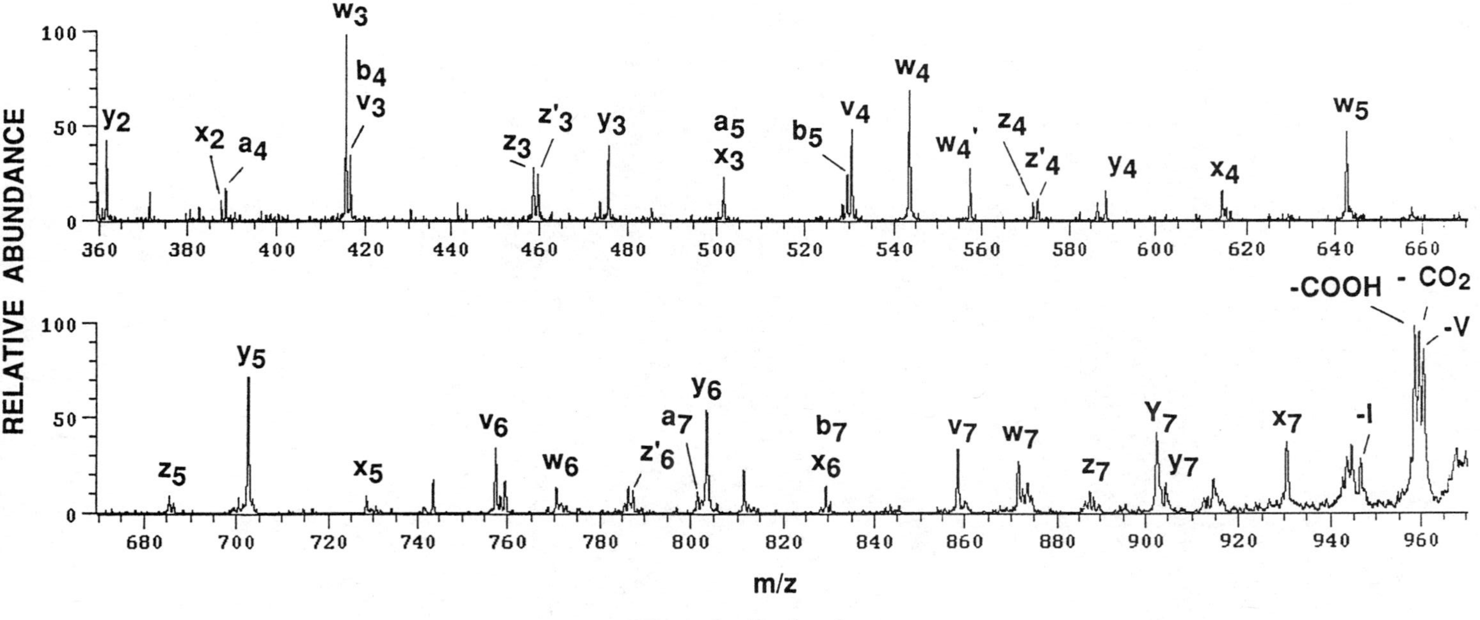

Figure 3. *Continued*

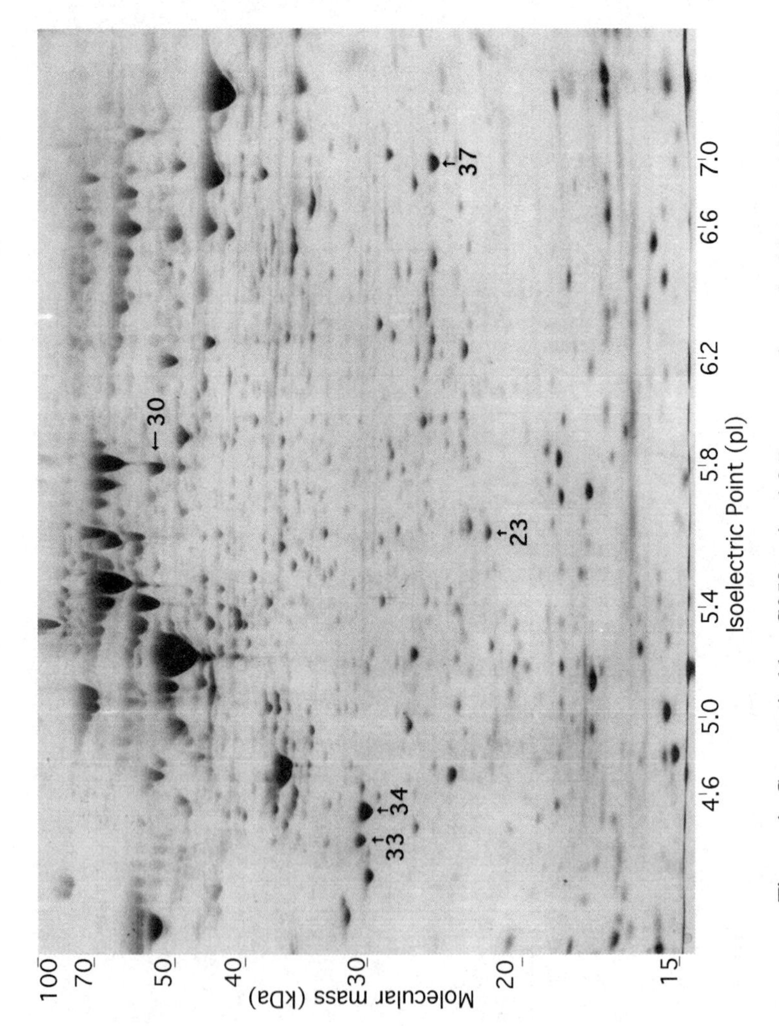

Figure 4. Coomassie blue G250-stained 2-D preparative gel of human A375 melanoma proteins.

which may be obtained, the spectra of the unseparated peptides from digestion of spot 23 is shown in Figure 5. The MALDI spectrum was recorded using α-cyano-4-hydroxycinnamic acid on a Fisons/VG TofSpec instrument in the linear mode. The database search identified two possible human protein matches, namely natural killer cell enhancing factor B and thiol-specific anti-oxidant protein. The sequences of these two possibilities are 93.4% identical. To increase our discriminating power, i. e., to determine unambiguously the identity of the protein(s) of interest, amino acid sequence information had to be obtained. After subsequent HPLC separation of this digest some components were subjected to so-called post source decay (PSD) analysis to gain partial amino acid sequence information using the MALDI instrument in the reflection mode. The microbore HPLC trace of this tryptic digest is shown in Figure 6. The PSD spectrum of fraction 18 (see Figure 6) is presented in Figure 7. Fragment ions expected from any particular known sequence can be calculated. From a comparison of the fragment ions observed and the known sequences of the two proteins the identity of this particular peptide was determined, as one derived from the thiol-specific anti-oxidant protein.

The necessity for more accurate mass measurement than that provided by linear MALDI-TOF mass spectrometers and for detailed sequence information is illustrated with proteins in spots 33 and 34 (see Figure 4). These proteins were digested with trypsin in the gel and the resulting peptides were separated by microbore reversed-phase HPLC. Using the linear MALDI-TOF method for measuring the peptides' molecular weight with external calibration, an HPLC fraction from both digests yielded an average MH^+ value of 1200 ±2 Da. LSIMS/high energy CID analysis of these fractions revealed two peptides differing by 1 Da (at m/z 1199.7 and 1200.6) corresponding to the blocked N-termini of two tropomyosin proteins, the non-muscle fibroblast type and the cytoskeletal type, respectively. The sequences are highly homologous. Only four amino acids are different in these two peptides; the fibroblast protein contains LNSL in positions 3-6, while the cytoskeletal protein features the sequence ITTI. Since Leu and Ile are isomeric amino acids with the same residue mass (isobaric), the mass difference between the two peptides is reduced two residues (NS vs TT) which may be distinguished in principle by PSD or low energy CID analyses. On the other hand, high energy CIDspectra are of advantage since they clearly indicate the difference even for the isobaric amino acid pairs (Figs. 8 and 9). Scheme 1 shows the correct assignment of the Leu/Ile positions from the side chain fragmentation (*6*).

Finally, a note of caution is necessary about the common occurrence of adventitious covalent modifications encountered from the gel matrix and/or digestion in the presence of urea which could interfere with database searches using only moleucular mass values. The peptide presented in Figure 10 was a candidate possibly originating from an unknown protein, since its measured molecular mass did not match any tryptic peptide for proteins in the database searched. However, upon interpretation of its high energy CID spectrum (see Figure 10) it became clear that this tryptic peptide was modified in two positions: by N-terminal carbamolyation from urea present in the digestion buffer (*6*) and by acrylamide alkylation of the cysteine thiol (*17*).

Our protocol has several advantages. It is inherently able to provide reliable primary sequence when a protein is not in a database (or when the database has errors), to carry out database searches seeking protein homology, to deal with non-specific proteolytic cleavage, to adjudicate the nature and origin of covalent modifications of peptides, and to detect and identify the presence of multiple co-migrating components. As outlined above and described in detail elsewhere (*16*), the procedure requires pooling of some 4-8 gel spots which stain well with Coomassie blue. Improving the sensitivity still further is an on-going effort.

Recently, we have coupled a capillary HPLC with a continuous flow inlet for the LSIMS source of our 4-sector mass spectrometer such that on-line recording of

Spot 23 Melanoma protein tryptic digest
MALDI Spectrum (linear mode)
Sample: 1/500th of total digest
Matrix: α-cyano-4-hydroxycinnamic acid
880.6
1211.9
1864.1
1930.9
100%
98
96
94
92
90
88
86
84
82
80
78
76
74
72
70
68
66
64
62

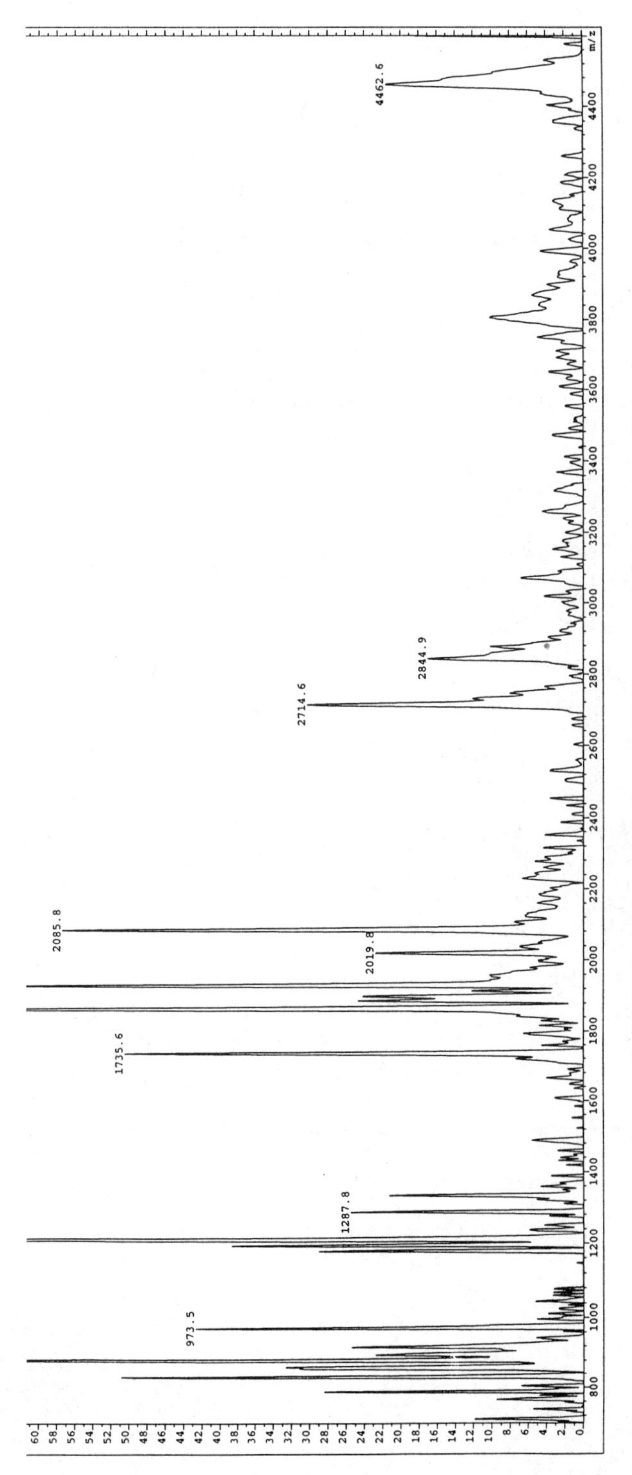

Figure 5. MALDI spectrum of the peptides extracted from an in-gel tryptic digest of spot 23 in Figure 4. An aliquot of the digest was diluted 1:10 with 50% ACN/0.1% TFA. After co-crystallizing the diluted digest with α-cyano-4-hydroxycinnamic acid, the crystals were rinsed twice with cold water to remove residual digestion buffer (25mM ammonium bicarbonate). 1/500 th of the total digest was loaded on the MALDI target.

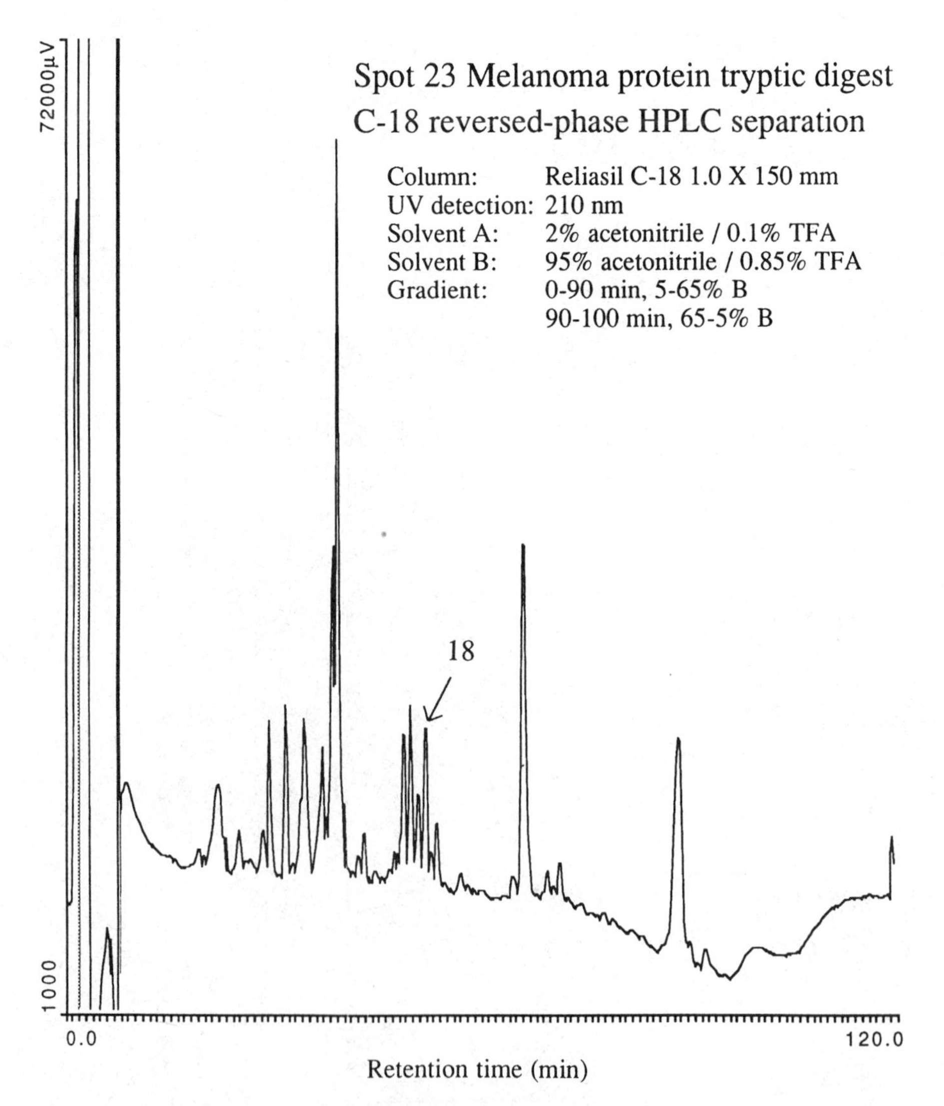

Figure 6. C-18 reversed-phase HPLC separation of the peptides extracted from an in-gel tryptic digest of spot 23 (Figure 4).

CID spectra can be achieved. The in-situ gel digest extract is first washed with water by concentrating the digest onto a small pre-column (300 micron i. d. x 5 mm, C18 packing) to remove extraneous salts which suppress LSIMS ionization of peptides. This pre-column is then used to "inject" onto the capillary for LC-CID analysis. Such an analysis was carried out on the tryptic digest of melanoma spot 30 (see Figure 4). The capillary HPLC trace is shown in Figure 11. Representative CID spectra from this run are presented in Figure 12 and 13. From interpreting these CID spectra and searching the database, it was established that gel spot 30 contained protein disulfide isomerase.

Selective detection of covalent protein modifications.

LC/ESIMS experiments yield the peptide molecular mass composition of the components in any given protein digest during their separation and elution. This method is also well suited for selectively detecting digest components possessing certain particular structural features, such as carbohydrates (*18, 19*), phosphate esters (*18*), or xenobiotically modified amino acid residues (see discussion below). This selective detection can be applied for all chemical structures that give diagnostic ion(s) under collisional activation conditions. For this purpose either in-source fragmentation is induced by increasing the skimmer- (cone/orifice) voltage in the electrospray source, or CID analysis is carried out in a tandem mass spectrometer while monitoring mass values characteristic of the moiety in question. SIM and molecular mass determination can be carried out in a single LC/ESIMS experiment, following Carr's "orifice-stepping" method (*18*). In this type of experiment the orifice voltage is set at a high value while scanning the low mass range for diagnostic fragment ions and is then decreased to the normal value once m/z 500 (or other appropriate m/z value) is reached during the scan.

This technique is well suited for detecting the glycosylated components in bovine fetuin tryptic digest as shown in Figure 14. The UV chromatogram is recorded in Panel a, the total ion current (TIC) in Panel b and four fragment ions characteristic of glycosylation, namely m/z 204 ($HexNAc^+$, i. e., oxonium ion of any N-acetylhexos-amine), 292 ($Neu5Ac^+$, i. e., oxonium ion of neuraminic acid), 366 ($Hex\text{-}HexNAc^+$, i. e., oxonium ion of a hexosyl-N-acetylhexosamine disaccharide), and 657 (oxonium ion of a Neu5Ac-Hex-HexNAc trisaccharide) in Panels c, d, e, and f, respectively. Thus, it is readily apparent that the digest contained seven glycopeptide-containing fractions. From the molecular mass values observed, it was established that O-linked glycopeptides eluted in fractions 3 and 4, while the other fractions contained N-linked glycopeptides (19). For example, Fraction 1 corresponds to the Asn^{138}-linked glycopeptide, $K^{126}LCPDCLLAPLNDSR^{141}$, with its cysteine sulfhydryl functions alkylated by ethylpyridyl groups and bearing a series of different oligosaccharides, while components in Fraction 2 were tentatively identified as the same glycopeptides but with free cysteine thiols. This technique was then used to detect all the ethylpyridyl-cysteine containing peptides by monitoring an ion diagnostic of this structure (m/z 106) (*20*) as shown in Figure 14, Panel g. It can readily be seen from the elution position of peaks 1 and 2 in Figure 14 that the component represented by peak 1 is in fact vinylpyridylated, while that from peak 2 is not. So peaks 1 and 2 in Panels c-f represent the same N-linked glycopeptide with residue Asn-138. Mass spectrometric structural elucidation of glycosylation of bovine fetuin has been presented in detail elsewhere (*19*). In addition, the use of sequential glycosidase digestions and repetitive LC-ESIMS analysis yields more detailed information about the structures present at a given site. We have described this approach recently in connection with the analysis of lecithin-cholesterol acyltransferase (LCAT) (*21*).

Finally, this technology was employed effectively in detecting drug-protein adducts of tolmetic glucuronide/human serum albumin (HSA) adducts (*22, 23*).

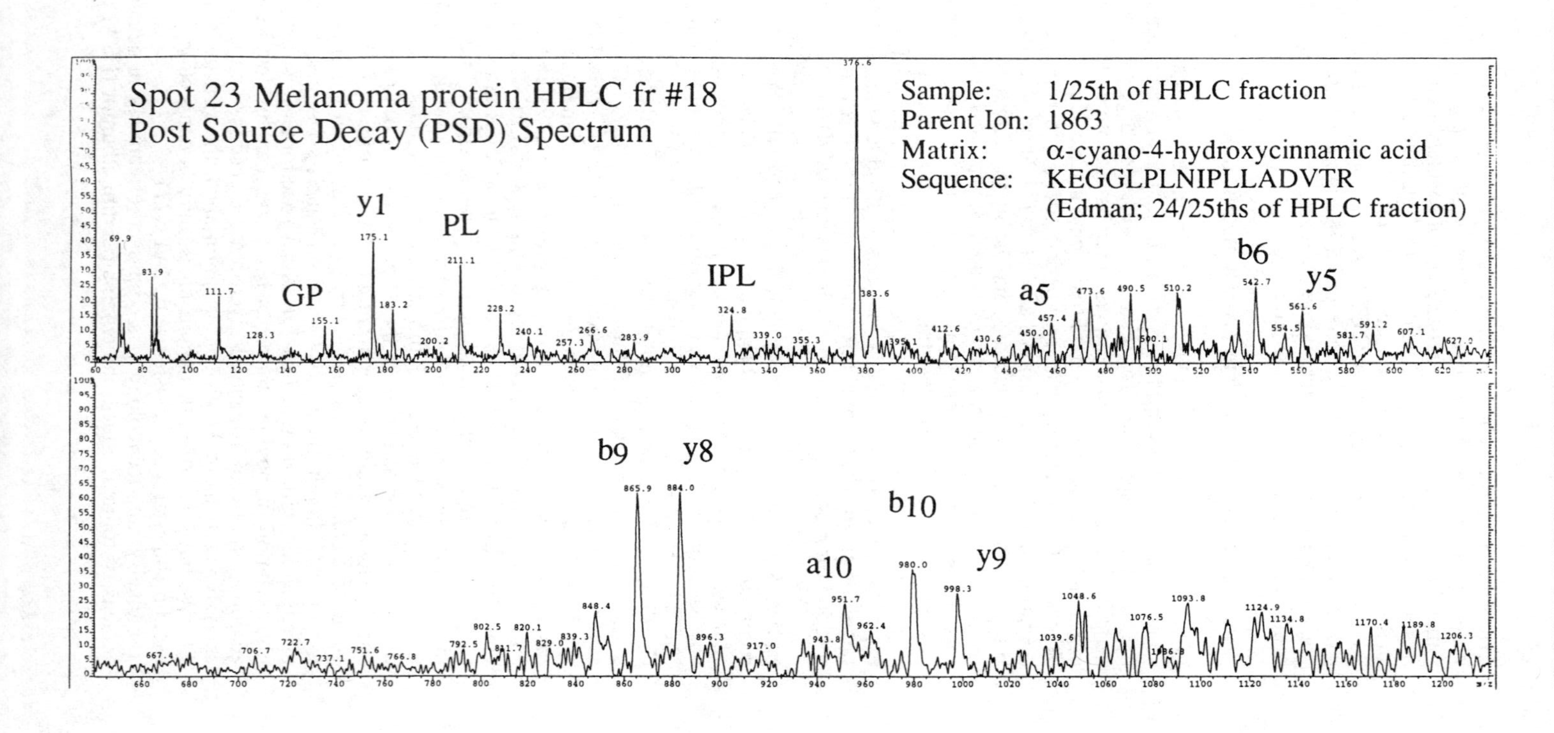

Spot 23 Melanoma protein HPLC fr #18
Post Source Decay (PSD) Spectrum
Sample: 1/25th of HPLC fraction
Parent Ion: 1863
Matrix: α-cyano-4-hydroxycinnamic acid
Sequence: KEGGLPLNIPLLADVTR
(Edman; 24/25ths of HPLC fraction)
GP
y1
PL
IPL
a5
b6
y5
b9
y8
a10
b10
y9

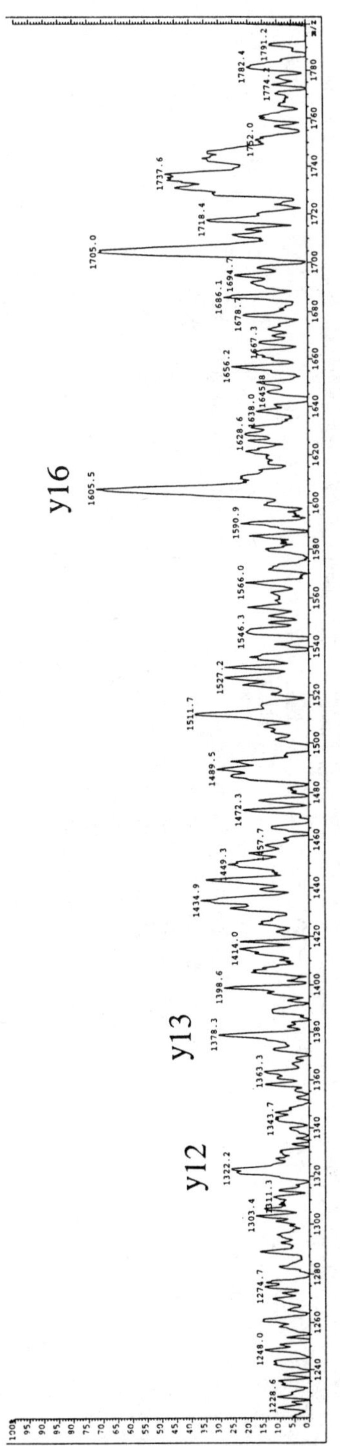

Figure 7. Post Source Decay spectrum of fraction #18 collected from the HPLC separation shown in Figure 6. Limited sequenced information is present. Ions are labeled as in Figure 8. The remainder of the fraction was sequenced by Edman degradation, yielding the sequence; KEGGLPLNIPLLADVTR, from thiol-specific antioxidant protein.

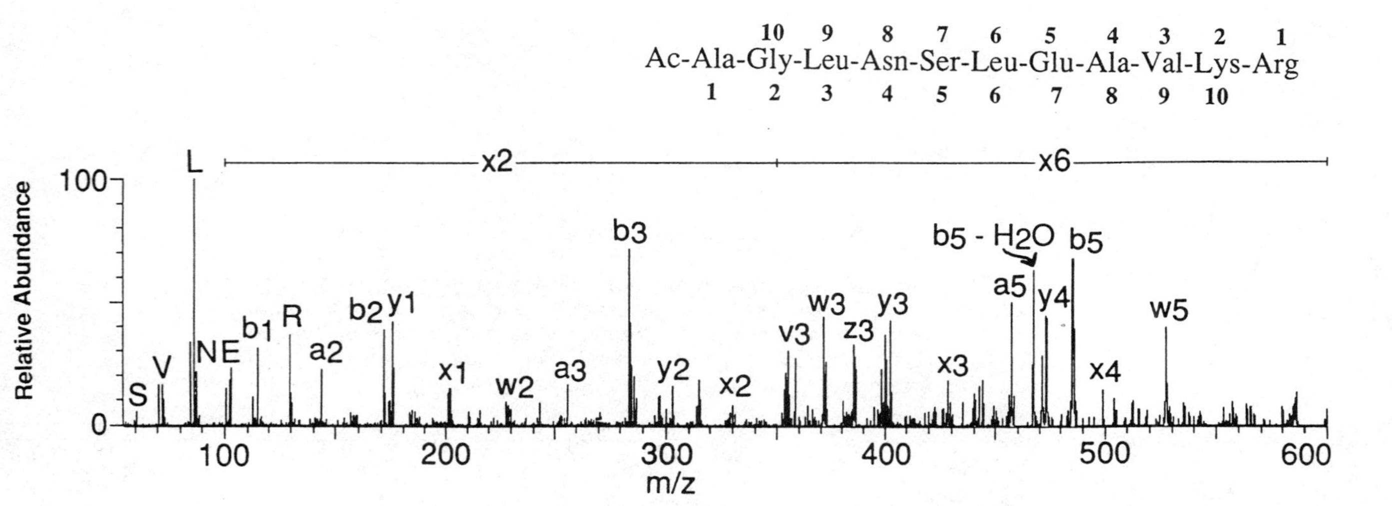

Figure 8. High-energy CID spectrum of fibroblast non-muscle tropomyosin N-terminal tryptic peptide from spot 33 (Figure 4). Immonium ions corresponding to individual amino acid residues are denoted by the single-letter code. Peptide backbone cleavage ions formed by fragmentation associated with charge retention at the C-terminus are denoted by **x**, **y**, and **z** and at the N-terminus are denoted by **a** and **b**. Ions formed by side-chain fragmentation are denoted as **v** and **w**. [Reprinted with permission (*16*).]

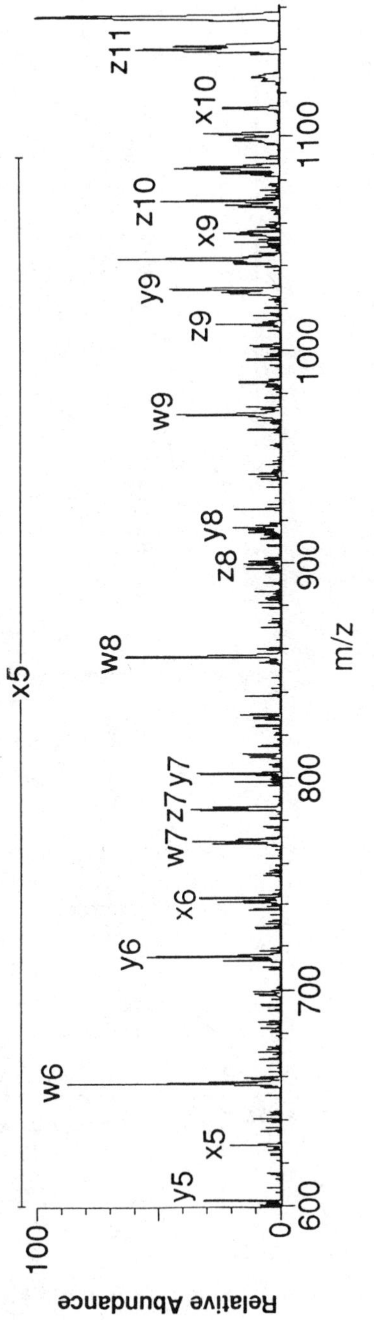

Figure 8. *Continued*

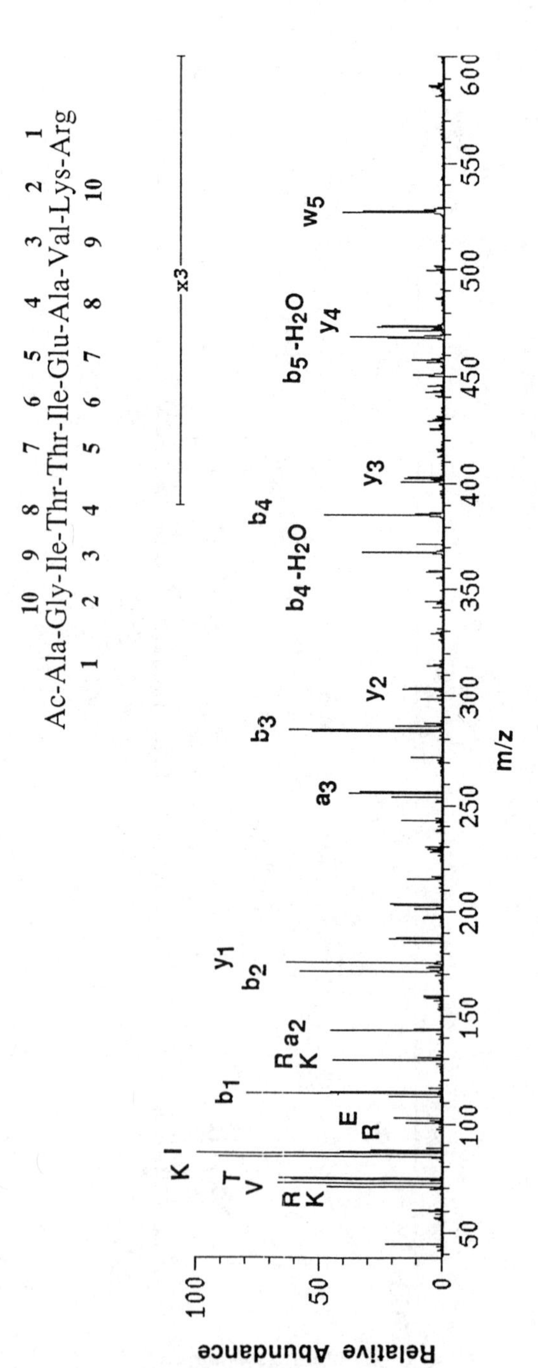

Figure 9. High-energy CID spectrum of cytoskeletal tropomyosin N-terminal tryptic peptide from spot 34 (Figure 4). Ions are labeled as in Figure 8.

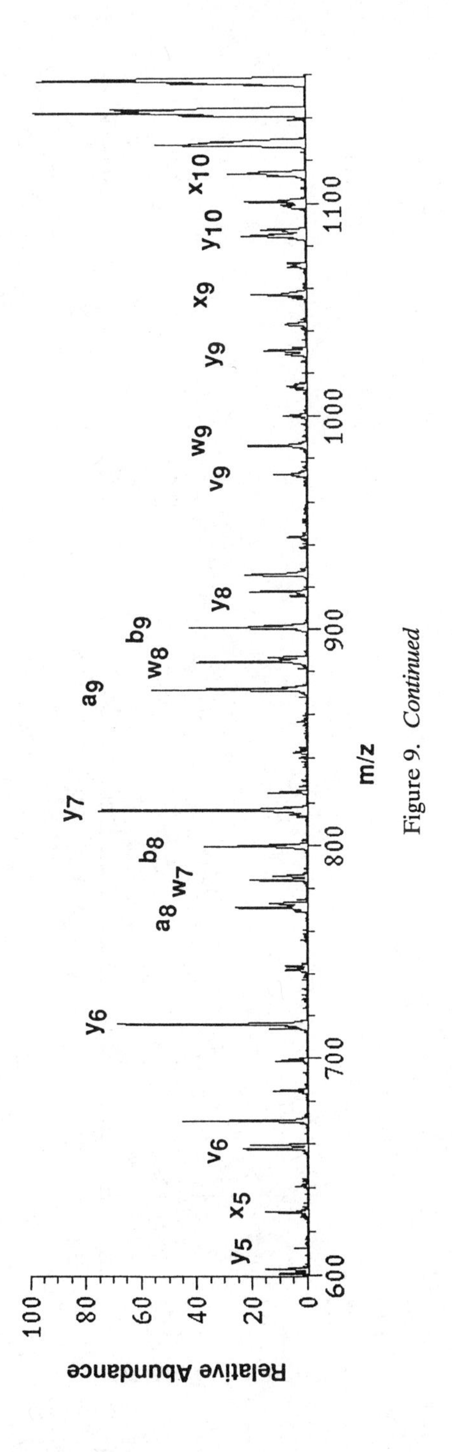

Figure 9. *Continued*

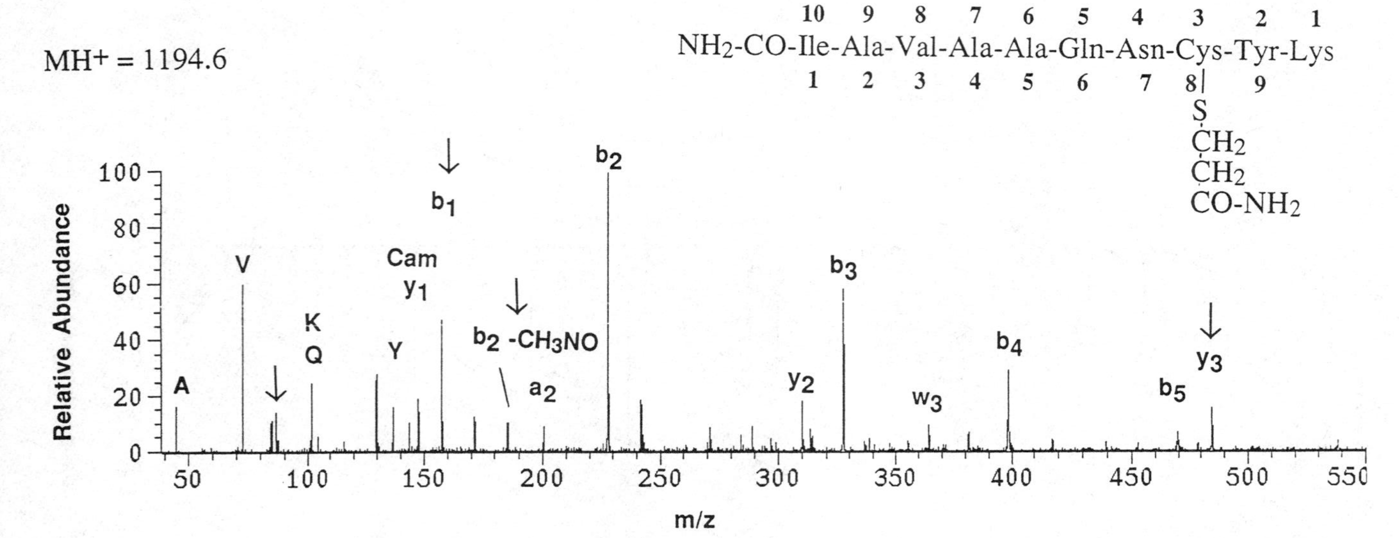

Figure 10. High-energy CID spectrum of a peptide from triosephosphate isomerase in spot 37 (Figure 4). The peptide is N-terminally carbamoylated (chemical modification resulting from urea present in the digest buffer) and contains an acrylamide modified cysteine (chemical modification occurring in the PAGE matrix). Ions are labeled as in Figure 8.

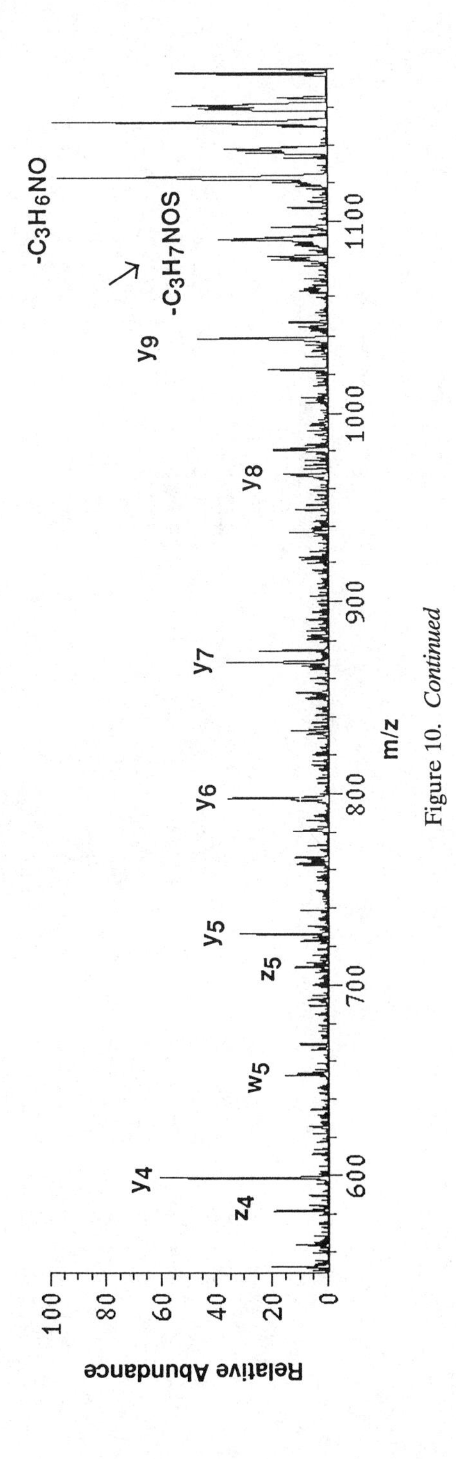

Figure 10. *Continued*

Spot 33 H_3CCO-A-G-L-N-S-L-E-A-V-K-R MH^+ = 1199.7 Da

Spot 34 H_3CCO-A-G-I-T-T-I-E-A-V-K-R MH^+ = 1200.6 Da

	y ion series		w ion series	
	acAGITTIEAVKR	acAGLNSLEAVKR	acAGITTIEAVKR	acAGLNSLEAVKR
1	175.1	175.1	73.0	73.0
2	303.2	303.2	229.1	229.1
3	402.3	402.3	371.2	371.2
4	473.3	473.3	456.3	456.3
5	602.4	602.4	527.3	527.3
6	715.4	715.4	670.4	656.4
7	816.5	802.5	783.5	769.5
8	917.5	916.5	884.5	856.5
9	1030.6	1029.6	985.6	970.5
10	1087.6	1086.6	---	---

Scheme 1. Comparison of Sequence Ions in Figures 8 and 9.

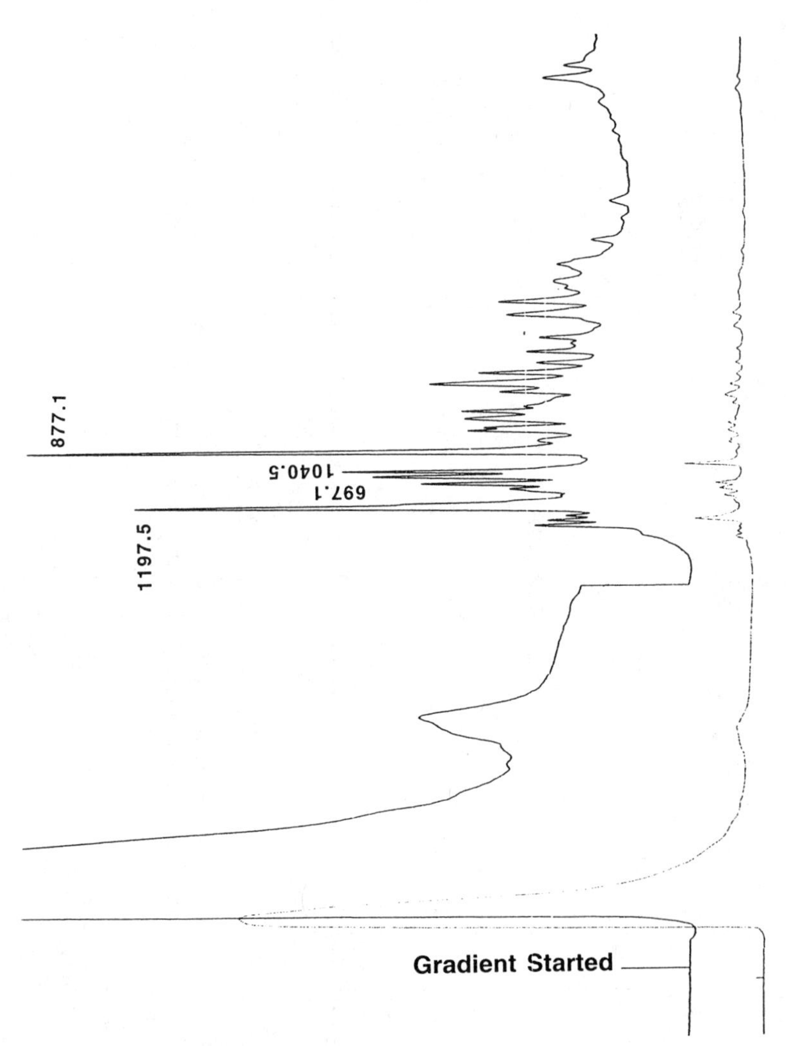

Figure 11. UV chromatogram (l = 215 nm) of tryptic digest capillary HPLC separation of melanoma protein isolated from 2-D gel spot A-30. Monoisotopic measured masses (Da) of peptides are labeled above their respective chromatographic peaks.

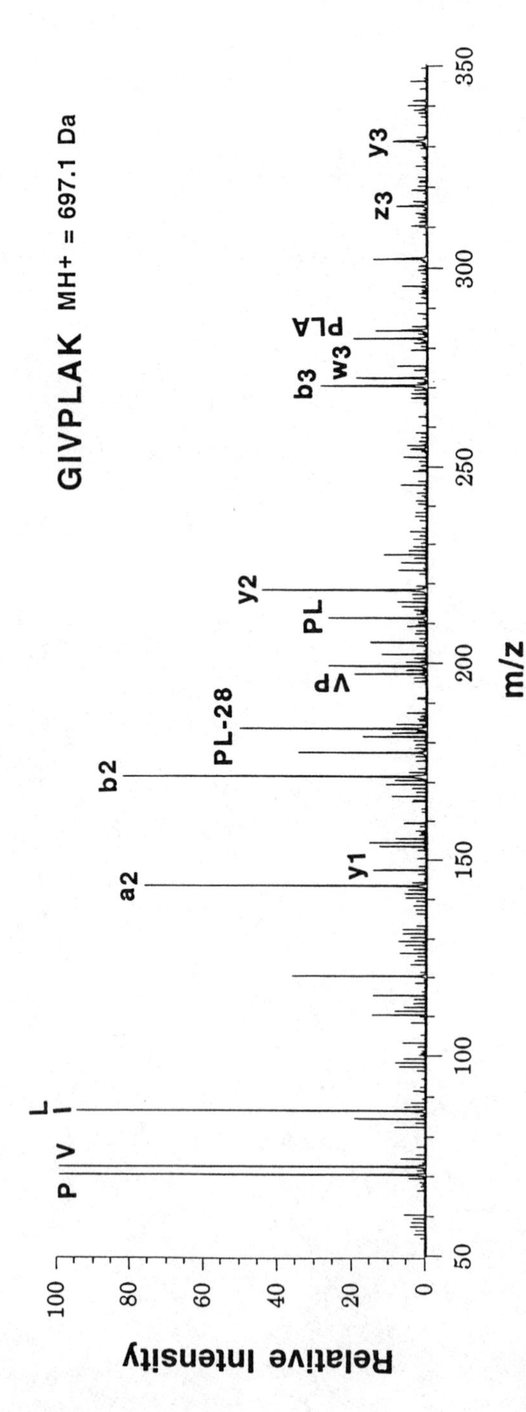

Figure 12. High energy CID spectrum of GIVPLAK, MH^+ = 697.1 Da, corresponding to residues 76-82 of human protein disulfide isomerase.

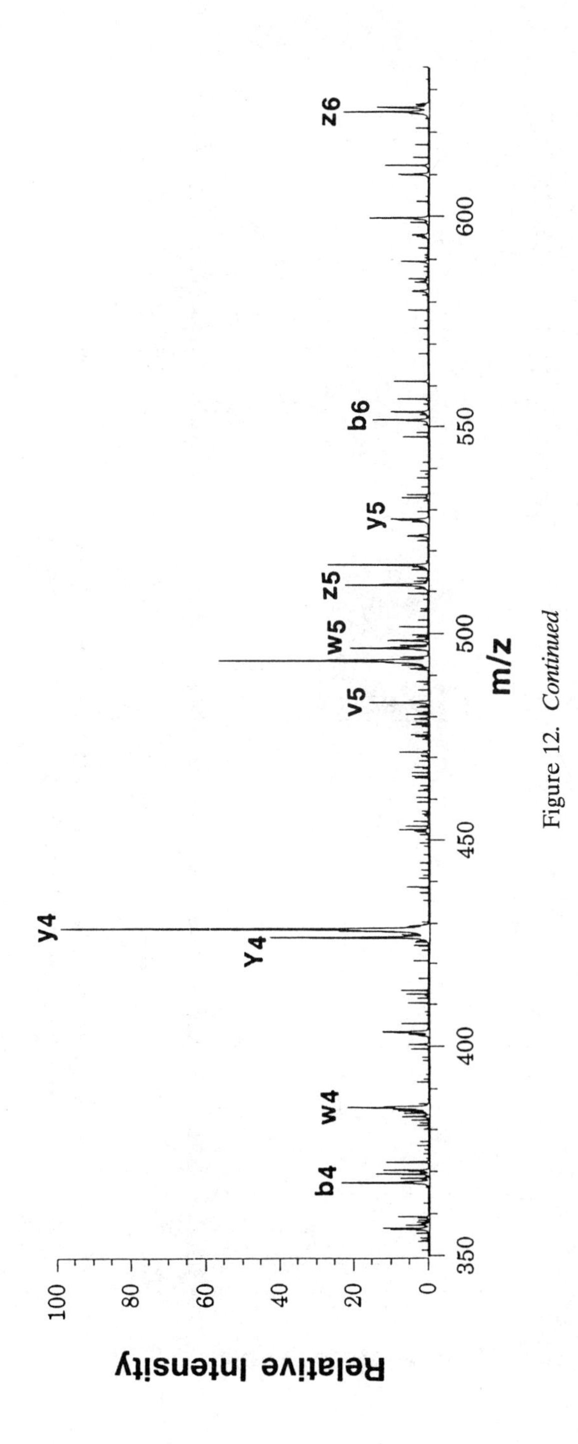

Figure 12. *Continued*

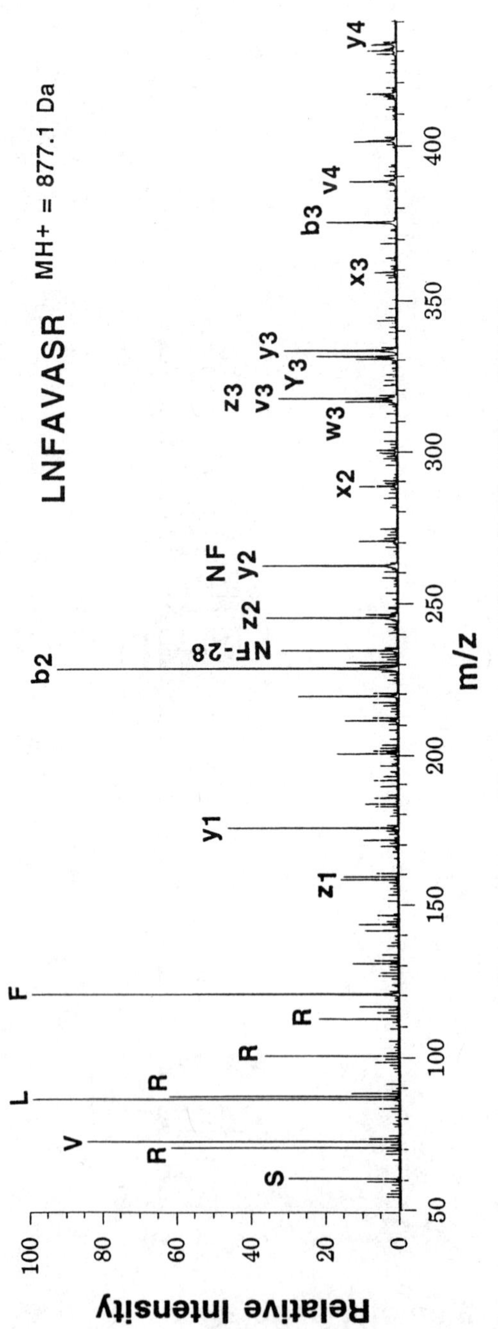

Figure 13. High energy CID spectrum of LNFAVASR, MH^+ = 877.1 Da, corresponding to residues 297-304 of human protein disulfide isomerase.

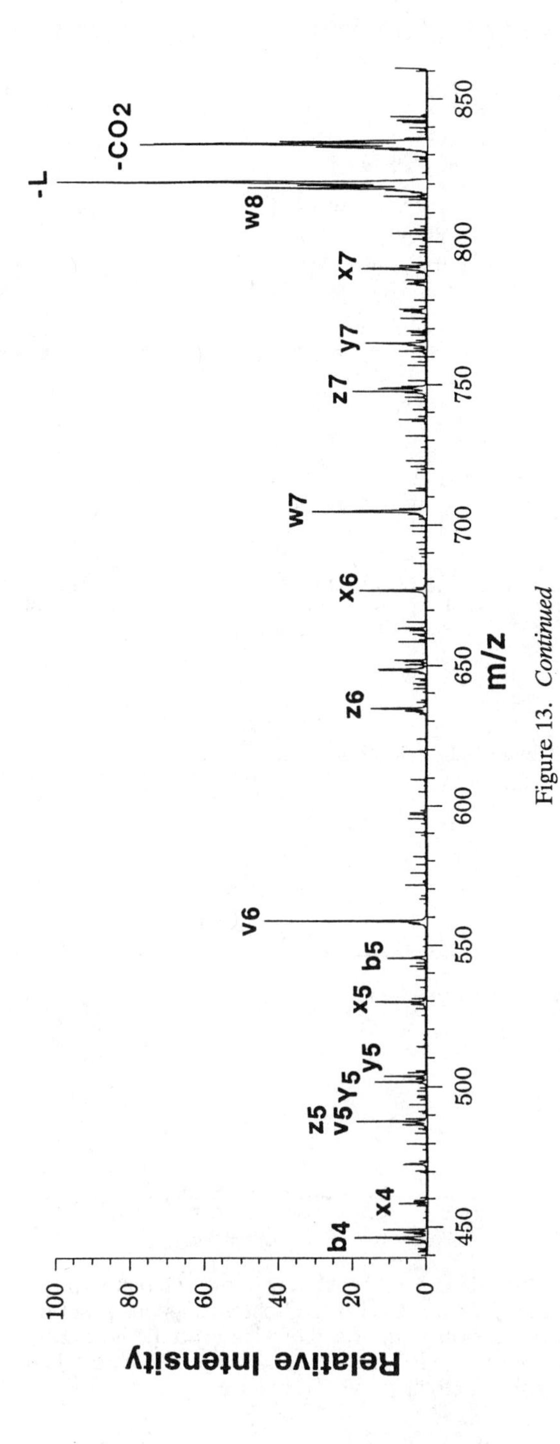

Figure 13. *Continued*

LC/SIM Analysis of Bovine Fetuin Tryptic Digest

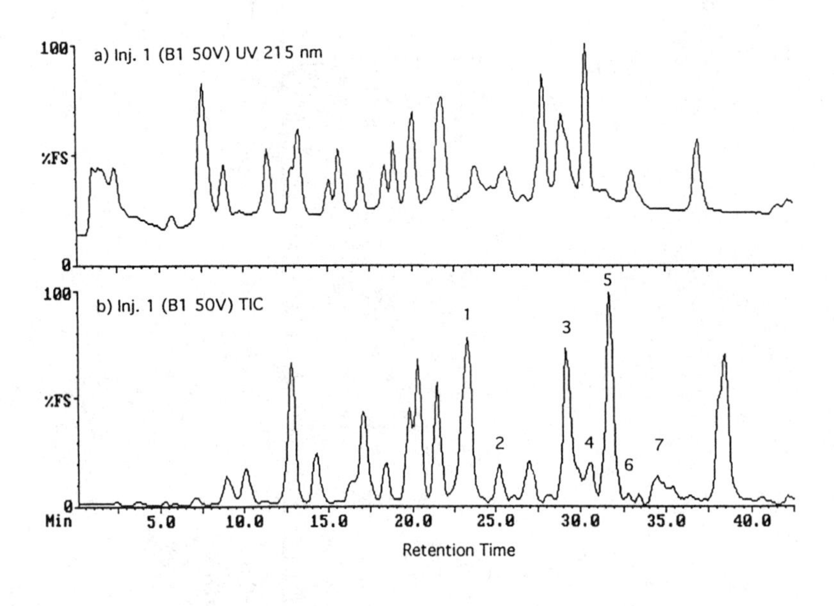

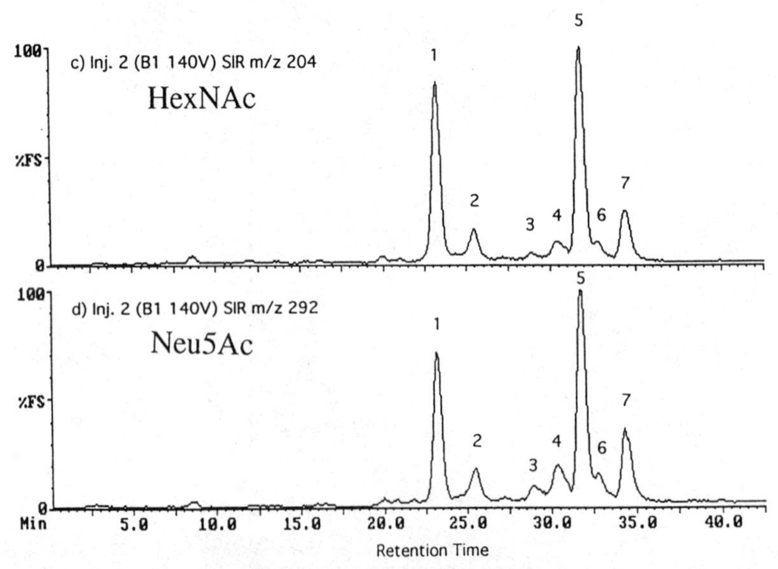

Figure 14. LC/ESIMS and SIM analysis of bovine fetuin tryptic digest. Panel a: absorbance at 215 nm. Panel b: TIC chromatogram. Panel c, d, e, and f show the SIM chromatogram of carbohydrate specific ions at m/z 204, 292, 366 and 657, respectively. Panel g shows the SIM chromatogram for m/z 106, characteristic of ethylpyridyl-Cys residues.

LC/SIM Analysis of Bovine Fetuin Tryptic Digest

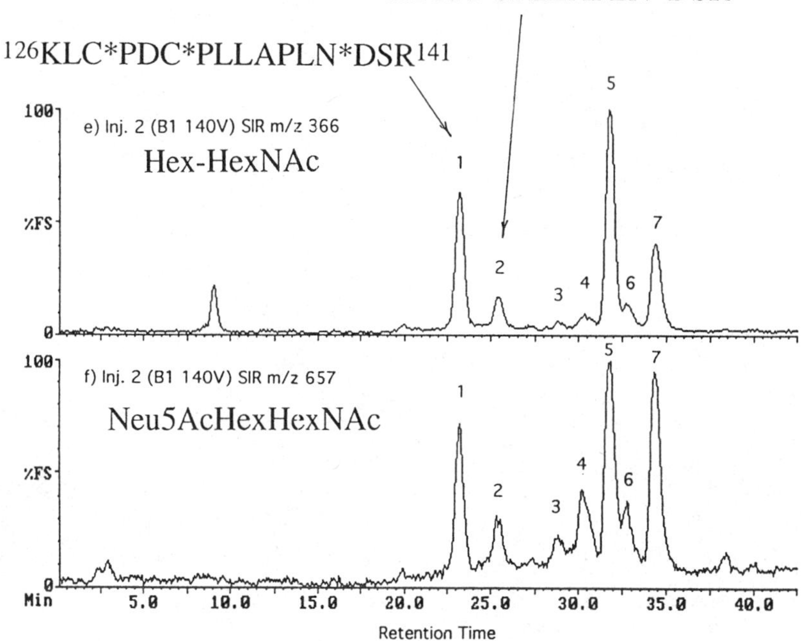

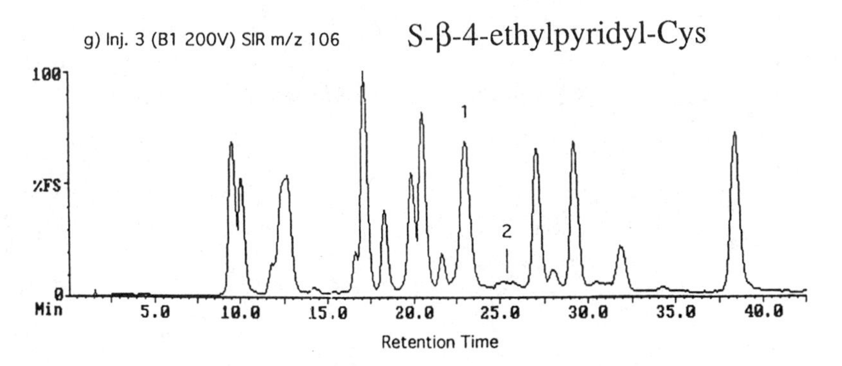

Figure 14. *Continued*

Diagnostic Fragment Ions of Tolmetin-Adducts

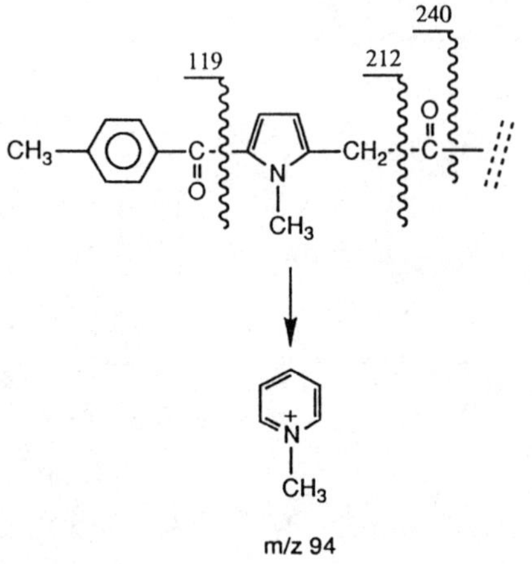

240

119 212

m/z 94

Scheme 2. Structures of moiety specific ions for tolmetin.

The moiety specific ions for tolmetin are shown in Scheme 2. Monitoring fragment ions at m/z 94 and 119 in an LC/ESIMS experiment identified five components which were modified during the incubation of HSA with the tolmetin metabolite (data not shown). The power of high energy CID analysis is well illustrated by the site specific detection and delineation of isomeric alkylation sites in the same peptide sequence resulting in two chromatographically unresolved isobaric components of the albumin digest (see Figure 15).

Capillary HPLC-ESI-magnetic sector multichannel array detection (EBE-MCA) instrumentation.

Use of a magnetic sector mass analyzer provides inherent advantages over those obtainable using quadrupole analyzers in both resolution and mass measurement accuracy (*24*). A resolving power of >20,000 has been demonstrated on ESI/sector instruments, and there does not appear to be any fundamental limitation beyond the inherent resolving power of the analyzer being used. Samples that give little or no useful information at a resolution of 500-1000 due to microheterogeneity or mixture complexity can be successfully analyzed at higher resolution with relative ease. The charge state of a molecule being analyzed by ESIMS is most commonly determined by finding a series of peaks corresponding to two or more adjacent charge states. However, when components are represented by a single charge state (or additional peaks are too weak to detect reliably), it is essential to be able to resolve the isotopic multiplet pattern to determine the charge state of the ion, and thus molecular mass of the compound. It is quite common for peptides resulting from tryptic digests, typically with masses<2000 Da, to give a "one peak" ESIMS spectrum. However, glycopeptides, which are typically larger, also frequently exhibit this property (*24*).

A second important advantage of higher resolution is the capability of sorting through and separating the ions generated from a complex mixture, such as four or five co-eluting peptides in an HPLC peak from enzyme digest of a moderately large protein. Computer algorithms rely on the identification of a number of tryptic fragment masses, or more efficiently on fewer peptides when the masses are more accurate (*26-28*). The ability to employ an array detector is one of the advantages of magnetic sector mass analyzers (*4, 6, 29*). Using the focal plane array detector, one can routinely achieve subpicomole detection limits, reaching into the attomole range in favorable cases.

Figures 16-19 show results from an LC/ESIMS experiment performed on a VG AutoSpec SE instrument with 50 fmoles of bovine fetuin tryptic digest injected. Figure 16 shows the TIC chromatogram in this experiment. While the spectrum of the peptide IPLDPVAGYK (Fig. 17) shows that the array detector delivers the required resolution even at this sample level, the spectrum of the glycopeptides (Fig. 18) again illustrates the impressive sensitivity of this method. The major glycoform labeled in this spectrum represents approximately 30 fmoles of the sample, according to NMR and previous ESIMS studies (19). As described earlier, in-source fragmentation of peptides can be induced by increasing to cone-voltage in the electrospray source. The spectra generated contain ion types which are qualitatively similar to the low energy CID spectra obtained using triple quadrupole mass spectrometers. Using this technique in conjunction with on-line HPLC separation on capillary columns provides a quick and sensitive method for obtaining significant sequence information from peptides in proteolytic digests. Figure 19 shows the cone-voltage induced fragmentation spectrum of 500 fmoles of a bovine fetuin tryptic acquired during a capillary LC/ESIMS run. The observed resolution (see insert) at this sample level is far superior to that demonstrated for quadrupole analyzers. This increase in resolution is important for assigning charge

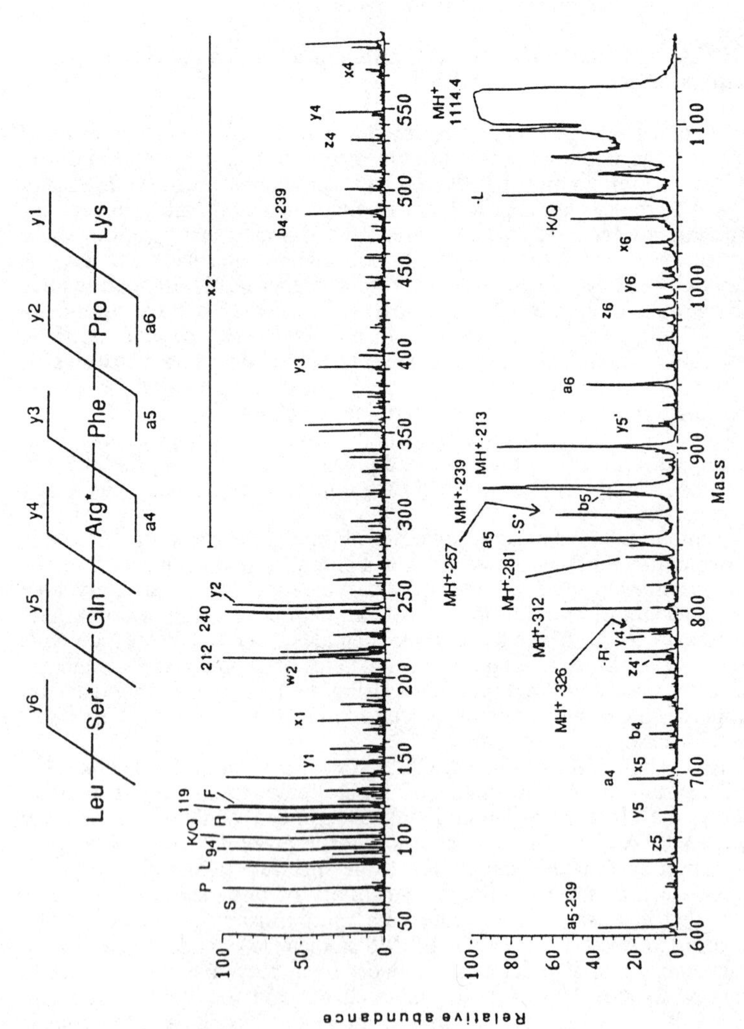

Figure 15. High energy CID spectrum of Leu-Ser-Gln-Arg-Phe-Pro-Lys which established that the tolmetin moiety is bound to Ser and Arg in an isobaric isomeric mixture of the two modified peptides with the same amino acid sequence.

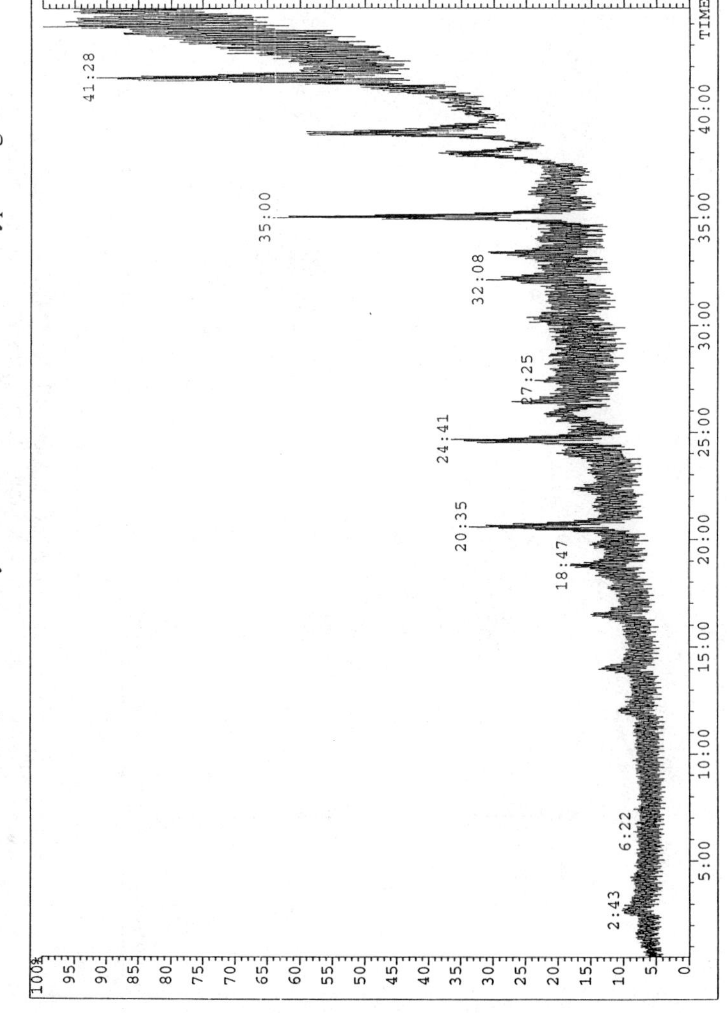

Figure 16. Total ion current trace obtained from the injection of 50 femtomoles of fetuin tryptic digest on a 180 micron I. D. capillary LC column interfaced to the ESI source of an AutoSpec SE with array detector.

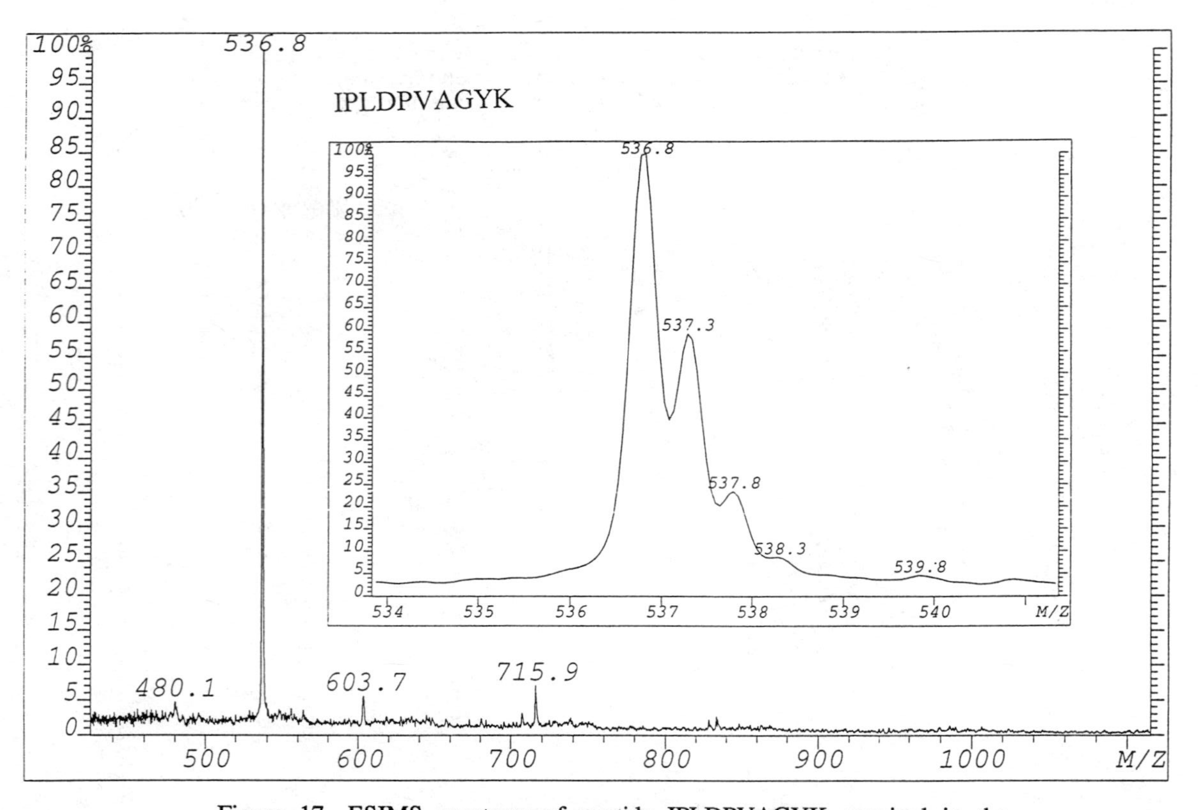

Figure 17. ESIMS spectrum of peptide IPLDPVAGYK acquired in the LC/ESIMS experiment shown in Figure 16. Insert shows sufficient resolution to establish ion at mass 537 as a doubly charged species.

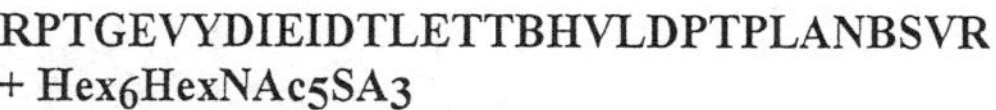
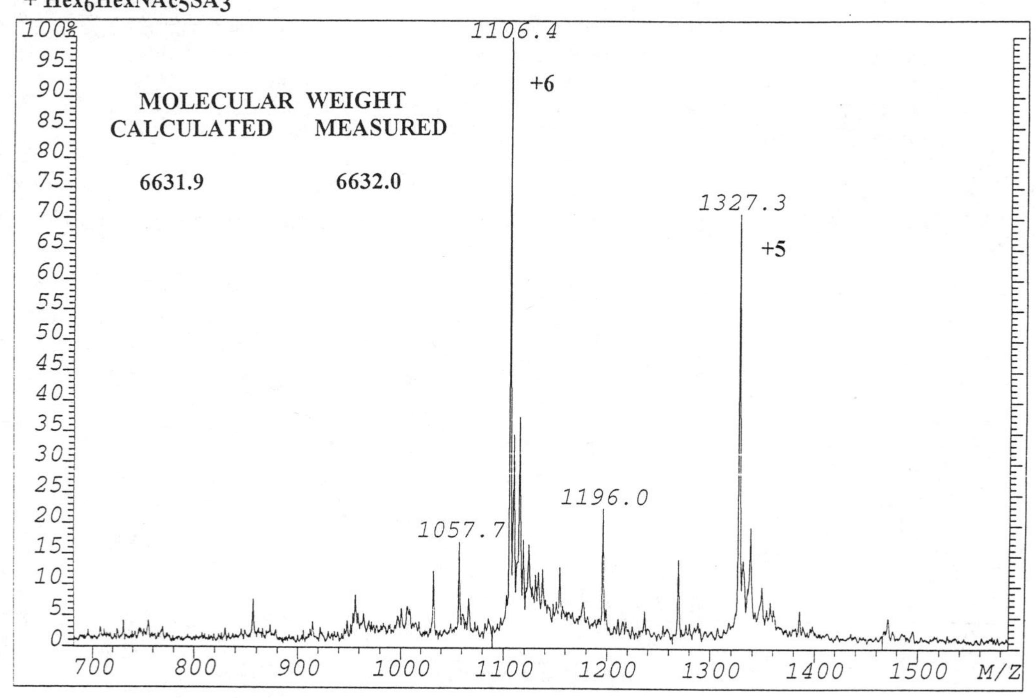

Figure 18. ESIMS spectrum of fetuin glycopeptides obtained in the LC/ESIMS experiment shown in Figure 16. The +6 and +5 charged state of the most prevalent isoform are labeled. The formula of this isoform is also shown along with the calculated and measured masses.

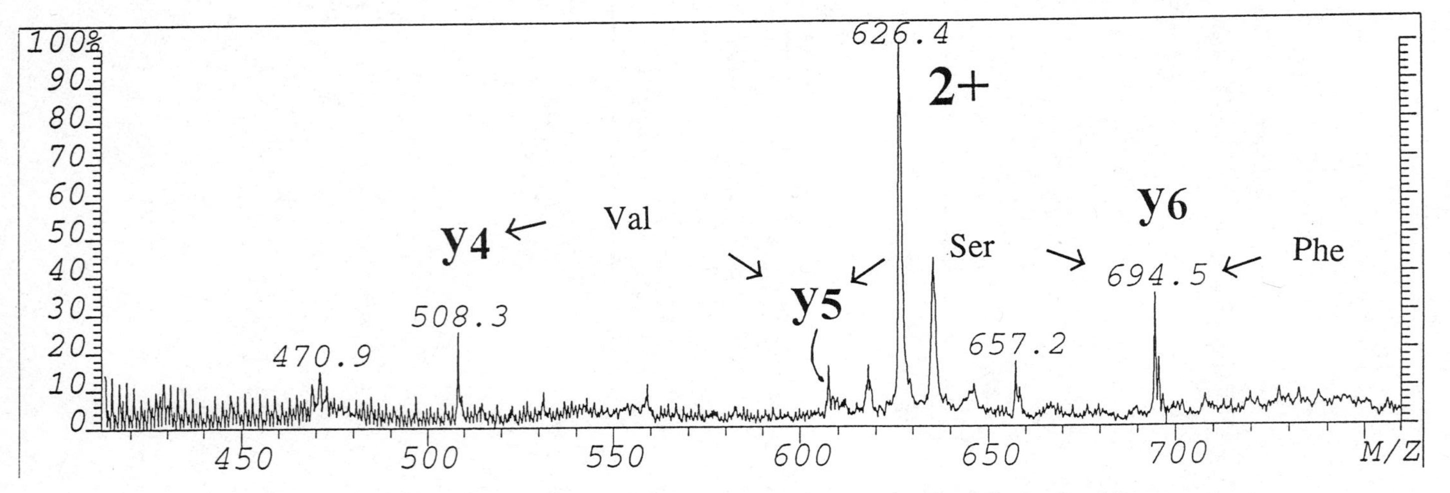

Figure 19. Cone-voltage induced fragmentation spectrum of fetuin peptide <QDGQFSVLFTK (<Q stands for pyroglutamic acid) acquired in an LC/ESI-MS experiment from injection of 500 femtomoles of a tryptic digest. Insert shows fragment ion $\mathbf{y_9}$ with sufficient resolution to determine charge state as single charged.

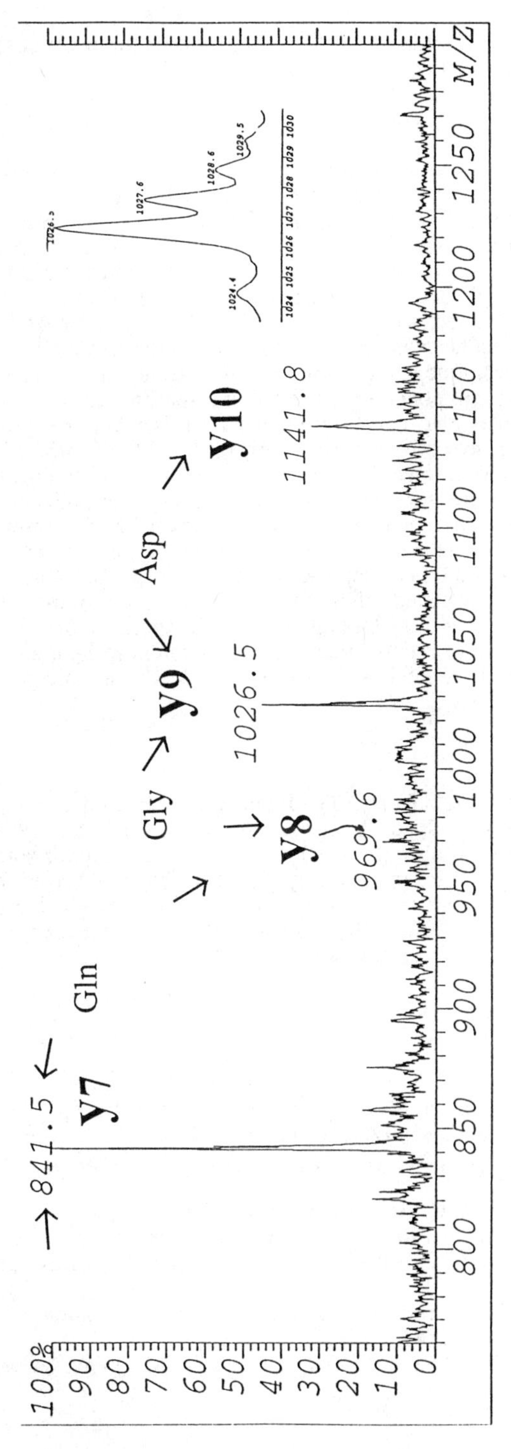

Figure 19. *Continued*

states of fragment ions, where charges of one, two or three are possible and also produces better mass accuracy.

Conclusions.

The tools are now in hand to characterize complex mixtures of proteins reliably. This involves use of a combination of techniques including 2-D gel electrophoresis, in-gel digestion, HPLC-separation and the powerful mass spectrometric techniques developed only recently. Based on our recent experience, MALDI-TOF instruments with reflectrons can be employed to obtain accurate molecular masses of digest components with high detection sensitivity and partial amino acid sequences derived from PSD spectra. On-line LC/ESIMS experiments yield a more comprehensive molecular mass composition, due to the less discriminative nature of this ionization technique, and also provide the means for selectively identifying fractions containing components with structural features of interest. In addition, amino acid sequences also can be obtained by this technique, either by cone-voltage induced in-source fragmentation or by on-line LC/ESIMS/MS experiments. LSIMS-high energy CID spectra yield usually the most comprehensive sequence information and thus, are the most suitable for primary sequence determination for cloning previously uncharacterized proteins. However, the sample quantities required for this technique are normally higher than that for the two other methods. However, very recent studies employing the EBE-orthogonal acceleration-TOF geometry have established that unknown peptides may be sequenced unambiguously from their MALDI high energy CID spectra (*30*).

Further evaluation of relative merits of the different techniques, especially the MALDI-PSD and CID spectra as well as LC/ESIMS cone-voltage fragmentation applications are underway.

Acknowledgments.

We thank Jonathan Pugh from VG Organic for carrying out the LC/ESIMS experiments on the AutoSpec SE instrument. We are indebted to Drs. Andreas Ding and Patrick Schindler for performing the SIM experiments. The work presented in this paper was supported by grants from the NIH National Center for Research Resources (grant number RR 01614), NSF Biological Instrumentation Program (grant number DIR-8700766) and National Institute of Environmental Health Sciences (grant number ES 04705).

Literature cited.

1. Barber, M.; Bordoli, R. S.; Sedgwick, R. D. In *Soft Ionization Mass Spectrometry*; Morris, H. R., Ed.; Heyden: London, 1981; pp. 137.
2. Aberth, W.; Straub, K. M.; Burlingame, A. L. *Anal. Chem.* **1982,** *54,* 2029.
3. Morris, H. R.; Panico, M.; Taylor, G. W. *Biochem. Biophys. Res. Com.* **1983,** *117,* 299.
4. Biemann, K.; Scoble, H. A. *Science* **1987,** *237,* 992.
5. Burlingame, A. L. In *Biological Mass Spectrometry: Present and Future*; Matsuo, T.; Caprioli, R. M; Gross, M. L.; Seyama, Y., Eds.; John Wiley & Sons: Chichester, 1987; p. 147.
6. Medzihradszky, K. F.; Burlingame, A. L. *Methods: A Companion to Methods in Enzymology* **1994,** *6,* 284.
7. Whitehouse, C. M.; Dreyer, R. N.; Yamashita, M.; Fenn, J. B. *Anal. Chem.* **1985,** *57,* 675.

8. Tanaka, K.; Waki, H.; Ido, Y.; Akita, S.; Yoshida, Y.; Yoshida, T. *Rapid Commun. Mass Spectrom.* **1988,** *2*, 151.
9. Karas, M.; Hillenkamp, F. *Anal. Chem.* **1988,** *60*, 2299.
10. Jones, J. L.; Dongre, A. L.; Somogyi, A.; Wysocki, V. H. *J. Am. Chem. Soc.* **1994,** *116*, 8368.
11. Spengler, B.; Kirsch, D.; Kaufmann, R. *Rapid Commun. Mass Spectrom.* **1991,** *5*, 198.
12. Hunt, D. F.; Zhu, N. Z.; Shabanowitz, J. *Rapid Commun. Mass Spectrom.* **1989,** *3*, 122.
13. *Mass Spectrometry in the Biological Sciences*; Burlingame, A. L.; Carr, S. A., Eds.; Humana Press: Clifton, NJ, 1995.
14. Burlingame, A. L.; Boyd, R. K.; Gaskell, S. J. *Anal. Chem.* **1994,** *66*, 634R.
15. Brown, J. D.; Hann, B. C.; Medzihradszky, K. F.; Niwa, M.; Burlingame, A. L.; Walter, P. *EMBO J.* **1994,** *13,* 4390.
16. Clauser, K. R.; Hall, S. C.; Smith, D. M.; Webb, J. W.; Andrews, L. E.; Tran, H. M.; Epstein, L. B.; Burlingame, A. L. *Proc. Natl. Acad. Sci. USA* **1995,** *92*, 5072.
17. Hall, S. C.; Smith, D. M.; Masiarz, F. R.; Soo, V. W.; Tran, H. M.; Epstein, L. B.; Burlingame, A. L. *Proc. Natl. Acad. Sci. USA* **1993,** *90*, 1927.
18. Huddleston, M. J.; Bean, M. F.; Carr, S. A. *Anal. Chem.* **1993,** *65*, 877.
19. Medzihradszky, K. F.; Maltby, D. A.; Hall, S. C.; Settineri, C. A.; Burlingame, A. L. *J. Am. Soc. Mass Spectrom.* **1994,** *5*, 350.
20. Falick, A. M.; Hines, W. M.; Medzihradszky, K. F.; Baldwin, M. A.; Gibson, B. W. *J. Am. Soc. Mass Spectrom.* **1993,** *4*, 882.
21. Schindler, P. A.; Settieri, C. A.; Collet, X.; Fielding, C. J.; Burlingame, A. L. *Protein Sci.* **1995,** *4*, 791-803.
22. Ding, A.; Ojingwa, J. C.; McDonagh, A. F.; Burlingame, A. L.; Benet, L. Z. *Proc. Natl. Acad. Sci. USA* **1993,** *90*, 3797.
23. Ding, A.; Zia-Amirhosseini, P.; McDonagh, A. F.; Burlingame, A. L.; Benet, L. Z. *Drug Metab. Dispos.* **1995,** *23*, 369.
24. Chapman, J. R.; Gallagher, R. T.; Mann, M. *Biochem. Soc. Trans.* **1991,** *19*, 940.
25. Carr, S. A.; Huddleston, M. J.; Bean, M. F. *Protein Sci.* **1993,** *2*, 183.
26. Mann, M.; Hojrup, P.; Roepstorff, P. *Biol. Mass Spectrom.* **1993,** *22*, 338.
27. Pappin, D. J.; Hojrup, P.; Bleasby, A. J. *Current Biology* **1993,** 327.
28. Henzel, W. J.; Billeci, T. M.; Stults, J. T.; Wong, S. C.; Grimley, C.; Watanabe, C. *Proc. Natl. Acad. Sci. USA* **1993,** *90*, 5011.
29. Evans, S. *Methods Enzymol.* **1990,** *193*, 61.
30. Medzihradszky, K. F.; Bateman, R. H.; Green, M. R.; Adams, G. W.; Burlingame, A. L. *J. Am. Soc. Mass Spectrom.,* submitted.

RECEIVED September 12, 1995

Chapter 25

Analysis of Biomolecules Using Electrospray Ionization–Ion-Trap Mass Spectrometry and Laser Photodissociation

James L. Stephenson, Jr.[1], Matthew M. Booth, Stephen M. Boué, John R. Eyler, and Richard A. Yost

Department of Chemistry, University of Florida, P.O. Box 117200, Gainesville, FL 32611–7200

The technique of photo-induced dissociation (PID) offers several advantages over that of collision-induced dissociation (CID) for structural analysis, including a well defined energy deposition process and no direct competition between ion dissociation and resonance ejection (typically observed in trapping instruments). Comparisons between the collisional and photon fragmentation processes for peptides, oligosaccharides and RNA dimers demonstrate the applicability of PID for structural elucidation of biological molecules. The instrumentation employed consisted of an rf-only octopole for ion injection of electrosprayed ions into a quadrupole ion trap. The ring electrode was modified with three mirrors to increase the photoabsorption pathlength for the PID process. The photoabsorption cross sections (IR) for the three compound classes studied followed the order: oligosaccharides (C—O—C ether linkage) > RNA dimers (phosphodiester linkage) > peptides (amide linkage).

Over the last decade, perhaps the most important advancement in quadrupole ion trap mass spectrometry has been that of tandem mass spectrometry or MS^n for the structural elucidation of polyatomic ions. The most frequently used method for the activation of these ions has been collisional activation or what is customarily called collision-induced dissociation (CID) (*1*). The major factors contributing to the success of CID experiments in the quadrupole ion trap mass spectrometer (QITMS) include the ability to perform tandem-in-time as opposed to tandem-in-space MS/MS experiments (*2*), the efficient conversion of parent ions to product ions (typically 10-50%) (*3*), and most importantly the high collision cross sectional area observed for a typical CID experiment (on the order of

[1]Current address: Oak Ridge National Laboratory, Building 5510, Mail Stop 6365, Oak Ridge, TN 37831–6365

0097–6156/95/0619–0512$20.25/0

10 to 200 Å^2). These advantages arise in part because in the quadrupole ion trap, uniquely amongst tandem mass spectrometers, kinetic energy is imparted to the parent ions between (rather than before) collisions.

In this chapter we discuss an alternative to collisional activation for tandem MS experiments, which utilizes the photon-absorption process for the activation of polyatomic ions (produced by electrospray ionization) in the QITMS. Until recently, photo-induced dissociation (PID) or photodissociation has been used almost exclusively in fundamental studies of gas-phase ions in physics and chemistry. Photo-induced dissociation is the next most frequently used method for activation of polyatomic ions after collisional activation. The range of internal energies present after the photon absorption process are much narrower than those obtained with collisional energy transfer. Therefore, the usefulness of PID for the study of ion structures is greatly enhanced. However, the reduced absorption cross sections observed with photodissociation (10^{-2} Å^2) compared to those of collision-induced dissociation (10 to 200 Å^2) can limit this technique for analytical applications. The recent availability of higher powered lasers over a wider range of wavelengths should provide greater flexibility for photodissociation as a routine analytical technique (*1,4*). In this study, extended irradiation of trapped ions and a multipass optical configuration make PID practical for analytical studies.

The process of photodissociation for a positive ion can be described by the following equation:

$$A^{+} \underset{\text{relaxation}}{\overset{nh\nu}{\rightleftharpoons}} A^{+*} \underset{\text{dissociation}}{\rightarrow} P^{+} + N \qquad (1)$$

where $\mathbf{A^+}$ is the ion of interest, **n** is the number of photons absorbed, $\mathbf{h\nu}$ is the photon energy, $\mathbf{A^{+*}}$ is the excited state, and $\mathbf{P^+}$ represents the product ion (with loss of neutral **N**). For photodissociation to occur several prerequisites must be met. The most important criteria include the absorption of photons with energy $\mathbf{h\nu}$, the existence of excited states above the dissociation threshold, a slow relaxation rate compared to the rate of photon absorption (multiphoton processes), and dissociation rates which are fast on the time scale of the type of mass spectrometer employed (*1,4*).

The information obtained from a photodissociation experiment can address a variety of gas-phase chemistry issues. One of the most important issues is the difference observed in fragmentation spectra between PID and CID. Since the range of internal energies after the activation step is much narrower in PID, typically the dissociation process proceeds via the fragmentation pathway with the lowest activation energy (especially for visible and infrared wavelengths) (*5-7*). Ion energy studies by Louris et al. using ultraviolet-visible (UV-Vis) wavelengths demonstrated the higher energy deposition available for the photodissociation process compared to the CID process (e.g. higher m/z 91 to m/z 92 branching ratios from the molecular ion of n-butyl benzene) in the quadrupole ion trap (*8*).

In addition, wavelength-dependent photodissociation spectra can be obtained as long as the internal energy of the excited ion population is above the dissociation threshold for the wavelength of interest. The photodissociation spectrum can then be compared with the vuv absorption spectrum of the neutral molecule, provided that sufficient transition intensity exists between the ground state neutral and the excited state ion. The information obtained from these experiments may constitute as a fingerprint in the determination of ion structures (*6,7*).

In trapping instruments, photo-induced ion fragmentation (i.e. product ion relative abundances) can be measured as a function of irradiance time. These data can be used effectively to distinguish isomeric ion structures in the gas-phase. Kinetic energy release data can also be used (below 100 mV) to add important ion fragmentation information (*9*). Perhaps some of the more interesting studies utilizing PID and trapping instruments involve ion trajectory studies, or what is frequently termed ion tomography. Several recent studies by Williams et al. have demonstrated the ability of PID in the UV-Vis range to map the instantaneous trajectories of ions stored in the quadrupole ion trap (*10,11*). Analogous studies by Lammert et al. have successfully used ion tomography to characterize the frequency of ion motion as well as the observed "mass shifts" caused by ion population effects in the trap (*12*).

Initial reports of PID combined with ion cyclotron resonance (ICR) mass spectrometry first demonstrated the usefulness of PID as an analytical tool. These early studies using photodissociation for structural studies of biological species in the ICR employed a wide range of wavelengths from infrared (IR) to UV (*13-15*). Williams et al. reported photodissociation efficiencies on the order of 100% for the peptide alamethicin using 193 nm light when the ions were confined to the beam path. Corresponding CID experiments involved pulsed-gas introduction of collision gas, which makes the MS/MS experiment much more complicated and produces dissociation efficiencies on the order of 15% which is significantly lower than for photodissociation (*14*). Comparison studies of photodissociation and surface-induced dissociation (SID) of porphyrins in the ICR by Castro et al. yielded higher dissociation efficiencies for the photodissociation process. Longer irradiance times (i.e., with no parent ion selection after the first MS/MS step), produced more fragment ions at higher abundances than those obtained with SID. At very long irradiance times (both UV and UV-Vis wavelengths), the new fragment ions produced were at the expense of diagnostically significant ions (*15*).

By combining the technique of electrospray ionization with that of PID for the structural elucidation of biological species, a unique and powerful tool can be developed to solve the more difficult problems faced by the analytical biochemist. Recently, infrared multiple photon dissociation (IRMPD) has been reported in the ICR for photodissociation of biological ions generated by electrospray ionization. Little et al. demonstrated the capability of IRMPD to obtain sequence information for peptides/proteins and oligonucleotides (*13*). The IRMPD of carbonic anhydrase produced fragment ions similar, but with valuable additions, to fragmentation information obtained by other methods (e.g., CID and SID). Optimization of irradiance times varied widely for peptides/proteins (from 50 to over 300 ms), indicating a greater range of ion stabilities than originally believed from CID data. Irradiance times for oligonucleotides (negative ion mode) were significantly less (e.g., 10-30 ms) than those of the peptide/protein experiments. This could be attributed to the photon resonance of the P—O stretching frequency. More importantly, IRMPD was shown to have greater selectivity, have less mass discrimination, and could dissociate much more stable ions than the corresponding CID process (*13*).

Here we report the development of a novel electrospray ion injection system combined with continuous wave (cw) and pulsed IR lasers for the analysis of peptides/proteins, carbohydrates, and oligonucleotides. Over the last eight years, ion injection into the quadrupole ion trap has become one of the most popular areas of research in mass spectrometry. Since the original report of ion injection using an external electron ionization (EI) ion source by Louris et al. (*16*), a myriad of different ion sources have been interfaced with the ion trap. These include but are not limited to EI/chemical ionization (CI) (*16-18*), fast atom bombardment (FAB) (*19*), particle beam (*20*), thermospray (*21*), electrospray (*22-26*), glow-discharge (*27-29*), atmospheric pressure (*30,31*), inductively coupled plasma (ICP) (*32*), laser desorption (*33-37*), and super critical fluid (*38*) sources. In all the aforementioned literature, every external ionization source has utilized some form of dc lens system for ion injection into the ion trap. Although a dc lens system may be the simplest to design and construct for an ion injection system, it is not necessarily the easiest and most efficient way to transfer ions from an external source to the ion trap analyzer. This is especially true for high pressure ionization sources such as electrospray or atmospheric pressure ionization where a large number of collisions between ionized species and gas-phase neutrals can occur, effectively scattering a large portion of the ion signal. Since dc lenses serve only to focus the ion beam and not to "recapture" scattered ions for ion injection, an alternative method which could recapture scattered ions and focus them into appropriate trajectories for ion injection would be desirable.

One technique which fulfills this requirement is the rf-only multipole. RF-only multipoles have been used extensively in both analytical and physical mass spectrometry (*39,40*). The ability of these devices to focus ions in a high pressure environment can be understood by examining the forces exerted on a given ion population as it moves through an rf-only device. As an ion is displaced from the center axis of the device, the restoring force acting upon that ion (in an rf-only multipole with 2n electrodes) is proportional to the $(n-1)^{th}$ power of that displacement, where n=2,3, and 4 for a quadrupole, hexapole, and octopole, respectively (*41*). Therefore, as ions are displaced from the center of the rf-only device due to collisions with neutral molecules, the restoring forces recapture the displaced ions and successfully transfer them from a high pressure region (e.g., electrospray ion source) to a lower pressure region (e.g., ion trap analyzer) with minimal scattering losses. Based on the above discussion, the octopole would be the logical choice for an ion injection device due to the greater restoring forces for the n=4 case.

Previously, the rf-only octopole has been used as an ion-molecule reaction cell (*42,43*), as a collision cell in tandem quadrupole instruments (*44*), as an ion injection device for triple quadrupole instruments (*45*), and as a method for determining ion-molecule reaction cross sections and energetics (i.e., translational energy dependence, product branching ratios, collision-cross sections) of ion-molecule reactions (*46-49*). Other rf-only devices have been designed and used (e.g., hexapoles and quadrupoles) for the aforementioned purposes and have been employed as ion transmission devices (*50-53*).

As mentioned previously, the main reason for the lack of analytical publications on PID is the low photoabsorption cross section (10^{-2} Å^2) observed for most organic ions (*1*). In order to overcome this inherent disadvantage, we have developed a modified ion trap ring electrode which facilitates an increase in the photoabsorption pathlength for the PID experiment. The modified optical arrangement, as originally described by White (*54*), increased the optical pathlength and thus the amount of photon absorption obtained in the IR region. The first application of this technique in mass spectrometry was for the ICR cell as described by Watson et al (*55*). The White-type cell was used for the study of resonance-enhanced two-laser infrared multiphoton dissociation of gaseous perfluoropropene cations, protonated diglyme cations, gallium hexafluoroacetylacetonate anions and allyl bromide cations (*55,56*). Note that ion traps such as these offer the opportunity for extended irradiation of trapped ions, helping to overcome the disadvantage of low photodissociation cross sections.

For the quadrupole ion trap, the White-type design was constructed by mounting three spherically symmetric concave mirrors in the radial plane of the ring electrode. The gain in PID efficiency obtained with this arrangement was directly related to the eight laser passes across the ring electrode which produced a concurrent increase in the photoabsorption pathlength. A detailed description of the modified ring electrode, theoretical considerations (using CO_2 lasers), and performance characteristics have been published previously (*57,58*). A brief summary of these results are presented in the experimental section of this chapter.

One major advantage of photodissociation (in the quadrupole ion trap) is that the kinetic energy of an ion does not have to be converted into the necessary internal energy to effect fragmentation. In addition, several tuning parameters, including helium buffer gas pressure, ion frequency, and ion population, can further complicate the single-frequency CID experiment even for an experienced user. Even when broadband excitation techniques are employed to minimize ion frequency and population effects, dissociation efficiencies observed are somewhat lower than the corresponding single-frequency experiments and can be significantly lower than photodissociation experiments when a sufficient photoabsorption cross-section exists for the ions of interest.

In figure 1 is shown the effect of ion population on the dissociation efficiency of protonated 12-crown-4 ether. Ionization time (shown on the x-axis) is directly proportional to the number of ions in the trap. For the case of the single-frequency CID experiment (where in this example, the tuning parameters were optimized for a low number of ions stored in the ion trap), a significant decrease in photodissociation efficiency was observed as the ionization time was increased and hence the ion population increased. Typically, an increase in ion population results in a shift of the fundamental frequency of ion motion to lower frequency, thus resulting in a decrease in dissociation efficiency for the given CID tune parameters (see figure 1) (*59*). In the case of broadband excitation where a whole range of frequencies are excited over a given time period, the dissociation efficiency is independent of ion population. However, a decrease is typically observed in dissociation efficiency compared to that of the optimum single-frequency results.

When the results of the two CID methods are compared to photodissociation data for the same protonated 12-crown-4 ether, the PID efficiency seen is independent of ion population. However, the overall dissociation efficiency is substantially higher (approximately 100% compared to 65% for the optimum single-frequency tune and 45% for the broadband excitation) than either CID technique (see figure 1). These results can be attributed to two factors: (1) the increased photoabsorption pathlength of the multipass ring electrode, which compensates for the reduced photoabsorption cross-sections of organic species; and (2) the elimination of collisions for transfer of translational energy to

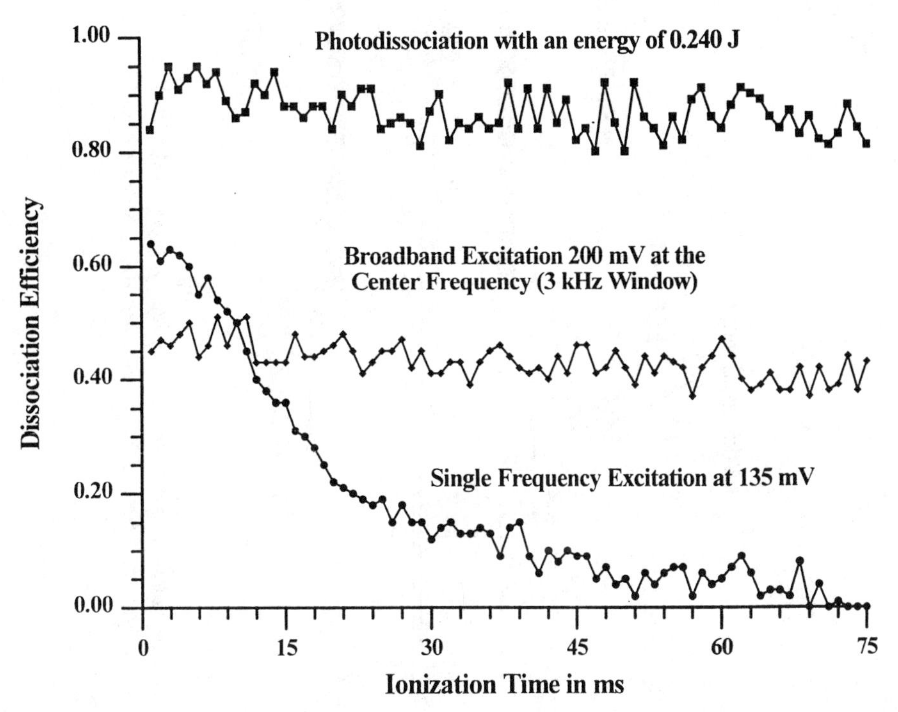

Figure 1. Evaluation of the various MS/MS techniques used with the quadrupole ion trap. The dissociation efficiency of photodissociation, single frequency CID, and broadband excitation are plotted versus ionization time for 12-crown-4 ether. The ionization time is directly proportional to the trapped ion population.

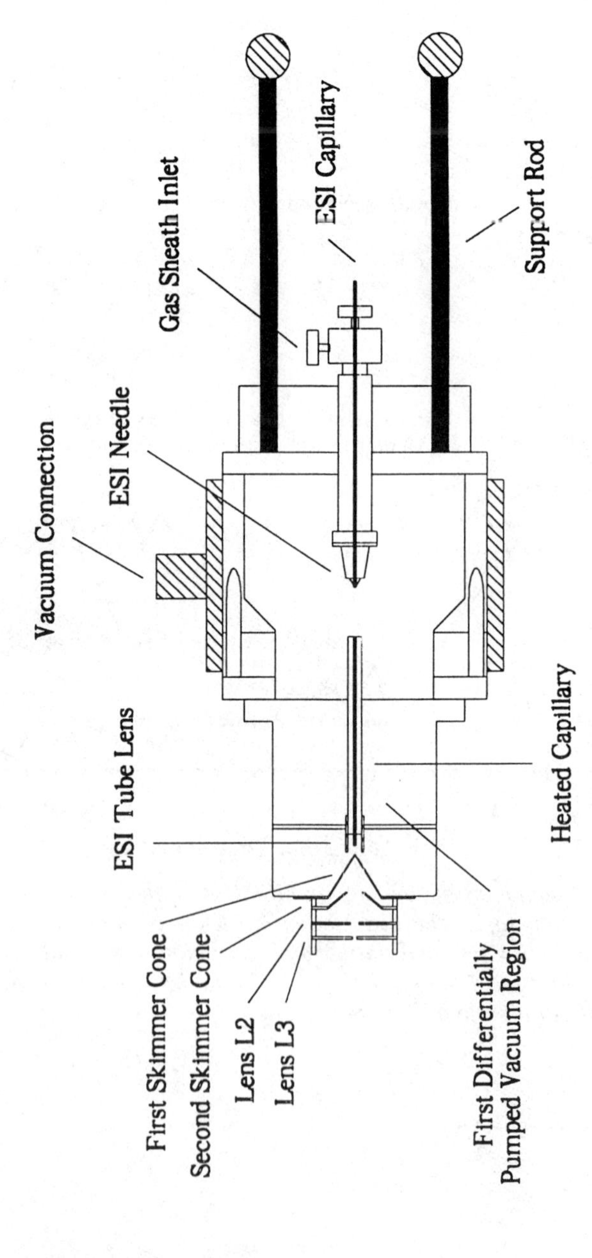

Figure 2. Analytica electrospray ionization source equipped with a heated capillary (as modified by Mark Hail at Finnigan MAT, San Jose, CA).

vibrational energy, where ion stability can cause competition between resonance ejection and CID during an MS/MS experiment.

Experimental

The electrospray ionization source used with the ESI/ion trap system was a standard Analytica (Analytica Inc., Branford, CT) source equipped with a stainless steel heated capillary (modification performed at Finnigan MAT) as seen in figure 2. The purpose of the heated capillary is to help droplet desolvation for the electrospray ionization process. The heated capillary was 0.308" in diameter by 4.580" long with an i.d. of 0.020". Temperature control of the heated capillary was by an Omega 6000 (Omega Engineering, Stamford, CT) series temperature controller with a 220 V output, previously used for controlling the Finnigan MAT ITMS manifold temperature. The 220 V output was stepped down via a transformer to 24 V (< 2 amp output current). Typical operating temperatures were in the 170° to 215 C range. For this temperature range, little if any thermal degradation or induced fragmentation of the $[M+nH]^{n+}$ (where n=1, 2, 3...) ion(s) was observed.

To facilitate coupling of the rf-only octopole to the ESI source, several design changes were made. First, to simplify the ion optics and obtain high ion transfer efficiencies, the second skimmer cone, lens L2 and lens L3 were removed from the electrospray head (see figure 3). An adaptor ring was then constructed from stainless steel to extend the only remaining skimmer cone 0.250" forward. The adaptor ring was mounted between the base plate of the electrospray head and the skimmer cone, employing o-ring seals between the skimmer cone and base plate assemblies. The alignment tool used to determine the exact distance from the heated capillary exit to the skimmer cone was then modified to compensate for the stainless steel adaptor ring to maintain a constant distance between the heated capillary and the skimmer cone of 3.5 to 4.0 mm. The electrospray needle was also moved forward to compensate for the adjusted heated capillary position. In figure 3 is shown the modified electrospray source described above.

The tube lens located at the end of the heated capillary (see figure 3) was used to gate ions into the rf-only octopole (via the skimmer cone). Control of the voltages applied to the tube/gate lens was accomplished by using the gate control circuit from the ITMS electronics. The circuit was modified so that both positive and negative ions could be gated efficiently. For the analysis of positive ions, a variable positive voltage is applied to the tube/gate lens (10-120 V) to focus the ion beam. To control the pulse width of the beam, a -180 V potential is applied to the tube/gate lens to stop ion transmission. For negative ions, a variable negative voltage (-10 to -100 V) is used for ion focusing and a +180 V signal is used to stop ion transmission (e.g., control the pulse width).

All electrical connections (capillary heaters/temperature sensors, capillary offset voltage) for the ESI source were made through a 6" Conflat flange (equipped with an Amphenol connector) located to the left of the ion source (see figure 5). High voltage for the electrospray needle and drying gas were controlled manually by an external power/gas distribution unit (Finnigan MAT, San Jose, CA). The electrospray source was pumped by two 500 L/min rotary pumps (model UNO 016B, Balzers Inc., Hudson, NH). A Harvard

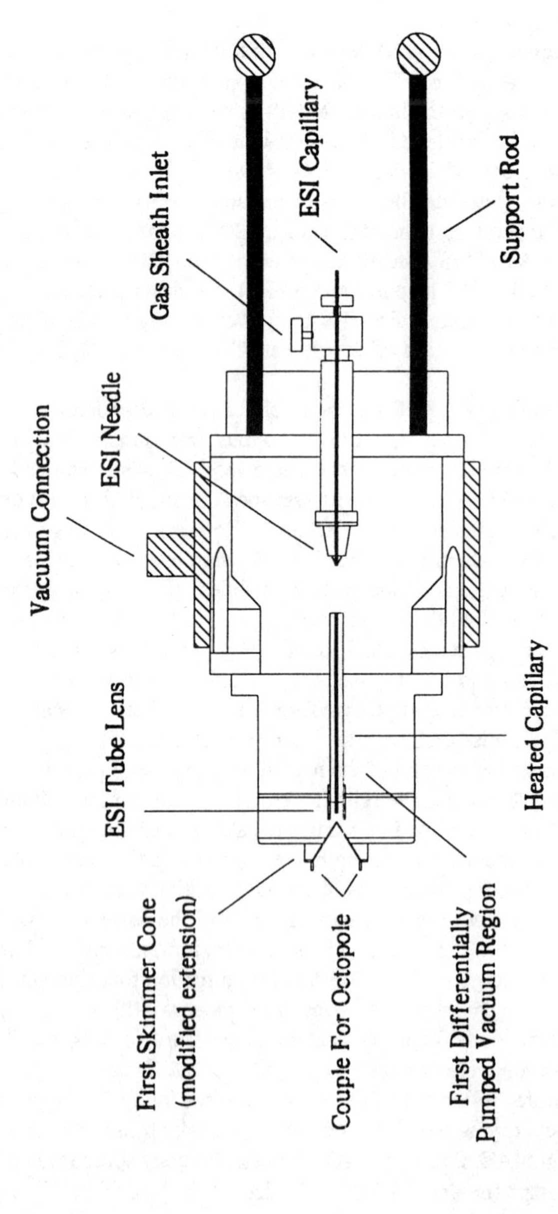

Figure 3. The modified Analytica electrospray ion source showing the elimination of the second skimmer cone, lens L2 and lens L3. Also shown is the stainless steel adaptor ring used to extend the first skimmer cone region 0.250" forward for coupling to the rf-only octopole.

Apparatus 22 syringe pump (South Natick, MA) was used for the direct infusion of samples into the electrospray source.

All flow rates were 3 μL/min unless otherwise stated. All proteins were sprayed (positive ion mode) in a 50:50 methanol/water solution containing 0.1% acetic acid. Carbohydrate samples (positive ion mode) were also sprayed in a 50:50 methanol/water solution with 5 mM ammonium acetate. Oligonucleotides samples (negative ion mode) were prepared in a 70:30 methanol/water solution.

The QITMS instrument employed in these studies was built and designed at the University of Florida. The system consists of a standard Finnigan MAT (San Jose, CA) ITMS ion trap analyzer mounted into a differentially pumped vacuum manifold (two 500 L/s turbo pumps). The electronics obtained from Finnigan MAT were modified to extend (and optimize) the mass range of the standard system (normally 650 u). Software used for integrated system control of the CO_2 laser and QITMS system (ICMS) was developed in our laboratory (*60*). The ion injection system utilized an rf-only octopole ion guide for the efficient transfer of the electrosprayed ions from the source region to the ion trap analyzer. This ion injection system is based on the design of Schwartz et al. (*61*). The major advantages of rf-only devices include the efficient transport of ions from a high pressure region (ESI source) to a region of lower pressure with minimal scattering losses, ease of use (e.g. no lenses to tune), and the ability to inject ions of different m/z ratios all at the same energy (*41*). The choice of an rf-only octopole over that of a corresponding hexapole or quadrupole can be explained by examining the effective trapping potential (equation 2) of any rf-only device:

$$V_{eff} = \frac{e^2 V_{0\text{-}P}}{4m\omega^2 r_0^2}\left(\frac{r}{r_0}\right)^{2n-2} \tag{2}$$

where $\mathbf{V_{eff}}$ is the effective trapping potential, **e** the charge on the ion, $\mathbf{V_{0\text{-}P}}$ the zero-to-peak rf voltage applied to the multipole, **m** the mass of the ion, **ω** the rf frequency applied to the multipole, $\mathbf{r_0}$ the inscribed radius of the multipole, **r** the displacement of an ion from the center axis of the multipole, and **2n** the number of poles present. In figure 4 is shown a plot of effective trapping potential versus ion displacement from the central axis of three different rf-only devices (quadrupole n=2, hexapole n=3, and octopole n=4). The octopole parameters used are seen in table 1. These parameters are the actual design specifications of the rf-only octopole developed in our laboratory. All parameters for the three multipole systems in table 1 are constant except for the value of n. The plot shown in figure 4 is for the +3 charge state of bovine insulin. The radial potential of the octopole has steep repulsive walls which approximate the ideal case of a square well. This means a large number of ions of differing m/z ratios can be transmitted to the ion trap at a constant translational energy (due to the large "flat bottom" portion of the well, where ion kinetic energies have only small perturbations), which yields higher transmission and trapping efficiencies. For the quadrupole and hexapole, the radial potential is more triangular in shape and, therefore, not as many ions can be transferred at constant energy, leading to a decrease in ion injection efficiency. This phenomenon can be explained by the dependence of the radial potential on the normalized ion displacement $(r/r_0)^{2n-2}$. The effective radial

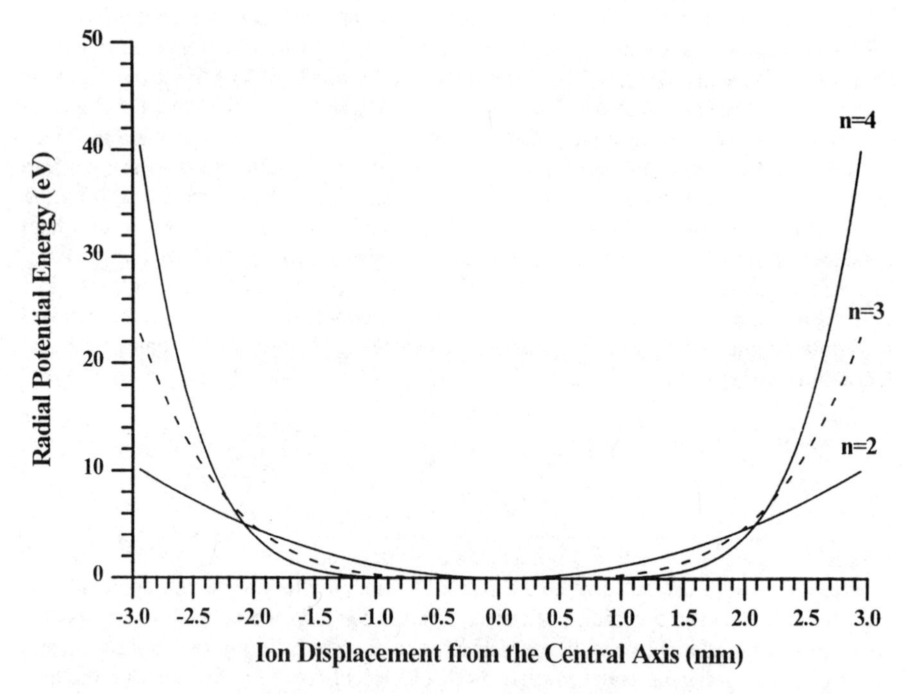

Figure 4. Graph of the trapping potential function (radial potential energy versus ion displacement) for a quadrupole (n=2), hexapole (n=3), and octopole (n=4). The curves represent the +3 charge state of bovine insulin (average molecular weight 5733 g/mol) where V_0 is 100 V, ω is 1.659 MHz, and r_0 is 2.94 mm for each multipole device.

potential in an octopole is proportional to r^6, which provides a large trapping volume for the ions of interest (*41,52*). In the case of the other extreme where n=2 for the quadrupole, the radial potential is proportional to r^2 with a maximum trapping energy one-fourth that of the octopole. Other advantages of the rf-only octopole include the ability to shield low energy ions from stray electric fields (or contact potentials) and reduced effects from space charging (*41,52*).

Table 1. Parameters for the radial potential energy equation.

Device	*n*	*e*	*ω (MHz)*	*r_0 (mm)*	*V_0 (V)*	*m (g/mol)*
Quadrupole	2	+3	1.659	2.94	100	5733
Hexapole	3	+3	1.659	2.94	100	5733
Octopole	4	+3	1.659	2.94	100	5733

For our instrument the inscribed radius, r_0, is 2.94 mm, with a corresponding rod radius (a_0) of 1 mm. This gives an a_0/r_0 ratio of 0.34 which closely approximates the ideal case for an octopolar field generated by round rods. The octopole is 30 cm long, and is operated at a frequency of 1.659 MHz and a maximum voltage output of 2000 $V_{0\text{-}P}$. Electronics used to control the rf power were from a Finnigan MAT 4500 single quadrupole mass spectrometer.

A diagram of the complete ESI ion injection system can be seen in figure 5 (which shows the complete instrument configuration). Ions first were passed through the heated capillary region to an ion gate located at the exit of the heated capillary. This gate was operated with a variable positive voltage to pass positive ions, and a variable negative voltage to pass negative ions. Ion gating was turned off by switching the voltage to ± 180 V (opposite polarity to the ions of interest). The ions were then passed through the grounded skimmer cone attached to the ESI source. The end of the octopole rods were placed within 0.020 inches of the skimmer cone. The ions then traveled through the differentially pumped regions and into the ion trap analyzer where the opposite end of the octopole was placed 0.050 inches away from the modified entrance endcap. By placing the octopole very close to both the ESI skimmer cone and the ion trap analyzer, the effects of ion entry/exit angle dependence and ion displacement from the center axis upon ion injection were minimized. Ions were then trapped by the rf field applied to the ring electrode; mass isolation was accomplished by a series of forward and reverse scans, followed by CID (single frequency) or PID experiments. Ions were then ejected via a mass-selective instability scan with resonance ejection, passing through a tube lens on their way to a 20 kV off-axis-dynode detector. Ion injection times range from 1 to 10 ms for peptide samples (approximately 300-900 fmol/µL), 10 to 20 ms for DNA samples (approximately 5-10 pmol/µL), and 50 ms for carbohydrate samples (approximately 20 pmol/µL).

For the PID experiments (using either a pulsed or continuous wave CO_2 laser), the IR laser beam entered through the ZnSe window (figure 5) and was attenuated by a 0.3 cm (1/8") entrance aperture on the ring electrode of the ion trap. The ring electrode was

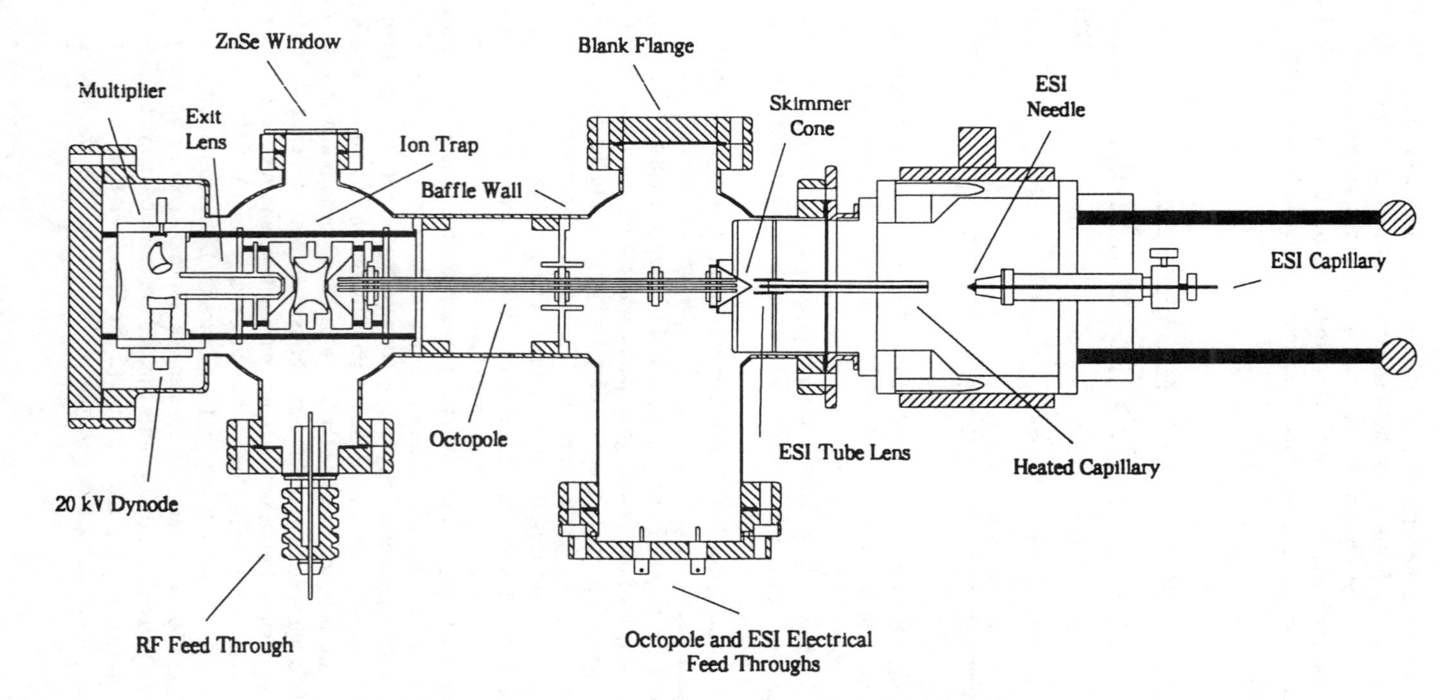

Figure 5. A simplified schematic diagram of the ESI/ion trap system including the electrospray source, octopole assembly, ion trap analyzer and detector assembly.

modified by incorporation of three polished stainless steel spherical concave mirrors (radius of curvature = 2.0 cm) mounted on the inner surface of the ring, as shown in figure 6. The relative photon density (assuming constant intensity across the attenuated beam width) observed in the radial plane of the ring electrode can be seen in figure 6. The most critical adjustment of the mirror system was the separation of the centers of curvature of the mirrors labeled A and B. This separation distance determines the number of beam transversals across the ring electrode: 4,8,12, or any other multiple of 4. For the multipass ring electrode described here, the separation distance between mirrors A and B was set for eight laser passes across the radial plane of the ring electrode. The mirrors were mounted on the ring electrode such that the centers of curvature of mirrors A and B were on the front surface of mirror C, and the center of curvature of mirror C was halfway between mirrors A and B. This method of mirror alignment establishes a system of conjugate foci on the reflecting surfaces of mirrors A,B, and C. Consequently, light leaving the surface of mirror A is focused by mirror C on the surface of mirror B, and the light leaving mirror B is then focused back to the original point on mirror A. Similarly, any light leaving mirror C and going to either mirror A or B is focused back to mirror C at some point offset from the original one (see figure 6) (*54*).

This technique for extending the optical pathlength in restricted volumes has many advantages over previous designs which incorporate flat mirror systems or a spherical mirror and a truncated prism scheme (*54*). One advantage is the ease in making adjustments since all tolerances with the exception of the horizontal angles of mirrors A and B are usually quite large. Another advantage is that light losses on mirror surfaces are kept to a minimum. Since there are only two reflections (at normal incidence), spots, dust or pinholes on mirrors A and B have a much less serious effect since the light from any point on the mirror surface always goes back to the same point; therefore, if there is a spot on the mirror surface, the light falling on the spot is lost but on the second reflection from that mirror no more light is lost. Yet another advantage is that there is only one transmission of light through an entrance aperture; no light travels through any glass or other optical material where losses due to reflection can occur (*54*).

Each mirror (and its mounting bracket) was constructed from a single piece of stainless steel. At one end of each piece, the radius of curvature was cut (r = 2.0 cm) and the surface was highly polished. The three mirrors were mounted into precision-drilled holes in the ring electrode, positioned such that the alignment was automatic; no realignment has been needed since the original assembly of the multipass ring electrode (36 months). A small machine screw was used to hold each mirror-mirror mount assembly in place on the ring electrode. Laser alignment was set such that the center portion of the 1-cm Gaussian beam profile was transmitted through the entrance aperture, thus yielding high photodissociation efficiency. With efficiencies already greater than those published for previous designs on a ion trap or ICR cell, condensing the beam down to 0.3 cm was considered less important than examining the numerous ways in which gas-phase ion chemistry can be studied via IRMPD with this unique design. Furthermore, with the beam focused to 0.3 cm, damage could possibly occur on the surface of the ion trap mirror when the laser is tuned to a strong IR emission line (e.g., 10.59 μm). With continued use over a 36-month period, no degradation of the ion trap mirror surfaces has been observed with the unfocused laser beam (*54,57,58*).

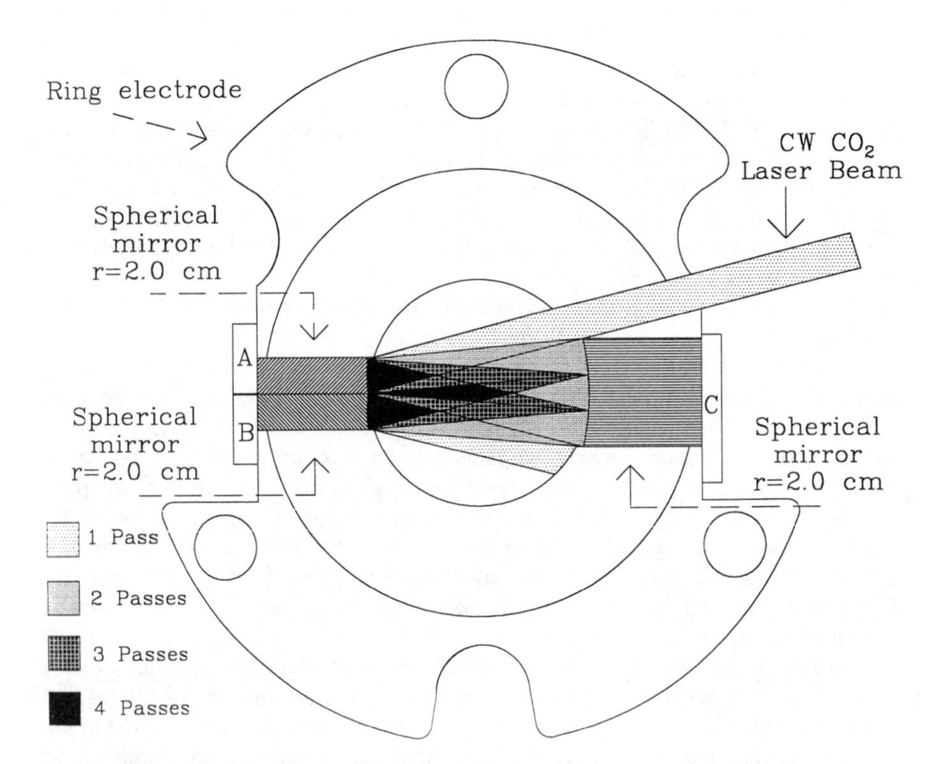

Figure 6. Modified ring electrode for multipass IRMPD experiments. Mirror positions and the eight laser passes across the radial plane of the ring electrode, along with approximate photon density in the radial plane of the ring electrode, are shown. The positions of mirrors A and B determine the number of laser transversals across the radial plane of the ring electrode.

Two different CO_2 lasers were employed for the photodissociation experiments described in this chapter. The first was a continuous wave CO_2 laser (Apollo Model 570) line-tunable over a wavelength range of 1099-924 cm^{-1} with a beam size of approximately 1.0 cm. The maximum power obtainable was 50 W at 944 cm^{-1}. The second laser was a pulsed CO_2 laser (Lumonics Series TE-860-4 Excimer, Ottawa, Ontario, Canada) capable of a 3 J pulse (at 10.60 μm) with an approximate 3 ns pulse width. The maximum repetition rate for full power operation of the laser was 20 Hz. Each laser required a different optical set-up depending on the type of photodissociation experiment employed. For optics mounted on the instrument table, two compensators were used to damp vibrations generated by the two 500 L/s turbomolecular pumps.

The vacuum-air interface for both the pulsed and cw lasers consisted of a 2.75" flange modified to accept a 1.5" ZnSe window (Melles Griot, Irvine, CA). The vacuum seal was made by two teflon rings placed on either side of the ZnSe window.

The photodissociation set-up for the cw laser can be seen in figure 7. A 13" x 20" optical table was constructed and used for mounting the various ion optics. The 1 cm diameter unfocused beam was passed through a beam selector (designed to pass IR radiation) and reflected at a 90° angle off a gold plated mirror. The beam then entered the mass spectrometer through the ZnSe window. A helium-neon (632.8 nm) laser (model 05 LLR 851, 5 mW power output, from Melles Griot, Irvine, CA) was used to expedite the alignment of the cw CO_2 laser. The helium-neon laser was placed on top of the instrument table and aimed directly at the 1" beam selector (Melles Griot, Irvine, CA) which reflected the 632.8 nm light and passed IR (9.0 to 11.0 μm) radiation. This greatly simplifies the alignment process since the helium-neon laser does not need to be placed in line with the cw CO_2 laser for beam alignment.

For the case of the pulsed laser (as seen in figure 8), the beam selector was not used since the beam shape from the laser (approximately a 1" x 1.25" square beam) exceeded the dimensions of the beam selector. As mentioned, the beam was first reflected at a 90° angle off the surface of a gold plated mirror. To focus the beam down to the appropriate size to pass through the entrance aperture of the multipass ring electrode, a convex focusing lens with a focal length of 4" was placed in line with the beam just in front of the ZnSe window (see figure 8). This lens focused the beam down to a 0.2 cm spot size. A coarse alignment was obtained by placing the helium-neon laser in line with the laser beam. Fine adjustment of the alignment was performed by placing burn-paper at the entrance aperture of the multipass ring electrode to determine spot size and position. The major disadvantage of using the pulsed laser is in the alignment procedure, since the instrument must be vented to ensure proper beam position.

Results and Discussion

Peptides and Proteins - CID. Collision-induced dissociation has seen a great deal of development for biological applications in mass spectrometry (*62-74*). Applications have ranged from determination of side chain fragmentations of leucine and isoleucine for isomer differentiation to determination of glycopeptide structures using LC/MS/MS in triple quadrupole instruments (*71-73*). Ion trap mass spectrometry (employing CID) has also

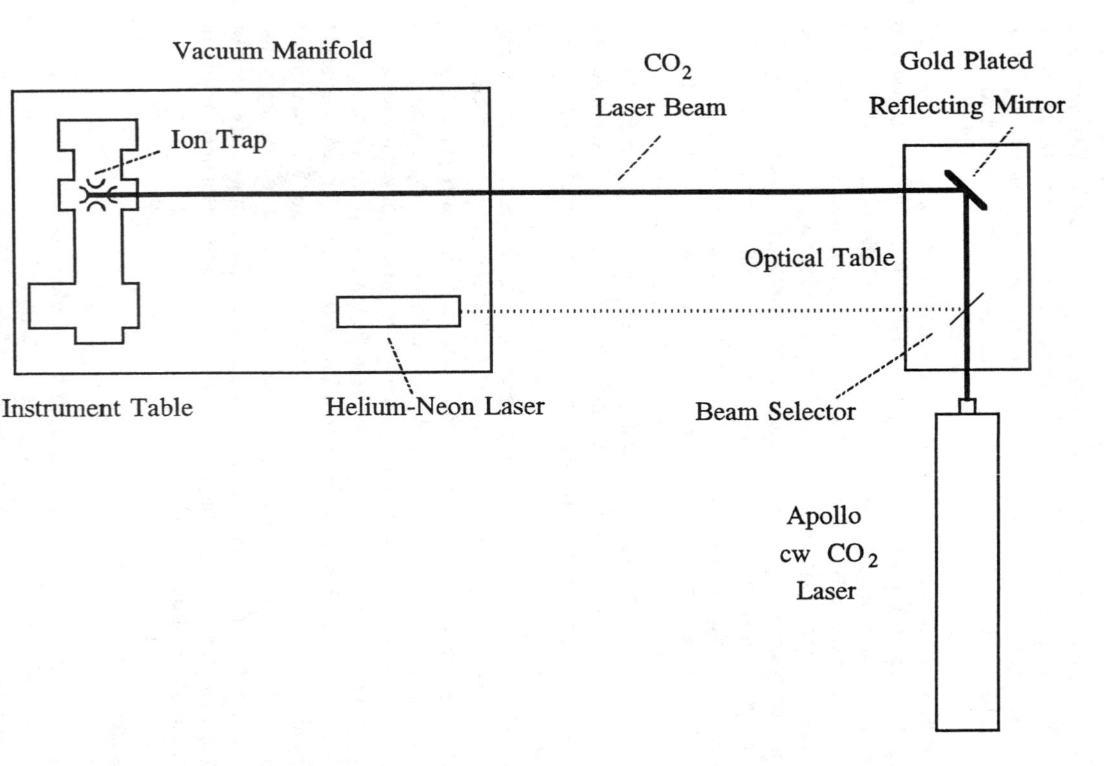

Figure 7. Photodissociation set-up for the continuous wave-laser. The beam selector allows 10.6 μm radiation to pass directly through to the gold plated reflecting mirror, while the output of the Helium-Neon laser (632.8) is reflected off the front surface of the beam selector.

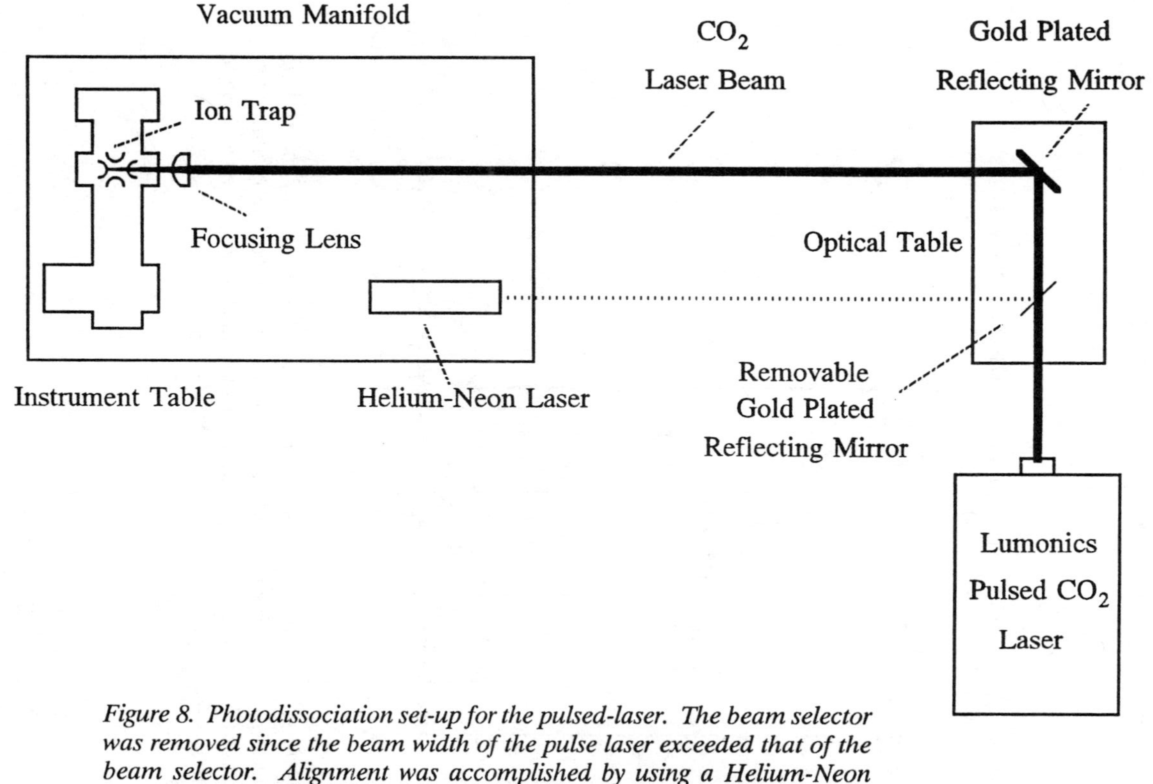

Figure 8. Photodissociation set-up for the pulsed-laser. The beam selector was removed since the beam width of the pulse laser exceeded that of the beam selector. Alignment was accomplished by using a Helium-Neon laser and an additional gold plated turning mirror. A focusing lens was added to make the beam diameter 0.2 cm as it entered the multipass ring electrode.

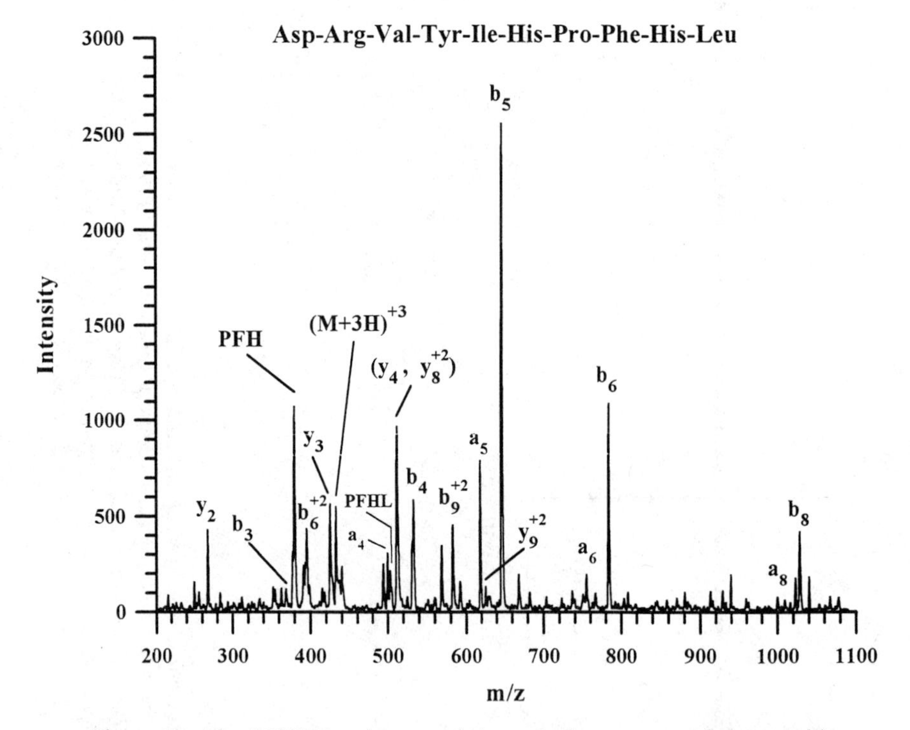

Figure 9. The MS/MS spectrum of the +3 charge state of the peptide angiotensin I. Observed fragmentation includes both doubly and singly charged b and y ions, singly charged a ions, and some sequence specific internal fragment ions (due to the presence of proline).

seen rapid development in the analysis of biological materials, particularly peptides and proteins (*70*).

To evaluate the MS^n capabilities of the ESI/ion trap instrument (via CID), a series of tandem mass spectrometry experiments was conducted on the peptide human angiotensin I, a decapeptide (DRVYIHPFHL), which is a pressor substance. Human Angiotensin I is converted to its active form (angiotensin II) by cleavage of the HL residues from the carboxy terminus. The human angiotensin I was obtained from ICN Pharmaceuticals (Costa Mesa, CA), with 700 fmol/µL standards dissolved in a 50:50 methanol:water solution with 0.1% acetic acid. The ESI spectrum of angiotensin I exhibits an intense, triply-charged ion at m/z 433, which was mass-isolated using the forward/reverse scan isolation technique (*75*)

Table 2. Instrumental parameters for the MS/MS of the triply-charged m/z 433 ion of human angiotensin I.

Parameter	*value*
q_z	0.3
MS/MS Single Frequency	118 kHz
Resonance Excitation Amplitude	1.5 V_{P-P}
Resonance Excitation Time	20 ms
He Buffer Gas Pressure	1.0×10^{-4} torr (uncorr.)

Figure 9 shows the CID spectrum of the triply-charged m/z 433 ion of angiotensin I. The MS/MS parameters for the CID of the triply-charged m/z 433 ion are shown in table 2. The two major types of fragment ions observed were b ions (charge retention on the carboxy terminus) and y ions (charge retention of the amino terminus). This nomenclature follows that originally developed by Reopstorff and Fohlman (*74*). The primary mechanisms of formation for these ions is shown in figure 10. In general, formation of either b or y type ions involves cleavage of the amide bond along the peptide backbone (see figure 10). The proton attached to the amide nitrogen is typically free to move along the peptide chain. Formation of b type ions occurs by β-cleavage and charge migration to the carbonyl end of the amino terminus fragment. The b ion in figure 10 is termed b_2 because it contains two amino acid residues from the original peptide structure (counting from the amino terminus). The complementary reaction (which involves charge retention on the amide nitrogen) for the formation of y type ions is characterized by a hydrogen atom migration, typically from the α-carbon. The y ion in figure 10 is termed y_3 because it contains three amino acid residues from the original peptide structure (counting from the carboxy terminus) (*76-78*).

The interpretation of the MS/MS spectrum in figure 9 can be explained by the charge localizations associated with the triply-charged parent ion at m/z 433. Much of the fragmentation observed for angiotensin I follows that of a low-energy CID spectrum from

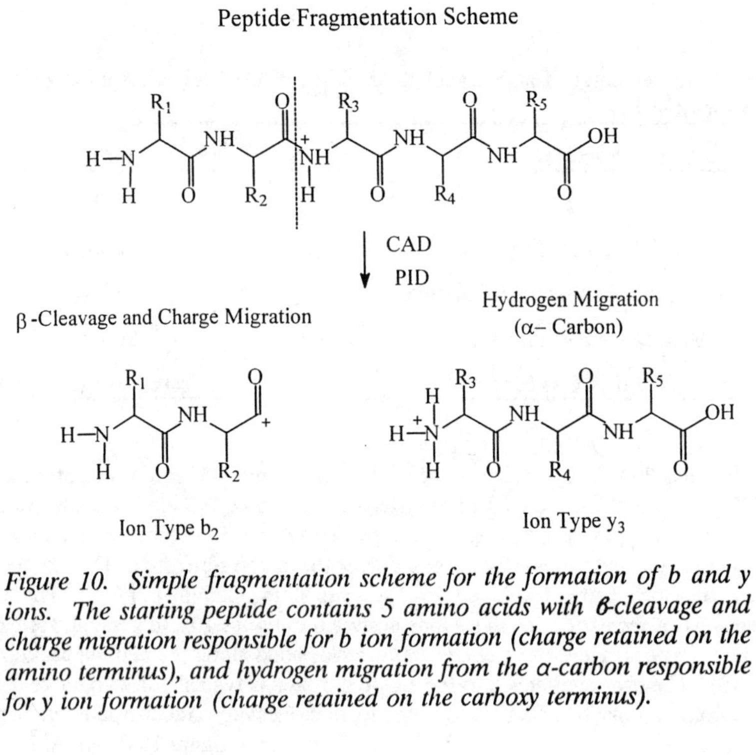

Peptide Fragmentation Scheme

R1 O R3 O R5
H—N NH +NH NH NH OH
H O R2 H O R4 O

CAD
PID

β-Cleavage and Charge Migration

Hydrogen Migration
(α– Carbon)

R1 O
H—N NH +
H O R2

Ion Type b_2

R3 O R5
H
H—+N NH NH OH
H O R4 O

Ion Type y_3

Figure 10. Simple fragmentation scheme for the formation of b and y ions. The starting peptide contains 5 amino acids with β-cleavage and charge migration responsible for b ion formation (charge retained on the amino terminus), and hydrogen migration from the α-carbon responsible for y ion formation (charge retained on the carboxy terminus).

a triple quadrupole mass spectrometer as reported by Boyd and co-workers (*79*). The fragmentation indicates that protons are located on Arg^2 and His^9, with the third proton free to move along the peptide backbone between these two residues. This leads to the large amount of acylium ion type mid-range fragmentation observed (b_4, b_5, b_6, and b_8). The corresponding series of a ions (a_4, a_5, a_6, and a_8) arise from the loss of neutral carbon monoxide from the b series of ions. Some doubly-charged ions are also observed (b_6^{+2}, b_9^{+2} and y_9^{+2}), associated with basic residues His^6, His^9 and Arg^2, respectively. A complementary series of y ions also appear (y_2, y_3, y_4) in the spectrum with an unusually high intensity observed for the y_3 ion (proline effect) (*79*). Attempts of MS^3 experiments on the y_3 ion to verify its structure produced no observable fragmentation due to the stability of the ion in the gas phase. In addition, no a_7-b_7 pair of ions is observed due to the proline effect (*79*).

The peaks labeled PFH and PFHL in figure 9 are internal sequence y type ions, driven by internal cleavage of the Pro^7 residue. These peaks arise due to the high basicity of proline in the gas-phase. Of all the basic residues, proline has the highest proton affinity, even exceeding those of arginine and lysine. The mechanism of formation for these ions follows the same convention as y type ions with charge retention on the amide nitrogen and subsequent hydrogen ion migration to the highly basic proline residue (see figure 11). Internal sequence type ions have been observed extensively with any peptides that contain proline residues in ion trap CID spectra. The high ion abundance of these peaks is probably due to the timescale of the ion trap MS/MS experiment, which takes place on the millisecond as opposed to the microsecond timescale found in either magnetic sector or triple quadrupole instruments. Over the millisecond timeframe, ions in the trap will tend to go towards their lowest energy state, if given the opportunity.

MS^3 experiments were performed on both the b_5 and b_9^{+2} ions from the MS/MS spectrum. In figure 12 is shown the MS^3 spectrum obtained from the b_5 ion of angiotensin I. The b_5 ion was mass isolated using forward and reverse scans and fragmented at a q_z of 0.3 (amplitude=2.0 $V_{0\text{-}P}$). Typically, singly-charged b ions fragment easily to produce lower mass b ions and some a ions as seen in figure 12. There is a great deal of interest in generating a large intensity of high-mass b ions, to evaluate MS^n techniques for sequencing peptides one residue at a time. From figure 12 the formation of the a_5, b_4, and a_4 verifies the loss of isoleucine from the b_5 fragment.

MS^3 data from the fragmentation of the doubly-charged b_9^{+2} ion produced a series of singly-charged higher m/z b ions, as shown in figure 13. It is interesting to note little or no complementary singly-charged y ions were observed. However, there was a series of four unknown peaks (labeled with question marks in figure 13), which may be possible cyclization rearrangements driven by the presence of the Pro^7 residue and the long reaction times associated with ion trap experiments. Further MS^3 experiments utilizing labeled standards would be necessary to verify these ion fragmentations.

Attempts using MS^3 to determine the possibility of a y_8^{+2} in the y_4 peak (both at m/z) in figure 9 yielded no interpretable data which could either confirm or deny the presence of this ion. Part of the reason (as mentioned earlier) is the high stability of the y ion series in general.

Human Angiotensin I - PID. A continuous wave (cw) photodissociation set-up was employed as described in the experimental section. The m/z 433 ion (triply-charged) of

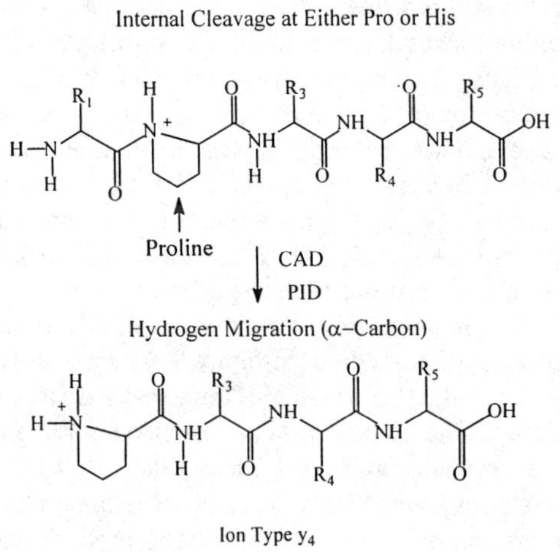

Internal Cleavage at Either Pro or His
Proline
CAD
PID
Hydrogen Migration (α–Carbon)
Ion Type y4

Figure 11. Mechanism for formation of internal y type ions from peptides which have proline within the peptide backbone (e.g. not located on a terminus).

Asp-Arg-Val-Tyr-Ile (b5 fragment)
Intensity
m/z
a4
b4
a5
b5

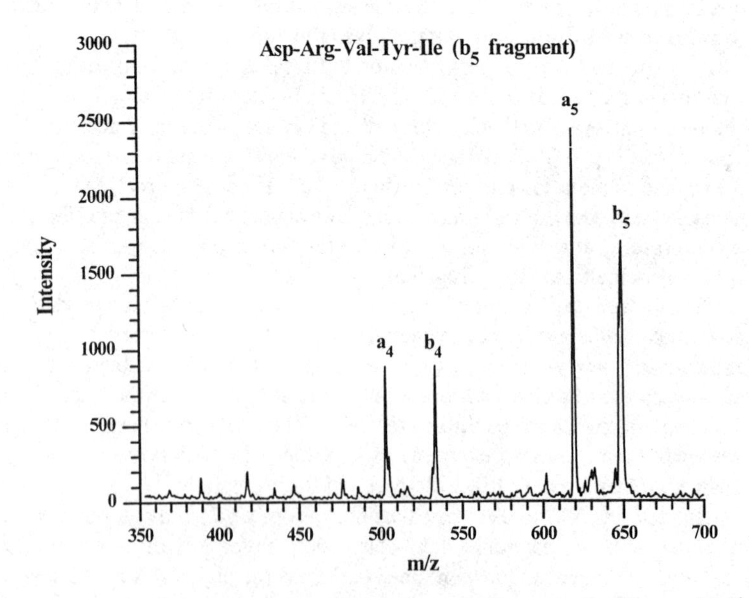

Figure 12. MS^3 *spectrum of the* b_5 *ion from angiotensin I. The predominant ions in the spectrum are the next lowest b ion in the series* (b_4) *and the corresponding a ions which arise from the loss of carbon monoxide from the acylium portion of the b ion.*

angiotensin I was mass-selected in the ion trap (using forward/reverse scans) as described previously. The helium buffer gas pressure was initially $1.0x10^{-4}$ torr (uncorrected). For this buffer gas pressure, no fragmentation was observed with the continuous wave CO_2 laser (for irradiation times up to 400 ms). This indicated that the large number of collisions of the triply-charged parent ion with helium buffer gas sufficiently deactivated (via collisional cooling) the parent ion, preventing the multiple photon absorption process from reaching the dissociation threshold energy. When the buffer gas pressure was reduced to $2.0x10^{-5}$ torr, the appearance of a photodissociation spectrum (85 ms irradiance time) was observed (see figure 14).

The fragmentation observed included the mid-range b fragments indicative of the +3 charge state (b_5 and b_6), the internal y ion series generated by the presence of Pro^7 (PFH and PFHL fragments), and the y_4-y_8^{+2} peak(s). These fragment ions compare well with what was generated using single-frequency CID experiments discussed above (see figure 9). One major difference observed between PID and CID was the presence of a b_9 ion in the PID spectrum. The singly-charged b_9 ion is of particular importance since it could be mass-isolated and a series of MS^n studies performed, where individual amino acids from the peptide chain could be sequenced in succession by either further PID or CID experiments. Generating high mass b ions is of particular importance for accomplishing a "ladder sequencing" experiment (removal of one consecutive amino acid from the peptide chain for each dissociation step) in order to verify peptide sequences in mass spectrometry and Edman degradation.

One negative aspect of this experiment is the reduced pressure of the helium buffer gas required for the appearance of the photodissociation spectrum. Improved fragmentation efficiencies could be observed for the IRMPD process if helium buffer gas pressures during the laser irradiation period were reduced to below $\approx 4.0x10^{-6}$ torr; however, significant reduction of pressure in the ion trap analyzer not only adversely affects peak shape and resolution, but more importantly reduces ion injection efficiency and thus sensitivity. For the case of the angiotensin I studies, ion injection times were increased from the 1 to 3 ms range (when operating with a helium buffer gas pressure of $1.0x10^{-4}$ torr) to the 68 to 75 ms range with the reduced operating pressures (e.g., $2x10^{-5}$ torr). The use of a pulsed-valve to separate the ion injection, photodissociation, and detection events in time from the helium pulse should significantly improve photodissociation efficiency without sacrificing sensitivity (*57*).

Gramicidin D - PID. The second peptide investigated using IRMPD was the antibiotic gramicidin D. Photodissociation was first observed for the singly-charged ion of gramicidin D with a helium buffer gas pressure of $1.7x10^{-5}$ torr and an irradiance time of 85 ms. Ion injection conditions were set such that the majority of the signal observed was that of the singly-charged species, with only a small portion of the doubly-charged ion stored for instrument calibration purposes. In figure 15 is shown the corresponding photodissociation spectrum.

The major fragment ion observed was a rearrangement loss of the modified carboxy terminus of the peptide. The $[M-61]^+$ ion at m/z is formed by hydrogen migration from somewhere along the peptide backbone (preferentially on the α carbon relative to the tryptophan residue in position 15), and subsequent cleavage of the amide bond between the

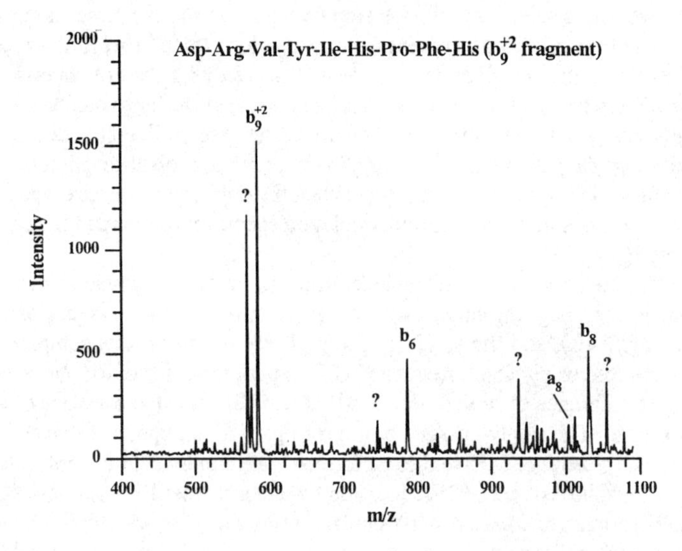

Figure 13. MS^3 spectrum of the b_9^{+2} ion from angiotensin I. The predominant ions observed are the higher m/z ions of b_6 and b_8. The unknown peaks may correspond to cyclization for the peptide backbone structure, however without proper labeling experiments this cannot be confirmed.

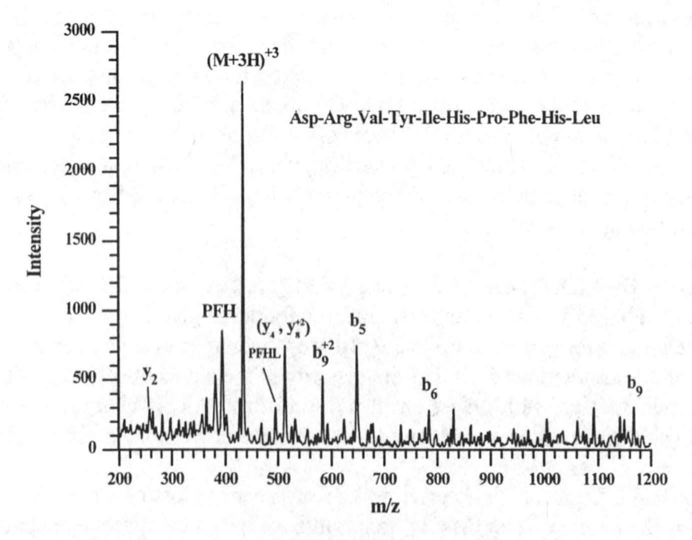

Figure 14. IRMPD of the triply charged m/z 433 ion of human angiotensin I. The peaks at PFH and PFHL represent internal y sequence ions generated by the presence of proline at residue 7.

modified carboxy terminus and the tryptophan residue in position 15. The two peaks to the immediate right of the $[M+H]^+$ ion are the sodium and potassium adducts of gramicidin D. Photodissociation of these species yielded no loss of the modified carboxy terminus. The appearance of the $[M-61]^+$ ion as the major fragment peak in the spectrum indicates that protonation occurs preferentially on the modified carboxy terminus of the peptide (e.g., the most basic site) where charge-site-driven fragmentation results. The only other structurally relevant peak observed was that of the b_{14} ion. As seen with the previous IRMPD experiment, a large sequence-specific b ion is produced which could be used to obtain ladder sequence information to verify peptide sequences.

Carbohydrates and Oligosaccharides. One of the biggest challenges in biochemistry is the identification and sequence analysis of complex carbohydrates and oligosaccharides. Perhaps the most difficult problem in carbohydrate chemistry is the investigation of biologically active binding-recognition systems. The biologically active portion of a carbohydrate or oligosaccharide can also be linked to a variety of other biochemically important compounds to form glycoproteins, peptidoglycans, lipopolysaccharides, and glycosphingolipids. To determine the structure of the oligosaccharide, or oligosaccharide portion of a complex biomolecule, several steps are necessary to establish structure: (1) determination of the reducing and nonreducing ends of branched or straight chain oligosaccharides, (2) determination of monosaccharide ratios and ring sizes, (3) determination of the internal sequence of individual monosaccharides, (4) determination of branching points (if any), (5) anomeric data relating to linkage type and bonding, and (6) presence of modified sugars possibly containing acyl groups, phosphates, sulfates, pyruvates, cyclic acetals, or taurine moieties. To obtain the aforementioned information, a myriad of techniques are employed, including: wet chemical, enzymatic, antibody or lectin affinity, chromatography (thin layer, gas, liquid, paper), nuclear magnetic resonance, circular dichroism, and mass spectrometry (*76-80*).

The major contribution of mass spectrometry to carbohydrate analysis has been high sensitivity (nanomole to femtomole) studies which have included selective detection of glycopeptides in protein digests, multi-residue confirmation of aminoglycoside antibiotics, sequencing of cationized carbohydrate antibiotics, identification and linkage position determination of reducing ends in oligosaccharides, and the structural analysis of steroidal oligoglycosides (*70,85-88*). Collision-induced dissociation studies needed to help determine oligosaccharide sequence information have also been successfully performed using FAB and electrospray ionization techniques for generation of abundant parent ion populations (*89-91*).

In this section, an alternative approach to traditional CID for determining carbohydrate structure is presented. Photodissociation in the IR region is a natural choice for cleaving glycosidic bonds, identifying ring structure, or possibly determining "post translational" modifications to monosaccharide residues. This is due mainly to the high photon absorption cross section seen for C—O—C linkages. Although the appropriate derivatization chemistry and enzymatic data are needed to help determine internal oligosaccharide sequence information, the use of photodissociation in combination with electrospray ionization can provide a basis for tackling the more difficult problem of anomeric confirmation in the gas-phase.

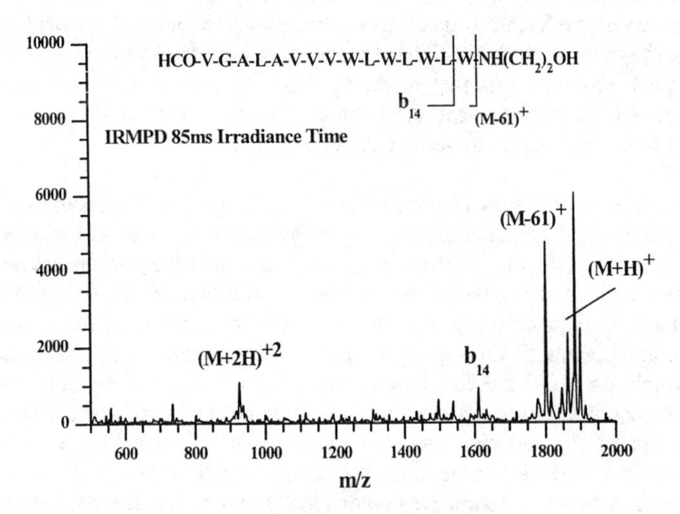

Figure 15. IRMPD of the singly charged $[M+H]^+$ ion of gramicidin D. The $[M+2H]^+$ ion is present for mass calibration purposes. The two peaks just above the $[M+H]^+$ ion represent the sodium and potassium adducts of gramicidin D respectively. The peak at $[M-61]^+$ represents cleavage of the modified carboxy terminus with subsequent hydrogen ion migration.

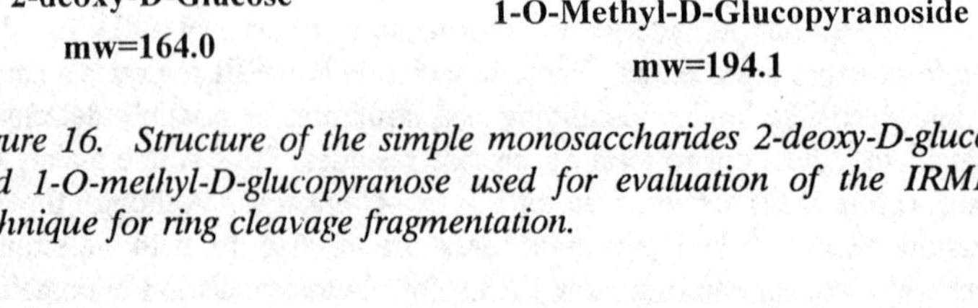

2-deoxy-D-Glucose

mw=164.0

1-O-Methyl-D-Glucopyranoside

mw=194.1

Figure 16. Structure of the simple monosaccharides 2-deoxy-D-glucose and 1-O-methyl-D-glucopyranose used for evaluation of the IRMPD technique for ring cleavage fragmentation.

The first portion of this section discusses the determination of the number of hydroxy substituents on monosaccharide residues. The initial studies performed were done using a solids-probe and ammonia chemical ionization (CI), in order to test the feasibility of the photodissociation technique. Original data comparisons with CID showed increased dissociation efficiencies and reduced analysis times for IR photodissociation experiments. The second portion of this section covers the use of photodissociation for cleaving the glycosidic bonds of straight-chain oligosaccharides (e.g., ammonium adducts of raffinose and stachyose) using the electrospray ion trap instrument.

Monosaccharide Cleavage - CID/PID. The first step in obtaining relevant carbohydrate sequence data employing photodissociation is understanding the ring fragmentations of simple monosaccharides subunits. The goal was to determine the number of ring substituents on an individual monosaccharide. In this section the ring dissociations of 2-deoxy-D-glucose and 1-O-methyl-D-glucopyranoside (structures shown in figure 16) are discussed and served as a feasibility study for more advanced applications (e.g., glycosidic bond cleavages). The samples were introduced via a solids probe into a standard Finnigan MAT ITMS set up for photodissociation experiments. Sample ionization was accomplished by ammonia CI with reaction times on the order of 65 to 90 ms and ammonia pressures of $\approx 5.5 \times 10^{-6}$ torr. For CID experiments, the helium buffer gas pressure was 1.0×10 torr (uncorrected). To maximize photodissociation efficiency, the ion trap was operated without helium buffer gas for these studies.

In figure 17 (top) is shown the CID mass spectrum of the ammonium adduct of 2-deoxy-D-glucose (m/z 182). The dissociation efficiency of the ammonium adduct ion was 90%. The major peak produced at m/z 164 involves the loss of neutral H_2O from the 2-deoxy-D-glucose ring. The peak at m/z 147 also showed the loss of neutral H_2O with an additional loss of ammonia. Low intensity ions at m/z 146 $[M+NH_4-2H_2O]^+$ and m/z 129 $[M-2H_2O-NH_3]^+$ indicated the loss of a second neutral H_2O from the monosaccharide ring. To determine the presence of other ring substitutions (e.g., more —OH or other functional groups) a third sequential step of MS was performed on the m/z 164 ion $[M+NH_4-H_2O]^+$.

The MS^3 spectrum of the ammonium adduct of 2-deoxy-D-glucose is also shown in figure 17 (bottom). Due to the stability of the m/z 164 ion, dissociation efficiencies were only ≈13%. Diagnostic ions indicating the loss of one and two neutral H_2O molecules from the ring at m/z 147, 146, and 127 were present in the MS^3 spectrum as well as the MS^2 spectrum. An additional ion at m/z 111 $[M+NH_4-3H_2O-NH_3]^+$ signified the presence of a third hydroxy substituent on the monosaccharide ring. There was no indication of cleavage of the hydroxyl group attached to position 6 of the pyranose ring, nor was there any evidence for the loss of methanol (e.g., cleavage of the C5—C6 bond) from the pyranose ring.

In the IRMPD spectrum of the ammonium adduct of 2-deoxy-D-glucose (figure 18), the m/z 164, 147, 146, 129, and 111 peak (indicating the losses of one, two and three hydroxyl moieties from the pyranose ring) were all present in a single MS/MS experiment. This was due to the formation of product ions from the m/z 182 parent species during the laser irradiance time (85 ms). As the first sequential products were formed, they could undergo further dissociation to yield more fragment ions, since there was no mass isolation step or filtering during the laser irradiance period. The dissociation efficiency of the m/z

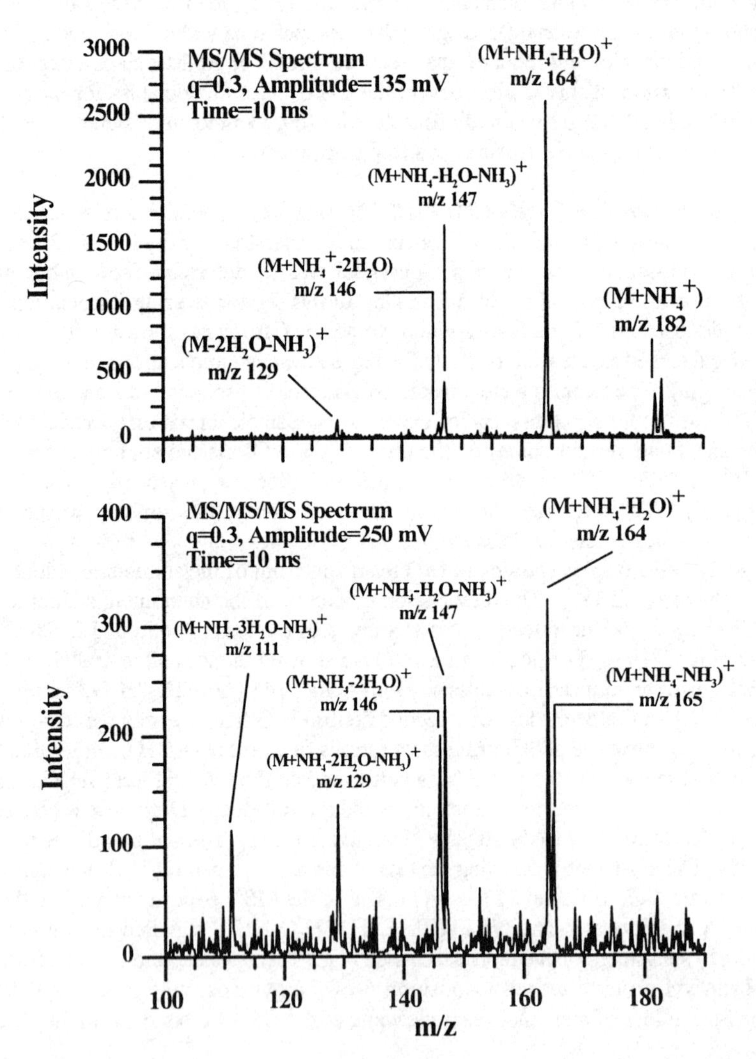

Figure 17. The MS^2 and MS^3 IRMPD spectra of the ammonium adduct of 2-deoxy-D-glucose. In the first experiment (top) the $[M+NH_4^+]$ ion at m/z 182 was mass isolated and fragmented at a q of 0.3. For the MS^3 experiment (bottom) the first sequential product ion from the MS^2 process at m/z 164 was mass isolated and fragmented at a q of 0.3. A significant reduction in CID efficiency marked the MS^3 spectrum.

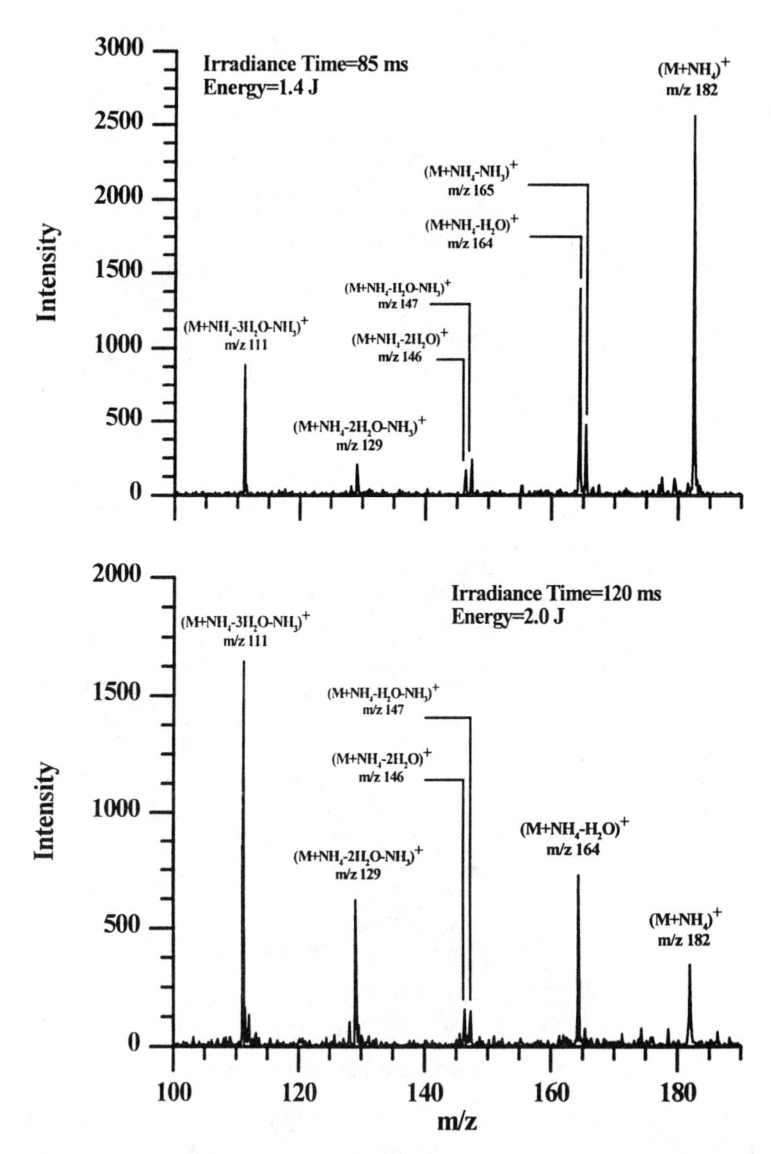

Figure 18. IRMPD spectra of the ammonium adduct of 2-deoxy-D-glucose at 85 ms and 120 ms irradiance time. High dissociation efficiencies were observed for IRMPD, since the kinetic energy of ion motion does not need to be converted to the internal vibrational energy needed for fragmentation. The number of hydroxyl substituents on the ring can be determined much easier with IRMPD, since dissociation efficiency is independent of ion population.

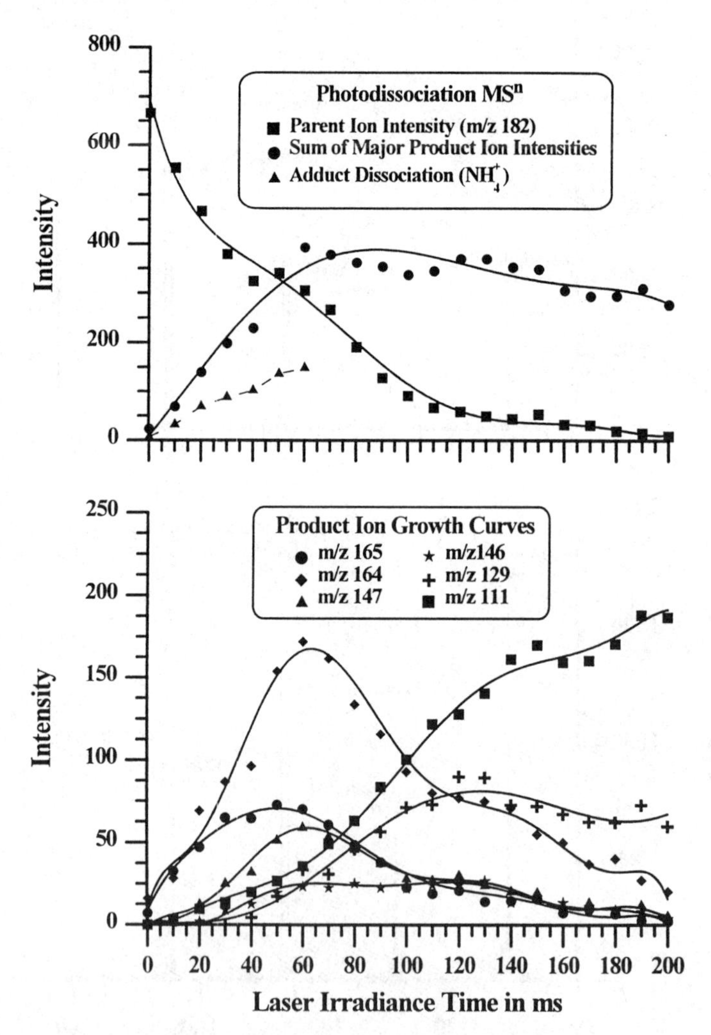

Figure 19. Consecutive reaction curves for the IRMPD of the ammonium adduct of 2-deoxy-D-glucose. The top half of the figure shows the sum of the major product ion intensities, plus the presence of the direct dissociation of the complex to showing the ammonium species at m/z 18 (dashed line). The bottom portion of the figure represents the individual curves comprising the major product ion intensity curve in the top half of the figure.

182 ammonium adduct was ≈100%. As the laser irradiance time was increased (figure 18), this essentially drove the series of consecutive reactions for the ammonium adduct of 2-deoxy-D-glucose to its final product ion, the m/z 111 species $[M+NH_4-3H_2O-NH_3]^+$.

The consecutive reaction curves in figure 19 (top) for the IRMPD of the ammonium adduct of 2-deoxy-D-glucose (m/z 182) show that only 61% of the fragment ions produced from the photodissociation process are structurally significant ions; the remaining 39% represent the direct dissociation of the ammonium adduct complex to form the ammonium ion and corresponding 2-deoxy-D-glucose neutral. The ammonium ion intensity curve in figure 19 was incomplete due to the complex series of back reactions observed for the NH_4^+ ion with neutrals present in the trapping region during the laser irradiance period (e.g., longer irradiance times correspond to longer reaction times for the dissociated NH_4^+ ion). The trace is present only to confirm the presence of this dissociation pathway; the experimental conditions were altered in a separate experiment to efficiently trap the m/z 18 NH_4^+ ion.

The bottom half of figure 19 shows the individual structurally diagnostic portion of the consecutive reaction curves. It is clearly seen that a series of complex consecutive and competitive reactions occur during the photodissociation process. The presence of the competitive reaction curves (e.g., m/z 164 and 165) could involve reactions where the critical energy of the two reaction channels are equivalent.

Although the types of fragment ions produced by CID and PID in this example were the same (which might be expected since both are low-energy processes in this case), the time involved in producing each spectrum was radically different. For the case of single-frequency CID, it is critical that a constant ion population is present so that re-tuning of the instrumental parameters involved for the CID process (typically 20 minutes, even for an experienced operator) need not occur (see figure 1). Also, in order to gain the same information obtained from the IRMPD experiment in figure 18, a third stage of mass analysis must be performed which means that the efficiency of the MS^3 step is directly influenced by the parent ion population frequency shifts and translates to a decreased MS/MS efficiency. This can drastically affect the formation of the parent species in the MS^3 process. In addition, the time required for tuning of the ion trap for an MS^3 process can take on the order of 30 minutes. More importantly, for the case of real-time sample concentration profiles observed in gas or liquid chromatography, single frequency CID experiments could produce a limited amount of structural information due to shifting peak concentration profiles. For the IRMPD process, there is an initial time investment for turning on and tuning the laser; however, since the dissociation event is independent of ion population (it only depends on the ability of functional groups to absorb photons), MS/MS experiments with photodissociation can be carried out for real-time chromatographic analysis. In addition, "multiple" stages of mass spectrometry for the IRMPD process can be obtained by simply increasing the irradiance time for the parent/product ion species. The one potential drawback of the IRMPD process could be the formation of low-intensity structurally significant ions which are never observed because the laser irradiation period is set such that they immediately absorb enough photons to dissociate quickly to the next sequential product ion.

For the other monosaccharide investigated in this study (1-O-methyl-D-glucopyranoside), the pyranose ring contained a methoxy substitution at the anomeric carbon

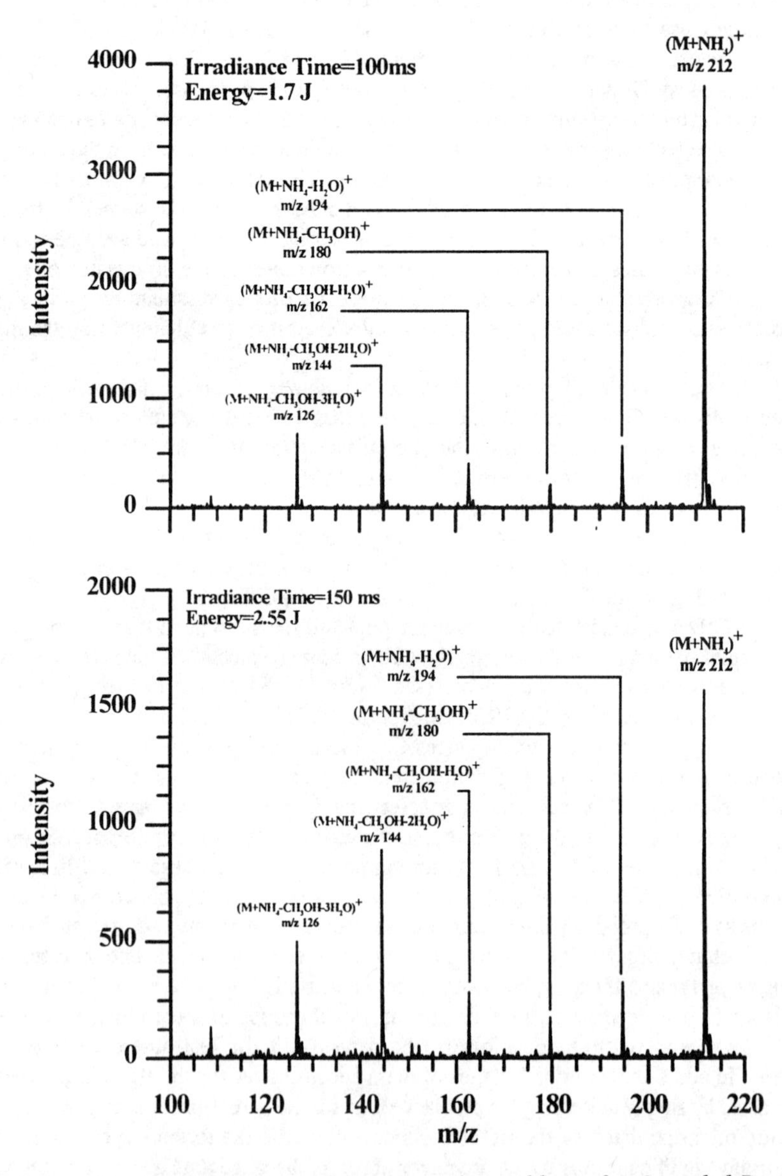

Figure 20. IRMPD spectra of the ammonium adduct of 1-O-methyl-D-glucopyranoside at irradiance times of 100 and 150 ms, respectively. The peak at m/z 180 is due to the loss of methanol, indicating the presence of a modified group on the ring structure of the monosaccharide.

(position 1). The fragmentation of this compound was examined to investigate the ability of IRMPD to cleave substituted groups on the pyranose ring structure (naturally occurring as well as permethylated derivatives). The case of the methoxy substituent is the simplest case, which serves quite well for evaluation purposes. The conditions for instrument operation were the same as described above for 2-deoxy-D-glucose.

In figure 20 is shown the IRMPD spectrum of 1-O-methyl-D-glucopyranoside. For an irradiance time of 100 ms, structurally diagnostic ions at m/z 194 $[M+NH_4-H_2O]^+$ indicating the loss of one hydroxyl group, m/z 180 $[M+NH_4CH_3OH]^+$ indicating the loss of the substituted methoxy group, m/z 162 $[M+NH_4-CH_3OH-H_2O]^+$ showing the loss of methoxy and one hydroxyl, m/z 144 $[M+NH_4-CH_3OH-2H_2O]^+$ showing the loss of methoxy and two hydroxyl groups, and m/z 126 $[M+NH_4-CH_3OH-3H_2O]^+$ marking the methoxy loss minus 3 hydroxyl groups (e.g., with m/z 126, all ring substituents have been removed) are observed.

Although photodissociation has been shown to remove all substituents from the pyranose ring structure to aid in ring identification, it does not indicate the substitution position for each individual loss. Perhaps the most important information provided by photodissociation of the ammonium adducts of monosaccharides will be the identification of "post-translationally" modified substituents.

Raffinose - CID/Pulsed PID. To determine the applicability of photodissociation to sequencing oligosaccharides, two straight chain oligosaccharides, raffinose (O-α-D-galactopyranosyl[1-6]-α-D-glucopyranosyl-β-D-fructofuranoside) and stachyose (α-D-galactopyranosyl-[1-6]-α-D-galactopyranosyl-[1-6]-α-D-glucopyranosyl-[1-2]-β-D-fructofuranoside), were chosen for study. In figure 21 are shown the structures of these two compounds. This section will focus on the photodissociation and fragmentation of raffinose, an oligosaccharide typically used in tissue culture media.

A 20 pmol/μL solution of raffinose was prepared in a 50:50 methanol:water solution with 5 mM NH_4OH. Samples were directly infused through the electrospray interface as described previously in the experimental section. For photodissociation experiments, a pulsed CO_2 excimer laser was employed. The laser was capable of an approximate 3-5 ns pulse with an energy of 1.1 J (at 944 cm^{-1}). Due to the quick pulse of laser energy and the high photon absorption cross section for the C—O—O stretch, the ion trap was operated with buffer gas at a pressure of $1.0x10^{-4}$ torr (uncorrected) for both CID and photodissociation experiments. The major (parent) ion produced from the electrospray process was the ammonium adduct of raffinose at m/z 522.

The fragmentation nomenclature employed for the spectral interpretation of the raffinose and stachyose data was that of Domon and Costello (*92*) In figure 22 is shown a simple fragmentation of a model disaccharide with cleavage points following the Domon and Costello nomenclature. Charge retention on the reducing portion of the sugar is indicated by the fragments X (ring opening cleavage), Y and Z. For charge retention on the nonreducing terminus of the sugar, the fragments are designated A (ring opening cleavage), B, and C. The superscripts in figure 22 indicate the bond cleavage positions, while the presence of a subscript identifies the residue number. Branches are labeled α, β, γ.... where α is the branch with the highest mass (*92*).

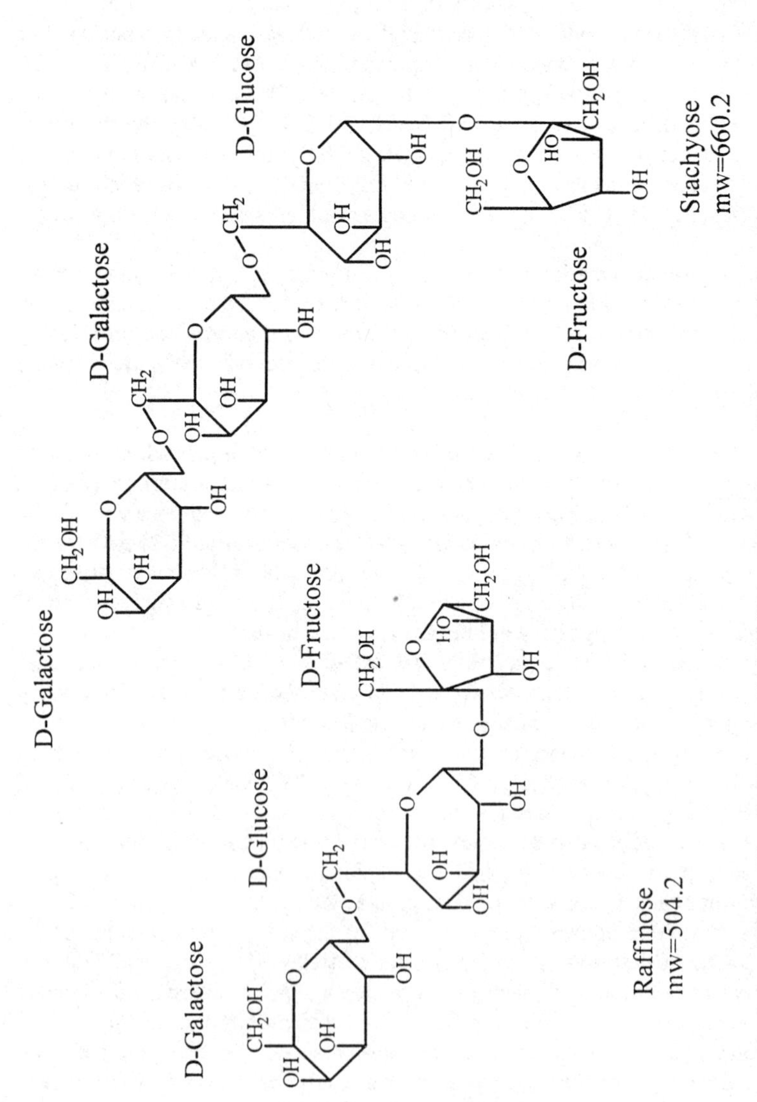

Figure 21. Structures of the oligosaccharides raffinose (O-α-D-galactopyranosyl[1-6]-α-D-glucopyranosyl-β-D-fructofuranoside) and stachyose (α-D-galactopyranosyl-[1-6]-α-D-galactopyranosyl-[1-6]-α-D-glucopyranosyl-[1-2]-β-D-fructofuranoside).

In figure 23 are shown both the CID and PID (pulsed CO_2 laser) spectra of the ammonium adduct of raffinose. The major product ion from the single-frequency CID process was m/z 505 $[M+H]^+$, arising from loss of neutral ammonia from the adducted species. Although this ion gave no structural information, it could be used to help confirm molecular weight. The peaks at m/z 343 and 326 indicate the loss of one of the terminal monosaccharide groups (D-fructose from the reducing end or D-galactose from the nonreducing end), with the peak at m/z 326 showing an additional loss of ammonia. Assignment of these ions as either $B_{2\alpha}$ or $Z_{2\alpha}$ is difficult since the neutral loss of either D-galactose or D-fructose has the same mass. However, from a biochemical standpoint, the most labile bond in the system is the glucopyranosyl-β-D-fructofuranoside bond, which would indicate that the ions formed in this mass spectrum are B type ions with charge retention on the nonreducing terminus. These results would also be consistent with other low-energy CID results of straight chain oligosaccharides reported in the literature (*80*). In figure 24 is found the fragmentation analysis for raffinose.

Also shown in figure 23 is the photodissociation spectrum (2 laser pulses) of the ammonium adduct of raffinose. Compared to the CID spectrum, the PID spectrum produced only a small $[M+H]^+$ peak, with the majority of the ion signal producing structurally diagnostic ions. The two largest ions in the spectrum were the m/z 326 and 343 ions indicative of the loss of a terminal monosaccharide (as seen in the CID spectrum). Also present were a series of peaks indicating successive losses of H_2O at m/z 308, 290, and 272 from the "$B_{2\alpha}$" ion at m/z 326. An additional peak at m/z 164 represents the cleavage of the internal monosaccharide residue (D-glucose) to form either the $B_{1\alpha}$ or $Z_{1\alpha}$ ion. The structure of the peak at m/z 365 cannot be assigned at this time. For determination of an actual unknown sample, identification of the internal sequence ions of an oligosaccharide would be extremely difficult using mass spectrometry alone (even using the appropriate derivatization chemistry), since anomeric and linkage position data are stereospecific. It is important to point out that with photodissociation, the "entire sequence" of the trisaccharide was obtained without complicated instrument tuning procedures. In contrast, for CID, a third stage of mass spectrometry would have to be performed in order to obtain the same information (e.g., cleavage of the internal residue) as from the photodissociation experiment. This means that for photodissociation another sample could be analyzed immediately after the first, since no tuning of instrument parameters is needed to induce fragmentation (only an initial time investment at the beginning of the day's analysis).

Stachyose - CID/Pulsed PID. The next compound studied was the tetrasaccharide stachyose (see figure 21 for structure). Stachyose (α-D-galactopyranosyl-[1-6]-α-D-galactopyranosyl-[1-6]-α-D-glucopyranosyl-[1-2]-β-D-fructofuranoside) differs from raffinose by the one additional D-galactose on the non-reducing end and a [1-2] β linkage from glucose to furanose. Electrospray and instrumental conditions were the same as for raffinose. The sample, at a concentration of 20 pmol/μL, was infused directly through the ESI source.

In figure 25 is shown the single-frequency CID spectrum of the ammonium adduct of stachyose (m/z 684). The major fragment peaks in the spectrum were observed at m/z 505 and 488 ($B_{3\alpha}$ or $Z_{3\alpha}$), indicating the loss of the nonreducing terminal (D-galactose) or

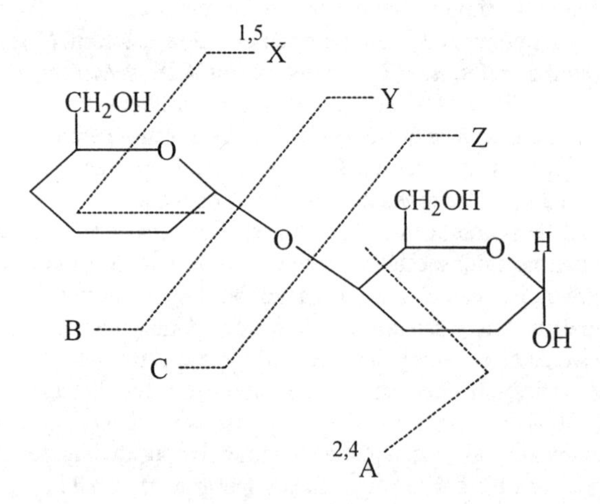

Figure 22. Carbohydrate fragmentation nomenclature as determined by Domon and Costello (92). Charge retention on the reducing terminal is indicated by the X, Y, and Z fragmentation. Charge retention on the nonreducing end is represented by the A, B, and C fragments. Ring opening cleavage fragments are X and A with superscripts denoting bond breaking points on the ring. Adapted from reference 92.

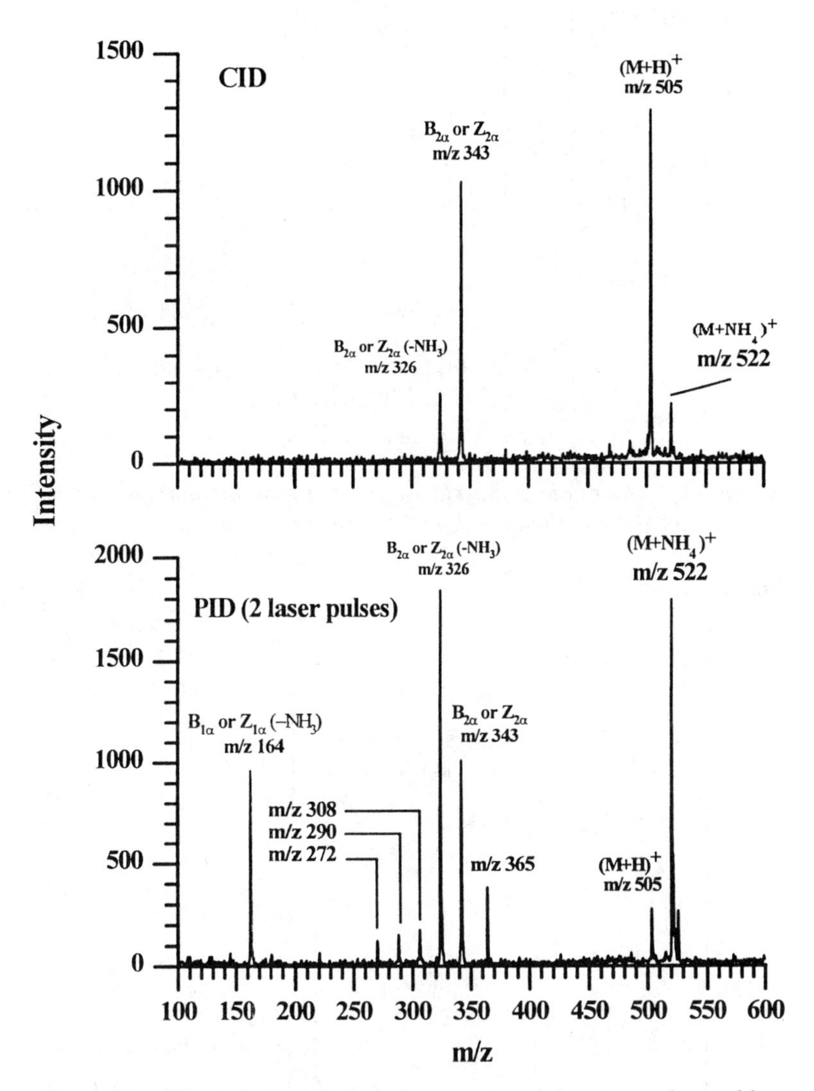

Figure 23. CID and photodissociation spectra of the ammonium adduct of raffinose. The major fragmentation observed in the CID spectrum is the $[M+H]^+$ at m/z 505 which essentially confirms the molecular weight, but provides no structural information. The photodissociation spectrum shows complete fragmentation information for the trisaccharide.

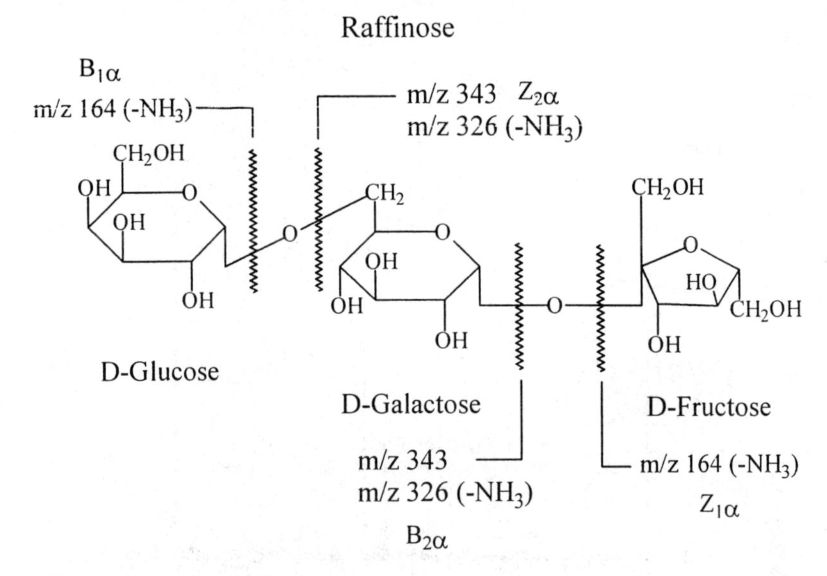

Figure 24. Glycosidic bond cleavages of the ammonium adduct of raffinose for the collisional and photon activation processes.

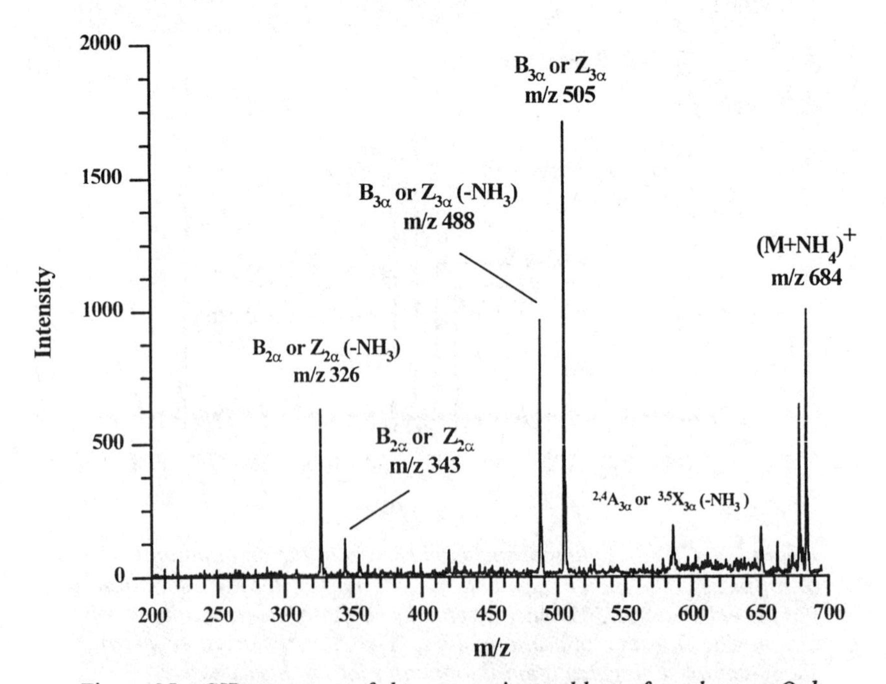

Figure 25. CID spectrum of the ammonium adduct of stachyose. Only the loss of two monosaccharide units is observed under CID conditions (one terminal and one internal residue). A small degree of ring cleavage on a terminal monosaccharide unit is observed by the presence of the $^{2,4}A_{3\alpha}$ or $^{3,5}X_{3\alpha}$ peak.

the D-fructose which has its reducing end bonded β-[1-2] to the glucopyranosyl moiety. The peak at m/z 488 shows the same cleavage with loss of ammonia. The peaks observed at m/z 343 and 326 indicate the loss of an internal sugar residue, either D-galactose or D-glucose ($B_{2\alpha}$ or $Z_{2\alpha}$). The small broad peak at approximately m/z 589 may indicate a ring opening reaction which cleaves in the 3 and 5 position of D-galactose to form the $^{3,5}X_{3\alpha}$ ion or cleaves in the 2 and 4 position of D-fructose to produce the $^{2,4}A_{3\alpha}$ ion. The broad peak width precludes direct mass assignment of this ion. The peak at m/z 680 was an occasional artifact peak observed, and depended on the amplitude of the resonance ejection frequency and single frequency CID signal. In figure 26 is shown the fragmentation scheme for the ammonium adduct ion of stachyose. As was the case for raffinose, additional stages of mass spectrometry are needed to obtain entire sequence information.

The photodissociation spectra for 1,2 and 3 laser pulses of the ammonium adduct of stachyose are seen in figure 27. After two laser pulses, the entire sequence of the tetrasaccharide is obtained, with the peak at m/z 164 indicating either the nonreducing terminus ($B_{1\alpha}$) ion or the blocked reducing end D-fructose ion ($Z_{1\alpha}$). For the three laser pulse spectrum, peaks at m/z 146 and 128 indicate the successive losses of H_2O. Interestingly enough, the small broad peak indicating ring opening cleavages at m/z 589 is not observed is the photodissociation spectra. However, an unknown peak at m/z 640 is observed in all three spectra (structure unknown). As was the case for raffinose, the [1-2] β linkage between the D-glucose and D-fructose is the most labile glycosidic bond in stachyose. Therefore, the sequential product ions in figure 27 probably represent charge retention on the nonreducing terminus, again consistent with previously reported low-energy CID data (*80*). Fragmentation patterns for the ammonium adduct of stachyose are found in figure 26.

Carbohydrate Antibiotics. Carbohydrate antibiotics are some of the most important compounds in all of biochemistry. In the next decade, an increasing emphasis will be placed on discovery of new carbohydrate antibiotics, as microorganisms continue to develop resistance to current antibiotic treatments. Over the years, one of the most important classes of carbohydrate antibiotics has been the macrolide antibiotics (*93*) Macrolide antibiotics all contain a large lactone ring (aglycone of 12 to 22 atoms) with no nitrogen atoms and few double bonds. Linked to the aglycone ring are one or more sugars (which can contain nitrogen). These sugar linkages to the aglycone ring are critical for biological activity. In figure 28 is shown the structure of the macrolide antibiotic erythromycin, with the aglycone moiety and two attached monosaccharides (D-desosamine and L-cladinose). The structural determination of the various functional groups comprising these compounds and other carbohydrate antibiotics has long been of interest (*85,86,89,93-105*).

Perhaps the most widely used macrolide antibiotic is erythromycin. This antibiotic is typically used to treat respiratory infections, and is particularly effective against Gram-positive bacteria such as those involved in streptoccal, staphyloccal, and pneumococcal infections (*93*). The mechanism of action involves binding of the antibiotic to the 50S subunit of the bacterial ribosome, thus blocking the action of peptidyl transferase in the peptide elongation process.

It is the purpose of this section to evaluate the ability of the photodissociation technique for structural elucidation of carbohydrate antibiotics. Studies were performed

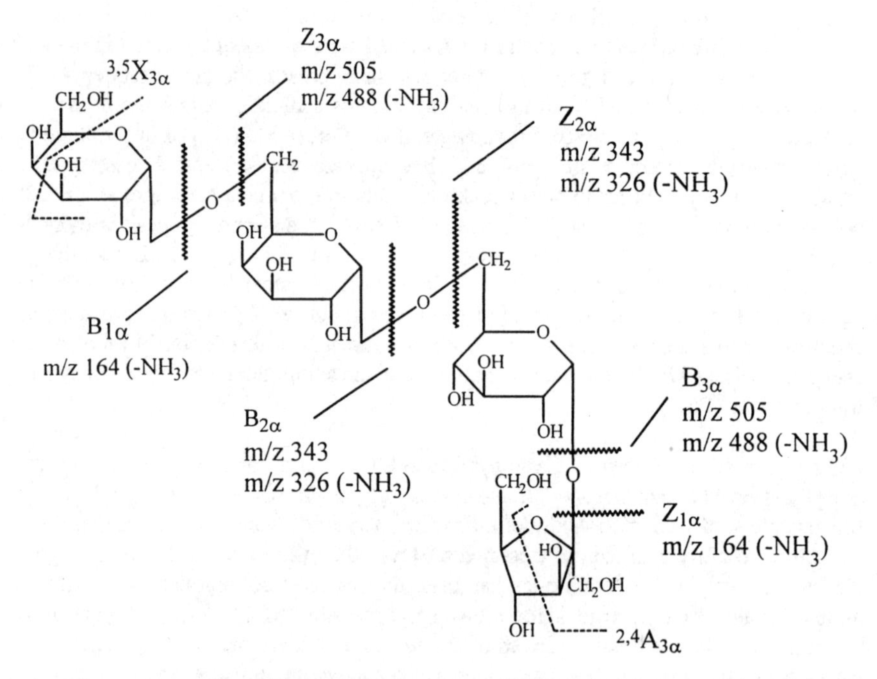

Figure 26. Glycosidic bond cleavages of the ammonium adduct of stachyose for the collisional and photon activation processes. The ring cleavages were only observed in the CID spectrum.

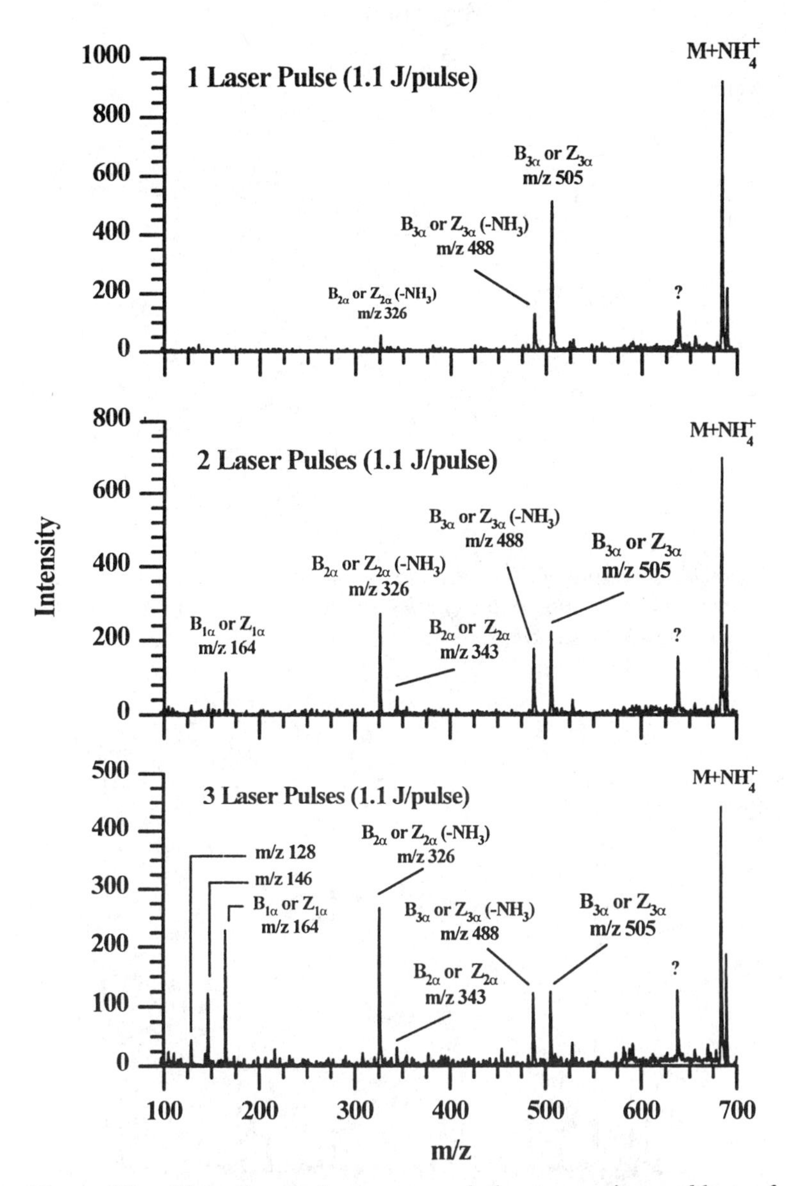

Figure 27. Photodissociation spectra of the ammonium adduct of stachyose for one, two, and three laser pulses. After two laser pulses, the "entire sequence" for stachyose was obtained. With the addition of a third laser pulse, ring fragmentations (e.g. loss of H_2O indicating the presence of the hydroxyl group) were observed for the terminal monosaccharide residue (m/z 128 and m/z 146).

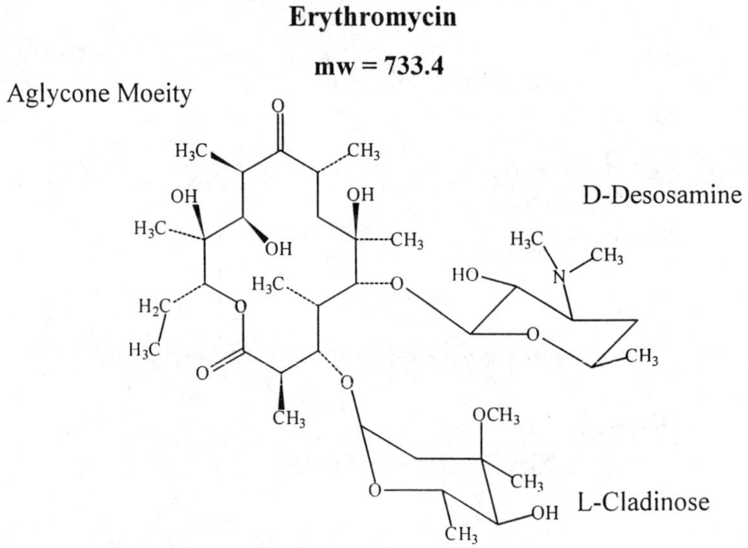

Erythromycin

mw = 733.4

Aglycone Moeity

D-Desosamine

L-Cladinose

Figure 28. Structure of the macrolide antibiotic erythromycin showing the lactone ring (aglycone moiety), the amino sugar D-desosamine and nonamino sugar L-cladinose.

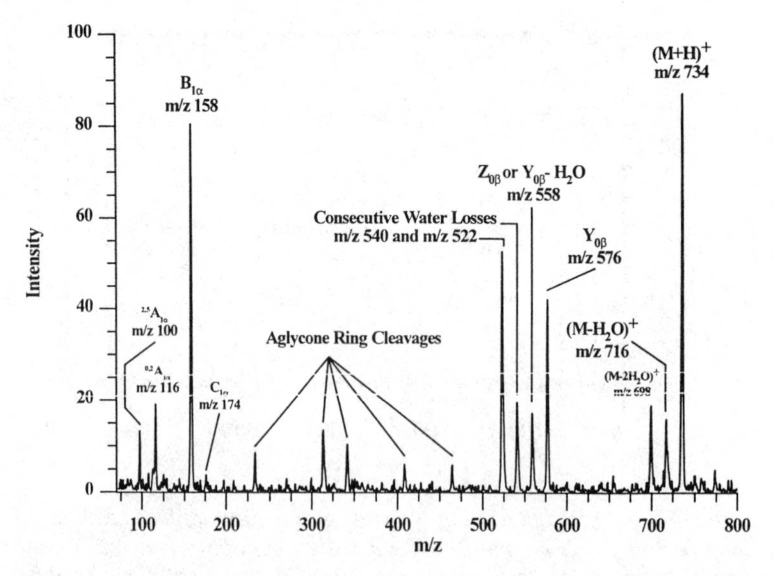

Figure 29. Photodissociation spectrum (7 laser pulses) of the protonated $[M+H]^+$ ion m/z 734 of erythromycin. The mechanism and composition of the fragments labeled aglycone ring cleavages are unknown. Many of the fragments observed (e.g. $C_{1\alpha}$ and $Y_{0\beta}$) correspond to data obtained using high energy CID as seen on magnetic sector instruments.

with protonated erythromycin as a model compound, in order to understand the fragmentations observed with basic macrolide antibiotics.

Macrolide Antibiotics - Erythromycin PID. The erythromycin standard was obtained from ICN Pharmaceutical (Costa Mesa, CA). Solutions were made up in 50:50 methanol:water with 0.1% acetic acid. The concentration of the electrospray standard was 15 pmol/μL. All electrospray and instrumental conditions were the same as described in the experimental section. The photodissociation set-up employed a pulsed CO_2 laser as discussed previously.

The preferential site of protonation is the tertiary amine group on the D-desosamine sugar. Therefore, it would be predicted that charge-site fragmentation should be driven from this portion of the ion. In figure 29 is seen the photodissociation spectrum (7 laser pulses) of protonated erythromycin. The low mass peaks at m/z 100 ($^{2,5}A_{1\alpha}$) and 116 ($^{0,2}A_{1\alpha}$) are charge-site-driven fragmentations for the ring opening of the D-desosamine sugar. Direct cleavage of the glycosidic bond on the D-desosamine side sugar (with accompanying hydrogen transfers) is indicated by the fragment (m/z 158) $B_{1\alpha}$. The possible appearance of a small peak at m/z 174 ($C_{1\alpha}$) could signify the transfer of two hydrogens from the amino sugar to aglycone-L-cladinose neutral. The unusual aspect of this fragment is the fact that it is typically observed only in the charge remote fragmentation process, indicative of a high-energy collision process. The formation of the m/z 116 ($^{0,2}A_{1\alpha}$) peak could arise from a two step elimination of $C_2H_2O_2$ from the m/z 174 ($C_{1\alpha}$) peak. (*85,105*)

Even more unusual is the presence of an intense peak at m/z 576 ($Y_{0\beta}$), indicating loss of the L-cladinose sugar from the aglycone ring (cleavage on the L-cladinose side of the glycosidic bond). The peak at m/z 558 could either be cleavage of the glycosidic bond on the aglycone side to form the $Z_{0\beta}$ ion, or loss of water from the m/z 576 ion. The peaks at m/z 540 and 522 are consecutive water losses occurring from the aglycone ring (indicating three hydroxyl substituents present). Around the protonated erythromycin region (m/z 734, 716 and 698) are observed two consecutive losses of water. This indicates that some form of charge migration or charge-remote fragmentation mechanism occurs since there is only one hydroxyl group present on D-desosamine (where protonation occurs on the tertiary nitrogen). In figure 30 are seen the fragmentation pathways observed for protonated erythromycin.

The photodissociation spectrum of protonated erythromycin bears a strong resemblance to its high-energy CID spectrum taken on a magnetic sector instrument (*85,105*). Protonated species under high-energy collisions give information on the monosaccharide residues present on the aglycone ring moiety. These fragmentations are typically some form of the charge-remote mechanism. One other interesting note is the formation of a series of ions (from m/z 200 to 500) in the photodissociation spectrum indicating ring opening reactions of the aglycone moiety. At this time reasonable neutral losses and peak assignments have not been made, and will require an extensive investigation. It is theorized that these ring opening cleavages are perhaps driven by the C—O—C ether group in the aglycone ring (with some form of accompanying charge migration or charge remote fragmentation mechanism). No ring fragmentations were observed in any previously published high-energy CID spectra. The peak observed at m/z 403 could be the loss of D-desosamine from the $Y_{0\beta}$ at m/z 576 to form the S type ion.

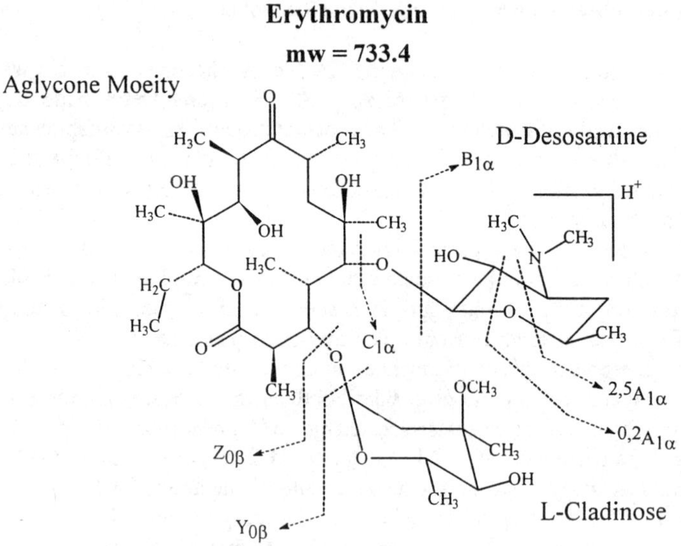

Erythromycin

mw = 733.4

Aglycone Moeity

D-Desosamine

L-Cladinose

$B_{1\alpha}$ $C_{1\alpha}$ $^{2,5}A_{1\alpha}$ $^{0,2}A_{1\alpha}$ $Z_{0\beta}$ $Y_{0\beta}$ H^+

Figure 30. Fragmentation of the macrolide antibiotic erythromycin obtained with photon activation (7 laser pulses).

Oligonucleotide Fragmentation Nomenclature

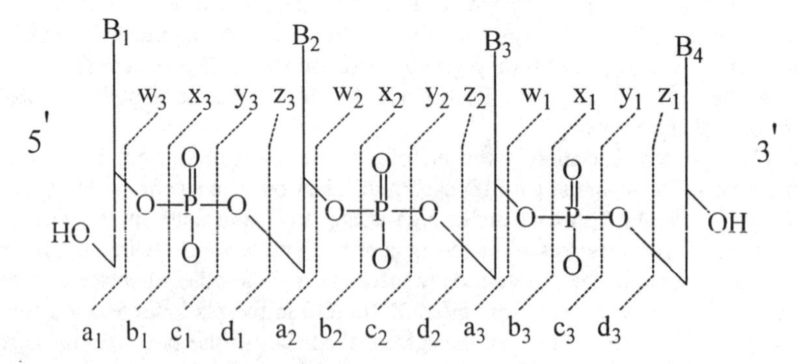

Figure 31. Oligonucleotide fragmentation scheme as defined by McLuckey et al. (109). The letter B_n represents the individual nucleoside bases, with position one defined from the 5' end. The w, x, y, and z fragments have the charge retained on the 3' end while the a, b, c, and d fragments have the charge retained on the 5' terminus. Adapted from reference 109.

Oligonucleotides. The first real advances in mass spectrometry for oligonucleotide analysis came with the advent of laser desorption and electrospray ionization methods (*106,107*). Recent successes in the field of electrospray have enabled ionization of oligonucleotides up to 76 base pairs (*108*). Typical electrospray spectra include sodium attachment to the multiply-charged negative ions of the oligonucleotides; this can complicate molecular weight analysis and has proved to be the single most limiting factor in electrospray mass spectrometry of this compound class (*108*). In addition, the weaker signals generated by the oligonucleotide anions can limit sensitivity of the electrospray technique (compared to the positive ion signal for peptides and proteins).

Perhaps the most difficult challenge of negative ion electrospray for the analysis of oligonucleotides is to obtain structural information from tandem MS/MS experiments. To date, only a handful of articles have appeared on the MS/MS of negatively-charged oligonucleotides (*13,109,110*). The main problem is in the area of data interpretation, where multiple charge states and adduct formation can complicate even the simplest of CID spectra. Previous reports have centered on the CID of small, multiply-charged negative ion DNA fragments (up to 8 base pairs) using electrospray ionization/ion trap mass spectrometry, and linker DNA 8-mers using FTICR and photodissociation (*13,109*). Both techniques show a great deal of promise for the analysis of larger oligonucleotides.

The fragmentation nomenclature employed for tandem MS/MS studies of oligonucleotides was developed by McLuckey et al. (*109*). Shown in figure 31 is a tetranucleotide with bases B_1 through B_4. Cleavage points along the phosphodiester backbone are indicated by the lower case letters a, b, c, and d for charge retention on the 5' end and w, x, y, and z for charge retention on the 3' end of the oligonucleotide. Subscripts indicate the number of bases contained in the fragment. Fragmentation of the nucleoside bases is represented by an upper case B, where B is the individual base A, G, C, T, or U. A subscript is assigned to the B symbol to represent the position of the base from the 5' end of the molecule. Bases are represented parenthetically to avoid confusion with the normal sequence terms (e.g., B_1(A)).

In this section are discussed the merits of photodissociation for DNA/RNA sequence analysis. Two RNA dimers were studied in order to determine the feasibility of the technique (pulsed-CO_2 laser) for future sequencing experiments.

RNA Dimers. Two RNA dimers adenyl adenosine (ApA) and adenyl cytidine (ApC) were used for the photodissociation experiments. The samples were obtained from the Core Biotechnology Facility at the University of Florida, and were purified to remove sodium salts. Sample concentrations were 2 pmol/μL in a 70:30 methanol:water solution. Electrospray conditions and instrumental parameters were set as described in the experimental section.

The lone parent species produced for the negative ion electrospray for the two dimers was a singly-charged negative ion $(M-H)^-$ at m/z 595 for ApA and m/z 571 for ApC (figure 32). Due to the desalinization process, no sodium adducts were observed. The negative charge is located on the phosphodiester bridge, as shown in figure 31. The maximum number of negative charges attached to any oligonucleotide is equal to the number of phosphodiester bridges present, plus any phosphate groups attached to the free 3' or 5' positions on the ribose sugar backbone. For a DNA tetramer (ApApApA) run on the electrospray ion trap, the highest charge state observed was that of $(M-3H)^{-3}$, indicating

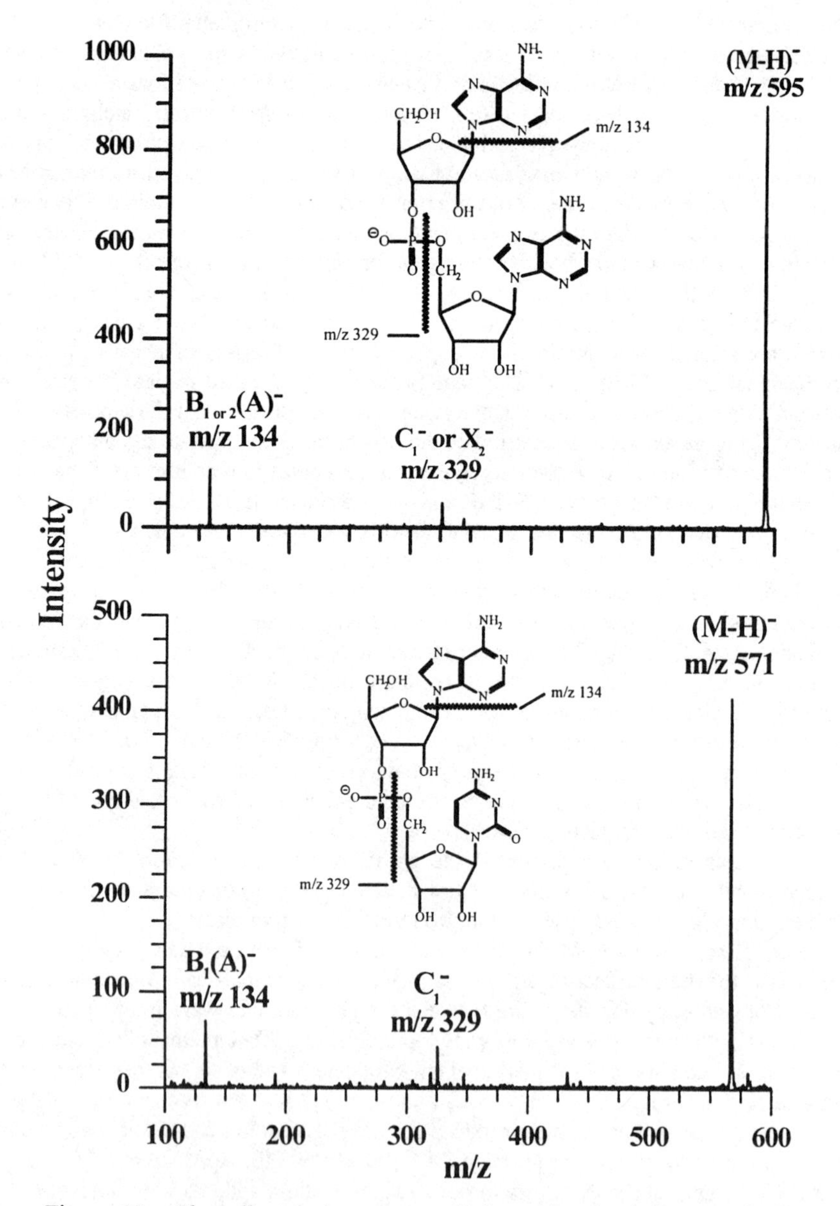

Figure 32. Photodissociation spectra (5 laser shots) of the negatively charged RNA dimers adenyl (ApA) adenosine (top) and adenyl (ApC) cytidine (bottom). The individual structures and fragmentations are also shown. Bond cleavage for the photodissociation process occurs at the phosphodiester bond P--O where there is a relatively high photoabsorption cross-section in the IR region.

the presence of the three phosphodiester bridges involved in the ionization process (data not shown).

In figure 32 are shown the photodissociation spectra, structure, and cleavage points of adenyl (ApA) adenosine and adenyl (ApC) cytidine. The photodissociation spectra represent 5 laser shots from the pulsed CO_2 laser. In the top spectrum is shown the fragmentation of ApA, with two characteristic peaks at m/z 134 and 329. The peak at m/z 134 represents loss of the adenine base from either position one or position two, therefore no subscript for the $B_n(A)$ peak can be assigned. The second peak at m/z 329 indicates direct cleavage of the phosphodiester bond (P—O), to form either a C_1^- ion (charge retention on the 5' side) or a X_1^- ion (charge retention on the 3' side).

The fragmentation of ApC is shown in the bottom portion of figure 32. As with the previous example, cleavage of the adenine base is observed and the fragment at m/z 134 can be assigned to the B_1^- peak. The second peak observed at m/z 329 represents cleavage of the phosphodiester bond to form the C_1^- peak containing the ribose sugar and cytosine base.

The types of ions observed in photodissociation spectra differ markedly from the CID of oligonucleotides (*109*). For CID spectra, preferential cleavage occurs at the C—O of the sugar (charge retention of the 3' end) to form w-type fragment ions. Complementary cleavages to form ions with a (a-B(A)) type fragmentation were also observed for oligonucleotides containing adenine base(s). The photodissociation spectra presented in this section show preferential cleavage at the P—O, which has a fairly high photoabsorption cross-section; there data corresponds well with that obtained using IRMPD in the FTICR (*13*).

Conclusions and Future Work

Much has yet to be learned about the photodissociation of electrosprayed ions. However, these studies have shown that by increasing the photoabsorption pathlength for the PID experiment, valuable structural information can be obtained on a variety of different compound classes. Currently we are in the process of verifying some of the more unusual fragmentations observed with PID using either labeled standards or MS^n experiments. As an overall observation, it was found that photoabsorption cross sections for the three compound classes studied follow the order: carbohydrates (C—O—C ether linkage) > DNA (phosphodiester linkage) > peptides/proteins (amide linkage). Future studies will focus not only on the fundamental aspects of photodissociation (e.g. quantitative measurements of photoabsorption cross-sections for multiply charged electrospray ions), but also on the extension of this technique to solve real-world problems (e.g. structural elucidation of glycopeptide antibiotics). Future PID studies will be directed towards the preferred cleavage at glycosidic bond linkages (C—O—C groups), using more appropriate laser wavelengths to achieve higher photoabsorption cross sections. These PID studies have the potential to provide significant structural information for oligosaccharide composition/heterogeneity, linkage analysis, and possibly identification of glycosylation sites on peptides and proteins in the gas phase.

Acknowledgements

The authors wish to acknowledge the Office of Naval Research, an ACS Analytical Division Fellowship (to JLS) funded by Procter & Gamble, and the University of Florida

Division of Sponsored Research for financial assistance with this project. Special thanks also go to Mr. Joseph A. Shalosky for the machine work done on this project, and to Mr. Scott T. Quarmby for his assistance with the rf circuitry.

Literature Cited

1. Busch, K.L.; Glish, G.L.; McLuckey, S.A. Reactions in MS/MS, in *Mass Spectrometry/Mass Spectrometry: Techniques and Applications of Tandem Mass Spectrometry*; VCH: New York, 1988; pp 87-90.
2. Johnson, J.V.; Yost, R.A.; Kelley, P.E.; Bradford, D.C. *Anal. Chem.* **1990**, *62*, 2162-2172.
3. McLuckey, S.A.; Van Berkel, G.J.; Goeringer, D.E.; Glish, G.L.; *Anal. Chem.* **1994**, *66*, 689A-696A.
4. Dunbar, R.C. in *Gas Phase Ion Chemistry, Vol. 2*; M.T. Bowers ed.; Academic Press: London, **1979**.
5. Dunbar, R.C. in *Gas Phase Ion Chemistry, Vol. 3*; M.T. Bowers ed.; Academic Press: London, **1984**.
6. van der Hart, W.J. *Mass Spectrom. Reviews* **1989**, *8*, 237-268.
7. van der Hart, W.J. *Int. J. Mass Spectrom. Ion Processes* **1991**, *118/119*, 617-633.
8. Louris, J.N.; Brodbelt, J.S.; Cooks, R.G. *Int. J. Mass Spectrom Ion Processes* **1987**, *75*, 345-352.
9. Dunbar, R.C.; Weddle, G.H. *J. Phys. Chem.* **1988**, *92*, 5706.
10. Hemberger, P.H.; Nogar, N.S.; Williams, J.D.; Cooks, R.G.; Syka, J.E.P. *Chem. Phys. Lett.* **1992**, *191*, 405-410.
11. Williams, J.D.; Cooks, R.G.; Syka, J.E.P.; Hemberger, P.H.; Nogar, N.S. *J. Am. Soc. Mass Spectrom.* **1993**, *4*, 792-797.
12. Lammert, S.A.; Cleven, C.D.; Cooks, R.G. *J. Am. Soc. Mass Spectrom.* **1994**, *5*, 29-36.
13. Little, D.P.; Speir, P.J.; Senko, M.W.; O'Connor, P.B.; McLafferty, F.W. *Anal. Chem.* **1994**, *66*, 2809-2815.
14. Castro, J.A.; Nuwaysir, L.M.; Ijames, C.F.; Wilkins, C.L. *Anal. Chem.* **1992**, *64*, 2238-2243.
15. Williams, E.R.; Furlong, J.J.P.; McLafferty, F.W. *J. Am. Soc. Mass Spectrom.* **1990**, *1*, 288-294.
16. Louris, J.N.; Amy, J.W.; Ridley, T.Y.; Cooks, R.G. *Int. J. Mass Spectrom. Ion Processes* **1989**, *88*, 97-111.
17. Pedder, R.E.; Yost, R.A.; Weber-Grabau, M. *Proceedings of the 37th ASMS Conference on Mass Spectrometry and Allied Topics*, Miami, FL, **1989**, 468-469.
18. Schwartz, J.C.; Cooks, R.G. *Proceedings of the 36th ASMS Conference on Mass Spectrometry and Allied Topics*, San Francisco, CA, **1988**, 634-635.
19. Soni, M.; Cooks, R.G. *Anal. Chem.* **1994**, *66*, 2488-2496.
20. Bier, M.E.; Hartford, R.E.; Herron, J.R.; Stafford, G.C. *Proceedings of the 39th ASMS Conference on Mass Spectrometry and Allied Topics*, Nashville, TN, **1991**, 538-539.
21. Kaiser, R.E.; Williams, J.D.; Schwartz, J.C.; Lammert, S.A.; Cooks, R.G. *Proceedings of the 37th ASMS Conference on Mass Spectrometry and Allied Topics*, Miami, FL, **1989**, 369-370.

22. Van Berkel, F.J.; Glish, G.L.; McLuckey, S.A. *Anal. Chem.* **1990**, *62*, 1284-1295.
23. Mordehai, A.V.; Hopfgartner, G.; Huggins, T.G.; Henion, J.D. *Rapid Commun. Mass Spectrom.* **1992**, *6*, 508-516.
24. Boué, S.M.; Stephenson, J.L.; Yost, R.A. *Presented at the 43rd ASMS Conference on Mass Spectrometry and Allied Topics*, Atlanta, GA, **1995**.
25. Doktycz, M.J.; Habibigoudarz, S.; McLuckey, S.A. *Anal. Chem.* **1994**, *66*, 3416-3422.
26. Boué, S.M.; Jones, J.A.; Yost, R.A. *Proceedings of the 42nd ASMS Conference on Mass Spectrometry and Allied Topics*, Chicago, IL, **1994**, 218.
27. Asano, K.G.; Glish, G.L.; McLuckey, S.A. *Proceedings of the 36th ASMS Conference on Mass Spectrometry and Allied Topics*, San Francisco, CA, **1988**, 636-637.
28. McLuckey, S.A.; Glish, G.L.; Asano, K.G. *Anal. Chem. Acta.* **1989**, *225*, 25-35.
29. Duckworth, D.C.; Barshick, C.M.; Smith, D.H.; McLuckey, S.A. *Anal. Chem.* **1994**, *66*, 92-98.
30. McLuckey, S.A.; Glish, G.L.; Van Berkel, G.J. *Proceedings of the 38th ASMS Conference on Mass Spectrometry and Allied Topics*, Tucson, AZ, **1990**, 512-513.
31. Chien, B.M.; Michael, S.M.; Lubman, D.M. *Anal. Chem.* **1993**, *65*, 1916-1924.
32. Koppenaal, D.W.; Barinaga, C.J.; Smith, M.R. *J. Anal. Atomic Spectrom.* **1994**, *9*, 1053-1058.
33. Louris, J.N.; Brodbelt-Lustig, J.S.; Kaiser, R.E.; Cooks, R.G. *Proceedings of the 36th ASMS Conference on Mass Spectrometry and Allied Topics*, San Francisco, CA, **1988**, 968-969.
34. Eiden, G.G.; Garrett, A.W.; Cisper, M.E.; Nogar, N.S.; Hemberger, P.H. *Int. J. Mass Spectrom. Ion Processes* **1994**, *136*, 119-141.
35. Booth, M.M.; Stephenson Jr., J.L.: Yost, R.A. *Proceedings of the 42nd ASMS Conference on Mass Spectrometry and Allied Topics*, Chicago, IL, **1994**, 693.
36. Yang, M.; Dale, J.M.; Whitten, W.B. *Anal. Chem.* **1995**, *67*, 1021-1025.
37. Dale, J.M.; Yang, M.; Whitten, W.B. *Anal. Chem.* **1994**, *66*, 3431-3435.
38. Pinkston, J.D.; Delaney, T.E.; Morand, K.L. *Anal. Chem.* **1992**, *64*, 1571-1577.
39. Yost, R.A.; Enke, C.G. *Anal. Chem.* **1979**, *51*, 1251A-1264A.
40. Teloy, E.; Gerlich, D. *Chem. Phys.* **1974**, *4*, 417-427.
41. Szabo, I. *Int. J. Mass Spectrom. Ion Processes* **1986**, *73*, 197-235.
42. Guettler, R.D.; Jones, G.C.; Posey, L.A.; Kirchner, N.J.; Keller, B.A. Zare, R.N. *J. Chem. Phys.* **1994**, *101*, 3763-3771.
43. Ervin, K.M.; Armentrout, P.B. *J. Chem. Phys.* **1985**, *83*, 166-189.
44. Syka, J.E.P. Szabo, I. *Proceedings of the 36th ASMS Conference on Mass Spectrometry and Allied Topics*, San Francisco, CA, **1988**, 1328-1329.
45. Hail, M.E. U.S. Patent
46. Shao, J.D.; Ng, C.Y. *Chem. Phys. Lett.* **1985**, *118*, 481-485.
47. Gerlich, D.; Disch, R.; Scherbarth, S. *J. Chem. Phys.* **1987**, *87*, 350-359.
48. Anderson, S.L.; Houle, F.A.; Gerlich, D.; Lee, Y.T. *J. Chem. Phys.* **1981**, *75*, 2153-2162.
49. Posey, L.A.; Guettler, R.D.; Kirchner, N.J.; Zare, R.N. *J. Chem. Phys.* **1994**, *101*, 3772-3786.
50. Miller, P.E.; Denton, M.B. *Int. J. Mass Spectrom. Ion Processes* **1986**, *72*, 223-238
51. Everdij, J.J; Huijser, A.; Verster, N.F. *Rev. Sci. Instrum.* **1973**, *44*, 721-725.

52. Hägg, A.; Szabo, I. *Int J. Mass Spectrom. Ion Processes* **1986**, *73*, 237-275.
53. Friedman, M.H.; Yergey, A.L.; Campana, J.E. *Rev. Sci. Instrum.* **1982**, *15*, 53-61.
54. White, J.U. *J. Opt. Soc. Am.* **1942**, *32*, 285-288.
55. Watson, C.H.; Zimmerman, J.A.; Bruce, J.E.; Eyler, J.R. *J. Phys. Chem.* **1991**, *95*, 6081-6086.
56. Peiris, D.L.; Cheeseman, M.A.; Ramanathan, R.; Eyler, J.R. *J. Phys. Chem.* **1993**, *97*, 7839-7843.
57. Stephenson, Jr., J.L.; Booth, M.M.; Shalosky, J.A.; Eyler, J.R.; Yost, R.A. *J. Am. Soc Mass Spectrom.* **1994**, *5*, 886-893.
58. Stephenson, Jr., J.L.; Booth, M.M.; Eyler, J.R.; Yost, R.A. submitted to *Rapid Comm. Mass Spectrom.*
59. Yates, N.A. *Methods for Gas Chromatography-Tandem Mass Spectrometry on the Quadrupole Ion Trap*, Ph.D. Dissertation University of Florida, **1994**.
60. ICMS Software, developed by Nathan A. Yates, Department of Chemistry, University of Florida, 1993.
61. Schwartz, J.C.; Bier, M.; Syka, J.E.P. personal communication.
62. Cox, A.L.; Skipper, J.; Chen, Y.; Henderson, R.A.; Darrow, T.L.; Shabanowitz, J.; Englehard, V.H.; Hunt, D.F.; Slingluff, C.L. *Science* **1994**, *264*, 716-719.
63. Hunt, D.F.; Michel, H.; Dickerson, T.A.; Shabanowitz, J.; Cox, A.L.; Sakaguchi, K.; Appella, E.; Grey, H.M.; Sette, A. *Science* **1992**, *256*, 1817-1820.
64. Hunt, D.F.; Henderson, R.A.; Shabanowitz, J.; Sakaguchi, K.; Michel, H.; Sevilir, N.; Cox, A.L.; Appella, E.; Englehard, V.H. *Science* **1992**, *255*, 1261-1263.
65. Henderson, R.A.; Michel, H.; Sakaguchi, K.; Shabanowitz, J.; Appella, E.; Hunt, D.F.; Englehard, V.H. *Science* **1992**, *255*, 1264-1266.
66. Ishikawa, K.; Nishimura, T.; Koga, Y.; Niwa, Y. *Rapid Comm. Mass Spectrom.* **1994**, *8*, 933-938.
67. Kilby, G.W.; Sheil, M.M. *Org. Mass Spectrom.* **1993**, *28*, 1417-1423.
68. Senko, M.W.; McLafferty, F.W. *Annu. Rev. Biophys. Biomol. Struct.* **1994**, *23*, 763-785.
69. Cox, K.A.; Williams, J.D.; Kaiser, R.E.; Cooks, R.G. *Biol Mass Spectrom.* **1992**, *21*, 226-241.
70. Huddleston, M.J.; Beam, M.F.; Carr, S.A. *Anal. Chem.* **1993**, *65*, 877-884.
71. Johnson, R.S.; Martin, S.A.; Bieman, K.; Stults, J.T.; Watson, J.T. *Anal. Chem.* **1987**, *59*, 2621-2625.
72. Johnson, R.S.; Martin, S.A. Bieman, K. *Int. J. Mass Spectrom. Ion Processes* **1988**, *86*, 137-154.
73. Baldwin, M.A. Natural Products Reports, **1995**, 33-44.
74. Reopstorff, P.; Fohlman, J. *Biomed. and Environ. Mass Spectrom.* **1984**, *11*, 601.
75. Kaiser, R.E.; Cooks, R.G.; Syka, J.E.P.; Stafford, G.C. *Rapid Commun. Mass Spectrom.* **1990**, *4*, 30.
76. Hunt, D.F.; Yates, J.R. III; Shabonowitz, J.; Winston, S.; Hauer, C.R. *Proc. Natl. Acad. Sci.* U.S.A. **1986**, *83*, 6233-6237.
77. Bieman, K.; Martin, S.A. *Mass Spectrom. Rev.* **1987**, *6*, 1-75.
78. Tang, X.J.; Boyd, R.K. *Rapid Commun. Mass Spectrom.* **1992**, *6*, 651-657.
79. Tang, X.J.; Thibault, P.; Boyd, R.K. *Anal. Chem.* **1993**, *65*, 2824-2834.
80. Laine, R.A. In *Methods in Enzymology, Volume 193, Mass Spectrometry* James A. McClosky ed., **1990**, Academic Press: San Diego, CA, pp 539-553.

81. Hellerqvist, C.G.; Sweetman, B.J. "Mass Spectrometry of Carbohydrates" in *Biomedical Applications of Mass Spectrometry*, vol. 34, John Wiley & Sons: New York, **1990**.
82. Kent, P.W. *Pestic. Sci.* **1994**, *41*, 209-238.
83. Reinhold, V.; Reinhold, B. *Anal. Chem.* **1995**, *67*, 1722-1784.
84. Kamerling, J.P.; Vleigenthart, J.F.G. in *Clinical Biochemistry, Principles, Methods and Applications: Mass spectrometry* A.M. Lawson, Ed., Walter de Gruyter: New York, **1989**, 177-244.
85. Florencio, M.H.; Despeyroux, D.; Jennings, K.R. *Org. Mass Spectrom.* **1994**, *29*, 483-490.
86. McLaughlin, L.G.; Henion, J.D.; Kijak, P.J. *Biol. Mass Spectrom.* **1994**, *23*, 417-429.
87. Garozzo, D.; Guiffrida, M.; Impallomeni, G. *Anal. Chem.* **1990**, *62*, 279-286.
88. Chen, Y.; Chen, N,; Li, H.; Zhao, F.; Chen, N. *Biomed. Environ. Mass Spectrom.* **1987**, *14*, 9-15.
89. Schneider, R.P.; Lynch, M.J.; Ericson, J.F.; Fouda, H.G. *Anal. Chem.* **1991**, *63*, 1789-1794.
90. Laine, R.A.; Pamidimukkala, K.M.; French, A.D.; Hall, R.W.; Abbas, S.A.; Jain, R.K.; Matta, K.L. *J. Am. Chem. Soc.* **1988**, *110*, 6931-6939.
91. Gu, J.; Hiraga, T.; Wada, Y. *Biol. Mass Spectrom.* **1994**, *23*, 212-217.
92. Domon, B.; Costello, C.E. *Glycoconjugate J.* **1988**, *5*, 397.
93. Corcoran, J.W.; Hahn, F.E. eds; *Antibiotics III: Mechanism of Action of Antimicrobal and Antitumor Agents*, Springer-Verlag: New York, **1975**, 396-479.
94. Roberts, G.D.; Carr, S.A.; Christensen, S.B. in *Proceedings of the 35th ASMS Conference on Mass Spectrometry and Allied Topics*, Denver, CO, **1987**, 933.
95. David, L.; Scanzi, E.; Fraisse, D.; Tabet, J.C. *Tetrahedron* **1982**, *38*, 1619.
96. Barbalas, M.P.; McLafferty, F.W. Occolowitz, J.L. *Biomed. Mass Spectrom.* **1982**, *10*, 258.
97. Cooper, R.; Unger, S.E. *J. Antibiot.* **1985**, *38*, 24.
98. Siegel, M.M.; McGahren, W.J.; Tomer, K.B.; Chang, T.T. *Biomed. Environ. Mass Spectrom.* **1987**, *14*, 29.
99. Holzman, G.; Ostwald, U.; Nickel, P.; Haack, H.J.; Widjaja, H.; Arduny, U. *Biomed. Mass Spectrom.* **1985**, *12*, 659.
100. Nelson, C.C.; McCloskey, J.A.; Isono, K. in *Proceedings of the 37th ASMS Conference on Mass Spectrometry and Allied Topics*, Miami, FL, **1989**, 724.
101. Curtis, J.M.; Bradley, B.; Derrick, P.J.; Sheil, M.M.; *Org. Mass Spectrom.* **1992**, *27*, 502.
102. Vincenti, M.; Guglielmetti, G.; Andriollo, N.; Cassani, G. *Biomed. Environ. Mass Spectrom.* **1990**, *19*, 240.
103. Edwards, D.M.F.; Selva, E.; Stella, S.; Zerilli, L.F.; Gallo, G.G. *Biol. Mass Spectrom.* **1992**, *21*, 51.
104. Straub, R.; Linder, M.; Voyksner, R.D. *Anal. Chem.* **1994**, *66*, 3651-3658.
105. Cerny, R.L.; MacMillan, D.K.; Gross, M.L.; Mallams, A.K.; Pramanik, B.N. *J. Am. Soc. Mass Spectrom.* **1994**, *5*, 152-158.
106. Karas, M.; Hillenkamp, F. *Anal. Chem.* **1988**, *60*, 2299.
107. Fenn, J.B.; Mann, M.; Meng, C.K.; Wong, S.F.; Whitehouse, C.M. *Science* **1989**, 264, 64.

108. Smith, R.D.; Loo, J.A.; Edmonds, C.G.; Barinaga, C.J.; Udseth, H.R. *Anal. Chem.* **1990**, *62*, 882-899.
109. McLuckey, S.A.; Van Berkel, G.J.; Glish, G.L. *J. Am. Soc. Mass Spectrom.* **1992**, *3*, 60-70.
110. Edmonds, C.G.; Barinaga, C.J.; Loo, J.A.; Udseth, H.R.; Smith, R.D. in *Proceedings of the 37th ASMS Conference on Mass Spectrometry and Allied Topics*, Miami Beach, FL, **1989**, 844.

RECEIVED November 14, 1995

Chapter 26

Atmospheric Pressure Ionization Liquid Chromatography–Mass Spectrometry for Environmental Analysis

Robert D. Voyksner

Analytical and Chemical Sciences, Research Triangle Institute, P.O. Box 12194, Research Triangle Park, NC 27709

Atmospheric Pressure Ionization-Liquid Chromatography/Mass Spectrometry (API-LC/MS) is discussed and reviewed as related to environmental analysis. The chapter covers the use of collision induced decomposition in the API interface to obtain structural information and applications of API-LC/MS for specific classes of environmentally relevant compounds. API-LC/MS determinations are demonstrated for dyes (sulfoninated, azo, cationic, and anthraquinone types), pesticides, herbicides, amines, nitro, hydroxy and carbonyl containing compounds. In particular, keypoints in API-MS operation such as pH, acceptable buffers and derivatizations are covered to achieve optimal sensitivity. Generally, the best sensitivity is achieved when conditions are used which ionize the compound in solution (prior to introduction to the API-MS).

Future trends in environmental LC/MS are discussed highlighting new capabilities in MS instrumentation and separations. API ion trap MS or time-of-flight-MS instruments offer capabilities to improve sensitivity, specificity and scanning rates. These capabilities are important when coupling API-MS techniques to chromatographic techniques such as capillary electrophoresis.

There are growing concerns over the world's environment. The air, water, and foods are highly scrutinized for organic and inorganic contaminants that can lead to health risks. Government legislation and EPA methods have led the way for regulations of wastes, and for establishing water and air quality. Legislation such as the Clean Water Act, Safe Drinking Water Act, Resource Conservation and Recovery Act (RCRA), Clean Air Act and Toxic Substance Control Act has lead to a series of methods using gas chromatography (GC), gas chromatography/mass spectrometry (GC/MS), liquid chromatography (LC) and most recently liquid chromatography/mass spectrometry (LC/MS) to monitor the environment.

0097–6156/95/0619–0565$12.00/0

Gas chromatography/mass spectrometry has been particularly important due to its capability to confirm or identify trace residues in complex matrices (*1-2*). It is routinely used in the identification and measurement of volatiles and semivolatiles in air, food stuffs, water and wastes. Even with intense monitoring programs there are major shortcomings in the characterization of environmental media, and in particular hazardous waste. Often analytical methodology based on GC and GC/MS can only account for a small portion of the carbon mass balance of oxidation products in air, pollutants in water or components in hazardous waste (*1*). Typically this unaccounted mass is summarized as polar, thermally liable or high molecular weight material unsuitable for GC based methods. While chemical derivatization methods have increased the potential of GC based methods for the detection of polar components, what is really needed is a complement to GC/MS, namely LC/MS.

The use of LC to separate nonvolatile, thermally unstable and high molecular weight compounds has been well demonstrated (*3-4*). Furthermore, the coupling of LC with MS has been reported more than 20 years ago, yet its use for environmental monitoring is still in it's infancy. This is partially due to the complexity, ruggedness and sensitivity limitations of the early interfaces.

The development of thermospray and particle beam LC/MS overcame some of the initial interface shortcomings, permitting their use on a more routine basis (*5-6*). Several methods have been developed to monitor chlorinated phenoxyacid herbicides, azo dyes, nitroso compounds and organophosphorus pesticides using particle beam and/or thermospray LC/MS (*7-10*). While these interfaces can be used routinely to solve environmental problems they still suffer from several severe drawbacks, which limit the specificity or sensitivity of the approach. For example, thermospray spectra often lack the fragmentation necessary for compound confirmation (*11*). This weakness can be overcome with the use of tandem MS which adds to the expense and complexity of the method (*11-12*). Particle beam lacks the ability to ionize nonvolatile compounds and often is insufficiently sensitive to conduct sub-ppb determinations (*9-13*). Solutions to these problems can possibly be achieved through refinements in these analytical techniques. However, it appears that the atmospheric pressure ionization (API) technique of electrospray and atmospheric pressure chemical ionization (APCI) superseded these developments, offering the required sensitivity and specificity for modern environmental analysis.

API-MS

API-electrospray is a technique pioneered by Dole and coworkers (*14*) and combined with mass spectrometry through research by Fenn and coworkers (*15-16*), has revolutionized mass spectrometry with the ability to ionize high molecular weight molecules and detect femtomole levels of material. Previous chapters in this ACS Symposium Series Volume have described API interfaces with the emphasis on the analysis of biological molecules, particularly high molecular weight moieties. In contrast, environmental analysis involves primarily low molecular weight polar, nonvolatile and thermally unstable compounds typically in the molecular weight range of 100-800. These compounds primarily exhibit single charge molecular ions under API conditions using ion evaporation ionization (*17-18*) or chemical ionization at atmospheric pressure (APCI) (*19-20*). While tandem MS can be used to provide structural information on the molecular ion species, the cost and instrument complexity of the technique has limited its routine use in environmental analysis. On the other hand, the use of collision induced decomposition (CID) in the API transport region can provide structural information. The CID in the API transport region occurs between elements in the interface such as the capillary-skimmer or between two skimmers. The extent of fragmentation is controlled by the voltage placed on these two elements (*21*). The process has also been called "up front CID" or "cone CID".

Obtaining CID spectra in the API transport region involves controlling the potential on the capillary exit for the system (Analytica of Branford, Branford, CT) used in our laboratory. An increase in voltage accelerates the ions which undergo multiple collisions with air or nitrogen (present from the counterflow or bath gas) resulting in an increase in internal ion energy. As the voltage is further increased the internal energy exceeds the bond energies resulting in fragmentation (Figure 1).

The internal energy imparted into a molecule has been shown to vary linearly with capillary voltage and over 16 eV of internal energy can be introduced into the molecule (*21*). With typical bond energies of about 1 eV, there is sufficient internal energy to cleave numerous bonds providing the structural information necessary for confirmation. For example, compare the mass spectra for carbofuran at a capillary voltage of 50 V and 120 V (Figure 2). The spectrum at 50 V consists essentially of $[M+H]^+$ ions. At 120 V the $[M+H]^+$ ions underwent CID to generate the product ions at m/z 165, 147, 123, 91 and 60. The CID spectrum generated in the API region is qualitatively similar to the product ion spectrum recorded on a tandem MS instrument (Figure 2C). The differences can be accounted for by differences in collision energies between the two techniques. Also the switching from a low (50 V) to a high (120 V) capillary potential can be performed quickly under computer control. This not only permits computer optimization of the CID voltage for better fragmentation, but allows "toggling" of the CID voltage on alternate scans or changing the CID voltage in sequence with the mass scan to achieve the best CID sensitivity for all fragmentations. Furthermore, the CID processes in the API transport is very efficient, with few losses from scatter or neutralization, as depicted by the near constant total ion current over the voltage range evaluated (Figure 1).

The use of CID capabilities in the API interface on a single mass analyzer instrument requires the introduction of a pure sample. Since there is no mass separation or isolation as is the case with tandem MS, sample purity relies on chromatography. The presence of coeluting LC peaks will result in a compound spectrum representing all components present and can make interpretation nearly impossible. This limitation seems to be out weighed by the simplicity, cost, ruggedness and sensitivity of LC/electrospray-MS using a single analyzer. This chapter will cover the use of LC/electrospray MS for determination of a variety of compounds of environmental interest including dyes, pesticides, herbicides, amine, hydroxy, and carbonyl compounds and hydrocarbons. In particular, the use of capillary columns, fast analysis times and future techniques employing ion trap mass spectrometry (ITMS) and time-of-flight mass spectrometry (TOF-MS) will be discussed.

Application of LC/API-MS to the Environmental Analysis of Dyes

The API techniques of electrospray or pneumatic assisted electrospray (ion spray) achieve the best sensitivity for compounds that are precharged in solution. For example, ionic species or compounds that can be protonated or deprotonated by adjusting pH are well suited for the ion evaporation ionization process in electrospray.

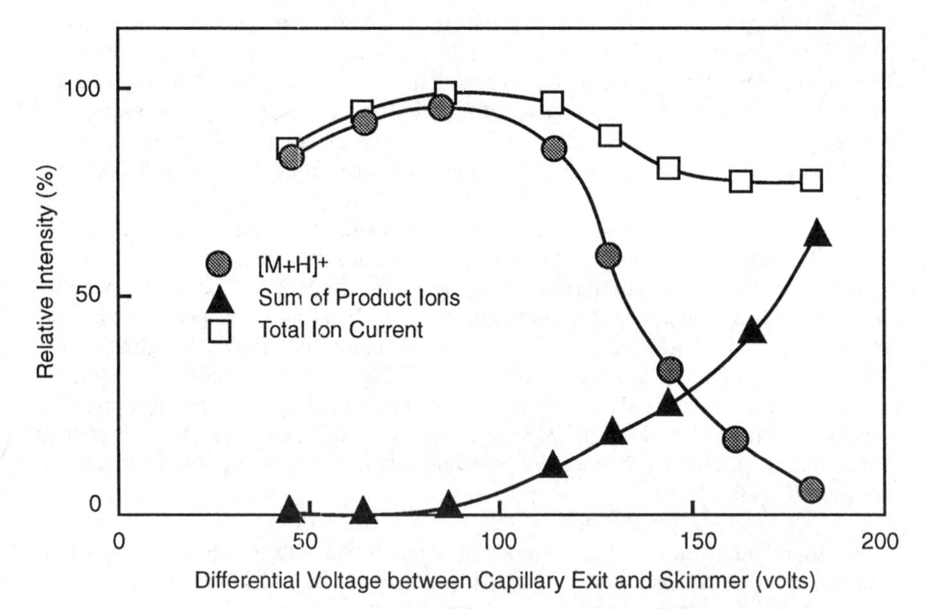

Figure 1. The relative abundance of the [M+H]$^+$ and the sum of the CID product ions for aldicarb at various capillary exit potentials.

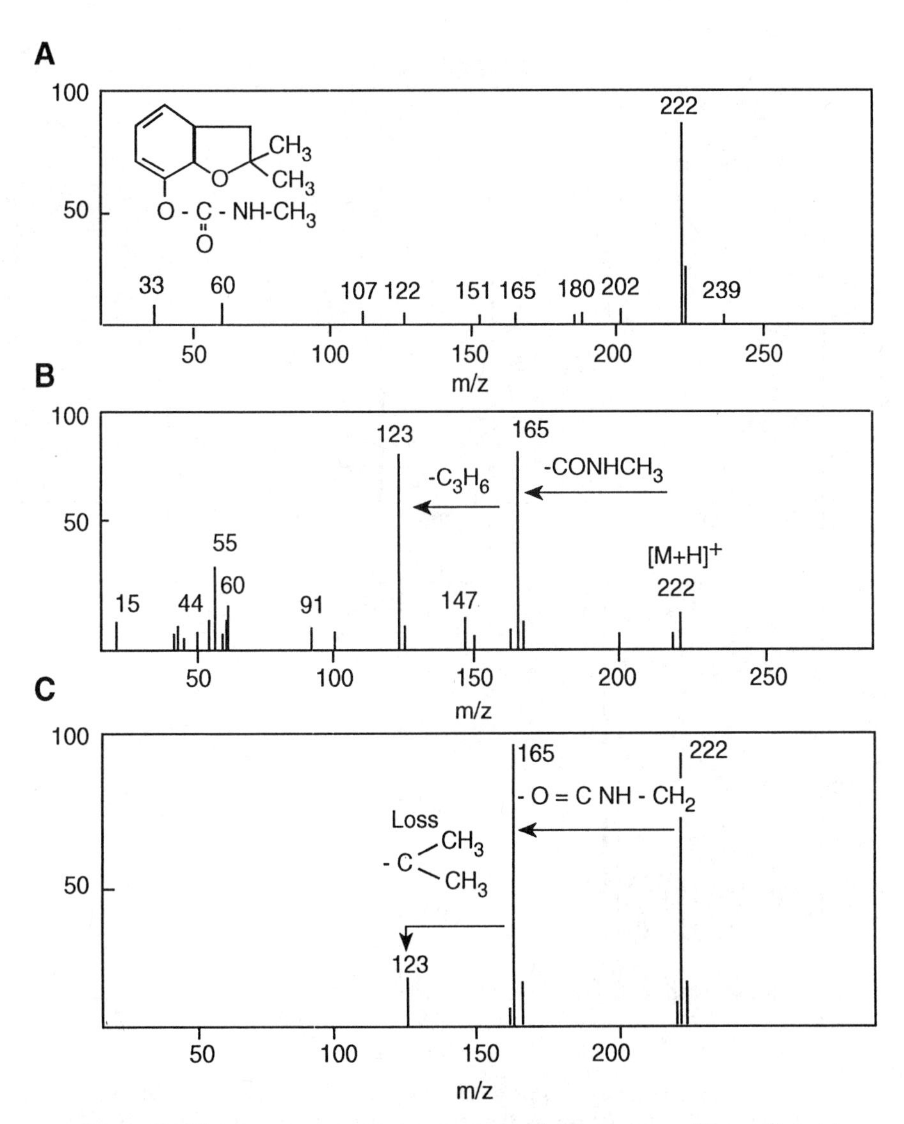

Figure 2. The electrospray mass spectrum of carbofuran at a capillary voltage of (A) 50 V (no CID); (B) 120 V (CID conditions), and (C) product ion spectrum of the $[M+H]^+$ ion of carbofuran (30 eV lab) on a triple quadrupole MS system.

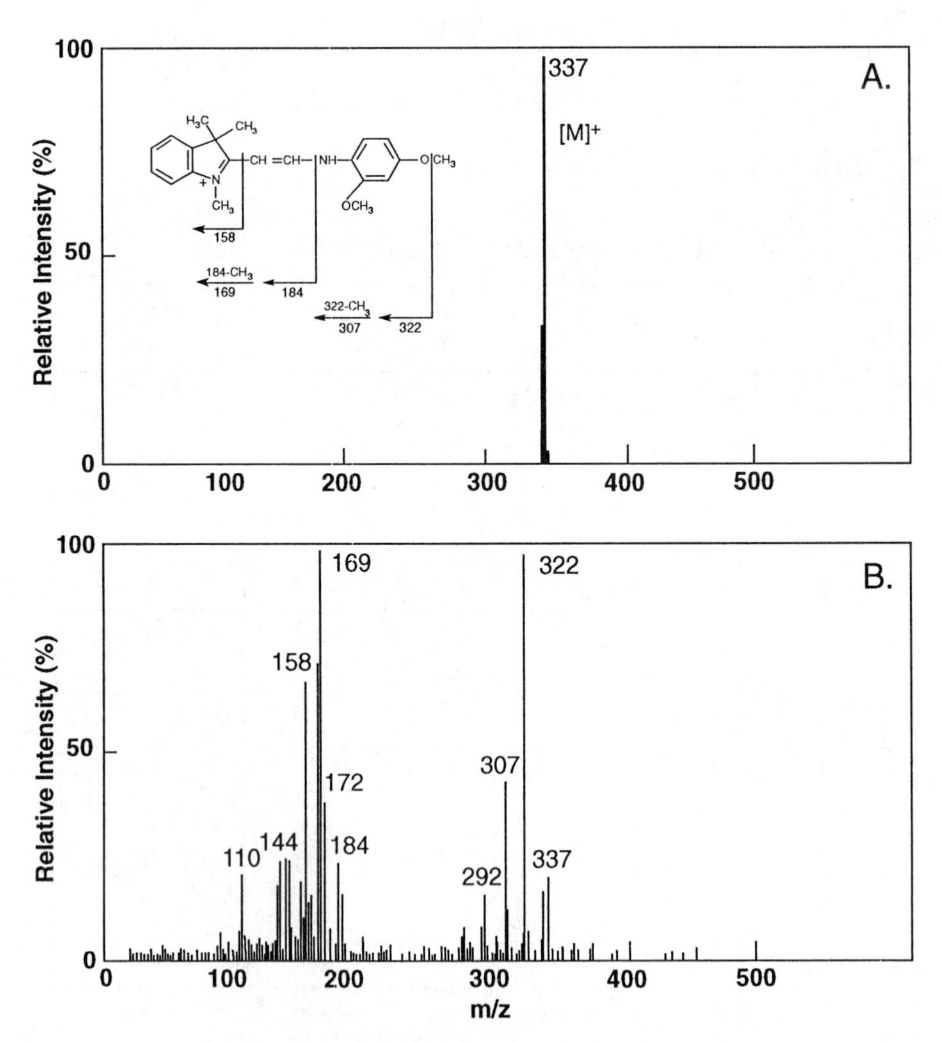

Figure 3. Positive ion electrospray mass spectra of Basic Yellow 11. (A) Non-CID conditions with the capillary exit at 120 V. (B) CID conditions with the capillary exit at 200 V.

For this reason some of the earlier environmental work centered on the analysis of sulfonated azo dyes. These dyes were problematic when analyzed by thermospray or particle beam LC/MS, contributing to the need for the development of an LC/MS analysis method. The ion spray or electrospray spectra of the sulfonated dyes consisted of an $[M-2Na]^-$ anion with little fragmentation (*22-26*). Di, tri and up to hexasulfonated azo dyes were analyzed, yielding multiply charged clusters with the highest charge state equal to the number of sulfonated groups present e.g., $[M-2Na]^{-2}$ for disulfonated, $[M-6Na]^{-6}$ for hexasulfonated (*26*). Structural information on these dyes could be generated in the API transport region, resulting in fragments from breaking the azo linkage with the charge remaining on the sulfonate. Also, the $[SO_3]^-$ ion at m/z 80 was a common CID fragment for these dyes. Tandem MS provided similar product ion mass spectral information from the molecular anion of these dyes.

The flexibility of the API-MS approach enabled coupling of separation techniques covering a wide range of flow rates. Separation performed on a C_{18} column at 2 mL/min, to analyze sulfonated azo dyes was accomplished using APCI (*22*). However, APCI usually resulted in a poorer response for polysulfonated dyes compared to electrospray due to the lack of sample volatility. Ion spray was used to analyze dyes separated on a 1.0 mm column at 40 μL/min flow rate (*22*). Also, capillary electrophoresis (CE) coupled to electrospray MS was used to separate sulfonated dyes, offering high resolution separation capabilities and high peak concentrations to achieve sub-picomole detection limits (*23-27*).

Cationic dyes have also been analyzed using positive ion detection API-MS, with great success (*28*). These precharged dyes are well suited for ion evaporation ionization analogous to the negative ion formation for sulfonated dyes. The mass spectrum of Basic Yellow II exhibits an $[M]^+$ ion at low voltages (non-CID). At a high voltage (200-240 V) on the capillary exit of an Analytica of Branford interface, numerous fragment ion for Base Yellow II are observed (Figure 3).

Azo dyes that are not ionic salts have been successfully analyzed by electrospray and APCI-MS (*26-28-29*). Dyes from the disperse and solvent classes result in optimal ion formation at low pH conditions (e.g., 1% acetic acid or formic acid), which results in sample protonation to form a cation in solution. These dyes exhibited an $[M+H]^+$ ion at low API-transport voltage (80-100 V), and structurally relevant fragment ions at higher voltages (e.g., capillary voltage of 160-200 V for CID). Several anthraquinone dyes have been analyzed by electrospray and APCI-MS (*29*). At low pH and CID voltage conditions, optimal $[M+H]^+$ ion currents could be detected. At higher CID voltage, fragment ions, primarily due to the loss of the alkyl side chains on the anthraquinone ring, were detected.

Pesticides/Herbicides. API-MS techniques of electrospray, ion spray and APCI have been used to determine a wide variety of pesticides and herbicides (*21,27,29-35*). Triazine, organophosphorus, carbamate, and chlorinated acid herbicides or pesticides have been the major classes analyzed by API-MS. The first three classes generate optimal ion signal using positive ion detection and low solution pH. The chlorinated acid herbicides and most organophosphorus compounds can be detected using negative ion detection by increasing the pH using ammonium hydroxide. Generally either mode of ionization yields molecular ions (e.g., $[M+H]^+$ for positive ion detection or $[M-H]^-$

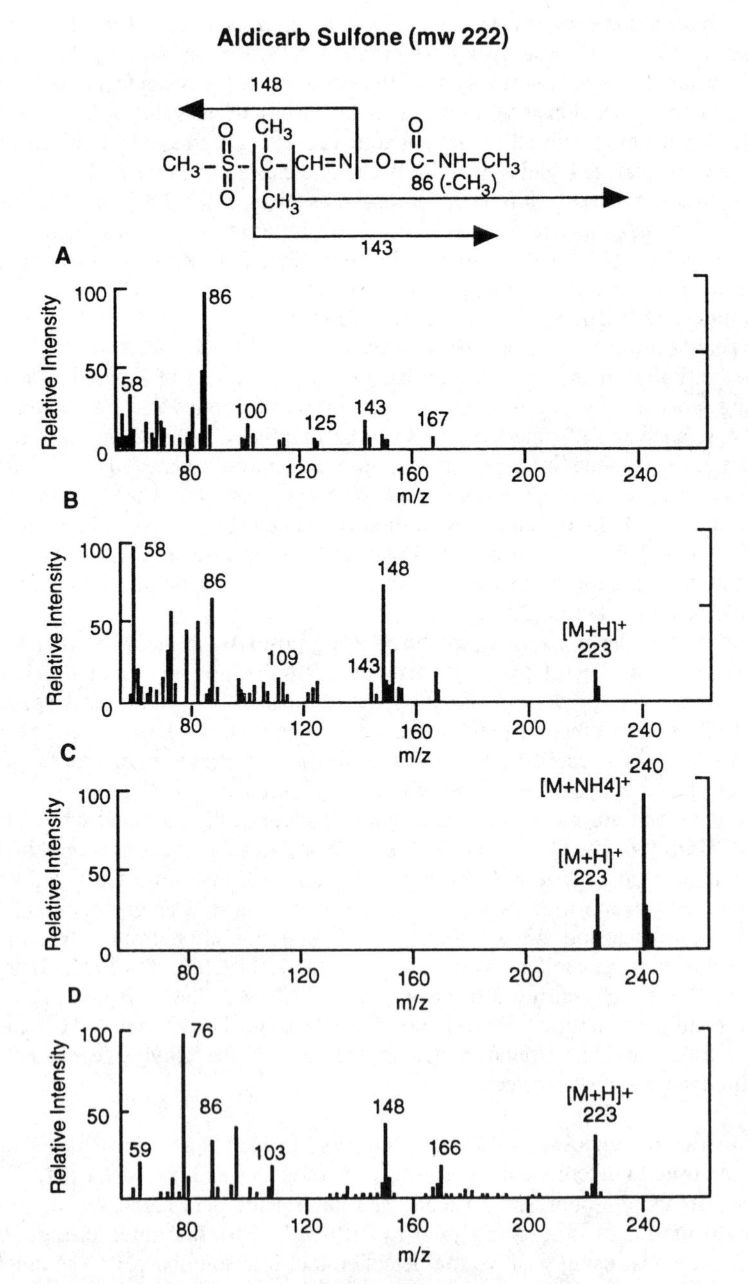

Figure 4. Comparison of mass spectra for aldicarb obtained by (A) 100 ng injected using particle beam EI, (B) 100 ng injected using particle beam methane positive CI, (C) 20 ng injected using thermospray and (D) 100 pg injected using electrospray at a capillary voltage of 90 V.

for negative ion detection). Structural information can be achieved by using a higher CID voltage in the transport region of the interface (Figure 2).

The analysis of these classes of pesticides have been previously accomplished by particle beam and thermospray LC/MS, therefore, the implementation of electrospray was to seek enhancements in sensitivity and/or specificity. A comparison of the particle beam electron ionization (EI) and chemical ionization (CI), thermospray and electrospray spectra in Figure 4 for aldicarb sulfone demonstrate the improvements from electrospray. The EI spectra from particle beam exhibits primarily fragment ions at m/z 86 and 143, which are common to aldicarb and its metabolites. The lability of the NO bond limits the usefulness of gas phase ionization techniques. On the other hand, thermospray lacks the structural information necessary for compound identification, showing only $[M+H]^+$ and $[M+NH_4]^+$ ions. Electrospray can provide the molecular ion and structurally relevant ions for confirmation by proper choice of API transport CID conditions. Furthermore, sensitivity of electrospray is far superior to these other interfaces, considering 100 pg was used to generate the electrospray spectrum compared to 20 ng for thermospray and 100 ng for particle beam spectra.

On-line LC-MS determinations of pesticides and herbicides have been reported using APCI, pneumatic electrospray and electrospray. Under APCI conditions, separation of organophosphorus compounds could be done using a 3.9 x 300 mm column at 1 mL/min (*30*). Pneumatic-assisted electrospray LC/MS has been demonstrated for the separation of selected carbamates on a 1 x 150 mm column or 0.32 x 150 mm perfusion column operating at a flow rate in the range of 50-100 μL/min (*32*) and for the separation of chlorophenoxy acid herbicides on a 3 x 125 mm column operating at 0.25 mL/min (*34*). Electrospray LC/MS was demonstrated to separate several carbamate pesticides using a capillary C_{18} column (0.32 x 150 mm) at 4 μL/min (*29*). Capillary electrophoresis - API-MS has been used to separate and detect chlorophenoxy acid herbicides (*23*) and sulfonylurea herbicides (*33*). The advantages of API-MS (APCI and ion spray MS) compared to particle beam and thermospray were compared for the analysis of carbamate pesticides (*35*). The comparison showed APCI resulted in the best detection limits (0.5-1 ng full scan) for the carbamates evaluated. Ion spray detection limits were 0.3-1.5 ng. Thermospray detection limits were 2-4 ng and particle beam detection limits were 100-240 ng (*35*).

Amines/Nitro Compounds. Electrospray or pneumatic assisted electrospray analysis of volatile and nonvolatile amines is best achieved at low pH solution conditions to ensure sample protonation. Mobile phase pH of 2-3 (acetic or formic acid) results in the best electrospray sensitivity, with the spectra primarily exhibiting $[M+H]^+$ ions under non-CID conditions in the ion transport region. APCI approaches also work well for amines that can be vaporized for gas phase ionization. The implementation of capillary LC combined with electrospray-MS was used to separate a mixture of 28 amines that are representative of the reduction products of azo dyes. Figure 5 shows the total ion current chromatogram for the separation of the amine mixture on a 0.32 x 150 mm C_{18} column operated at a flow rate of 6 μL/min in about 35 minutes. The chromatogram also indicates the variance in response factors for the amines, which were each present at the 32 ng/μL level (0.5 μL injected). The separation was sufficient to isolate most amines from one another so as to obtain CID

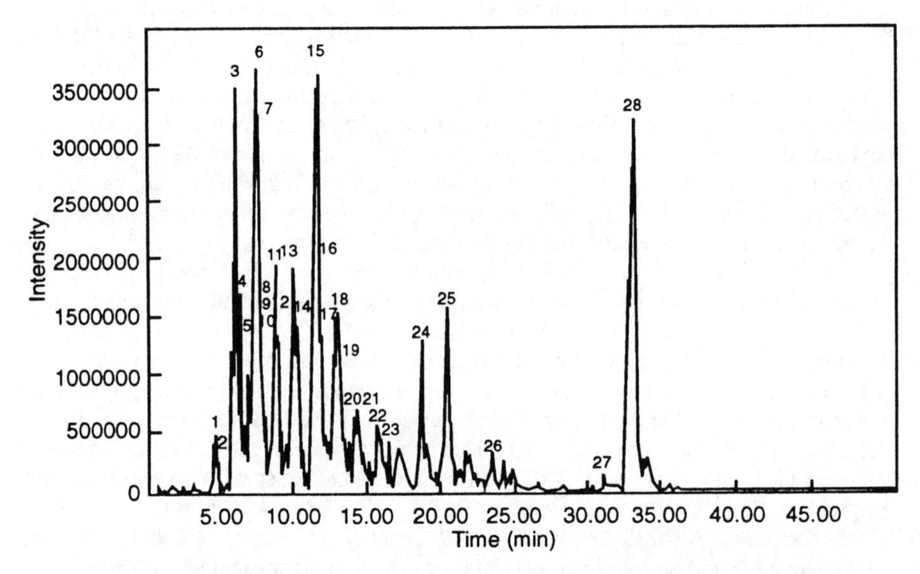

Figure 5. Capillary LC/electrospray-MS total ion current chromatogram for the separation of a mixture of aromatic amines. (Conditions: 0.32 x 150 mm C_{18} column, 5 μm particles, using a gradient of 40-90% acetonitrile (1% acetic acid) in 30 minutes at a flow rate of 6 μL/min. 16 ng of each amine was injected. The identities of the peaks are as follows: (1) aniline, (2) phenol, (3) m-phenylenediamine, (4) 2-fluoroaniline, (5) 4-fluoroaniline, (6) benzidine, (7) 2,4-dinitroaniline, (8) 3,3'-dimethoxybenzidine, (9) 2-methylaniline, (10) 2-methoxyaniline, (11) 3,3'-dimethylbenzidine, (12) 4-fluoro-2-methylaniline, (13) 4-nitroaniline, (14) ethylene dianiline, (15) 2,4-dimethylaniline, (16) 4-amino-3-nitrobenzonitrile, (17) 3,3,',5,5'-tetramethylbenzidine, (18) N,N,N',N'-tetramethylbenzidine, (19) methyl-mercaptoaniline, (20) thiochroman-4-ol, (21) 1-naphthylamine, (22) 4,4'-difluorobiphenyl, (23) 4-chloro-2-methylaniline, (24) 4,5-difluoro-2-nitroaniline, (25) 3,3'-dichlorobenzidine, (26) 2,6-dichloro-4-nitroanline, (27) 4,4'-diaminooctafluorobiphenyl, (28) diphenylamine.

spectra in the API transport region for confirmation. Nitro compounds also have been successfully analyzed with pneumatic assisted electrospray (*36*). The chlorinated nitroaromatic compounds showed the best sensitivity using negative ion detection, with structural information obtained by tandem MS.

Carbonyl and Hydroxy Compounds. The determination of oxygenated compounds of environmental interest proves a bit more challenging due to the relative difficulty for protonation or deprotonation to form ions in solution compared to the amines, acids or ionic dyes previously discussed. Hydroxy aromatic compounds have been analyzed by electrospray with positive and negative ion detection (*37*). These hydroxy aromatics exhibited $[M+H]^+$ and $[M-H]^-$ for the respective modes of detection. Often APCI is performed for the analysis of the compounds that can be transferred into the gas phase since the proton affinity of hydroxy compounds is sufficiently high to enable gas phase protonation with H_3O^+ (water CI).

Carbonyl compounds often lack a site for protonation or deprotonation to be well suited for electrospray. In this case, APCI approaches may prove superior for volatile and thermally stable molecules. Work in our laboratory to detect both volatile and nonvolatile carbonyl compounds has involved derivatization with 2,4-dinitrophenylhydrazine (DNPH) (*38*). This derivatization processes served several purposes:

(1) The formation of the DNPH derivatives helped trap volatile aldehydes and ketones which could not be sampled on Tenax cartridges.
(2) The DNPH derivatives provided a UV chromophore for LC/UV detection.
(3) DNPH has sites of protonation and deprotonation, so the addition of this moiety to a carbonyl compound would greatly improve electrospray MS response for this class of compounds.

The electrospray LC/MS ion chromatograms for the analysis of the DNPH derivatives of various aldehydes and ketones are shown in Figure 6. The $[M-H]^-$ ion for each derivative under non-CID conditions is shown. Negative ion detection (at pH 8) proved superior in sensitivity compared to positive on detection (pH 3) for these derivatives. The electrospray mass spectra of the standard aldehydes and ketones in Figure 6 exhibited the $[M-H]^-$ ion for the derivative [m/z = molecular weight of organic compound - 16 (for oxygen) + 196 (for molecular weight of DNPH) -1 (for the loss of H to form the anion)] at a low capillary voltage (120 V). A spectrum for the DNPH derivative of acetone is shown in Figure 7, as a representative spectrum for the DNPH derivative of aldehydes and ketones. At a high capillary exit voltage (140 V) several common fragment ions were detected at m/z 181 and m/z 169 (data not shown). These fragments were postulated to be $[(NO_2)_2C_6H_3N]^-$ and $[(NO_2)_2C_6H_5]^-$ from the DNPH derivative. No ions corresponding to the aldehyde or ketone were detected, therefore the fragment ion information was of limited use, although these ions can serve to screen for compounds that have undergone derivatization.

Hydrocarbons. Hydrocarbons and polynuclear aromatic hydrocarbons (PAHs) prove the most difficult to be ionized by electrospray due to the lack of sites of protonation or deprotonation. Furthermore, they have a low proton affinity limiting their ability to form ions by APCI. While most hydrocarbons and PAH's are volatile,

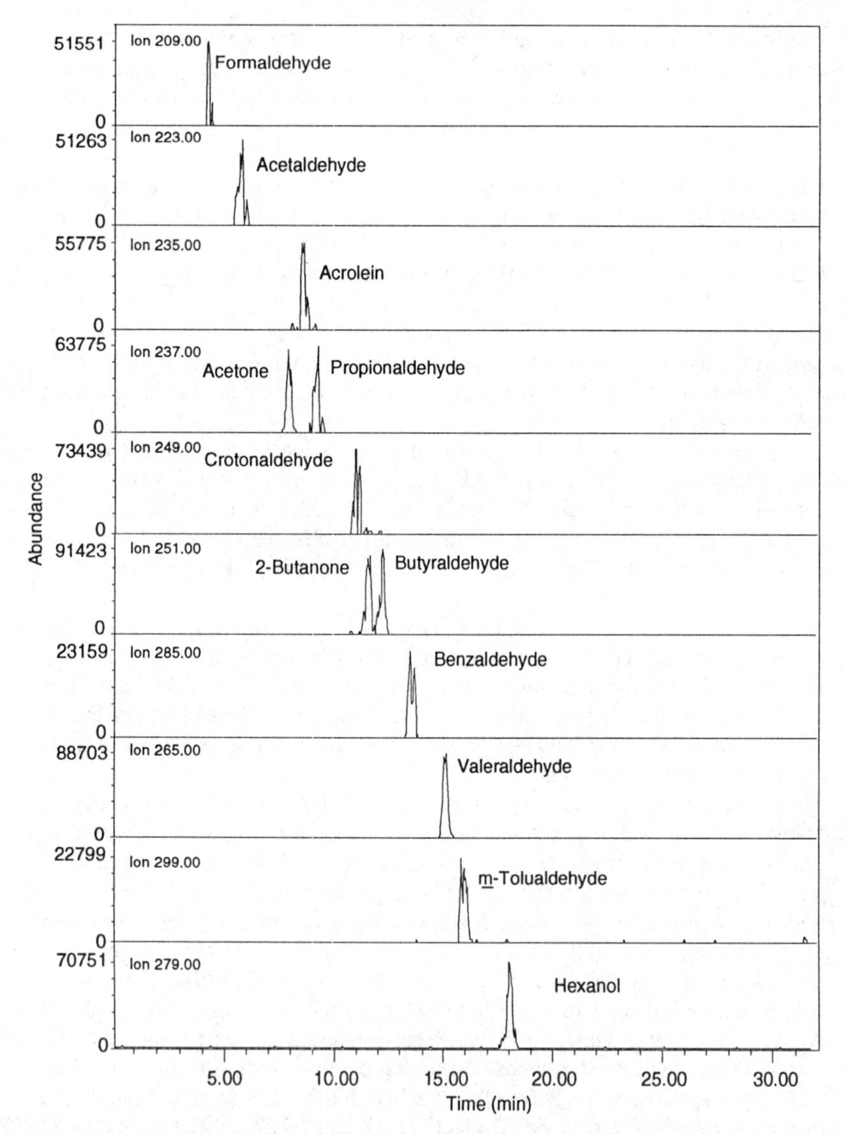

Figure 6. Ion chromatogram for the [M-H]⁻ ion of the DNPH derivatives of various aldehydes and ketones, including formaldehyde (m/z 209), acetaldehyde (m/z 223), acetone and propionaldehyde (m/z 237), acrolein (m/z 235), crotonaldehyde (m/z 249), and 2-butanone and butyraldehyde (m/z 251). The LC/MS separation was performed using a Nova Pak C_{18} 3.9 x 150 mm column with a gradient from 30% acetonitrile:10% THF in water (hold 3 min) to 60% acetonitrile in water in 10 min at 1.5 mL/min. A post-column 10:1 split reduced the flow of the electrospray interface which was mixed with 200 mM ammonium hydroxide at 0.1 mL/min to increase the pH. A 10 μL sample injection of a 60 ng/μL standard was performed. The electrospray system was operated in the negative ion mode.

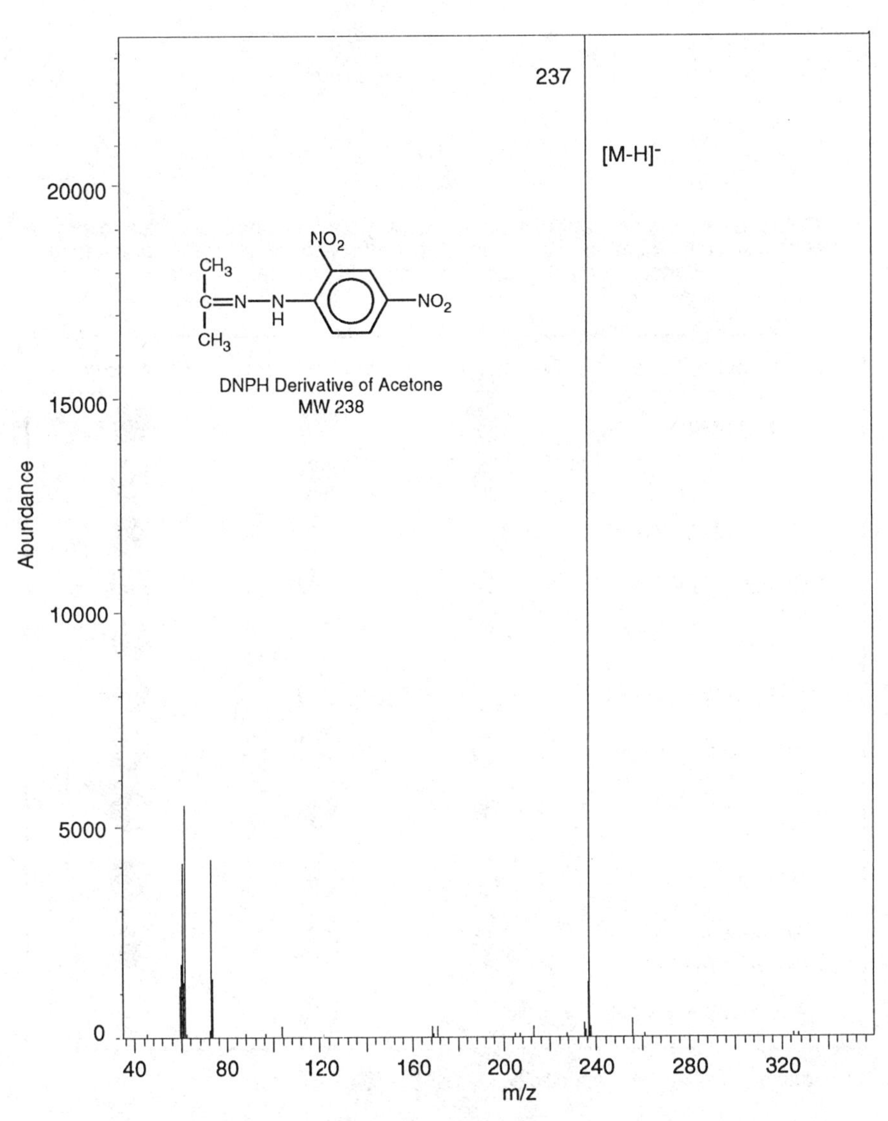

Figure 7. Electrospray negative ion mass spectrum of the DNPH derivative of acetone at a capillary exit voltage of 120 V (non-CID condition in the API transport). The ion at m/z 237 is the [M-H]⁻ ion for the acetone derivative and the signals below m/z 80 are mobile phase background ions. (See Figure 6 for LC/MS conditions.)

Table 1

Comparison of the capabilities of thermospray, particle beam and electrospray MS to provide the combination of sensitivity and specificity for the determination of various compounds classes of environmental interest

Compound Class	Thermospray	Particle Beam	Electrospray
Phenyl urea	◑	●	●
Triazines	◑	●	●
Organophosphorus	●	◑	●
Chlorinated acids	◑	◑	●
Azo dyes	◑	○	●
Sulfonated azo dyes	○	○	●
Anthraquinone dyes	◑	○	●
Alcohols, ketones, aldehydes	◑	●	◑
Nitro compounds	○	◑	●
Aromatic amines and nitro compounds	◑	◑	●
Hydrocarbons, PAH	○	●	○

Increasing sensitivity and specificity ○ ◑ ● (→ →)

certain species from these classes do not elute through a GC column due to limited volatility and thus require LC/MS. Electrospray has been demonstrated to ionize these non-polar species by charge exchange with solutions such as acetonitrile. For example, negative ion detection of fullerenes have been reported using electrospray-MS (*39*).

Conclusions and Future Trends

In conclusion, API-MS has the combination of sensitivity and specificity needed in the detection of nonvolatile and thermally unstable compounds of environmental concern (Table I). This combination is not achieved to the same degree using particle beam or thermospray LC/MS. While the electrospray ionization process is best suited to polar species or molecules with an acidic or basic site, less polar compounds (e.g., hydrocarbons, ethers, aldehydes, etc.) can be addressed using either APCI, derivatization with an easily ionizable functionality (e.g., DNPH) or solvent initiated charge exchange (e.g., fullerenes). Particle beam LC/MS is still valuable in the analysis of some of these less polar relatively volatile molecules. Furthermore, the versatility of API-MS for handling various column flow rates and mobile phases (using electrospray, pneumatic electrospray and APCI) makes it especially suitable for capillary LC and microsampling techniques and for using 2.1 or 4.6 mm i.d. capillary columns. These capabilities demonstrate that electrospray LC/MS will play a rapidly increasing role in monitoring our environment.

Future techniques in LC/MS will address improvements in sensitivity, specificity, analysis time and instrument cost. In particular, time-of-flight MS (TOF-MS) (*40-44*) and ion trap MS (ITMS) (*45-49*) show particular promise, especially when coupled to the rapid and high resolution separation technique of capillary electrophoresis (CE) (*50,51*). Both TOF and ITMS offer the ability to sample a higher percentage of the ions generated (higher duty cycle), therefore improving signal levels. TOF-MS instruments have employed orthogonal storage devices to improve the duty cycle to greater than 20-30%. Ion traps can accumulate ions in the analyzer to achieve similar duty cycles. These duty cycles are far superior to the 0.1% duty cycle achieved in a quadrupole MS scanning from 0-1000 a.m.u. This translates into the ability to obtain full scan spectra by ITMS or TOF-MS at levels that were obtained only by selected ion monitoring on a quadrupole instrument.

Secondly, the ITMS offers the ability to obtain CID spectra on a mass selected ion (MS/MS). For example, the ITMS product ion mass spectrum of the $[M]^+$ ion at m/z 337 for Basic Yellow 11 (Figure 8) exhibits several intense product ions, specific for this dye. While many of these product ions were observed for the CID of this dye in the API transport region (Figure 3B), additional specificity is achieved by mass selection prior to CID in the ITMS. This removes the need to completely resolve components with chromatography prior to generation of the product ion mass spectra which is not the case with CID in the API transport. Furthermore, tandem mass spectrometry capabilities can be achieved using post source decay on a TOF-MS (*52*).

Thirdly, scan times can be greatly increased, enabling a full scan mass spectrum to be acquired in 50 μs for a TOF-MS instrument or in 100 ms for an ITMS. These cycle times become increasingly important when considering rapid and high resolution chromatography. High resolution techniques like CE can only become viable with

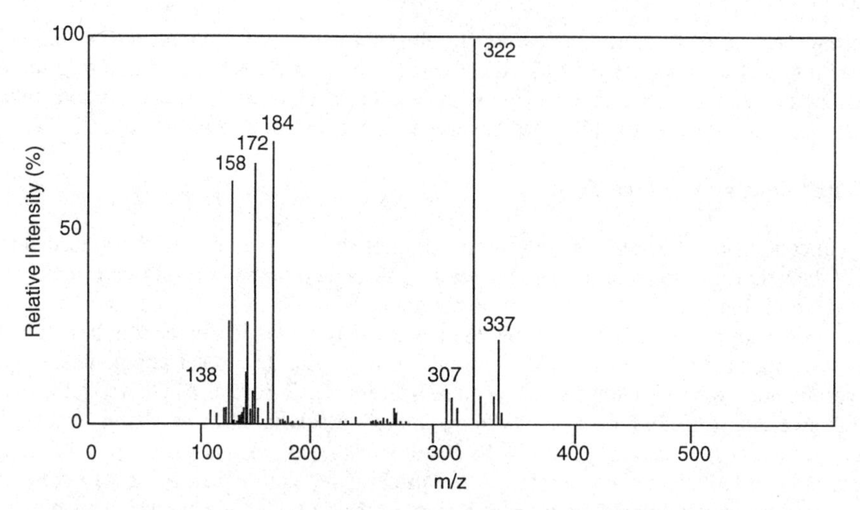

Figure 8. Electrospray-ITMS product ion mass spectrum of the $[M]^+$ ion for Basic Yellow 11. The ion at m/z 337 was isolated by a combination of Rf and dc voltages on the ring electrode and that ion was subjected to helium CID using a tickle voltage of 1.7 V for 30 ms at a q_z of 0.3.

sensitive detectors which can acquire a sufficient number of data points (mass spectra) to define a 1-3 s wide peak. To obtain 20 samples across a 1 s wide CE peak will require 50 ms cycle times which can easily be achieved by TOF-MS.

Finally, the cost of these instruments are nearly equivalent and potentially lower than quadrupole based MS analyzers. Obviously, the capabilities of the TOF-MS and ITMS combined with LC can offer significant advantages in monitoring the environment and could play a significant role in future environmental methods.

Acknowledgments

Although the research described in this article has been funded in part by the U.S. Environmental Protection Agency through Cooperative Agreement Number CR-819555 and Contract Number 68-02-4544 to Research Triangle Institute, it has not been subjected to agency review. Therefore, it does not necessarily reflect the views of the Agency. Mention of trade names or commercial products does not constitute endorsement or recommendation for use.

Literature Cited

1. Lesage, S. *J. Chromatogr.*, 1993, *642*, 65-74.
2. Holland, P. T. *Pure Appl. Chem.*, 1990, *62*, 317-336.
3. Barceló, D. *Chromatographia*, 1988, *25*, 928-936.
4. Grosser, Z. A.; Wren, J.F.; and Dong, M. W. *J. Chromatogr.*, 1993, *642*, 75-87.
5. Winkler, P. C.; Perkins, D. D.; Williams, W. K.; and Browner, R. F. *Anal. Chem.*, 1988, *60*, 489.

6. Blakely, C. R.; and Vestal, M. L. *Anal. Chem.*, 1983, *55*, 750-754.
7. U.S. Environmental Protection Agency. Methods for the Determination of Organic Compounds in Drinking Water. Supplement II (Method 8321). EPA-600/R-92-129, Washington, D.C., August 1992.
8. Behymer, T.; Bellar, T. A.; and Budde, W. L. *Anal. Chem.*, 1990, *62*, 1686-1690.
9. Jones, T. L.; Betowski, L. D.; Lesnik, B.; Chiang, T. C.; and Teberg, J. E. *Environ. Sci. Technol.*, 1991, *25*, 1880-1884.
10. Ho, J. S.; Bellar, T. A.; Eichelberger, J. W.; and Budde, W. L. *Environ. Sci. Technol.*, 1990, *24*, 1748-1751.
11. Voyksner, R. D.; and Cairns, T. Application of Liquid Chromatography - Mass Spectrometry to the Determination of Pesticide. In: Analytical Methods for Pesticides and Plant Growth Regulations, J. Sherma, ed., Academic Press, Orlando, Florida, 1989, pp. 119-164.
12. Voyksner, R. D.; McFadden, W. H.; and Lammert, S. A. *Applications of New Mass Spectrometry Techniques in Pesticide Chemistry.* J. D. Rosen, ed. John Wiley and Sons, New York, NY, 1987, pp. 247-258.
13. Voyksner, R. D.; Smith, C.; and Knox, P. *Biomed. Environ. Mass Spectrom.*, 1990, *19*, 523-534.
14. Dole, M.; Mack, L. C.; Hines, R. L.; Mobley, R. C.; and Ferguson, L. D. *J. Chem. Phys.*, 1968, *49*, 2240-2249.
15. Yamashita, M.; and Fenn, J. B. *J. Chem. Phys.*, 1984, *88*, 4451-4459.
16. Whitehouse, C. M.; Dryer, R. N.; Yamashita, M.; and Fenn, J. B. *Anal. Chem.*, 1985, *57*, 675-679.
17. Iribarne, J. V.; and Thompson, B. A. *J. Chem. Phys.*, 1976, *64*, 2287-2289.
18. Thompson, B. A.; and Iribarne, J. V. *J. Chem. Phys.*, 1979, *71*, 4451-4463.
19. Huane, G. C.; Wacks, T.; Conboy, J.J.; and Henion, J.D. *Anal. Chem.*, 1990, *62*, 713A-725A.
20. Covey, T. R.; Lee, E.D.; Bruins, A.P.; and Henion, J.D. *Anal. Chem.*, 1986, *58*, 1451A-1461A.
21. Voyksner, R. D.; and Pack, T. *Rapid Comm. Mass Spectrom.*, 1991, *5*, 263-268.
22. Bruins, A. P.; Weidolf, L. O. G.; Henion, J. D.; and Budde, W. L. *Anal. Chem.*, 1987, *59*, 2647-2652.
23. Lee, E. D.; Mück, W.; Henion, J. D.; and Covey, T. R. *Biomedical and Environmental Mass Spectrometry*, 1989, *18*, 253-275.
24. Straub, R.; Voyksner, R. D.; and Keever, J. T. *J. Chromatogr.*, 1992, *627*, 173-186.
25. Covey, T. R.; Bruins, A. P.; and Henion, J. D. *Organic Mass Spectrometry*, 1988, *23*, 178-186.
26. Lee, E. D.; and Henion, J. D. *Rapid Comm. in Mass Spectrom.*, 1992, *6*, 727-733.
27. Tetler, L. W.; Cooper, P. A.; and Carr, C. M. *Rapid Comm. in Mass Spectrom.*, 1994, *8*, 179-182.
28. Voyksner, R. D.; Betowski, L. D.; and Lin, H. -Y. Proceedings from the 42nd Annual ASMS Conference on Mass Spectrometry, May 29 - June 3, 1994, Chicago, IL, pg. 817.
29. Lin, H. -Y.; and Voyksner, R. D. *Anal. Chem.*, 1993, *65*, 451-456.

30. Kawasaki, S.; Ueda, H.; Hideo, I.; and Tadano, J. *J. of Chromatography*, 1992, *595*, 193-202.
31. Zhao, J.; Zhu, J.; and Lubman, D. M. *Anal. Chem.*, 1992, *64*, 1426-1433.
32. Voyksner, R. D. *Environ. Sci. Technol.*, 1994, *28*(3), 118-127.
33. Garcia, F.; and Henion, J. *J. of Chromatography*, 1992, *606*, 237-247.
34. Chiron, S.; Papilloud, S.; Haerdi, W.; and Barreló, D. *Anal. Chem.*, *67*, 1637-1643, 1995.
35. Pleasance, S.; Anacleto, J.F.; Bailey, M.R.; and North, D.H. *J. Am. Soc. Mass Spectrom.*, 1992, *3*, 378-397.
36. Hughes, B. M.; McKenzie, D. E.; and Duffin, K. L. *J. Am. Soc. Mass Spectrom.*, 1993, *4*, 604-610.
37. Galceran, M. T.; and Moyano, E. *J. of Chromatography A*, 1994, *683*, 9-19.
38. Raymer, J.H.; and Novotny, M.V. *Recent Developments in the Determination of Carbonyl Compounds in Biological Fluids and Tissues in Trace Analysis*, edited by J.F. Lawerence, *Vol. 3*, Academic Press Inc., Orlando, FL, 1984, 3-30.
39. Fujimaki, S.; Kudaka, I.; Sato, T.; Hiraoka, K.; Shinohara, H.; Saito, Y.; and Nojima, K. *Rapid Comm. Mass Spectrom.*, 1993, *7*, 1077-1081.
40. Sin, C. H.; Lee, E. D.; and Lee, M. L. *Anal. Chem.*, 1991, *63*, 2897-2900.
41. Boyle, J. G.; Whitehouse, C. W.; and Fenn, J. B. *Rapid Comm. Mass Spectrom.*, 1991, *5*, 400-405.
42. Verentchikov, A. N.; Ens, W.; and Standing, K. G. *Anal. Chem.*, 1994, *66*, 126-133.
43. Mirgorodskaya, O. A.; Shevchenko, A. A.; Chernushevich, I. V.; Dodonov, A. F.; and Miroshnikov, A. I. *Anal. Chem.*, 1994, *66*, 99-107.
44. Michael, S. M.; Chien, B. M.; and Lubman, D. M. *Anal. Chem.*, 1993, *65*, 2614-2620.
45. Todd, J. F. J. *Mass Spectrom. Rev.*, 1991, *10*, 3-52.
46. Nourse, B. D.; and Cooks, R. G. *Anal. Chim. Acta*, 1990, *228*, 1-21.
47. Cooks, R. G.; and Kaiser, Jr., R. E. *Acc. Chem. Res.*, 1990, *23*, 213-219.
48. Todd, J. F. J.; and Penman, A. D. *Intl. J. Mass Spectrom. Ion Process*, 1991, *106*, 1-20.
49. McLuckey, S. A.; Van Berkel, G. J.; Glish, G. L.; Huang, E. C.; and Henion, J. D. *Anal. Chem.*, 1991, *63*, 375-383.
50. Ewing, A. G.; Wallingford, R. A.; and Olefirowicz, T. M. *Anal. Chem.*, 1989, *61*(4), 292-303.
51. Smith, R. D.; Goodlett, D. R.; and Wahl, J. H. *Handbook of Capillary Electrophoresis*, 1994, Chapt. 8, 185-206.
52. Price, D.; Milnes, G.J. *Int. J. Mass Spectrom. Ion Processes*, 1990, *99*, 1-39.

RECEIVED July 21, 1995

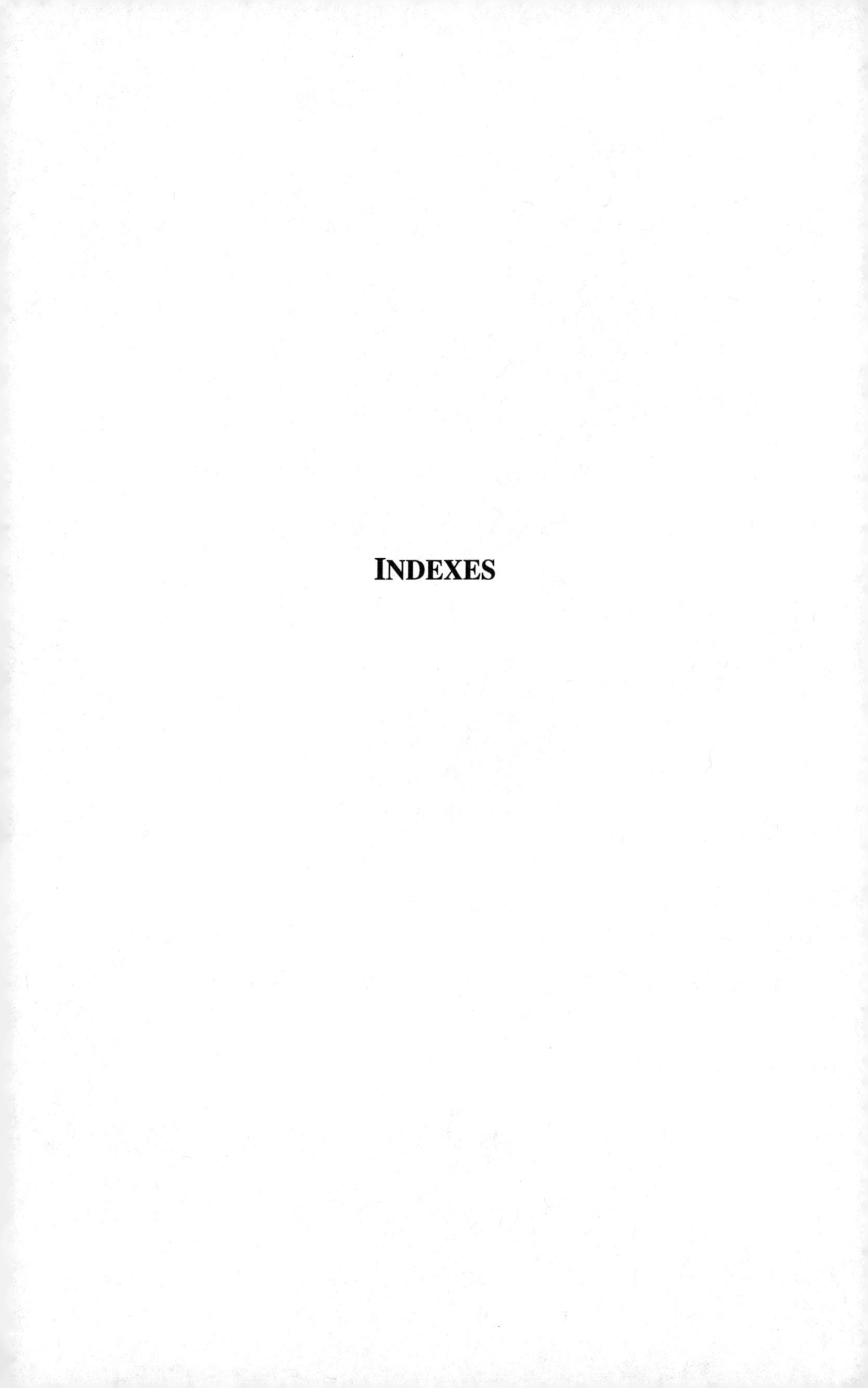

INDEXES

Author Index

Aebersold, Ruedi, 381
Affolter, Michael, 381
Altman, E., 149
Apffel, A., 432
Arnott, David P., 226
Auriola, S., 149
Banks, F. J., Jr., 60
Black, Gavin E., 81
Bonate, Peter L., 315
Booth, Matthew M., 512
Boué, Stephen M., 512
Burlingame, A. L., 472
Chakel, J., 432
Chen, Ruidan, 294
Cheng, Xueheng, 294
Clauser, K. R., 472
Cole, Richard B., 185
Cornpropst, J. David, 315
Covey, Thomas, 21
DeLong, Allyn F., 315
DeRoos, Paul C., 207
Derby, Patricia, 408
Eng, Jimmy K., 207
Engstrom, Jeffrey J., 166
Eyler, John R., 512
Fenn, J. B., 60
Fenselau, Catherine, 424
Fox, Alvin, 81
Gibson, Bradford W., 166
Gilbert, John D., 330
Gillespie, Todd A., 315
Hall, S. C., 472
Hancock, W. S., 432
Handwerger, Sandra, 106
Henzel, William J., 226
Hofstadler, Steven A., 294
Iden, Charles R., 281
Katta, Viswanatham, 408
Kelly, Michele, 424
Kerwin, James L., 244
Kim, Hee-Yong, 258
Krebs, Danielle L., 381
Lee, Mike S., 106
Lindsay, Thomas J., 315
Liu, Jinping, 106
Ma, Yee-Chung, 258
Maltby, D. A., 472
Martin, LeRoy B., 281
Masoud, H., 149
McCormack, Ashley L., 207
McLoughlin, Debra A., 330
Medzihradszky, K. F., 472
Melaugh, William, 166
Nohmi, T., 60
Olah, Timothy V., 330
Phillips, Nancy J., 166
Pucci, Michael J., 106
Pungor, E., Jr., 432
Quilliam, M. A., 351
Reinhold, Bruce B., 130
Reinhold, Vernon N., 130
Richards, J. C., 149
Rieger, Robert A., 281
Rohde, Michael F., 408
Rosell, J., 60
Ross, N. W., 351
Rudensky, Alexander Y., 207
Rush, Robert S., 408
Sadovskaya, I., 149
Samelson, Lawrence E., 381
Schwartz, Brenda L., 294
Shen, S., 60
Shipley, Lisa A., 315
Skaggs, Theresa G., 315
Smith, Richard D., 294
Snyder, A. Peter, 1
Souders, C., 432
Stephenson, James L., Jr., 512
Stults, John T., 226
Thibault, P., 149
Torres, M. Cecilia, 281
Udiavar, S., 432

Volk, Kevin J., 106
Voyksner, Robert D., 565
Walls, F. C., 472
Wang, Tao-Chin Lin, 258
Wange, Ronald L., 381
Watts, Julian D., 381
Withers, Stephen G., 365
Yates, John R., III, 207
Yost, Richard A., 512

Affiliation Index

Amgen, Inc., 408
Analytica of Branford, Inc., 60
Berlex Biosciences, 432
Boston University School of Medicine, 130
Bristol-Myers Squibb, 106
Fisons Instruments, 281
Genetech, Inc., 226
Hewlett-Packard Laboratories, 432
Lilly Research Laboratories, 315
Merck Research Laboratories, 330
NESTEC Ltd. Research Center, 381
National Institutes of Health, 258,381
National Research Council
of Canada, 149,351
Nohmi Bosai Ltd., 60
Pacific Northwest Laboratory, 294
Perkin Elmer-Sciex, 21
Research Triangle Institute, 565
Rockefeller University, 106
State University of New York—
Stony Brook, 281
U.S. Army Edgewood Research,
Development and Engineering Center, 1
University of British Columbia, 365,381
University of California—
San Francisco, 472
University of California—San Francisco
School of Pharmacy, 166
University of Florida, 512
University of Maryland—Baltimore
County, 424
University of New Orleans—Lakefront, 185
University of South Carolina School
of Medicine, 81
University of Washington, 207,224,381
Virginia Commonwealth University, 60

Subject Index

A

Acid–base catalyst identification in
glycosidases, labeling of *Cellulomonas fimi* exoxylanase/glucanase, 374*f*–378
Acid hydrolysis, lipid A, 196*f*–200*f*
Activated I cells, electrospray ionization
MS analysis of inducible protein
phosphorylation sites, 381–405
Active site nucleophile, identification,
367–375
Active site residues in glycosidases
acid–base catalyst identification, 374–378
active site nucleophile identification,
367–375
experimental description, 367
Aerospray ionization, description, 62
Affinity capillary electrophoresis,
binding studies, 121–127
Alzheimer's disease, xanomeline analysis
using electrospray ionization tandem
MS, 315–329
Amines, environmental analysis, 573–575
Analytical characteristics, electrospray
ionization process, 21–58
Analytical methodology
challenges, 432
multidimensional techniques, 433
Angiotensin I, analysis, 533,535,536*f*
Antibiotic(s)
carbohydrate, *See* Carbohydrate antibiotics
macrolide, 554–556

Antibiotic-resistant bacteria, identification of cytoplasmic peptidoglycan precursor, 106–128
Atmospheric pressure ion evaporation, description, 61
Atmospheric pressure ionization
 advantages, 333
 applications, 333
 use in heated nebulizer interface, 332–333
Atmospheric pressure ionization III MS, advantages, 332
Atmospheric pressure ionization LC–MS
 advantages, 580
 collisionally induced spectrum procedure, 567,568*f*
 comparison to other techniques, 578*t*,579
 description, 566–567
 environmental analysis
  amines/nitro compounds, 573–575
  carbonyl and hydroxy compounds, 575–577*f*
  dyes, 567,570–571
  hydrocarbons, 575,579
  pesticides/herbicides, 571–573
 experimental description, 565
 future work, 579–580
 internal energy vs. capillary voltage, 567,569*f*
Atmospheric pressure ionization tandem MS, simultaneous determination of drugs and metabolites, 344–347
Avidin complexes, electrospray ionization MS, 307–308
Azo dyes, analysis using atmospheric pressure ionization LC–MS, 571

B

Bacillus subtilis xylanases, active site nucleophile identification, 361–372
Bacteria, antibiotic resistant, identification of cytoplasmic peptidoglycan precursor, 106–128
Bacterial cell hydrolysate characterization
 LC–MS, 87,90–98
 LC–tandem MS, 81–104
Bacterial endotoxins
 cell wall regions, 186*f*,187
 function, 185,187
 structural characterization of lipid A component, 185–203
Bacterial lipopolysaccharides, toxicity mechanism, 187
Bacterial surfaces, use of electrospray ionization MS, 12–13
Base hydrolysis, lipid A, 199,201–203*f*
Binding, cytoplasmic peptidoglycan precursor in antibiotic-resistant bacteria, 121–127
Bioanalytical chemistry, quantitative, HPLC–atmospheric pressure ionization tandem MS, 330–348
Biological analysis of proteins, *See* Protein biological analysis
Biological applications, electrospray ionization MS, 12–16
Biological macromolecules
 advantages, 279
 electrospray ionization MS, 268,270–272
 experimental description, 267–268
 HPLC–electrospray ionization MS, 269,273,274
 incorporation of 22:6n3 to C-6 glioma cells, 275–277*f*
 polyunsaturated phospholipid turnover in mouse brain, 275,278–279
 quantitation, 275
 sensitivity, 269,272
 structural characterization, 472
 study methods, 267–268
Biological matrix phospholipids, analytical problems, 267
Biomolecule analysis
 CO_2 lasers
  continuous wave laser, 527,528*f*
  pulsed laser, 527,529*f*
 collisionally induced dissociation
  peptides and proteins, 527,530–534,536
  RNA dimers, 559
 collisionally induced dissociation–photodissociation, monosaccharide cleavage, 538*f*–545

Biomolecule analysis—*Continued*
collisionally induced dissociation–pulsed photodissociation
raffinose, 545–550
stachyose, 547,550–553
development, 515
electrospray ionization ion injection system, 523–525
electrospray ionization source, 518*f*,519
experimental procedure, 519–529
future work, 559
instrumentation, 519–529
ion population vs. dissociation efficiency, 516–517,519
ion sources, 515
ion trap ring electrode for photoadsorption path length increase, 56
modified electrospray ionization source, 519,520*f*
modified ring electrode, 525,526*f*
photodissociation
erythromycin, 554*f*–556
gramicidin D, 535,537,538*f*
human angiotensin I, 533,535,536*f*
RNA dimers, 556–559
trapping potential function for quadrupole, hexapole, and octopole, 521–523
use of radio frequency only multipole, 515
Biotin complexes, electrospray ionization MS, 307–308
Bovine adenosine deaminase, mass map of peptides from digestion, 429–430
Bovine fetuin tryptic digest, selective detection, 485,500–501*f*
N-(Bromoacetyl)glycosylamines, labeling of acid–base catalyst, 375

C

Capillary electrophoresis electrospray MS, structural characterization of lipopolysaccharides from *Pseudomonas aeruginosa*, 149–163
Capillary GC, advantages and disadvantages, 331
Capillary HPLC, comparison to capillary LC–MS for protein identification, 239
Capillary HPLC–electrospray ionization–magnetic sector multichannel array detection instrumentation, use for protein biological analysis, 503,505–510
Carbohydrate(s)
analysis, 537–553
characterization
GC–MS, 83
LC–tandem MS, 90,99,–103*f*
tandem MS, 83
importance of chromatographic characteristics in identification, 82
in whole bacterial cell hydrolysates, characterization using LC with electrospray tandem MS, 81–104
Carbohydrate antibiotics
photodissociation, 551,554–556
structures, 551,554*f*
Carbonyl compounds, environmental analysis, 575–577*f*
Cationic dyes, analysis using atmospheric pressure ionization LC–MS, 570,571
Cell's ability to respond to external stimuli, role of cell surface receptor complexes, 381
Cellulomonas fimi exoxylanase/glucanase, labeling of acid–base catalyst, 374*f*–378
Charged droplets, gas-phase ion formation from solute species, 60–79
Charged residue model of charged droplet formation
comparison to ion desorption model, 66–67,69,78–79
description, 65–66
experimental evidence
complex ions, 73–75
droplet evaporation rate, 76–77
metal cation behavior, 71,74*f*
peptides, 71–73
tetraalkylammonium ions, 68*f*–71
schematic representation, 64*f*,65
Chemical ionization, function, 90

Chemical ionization techniques, usage trends, 8–10
Chinese hamster ovary cells, characterization of recombinant glycoproteins, 408–422
Collisionally induced dissociation
advantages, 512–513
at low energy, structural characterization of prokaryotic glycans and oligosaccharides, 132–133
characterization of recombinant DNA derived glycoproteins, 461–464
complex protein mixture analysis, 472–510
deoxynucleotide sequence verification of multiply charged molecular ions in triple quadrupole instrument, 290*f*–292
O-deacylated lipooligosaccharides, 179–181
Collisionally induced dissociation fragmentation process, comparison to photodissociation, 513–559
Complex protein mixture analysis
protein primary sequence determination, 474–479*f*
strategy, 473–474
two-dimensional sodium dodecyl sulfate–polyacrylamide gel electrophoresis of melanoma proteins, 474,480–499
Complications, determination of molecular weights of proteins and peptides, 424–430
Concentration-sensitive detector, description, 43
Concentration sensitivity, flow rates in electrospray ionization, 45,49*f*
Conduritol epoxides, labeling of acid–base catalyst, 375
Core oligosaccharides, structural characterization of lipopolysaccharides from *Pseudomonas aeruginosa*, 150*f*,151
Covalent protein modifications
analysis using MS, 472–510
selective detection, 485,500–504
Cyclosporin A, charged droplet formation, 71–73
Cytoplasmic peptidoglycan precursor in antibiotic-resistant bacteria
binding studies, 121–127
experimental description, 111,113
future work, 125,128
instrumentation, 112*f*,113
previous studies, 107,111
structure analysis, 114–121

D

Data processing for atmospheric pressure ionization LC–tandem MS, automation, 346–347
Deoxynucleotide sequence, determination method, 283
Desorption, definition, 61–62
Diarrhetic shellfish poisoning, description, 351
Diarrhetic shellfish poisoning toxins and metabolites in plankton and shellfish
experimental description, 353–354
ionspray MS, 355–357
plankton sample analysis, 359–363
shellfish tissue analysis, 357–359
2,4-Dinitrophenylhydrazine, analysis using atmospheric pressure ionization LC–MS, 575–577*f*
Dinophysis species, shellfish contamination, 351
Dinophysis toxins
isolation of derivatives, 351–353
structures, 351,352*t*
Direct injection MS, characterization of underivatized carbohydrates, 83–86
DNA-derived glycoproteins, recombinant, characterization using hyphenated liquid-phase analyses–MS approaches, 432–469
Dole, Malcolm, use of electrically charged droplets as ion source for MS, 60–61
Drug(s), simultaneous determination using atmospheric pressure ionization LC–tandem MS, 344–347
Drug–oligonucleotide associations, electrospray ionization MS, 302–305*f*
Dyes, environmental analysis, 567,570–571

Dynamic range, electrospray ionization, 50–51

E

Edman degradation, protein amino acid determination, 226
Electrically charged droplets as source of ions
for MS, 60
in atmosphere, 61–62
Electrohydrodynamic ionization, description, 61
Electron ionization, function, 90
Electrospray ionization
amenable molecules
inorganic anions and cations, 54,56–57*f*
nonpolar neutral species, 51,54,55*f*
polar neutral species, 51–53*f*
species charged in solution, 51
application areas, 2
combination with photodissociation, 514
current status, 104
development, 2–3
dynamic range, 50–51
evolution, 1–2
factors affecting sensitivity, 50
flow rates
column diameter, 23,24*f*
concentration sensitivity, 45,49*f*
high flow range, 35,42–48
intermediate flow range, 33–41
low flow range, 25,27–33
mass sensitivity, 45
sprayer design, 23,25,26*f*
future improvements, 54,58
instrumentation, 22
interface with MS, 2
linearity, 50–51
LC separations, 81–82
names and abbreviations for techniques, 3–7
prerequisite, 22
usage, 22
usage trends, 8–10,12
use for protein biological analysis, 472–510
versatility, 22
Electrospray ionization–capillary HPLC–magnetic sector multichannel array detection instrumentation, use for protein biological analysis, 503,505–510
Electrospray ionization–collisionally induced dissociation–tandem MS, structural characterization of prokaryotic glycans and oligosaccharides, 130–146
Electrospray ionization ion trap MS, use for biomolecule analysis, 512–559
Electrospray ionization LC–MS
characterization of recombinant DNA derived glycoproteins, 432–469
structure analysis, 114–121
use in identification of N- and O-linked glycosylation sites, 433
Electrospray ionization MS
advantages
general, 282,424,430
structural determination
carbohydrate moieties, 409
lipid A, 188
structure–function analysis of bacterial lipooligosaccharides, 167,170
analysis of signal transduction pathways, 381–405
applications, 424
biological applications, 12–16
characterization
polyunsaturated phospholipid remodeling in mammalian cells, 267–279
recombinant glycoproteins from Chinese hamster ovary cells, 408–422
underivatized carbohydrates, 83–86
comparison to atmospheric pressure ionization LC–MS, 578*t*,579
future, 16–17
lipid metabolism of *Lagenidium giganteum* and hosts, 244–265
lipopolysaccharides from different pathogenic microorganisms, 151
noncovalent complexes of nucleic acids and proteins, 294–312
oligodeoxynucleotide analysis
degradation of oligomers containing 8-oxo-2'-deoxyguanosine, 287–289,291

Electrospray ionization MS—*Continued*
oligodeoxynucleotide analysis—*Continued*
deoxynucleotide sequence verification using collisionally induced dissociation of multiply charged molecular ions in triple quadrupole instrument, 290*f*–292
experimental description, 284–285
spectra, 285–288*f*
structural characterization of prokaryotic glycans and oligosaccharides, 130–146
structure and function determination of surface glycolipids in pathogenic *Haemophilus* bacteria, 166–183
use for structural characterization of biological macromolecules, 473
Electrospray ionization tandem MS
advantages, 315
identification of active site residues in glycosidases, 365–378
xanomeline analysis, 315–329
Electrospray nebulizer, variations for maximum performance, 25,26*f*
Electrospray tandem MS, characterization of underivatized carbohydrates, 83–86
Electrospray volatilization/ionization of oligodeoxynucleotides, ion formation, 282
Endotoxins
bacterial, structural characterization of lipid A component, 185–203
from *Pseudomonas aeruginosa* serotype O6, analysis of underivatized lipopolysaccharides, 149–163
Enterococcus faecalis, mechanism of glycopeptide resistance, 106–128
Environment, concerns about organic and inorganic contaminants, 565
Environmental analysis
atmospheric pressure ionization LC–MS, 565–580
electrospray ionization MS, 16
Environmental monitoring, use of LC techniques, 565–566
Erythromycin
analysis, 554*f*–556
structures, 551,554*f*
Exocyclic glycosyl epoxides, labeling of acid–base catalyst, 375
Exopolysaccharides, structural characterization, 133–134
Expressed sequence tag data bases, growth of partial complementary DNA sequence entries, 240

F

Fast atom bombardment MS
lipopolysaccharides from different pathogenic microorganisms, 151
use for oligodeoxygenated analysis, 282
Fast atom bombardment–secondary ion MS desorption techniques, usage trends, 8–10
Fatty acids, identification, 256–263,265
Flow rates in electrospray ionization
column diameter, 23,24*f*
concentration sensitivity, 45,49*f*
high flow range, 35,42–48
intermediate flow range, 33–41
low flow range, 25,27–33
mass sensitivity, 45
sprayer design, 23,25,26*f*
FRAGFIT, use in protein identification, 235,237,238*t*
Functions, surface glycolipids in pathogenic *Haemophilus* bacteria, 166–183

G

Gas chromatography (GC)–MS
characterization of carbohydrates, 83
comparison to LC–MS, 90,95–96
current status, 99,104
use for environmental monitoring, 565–566
Gas chromatography (GC)–tandem MS
advantages and disadvantages, 331
current status, 104

Gas-phase ions, formation from solute species in charged droplets, 60–79
Gene sequencing, comparison to electrospray ionization MS, 425
Glycans, structural characterization, 138–141
Glycerophosphoethanolamine, identification, 251–256
Glycerophospholipids, metabolism, 244–265
Glycolipids, surface, *See* Surface glycolipids in pathogenic *Haemophilus* bacteria
Glycopeptide(s), response, 433,436–437*f*
Glycopeptide oligomers, use of electrospray ionization MS, 12
Glycoproteins, recombinant DNA derived, characterization using hyphenated liquid-phase analyses–MS approaches, 432–469
Glycosidases
 active site residue identification, 366–378
 functions, 365–366
 identification of active site residues, 365–378
 mechanistic categories, 366
Gramicidin D, analysis, 535,537,538*f*
Gramicidin S, charged droplet formation, 71–73

H

Haemophilus, diseases caused, 166
Heated nebulizer interface, description, 332–333
Herbicides, environmental analysis, 571–573
3-[4-(Hexyloxy)-1,2,5-thiadiaxzol-3-yl]-1,2,5,6-tetrahydro-1-methylpyridine, *See* Xanomeline
High-performance capillary electrophoresis (HPCE), characterization of recombinant DNA derived glycoproteins, 432–469
High-performance LC (HPLC)
 characterization of recombinant DNA derived glycoproteins, 432–469
 development, 331–333
High-performance LC (HPLC)—*Continued*
 with atmospheric pressure ionization tandem MS for quantitative bioanalytical chemistry
  good laboratory practice, 347–348
  isotopes, 340–344
  LC–MS vs. LC–tandem MS, 334–339
  support of basic research, 347
High-performance LC (HPLC)–electrospray ionization MS
 analysis of T cell receptor ζ phosphorylation sites, 389,397–401*f*
 characterization of polyunsaturated phospholipid remodeling in mammalian cells, 267–279
High-performance LC (HPLC)–electrospray ionization tandem MS, identification of active site residues in glycosidases, 365–378
High-performance LC (HPLC)–MS, use for structural characterization of biological macromolecules, 473
High-performance LC (HPLC) separation, complex protein mixture analysis, 472–510
High pH anion-exchange LC, current status, 99
High pH anion-exchange LC–MS, carbohydrate characterization, 87,90–98
Human angiotensin I, analysis, 533,535,536*f*
Human lysosomal glucocerebrosidase, active site nucleophile identification, 371,373–375
Human urine and plasma, xanomeline analysis using electrospray ionization tandem MS, 323–328
Hydrocarbons, environmental analysis, 575,579
Hydroxy compounds, environmental analysis, 575–577*f*
Hydroxy fatty acids, metabolism, 244–265
Hyphenated liquid-phase analyses–MS approaches, characterization of recombinant DNA derived glycoproteins, 432–469

Hyphenated techniques, development, 433

I

Iminobiotin complexes, electrospray ionization MS, 307–308
Immobilized metal affinity chromatography–HPLC–electrospray ionization MS, ZAP–70 phosphorylation site analysis, 388–386
Immunoassay
 applications, 331
 requirements, 330–331
Immunological samples, analysis using microcolumn LC electrospray ionization tandem MS, 207–223
Immunoreceptor tyrosine based activation motif, description, 383
In-gel digestion, complex protein mixture analysis, 472–510
Inducible protein phosphorylation sites in activated T cells, analysis using electrospray ionization MS, 381–405
Infrared multiple photon dissociation, studies, 514
Inhibitor–protein complexes, electrospray ionization MS, 308–311*f*
Inorganic anions and cations, electrospray ionization, 54,56–57*f*
Interface conditions, observation of noncovalent associations, 295,297–298
Inverting mechanism, glycosidases, 366
Ion cyclotron resonance, combination with photodissociation, 514
Ion desorption, definition, 61–62
Ion desorption model of charged droplet formation
 comparison to charged residue model, 66–67,69,78–79
 description, 65–66
 experimental evidence
  complex ions, 73–75
  droplet evaporation rate, 76–77
  metal cation behavior, 71,74*f*
  peptides, 71–73
  tetraalkylammonium ions, 68*f*–71
Ion desorption model of charged droplet formation—*Continued*
 schematic representation, 64*f*,65
Ion formation, mechanisms, 283
Ion trap ring electrode for photoadsorption path length increase, use for biomolecule analysis, 515–516
Ions without droplet evaporation, evidence, 62–63,65
Ionspray interface, description, 333
Ionspray LC–MS, analysis of diarrhetic shellfish poisoning toxins and metabolites in plankton and shellfish, 351–363
Iribarne, charged droplets as source of ions in atmosphere, 61–62
Isotopes, HPLC–atmospheric pressure ionization tandem MS for quantitative bioanalytical chemistry, 340–344

L

Lactobacillus casei, mechanism of glycopeptide resistance, 106–128
3-*O*-Lactylglucosamine, *See* Muramic acid
Lagenidium giganteum
 description, 244–245
 development as mosquito control agent, 245
 lipid metabolism, 244–265
Laser photodissociation, use for biomolecule analysis, 512–559
Lecithin–cholesterol acyltransferase, selective detection, 48
Lentevirus proteins processed from *gag* preproteins, molecular weights, 425–426
Leuconostoc mesenteroides, mechanism of glycopeptide resistance, 106–128
Linearity, electrospray ionization, 50
Lipid(s), use of electrospray ionization MS, 13–14
Lipid A
 acid hydrolysis, 196–200
 base hydrolysis, 199,201–203*f*
 structural characterization, 141–145

Lipid A component in bacterial endotoxins, structural characterization, 185–203
Lipid A disaccharide, structural characterization, 158–159
Lipid metabolism of *Lagenidium giganteum* and hosts
advantages, 245
experimental description, 245,246
fatty acid and oxygenated fatty acid identification, 256–263,265
glycerophosphoethanolamine identification, 251–256
loss of phosphoethanolamine from precursor ion, 247,250*f*
positive ion MS
Culex tarsalis, 247,249*f*
Lagenidium giganteum, 247,248*f*
process, 245
resolution and sensitivity, 247,250*f*,264*f*,265
Lipooligosaccharides
description, 149
from different pathogenic microorganisms, structural characterization methods, 151
structure and function in pathogenic *Haemophilus* bacteria, 166–183
Lipopolysaccharides
structural characterization, 149–163
See also Surface glycolipids in pathogenic *Haemophilus* bacteria
Liquid chromatography (LC), use for environmental monitoring, 566
Liquid chromatography (LC)–electrospray ionization MS
complex protein mixture analysis, 472–510
selective detection, 485,500–504
Liquid chromatography (LC)–electrospray ionization tandem MS
advantages, 81–82
characterization of carbohydrates in whole bacterial cell hydrolysates, 82–104
xanomeline analysis, 315–329
Liquid chromatography (LC)–MS
carbohydrate characterization, 87,90–98
combination with HPLC, 331–333
comparison to
GC–MS, 90,95–96
LC–tandem MS, 334–339
cross-validation, 337–339
current status, 99
ionspray, 351–363
isotopes, 340–344
Liquid chromatography (LC)–pulsed amperometric detection, bacterial hydrolysate characterization, 90–98
Liquid chromatography (LC)–tandem MS
advantages for carbohydrate analysis, 83
carbohydrate characterization, 90,99–103
comparison to LC–MS, 334–339
current status, 99,104
isotopes, 340–344
protein identification from two-dimensional electrophoresis gels, 226–240
simultaneous determination of drugs and metabolites, 344–347
structure analysis, 114–121
Liquid matrix secondary ion ionization, use for protein biological analysis, 472–510
Liquid-phase analyses–MS approaches, hyphenated, characterization of recombinant DNA derived glycoproteins, 432–469

M

Macrolide antibiotics, photodissociation, 554–556
Major histocompatibility complex class I antigens recognized by specific I cells, identification methods, 208
Major histocompatibility complex class II alleles associated with rheumatoid arthritis, microcolumn LC electrospray ionization tandem MS, 213–214,216*t*,219*f*

Major histocompatibility complex class II presentation mutants, microcolumn LC electrospray ionization tandem MS, 215,217–218,220–223
Major histocompatibility complex molecules, presentation of peptide antigens, 207–208
Major histocompatibility–peptide complexes, processing mechanisms, 208,218*f*
Mammalian cells, characterization of polyunsaturated phospholipid remodeling using HPLC–electrospray ionization MS, 267–279
Mass-sensitive detectors, description, 45
Mass sensitivity, flow rates in electrospray ionization, 45
Mass spectrometry (MS)
  analysis of proteins, 472–510
  characterization of bacterial lipid A species, 187–188
  interface with electrospray ionization, 2
  lipopolysaccharides from different pathogenic microorganisms, 151
  use of electrospray ionization, 1–17
Mass spectrometry (MS)–liquid-phase analysis approaches, hyphenated, characterization of recombinant DNA derived glycoproteins, 432–469
Mass spectrometric analysis, common methods of sample ionization, 3,8–9*f*
Matrix-assisted laser desorption ionization, usage trends, 8–10
Matrix-assisted laser desorption ionization MS
  oligodeoxygenated analysis, 282
  protein biological analysis, 472–510
  structural characterization of biological macromolecules, 473
Matrix-assisted laser desorption ionization time of flight MS, characterization of recombinant DNA derived glycoproteins, 432–469
Melanoma proteins, two-dimensional sodium dodecyl sulfate–polyacrylamide gel electrophoresis, 474,480–499
Metabolites identification from drug substance, steps, 315
Microbial identification, methods, 82
Microcolumn LC electrospray ionization tandem MS
  class II alleles associated with rheumatoid arthritis, 213–214,216*t*,219*f*
  class II presentation mutants, 215,217–218,220–223
  experimental description, 208–213
  peptides isolated from subcellular fractions, 214–215,217*t*
Microcolumn reversed-phase LC electrospray ionization tandem MS, applications, 207
Molecular weights of proteins and peptides
  advantages, 424,430
  comparison of electrospray ionization to
    gene sequencing, 425
    polyacrylamide gel electrophoresis, 425
  example of spectrum, 428*f*,429
  experimental description, 424
  lentevirus proteins processed from *gag* preproteins, 425–426
  mass map of peptides from digestion of bovine adenosine deaminase with clostripain, 429–430
  processing problems, 427
  protein truncation due to collisions, 427,429
  purification problems, 426
Molecules
  biological, *See* Biological macromolecules
  electrospray ionization, 51–57
Molecules that form structurally specific noncovalent associations in solution, biological importance, 294
Mosquito pathogen, lipid metabolism, 244–265
Multidimensional techniques, development, 433
Multimeric proteins, electrospray ionization MS, 303,306–307
Muramic acid, taxonomic discrimination, 82

N

Nebulizer-assisted electrospray ionization techniques, advantages, 111
Neisseria, diseases caused, 166
Nitro compounds, environmental analysis, 573–575
Noncovalent complexes
  examples, 294–295
  nucleic acids and proteins
    avidin and streptavidin complexes with biotin and iminobiotin, 307–308
    contributions due to nonspecific solution associations, 299–300
    distinguishing specific and nonspecific noncovalent association
      condition, 300
      dissociation due to modification of solution conditions, 301
      gas-phase lability, 301
      relative abundance, 301
      sensitivity to structural modification, 301–302
      stoichiometry, 301
    experimental description, 295
    formation model, 295,296*f*
    future work, 310,312
    interface conditions for observation of noncovalent associations, 295,297-298
    multimeric proteins, 303,306–307
    oligonucleotide–drug associations, 302–305*f*
    previous studies, 294–295
    protein–inhibitor complexes, 308–311*f*
    solution conditions for retention of noncovalent associations, 298–299
  study methods, 294
Nonpolar neutral species, electrospray ionization, 51,54,55*f*
Nonspecific noncovalent association, distinguishing from specific noncovalent association, 300–302
Nucleic acid–protein complexes, electrospray ionization MS, 294–312

O

O-deacylated lipooligosaccharides, collisionally induced dissociation, 179–181
Oligodeoxynucleotide(s)
  analysis, 556–559
  analysis using electrospray ionization MS, 281–292
  analytical methods, 282
Oligonucleotide–drug associations, electrospray ionization MS, 302–303,304–305*f*
Oligosaccharide
  analysis, 537–553
  structural characterization, 134–138,140*f*
  use of electrospray ionization MS, 16
Oomycetes, description, 244–245
8-Oxo-2'-deoxyguanosine oligomer degradation, electrospray ionization MS, 287–289,291
Oxygenated fatty acids, identification, 256–263,265

P

Parasitic microorganisms, steps in interaction with other microorganisms, plants, insects, and mammals, 244
Particle beam LC–MS, use for environmental monitoring, 566
Particle beam MS, comparison to atmospheric pressure ionization LC–MS, 578*t*,579
Pathogenic bacteria from *Neisseria* and *Haemophilus*, diseases caused, 166
Pathogenic *Haemophilus* bacteria, structure and function determination of surface glycolipids, 166–183
Pathogenic microorganisms, steps in interaction with other microorganisms, plants, insects, and mammals, 244
Peptide(s)
  analysis, 527,530–534,536
  molecular weight determination, 424–430
  use of electrospray ionization MS, 16

Peptide(s) isolated from subcellular fractions, microcolumn LC electrospray ionization tandem MS, 214–215,217*t*
Peptide antigens, presentation using major histocompatibility complex molecules, 207–208
Peptide mapping
application of LC, 433
characterization of recombinant glycoproteins from Chinese hamster ovary cells, 411,414–419
limitations, 433
Peptide mass fingerprinting
protein identification from two-dimensional electrophoresis gels, 226–240
protein identification methods, 227,228*f*
proteins separated using two-dimensional electrophoresis, 227,229
Peptidoglycan-associated sugar monomers, use of electrospray ionization MS, 12
Peptidoglycan precursor in antibiotic-resistant bacteria, cytoplasmic, *See* Cytoplasmic peptidoglycan precursor in antibiotic-resistant bacteria
Pesticides, environmental analysis, 571–573
Phospholipid(s) from biological matrix, analytical problems, 267
Phospholipid bilayers, role in cell membrane characteristics, 267
Phosphorylation sites in activated T cells, analysis using electrospray ionization MS, 381–405
Photodissociation
applications, 513
combination with
electrospray ionization, 514
ion cyclotron resonance, 514
comparison to
collisionally induced dissociation fragmentation process, 513–559
surface-induced dissociation, 514
for positive ion, process, 513
information obtained, 513–514
Photoinduced dissociation, *See* Photodissociation
Photoinduced ion fragmentation, measurements, 513–514
Plankton, analysis of diarrhetic shellfish poisoning toxins and metabolites, 351–363
Plasma, xanomeline analysis, 323–328
Plasma desorption MS, structural characterization of lipid A component in bacterial endotoxins, 189–195,197
Polar neutral species, electrospray ionization, 51
Polyacrylamide gel electrophoresis, comparison to electrospray ionization MS, 425
Polyacrylamide gel electrophoresis–sodium dodecyl sulfate, analysis of melanoma proteins, 474,480–499
Polyatomic ions, activation method, 512
Poly(ethylene glycols), charged droplet formation, 78
Polynuclear aromatic hydrocarbons, analysis using atmospheric pressure ionization LC–MS, 575,579
Polyunsaturated fatty acids, function in neuronal membranes, 267
Polyunsaturated phospholipid remodeling in mammalian cells, characterization using HPLC–electrospray ionization MS, 267–279
Proceedings of the American Society for Mass Spectrometry Conferences on Mass Spectrometry and Allied Topics, function, 3
Processing pathways, study requirements, 208
Prokaryotic cell surface components, structural characterization, 133–134
Prorocentrum species, shellfish contamination, 351
Protein(s)
analysis, 527,530–534,536
molecular weight determination, 424–430
separation using two-dimensional electrophoresis, 227,229
use of electrospray ionization MS, 16

Protein amino acid sequence, determination method, 226
Protein biological analysis
capillary HPLC–electrospray ionization–magnetic sector multichannel array detection instrumentation, 503,505–510
complex protein mixtures, 473–499
future work, 510
selective detection of covalent protein modification, 485,500–504
Protein glycans, structural characterization, 138–141
Protein identification from two-dimensional electrophoresis gels using peptide mass fingerprinting
advantages of two-dimensional gels, 236
base peak profile, 234*f*,235
capillary HPLC vs. capillary LC–MS, 239
Coomassie-stained blot, 232–233,235
data-base availability, 236
detection limit, 226
experimental description, 229–232
false matches, 237,239
future prospects, 239–240
output from FRAGFIT searches, 235,237,238*t*
previous studies, 227,229
proteins identified, 235–237
Protein–inhibitor complexes, electrospray ionization MS, 308–311*f*
Protein mixture analysis, complex, *See* Complex protein mixture analysis
Protein–nucleic acid complexes, electrospray ionization MS, 294–312
Protein phosphorylation, analytical techniques, 383
Protein phosphorylation sites in activated T cells, analysis using electrospray ionization MS, 381–405
Protein primary sequence, analysis using MS, 472–510
Protein sequence data bases, protein identification problems, 226–227
Pseudomolecular ions, formation, 472
Pseudomonas aeruginosa
occurrence, 149
strain differences, 151
Pseudomonas aeruginosa lipopolysaccharides, structural characterization, 149–163
Pulsed amperometric detection, carbohydrate characterization, 82–83

Q

Quadrupole ion trap MS, advancement, 512
Quantitative bioanalytical chemistry
challenges, 330
HPLC–atmospheric pressure ionization tandem MS, 330–348
measurement methods, 330

R

Radio frequency only multipole, use in biomolecule analysis, 515
Radioimmunoassay, cross-validation, 338–339
Raffinose, analysis, 545–550
Rat urine, xanomeline analysis using electrospray ionization tandem MS, 318*f*–323
Recombinant desmodus salivary plasminogen activator, characterization using hyphenated liquid-phase analyses–MS approaches, 432–469
Recombinant DNA derived glycoprotein characterization
analytical strategy
collisionally induced dissociation, 461–464
HPLC fractionation of digest
HPCE, 465,467,469
HPLC, 465,466*f*
matrix-assisted laser desorption ionization time of flight MS, 468*f*,469
intact protein
description, 441,443
HPCE, 443,445*f*,447

Recombinant DNA derived glycoprotein characterization—*Continued*
analytical strategy—*Continued*
intact protein—*Continued*
HPLC, 443,444*f*,447
matrix-assisted laser desorption ionization time of flight MS, 443,446*f*,447
process, 440–442*f*
proteolytic digest
electrospray ionization LC–MS, 450*t*–455
HPCE, 447,448*f*
HPLC, 447,448*f*
matrix-assisted laser desorption ionization time of flight MS, 447,449*f*
two-dimensional maps, 455–461
characterization techniques, 433,435
procedure
electrospray ionization LC–MS, 439
HPCE, 439
HPLC, 435,438–439
matrix-assisted laser desorption ionization time of flight MS, 440
sample preparation, 440
Recombinant glycoproteins from Chinese hamster ovary cells
additional substituents, 419,421
direct analysis of intact proteins, 410–413*f*
expanded HPLC, 419,420*f*
experimental procedure, 409–410
future work, 421–422
peptide mapping, 411,414–419
5-α-Reductase inhibitor, determination using atmospheric pressure ionization LC–tandem MS, 344–346
Retaining mechanism, glycosidases, 366
RNA, use of electrospray ionization MS, 16
RNA dimers, analysis, 556–559
Rough-type lipopolysaccharides, description, 149

S

Sample preparation for atmospheric pressure ionization LC–tandem MS, automation, 346
Sensitivity
electrospray ionization MS of biological macromolecules, 269,272
influencing factors for electrospray ionization, 50
Serotype O6, heterogeneity of core and O chain, 159–163
Shellfish, analysis of diarrhetic shellfish poisoning toxins and metabolites, 351–363
Shellfish poisoning toxins and metabolites, diarrhetic, *See* Diarrhetic shellfish poisoning toxins and metabolites in plankton and shellfish
Sialic acid, determination of lipooligosaccharides, 171,173*f*,175
Signal recognition particle, sequence determination, 474,475–479*f*
Signal transduction
description, 381
influencing factors, 381–382
Signal transduction pathways
advantages, 403
experimental description, 383–387
inhibition of tyrosine activation and phosphorylation, 403–405
interpretation of two-dimensional phosphopeptide maps, 397,402–403
T cell receptor ζ phosphorylation sites, 389,397–401*f*
ZAP–70 phosphorylation sites, 388–396*f*
Simultaneous determination of drugs and metabolites
data processing automation, 346–347
sample preparation automation, 346
Single-stranded RNA, use of electrospray ionization MS, 16
Sites of inducible protein phosphorylation in activated T cells, analysis using electrospray ionization MS, 381–405
Sodium dodecyl sulfate–polyacrylamide gel electrophoresis, analysis of melanoma proteins, 474,480–499
Solute species in charged droplets, gas-phase ion formation, 60–79
Solution conditions, retention of noncovalent associations, 298–299

Solvent evaporation, mechanisms, 283
Species charged in solution, electrospray ionization, 51
Specific noncovalent association, distinguishing from nonspecific noncovalent association, 300–302
Spray formation, mechanisms, 283
Stachyose, analysis, 547,550–553
Streptavidin complexes, electrospray ionization MS, 307–308
Structural characterization
 biological macromolecules
  requirements, 473
  revolution, 472–473
 carbohydrate moieties attached to protein
  analytical problem, 408–409
  use of electrospray ionization MS, 409
 lipid A component in bacterial endotoxins
  acid hydrolysis, 196–200
  base hydrolysis, 199,201–203*f*
  experimental description, 188–189
  plasma desorption MS vs. electrospray ionization MS, 189–195,197
  structure, 186*f*,187
 lipopolysaccharides from *Pseudomonas aeruginosa*
  core oligosaccharide, 150–151,153–158
  experimental procedure, 151–153
  heterogeneity of core and O chain of serotype O6, 159–163
  lipid A disaccharide, 158–159
  regions, 149
  strategy, 153,157*t*
 prokaryotic glycans and oligosaccharides
  collisionally induced dissociation at low energy, 132–133
  experimental description, 130–131,146
  lipid A, 141–145
  oligosaccharides, 134–138,140*f*
  profile analysis using electrospray ionization MS, 131–132
  prokaryotic cell surface components, 133–134
  protein glycans, 138–141
Structure
 cytoplasmic peptidoglycan precursor in antibiotic-resistant bacteria, 114–121
 surface glycolipids in pathogenic *Haemophilus* bacteria, 166–183
Sugar, structure, 82
Surface glycolipids in pathogenic *Haemophilus* bacteria
 advantages of electrospray ionization MS, 167,170
 collisionally induced dissociation of *O*-deacylated lipooligosaccharides of *Haemophilus ducreyi*, 179–181
 conversion to *O*-deacylated forms using hydrazine treatments, 171,172*f*
 experimental description, 167,170
 future work, 183
 intact lipooligosaccharide analysis, 181,182*f*
 lipooligosaccharide glycoform analysis for *Haemophilus influenzae*, 174*f*–179
 sialic acid determination in lipooligosaccharides of *Haemophilus ducreyi*, 171,173*f*,175
 structures, 166–169*f*
Surface-induced dissociation, comparison to photodissociation, 514

T

T cell(s), induction of tyrosine phosphorylation, 382
T cell receptor complex, schematic representation, 382–383,384*f*
T cell receptor ζ phosphorylation sites, analysis using HPLC–electrospray ionization MS, 389,397–401*f*
Tandem MS
 characterization
  carbohydrates, 83
  underivatized carbohydrates, 86,88–89*f*
 lipid metabolism of *Lagenidium giganteum* and hosts, 244–265
 structural characterization of lipopolysaccharides from *Pseudomonas aeruginosa*, 149–163

Taxonomic differentiation, methods, 82
Taxonomic discrimination, muramic acid, 82
Thermospray ionization
 advantages, 332
 description, 62
Thermospray ionization LC–MS
 comparison to atmospheric pressure ionization LC–MS, 578*t*,579
 use for environmental monitoring, 566
Thomson, charged droplets as source of ions in atmosphere, 61–62
Tissue plasminogen activator, glycoforms, 433,434*t*
Tolmetic glucuronide–human serum albumin drug adducts, selective detection, 485,502–504*f*
Toxins, *See* Diarrhetic shellfish poisoning toxins and metabolites in plankton and shellfish
Two-dimensional electrophoresis, description, 227
Two-dimensional electrophoresis gels, protein identification using peptide mass fingerprinting, 226–240
Two-dimensional maps, characterization of recombinant DNA derived glycoproteins, 455–461
Two-dimensional sodium dodecyl sulfate–polyacrylamide gel electrophoresis, analysis of melanoma proteins, 474,480–499

U

Underivatized carbohydrates, analytical techniques, 83–89
Urine, xanomeline analysis, 318–328

V

Vancomycin
 binding with peptidoglycan precursor, 107,110*f*
 function, 107
 mode of action, 107,109*f*
 role of VanA in resistance, 107
 structure, 107,108*f*
Vancomycin-resistant bacteria, role of VanA in resistance, 107
Vestal, Marvin, thermospray ionization, 62

X

Xanomeline, structure, 316
Xanomeline analysis
 experimental description, 316–317,319
 future work, 329
 human urine and plasma, 323–328
 rat urine, 318–323
Xylanases, active site nucleophile identification, 369–371,372*f*

Z

ZAP–70, role in induction of full T cell receptor mediated response, 383
ZAP–70 phosphorylation sites, analysis using immobilized metal affinity chromatography–HPLC–electrospray ionization MS, 388–396

Production: Susan Antigone
Indexing: Deborah H. Steiner
Acquisition: Barbara E. Pralle
Cover design: A. Peter Snyder & Amy Hayes

Printed and bound by Maple Press, York, PA